Studies of
FOOD MICROSTRUCTURE

Based on programs organized by
S.H. Cohen, E.A. Davis, D.N. Holcomb and M. Kalab

Guest Editors
D.N. Holcomb and M. Kalab

Published by:
Scanning Electron Microscopy, Inc.
P.O. Box 66507
AMF O'Hare, IL 60666

Copyright © 1981, Scanning Electron Microscopy, Inc.
except for contributions in the public domain.

All rights reserved.

Individual readers of this volume and non-profit libraries acting for them are freely permitted to make fair use of the material herein, such as to copy an article for use in teaching or research. Permission is granted to quote from this volume in scientific works with the customary acknowledgment of the source. To print a table, figure, micrograph or other excerpt requires, in addition, the consent of one of the original authors and notification to SEM, Inc. Republication or systematic or multiple reproduction of any material in this volume (including the abstracts) is permitted only after obtaining written approval from SEM, Inc.; and, in addition, SEM, Inc. will require that permission also be obtained from one of the original authors.

Every effort has been made to trace the ownership of all copyrighted material in this volume and to obtain permission for its use.

The articles in this book are reprinted from the Journal:

    Scanning Electron Microscopy, Vol. 1979 Part III
    Scanning Electron Microscopy, Vol. 1980, Part III
    Scanning Electron Microscopy, Vol. 1981, Part III

The paginations from these publications are given at the top left of the first page of each article.

In quoting the papers in this book, it is strongly recommended that the original pagination be used in the following format
    Scanning Electron Microsc. year; part; p. no.

For information on availability of past SEM issues or other inquiries contact:

    Dr. Om Johari — phone 312-529-6677
    SEM, Inc.
    P.O. Box 66507
    AMF O'Hare, IL 60666, U.S.A.

ISBN: 0-931288-22-3   Library of Congress
    Catalog Card Number: 81-84080.

Printed in the United States of America

Editor   Om Johari
Associate Editors   Sudha A. Bhatt & Irene Corvin Pontarelli
Editorial Assistance   Joseph Staschke

ORGANIZERS AND MEMBERS OF EDITORIAL BOARD

Sam H. Cohen   U.S. Army Natick R & D Command, Natick, MA
Eugenia A. Davis   University of Minnesota, St. Paul, MN
David N. Holcomb   Kraft Inc., R & D, Glenview, IL
Milos Kalab   Agriculture Canada, Ottawa, Canada

SCANNING ELECTRON MICROSCOPY, INC.

President   John D. Fairing
Vice President   Robert P. Becker
Secretary-Treasurer   Om Johari

FOREWORD

Beginning in 1979, programs on food microstructure have taken place at annual Scanning Electron Microscopy (SEM) meetings. The papers presented in those programs have been published in the respective SEM volumes. Thirty-six of those papers have been compiled to form this book which will serve as a prelude to the forthcoming journal, *Food Microstructure*.

The range of titles in this compilation reflects the rapidly escalating interest in the area. This interest includes: the fundamental aspects of food microstructure, such as the molecular and colloidal forces which determine it; and the practical relations between food microstructure, processing, ingredient changes, shelf life, and consumer acceptability. The era in which food product development and improvement was as likely done by a chef as by a food scientist ended decades ago; solution of present day problems in the food industry requires increasingly sophisticated tools and approaches. Such modern approaches include microscopic examination of foods and ingredients and efforts to understand the factors which lead to the microstructures observed.

The papers in this volume have been divided into four general areas: I. general applications of microscopy in food sciences (5 papers); II. meat foods (7 papers); III. milk products (13 papers); IV. foods of plant origin (11 papers). In these four sections, leading scientists have demonstrated that microstructural studies of foodstuffs is invaluable. We believe that, significant though these contributions are, microscopic examination and understanding of foods is still in its infancy. It is our hope that this volume of "Studies of Food Microstructure" and the journal to follow will give impetus to future developments in the field.

S.H. Cohen   E.A. Davis   D.N. Holcomb   M. Kalab

August 31, 1981

REVIEWING PROCEDURE
AND
DISCUSSION WITH REVIEWERS

Each paper in this volume contains a Discussion with Reviewers. This discussion follows the text and should be read with the paper. Each paper submitted to SEM, Inc. for publication is reviewed by at least three, up to an average of five, reviewers. The reviewers are asked to separate their comments from their questions. The comments are useful in determining the acceptability of the papers as submitted. Although the comments require no written response, in several cases, the authors have included responses to comments, or to questions phrased from, or based on, comments (either as a result of editorial suggestions or on the author's own initiative). Based on these comments approximately 15% of the submitted papers were not accepted for publication; while almost all of the others were asked to make changes involving from minor to major revisions.

The questions, for the most part, originate as a result of statements included in our cover letter accompanying each paper sent to the reviewers. The reviewers are asked to suppose they are attendees at a conference where this paper, as written, is being presented, and then ask relevant questions which would occur to them resulting from the presentation. From the questions so asked, some are not included with the published paper because the authors attended to them by text revisions. In some cases, editorial and/or space considerations may exclude inclusion of all questions asked by reviewers. The authors are asked to prepare their Discussion with Reviewers section in a camera-ready format in accordance with detailed instructions which they are sent by SEM, Inc. In some instances the authors edit the questions and/or combine several similar questions from different reviewers to provide one answer. While all efforts are made to check that the questions in the printed version faithfully follow the views of the specific reviewer, the editors apologize, if in some instances, the actual meaning and/or emphasis may have been changed by the author.

The cover letter to the reviewers states:

"1. Your name will be conveyed to the author with your review UNLESS YOU ASK US NOT TO.

2. The questions published in the Journal will be identified as originating from you UNLESS YOU ADVISE OTHERWISE..."

In all cases sincere efforts are made to respect the reviewer's wishes to remain anonymous; however, in nearly 95% of the cases, the reviewers have given permission to be identified; so their names are conveyed to the authors and are included with the questions printed with each paper. An overall list of reviewers is provided in the opening pages of each SEM part. We apologize for any error/omissions which may occur.

Finally, readers are urged to be cautious regarding the weight they attach to the authors' replies, since the answers to the questions represent the authors' unchallenged views--except for minor editorial changes--the authors generally have the last word. Also, please consider that the questions were, in most all cases, relevant to the originally submitted paper, and they may not have the same significance for the revised paper published in this volume.

If you disagree with the results, conclusions or approaches in a paper, please send your comments, as a Letter to Editor, typed in a column format (each column is 4-1/8 inches wide and 11-1/2 inches long; i.e., 10.5 by 29.3 cm.). Your comments along with author's response will be published in a subsequent issue.

The editor gratefully thanks the authors and reviewers (see p. x-xv) for their contributions, invites your comments on ways to improve this procedure and seeks qualified volunteers to assist with reviewing papers in the future. (see p. xv)

ERRATA: Despite the best efforts of authors, reviewers and editors, errors may remain. Please help by pointing out errors that you notice. Please provide enough information to locate each error (volume, part, page, column, line, etc.) and indicate suitable correction.

TABLE OF CONTENTS

Editorial Board, SEM Inc. — iii
Foreword — iii
Discussion with Reviewers, Errata — iv
Reviewers List, Call for Reviewers — viii
Call for Papers — x

PART I - GENERAL PAPERS

SOME EXAMPLES OF SCANNING ELECTRON MICROSCOPY IN FOOD SCIENCE
    R.J. Carroll and S.B. Jones   (REVIEW PAPER)* — 1

PREPARATION OF FOOD SCIENCE SAMPLES FOR SEM   (REVIEW PAPER
    J.F. Chabot — 9

APPLICATION OF SCANNING ELECTRON MICROSCOPY FOR THE DEVELOPMENT OF MATERIALS FOR FOOD   (REVIEW PAPER)
    C.-H. Lee and C.K. Rha — 17

THE USE OF MICROSCOPY TO EXPLAIN THE BEHAVIOUR OF FOODSTUFFS - A REVIEW OF WORK CARRIED OUT AT THE LEATHERHEAD FOOD RESEARCH ASSOCIATION   (REVIEW PAPER)
    D.F. Lewis — 25

IDENTIFICATION OF FOREIGN MATTER IN FOODS
    J.T. Stasny, F.R. Albright and R. Graham — 39

PART II - MEAT FOODS

SCANNING ELECTRON MICROSCOPY IN MEAT SCIENCE   (TUTORIAL PAPER)*
    C.A. Voyle — 51

PREPARATION OF MUSCLE SAMPLES FOR ELECTRON MICROSCOPY
    H.D. Geissinger and D.W. Stanley   (TUTORIAL PAPER) — 61

THE EFFECT OF CATHEPTIC ENZYMES ON CHILLED BOVINE MUSCLE
    S.H. Cohen and L.R. Trusal — 73

MICROSCOPICAL OBSERVATIONS ON ELECTRICALLY STIMULATED BOVINE MUSCLE
    C.A. Voyle — 79

SCANNING AND TRANSMISSION ELECTRON MICROSCOPY OF NORMAL AND PSE PORCINE MUSCLE
    J.D. Cloke, E.A. Davis, J. Gordon, S.-I. Hsieh, J. Grider, P.B. Addis and C.J. McGrath — 87

IDENTIFICATION OF FAT AND PROTEIN COMPONENTS IN MEAT EMULSIONS USING SEM AND LIGHT MICROSCOPY
    F.K. Ray, B.G. Miller, D.C. Van Sickle, E.D. Aberle, J.C. Forrest and M.D. Judge — 99

MEAT EMULSIONS - FINE STRUCTURE RELATIONSHIPS AND STABILITY
    R.J. Carroll and C.M. Lee — 105

*See page vii.

PART III - MILK PRODUCTS, GELS, AND MAYONNAISE

SCANNING ELECTRON MICROSCOPY OF DAIRY PRODUCTS: AN OVERVIEW
    M. Kalab   (REVIEW PAPER)   111

ELECTRON MICROSCOPY OF MILK PRODUCTS: A REVIEW OF TECHNIQUES
    M. Kalab   (REVIEW PAPER)   123

MICROSTRUCTURE AND RHEOLOGY OF PROCESS CHEESE
    A.A. Rayan, M. Kalab and C.A. Ernstrom   143

ELECTRON MICROSCOPY AND SENSORY EVALUATION OF COMMERCIAL CREAM CHEESE
    M. Kalab, A.G. Sargant and D.A. Froehlich   153

MORPHOLOGICAL, ULTRASTRUCTURAL AND RHEOLOGICAL CHARACTERIZATION OF CHEDDAR AND MOZZARELLA CHEEZE
    M.V. Taranto, P.J. Wan, S.L. Chen and K.C. Rhee   163

MORPHOLOGICAL AND TEXTURAL CHARACTERIZATION OF SOYBEAN MOZZARELLA CHEESE ANALOGS
    M.V. Taranto and C.S.T. Yang   169

POSSIBILITIES OF AN ELECTRON-MICROSCOPIC DETECTION OF BUTTERMILK MADE FROM SWEET CREAM IN ADULTERATED SKIM MILK
    M. Kalab   179

A SCANNING ELECTRON MICROSCOPICAL INVESTIGATION OF THE WHIPPING OF CREAM
    D.G. Schmidt and A.C.M. van Hooydonk   187

A COMPARISON OF THE MICROSTRUCTURE OF DRIED MILK PRODUCTS BY FREEZE-FRACTURING POWDER SUSPENSIONS IN NON-AQUEOUS MEDIA
    W. Buchheim   193

SEM INVESTIGATION OF THE EFFECT OF LACTOSE CRYSTALLIZATION ON THE STORAGE PROPERTIES OF SPRAY DRIED WHEY
    M. Saltmarch and T.P. Labuza   203

EFFECT OF ACIDULANTS AND TEMPERATURE ON MICROSTRUCTURE, FIRMNESS AND SUSCEPTIBILITY TO SYNERESIS OF SKIM MILK GELS
    V.R. Harwalkar and M. Kalab   211

STRUCTURE OF VARIOUS TYPES OF GELS AS REVEALED BY SCANNING ELECTRON MICROSCOPY (SEM)
    V.E. Colombo and P.J. Spath   223

MICROSTRUCTURE OF MAYONNAISE AND SALAD DRESSING
    M.A. Tung and L.J. Jones   231

PART IV - FOODS OF PLANT ORIGIN

SCANNING ELECTRON MICROSCOPY OF SOYBEANS AND SOYBEAN PROTEIN PRODUCTS   (REVIEW PAPER)
    W.J. Wolf and F.L. Baker   239

SOYBEAN SEED-COAT STRUCTURAL FEATURES: PITS, DEPOSITS AND CRACKS
    W.J. Wolf, F.L. Baker and R.L. Bernard   253

AN SEM STUDY OF THE EFFECTS OF AVIAN DIGESTION ON THE SEED COATS OF THREE COMMON ANGIOSPERMS
    L.B. Smith   267

MICROSTRUCTURE OF TRADITIONAL JAPANESE SOYBEAN FOODS
    K. Saio   (REVIEW PAPER)   275

EFFECTS OF EXOGENOUS ENZYMES ON OILSEED PROTEIN BODIES
    R.D. Allen and H.J. Arnott   283

TANNIN DEVELOPMENT AND LOCATION IN BIRD-RESISTANT SORGHUM GRAIN
    P. Morrall, N.v.d.W. Liebenberg and C.W. Glennie   293

LIGHT MICROSCOPY OF PLANT CONSTITUENTS IN ANIMAL FEEDS
    J.G. Vaughan   (TUTORIAL PAPER)   299

THE RELATIONSHIP BETWEEN WHEAT MICROSTRUCTURE AND FLOURMILLING
    R. Moss, N.L. Stenvert, K. Kingswood and G. Pointing   305

SCANNING ELECTRON MICROSCOPY OF FLOUR-WATER DOUGHS TREATED WITH OXIDIZING AND REDUCING AGENTS
    L.G. Evans, A.M. Pearson and G.R. Hooper   313

STRUCTURAL STUDIES OF CARROTS BY SEM   (TUTORIAL PAPER)
    E.A. Davis and J. Gordon   323

THE MICROSTRUCTURE OF ORANGE JUICE
    G.G. Jewell   333

ABOUT SCANNING ELECTRON MICROSCOPY, INC.   222

SUBJECT INDEX   339

AUTHOR INDEX   341

ANNOUNCING: "FOOD MICROSTRUCTURE" - An International Journal   342

---

*EXPLANATION OF THE TYPES OF PAPERS IN THIS VOLUME:

<u>TUTORIAL</u>: Presentation of established material in teaching format emphasizing techniques.

<u>REVIEW</u>: A review of the chosen subject with emphasis on author's own work, placing it in context with relevant literature and putting the topic in perspective.

Volunteers to prepare tutorial, review papers or bibliographies should contact Om Johari (see page ii)

REVIEWERS LIST

The editors gratefully acknowledge the help of the following reviewers with the papers on Food Microstructure submitted during 1979, 1980 and 1981.

Abbott, M.T.	Campbell Soup Co., Camden, NJ
Amanthea, G.F.	Fraser Valley Milk Prod. Assoc., Vancouver, Canada
Amer, M.A.	Gay Lea Foods, Guelph, Canada
Anderson, M.	National Inst. Res. Darying, Reading, U.K.
Arnott, H.J.	Univ. Texas, Arlington
Ashraf, M.	Univ. Cincinnati, OH
Bair, C.W.	Frito-Lay, Inc., Irving, TX
Becker, R.P.	Univ. Illinois Med. Ctr., Chicago
Bridges, A.	John Labatt Ltd., London, Canada
Brisson, J.D.	Quebec Dept. Agriculture, St.-Foy, Canada
Brown, J.A.	McCrone Associates, Chicago, IL
Buchheim, W.	Bund. Milchforschung, Kiel, W. Germany
Bullard, R.W.	US Dept. Interior, Denver, CO
Buma, T.J.	Netherland Inst. Dairy Res., Ede
Buri, M.	Kraft Inc. R&D, Glenview, IL
Carpenter, D.E.	Kraft Inc. R&D, Glenview, IL
Carroad, P.A.	Univ. California, Davis
Carroll, R.J.	USDA Eastern Reg.Res.Ctr., Philadelphia, PA
Cassens, R.G.	Univ. Wisconsin, Madison
Chabot, J.F.	Cornell Univ., Ithaca, NY
Chatfield, E.J.	Ontario Res. Fndn., Mississauga, Canada
Christman, M.A.	Kraft Inc. R&D., Glenview, IL
Cohen, S.H.	US Army Natick R&D Command, MA
Croxdale, J.	Univ. Wisconsin, Madison
Davey, C.L.	Meat Ind. Res. Inst., Hamilton, New Zealand
Davis, E.A.	Univ. Minnesota, St. Paul
De Man, J.M.	Univ. Guelph, Canada
Draftz, R.G.	IIT Res. Inst., Chicago, IL
Dutson, T.R.	Texas A&M Univ., College Station
Ernstrom, C.A.	Utah State Univ., Logan
Fairing, J.D.	Monsanto Company, St. Louis, MO
Fuwa, H.	Osaka City Univ., Japan
Gallant, D.J.	INRA Nantes, France
Gaud, S.M.	Kraft Inc. R&D, Glenview, IL
Geissinger, H.D.	Univ. Guelph, Canada
Glabe, E.F.	Food Tech Lab, Chicago, IL
Gordon, J.	Univ. Minnesota, St. Paul
Harper, W.J.	Ohio State Univ., Columbus
Harwalker, P.A.	Agriculture Dept., Ottawa, Canada
Haslam, E.	Univ. Sheffield, U.K.
Holcomb, D.N.	Kraft Inc. R&D, Glenview, IL
Holmes, L.G.	US Army Natick Res. Labs., MA
Hood, L.F.	New York State College of Agri., Ithaca, NY
Horner Jr., H.T.	Iowa State Univ., Ames
Hoseney, R.C.	Kansas State Univ., Manhattan, KS
Ilker, R.	General Foods Corp., Tarrytown, NY
Jewell, G.G.	Cadbury Schweppes Ltd, Birmingham, U.K.
Johari, O.	Scanning Electron Microscopy, Inc., Chicago, IL
Jones, S.B.	USDA Eastern Reg. Res. Ctr., Philadelphia, PA
Kalab, M.	Agriculture Dept., Ottawa, Canada
Karel, M.	Massachusetts Inst. Technology, Cambridge, MA
Keeney, P.G.	Pennsylvania State Univ, Univ. Park.
Knoop, A.-M.	Bund. Milchforschung, Kiel, W. Germany
Knutson, C.A.	USDA Northern Reg. Res. Lab., Peoria, IL
Krishnamurthy, R.G.	Kraft Inc., R&D, Glenview, IL
Kvenberg, J.E.	USDA Food & Drug Adm., Washington, DC
Lineback, D.R.	North Carolina State Univ., Raleigh
Loh, J.L.	General Foods Corp., Terrytown, NY
Lott, J.N.	McMaster Univ., Hamilton, Canada
Lowrie, R.A.	Univ. Nottingham, U.K.
MacRitchie, F.	CSIRO Wheat Res. Unit, N.Ryde, Australia
McGrath, P.P.	USDA Food & Drug Adm., Bethesda, MD

Moore, J.A.	DHHS/USPHS/NIH, Research Triangle Park, NC
Moss, R.	Bread Res. Inst., N. Ryde, Australia
Murphy, J.A.	Southern Illinois Univ., Carbondale, IL
O'Brien, T.P.	Monash Univ., Clayton, Vic., Australia
Oles, J.G.	Kraft Inc. R&D, Glenview, IL
Patton, S.	Pennsylvania State Univ., Univ. Park, PA
Pesheck, P.S.	Pillsbury Co., Minneapolis, MN
Peterson, R.L.	Univ. Guelph, Canada
Pomeranz, Y.	USDA Grain Marketing Res. Lab., Manhattan, KS
Postek, M.T.	AMRAY Inc., Bedford, MA
Prentice, J.H.	Natl. Inst. Res. Dairying, Reading, U.K.
Prescott, H.E.	General Foods Corp., Tarrytown, NY
Price, M.L.	Battelle Labs, Columbus, OH
Ray, F.K.	Oklahoma State Univ., Stillwater
Rha, C.K.	Mass. Inst. Technology, Cambridge
Rockland, L.B.	Chapman College, Orange, CA
Roetman, K.	DOMO Melkproduktenbedrijven, Beilen, W. Germany
Rooney, L.W.	Texas A&M Univ., College Station
Ruegg, M.	Fed Dairy Res Inst, Liebefeld, Switzerland
Sachs, I.B.	USDA Forest Products Lab., Madison, WI
Sahasrabudhe, M.R.	Agriculture Canada, Ottawa, Canada
Saio, K.	Natl. Food Res. Inst., Ibaraki, Japan
Sargant, A.G.	Silverwood Industries Ltd., London, Ontario, Canada
Schmidt, D.G.	Netherland Inst. Dairy Res., Ede
Smith, L.B.	Univ. Texas, Arlington
Stanley, D.W.	Univ. Guelph, Canada
Steere, R.L.	USDA Beltsville Agri. Res. Ctr., MD
Sullins, R.D.	Ralston Purina Co., St. Louis, MO
Swatland, H.J.	Univ. Guelph, Ontario, Canada
Taranto, M.V.	ITT Continental Baking Co, Rye, NY
Thurston, E.L.	Texas A&M Univ., College Station
Tung, M.A.	Univ. British Columbia, Vancouver, Canada
Ueda, S.	Kyushu Univ., Fukuoka, Japan
Varriano-Marston, E.	Kansas State Uni., Manhattan
Vaughan, J.A.	Queen Elizabeth College, London., U.K.
Voyle, C.A.	Meat Res. Inst., Bristol, U.K.
Walstra, P.	Univ. Wageningen, Netherlands
Wolf, W.J.	USDA Northern Reg. Res. Lab., Peoria, IL

CALL FOR REVIEWERS

The contribution of reviewers to the quality of this publication and our meetings is tremendous. We find suitable reviewers from the suggestions we receive from the authors, our advisors, and from our past contacts. We will welcome your suggesting your own name or other's names (along with full mailing address) as reviewers.

Important Note: The time restrictions we work under require that each reviewer returns his review (along with the manuscript sent) within a set time from its receipt. *Please do not commit yourself if you feel that you cannot respond within this time frame; while we are grateful for your desire and efforts to help us, the reviewers who do not respond in time, in fact, seriously hamper our efforts.*

CONTACT: Om Johari, SEM Inc., P.O. Box 66507, AMF O'Hare, IL 60666, USA
Phone - 312-529-6677

CALL FOR PAPERS

Papers for Food Microstructure should deal with microscopy and microanalysis of foods, feeds and their ingredients. Studies on beverages, fruits, vegetables, cereals, meat, seafood, milk products, edible oils and fats, etc. are welcome. Experimental techniques may include any type of microscopy (scanning electron, transmission electron, or light microscopy), x-ray microanalysis or related microscopical/microanalytical methods.

TYPES OF PAPERS

Papers can be offered either for publication only (see below) or **publication and oral presentation**. **Only papers submitted for publication and having a reasonable chance of acceptance can be orally presented.**

The following types of papers may be submitted:

A. Contributed Papers: Present new unpublished findings as in a **paper or note** submitted to any other professional journal.

B. Review Papers: Include an extended literature review and bibliography (including author's own work), **emphasize author's new unpublished findings** and in an extended discussion put the topic of review in proper perspective.

C. Tutorial Papers: Contain an organized comprehensive review and bibliography of ALL relevant published material presented in a teaching manner.

The above definitions must be carefully adhered to, since reviewers are asked to judge papers accordingly. *Space restrictions are not imposed on papers; shorter but complete papers are welcome.*

PAPER FOR PUBLICATION ONLY

Authors with papers ready for publication can offer them at any time by submitting four (4) copies of their papers. The cover letter must include information sought by items 1, 3a, 4 and 5 of the "Procedure for Offering Papers".

PROCEDURE FOR OFFERING PAPERS

To offer a paper, the prospective author(s) must first submit a **Letter of Intent** giving:

▶[1] (a) Short but representative title, (b) type of paper (see above), and (c) name, **phone number**, and mailing address of the person to contact.

▶[2] **Fifty word summary of the proposed paper.** This summary is used to organize papers for different sessions and may also be published.

▶[3] (a) **Each Letter of Intent must contain the following statement: The Publication Requirements described in this announcement have been carefully read, understood, and will be followed.** (b) Authors are encouraged to discuss papers with organizers of this program, however, each Letter of Intent must be sent to the SEM office, at P.O. Box 66507, AMF O'Hare, IL 60666.

▶[4] Names and full mailing addresses (and, if possible, phone numbers) of **four persons competent to review the complete paper** based on the summary. **This information is extremely important and must be carefully prepared and included in your Letter of Intent.** Please note: (a) Suggested reviewers must neither be from author's current or recent affiliations, nor co-workers; (b) preferably, suggested reviewers should be amongst active researchers in the field of the paper (e.g., whose work is being extensively referenced or commented on); (c) authors are neither expected to personally know nor contact the suggested reviewers. From the names supplied by authors and the organizers, the SEM office will select and contact the most suitable reviewers **irrespective of their geographical location.** It is in the author's interest to provide reviewer names since it will significantly hasten the selection of reviewers.

▶[5] FOOD MICROSTRUCTURE will be a copyrighted publication. If part, or all of the material has been, is being, or will be published and/or submitted elsewhere, details must be included in the Letter of Intent. However, oral presentation of a paper at some other meeting as well as publication in the form of *unreviewed* abstract (e.g., in proceedings, non-English publications, etc.) does not preclude consideration of a paper by Food Microstructure.

▶[6] All authors intending to present their papers are expected to register for the meetings, (see registration form). In case of hardship, an author can request a form to apply for registration fee waiver. Except for certain special requirements (e.g., color photographs etc.), no page charges will be imposed on the authors of papers in Food Microstructure.

PUBLICATION REQUIREMENTS & PROCEDURE

For oral presentation at the 1982 meetings, **a paper must be first submitted for publication in the Food Microstructure.** Each paper is intensely reviewed by at least three, and often more, referees. Therefore, an invitation or an acceptance of a Letter of Intent does not guarantee acceptance of the paper by its reviewers.

All authors (particularly those from outside of the United States) are urged to submit their Letters of Intent and papers as soon as possible. A paper number, and instructions to prepare the paper will be sent immediately upon the receipt of the Letter of Intent. Four copies of this paper, each with its glossy photographs, should be submitted in a regular typed format (on 8½x11" or similar size standard paper). After completion of the review phase, the authors will be required to supply their final manuscript in a camera-ready format on special model sheets which SEM will supply. In addition to all the text, the final manuscript will also have to contain author's publishable responses to questions raised by the paper's reviewers (see for example, discussion with reviewers in past SEM volumes).

The deadlines for submission are: (a) Letter of Intent — November 1, 1981, and (b) Paper — Jan. 15, 1982.

Since papers for publication in the Food Microstructure can be offered at any time, Letters and papers can be submitted after the above deadlines. All such papers will also be processed for publication; however, late submission of a paper will jeopardize its possibility for oral presentation. **Papers offered earlier in 1981 will be published in the first issue of Food Microstructure if the processing can be completed on time.**

Immediately upon the receipt of a paper intended for oral presentation, one of the organizers of this program will be asked to initially assess its probable acceptance. Following this assessment, **a letter regarding the oral presentation of the paper will be sent to the author.** Final decision on the publication of a paper will depend on its reviews.

All internal clearances and approvals as well as permissions to use any copyrighted material are the responsibility of the authors and should be so obtained that the complete paper can be submitted on time.

Publication of a paper in Food Microstructure will involve (on the part of the authors, reviewers and editorial staff) considerable effort. All authors should therefore be prepared to submit their papers, manuscript and responses to reviewers as described above. AUTHORS MUST NOT COMMIT THEMSELVES UNLESS THEY ARE CONFIDENT THAT THEY CAN MEET THE DEADLINES AND THAT THEIR WORK WILL STAND UP TO CAREFUL INTENSIVE SCRUTINY BY THEIR COLLEAGUES IN THE SCIENTIFIC COMMUNITY.

TRAVEL SUPPORT FOR 1982 MEETINGS

Authors offering tutorial or review papers may apply for travel support. Authors whose papers make significant contributions to 1982 program may apply for a **Presidential Scholarship.** Submit a **complete Letter of Intent** as described above and *include the extent of subsidy desired* (limited to $300 for travel within North America and $500 elsewhere). Scholarship applicants should also submit a recommendation letter (preferably from someone associated with SEM Inc. activities). The decision to support travel will be made in consultation with the organizers of this program following established guidelines.

The promise of travel support (to be confirmed in writing by Nov.-Dec.) will be automatically considered withdrawn, if (a) the full paper was not submitted on time (January 15, 1982), or (b) if the paper was not acceptable to reviewers or editors, or (c) if the paper was not presented by the person promised support.

SOME EXAMPLES OF SCANNING ELECTRON MICROSCOPY IN FOOD SCIENCE

R. J. Carroll and S. B. Jones

Eastern Regional Research Center
Agricultural Research, Science and Education Administration
U. S. Department of Agriculture
600 East Mermaid Lane
Philadelphia, PA 19118

Abstract

The scanning electron microscope was applied to a variety of problems involving agricultural research at our Research Center. The investigations described include: (a) host-pathogen interactions in potato, (b) cracking of maturing cherries, (c) changes in meat structure as a consequence of thermal and mechanical stress.

Each study required specific fixation, dehydration and drying procedures depending on the physical characteristics and the nature of the structural information desired.

The study of fungal interaction with potatoes required freeze drying as well as chemical fixation to preserve fungal-potato cell relationships. The studies on meat depended on maintaining fiber-connective tissue orientations that resulted from heating and tensile stresses. A minitensile stage stressed the samples, and a special clamping holder maintained the specimens in a stress position through all stages of sample preparation and observation. The cherry tissue integrity was maintained by use of freeze-drying techniques.

Application of scanning microscopy to food and food products often requires ingenuity and expertise to retain structural relationships and thus gain an insight into functional properties of these commodities.

Research supported in part by U. S. Army Natick Development Center Project AMXRED 73-161.

KEY WORDS: Potato, Phytophthora infestans, Cherry Cracking, Ethyl Oleate, Meat, Tensile Stress, Connective Tissue, Meat Tenderness, Muscle

Introduction

The use of the scanning electron microscope (SEM) in the determination of structural changes in a wide variety of food and food products is finding increased applications. This use is governed for the most part by the availability of reliable procedures for adequate preservation of structures of the food and food products under investigation. This preservation of food structure depends on (a) methods of isolation of specimens, (b) fixation, (c) dehydration, (d) physical characteristics of the specimen, (e) structure of interest, and (f) instrumentation conditions.

For most studies, the fixation of choice is glutaraldehyde or glutaraldehyde-paraformaldehyde (after Karnovsky)[1] at the temperature, osmolarity, pH, and buffer type appropriate for each food sample. Dehydration, usually with ethyl alcohol, followed by critical point drying with carbon dioxide, results in preservation of structure. At times, better structure retention is obtained by freeze drying, depending on the food specimen.

The cryofracture technique developed by Humphreys et al.[2] is one of the better methods which can be used to obtain internal surfaces with good structural features and which minimizes cutting artifacts and eliminates ice crystal damage.

As an introduction to this Workshop on "The Scanning Electron Microscope in the Food Sciences," we would like to present the results of our SEM research which give information on the structures of selected foods. These investigations include: (1) host-pathogen interactions in potato tuber and leaf involving the fungus Phytophthora infestans, (2) use of ethyl oleate to reduce cracking of sweet cherries, and (3) effects of thermal and mechanical stresses on meat texture.

Each of these studies required a different approach to preserve food structures to obtain the desired information with the scanning electron microscope.

Instrumentation and Methods

Details of each sample preparation procedure are discussed under each subject. Some samples were freeze-dried, others were fixed, dehydrated in ethyl alcohol and critical point dried; then the samples were mounted on copper stubs. A JEOL 50-A* scanning electron microscope, operating at 10-20 kV, was used in these investigations.

Fungal Interaction with the Potato

The fungus *Phytophthora infestans* causes Late Blight in potato plants. In the field, fungal spores fall onto the soil and infect tubers through wounds or natural openings. Tubers then rot while in storage or in market. Late Blight is thus both a field and a post-harvest agricultural problem.

Certain varieties of potato are resistant to one or more races of the fungus. Our research has been directed toward investigating the ultrastructural differences between the fungus interacting with susceptible and resistant cultivars and complements other research in our Center on the biochemistry of stress metabolites produced in cells of potato plants. We have documented structural aspects of the growth and development of *P. infestans* on both susceptible and resistant tubers.[3] Our current work utilizes both transmission electron microscopy (TEM) and SEM for the examination of leaves and tubers.

SEM is ideal for the examination of surfaces of leaves and tuber slices that have been inoculated with sporangia or zoospores of *P. infestans*. Inoculated leaves were incubated at 12 C and sampled at time intervals ranging from a few hours to four days, at which time growth at the leaf or tuber-slice surface was easily visible without magnification. Samples of leaf were removed and fixed in 3% glutaraldehyde in 0.025M sodium cacodylate, pH 6.0, for microscopic investigations.

Specimens for SEM examination were dehydrated in ethanol and critical point dried. Fig. 1 shows the surface of a tuber slice with fungal hyphae growing across the cells. Starch granules can be seen on the floor of the tuber cells. Most of the starch was washed out before inoculation to make visualization of the fungal structures easier.

Both zoospores and sporangia of *P. infestans* can germinate, depending on conditions of temperature and humidity. Fig. 2 shows a sporangium which has germinated on a leaf surface. The germinating hypha extends across an open stomata. This germination structure will not necessarily infect the leaf, because the ability of the structure to penetrate tuber cells depends on a specific interaction between host and pathogen. On the leaf surface, hyphae emerging from sporangia have been observed to grow across the surface. We have not observed any hyphae entering stomata or any conclusive observations up to this time of direct penetrations of the cuticle. Infections must occur in leaves thus inoculated because after approximately four days the zoosporangiophores (the sporangia-bearing hyphae) emerge from stomata (Fig. 3). Each zoosporangiophore terminates in a characteristic bulb-like structure. Sporangia easily disengage and can be seen resting on the leaf surface in the micrograph. Only after the fungus has ensured the completion of its life cycle does the destruction of the leaf commence. Then the leaf "withers and dies."

Cherry Surface Alteration

SEM provides a way to view the surfaces of raw agricultural products. One example of such an application is a study in which cherries were examined for alterations in surface wax.

Sweet cherries in the orchard often crack just before harvest. The cause has been shown to be water absorbed through the cuticle,[4] with much damage occurring within a few hours after a rainstorm.

Incidence of cracking can be reduced by spraying the fruit with a water emulsion of ethyl oleate (EO).[5] Fruit treated with EO has been examined to determine what structural changes have occurred. Tangential sections (5x5x2) of untreated and EO-treated cherries were cut with a razor blade. Specimens were rinsed with deionized water to remove juice, frozen in liquid nitrogen, and dried overnight in a vacuum evaporator. Fixation was omitted because the fixative containing glutaraldehyde and paraformaldehyde produced artifacts in the waxy coating of the cherries. Surface wax in untreated fruit formed a continuous layer over the cuticle which sometimes partially or wholly obstructed stomata (Fig. 4). Cherries which had been dipped in 4% EO appeared to have little or no wax on the cuticle or around stomata (Fig. 5). The surface was similar to that of a cherry which had been dewaxed with chloroform. Cherries sprayed with EO in the field showed no visible effects over most of the surface examined. However, on some sprayed samples, surface wax appeared to have been moved about and formed into flow patterns clearly different from the usual appearance of the wax (Fig. 6). The effect of the EO may be to redistribute the wax on the surface, permitting enhanced vapor exchange in certain areas of the cuticle. It is unlikely that spraying with EO removed wax clogging stomatal pores because, even after the cherries were dipped in EO, wax in stomatal opening often remained and appeared to be more resistant to solubilization than cuticular wax.

Meat Structure

An important desired characteristic of meat is tenderness, a quality easy to determine at mealtime but difficult to predict by any objective measurement. Much information is now available concerning the factors contributing to toughness or tenderness. Toughness can be separated into two components: one arising from

---

* Reference to brand or firm name does not constitute endorsement by the U. S. Department of Agriculture over others of a similar nature.

Examples of SEM in Food Science

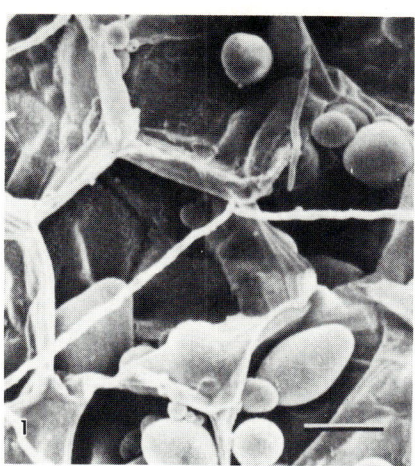

Fig. 1 Potato tuber cells with hyphae of Phytophthora infestans. Some starch granules are present. Scale marker (S.M.) = 40 μm.

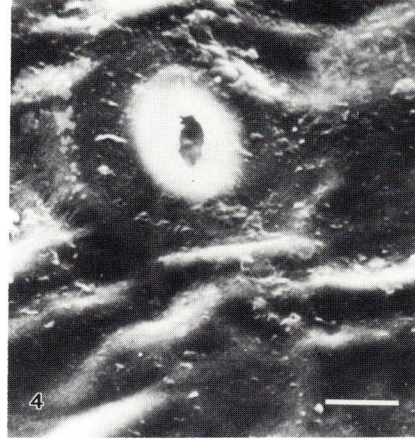

Fig. 4 Surface of cherry fruit. Wax covers the cuticle surface nearly closing stomata. S.M. = 10 μm.

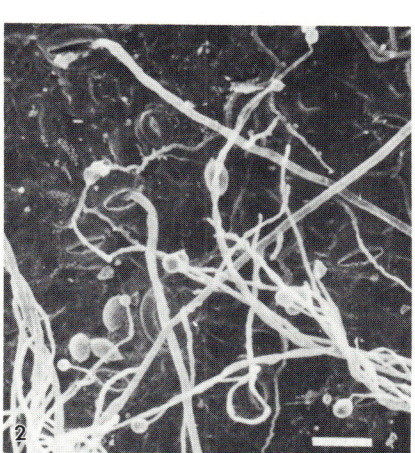

Fig. 2 Germinating hypha of Phytophthora infestans on potato leaf. S.M. = 12.5 μm.

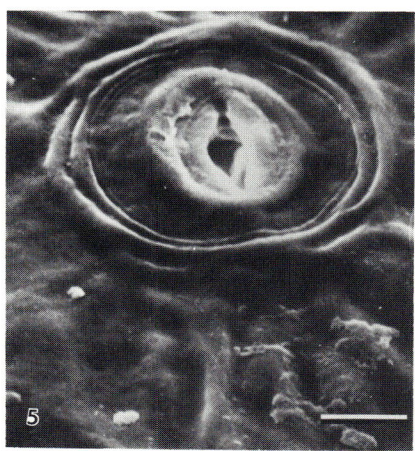

Fig. 5 Cherry treated with 3% EO in laboratory. Most wax removed. S.M. 15 μm.

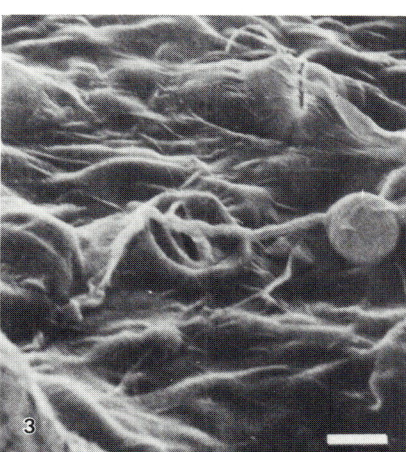

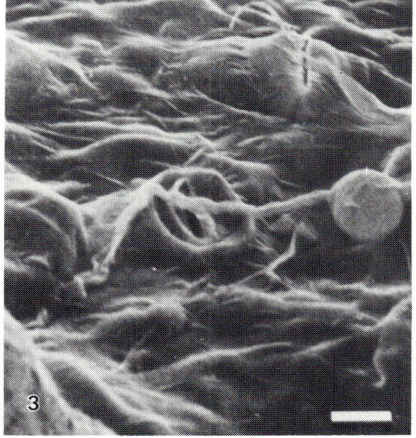

Fig. 3 Potato leaf surface with hyphae exiting from stomata. S.M. = 33 μm.

Fig. 6 Surface of cherry sprayed with 3% EO in the field. Wax formed flow pattern. S.M. = 20 μm

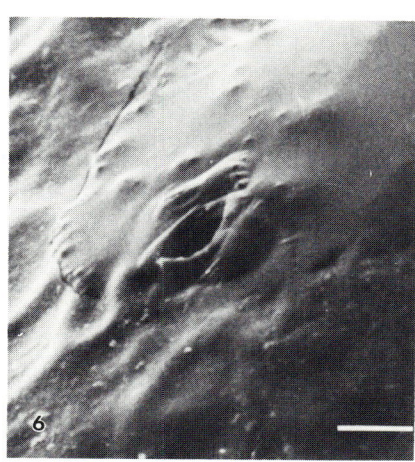

actomyosin interaction, and the other from background or connective tissue contributions.[6]

In an effort to elucidate the structural bases of tenderness and/or toughness and, hopefully, to find a foundation for an objective measurement of tenderness, scanning electron microscopy was used to examine alterations of meat structure as influenced by heating[7] and applied tensile stress.[8] Bovine semitendinosus (eye-of-round) was obtained either commercially or from aged hanging carcasses. A standard sample preparation procedure was developed to minimize introduction of surface artifacts.[9] The preferred fixative was a solution which contained 2.5% glutaraldehyde and 2% paraformaldehyde buffered in 0.07M phosphate to the pH of the sample.[1] Meat samples were clamped to prevent contraction during fixation. Ethyl alcohol dehydration, coupled with the cryofracture technique of Humphreys,[2] followed by critical point drying, gave the best preservation of meat structures. For cooked meat studies, 6 mm thick slices of the meat were placed in water within a polyethylene bag, and heated in a water bath at the specified temperature for 45 minutes.

A special tensile stage was designed and built to carry out dynamic stress studies. Stressed samples were placed in a clamping holder for observation in the SEM. All of the additional specimen preparation steps (fixation, dehydration, and critical point drying) were performed with the stressed tissue in the clamping holder. All stress experiments were monitored by a stereo microscope equipped with a video recording system. Specific areas of the stressed tissue were compared with similar areas of tissue examined with the SEM.

Fig. 7 shows in cross fracture the main components of meat tissue. The fiber bundle is surrounded by a perimysial connective tissue (P). The endomysium (E) is the sheath-like connective tissue surrounding the individual muscle fibers.

Structural changes in meat tissue heated to 50, 60, and 90 C were determined.[7] The structures examined included the sarcomere, endomysial connective tissue, and the sarcolemma. In Fig. 8 the tissue which had been heated to 50 C appears similar to unheated meat, with the fibers well separated and the endomysium free. The tissue which had been heated to 60 C became compact (Fig. 9). The fracture plane passed through and around the fiber, and the endomysium was in close contact with the muscle fibers accompanied with the deposition of particulates. Heating to 90 C produced most drastic changes. The endomysium was quite congealed, and the sarcolemma was granular.

Heating to 50 and 60 C did not alter the sarcomere length as compared to that of unheated tissue. The sarcomeres were about 1.9 µm long. In tissue heated to 60 C, congealed material was observed, particularly at the Z-line region. In tissue heated to 90 C, however, the sarcomeres were shortened to about 1.4 µm, the Z-line material was detached, and particulates were observed at the Z-line (Fig. 10). Denaturation of the connective tissue, accompanied by the formation of rigid protein blocks, corresponds respectively to the connective tissue and the actomyosin complex as they influence meat tenderness. The changes in the endomysium at 60 C (denaturation of connective tissue) probably help to make the meat more tender. Conversely, the shortening of the sarcomeres and the "rigid blocks" of the actomyosin complex probably make the meat less tender. The ultimate tenderness in meat is in part the result of opposing changes in the two major meat components. These observations of the changes induced in meat tissue structures may help in the elucidation of their effects on meat tenderness.

Tensile stress, both parallel and perpendicular to the meat fiber axis, was applied to unheated and heated meat tissue, to ascertain the weakest and the strongest structures in meat.

The muscle tissue samples were stressed on a ministage (Fig. 11) and observed with a stereo microscope. Applications of tensile stress parallel to the fiber axis caused initial rupture of the endomysium muscle fibers accompanied by strand formation. Fig. 12A shows the strands as observed with a light microscope, and Fig. 12B shows a similar area observed with the SEM. In the lower part of the micrograph, the muscle fibers have ruptured and the strands appear taut, while some of the strands have broken and coiled back. The strands appear to originate from the perimysial connective tissue.

Raw and heated (90 C) meat samples were stressed either parallel or perpendicularly to the fiber axis almost to rupture. Interior surfaces of the tissue adjacent to the rupture point were obtained by the cryofracture technique. Fig. 13 shows the alignment of the perimysial connective tissue fibers in uncooked meat after parallel stress was applied. Similar results were obtained with heated tissue. Fig. 14 shows the effect of stress perpendicular to the fiber axis in tissue heated to 90 C. The initial rupture occurred at the junction of the endomysial-perimysial connective tissue.

Orientation of the strands originating from the perimysial connective tissue took place after application of considerable stress. The apparent elasticity of the strands resulted from the orientation of the random connective tissue fibers which took up the stress after rupture of the endomysial-perimysial connective tissue junction.[10] When perpendicular stress was applied, the muscle fiber-endomysium complex was the stronger structure. These strands were identified as collagen with light microscopy staining techniques. The strands were birefringent when viewed under polarized light. Small amounts of elastin were also observed in association with blood vessels.

Transmission electron microscopy of dispersed perimysial connective tissue showed a random network of collagen fibers with the typical banding pattern. Perimysial connective tissue heated to 90 C and viewed in the TEM was highly denatured as expected. Approximately 10% of the fibrils, however, still retained the banding pattern of native collagen. This result was unexpected since collagen denatures in the range of 58-63 C.[11]

The tensile properties of meat tissue depend considerably on the nature and extent of the perimysial connective tissue and its reaction to

Examples of SEM in Food Science

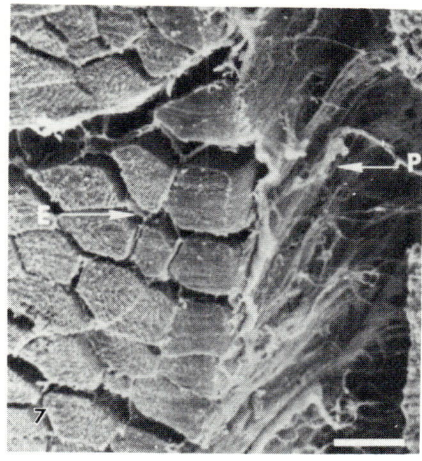

Fig. 7 Raw bovine semitendinosus. Fibers are surrounded by endomysial (E) collagen. The perimysium (P) encloses fiber bundle. S.M. = 33 μm.

Fig. 10 Sarcomeres of meat heated to 90 C. Z-disc (Z) material detached from sarcomere. Protein block broken away at unmarked arrow. S.M. = 1 μm.

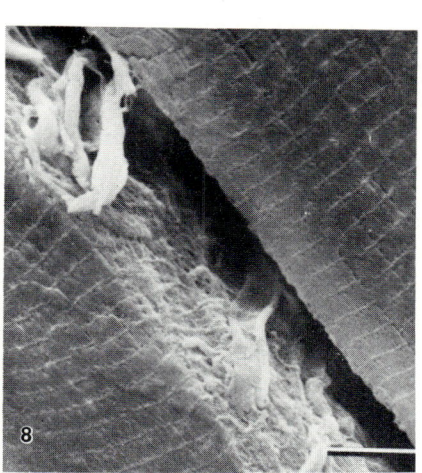

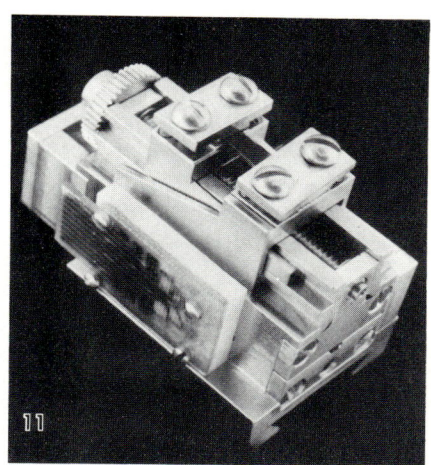

Fig. 8 Endomysial collagen between muscle fibers heated to 50 C shows no change from raw control. S.M. = 4.5 μm.

Fig. 11 Minitensile stage designed to operate outside the SEM.[8]

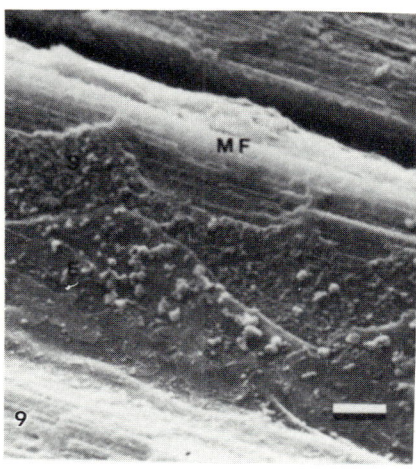

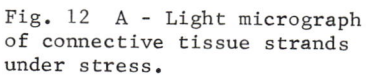
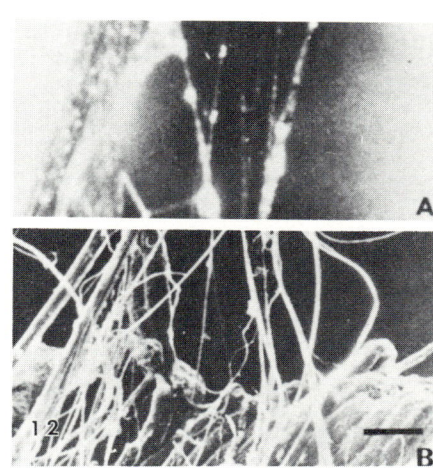

Fig. 9 Heating at 60 C endomysium (E) congealed and sarcolemma (S) became granular. Myofibrils (MF) remained intact. S.M. = 5 μm.

Fig. 12 A - Light micrograph of connective tissue strands under stress.
B - SEM of area similar to A, strands under stress and ruptured muscle fibers. S.M. = 45 μm.

5

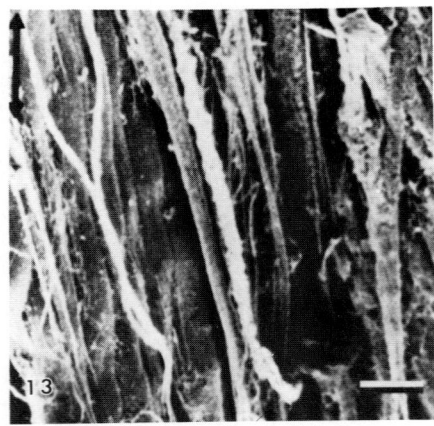

Fig. 13 Fractured interior surface of raw semitendinosus stressed parallel to fiber axis. The perimysium fibers are oriented in the direction of applied stress. (Arrow upper left.) S.M. = 30 µm.

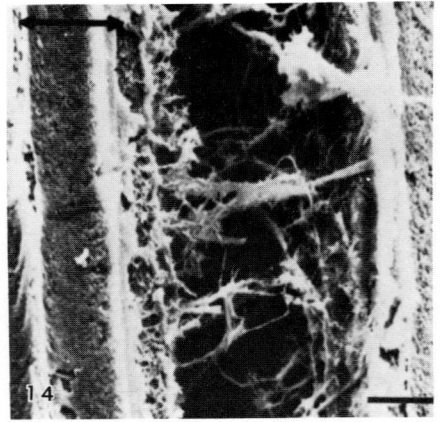

Fig. 14 Fractured interior surface of heated (90 C) semitendinosus stressed perpendicular to the fiber axis. (Arrow upper left.) Stress on the perimysium fibers is evident; muscle fibers undisturbed. S.M. = 20 µm.

thermal stress. Such tensile properties, in turn, have a significant effect on both the subjective and objective evaluation of meat tenderness.

Future Directions

The strength of SEM for food science studies is its magnification range and large depth of field. The images of SEM often give information directly relating to textural properties of foods in a unique way and should be useful in assessing and predicting consumer appeal and in developing processed foods.

Operationally, improvements should be developed in techniques of sample handling which will preserve and protect the features of the viewed surface. Cryofracture techniques are useful and can be used routinely for the examination of a matrix structure such as muscle.

A potentially useful technique is low-temperature SEM of unfixed, frozen biological specimens. This technique avoids the drawbacks of chemical fixation, dehydration, and drying. Although different artifacts are introduced, cryofreezing may give an insight into structure determinations of lipids and lipid-protein systems that are difficult to preserve by conventional procedures. The cryo technique will also serve as a check on possible artifacts introduced by normal procedures. The use of a cryofracture device will expose additional sample surfaces for characterization at different levels within the sample without thawing. Use of a metal deposition device will permit specimen observation at higher accelerating voltages while minimizing specimen charging.

The use of energy dispersive X-ray analysis in conjunction with the SEM should provide needed information regarding chemical changes in food and food products associated with aging, processing, packaging, and other variables. This powerful technique gives elemental information related to chemical changes in food tissues. Combined with the SEM structural observations, it should permit food investigators to obtain a more comprehensive characterization of food and food products.

References

1. M. J. Karnovsky. A formaldehyde-glutaraldehyde fixative of high osmolarity for use in electron microscopy. J. Cell Biol. 27, 1965, 137A.
2. W. J. Humphreys, B. O. Spurlock, and J. S. Johnson. Critical point drying of ethanol-infiltrated, cryofractured biological specimens for scanning electron microscopy. SEM/1974, IIT Research Institute, Chicago, IL 60616, 275.
3. S. B. Jones, R. J. Carroll, and E. B. Kalan. A scanning electron microscope study of the host-pathogen interaction of Phytophthora infestans with potato tissue. SEM/1974, IIT Research Institute, 397-404.
4. D. C. Davenport, K. Uria, and R. M. Hagen. Antitranspirant film: curtailing intake of external water by cherry fruit to reduce cracking. HortScience 7, 1972, 507-508.
5. W. O. Harrington, C. H. Hills, S. B. Jones, et al. Ethyl oleate sprays to reduce cracking of sweet cherries. HortScience 13, 1978, 279-280.
6. B. B. Marsh and N. G. Leet. Studies in meat tenderness. 3. The effects of cold shortening on tenderness. J. Food Sci. 31, 1966, 450-459.
7. S. B. Jones, R. J. Carroll, and J. R. Cavanaugh. Structural changes in heated bovine muscle: scanning electron microscope study. J. Food Sci. 42, 1977, 125-131.
8. R. J. Carroll, F. P. Rorer, S. B. Jones, et al. Effect of tensile stress on the ultrastructure of bovine muscle. J. Food Sci. 43, 1978, 1181-1187.
9. S. B. Jones, R. J. Carroll, and J. R. Cavanaugh. Muscle samples for scanning electron microscopy: preparative techniques and general morphology. J. Food Sci. 41, 1976, 867-873.

10. D. W. Stanley, L. M. McKnight, W. G. S. Usborne, et al. Predicting meat tenderness from muscle tensile properties. J. Text. Stud. 3, 1972, 51-68.
11. P. E. McClain, G. J. Creed, E. R. Wiley, et al. Effect of postmortem aging on isolation of intramuscular connective tissue. J. Food Sci. 35, 1970, 258-259.

Discussion with Reviewers

R. G. Cassens: Does SEM of foods result in any well known or easily recognized artifacts?
Authors: Yes, food samples for the SEM require extreme care in all stages of sample preparation. One of the more common artifacts is shrinkage. Clamping of the meat tissue during fixation and dehydration kept shrinkage to a minimum. In addition, the loss of lipid from fat cells in meat samples processed by critical point drying is observed. Adequate metallic coating of the sample is essential to minimize charging in the SEM.

J. F. Chabot: Did you identify a structural component to resistance or susceptibility to infection by P. infestans?
A-M. Knoop: Did you find structural differences between races of fungi or between susceptible and resistant potatoes?
Authors: Work on structural relationships and host-pathogen interactions is being actively studied in our laboratory. In TEM studies of infected susceptible tuber cells, new structural features were observed in the region where cell penetration occurred. The significance of these structures and their frequency in resistant, as well as susceptible interactions, is not clear at this time.

J. F. Chabot: Why were penetrations of the leaf cuticle by hyphae not seen?
A-M. Knoop: Have you an explanation for the fact that no entering of hyphae into the stomata and into the cuticle of leaves was to be seen in your SEM work? Was the number of zoospores too small or were the intervals of sampling too large?
Authors: Penetration through the cuticle was expected, and it is not clear why it wasn't observed. It is possible that the geometry of direct penetration makes it difficult to detect by SEM. In general, entrance via stomata is much more readily documented although no such penetration was observed in this case. In this study, sporangia rather than zoospores were used; these sporangia germinate and presumably penetrate. In order to answer these important questions, work is continuing using both sporangia and zoospores and other sampling intervals.

J. F. Chabot: Is there other evidence that wax redistribution is important in preventing cracking of cherries in light of the lack of a major change observable in the SEM? What is the effect of ethyl oleate in stomatal function?
Authors: This is the only evidence involving rearrangement of surface wax and is offered as suggestive, not conclusive. These findings are subject to the qualifications imposed by random sampling as well as random deposit of spray on cherry surfaces. Nevertheless, the wax flow patterns were observed only on sprayed cherries. Although stomatal function was not measured in this study, the microscopy observations indicated that EO application by spraying had no effect on stomatal wax. Text reference 9.

R. J. Cassens: Why does heating to 90 C cause shortening of sarcomeres?
Authors: Application of heat to meat tissue causes (a) unfolding of peptide chains, (b) aggregation of protein molecules, (c) release of water, and (d) increased rigidity of physical structure. The combination of these factors results in a decrease in sarcomere length. (E. Laakkonen. Factors affecting tenderness in meat. Advances in Food Research, C. O. Chichester, Ed. Academic Press, New York, N. Y. 1973. 20, p. 313)

A-M. Knoop: In which way do the heat induced changes of the meat structure influence the meat tenderness?
Authors: When meat is heated above 60 C, the connective tissue components begin to denature, making the meat more tender. As the temperature is increased above 60 C, the induced changes in the actomyosin components (denaturation, aggregation, and loss of water binding capacity) cause the formation of "rigid blocks" making the meat less tender. The relative amounts of connective tissue, method of cooking, origin of muscle, age of animal, etc., all contribute to the ultimate toughness and/or tenderness of the meat tissue.

D. N. Holcomb: Can the authors estimate the magnitude of forces developed with their mini-stage? Microstructure-force correlations might be important.
A-M. Knoop: How large was the stress applied to the muscle fibers? Was the effect of stress different for heated and raw meat and how large were the differences?
Authors: Strain gauges were not incorporated into the minitensile stage. Therefore, it is not possible to determine the magnitude of the stress or microstructure-force relationships. The stage is driven by a D.C. power supply and the current was held constant as the voltage was increased to apply the stress at a more or less constant rate. A rough estimate of breaking force can be obtained from the relative voltage required to break the sample. In raw meat about twice the applied voltage was required to rupture parallel to the fiber axis than perpendicular to the fiber axis. Heated meat samples required half the voltage required to rupture raw meat in both instances. These estimates are in agreement with unpublished Instron tensile results on larger samples of unheated meat (cross-section = 6.35 mm x 12.7 mm). The tensile strength averaged 10.7 gm/mm^2 for parallel and 5.2 gm/mm^2 for the perpendicular stressed sample.

Additional discussion with reviewers of the paper "Preparation of Food Science Samples for SEM" by J. F. Chabot continued from page 16.

S. H. Cohen: Although meat is the predominant tissue of interest from the viewpoint of human consumption, fish is also very important. Therefore, what preparative techniques might be used for fish?
Author: In general several different procedures should be used, when practical, on any material in order to appreciate the types of alterations that might be induced. As an animal tissue with protein as a major constituent, conventional fixation, dehydration, and critical point drying would provide the opportunity to compare SEM work with published work on fish muscle structure using TEM. Cryofracturing has been used with success on mammalian muscle and is also an alternative.

K. Saio: I agree with your opinion that the best treatment is no treatment for SEM observation. However, the foods which contain a large amount of lipid are not easy to coat without fixation. Don't you have something advisable for such like foods?
Author: In my experience freeze drying does not yield stable preparations of cheese and fixation may be necessary. In cheeses with a high lipid content, lipid extraction, as extreme as it seems, may be necessary and has yielded useful information on the protein matrix.

D. N. Holcomb: It is indicated that both critical point drying and freeze drying cause "major alterations" in cheese structural detail. Are there available micrographs of "unaltered" cheese, or how is it ascertained that major alterations have occurred?
Author: Workers on cheese often employ fixation and extraction procedures before freeze drying or critical point drying. The solvents used in critical point drying can extract lipids. Kalab[41] illustrated some types of artifacts produced by freeze drying in milk products. These were especially noticeable in products such as yogurt that contained added starch. There are few available micrographs on unaltered ripe cheese because of the difficulty of working with lipid-rich material.

PREPARATION OF FOOD SCIENCE SAMPLES FOR SEM

J. F. Chabot

Section of Ecology and Systematics
Langmuir Laboratory
Cornell University
Ithaca, NY 14850

Abstract

Food scientists encounter many of the same problems as biologists in preparing material for scanning electron microscopy. Several methods for drying biological samples are in common use: air drying, freeze drying, and fixation and chemical dehydration followed by critical point drying. Many food samples can be satisfactorily examined after air drying, since this is their natural state. Freeze drying has gained acceptance in food preparation, and is therefore also of merit. However, methods that require fixation and alcohol or acetone dehydration must be approached with caution. Aqueous fixatives alter the structure of most low-moisture foods. Conventional fixation preserves protein and lipid; most plant-derived foods have abundant carbohydrates that are integral structural components, and these may be destroyed during chemical dehydration. Critical point drying is useful on these materials primarily to understand the type of changes that occur in preparation for transmission electron microscopy. Freeze drying also produces artifacts, mainly ice crystal damage, but since the damage is so obvious there is less temptation to misinterpret the results.

KEY WORDS: Food, Starch, Meat, Biological Specimen Preparation, Bread, Dehydration

Introduction

Food scientists work with an array of materials many of which are biological in origin. Biologists have the goal of understanding the living state. The goal of the food scientist is to understand structural features of a material that are important in its functional role in a food. Despite rapid advances in instrumental design of scanning electron microscopes persons working with material of biological origin find their work and resolution limits set by specimen preparation problems. Biological substances in general present two problems for scanning electron microscopy (SEM)--water must be removed without destroying structural relationships and the specimen must be made conductive.

Food science samples run the gamut of biological material from plant to animal, from materials as hard as eggshells to as formless as unbeaten egg white, from dry seeds and bran to liquid cake batter, from microbes to loaves of bread. A small proportion of specimens are nonbiological, such as filters used in ultra-filtration and purification studies. Because of the diversity of material encountered in a food science electron microscopy facility, it is necessary to understand the specimen preparation methods thought best by specialists working on a variety of biological tissue types. Running every sample through a routine procedure such as fixation, dehydration, critical point drying as used in a laboratory concentrating on animal tissue is not a reasonable practice. On the other hand when a food scientist does go into the literature on biological specimen preparation, immediately he or she encounters a variety of good but highly critical reviews of each and every method[1,2,3]. The commonly used procedures of dehydration and coating have been thoroughly presented as tutorials to illustrate their use, and thoroughly criticized as to the artifacts they produce.

The aim of this review is to examine some of the methods used in food research and to point out some of the problems encountered and results obtained, particularly with reference to fixation and dehydration procedures commonly used for both scanning and transmission electron microscopy. Reviews on the general use and theoretical aspects of biological specimen preparation are available[3,4,5,6]. Pomeranz[7] provides an introduction to the use of SEM in food science and an indication of areas where it is proving useful. The focus here will be on methods of preparation as they affect the results.

A large number of foods are derived from microorganisms, plant and animal tissues with little or no processing and have organelles, cells, cell walls, and tissue structure corresponding to the living state. A second category of foods are those derived from biological material such as seeds and grains that may be in a dormant state. Many cereal products are characterized by low moisture content at many periods during processing into food items. A third general category of foods for the purposes of this review are those that are liquids or gels. Examples of preparation procedures for each type will be presented, with an emphasis on the latter two.

Methods of Drying

Air Drying

Complicated methods of removing water from soft specimens that require chemicals or specialized equipment have been developed all to the end of avoiding air drying. At the air-water interface forces develop that lead to shrinkage, flattening, and alteration of surface details. However, in a food science laboratory air drying is often the method of choice, not because the surface distortion is unimportant, but because this is the natural state of many food items. The same applies to even more drastic treatments such as oven, drum, or spray drying. The SEM is put to its most straightforward use when used to examine food that is normally dry. This includes items such as grains, flours, starch granules, and processed products that are dry in their final form. If the problem being investigated is to evaluate the structure of such materials, no artifacts are being introduced by the microscopist.

Freeze Drying

Freeze drying avoids a liquid-air interface by placing a frozen specimen under vacuum until sublimation has resulted in water removal. Two steps in freeze drying are considered important [5,6]. The first is rapid freezing; the second is that the temperature of the material must be kept below the recrystallization point of ice during the sublimation process. The importance of rapid freezing to reduce ice crystal damage has been reviewed with particular detail to the freeze-etching technique. High enough freezing rates can be practically obtained only with very small samples. For freeze etch of larger samples a cryoprotectant is often used, an option that is not acceptable for SEM. Under most conditions some ice crystal disruption of specimens that are of the size used for SEM is inevitable. Freeze driers have become essential equipment items in many food science laboratories because of this method's superiority in product processing. Despite its drawbacks, freeze drying is often a feasible method for SEM preparation.

Critical Point Drying

Critical point drying theory and practice are the subjects of a number of reviews[4]. This method has resulted in impressive micrographs of the surface structure of delicate cells. Severe shrinkage problems have become recognized[8], but the method remains popular because the images are more satisfying than samples that have the types of distortion produced by air or freeze drying. For studies correlating SEM and TEM images, this is the method of choice because the tissue is put through the same series of fluids for most of the preparation.

Miscellaneous Methods

Other drying methods such as air drying or freeze drying from a volatile solvent[9] or cryofracturing followed by critical point drying[10] are also used occasionally.

SEM of Limited Water Systems

Bread

Many important food items derived from cereals have a low moisture content. Bread and other wheat products are such an important part of the diet that these materials have long been the subject of chemical and microstructure studies. Wheat kernel structure, flour components, and dough formation have been studied using light microscopy, transmission electron microscopy of thin sections, and scanning electron microscopy[11,12,13,14]. Only wheat flour produces the type of loaf that is now expected for bread. Both gluten film forming proteins and starch granule size and morphology are important, since substitution of other starch or proteins drastically alters crumb structure. An understanding of the precise functional relationships between all the components in a loaf of bread is important not only from a need for basic knowledge of foods, but also in aiding attempts at adding other nutrients such as protein supplements, or increasing the fiber content[15,16].

The complex interaction between two types of starch grains, and proteins, lipids, yeast, and water during the conversion of flour to dough involves changes in morphology of protein bodies, from spherical structures in the seed, to elongate strands of protein after hydration. Mechanical dough development, with stretching of gluten strands over starch and baking with expansion of gases, produces air cells lined with a nearly continuous film of protein.

Bread provides an excellent system for testing methods of preparation for the scanning electron microscope for many food samples of low moisture, high starch content, because some of the changes are so obvious. The types of artifacts that result when flour, dough, and bread samples were prepared for light microscopy were recognized long ago[17]. Recent work using electron microscopy has reaffirmed problems in investigating dough structure[18,19,20,21]; there have been few published pictures on bread structure.

If an aqueous fixative is employed in hopes of stabilizing structures for dehydration (and embedding for TEM), immediate changes occur.

White bread cubes, transferred to a vial of glutaraldehyde or glutaraldehyde/formaldehyde solutions, or plain water, swell rapidly, imbibing water. Starch in particular can be seen to swell and separate from the matrix. These changes can be seen with the unaided eye and with wet mounts for light microscopy. The separation of starch and protein is also well known from reconstitution experiments. To separate starch and gluten a dough ball is washed under a stream of water. Starch washes out, leaving the gluten.

Figures 1-7 illustrate the structure of white bread prepared by several commonly used methods. Surfaces were exposed by tearing or fracturing dry pieces, which were mounted on aluminum stubs using double-sided tape or small amounts of silver paint. After sputter coating the samples were examined in an AMR 1000 at 10 or 20 kV. The source of bread was a commercial white sponge type used because of the uniformity of air cells.

Bread that was oven, air, freeze, or vacuum dried had air cell walls with a nearly continuous gluten film coating starch granules (Figs. 1-3). Fixation with glutaraldehyde and postfixation with osmium tetroxide in buffers, followed by several methods of dehydration all resulted in profound changes in structure of bread crumb. Gas cells were still recognizable in most areas of samples, but walls were distorted. Most important was the fragmentation of the gluten film with the formation of strands and the exposure of starch granule surfaces. All material that had been exposed to ethanol, whether air dried, freeze dried, or critical point dried, showed a similar disruption of structure. Bread fixed and postfixed but freeze dried from distilled water was different (Fig. 7) with a more open, less condensed, structure. Protein was filamentous. Starch granules were larger and had more wrinkles in the surface.

The conclusion from a study of fixation conditions on bread morphology was that procedures commonly used for TEM caused profound changes. These changes were most important in precisely the area that seems to have attracted the most attention, the interaction between starch and protein. Most interpretations of such interaction, whether in bread, dough, or the grain, are probably invalid. Fixation induced significant changes in protein. Starch granules increased in size in aqueous solvent, and decreased past the original size in ethanol dehydration steps. The starch protein connections that were found at the end of treatment are largely the result of shrinkage during alcohol dehydration and are not uniform.

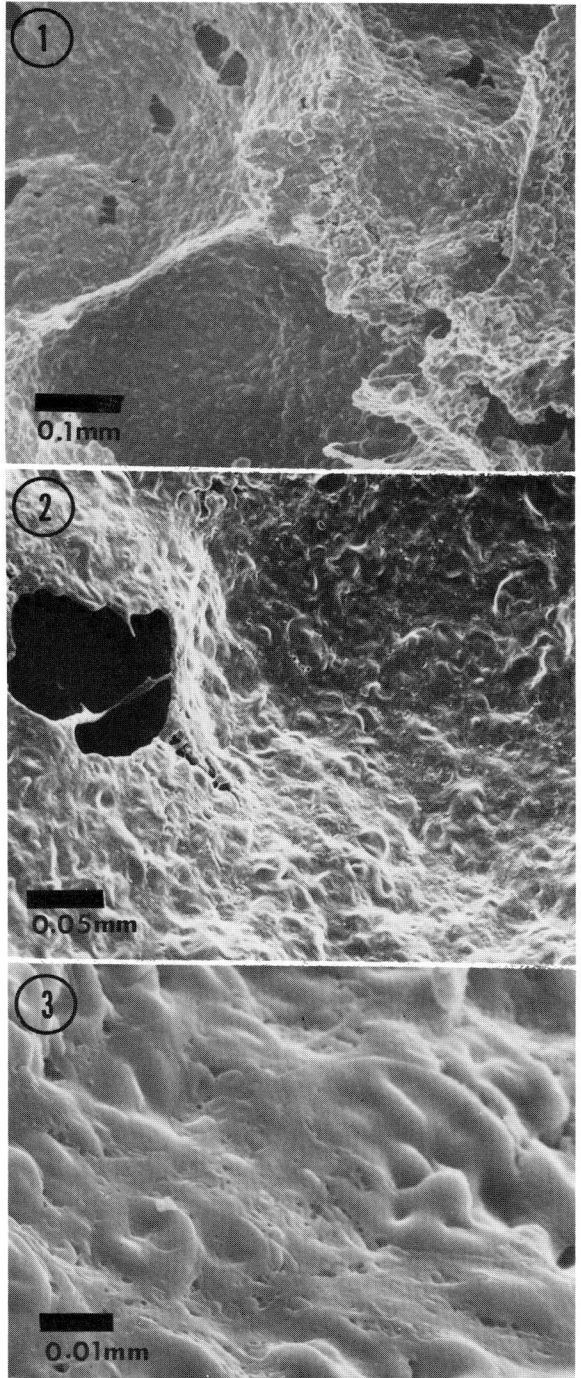

Figs. 1, 2. Fresh bread, surface exposed by tearing, mounted with silver paint, placed in vacuum for 5 min, sputter coated with gold/palladium. Air cells have a nearly continuous film. Silhouettes of starch granules are seen under the gluten film. Fig. 1 shows several air cells and sections of thin walls. Fig. 2 is a higher magnification illustrating continuous film forming coating of granules and making up surface of air cell wall. Large and small holes connect one cell to another.

Fig. 3. Bread frozen in liquid N_2, freeze dried. High magnification shows air cell wall with gluten film coating starch grains of several sizes. Small holes are present.

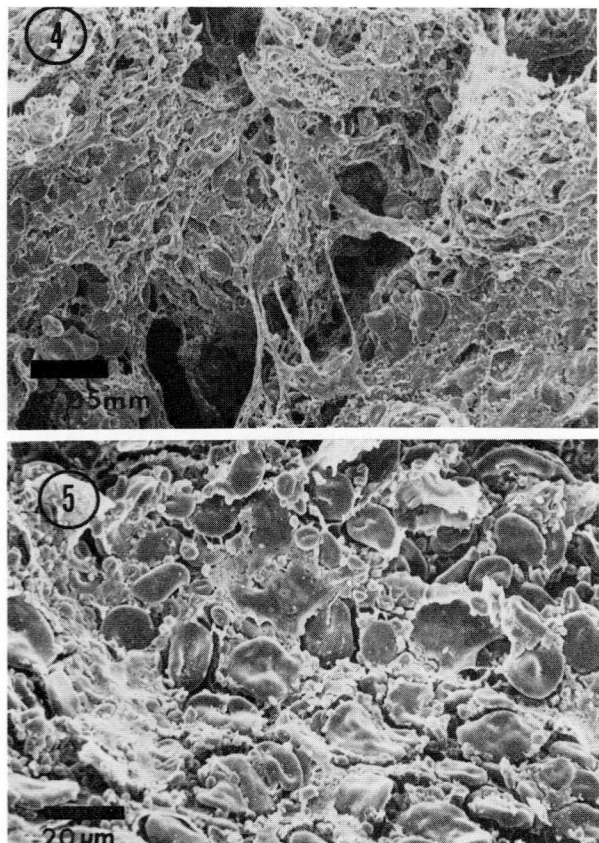
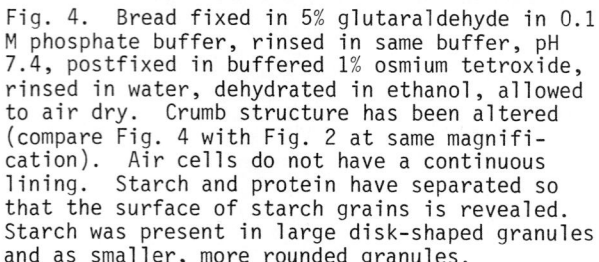
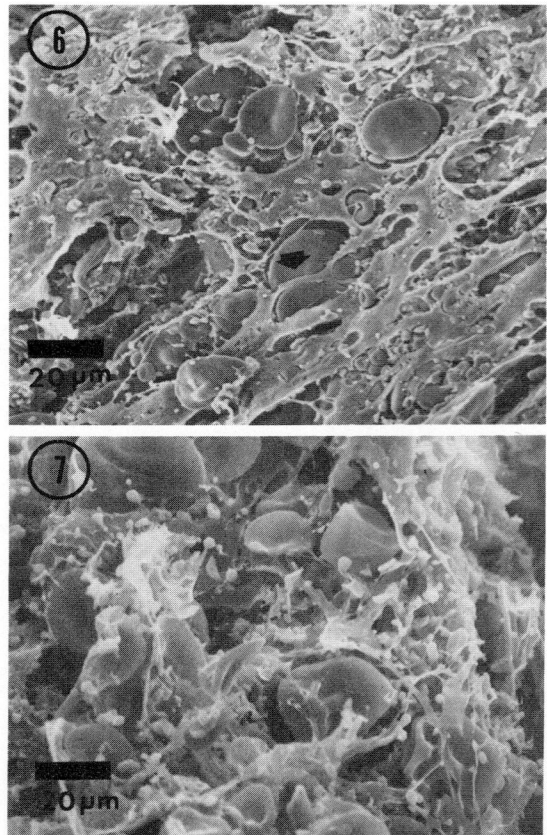

Fig. 4. Bread fixed in 5% glutaraldehyde in 0.1 M phosphate buffer, rinsed in same buffer, pH 7.4, postfixed in buffered 1% osmium tetroxide, rinsed in water, dehydrated in ethanol, allowed to air dry. Crumb structure has been altered (compare Fig. 4 with Fig. 2 at same magnification). Air cells do not have a continuous lining. Starch and protein have separated so that the surface of starch grains is revealed. Starch was present in large disk-shaped granules and as smaller, more rounded granules.

Fig. 5. Bread fixed in glutaraldehyde and osmium tetroxide and dehydrated in ethanol as above, but critical point dried, using ethanol/CO_2. The same distortion of crumb structure and separation of protein and starch components were noted.

Fig. 6. Bread fixed as above, dehydrated in ethanol and freeze dried, is very similar to Fig. 5. Note strands of protein over large starch granule surface (arrow).

Fig. 7. Fixed and postfixed as above but not exposed to alcohol. Freeze dried from distilled water rinse. Protein was divided into fine filaments that formed an open weblike structure. Starch granule surfaces were slightly wrinkled. Crumb structure was much more open and spongy than in ethanol dehydrated specimens.

Rather than assume a totally pessimistic note on the goal of obtaining high resolution micrographs on dough and bread structure it should be pointed out that the experimentally fixed samples reveal starch granule morphology much more clearly than the samples never exposed to water. As long as artifacts are clearly understood and reproducible, methods can be used for the information that is provided. What this study should reinforce is that for many cereal products, the least preparation is the best. Samples that are not moist cannot be treated with aqueous fixatives without alteration. Finally, glutaraldehyde, formaldehyde, and osmium tetroxide may be the best available fixatives at present for animal protein and lipids, but they do not preserve structure when starch is a prime component, as it is in many foods from plants.

Seeds

Seed structure has been studied to characterize species, to screen for genetic lines with particular nutritional features, and to study processing conditions[7]. In general grains and seeds are dry enough that splitting with a razor blade and coating is sufficient preparation. Freeze drying is used for wetter preparation. Again, treating with aqueous solutions may cause rapid and significant alterations[22].

Texturized Protein

Textural properties of many foods are important in consumer acceptance. The relationship between microstructure and texture has prompted an increasing effort at relating scanning electron microscope images to other measures[23]. These studies often rely on dry products for both types of measurements, thus direct comparisons are possible. Extruded protein products from several seeds and types of processing conditions needed to generate a specific structure are types of studies that benefit from SEM but present difficulty in doing TEM in parallel[24,25].

Starch

Many foods derived from plants have starch as a major constituent. Starch consists of glucose polymers tightly packed together in distinct units termed granules or grains which have a size and shape characteristic for each plant species. SEM has been particularly helpful in starch studies supplementing the light microscope in revealing details about granule morphology and development[26,27,28]. Light passes through the granules with complex interactions so it is not always possible to resolve surface versus interior features using light microscopy. Ungelatinized granules are relatively resistant to drying; air drying after extraction from the material is a normal state for starch so SEM on this type of specimen is simply a matter of attaching the particles firmly to stubs and coating in a vacuum evaporator or sputter coater. For studies on the surface of granules in air-dried products the initial drying conditions are appropriate. In experiments using aqueous extractions or enzyme incubations, care must be taken in washing off buffer salts that might form crystalline deposits on the surface.

SEM has been used to evaluate enzyme action of amylases from different sources--plant, pancreatic, fungal--on starch granules[29,30,31]. An important finding from this type of study is the realization that attack is not uniform over the surface; only very susceptible granules show extensive exocorrosion. Most uncooked granules develop small surface pits after being exposed to amylase enzymes. With time the pits penetrate toward the granule center. Granule interiors are often more susceptible to enzyme action than the outer surface. The surface pits get larger and more numerous with time and eventually the granule is sufficiently hydrolyzed that it falls apart leaving only fragments. The pitting pattern revealed that some areas of granules such as the crease along the side of large wheat starch granules are more susceptible than other areas. These types of studies provide insights into results on enzyme digestion experiments. The enzyme does not work on a uniform soluble substrate, but on a substance that is insoluble, has a definite geometry, and has areas of high and low enzyme susceptibility. These factors become more complex in natural systems in which some cellular structure is present.

Liquid Food Systems

At the other extreme from low-moisture products are those that at some stage have such a high water content they flow as liquids. These systems have received less study, but there is increased interest in relating viscosity and gel properties to structure of the solid components. Two major areas are gelatinized starch and dairy products.

Gelatinized Starch

When starch granules are heated in water they gelatinize; they absorb water, increase in volume, change shape, and lose birefringence. Gelatinization has been studied by a large number of physical methods and can be seen using a hot stage on a light microscope. During heating granules rapidly swell, and for a time resemble balloons but retain a distinctive morphology so that for example tapioca can be distinguished from corn. Gelatinized granules are delicate and during prolonged heating or mechanical agitation will fall apart or collapse losing a distinctive shape. Despite this delicate structure, it is the gelatinized form of starch granules that is of great importance in many foods. The formation of cake and bread structure and its stability is directly related to gelatinized starch. In liquids viscosity maintenance is also related to the retention of a swollen sac morphology by starch granules. The food industry often covalently cross-links the glucose chains to increase the stability of gelatinized granules used in systems in which viscosity maintenance is essential.

Despite the well-known importance of gelatinized granule morphology in the functional properties of many foods and the many successful studies using SEM on ungelatinized starch, such studies on gelatinized granules have been few and have raised a number of questions that are not yet resolved. Hill and Dronzek[32] reported that during early stages of gelatinization a film of starch coated the intact granules. During later stages of heating they found no three-dimensional starch granule structure. However, they centrifuged the starch suspension; this step may have destroyed the swollen granules. Two studies[33,34] used freeze drying of dilute suspensions of gelatinized starch and found that granules did maintain a three-dimensional structure, and that a fibrillar exudate was present outside the granule walls in many preparations. While recognizing that freezing could have altered the morphology of both the granule ghosts and the exudate, light microscopy of wet mounts confirms the general shape of granules and the presence of amylose in between the granules. Critical point drying of gelatinized granules results in shrinkage[35].

Light microscopy remains the best check on SEM preparation as far as simplicity and ease for many liquid food systems. At least the most drastic alterations can be monitored. For example, it is not difficult to monitor the alterations induced in gelatinized granules during preparation for transmission electron microscopy. During fixations with glutaraldehyde and osmium tetroxide little or no changes occur in the granules. However, during dehydration the granules begin to shrink at about 80% alcohol or acetone. Slow dehydration or small steps result only in slow shrinkage but the net result at 100% organic solvent is the same, a drastically altered granule.

The only promising method for high resolution of gelatinized granules in water is the freeze-etch technique[36] in which small drops of slurry are rapidly frozen in liquid N_2 cooled Freon and a replica of a fractured surface is examined. There remain some differences to be explained in the structure of gelatinized starch as viewed with these two methods. Slurries having a high starch concentration may form a sponge-like structure during freeze drying for SEM and individual granules cannot be discerned[37].

In conclusion, air drying ungelatinized starch granules is sufficient to preserve the type of major information on granule size, morphology, and hydrolysis patterns although undoubtedly some more delicate changes are not seen that can be found using freeze etching[38]. The only method of preparation for gelatinized granules in a liquid medium is freeze drying. Gelatinized granules contain water by hydrogen bonds that is a major part of their structure. Removing this water with dehydrating chemicals results in shrinkage. During the shrinkage, granules do not return to their native morphology. In general critical point drying is not a feasible method.

Another point which should be made on the study of gelatinized starch is that the granules are often difficult to distinguish in baked goods[39]. This has led to a tendency for granules to be isolated by washing the material with water[40]. However, many starches that have been heated and partially gelatinized retain a capability for further swelling if added to a surplus of water. In order to understand the relationship between starch granule structure and baked goods structure the system must be examined with the materials undisturbed.

Dairy Products

Milk and cheese present serious technical problems in preparation for SEM or TEM[41,42]. The volume of solids may be small. The resulting matrix is so open that continuous conductivity is hard to achieve. Studies on the microstructure of skim milk powders as related to drying conditions are fairly direct by SEM[43], but then cannot be directly compared to TEM images. These types of dry materials present problems similar to starch-containing food. If the product is already dry, it can be examined directly with confidence, but it is difficult to do parallel work using other techniques that require embedding. If the product is liquid, freezing produces large ice crystals so that the freeze-dried structure reflects ice crystal pattern in a dilute protein solution.

Cheese is a complex, delicate material that presents some problems related to the high lipid content. Drastic treatments, such as lipid extraction or trypsin hydrolysis of proteins termed etching, before freeze drying or critical point drying still left sufficient structural detail to be of value in studies of the effect of manufacturing conditions[44]. Eino et al.[45] preferred critical point drying to freeze drying, but it is difficult to judge the relative merits of the two methods because of the major alterations by both.

Animal Tissues

Muscle

Food scientists collaborating on nutrition studies may find themselves looking at many types of animal tissue, but from the viewpoint of human consumption muscle is the predominant tissue of interest. Jones et al.[46] discussed methods of preparing muscle tissue for both scanning and transmission electron microscopy and some of the problems with usual methods. Shrinkage during water removal, changes due to cutting, or abrasion of the surface were common problems. Ethanol freeze fracture of previously fixed muscle followed by critical point drying or air drying[10] circumvented some problems. This method had the advantage that it was similar to typical transmission preparation procedures so that direct comparisons could be made with abundant published work on TEM of muscle. Cryofracturing before fixation, dehydration, and critical point drying has been used in other studies on meat products[47]. Freeze drying after cryofracture was also used, but ice crystal damage can be substantial. As an independent method of observation Jones et al.[46] studied shrinkage during fixation and dehydration with the light microscope. Minimal shrinkage occurred during each fixation step. At 85% ethanol, shrinkage which was discernible by eye was present. It is surprising in light of this that the freeze-dried sample had more shrinkage than the critical-point-dried material that had been treated with ethanol.

Despite some remaining problems the cryofracture technique has provided information in a variety of studies on muscle that have related structural factors of muscle tissue components to meat quality[48].

Conclusions

It has been said before that the best treatment is no treatment. Fortunately for many food samples this is possible. Many of the items studied are sufficiently dry for insertion into a vacuum and must only be coated with a conducting layer. But for correlations with other methods of microscopy, especially when high resolution from thin sections is desired, treatments with fixatives and dehydrating agents become necessary prior to embedding. SEM shows very clearly how disruptive these procedures can be in limited water systems.

Foods that have a high water content can be fixed with no difficulty, but shrinkage during dehydration can be catastrophic. Freeze drying results in less shrinkage, but ice crystal formation distorts delicate features. Critical point drying, now a routine method for many animal tissues, is only rarely called for in food science. It produces fine images, but may induce false confidence.

Scanning electron microscopy is an important technique in food science, but its true value rests on scientists learning to control avoidable artifacts and recognizing the range of effects produced by different processing and preparation procedures. Conducting parallel studies using transmission electron microscopy on thin sections was one recommended way to monitor problems[49]. Using a variety of preparation procedures for SEM could also reveal problems. Food scientists regularly encounter materials that are low in moisture content and are very susceptible to the addition of water, so correlative TEM on thin sections turns out to be more difficult than in animal or plant work. Freeze etching provides an ideal method of high resolution on food structure but requires expensive equipment. For most studies the food scientist is advised to use visual examination, the light microscope, and common sense in preparing material for scanning electron microscopy.

References

1. Boyde, A., Do's and don'ts in biological specimen preparation for the SEM. In SEM/1977/I, IIT Research Institute, Chicago, IL, 60616, pp. 683-690.
2. Hayes, T. L., Scanning electron microscope techniques in biology. In Advanced techniques in biological electron microscopy, J. K. Koehler (ed.), Springer-Verlag, New York, 1973, pp. 153-214.
3. Boyde, A., Biological specimen preparation for the SEM--an overview. In SEM/1972, IITRI, pp. 257-264.
4. Cohen, A. L., A critical look at critical point drying--theory, practice and artefacts. In SEM/1977/I, IITRI, pp. 525-536.
5. Nermut, M. V., Freeze-drying for electron microscopy. In Principles and techniques of electron microscopy; biological applications, Vol. 7, M. A. Hayat (ed.), Van Nostrand Reinhold Co., New York, 1977, pp. 79-117.
6. Boyde, A., and C. Wood, Preparation of animal tissues for surface-scanning electron microscopy. J. Microscopy 90, 1969, 221-249.
7. Pomeranz, Y., Scanning electron microscopy in food science and technology. Adv. Food Res. 22, 1976, 205-307.
8. Boyde, A., E. Bailey, S. J. Jones, et al., Dimensional changes during specimen preparation for scanning electron microscopy. In SEM/1977/I, IITRI, pp. 507-518.
9. Humphreys, W. J., Drying soft biological tissue for scanning electron microscopy. In SEM/1975, IITRI, pp. 708-714.
10. Humphreys, W. J., B. O. Spurlock, and J. S. Johnson, Critical point drying of ethanol-infiltrated, cryofractured biological specimens for scanning electron microscopy. In SEM/1974/I, IITRI, pp. 275-282.
11. Bechtel, D. B., Y. Pomeranz, and A. de Francisco, Breadmaking studied by light and transmission electron microscopy. Cereal Chem. 55, 1978, 392-401.
12. Khoo, U., D. D. Christianson, and G. E. Inglett, Scanning and transmission microscopy of dough and bread. Bakers Digest 49(4), 1975, 24-26.
13. Sandstedt, R. M., L. Schaumburg, and J. Fleming, The microscopic structure of bread and dough. Cereal Chem. 31, 1954, 43-49.
14. Simmonds, D. H., The ultrastructure of the mature wheat endosperm. Cereal Chem. 49, 1972, 212-222.
15. Evans, L. G., T. Volpe, and M. E. Zabik, Ultrastructure of bread dough with yeast single cell protein and/or emulsifier. J. Food Sci. 42, 1977, 70-74.
16. Pomeranz, Y., M. D. Shogren, K. F. Finney, et al., Fiber in breadmaking - effects on functional properties. Cereal Chem. 54, 1977, 25-41.
17. Burhans, M. E., and J. Clapp, A microscopic study of bread and dough. Cereal Chem. 19, 1942, 196-216.
18. Barlow, K. K., M. S. Buttrose, D. H. Simmonds, et al., The nature of the starch-protein interface in wheat endosperm. Cereal Chem. 50, 1973, 443-454.
19. Aranyi, C., and E. J. Hawrylewicz, Application of scanning electron microscopy to cereal specimens. Cereal Sci. Today 14, 1969, 230-233, 253.
20. Crozet, N., Ultrastructural changes in wheat-flour proteins during fixation and embedding. Cereal Chem. 54, 1977, 1108-1114.
21. Varriano-Marston, E., A comparison of dough preparation procedures for scanning electron microscopy. Food Tech. 31(Oct), 1977, 32-36.
22. Buttrose, M. S., Rapid water uptake and structural changes in imbibing seed tissues. Protoplasma 77, 1973, 111-122.
23. Stanley, D. W., and M. A. Tung, Microstructure of food and its relation to texture. In Rheology and texture in food quality, J. M. de Man, P. W. Voisey, V. Rasper, and D. W. Stanley (eds.), Avi Publ. Co., Westport, CN, 1976, pp. 28-78.
24. Taranto, M. V., and K. C. Rhee, Ultrastructural changes in defatted soy flour induced by nonextrusion texturization. J. Food Sci. 43, 1978, 1274-1278.
25. Taranto, M. V., G. F. Cegla, and K. C. Rhee, Morphological, ultrastructural and rheological evaluation of soy and cottonseed flours texturized by extrusion and non-extrusion processing. J. Food Sci. 43, 1978b, 973-979.
26. Hall, D. M., and J. G. Sayre, A scanning electron-microscope study of starches, Part I, Root and tuber starches. Textile Res. J. 39, 1969, 1044-1052.
27. Hall, D. M., and J. G. Sayre, A scanning electron microscope study of starches II, Cereal starches. Textile Res. J. 40, 1970a, 257-266.
28. Evers, A. P., Scanning electron microscopy of wheat starch, III, Granule development in the endosperm. Starch 23, 1971, 157-162.

29. Gallant, P. D., A. Derrien, A. Aumaitre, et al., Degradation in vitro de l'amidon par le suc pancreatique. Starch 25, 1973, 56-64.
30. Lineback, D. R., and S. Ponpipom, Effects of germination of wheat, oats, and pearl millet on alpha-amylase activity and starch degradation. Starch 29, 1977, 52-60.
31. Takaya, T., Y. Sugimoto, E. Imo, et al., Degradation of starch granules by alpha-amylases of fungi. Starch 30, 1978, 289-293.
32. Hill, R. D., and B. L. Dronzek, Scanning electron microscopy studies of wheat and corn starch during gelatinization. Cereal Sci. Today 17, 1972, 266.
33. Miller, B. S., R. I. Derby, and H. B. Trimbo, A pictorial explanation for the increase in viscosity of a heated wheat starch-water suspension. Cereal Chem. 50, 1973, 271-280.
34. Chabot, J. F., L. F. Hood, and J. E. Allen, Effect of chemical modifications on the ultrastructure of corn, waxy maize, and tapioca starches. Cereal Chem. 53, 1976, 85-91.
35. Hood, L. F., A. J. Seifried, and R. Meyer, Microstructure of modified tapioca starch-milk gels. J. Food Sci. 39, 1974, 117-120.
36. Allen, J. E., L. F. Hood, and J. F. Chabot, Effect of heating on the freeze-etch ultrastructure of hydroxypropyl distarch phosphate and unmodified tapioca starches. Cereal Chem. 54, 1977, 783-793.
37. Berghofer, E., and H. Klaushofer, Untersuchung gefriergetrockhater Stärkekleister und Stärkeschwämme im Rasterelektronenmikroskop. Starch 28, 1976, 113-121.
38. Chabot, J. F., J. E. Allen, and L. F. Hood, Freeze-etch ultrastructure of waxy maize and acid hydrolyzed waxy maize starch granules. J. Food Sci. 43, 1978, 727-730, 734.
39. Stevens, D. J., The role of starch in baked goods, Part I, The structure of wafer biscuit sheets and its relation to composition. Starch 28, 1976, 25-29.
40. Hoseney, R. C., D. R. Lineback, and P. A. Seib, Role of starch in baked goods. Bakers Digest 52, 1978, 11-14, 16, 18, 40.
41. Kalab, M., Milk gel structure, VIII, Effect of drying on the scanning electron microscopy of some dairy products. Milchwissenschaft 33, 1978, 353-358.
42. Kalab, M., D. B. Emmons, and A. G. Sargant, Milk-gel structure, IV, Microstructure of yoghurts in relation to the presence of thickening agents. J. Dairy Res. 42, 1975, 453-458.
43. Saito, Z., Electron microscopic and compositional studies of casein micelles. Neth. Milk Dairy J. 27, 1973, 143-162.
44. Eino, M. F., D. A. Biggs, D. M. Irvine, et al., A comparison of microstructures of cheddar cheese curd manufactured with calf rennet, bovine pepsin, and porcine pepsin. J. Dairy Res. 43, 1976, 113-115.
45. Eino, M. F., D. A. Biggs, D. M. Irvine, et al., Microstructure of cheddar cheese: sample preparation and scanning electron microscopy. J. Dairy Res. 43, 1976, 109-111.
46. Jones, S. B., R. J. Carroll, and J. R. Cavanaugh, Muscle samples for scanning electron microscopy: preparative techniques and general morphology. J. Food Sci. 41, 1976, 867-873.
47. Theno, D. M., D. G. Siegel, and G. R. Schmidt, Meat massaging: effects of salt and phosphate on the ultrastructure of cured procine muscle. J. Food Sci. 43, 1978, 488-498.
48. Carroll, R. J., F. P. Rorer, S. B. Jones, et al., Effect of tensile stress on the ultrastructure of bovine muscle. J. Food Sci. 43, 1978, 1181-1187.
49. Clark, J. M., and S. Glagov, Evaluation and publication of scanning electron micrographs. Science 192, 1976, 1360-1361.

Discussion with Reviewers

S. H. Cohen: You state that glutaraldehyde, formaldehyde, and OsO4 aren't very satisfactory for starch fixation. What, then, would you suggest as an alternative protocol to these fixatives?
Author: Freeze etching appears to be the best method for high resolution TEM studies on starch. For SEM freeze drying is the best method. Ungelatinized starch granules are relatively stable. Gelatinized starch cannot be exposed to either aqueous solutions or dehydrating chemicals without alterations.

D. N. Holcomb: Could one avoid artifacts on drying food specimens by viewing frozen specimens directly in the manner of Pawley et al. (in SEM/1978/II, pp. 683-690)?
Author: Observation of specimens stabilized by freezing could be important in future work in SEM because it will avoid drying artifacts. However, literature on freeze etching indicates that ice crystal damage is difficult to avoid in large samples without cryoprotectants.

K. Saio: I observed differences between sputter and vacuum evaporator coatings. The sputter coating seems to be more destructive for fine and fragile structures (like lipid granules). Would you please refer to the varieties (instrument, materials) and methods (thickness, temperature) usable for the wide variety of foods in your review?
Author: The relative merits of vacuum evaporator coating versus types of sputter coaters have been mentioned in several reviews on biological structures. Lipids are sensitive to heat damage that can be difficult to avoid in diode sputter coaters. Each step in the preparation of a sample for SEM is important. My emphasis in this review was on dehydration procedures because the alterations at this stage can be so profound. It is difficult to specify exact procedures for a wide variety of foods, some are robust, others almost defy our present expertise. Even a simple parameter such as film thickness must be varied to suit a particular specimen, depending on composition, surface irregularity, and porosity.

For additional discussion see page 8.

APPLICATION OF SCANNING ELECTRON MICROSCOPY FOR THE DEVELOPMENT OF MATERIALS FOR FOOD

C.-H. Lee and C.K. Rha

Department of Nutrition and Food Science
Massachusetts Institute of Technology
Cambridge, MA 02139

Abstract

Understanding the physical and structural properties of biological materials from the three dimensional structure exhibited by SEM provides the means to develop an effective isolation or purification of food components. SEM study also provides the means for characterizing the physical properties and textural attributes of food ingredients.

This paper illustrates applications of SEM for evaluation of the procedures for the isolation of protein and the characterization of the physical properties of the protein aggregates. The SEM studies for the structural characteristics as a result of processing of food materials, with spray-drying and freeze drying as examples, are also discussed. The mechanism of structure forming in protein semisolids upon freeze drying is discussed on the basis of SEM studies.

KEY WORDS: Food Material Development, Dehydrated Protein Structure, Spray-Dried Structure, Freeze-Dried Structure, Structure Property Relationship

Introduction

Since commercial SEM became available in 1965, the three dimensional microstructure of materials has been used to explain and predict the physical and chemical behavior of biological materials. Its use is increasing rapidly in the field of food science and technology. The stereostructure and geometry of materials shown by SEM can be used as the basis for the selection and identification of the potential utilization of food resources, and for the process optimization and quality evaluation of manufactured food. The spherical protein bodies in legumes[1,2], the casein micelles in milk[3,4], and the starch granules in seeds and roots[5,6,7], have been characterized by SEM micrographs. The visualization by SEM of their changes in organization upon different treatment provides better understanding of the physical and rheological properties of food materials, and serves to improve food processing methods.

Freezing and dehydration are important procedures for food processing, and the resultant microstructure has important implication to the physical and functional properties of food materials[8]. This paper discusses the SEM studies of the effect of dehydration and freezing on the structural characteristics of liquid and semisolid food. In addition, the relationship between the structural characteristics and the functional properties of these food materials is discussed as it relates to the use of SEM in process optimization.

Microstructure Resulted From Dehydration

Spray drying has been used widely for the dehydration of liquid food. The process involves forming atomized liquid droplets and hot air dehydration. The process results in the hollow spherical particles. Wolf and Baker[2] studied the three dimensional structure of spray-dried commercial soybean protein isolates, and noticed a marked distinction between isoelectric precipitated protein isolates and the sodium proteinates. The isoelectric isolates contained particles from 2 μm to more than 40 μm in diameter that were nearly spherical and had rough surfaces. The sodium proteinate had hollow spheres and partially collapsed spheres with smooth surfaces.

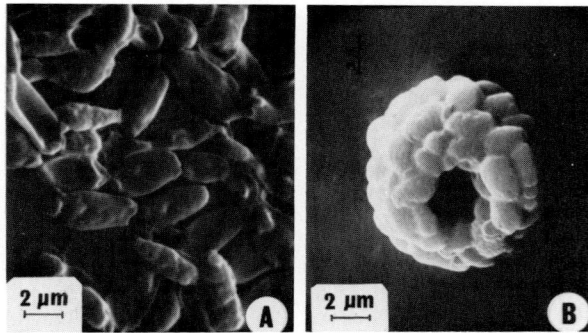

Figure 1. *Candida lipolytica* yeast as observed through SEM, (A) Fresh yeast, (B) Spray-Dried Yeast.

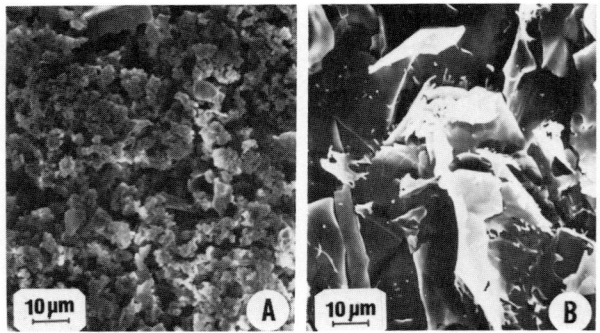

Figure 2. Freeze-dried soybean isolates as observed through SEM, (A) Isoelectric type isolate, (B) Neutralized sodium proteinate.

The differences in the surface characteristics could be explained by the fact that the isoelectric isolates were dried from relatively insoluble dispersions resulting in rough particle surfaces, whereas the sodium proteinates were dried from more soluble form that yields smooth, partially hollow particles. The partially collapsed sphere of soybean protein isolate was also reported by Pomeranz et al[9]. Hollow and partially collapsed spheres were observed in spray-dried calcium caseinate[10], and also in materials other than protein such as coffee[11]. The implosion of casein droplets and uneven shrinkage of the casein have been suggested as causes of the collapsed appearance[12].

Recently, we showed the three dimensional microstructure of single cell protein[13]. Figure 1 is the SEM micrographs of fresh and spray-dried *Candida lipolytica*. It shows that in commercial spray dried yeast, the ellipsoidal individual cells are aggregated to form a hollow sphere. The rough surfaces of the spray-dried yeast clumps were similar in appearance to the spray-dried isoelectric soybean protein isolates[2].

The particle structure resulting from freeze drying of liquid food is quite different from that resulting from spray drying. Figure 2 shows the SEM micrograph of freeze-dried soybean protein isolates; Figure 2A is the isoelectric type isolate and 2B the neutralized sodium proteinate. The protein suspensions were frozen in liquid nitrogen and freeze-dried. The isoelectric type isolate contained particles from 2 μm to 10 μm in diameter and had irregular shape with rough surfaces. The sodium proteinates showed a large sheet-like entity with smooth surfaces. The freeze-dried particles of isoelectric type isolate are relatively insoluble in water, while the freeze-dried sodium proteinate is readily soluble. Therefore, the solubility and other physical properties of the dehydrated proteins may be predicted based on the appearance in SEM Micrograph.

Evaluation of Protein Extraction Process

The three dimensional structure of protein bodies in soybean cotyledon was first demonstrated by Wolf[1] using SEM micrograph. Wolf and Baker[2] further studied the morphology of the cytoplasmic network and the removal of protein bodies from the cell wall structure. The contents of the mechanically fractured cotyledon cells were easily washed out with water. This behavior explained why aqueous extraction of protein was possible in soybeans. The protein could be extracted by swelling soybeans in water, mechanically grinding and then heating. The mechanical grinding is undoubtedly a key step because once the cell wall is ruptured, extraction is very rapid and complete[2].

The extraction of protein from spray-dried yeast *Candida lipolytica* was studied by the authors[14]. The spherical yeast clumps shown in Figure 1 were stable in water, acetone or alcohol and combinations of those solvents.

Figure 3A shows that the cell clumps were hydrolyzed indiscriminately by 6N HCl at 55°C for 2 hr. This treatment extracted 64% of the total cellular protein. The extracted proteins were hydrolyzed and could not be precipitated at the isoelectric point. Figure 3B shows that the

Figure 3. Spray-dried *Candida lipolytica* (A) 6N-HCl at 55°C for 2 hr. (B) 0.1N-NaOH at 25°C. The bar designates 2μm.

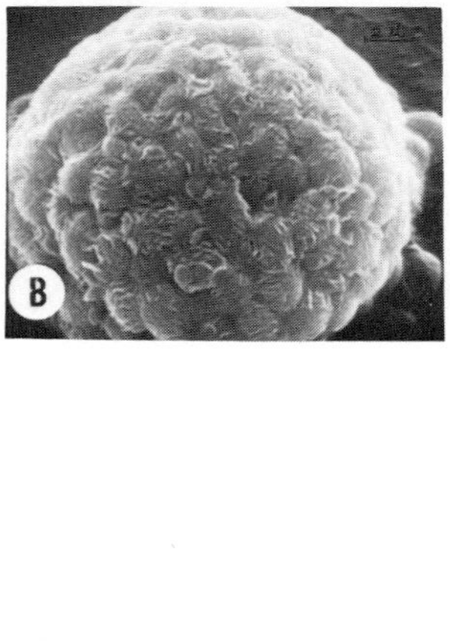

Figure 4. Spray-dried *Candida lipolytica* (A) Homogenized in H2O (B) Homogenized in 0.1 N-NaOH. The bar designates 5μm.

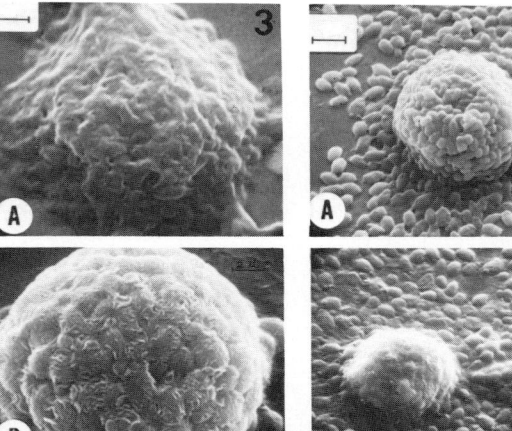

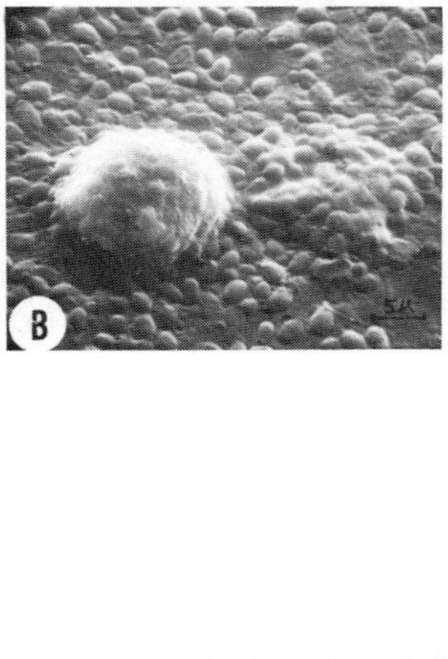

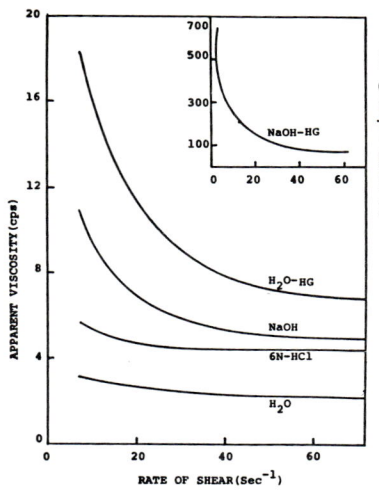

Figure 5. Viscosity of spray-dried *Candida lipolytica* in different suspension media. HG designates homogenization.

treatment in 0.1N NaOH solution at 25°C for 1 hr. caused the cell wall to wrinkle while most of the cell clumps were retained. Only 20% of the total cellular protein was extracted by this treatment.

Figure 4 indicates that an effective disintegration of cell clumps has been achieved by high pressure homogenization (9000 psi) in either water or 0.1 N NaOH suspension. However, the amounts of extracted protein by the homogenization in water were only 11% of the total cellular protein while alkaline homogenization extracted 93% of the cellular protein. The SEM micrograph of alkaline homogenized cells (Figure 4B) shows a large amount of cellular extract outside the cells. The original shape of most of the individual cells was still retained without fragmentation after the homogenization. This study indicated that the alkaline treatment weakened the cell wall structure and facilitated cell wall cracking during homogenization. In addition, the alkali had a solubilization effect on the spray-dried yeast protein.

The changes in the cell clump particle size as observed by SEM suggested changes in the flow behavior of the cell suspension. Figure 5 shows

Figure 6. Sedimentation property of spray-dried *Candida lipolytica*.

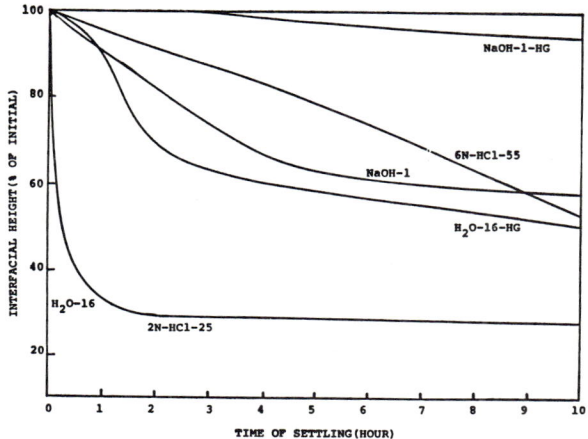

the apparent viscosity of the cell suspensions in the extraction medium. The increase in surface area of the cell caused by the disintegration of clumps, as evidenced in Figure 4, led to the increase in the apparent viscosity after homogenization. The high apparent viscosity obtained by homogenization in alkaline solution could mainly be attributed to extracted cellular proteins.

The SEM microstructure of the spray-dried yeast was used to illustrate the effect of particle size on the sedimentation property[14]. Figure 6 shows that the spray-dried yeast cell clumps (Figure 1B) in water sediment much faster than the disintegrated individual cells in water after homogenization (Figure 4A). The marked distinctions in the sedimentation pattern between cells treated with acid or alkali may reflect the different mechanism of cellular hydrolysis. The very slow sedimentation of the cell homogenized in alkaline solution can be explained by the collapsed cell structure observed by SEM (Figure 4B), where the higher density cellular plasma is depleted and subsequently replaced by suspending water.

Microstructure of Freeze-Dried Semisolid

The structural changes of semisolid food during dehydration processes have a special interest because of the high water content and weak structure. In many semisolid systems, the structural changes derived in dehydration are drastic enough that the dehydrated material is irreversibly altered and can not be reconstituted to the original structure of the semisolid. Freeze drying has found most use in food processing because of its ability to preserve the physical and functional properties of food materials[15]. However, the freezing and freeze-drying methods normally used in food processing can cause irreversible structural changes in many protein and other hydrocolloid systems.[16,17,18,19] Different structural features can also be observed for identical samples depending upon the freezing and dehydration conditions[20,21]. Furthermore, similar microstructures upon freeze drying of quite different types of materials have been published. For example, the three dimensional network structure of hyaluronic acid gel[22] was very similar to the microstructure of freeze dried gelatin gel.[6] Similar microstructure to a Japanese frozen and air-dried soybean protein curd[23] was observed in the thermally induced rapeseed protein gel upon freeze-drying[24].

The structural changes upon freezing are mainly attributed to the mechanical stresses exerted by the growth of ice crystals and the chemical denaturation of the material due to the concentration of salt in the unfrozen water layer. In plant tissues, the nucleation and growth of ice crystals proceeds in the continuous extracellular liquid phase followed by nucleation and growth of ice crystals in individual supercooled cells[25]. In a system such as a protein semisolid which does not contain cellular compartments, the growth of ice crystals will extend over a longer range than in the plant tissue. The water content of semisolid becomes an important factor determining the growth of ice crystals for a given freezing

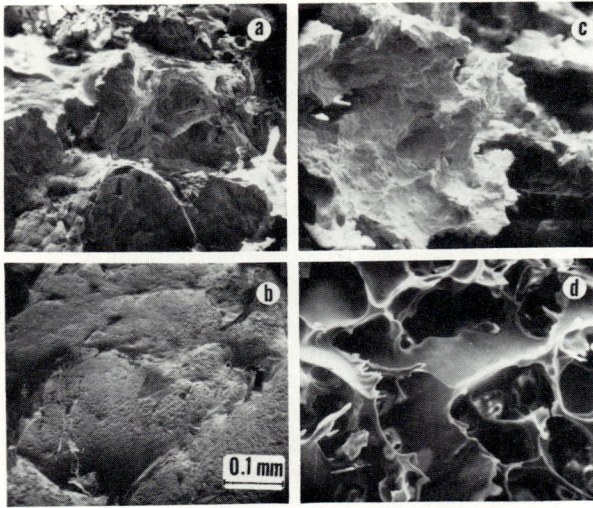

Figure 7. Freeze-dried yeast protein aggregates as observed through SEM (a) Glutaraldehyde fixed, isoelectric point precipitates (b) Glutaraldehyde fixed, calcium precipitates (c) Unfixed, isoelectric precipitates (d) Unfixed, calcium precipitates. All are in the same magnitude.

Table 1. Physical properties of yeast protein aggregates

Samples[a]	W.H.C.[b]	Hardness (Kg force)	Springiness
Fresh			
YPC-I	3.3	1.9	1.7
YPC-Ca	4.2	0.8	1.6
Freeze-Thawed			
YPC-I	2.6	3.1	1.8
YPC-Ca	2.7	2.8	2.3

[a] YPC is yeast protein concentrate. I and Ca indicate that the protein was precipitated at isoelectric point or by calcium respectively.

[b] W.H.C. is water holding capacity of yeast protein aggregates in gH_2O/g protein.

condition.

Tsintsadze et al.[17] studied the SEM microstructure of freeze-dried yeast protein aggregates. Figure 7 shows the differences in the freeze-dried microstructure between glutaraldehyde fixed (Figure 7a and b) and unfixed yeast protein aggregates (Figure 7c and d). The isoelectric precipitates (sample a and c) contained 77% water and the calcium precipitates (sample b and d) had 81% water. Both were frozen at -20°C and freeze-dried. Fixation of protein gel with glutaraldehyde reduces the structural distortion due to the growth of ice crystals during freezing[15]. The difference in the freeze-dried microstructure between glutaraldehyde fixed and unfixed yeast protein aggregates was not noticeable in the isoelectric precipitated aggregates, while a remarkable difference was observed in the calcium precipitated aggregates.

Table 1 shows that a large amount of water was lost after freeze-thawing of the yeast protein aggregates. The network structure became firmer after freeze-thawing and the springiness increased. The loss of water and the increase in the hardness and springiness after freeze-thawing were much higher in calcium precipitated than in isoelectric precipitated aggregates.

The effect of water content on the structural changes during freeze drying is evident upon comparing the above mentioned results with a simulated cheese product which contains 51% of water with 20% calcium/sodium caseinate and 18% vegetable oil[26]. Figure 8A is the freeze-dried microstructure of glutaraldehyde fixed simulated cheese and Figure 8B is the microstructure of unfixed simulated cheese. Both were frozen at -20°C and freeze-dried. The microstructure of these two samples are similar except that the glutaraldehyde fixed sample has more detailed fine structure, which is partly due to the loss of some chemical constitutents from cheese during fixation. The difference in SEM microstructure between the fixed and unfixed simulated cheeses was not as great as that experienced by the single cell protein curds.

In a SEM study with bread dough containing only 45% water, Varriano-Marston[20] found that the chemical fixation prior to freezing and dehydration was not necessary. In fact, chemical fixation caused structural distortion during the specimen preparation for SEM. These studies proved the important role of water content to the structure of freeze-dried semisolids.

The differences in the freeze-dried structure was also observed by changing the coagulation method of protein. Lee and Rha[27] studied the effect of heat treatment and coagulants on the microstructure of freeze-dried soybean protein. Figure 9 shows the glutaraldehyde fixed structure of freeze-dried soybean protein aggregates. Unheated soybean proteins precipitated at the isoelectric point or with calcium had globular particles (Figure 9c and d). Figure 10 shows the freeze-dried structure of the same protein aggregates without glutaraldehyde fixation. The isoelectric point precipitated aggregates from

Figure 8. Freeze-dried imitation cheese as observed through SEM, (A) Glutaraldehyde fixed (B) Without fixation.

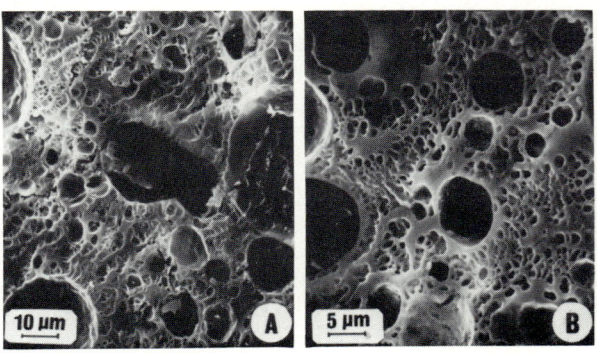

SEM for Development of Materials for Food

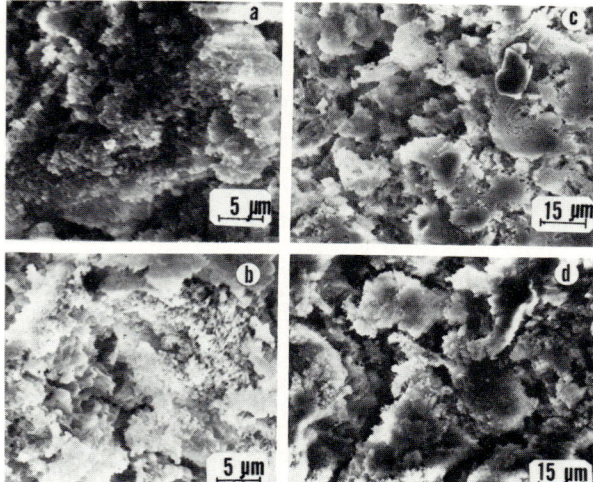

Figure 9. Freeze-dried soybean protein aggregates fixed with glutaraldehyde as observed through SEM (a) Unheated protein, isoelectric precipitates (b) Unheated protein, calcium precipitates (c) Heated protein, isoelectric precipitates (d) Heated protein, calcium precipitates.

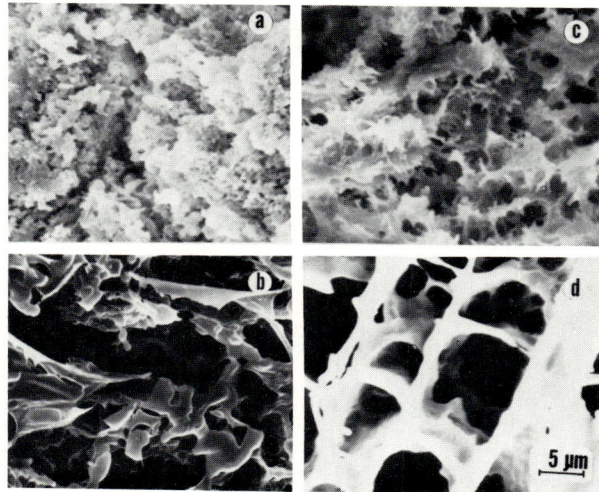

Figure 10. Soybean protein aggregates freeze-dried without glutaraldehyde fixation as observed through SEM. Sample designations are the same as in Figure 9. All are in the same magnitude.

unheated soybean protein retained the globular particles (Figure 10a), while the calcium precipitated aggregates formed a sheet-like structure (Figure 10b). The three dimensional network structure of heated protein precipitates became extensively enlarged upon freeze drying (Figure 10c and d), especially for the calcium precipitates with a well defined honeycomb-like structure.

The influence of heat denaturation and coagulants on the microstructure of freeze-dried protein aggregates indicates that the dimensions of the molecules or particles and the mechanisms of intermolecular interactions play important roles in the structure formation. The

Figure 11. Illustration of structure formation mechanism upon freeze drying.

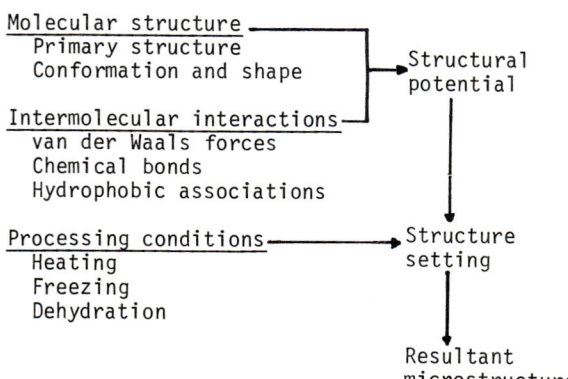

microstructure resulting from freeze drying of a semisolid is primarily determined by the size and shape of ice crystals which are formed during freezing and freeze drying. The amount of water in semisolid is dependent upon the water binding property of molecules and the development of network structure where the entrapped water is occluded. The particle dimensions and intermolecular interactions will exert the major constraint against the growth of ice crystals in semisolid. Therefore, a conclusion can be made that the structural changes in protein semisolid upon freeze drying depend on the molecular structure, particle dimension and the interaction forces acting between particles. The integrated effect of these factors can be represented as the structural potential of a semisolid system.

This conclusion results in a working hypothesis that the microstructures of freeze-dried food, and thus, their texture, can be manipulated by selection of raw materials having certain molecular dimensions, and introduction of a major interaction to create the structural potential, as shown in Figure 11. The structural potential in a system determines the size and shape of ice crystals, and thus controls the microstructure for a given freeze drying condition.

Concluding Remarks

SEM is a useful instrument for studying the microstructure of a variety of food materials. It shows the three dimensional microstructure which explains the physical and functional properties of food materials, changes in the microstructure due to processing and thus offers interpretation for the mechanism of the process. The use of SEM for the study of intact structure of semisolid is complicated because of the difficulties in the preparation of the specimen free of artifacts. However, changes in the microstructure upon the processing of semisolid foods can be studied with SEM by selecting a reference condition. For instance, the glutaraldehyde fixed protein semi-solid can be used as a reference for the study of the effect of the growth of ice crystals on the microstructure of freeze-dried sample. The most interesting aspect of SEM study is that by introducing a specific well-defined interaction mechanism to a model system, it is possible to show the magnitude of contribution of the interaction to the specific structure. Such information will provide the basis for the manipulation of microstructure and texture of fabricated foods in which SEM will find the most creative use.

References

1. Wolf, W.J. 1970. Scanning electron microscopy of sobyean protein bodies, J. Am. Chem. Soc. 47, 107-108.
2. Wolf, W.J. and Baker, F.L. 1975. Scanning electron microscopy of soybeans, soyflours, protein concentrates and protein isolates, Cereal Chem., 52, 387-396.
3. Kalab, M. and Harwalkar, V.R. 1973. Milk gel structure, I. Application of scanning electron microscopy of milk and other food gels, J. Dairy Science, 56, 835-842.
4. Kalab, M. and Harwalkar, V.R. 1974. Milk gel structure, II. Relation between firmness and ultrastructure of heat-induced skim-milk gels containing 40-60% total solids. J. Dairy Res., 41, 131-134.
5. Rockland, L.B., Jones, F.T. and Hahn, D.M. 1977. Light and scanning electron microscope studies of drybeans: extracellular gelatinization of lima bean starch in water and a mixed salt solution, J. Food Sci., 42, 1204-1212.
6. Evans, L.G., Volpe, T. and Zabik, M.E. 1977. Ultrastructure of bread dough with yeast single cell protein and/or emulsifier, J. Food Sci., 42, 70-74.
7. Silva, H.C. and Luh, B.S. 1978. Scanning electron microscopy studies on starch granules of red kidney beans and bean sprouts, J. Food Sci, 43, 1405-1408.
8. Stanley, D.W. and Tung, M.A. 1976. Microstructure of food and its relation to texture in Rheology and Texture in Food Quality, ed. J.M. de Man, P.W. Voisey, V.F. Rapser and D.W. Stanley, AVI, Westport, Con., 28-78.
9. Pomeranz, Y., Shogren, M.D. and Finney, K.F. 1977. Flour from germinated soybeans in high-protein bread, J. Food Sci., 42, 824-827.
10. Buma, T.J. and Henstra, S. 1971. Particle structure of spray-dried milk products as observed by a scanning electron microscope, Neth. Milk Diary J., 25, 75.
11. Gejl-Hansen, F. and Flink, J.M. 1976. Application of microscopic techniques to the description of structure of dehydrated food systems. J. Food Sci., 41, 483-489.
12. Buma, T. and Henstra, S. 1971. Particle structure of spray-dried caseinate and spray-dried lactose as observed by a scanning electron microscope, Neth. Milk Diary J. 25, 278.
13. Tsang, S., Lee, C.H. and Rha, C.K. 1979. Disintegration of cell wall and extraction of protein from Candida lipolytica, J. Food Sci., 44, 97-99.
14. Lee, C.H., Tsang, S.K., Urakabe, R. and Rha, C.K. 1979. Disintegration of the dried yeast cell and its effect on protein extractability, sedimentation property and viscosity of the cell suspension, Biotech. Biogeng., 21, 1-17.
15. Boyde, A. and Echlin, P. 1973. Freeze and freeze-drying - a preparative technique for SEM, in Scanning Electron Microscopy/1973, eds., O. Johari and I. Corvin, IIT Research Institute, Chicago, Ill., 759-766.
16. Halberstadt, E.S., Henish, H.K., Nickl, J. and White, E.W. 1969. Gel structure and crystal nucleation, J. Colloid Interface Sci., 29, 469-471.
17. Tsintsadze, T.D., Lee, C.H. and Rha, C.K. 1978. Microstructure and mechanical properties of single cell protein curd, J. Food Sci., 43, 625-630.
18. Torres, D.A., Schwartzberg, H.G., Peleg, M. and Rufner, R. 1978. Textural properties of amylose sponges, J. Food Sci., 43, 1006-1009.
19. Gejl-Hansen, F.A. 1977. Microstructure and stability of freeze-dried solute containing oil-in-water emulsions, Ph.D. Thesis, MIT, 155-183.
20. Varriano-Marston, E. 1977. A comparison of dough preparation procedures for scanning electron microscopy, Food Technol. 31(10), 32-36.
21. Eino, M.F., Biggs, D.A., Irvine, D.M. and Stanley, D.W. 1976. Microstructure of cheddar cheese: sample preparation and scanning electron microscopy, J. Dairy Res., 43, 109-115.
22. Seller, P.C., Dowson, D. and Wright, V.W. 1971. The rheology of synovial fluid, Rheol. Acta, 10, 2-7.
23. Matsumoto, S. 1975. Recent progress in food texture research in Japan, J. Texture Stud., 5, 373-398.
24. Gill, T.A. and Tung, M.A. 1978. Thermally induced gelation of the 12S rapeseed glycoprotein, J. Food Sci., 43, 1481-1485.
25. Brown, M.S. and Reuter, F.W. 1974. Freezing of nonwoody plant tissues, III. Videotape micrography and the correlation between individual cellular freezing events and temperature changes in the surrounding tissue, Cryobiology, 11, 185-191.
26. Petka, T.E. 1976. Novel Caseinate-effects natural functional properties for imitation cheese, Food Product Development, 10(10), 26-27.
27. Lee, C.H. and Rha, C.K. 1978. Microstructure of soybean protein aggregates and its relation to the physical and textural properties of the curd, J. Food Sci., 43, 79-84.

Discussion With Reviewers

V.R. Harwalker: What was the reason for drying the spray-dried suspensions of yeast cells in the air?
Authors: Because the yeast sample used was grown in hydrocarbon medium, use of organic solvent in the preparation for SEM specimen was avoided. Although the cell clumps did not disperse in acetone or alcohol, some indentation appeared on the clumps when treated in the organic solvents. The original spray-dried cells and those suspended in water and subsequently dried in air showed little difference in the microstructure.

S.H. Cohen: What isoelectric focusing technique did you use? Did varying the pH change the morphology of the curd?
Authors: The isoelectric precipitation was carried out by adding dilute HCl solution to the neutralized protein dispersion. The effect of pH on the structure of protein aggregates was not studied specifically. Isoelectric precipitated soybean protein aggregates had pH 4.5 and the calcium precipitated one had pH 5.5. However, the pH difference could not be considered to have a major influence on the observed microstructure changes. That is because the major interactions responsible for the structure of isoelectric and calcium precipitates were different and this difference was considered to be the predominating factor (see text reference 27).

R.C. Hoseney: The effect of calcium precipitation on the observed structure changes was assumed to depend on the aggregation mechanism. Could those differences also be explained as an effect of the salt on the thawing that occurred during your freeze-drying?
Authors: No thawing occurred in the freeze-drying process we used. The freeze-thawed sample was examined specifically to study and characterize the effect of freeze-thaw on the physical and mechanical properties. The concentration of salt in the frozen water layer upon freezing and thawing of semisolid would certainly influence the microstructure as mentioned in the text.

K. Saio: Why didn't you use freeze drying after immersing in liquid-N on yeast curds, in spite of using it on soybean curds?
Authors: The studies with yeast curd were aimed specifically at the structure resulting from freezing and thawing and the resultant changes in the mechanical properties. Therefore, conditions similar to industrial processes were used. In soybean curd, the study was aimed more directly on the structure of aggregates as formed by different mechanisms, therefore to preserve the structure liquid nitrogen was used as a freezing medium.

K. Saio: According to your photographs of soybean curd fixed or/and unfixed with glutaraldehyde, their SEM-image gave a honeycomb-like structure. How many minutes does it take for freeze-drying them? Did you use a high speed freeze-drying?
Authors: The soybean protein curd having a thickness of 0.5 cm was frozen in liquid nitrogen for rapid freezing. The sample was freeze-dried at the condenser temperature -55°C. It is well known that the microstructures of frozen or freeze-dried protein curds depend on the freezing rate. For this reason, glutaraldehyde was used to fix the protein curd. The effect is shown clearly in Figures 9 and 10. The differences in the structures between acid and calcium precipitated curds are also shown by the photographs provided by Dr. Saio given below. The photograph on the left shows the microstructure of freeze-dried soybean protein curds made by isoelectric precipitation. (Bar designates 10μm)

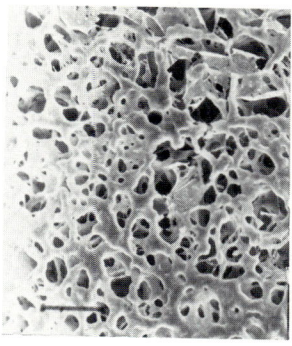 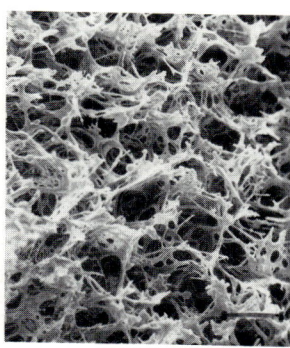

The photograph on the right shows the microstructure of freeze-dried soybean protein curds made by calcium precipitation. The isoelectric precipitated soybean protein curd shows expanded round cavities, whereas calcium precipitated curd has a more extended filament structure. These pictures are very similar to those of corresponding samples shown in Figure 10c and 10d in the text. For more details, refer to Gallant, D.J., Saio, K. and Ogura, K., 1976. The observation of high-moisture foods with a scanning electron microscope using the cryo-techniques, JEOL News, 14(10), 6-10.

K. Saio: You discussed the globular particles on the SEM-images of soybean curds. The globular particles of soybean 7S and 11S globulins have quite minute size, 11S is less than 100A in diameter. Such a small molecule can never be observed on SEM-images. If you find a sort of globular structure on the SEM-images of not-heated proteins, they might be aggregates of proteins. What do you think about it?
Authors: We agree that the globular particles should be aggregates of proteins but not the original protein body of either 7S or 11S.

Additional discussion with reviewers of the paper "...Foreign Matter in Foods" continued from page 50.

D.N. Holcomb and M. Buri: Isn't it true that the FDA Technical Bulletin No. 2 published in 1977 is not considered to be complete, and as useful as the Technical Bulletin No. 1?
Authors: FDA Technical Bulletin No. 1 addresses the total area of inspection and analysis, whereas FDA Technical Bulletin No. 2 is, more specifically, a training manual for analytical entomology in the food industry, as its title suggests.

D.N. Holcomb and M. Buri: When you talk of insect fragments, what size are you referring to?
Authors: In general, the sizes of the insect fragments will vary depending on what point in the processing of the food the insect or insect part contaminated the food. Likewise, as mentioned previously, the size of the sieve used to collect such fragments limits what size particles are available for further analysis. Particle sizes smaller than the sieve will not be retained and will go undetected.

D.N. Holcomb and M. Buri: What length rodent hair is meant in discussion of pest fragments? Most hairs found in processed foods, in our experience, have usually been 2mm and less in length.
Authors: Likewise, this has been our experience.

D.N. Holcomb and M. Buri: How can you tell whether an insect fragment, approximately 1mm in size, came from a raw material or from the processing area if the insect fragment is found in a finished product? In many cases the raw material is "processed".
Authors: Contaminations with insects or insect parts are possible at any stage from the raw material to the consumer. Prevailing conditions of the raw material before and during processing would provide important clues to the origin of similarly sized insect fragments. Likewise, the type of processing as well as the structural and color changes anticipated in insect fragments would be useful to determine whether the product was contaminated in the early or late stages of processing. During milling operations, shearing or crushing action can occur and characteristic particle sizes and structural pattern can result which can subsequently be detected by either LM and/or SEM. Additionally, color changes resulting from bleaching and/or leaching of insect parts during processing would aid in the determination of the point of origin of the contaminant during processing.

D.N. Holcomb and M. Buri: Isn't it rather difficult to determine who is responsible for contamination in a product unless it is a part of a machine, a piece of glass with identifiable structures, etc? Many packages are not tamper proof.
Authors: Discovery of a contaminant in a packaged food product usually requires opening of the package, unless of course, the package was previously damaged and the contents were accidentally exposed. In the case of tampering and deliberate additions of contaminants to packaged foods, often times, simulated conditions and elaborate controls must be used to prove the point of origin of the food contaminant.

D.N. Holcomb and M. Buri: If a 1.0% diet level of asbestos had no deleterious effect on hamsters, do we know at what level the asbestos would be carcinogenic?
Authors: What, if any, dose levels of ingested asbestos greater than 1.0% would be carcinogenic to hamsters is not presently known.

D.N. Holcomb and M. Buri: If most of the glass particles originating in a manufacturing plant were 0.1 to 0.4mm in size, wouldn't detection by the consumer be difficult, if not impossible?
Authors: The detection of glass particles in a consumer food product would depend on particle size and concentration, as well as the mode of particle recognition (sight vs. texture) by the consumer. One or more glass particles may be present and escape either or both visual or textural detection. Since most solid and some semi solid foods are chewed during the ingestion process, the coarse texture and sharp edges usually associated with glass particles could be detected and recognized, and perhaps associated with concomitant physical damage to teeth and tissues within the oral cavity.

THE USE OF MICROSCOPY TO EXPLAIN THE BEHAVIOUR OF FOODSTUFFS — A REVIEW OF WORK CARRIED OUT AT THE LEATHERHEAD FOOD RESEARCH ASSOCIATION

D.F. Lewis

Microscopy Section, Leatherhead Food Research Association,
Randalls Road, LEATHERHEAD, SURREY KT22 7RY, ENGLAND

Abstract

Microscopy has been used to investigate a wide range of foodstuffs and many food properties have been shown to be related to the structures found by microscopy. The work carried out at the Leatherhead Food Research Association has demonstrated that bloom on chocolate is formed by crystals projecting out from the surface of the bar. It has also been shown that bloom formation is associated with polymorphic changes in cocoa butter and that these changes may be found through the bar and not just at the surface. Work on gelatin gels has shown that the number and thickness of fibres in the gel increase with ageing and that it is possible to obtain collagen-like fibrils from a gelatin gel. Work on pectin has shown that the build up of the gel occurs in four stages. Investigations on starch gels have revealed that gel firmness is related to both the state of starch grains and the extent of a network formed from material extracted from the grains. In the field of meat products the action of salt and polyphosphate has been shown to be one of dispersing myofibrillar proteins. The retention of fat in meat products has been related to the nature of connective tissue and the crystalline state of the fat within cells. The review covers a period from 1921 to the present day.

KEY WORDS: Food Microscopy; Chocolate; Cocoa Butter; Fats; Pectin Gels; Starch Gels; Gelatin Gels; Meat Products; Fatty Tissue; Adipose Tissue.

Introduction

The origins of the Leatherhead Food Research Association (RA) go back to 1919 and the first reported use of microscopy to explain food behaviour was in 1921. During the very early days the microscopy was carried out using microscopes belonging to individual members of staff as the following extract from the journal Research in July 1922 illustrates:

"The Director desires to acknowledge with thanks the generous gift of a microscope to the Association's laboratories by a member who wishes to remain anonymous. Up to the present the Association has not possessed a microscope of its own, those in use being the property of members of the staff; consequently this gift will prove a valuable addition to the equipment of the laboratories".

Since those early days the Food RA has produced over sixty reports in which microscopy features as an explanative tool. This total does not include the use of the microscope for bacteriological examinations.

A wide range of food products have been scrutinised by microscopy at the Food RA and these can be grouped as follows:

1. Chocolate and Fat based products
2. Toffee and Sugar based products
3. Gel based products
4. Fruit and Vegetable based products
5. Meat and Fish based products

For the purpose of this paper three of these groups have been selected, namely Chocolate and Fats; Gels; and Meat Products. In presenting work in these three areas the aim is to demonstrate the important contributions to the understanding of food behaviour to be gained via microscopical observations and to point out how this understanding can aid the food technologist.

Investigation of foodstuffs by microscopy uses essentially biological specimen preparation techniques although in some cases specially developed techniques are needed.

Part I Chocolate and Fat Based Products

Chocolate was one of the first materials to be studied microscopically and the particular aspect to be considered was bloom formation. Bloom is a defect in the surface appearance of chocolate which appears as a white powdery deposit. It is now known to be associated with changes in the polymorphic forms of cocoa butter.

Campbell and Clothier (1921) used light microscopy to study the effect of the rate of cooling on cocoa butter and the development of crystals in cocoa butter, Verberine, illipe, Nucocos, Extra Nucoa and palm kernel fat during slow cooling. From their observations they concluded:
i) that bloom was due to fat crystals on the surface of chocolate
ii) that during slow cooling fats tended to form colonies of fat crystals
iii) that cocoa butter formed crystals below its measured "melting point"
iv) that solid particles such as cocoa solids or charcoal did not affect the formation of colonies during slow cooling
v) that stirring during cooling tended to reduce the incidence of spots or colonies.

This work was followed a year later with more observations on these fats (Campbell, Hinton and Heather, 1922). The authors prepared samples of fats at 60°C on microscope slides and then observed the effects of rapid cooling to 0°C or −18°C, re-warming to 23°−24°C and storing at that temperature and of warming on a microscope hot stage.

They observed that cocoa butter was 'microcrystalline' on rapid cooling from 60°C and that this microcrystalline state was an unstable form at temperatures above 0°C. Cocoa butter in the microcrystalline state produced colonies of large crystals within 1 hr at 23°−24°C. They also showed that the melting point of a fat was dependent on its state of crystallisation.

They concluded that bloom formation on chocolate occurred when cocoa butter changed from an unstable state to its most stable state and that it could therefore be prevented by producing a chocolate bar with the fat crystals in the most stable state.

Further observations on bloom were made by Campbell and Clothier (1923). They observed that bloom could be seen in the microscope as four different forms:
i) Irregular extruding crystals
ii) Boundary crystals of closely packed minute groups
iii) Small isolated groups or tufts
iv) Scales of fat, each scale being at an appreciable interval from its neighbours.

The culmination of these three reports were some recommendations as to how to process chocolate to minimise bloom formation. These recommendations were considered by members of the Association and published as an agreed scheme (Anon, 1923). The main points of the recommendations were that chocolate should be grained before cooling; that cooling should be carried out slowly with stirring to 27°−28°C before coating and that after coating the chocolate should be allowed to cool fairly slowly.

The concepts of fat crystallisation in chocolate are now widely accepted and present day tempering regimes are designed with much consideration of fat crystallisation. From records of the discussion of the above work, however, it appears that these observations about different crystal states and forms came as rather a surprise to the confectioners of the day and led to a reappraisal of their procedures. This is a recurring feature of much of the microscopy carried out at the Food RA.

The influence of 'nutty' centres on chocolate was studied by Campbell and Hallas (1936). They prepared samples of unstable (microcrystalline) cocoa butter and grained cocoa butter on microscope slides and observed the effect of Brazil nut oil on these preparations. In both cases a change in crystallinity occurred, resulting in the production of large fat crystals, as the oil diffused into the chocolate. They concluded that oil diffusion from 'nutty' centres was a cause of bloom on chocolates. They suggested that coating the nutty centres with a gelatin/sugar syrup layer would prevent this sort of bloom formation. The effectiveness of this treatment was subsequently demonstrated.

Chocolate and fat systems were one of the first areas to be investigated by electron microscopy at the Association. A method of preparing samples was described by Jewell, Saxton and Meara (1969) and Jewell and Meara (1970). The method essentially consisted of freezing the sample in liquid nitrogen and preparing a replica of the frozen sample. Methods for removing liquid fat with detergents so as to allow better observations of fat crystals were also described.

The method for the electron microscopy of fat was applied to chocolate by Jewell (1972). By this time work at the Association and elsewhere had established that cocoa butter could be made to exist in six different polymorphic forms. These can be conveniently arranged in ascending order of melting point and named from I to VI as in the following table 1.

Table 1

Form	Melting point °C
I	17.3
II	23.3
III	25.5
IV	27.5
V	33.8
VI	36.4

Examination of normal chocolate showed that the cocoa butter in the chocolate consisted of a mosaic of fairly small fat crystals. This description applied to both the surface and fracture face of the chocolate (see Figures 1a and 2a). Examining bloomed chocolate showed the presence of large finger like projections of fat crystals from the surface of the bar (Figure 1b). The fracture face of the bloomed chocolate also showed changes in the fat crystal structure; instead of a mosaic of small crystals the fat now existed as larger, elongated crystals (Figure 2b). Hence the surface phenomenon of fat bloom on chocolate was shown to be connected with a more general recrystallisation in the chocolate bar.

Samples of pure cocoa butter which had been given temperature cycles designed to produce specific polymorphic forms were also examined by electron microscopy. The identity of the polymorphic forms was confirmed by x-ray diffraction studies.

The surface of cocoa butter prepared to contain a mixture of forms I and II showed little crystallinity although lamellae were seen in the fracture face of this sample. Form III showed a sparsely crystalline surface with some tubular crystals lying on the surface. Form IV (Figure 3a) showed a tightly packed mass of smallish needle like crystals. Form V (Figure 3b) is the 'normal' form for chocolate and this was seen to consist of fairly large rectangular crystals;

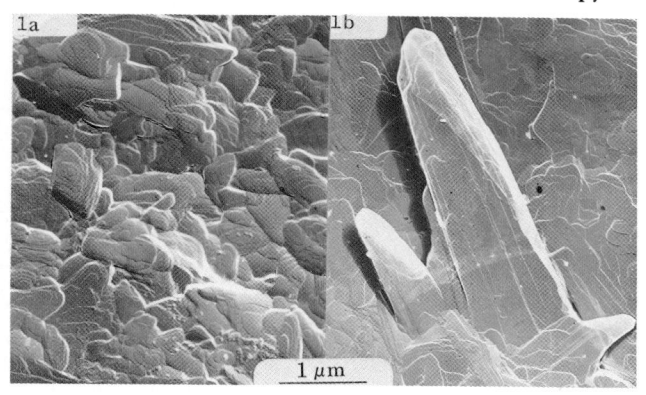

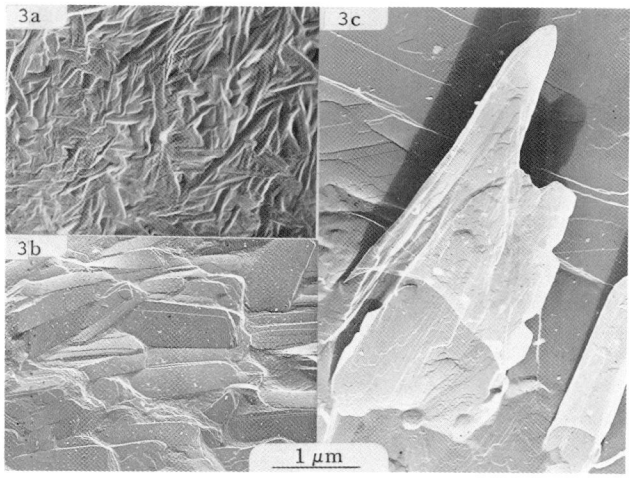

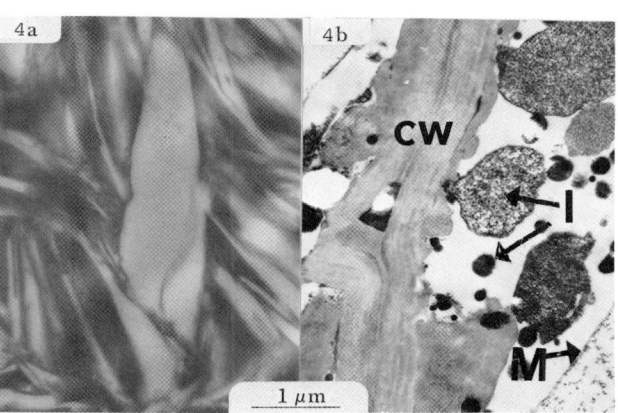

Captions to Figures 1-4

Figure 1a. Normal Chocolate surface replica. 1b. Bloomed Chocolate surface replica. Note that the normal chocolate has small flat crystals on the surface whilst the bloomed sample has large crystals protruding from the surface. Figure 2a. Normal Chocolate fracture face replica. 2b. Bloomed Chocolate fracture face replica. Note small crystals in normal chocolate but large needle or plate like crystals in bloomed chocolate. Figure 3a. Cocoa butter form IV polymorph. 3b. Cocoa butter form V polymorph. 3c. Cocoa butter form VI polymorph. All replicas. Note similarity of form VI crystals to bloomed chocolate surface. Figure 4a. Thin section of cocoa butter. 4b. Thin section of chocolate. Note clear crystals of solid fat in dark sea of osmium fixed liquid fat in cocoa butter. Chocolate section shows cocoa solids. cw — Plant Cell Wall, M — Membrane, I — Cell inclusions.

---

however, if the cocoa butter was stirred the crystals were smaller and started to resemble the mosaic structure seen in chocolate samples. Form VI (Figure 3c) was produced by keeping cocoa butter (as form V) at 26°C for 12 weeks; after this treatment bloom-like crystals were found on the surface of the cocoa butter and the crystals in the bar were now larger and elongated. Thus bloom formation was related to the polymorphic change from Form V to Form VI. Electron microscopy revealed small bloom crystals on a bar of cocoa butter (form V) after only 3 days at 26°C and by 7 days well developed bloom crystals were observed. Bloom on the surface of Brazil nut chocolates was examined and found to be morphologically similar to normal bloom.

Replica techniques for examining chocolate are good for revealing the fat crystal structure of cocoa butter but are not as good at showing the relationship between the various components in a bar of chocolate. Consequently, techniques for embedding and thin sectioning chocolate were investigated by Cruickshank (1976). He concluded that fixation in osmium tetroxide followed by dehydration and infiltration with the water soluble resin Durcupan gave satisfactory results if methyl nadic anhydride (MNA) was included with the resin. Using this approach, fat crystals were seen as clear areas whilst liquid fat stained darkly with osmium (Figure 4a). Milk particles, cocoa particles and the shapes of extracted sugar crystals could be recognised and high levels of lecithin gave rise to membranous structures (Figure 4b) in the fat.

The thin-sectioning techniques have recently been used to complement the replica technique and scanning electron microscopy in studying the role of lecithin and polyglycerol polyricinoleate (PGPR) in chocolate. This work has showed that lecithin tends to produce a more complete covering of sugar crystals by fat layers whilst PGPR seems to affect the relationship between fat and milk crumb particles. This difference in site of action accounts for the greater effect of lecithin in reducing the viscosity of chocolate. It also appears that the fat crystals were smaller in laboratory prepared milk chocolate than in laboratory prepared plain chocolate.

The effect of tempering on cocoa butter has also been studied. The electron microscopy of cocoa butter can be related to its state of temper, differential scanning calorimetry and x-ray diffraction data. It is concluded that the aim of tempering chocolate is to produce in the bar a network of small, fine fat crystals. If this is achieved, then the bar tends to be more stable since there is little liquid fat available to allow crystal growth and there is little available space for crystal growth. Also diffusion of liquid fat (from say, a nut centre) becomes more difficult. Slow crystallisation conditions tend to produce large Form VI crystals and with it bloom crystals. It should be pointed out that Form VI crystals in themselves were not necessarily harmful but that bloom is produced by conditions allowing slow crystal growth. They predicted that if a method of producing a fine network of small form VI crystals in a bar could be found then this would produce a stable chocolate.

Hence it can be seen that light microscopy suggested that tempering was necessary to produce a stable chocolate bar and that electron microscopy has helped to refine the theory of tempering.

In addition to bloom caused by fat crystals chocolate is also susceptible to sugar bloom. This phenomenon was investigated by Campbell (1925), Campbell and Clothier (1928) and Nicol (1940). These reports described the microscopical appearance of sugar bloom as being small, but complete sugar crystals. They confirmed ideas that this type of bloom could be caused by the presence of a film of moisture on the surface of the chocolate. Sugar could then dissolve in this film and subsequently crystallise out as small crystals. The main methods suggested for controlling this type of bloom were to store at fairly low relative humidity and to use suitable packaging materials. The factors within the chocolate which contributed to a susceptibility to sugar bloom were:
i) Low fat content
ii) Small particle size for sugar
iii) High air content in chocolate.

The low fat content and small sugar particle size encouraged the absorption of moisture by allowing a larger area of sugar to be uncoated at the surface of the bar.

In the case of air bubbles it appeared that crystallisation could occur within the air bubbles which might then become exposed at the surface. Air content did not affect the moisture uptake of the chocolate. High air contents were associated with using untempered chocolate or operating enrobers at too low a temperature.

A number of techniques including light and electron microscopy were used to explain the behaviour of different fats in incorporating air in cake batters.

The initial observations demonstrated that fats which incorporated air well had small uniform fat crystals present in small clusters of fairly uniform size. The fats which did not incorporate air well had rather larger crystal clusters. Electron microscopy demonstrated the presence of a crystal lattice at the edge of the air cells and naturally small crystals fit more readily into this lattice than larger ones. Electron microscopy also showed the poor performance fats to have larger crystals. A few fat samples did not follow the simple crystal size vs performance relationship. These were re-arranged lards which performed quite well but had large crystals when visualised by light and electron microscopy. When viewed by electron microscopy the crystals in these samples were seen to have 'stress' marks and an examination of these samples after creaming with sugar showed that the crystals broke down — presumably at these stress marks. With the normal fats the crystals appeared to survive creaming more or less intact. This work has been published by Berger, Jewell and Pollitt (1979).

The specimens shown in Figures 1-4 were prepared by the techniques described below.
a) Replicas. Replicas of both the moulded surface and a fractured face of the bars were prepared. The bars were fractured either at ambient temperature or −190ºC. The fractured surfaces and moulded surfaces were frozen in liquid nitrogen and then stored under liquid nitrogen until required. The samples were then transferred rapidly to a precooled brass block in the vacuum coating unit. When a vacuum of better than 5×10^{-4} torr had been achieved, a thin layer of platinum was first applied from a source at an angle of 45º to the chocolate surface. After this a thicker layer of carbon was evaporated from a carbon arc source directly above the sample. The replica was released from the substrate by immersion in diethyl ether. The remaining traces of chocolate were removed by treatment with a solution of biological detergent for up to one week. Pieces of replica were then briefly rinsed in water and left in sodium hypo-chlorite solution for one to two hours. The pieces were finally rinsed in water once more before being picked up on copper electron microscope grids for examination.

b) Sections. The technique for thin sections was developed as a result of prolonged experimentation with different embedding resins and fixation and dehydration procedures (Cruickshank, 1976). The final processing schedule was as follows:
1) Small pieces of chocolate of less than 1 mm^3 were fixed in 2% Palades fixative for 3 weeks at 4ºC.
2) The samples were dehydrated by immersion in Durcupan resin. The resin was changed after half an hour and then once more after a further hour had elapsed, the total dehydration time being 2½ hours. Dehydration was carried out at 4ºC.
3) Samples were then transferred to the complete embedding mixture for 1 hour and into fresh mixture overnight at 0ºC.
4) The chocolate pieces were finally embedded in polythene capsules in fresh embedding mixture and polymerised at 60ºC for 48 hours.
5) The polymerised blocks were aged for a further two or three days at room temperature before being sectioned.

The resin recipe was as follows: Durcupan resin 5.25 g, Dodecenyl succinic anhydride (DDSA) 9.30 g, 2,4,6-tri-(dimethylamino methyl) phenol (DMP 30) 1.0 g, methyl nadic anhydride (MNA) 2.0 g, and dibutyl phthalate 0.3 g.

Part II Gel Based Products

The behaviour of starches was observed microscopically with respect to lemon curd production by Kerlogue and Campbell (1933). They observed maize starch after various degrees of processing and showed that sugar syrup helped to stop starch grains from breaking down. Higher sugar levels gave greater protection to the grains. They also found that the time of addition of citric acid to the mixture in lemon curd production was important. Citric acid promoted the breakdown of starch grains. If it was added too early in the boiling process then considerable starch grain disintegration occurred and this produced fat separation in the product. On the other hand if the citric acid was added at the end of boiling the starch grains remained largely intact and this produced a product with a soft consistency, liquid separation and loose cracks or flakes in the curd. Hence citric acid needed to be added about half way through the boiling. This produced highly swollen starch grains with some disintegration and results in a high viscosity to reduce fat separation but also

Food Microscopy at the Leatherhead Food RA

Captions to Figures 5-9

Figure 5a. 5% low bloom gelatin matured at 10°C for 5 hrs. 5b. 5% high bloom gelatin matured at 10°C for 5 hrs. 5c. 5% low bloom gelatin matured at 10°C for 48 hrs. All freeze-etched replicas. Note that a fibrous network has developed very much earlier in the high bloom strength gelatin gel. Figure 6. All are negative stained preparations from gelatin gels which have been disrupted using trypsin. 6a. Acid prepared gelatin treated 1 hr with trypsin. 6b. Lime prepared gelatin treated 1 hr with trypsin. 6c. Acid prepared gelatin treated 1 week with trypsin. Note increased size of spindles with prolonged trypsin treatment and different nature of aggregates in acid and lime prepared gelatins. A — aggregates. Figure 7a. 1.2% pectin gel made with sucrose to 68% total solids and pH 3.0 fixed in uranyl acetate at 90°C. 7b. 1.2% pectin gel made with sucrose to 68% total solids and pH 3.0 fixed in uranyl acetate at 70°C. 7c. 1.2% pectin gel made with sucrose to 68% total solids and pH 3.0 fixed in uranyl acetate at 25°C. All are thin-sectioned. Note aggregation of pectin to form a network as the gel cools. Figure 8a. 1.2% pectin sol made with sucrose to 68% total solids and pH 4.0 fixed in uranyl acetate at 90°C. 8b. 1.2% pectin sol made with sucrose to 68% total solids and pH 4.0 fixed in uranyl acetate at 25°C. Both are thin-sectioned. Note that pectin only aggregates as clusters of fibrils and does not appear as a continuous network. Figure 9. 1.2% pectin gel made with glucose syrup to 68% total solids and pH 2.7, fixed at 25°C in uranyl acetate. Note that the pectin has formed large clumps rather than a continuous even network.

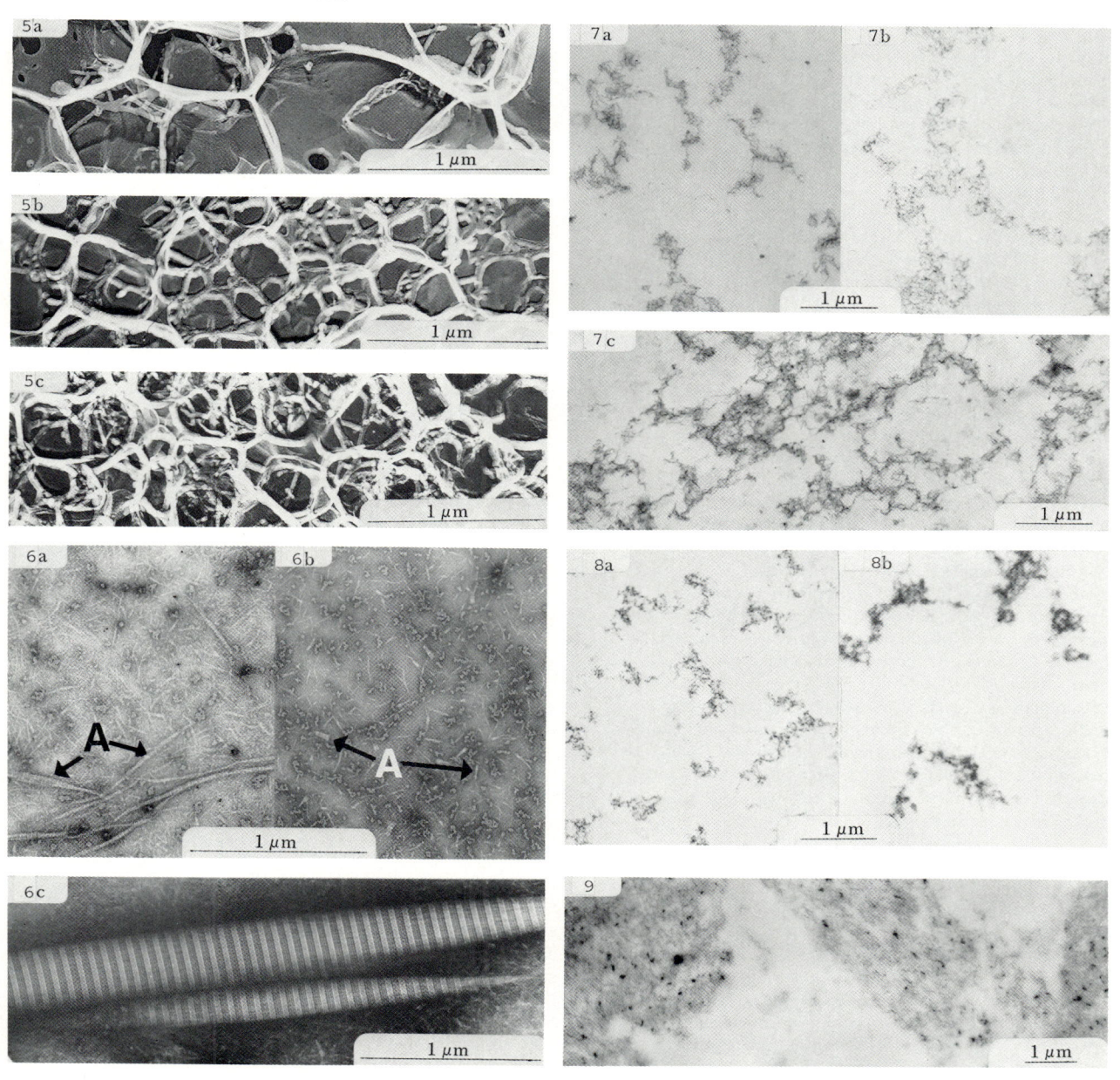

avoids syneresis which would cause liquid separation.

Smith and Hallas (1943) observed the behaviour of rice, potato, sago, maize, wheat and tapioca starches cooked in water to various temperatures. They related the microscopical state of the starch grains to gel rigidity and syneresis. They noted that the point of maximum rigidity tended to occur at about the point of maximum swelling of the granules, just as the granules were starting to break down. This applied to all the starches studied although the temperature of maximum swelling/rigidity varied according to the type of starch. As some rigidity was maintained after the starch grains had broken down they considered that not all the rigidity was due to the grains themselves, but that some was due to their contents extracted during cooking.

Further work on starches and lemon curd was reported by Smith (1949). He considered a starch gel as two phases with amylose as the binding layer and the amylopectin remains of the starch grains as filler. As gelatinisation proceeds the starch grains swell to a point where they start to collapse and break down. After the grains have broken down the rigidity is produced almost entirely by the amylose fraction. Syneresis is greatest where most intact grains are present.

The best condition for lemon curd production is to have half the starch grains fully swollen with the rest collapsed and broken. In this way syneresis is avoided as is fat separation. For a bakery curd the starch should be less gelatinised to allow for further cooking. An appendix to Smith's report (1949) describes a method for examining extent of starch gelation in a lemon curd.

The role of rusk in meat paste was also studied microscopically by Smith in Harvey and Ray (1950). It was concluded that rusk behaved substantially like normal wheat starch and again the state of starch grains was related to the fat separation and shrinkage in the meat paste.

Gelatin has been studied by electron microscopy at the Association. Acid pigskin gelatin gels were studied using freeze-etching and negative staining.

Freeze-etching indicated that gelatin gels were composed of a network of fibrous material which entraps the aqueous phase. The fibres in this network were usually about 200 Å in diameter and were composed of smaller fibrils about 35–40 Å in diameter. As the gel cooled and matured the fibre network became more tightly packed. It was proposed that an initial network was set up as the gel set and that this network became more extensive as the gel matured.

Examination of gelatins of different bloom strength showed that gelatin of high bloom strength layed down more fibres during the early stages of maturation. After 48 hrs maturation the high and low bloom strength gels had similar amounts of fibre but the fibres in the high bloom gelatin were rather thicker (Figures 5a, b and c).

In order to examine the gels by negative staining the gels were disrupted either by ultrasonic treatment or by digestion with trypsin. When the gels were disrupted with ultrasonics only very fine fibrils were seen, but in the enzyme disrupted preparation some of these fibrils had aggregated into large spindles which had the characteristic collagen staining pattern (Figure 6c). This observation showed that some of the gelatin molecules could reform into a collagen-like structure and it was considered that this mechanism was involved in the formation of the fibre network in gelatin gels. In other words the build up of the fibres in the gelatin gel was considered to be an ordered process based on the collagen structure rather than a random association of α-chains.

Negative staining procedures were applied to acid and lime processed gelatins. In this study it was found that the size of the banded spindles increased as longer times of trypsin treatment were used to disrupt the gel (Figures 6a, b and c). It was also found that the spindles could be obtained by holding the disrupted gel at its melting point in water without enzyme being present. Again there was some indication that the spindles increased in size with prolonged treatment.

A difference was noted in the negative staining pattern between acid and lime processed gelatins (Figures 6a and b). The acid processed gelatins produced banded spindles whilst limed gelatins tended only to produce small banded segments. This suggested that the mechanism of gel production was different in lime processed gelatin than in acid processed gelatin.

Work on gel structures has continued with a consideration of pectin gels. High methoxy pectin was studied and as this gels at low pH and in the presence of high sugar levels considerable problems of specimen preparation were encountered. A method involving fixation of the gel with uranyl acetate was developed and used to study the process of gel formation in pectin gels. Other methods were used to confirm specific observations.

The formation of a pectin gel on cooling from boiling point was seen as occurring in four stages. Firstly at high temperatures pectin forms into fibrils. The fibrils then aggregate into bundles up to 100 nm in diameter (Figure 7a). The third stage of aggregation occurs around the setting point of the mixture when the bundles start to join up to form a continuous network (Figure 7b). Finally more pectin is incorporated into this network to produce a firmer gel (Figure 7c).

When gelation is prevented because the pH is too high then the first two stages of aggregation still occur but the bundles do not join together to form a network (Figures 8a and b). Hence it seems likely that the ability of the system to gel or not is not explained by purely conformational change at the molecular level but is possibly dependent on the density of charged groups within the bundles.

If sucrose is replaced by glucose syrup then the gelling ability of the pectin is altered. Pectin/sucrose mixtures will gel at pH values of about 3.4 or less and over the range pH 2.8 to 3.4 firmer gels are formed at lower pH values. Pectin/glucose syrup mixtures will also gel at pH 3.4 or thereabouts but the gel becomes weaker at lower pH values. Electron microscopy was used to study this effect. It was found that the first two stages of pectin aggregation into fibrils and clusters occurs very rapidly in the presence of glucose syrup. Hence all the available pectin is incorporated into the fibrils and clusters whilst the system is still being stirred. The aggregation of clusters into a network cannot occur and the final system consists of many quite large and densely packed clusters which cannot develop into a network (Figure 9). This effect is intensified by lower pH values, Consequently glucose syrup produces very poor gels at lower pH values; this problem can be overcome experimentally by not stirring the system after buffering or by using a buffer which gives a slow pH fall (eg succinic anhydride) when firmer gels are produced.

Food Microscopy at the Leatherhead Food RA

Captions to Figures 10-13

Figure 10. All are scanning electron microscope preparations made by solvent dehydration and vacuum drying. All are gels formed after 2 hrs at 20°C after different cooking periods in a Brabender Amylograph. 10a. 10% maize starch cooked to before peak viscosity. 10b. 10% maize starch cooked to peak viscosity. 10c. 10% maize starch cooked beyond peak viscosity. Figure 11. Samples cooked as for Fig. 10 but fixed for transmission electron microscopy and thin-sectioned. 11a. 10% maize starch before peak viscosity. 11b. 10% maize starch at peak viscosity. 11c. 10% maize starch beyond peak viscosity. Note in this figure and Figure 10 the extraction of amylose and loosening of starch grain structure as cooking proceeds. Figure 12. Samples prepared as for Figure 10. 12a. 10% acid thinned maize starch at peak viscosity. 12b. 10% acid thinned maize starch plus 0.5% locust bean gum at peak viscosity. Figure 13. Samples prepared as for Figure 11. 13a. 10% acid thinned maize starch at peak viscosity. 13b. 10% acid thinned maize starch plus 0.5% locust bean gum at peak viscosity. 13c. 10% acid thinned maize starch plus 0.5% locust bean gum beyond peak viscosity. Note how the fragmentation of the acid thinned grains is delayed by the addition of locust bean gum to the mixture. In Figures 12 and 13 G — starch grain, M — cement matrix.

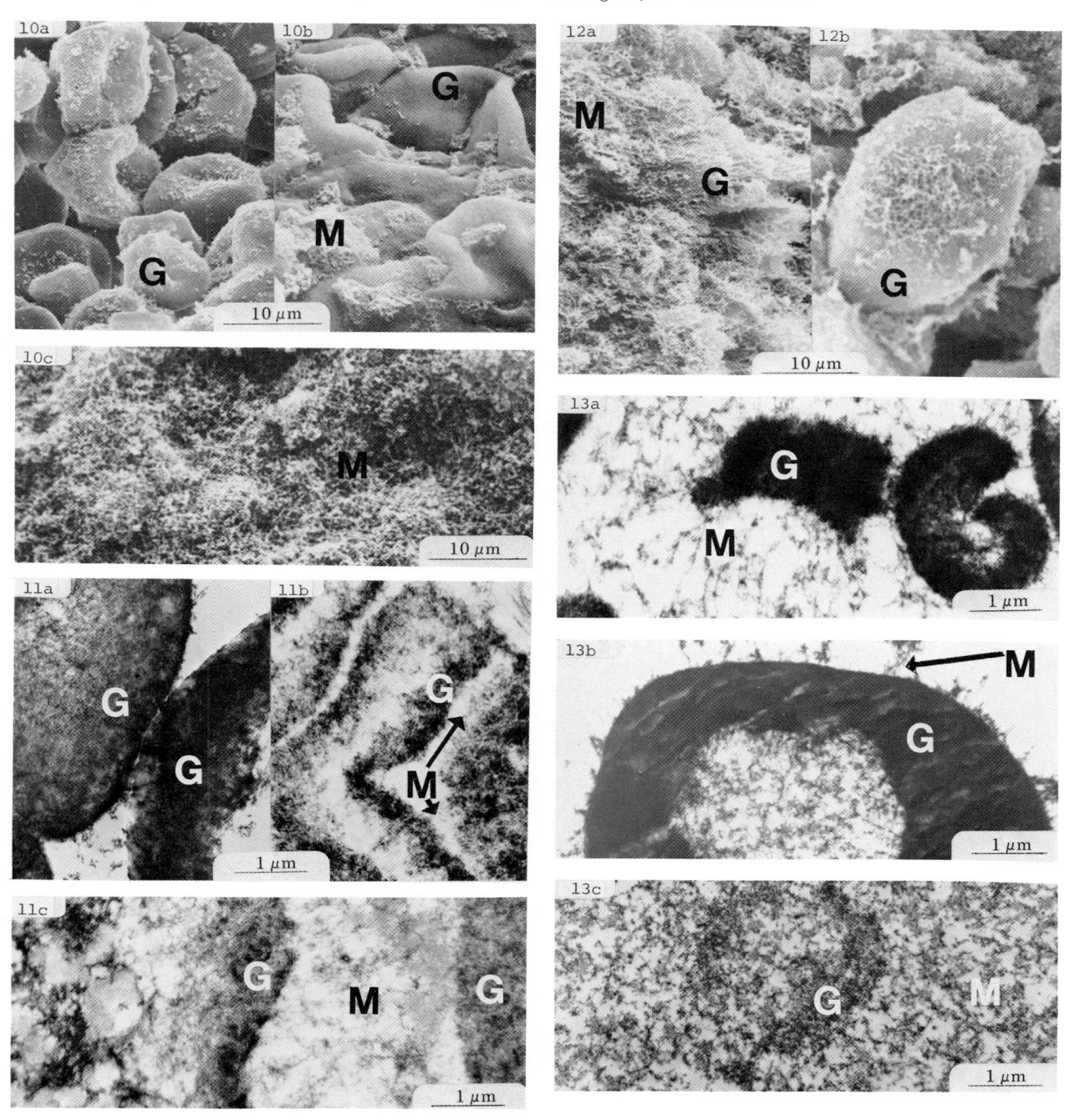

Recently work has been carried out on the electron microscopy of starch gels using both scanning electron microscopy and transmission electron microscopy. The starches examined were maize, acid thinned maize, and potato. Three different cooking procedures were used. These were: 1) in a Brabender Amylograph — taking samples at various stages during the cooking process, 2) in cans at 125°C or 140°C and, 3) at 95°C using different stirring procedures. Scanning and transmission electron microscopy (thin-sectioning and freeze-etching) were used to follow changes in structure.

As previously reported by Smith (1949) starch gels can be considered as two fractions, amylose acting as cement and swollen starch grains acting as filler. The nature of both these components influences the texture of the gel.

When normal maize starch is cooked in a Brabender Amylograph the starch grains swell and amylose is extracted from the grains. The viscosity of the paste increases to a maximum during cooking and then declines. The maximum rigidity of the gel formed occurs at about the point of maximum viscosity during cooking. In microscopical terms this point of maximum gel strength and viscosity occurs when the starch grains are fully swollen and virtually all the amylose has been extracted from them. Cooking beyond this point weakens the starch grains and consequently a weaker gel results (Figures 10a, b and c and 11a, b and c).

The effect of acid thinning on maize starch is to cause the starch grains to fragment during cooking and hence a very much smaller rise in viscosity is noted. A second effect is that the network produced from the extracted amylose is clumped (Figures 12a and 13a). The consequence of these two effects is to produce a weaker gel.

If the acid thinned starch was cooked with locust bean gum then the grains tended to hold together during the early stages of cooking and in consequence more viscosity developed (Figures 12b and 13b). This effect was short lived and the viscosity fell off sharply as cooking proceeded (Figure 13c). Locust bean gum also suppressed the extraction of amylose during the early stages of cooking so that a weaker gel resulted. After longer cooking however the locust bean gum seemed able to act as an extender for the gel network and so a firmer gel resulted.

During canning normal maize starch produces a weaker gel after cooking to 140°C than 125°C.

Microscopical examination led to the following explanation. At 125°C normal corn starch consists of very swollen starch grains surrounded by much extracted amylose so giving a fairly firm gel. Heating to 140°C causes the starch grains to break down and the amylose fraction is unable to bind together all the fragments so that a weak gel with a clumped network is formed.

As with chocolate it can be seen that microscopy has contributed to the understanding of mechanisms involved with gel formation. A common feature of the gels examined has been the presence of organised structures in the region of 100 Å to 1000 Å.

The following methods of specimen preparation were used for Figures 5-13.

1. Gelatin

a) Freeze-etching. Small samples (~ 1 mm cube) were cut and placed on a small freeze-etching stub. The sample was frozen by immersion in liquid Freon 12 at −150°C and the samples stored in liquid nitrogen. For preparation the samples were rapidly transferred to the cold stage (−150°C) of an NGN FE 600 freeze-etching device. A vacuum of 4.10^{-6} Torr or better was obtained and the specimen warmed to −100°C. The specimen was now fractured with a microtome at −194°C and after allowing etching to proceed for 3 minutes a carbon/platinum replica was made of the fractured surface. The specimen was removed from the freeze-etching device, allowed to thaw and then the replica was removed from the sample with concentrated hydrochloric acid.

b) Negative staining. A sample of the gel was partially disintegrated by mechanical means then given a mild enzyme treatment to provide a suitable preparation for negative staining. 1g of the gel was treated with 5 ml of trypsin solution (0.2 mg/ml) and held at 25°C until the gel was sufficiently broken down.

Tap water was used in preference to distilled water for the preparation of gels and trypsin solution as more consistent results were obtained. Trypsin activity is optimum over the range pH 7-8. The pH of the laboratory distilled water was 6.3 compared with that of tap water of 7.7. It was concluded that the use of tap water facilitated the enzymic breakdown of the gels and therefore resulted in preparations which were more suitable for negative staining.

A portion of the digest was mixed with an equal amount of the negative stain. 2% ammonium molybdate buffered with 0.5% ammonium acetate was used. A thin film of the mixture was deposited on a collodion-covered grid and allowed to dry.

2. Pectin

a) Thin sectioning. Samples of 2−3 mm diameter were taken from the gel and placed in 5% uranyl acetate solution. The uranyl acetate was adjusted, using acetic acid, to be of similar pH to the gel and was kept at the same temperature as the gel at the time of sampling. The sample was left in the uranyl acetate for 30 minutes and then washed in water for 2−3 hours. The samples were cut into ~ 1 mm cubes and placed in a 1:5000 w/v solution of ruthenium red in water. The samples were left in ruthenium red overnight or longer and the solution was gradually decolorised as the cubes of pectin took up stain. The cubes were washed in water for 1−2 hours and then dehydrated through a graded alcohol series starting with 70% ethanol and working up to absolute alcohol. Approximately 30 minutes was allowed in each alcohol. The samples were cleared with propylene oxide, infiltrated and embedded with Epon 812 resin and sectioned.

A second method of fixation was used in a few cases to compare with uranyl acetate fixation. This involved placing the samples directly into 70% ethanol for 2 hours followed by overnight treatment with a 1:2000 solution of ruthenium red in 70% ethanol. The samples were then dehydrated and embedded as above. This method could not be used for samples taken at 90°C and did not always work with samples at 70°C.

3. Starch Gels

a) Scanning electron microscopy (SEM). Cubes (~ 5 mm sides) were dissected from the samples and were dehydrated using either a graded series of ethanol or two changes of 2,2,-dimethoxypropane (2,2,DMP) acidified with 1 drop of concentrated hydrochloric acid per 50 ml of 2,2, DMP. After dehydrating the samples were placed in an Edwards 12E6 coating unit. After drying the cubes were fractured and attached, fractured face upwards, to SEM stubs with double-sided adhesive tape. The samples were coated with a conducting layer of gold using a Nanotech sputter coater and examined in a Cambridge Stereoscan SIIA scanning electron microscope operated at 20 kV.

Samples of some starches were also prepared by freeze-drying. In this case small pieces of the gels were frozen in liquid nitrogen and freeze-dried at $-50°C$ and 10^{-5} torr. using an NGN FE600 freeze-etching device. The samples were fractured, coated and examined in the same way as the solvent dried samples.

Transmission electron microscopy (TEM)

b) Thin-sectioning. One millimetre cubes of the starch gels were first treated with 1% periodic acid (normally in water but some samples dissolved and for these 1% periodic acid in 70% ethanol was used) for 1 hour. The samples were washed 3 times in water for 15 minutes each and then immersed in 1% thiocarbohydrazide in 5% acetic acid for 1 hour. The samples were again washed with water, stained with 1% osmium tetroxide for 3 hours or overnight, washed 3 times with water and dehydrated either with graded ethanol solutions or 2,2, DMP (see SEM method). After dehydrating with ethanol the starches were cleared with propylene oxide and infiltrated and embedded with resin. Samples dehydrated with 2, 2, DMP were infiltrated directly with resin. Epon 812 resin was used for earlier specimens and Spurr low viscosity resin used for later ones.

c) Freeze-etching. A few specimens were prepared using the freeze-etching technique. One millimetre cube samples were placed on gold freeze-etching stubs with a small amount of water. The stubs were frozen in liquid Freon and transferred to the stage of an NGN FE600 freeze-etching unit. The specimen was maintained at $-100°C$ whilst a vacuum of 4.10^{-6} torr was applied. The specimen was fractured with a knife at $-196°C$, allowed to etch for 3 minutes and a carbon/platinum replica of the surface prepared. The replica was released and cleaned using hydrochloric acid.

Part III Meat Products

The application of electron microscopy to study meat processing was first carried out at the Association by Lewis (1974). The effect of salt solutions on pork meat structure was studied using thin-sectioning and freeze-etching techniques. The observations showed a marked change in the organisation of the myofibrillar proteins. Meat consists of long tube-like cells, or fibres, $10-100~\mu m$ in diameter. The muscle cells are characterised by striations which run across the cells when viewed under the light microscope. These striations are caused by the arrangement of the myofibrillar proteins into sarcomeres. The sarcomere can be considered as the basic unit of muscle cells and changes in the arrangement of proteins in the sarcomere have a large effect on the behaviour of meat on processing. The structure of the sarcomere is illustrated in Figure 14a.

The effect of salt is to disrupt the sarcomere particularly in the myosin rich 'A'-band region, protein is dispersed from the sarcomeres and this enables the meat to retain more moisture.

The effect of salt and polyphosphate in combination and the effect of heating pork meats after treatment with brines was also studied by Lewis (1974).

When electron micrographs of meat (24 hrs post mortem) are compared with similar preparations of freshly excised muscle the meat generally appears with less clearly defined features than the muscle. This is probably due to some precipitation of sarcoplasmic proteins within the sarcomeres. Figures 15a and b show the appearance of raw meat. After treating the meat with polyphosphate the myofibrillar proteins within the sarcomeres become more clearly defined (Figure 16a). When salt and polyphosphate are used in combination very much more disruption is produced than with salt or polyphosphate alone (Figures 16b and c).

The synergistic behaviour of salt and polyphosphate is well known in technological terms as promoting water retention in meat. The explanation of polyphosphate action has been that it mimics adenosine triphosphate and dissociates actomyosin into actin and myosin. The electron micrographs of salt and polyphosphate treated meats, however, show that the regions of the sarcomeres most affected are those where actin and myosin would be expected not to be complexed as actomyosin (Figure 16c). In particular the H-zones tend to be most dispersed.

The effect of heat on meat structure is quite marked. In general proteins are coagulated in situ and some shrinkage of all elements is observed. Where meat is heated without treatment with salts a loss of fine structure is seen. Thus the actin and myosin filaments are lost and the A and I bands are seen as areas of coagulated protein (Figure 17a). Heating after treatment with a salt solution reveals a loss of protein from the actin-rich I band (Figure 17b) whilst heating after treatment with both salt and polyphosphate produces a pattern of narrow darkly stained bands (Figure 17c). Examination of the spacing of the dark bands indicates that they originate from the region of overlap between the A and I bands; this region is where actin and myosin are most likely to form actomyosin in meat. Thus the most resistant region of the sarcomere even after treatment with salt and polyphosphate seems to be where actomyosin is formed.

It seems from the microscopical evidence therefore that the proposed mechanism of polyphosphate dissociating actomyosin is not the most important feature. An alternative explanation which is supported by the microscopical observations is that polyphosphate can remove and precipitate sarcoplasmic protein from the actin and myosin filaments and hence allow the salt to disperse these proteins more effectively. On heating the dispersed myofibrillar proteins form a gel which retains water within the meat structure.

Similar structures have been seen in commercial ham products. These results are presented diagrammatically in Figure 14b. This work has been reported generally by Lewis (1979).

The behaviour of fatty tissue and its interaction with lean meat in the production of comminuted meat products has also been studied microscopically at the Association.

Evans and Ranken (1975) measured the water and fat losses on cooking lean beef and pork fat, with and without added water and salt, after various times of comminution. They concluded that the major factor controlling fat loss in these systems was the extent of the damage to the cellular fatty tissue structure. Different fatty tissues behaved differently on processing, soft fats (jowl fat) being more resistant to damage than hard fats (flare fat). Once fat had been released from the cells it could be retained by a matrix of meat protein but there was little evidence of true

Caption to Figure 14.

Figure 14a. Diagram showing location of proteins within a single sarcomere of a muscle cell. 14b. Diagrammatic representation of the changes produced in the sarcomere structure by combinations of salt, polyphosphate and heat treatment.

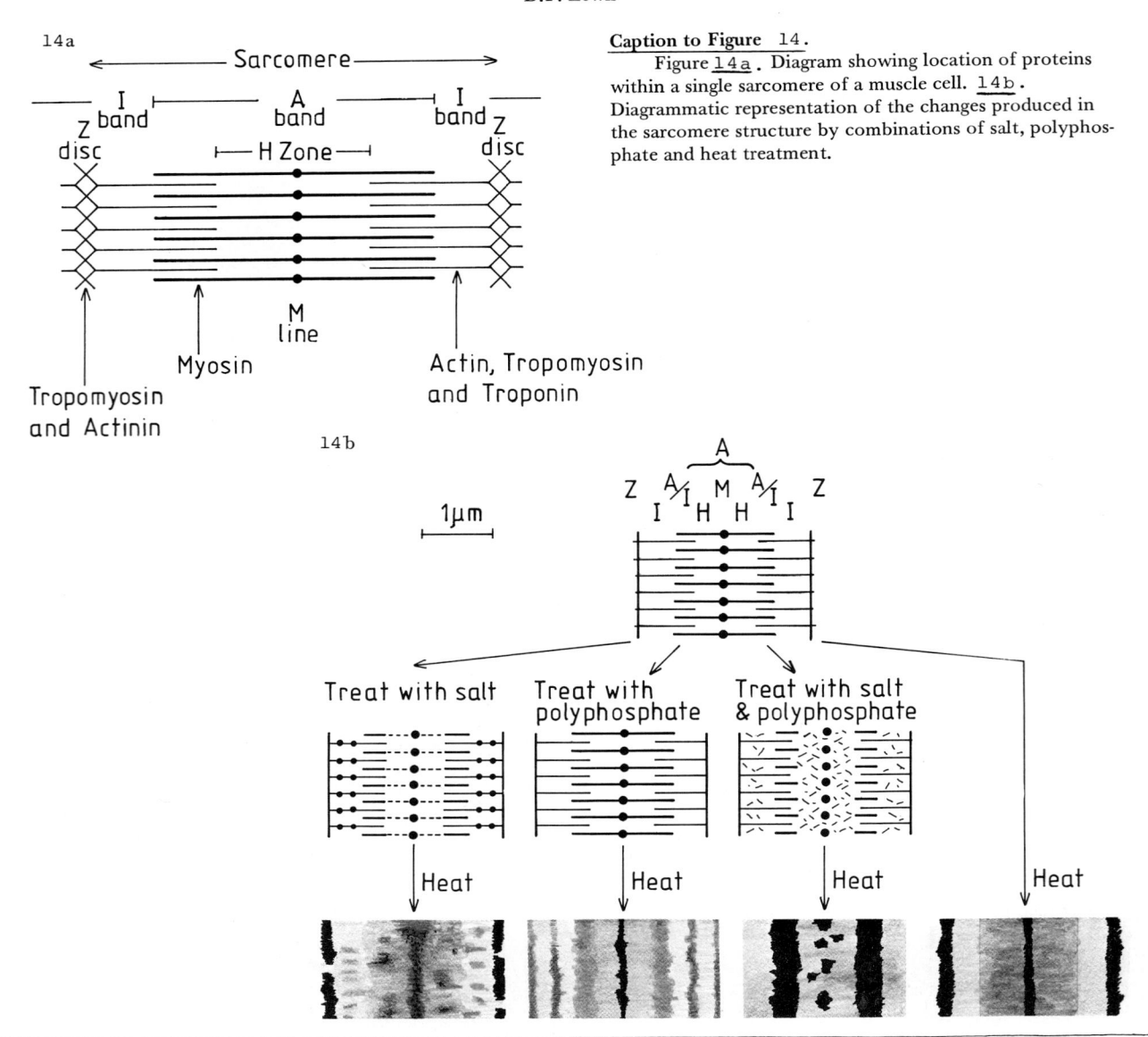

emulsification of fat. The consequence of this observation was that in manufacturing a coarsely comminuted product the conditions should be such as to minimise fatty tissue breakdown in order to reduce fat loss.

The structure of fatty tissues was studied by electron microscopy by Lewis (1979). Thin-sectioning techniques using long term osmium tetroxide fixation were used along with freeze-etching to study both the lipid and non-lipid parts of pork adipose tissue from different anatomical locations.

It was found that fatty tissue consisted of cells with connective tissue 'cell walls'. The cell contents were almost entirely lipid in nature. Differences in both the lipid and non-lipid parts of the fatty tissue structure were found between the different fats. Hard fats (flare or mesenteric fat) had highly crystalline fat within the cells (Figure 18b), thin connective tissue cell walls with poorly organised collagen fibres (Figure 19b) and tended to have quite large cells. This fat had a high fat loss after mincing and cooking. Soft fats (jowl or head fat) had much less crystalline fat (Figure 18a), smaller cells and thicker and more organised cell walls (Figure 19b). Intermediate fats (leg or shoulder fat) generally had features part way between the hard and soft fats; however they often showed a specific crystalline fat layer at the edge of each cell.

This work has been followed up by observations on the behaviour of fatty tissue on processing. This work was mainly light microscopy and involved using a stereological counting procedure to assess the extent to which different features were produced on processing.

Five areas of each sample were photographed and prints were produced on a format 15 cm x 10 cm and a final magnification of 60–100 times. A transparent film with 100 crosses evenly arranged in ten rows and contained in a

Food Microscopy at the Leatherhead Food RA

Captions to Figures 15-19

Figure 15a. Raw pork meat prepared by thin-sectioning after glutaraldehyde fixation. 15b. Raw pork meat prepared by freeze-etching after soaking in 30% glycerol. Note myosin rich A-bands (A), actin rich I-bands (I), Z-lines (Z) and M-lines (M) in the sarcomeres of the thin-sectioned preparation. Note also the looser arrangement of myofibrillar proteins when visualised by freeze-etching. Figure 16. Samples of pork meat prepared by thin-sectioning following soaking overnight in the following solutions. 16a. 1% sodium tripolyphosphate. 16b. 4% sodium chloride. 16c. 4% sodium chloride plus 1% sodium tripolyphosphate. Note the clearing of myofibrillar proteins by polyphosphate and enhanced dispersion of proteins by salt and polyphosphate together. Figure 17. Samples of pork meat cooked to 75°C after soaking overnight in the following solutions. 17a. Water. 17b. 4% sodium chloride. 17c. 4% sodium chloride plus 1% sodium tripolyphosphate. All prepared by thin-sectioning. Note the coagulation of proteins on heating. Also the dark bands (D) in the salt and polyphosphate treated meat which correspond to the regions of the sarcomere where actomyosin is likely to be. Figure 18. Samples of adipose tissue thin-sectioned following long term osmium tetroxide fixation. 18a. Pork jowl fat. 18b. Pork flare fat. Note more clear fat crystals in flare fat. Figure 19. Samples of adipose tissue thin-sectioned following short osmium tetroxide fixation. 19a. Pork jowl fat. 19b. Pork flare fat. Note the more organised connective tissue in the jowl fat.

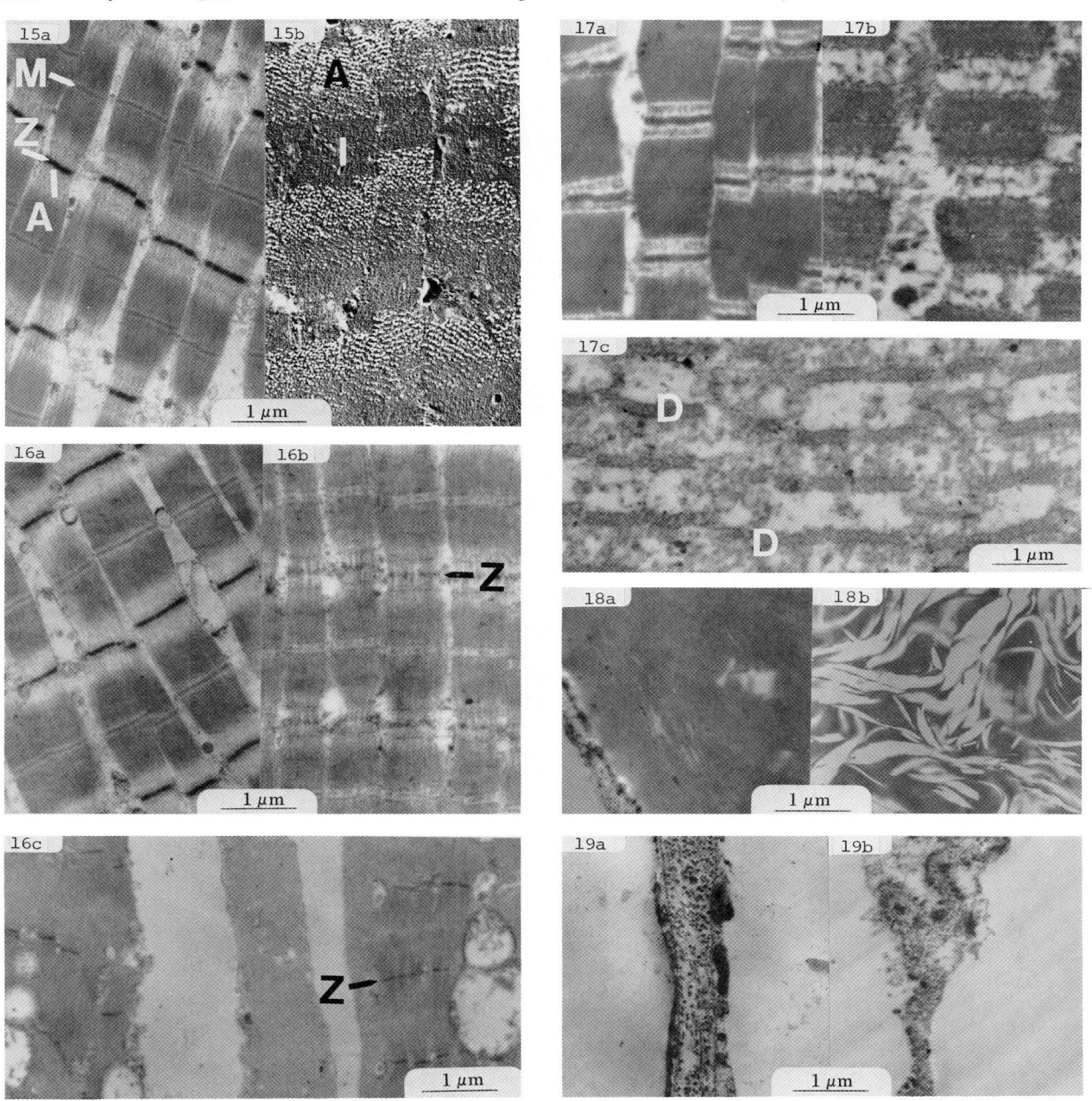

square 7 cm x 7 cm was placed over the print. Any feature touching the centre of a cross was recorded in one of five categories: Cell Groups, Isolated Cells, Broken Cells/ Connective Tissue, Free Fat and Empty Space. After counting the empty space category was discounted and the other features increased proportionately so that they totalled 100%.

In order to observe changes in components on cooking it is necessary to correct the figures obtained after cooking to allow for the cooking loss. The procedure for this was as follows:

The assumption was made that the stereological percentage of a component was equivalent to the weight percentage of that component. After cooking, if a sample gave a count of z% of a particular component it was assumed that 100 g of cooked sample contained z g of that component. On cooking the samples lost weight: thus 100 g of uncooked sample would give rise to y g after cooking. Hence the weight of a component derived from 100 g of uncooked sample would be $\frac{y.z.}{100}$ g. This is the figure corrected to allow for cooking loss. If the uncooked sample contained x% of the component, the percentage loss of the component can be calculated as $\frac{x - (y.z./100)}{x} \times 100\%$

Where x = percentage of the component deduced in the uncooked sample.
y = the weight of sample remaining after 100 g of sample had been cooked.
z = percentage of the component deduced in the cooked sample.

As fatty tissue is comminuted in a bowl chopper the fatty tissue structure tends to break down into the following components:
1) Groups of fatty tissue cells
2) Single isolated fat cells
3) Free fat liberated from fat cells
4) Broken cells and connective tissue.

The ratio of these components was measured at various stages of chopping with and without water and salt, and chopping the fat frozen and unfrozen using hard and soft back fat.

In general it was found that the fatty tissue broke down into groups of cells early in the chopping process and these broke down to produce free fat later. The conditions of chopping which affected the state of the fat cell contents and connective tissue were also found to affect the manner in which the fatty tissue was broken down.

Some work has been carried out at the Association to study mechanisms of retaining fat which has been removed from fat cells. In particular the stability of cold formed fat, water and protein 'emulsions' has been considered.

Light microscopy was used to study soya isolate, water and fatty tissue mixtures prepared in a laboratory blender.

It was observed that the 'free fat' was retained by a network of soya protein which formed on cooking. The state of this network was influenced by the presence of salt and the time of addition of salt.

Thus the ability to retain free fat by a protein isolate seems to be related to the extent to which the protein could be dispersed and its ability to form a network on cooking.

Salt improved the ability to produce a network but hindered the dispersion of the soya isolate.

Fat separation in meat pastes has also been studied microscopically.

Examination of commercial meat pastes microscopically showed that they relied on either a starch matrix or a caseinate soya matrix to retain the fat whilst maintaining 'spreadability'. In meat paste the meat is cooked before comminution with the fat so that the meat proteins are coagulated and do not contribute very much to the matrix. Instead the meat pieces serve to interrupt the continuous matrix and so help the paste to spread. In the paste mixture before it is finally heat processed none of the components form a continuous matrix. In order to prevent the paste from separating out while cooking in the jars the fat must be distributed finely through the mixture. On cooking the starch takes up water and swells to become the continuous phase and immobilise the fat.

Water and fat binding in meat and meat product is a complex subject and there are many mechanisms which may operate. Mostly the mechanisms are structural rather than chemical and the key to their understanding involves the use of all types of microscopy. In the case of meat products the work has involved quite close liaison between the microscopists and the meat technologists and it has been possible to turn the mechanistic theory to practical use in producing stable meat products.

The methods used to prepare specimens for Figures 15 - 19 are given below.

Meat and Fat Samples

a) **Thin-sectioning.** The pieces were placed in 3% glutaraldehyde in 0.1M cacodylate buffer for 1—2 hours. The samples were rinsed overnight in cacodylate buffer, or distilled water. Depending on the aim of the investigation the fatty tissue samples were then placed in 1% osmium tetroxide solution buffered to pH 7.2 with a phosphate buffer for a time upwards of 2 weeks at 0°—5°C. For meat samples or where the investigation was aimed at observing changes in the interstitial connective tissue, no osmium fixation was used. In the fatty tissue samples which had not been fixed with osmium the dehydration, cleaning and infiltration times were extended to encourage fat extraction before embedding. Thus dehydration through an alcohol series was over a period of 4 hours, clearing was in propylene oxide for 1 to 2 hours and infiltration was overnight at 40°C in a 50:50 propylene oxide/Epon mixture. Samples which had been fixed for a long time in osmium were dehydrated over a period of 2 hours, cleared in propylene oxide for 30 minutes and infiltrated overnight in a 50:50 propylene oxide/ Epon mixture at room temperature. All samples were embedded in Epon 812 which was cured overnight at 60°C. After embedding, thin sections were cut and these were generally stained with 4% uranyl acetate for 30 minutes at 60°C followed by 0.5% lead citrate in 0.2N sodium hydroxide for 5 minutes to improve contrast.

b) **Freeze-etching.** A few samples were freeze-etched. The cubes were placed in 30% glycerol solution until used; where a long delay was anticipated 1% glutaraldehyde was added to the glycerol solution. For freeze-etching the samples were placed on gold discs, frozen in liquid Arcton 12 and transferred to the freeze-etching device. An NGN FE600 was used and the specimen etched for 3 minutes at a temperature of —100°C and a vacuum better than 4×10^{-6}

torr after fracturing. A carbon-platinum replica of the etched surface was produced. After removing the specimen from the freeze-etching device the replica was cleaned using alcoholic caustic soda.

Conclusions

If microscopy has a role to play in food science and technology then its findings have to be applicable in practice. Over many years the Research Association has used microscopy and applied the results.

The early work on chocolate showed how an understanding of the nature of crystal formation in cocoa butter led to the suggestion of a 'tempering' regime to produce more stable chocolate. Later work illustrated more precisely the nature of bloom on chocolate and re-emphasised the need to obtain the right crystal structure in chocolate.

On the subject of gels, faults in lemon curd were related to structural changes in the starch and different processes were devised to produce the desired structure in the starch and hence produce an acceptable lemon curd. Later work has outlined the mechanism of gelatin gel formation and maturation and revealed some differences between acid and lime processed gelatins. The study of pectin gels has shown a number of stages to be involved in forming a gel and shown how changes in conditions can affect the development of structure at these stages and hence change the properties of the gel. Returning to starch gels electron microscopy is now pointing the way to better control of gel texture by allowing the components of the gel to be controlled.

With meat products consideration of the structures involved in holding water and fat within the products and the way that processing conditions affect the formation of these structures has led to a better understanding and control of manufacturing conditions.

Similar observations have been made at the Food RA in the fields of fruit and vegetable products and toffee and sugar based products. In this way microscopy has and is contributing to food technology.

References

Anon. (1923). Prevention of bloom. *Research* No. 12, Suppl. No. 3.

Berger KG, Jewell GG, Pollitt JM. (1979). Oils and fats. In Food Microscopy, ed Vaughan JG, publ Academic Press London, New York and San Francisco, pp. 445–497.

Campbell LE. (1925). Sugar bloom on chocolate I. *Research Records* No. 7.

Campbell LE, Clothier GL. (1921). Bloom and allied defects of the surface appearance of chocolate. *Research* No. 8, Suppl. No. 1.

Campbell LE, Clothier GL. (1923). Bloom and allied defects of the surface appearance of chocolate. *Research* No. 11, Suppl. No. 1.

Campbell LE, Clothier GL. (1928). Sugar bloom on chocolate II. *Research Records* No. 16.

Campbell LE, Hallas CA. (1930). Bloom on chocolate containing nutty centres. *Research Records* No. 27.

Campbell LE, Hinton CL, Heather JR. (1922). Crystallisation of solid fats. *Research* No. 10, Suppl. No. 1.

Cruickshank DA. (1976). Electron microscopy of chocolate Part II, Thin sectioning. *Leatherhead Food RA Technical Circular* No. 611.

Evans GG, Ranken MD. (1975). Fat cooking losses from non-emulsified meat products. *J. Fd Technol.,* **10** (1), pp. 63–72.

Harvey HA, Ray N. (1950). Separation and shrinkage of meat and fish pastes. *Leatherhead Food RA Research Report* No. 24.

Jewell GG. (1972). Some observations on bloom on chocolate. *International Chocolate Revue,* 27 (6), pp. 161–162.

Jewell GG, Meara ML. (1970). A new and rapid method for the electron microscopic examination of fats. *J. Am. Oil Chem. Soc.,* 47, p. 535.

Jewell GG, Saxton CA, Meara ML. (1969). The structure of fats. A preliminary examination with the aid of the electron microscope. *Leatherhead Food RA Technical Circular* No. 416.

Kerlogue RH, Campbell LE. (1933). Defects of lemon curds or cheese. *Research Records* No. 39.

Lewis DF. (1974). An electron microscope study of the factors affecting the structure of meat. PhD Thesis, University of Leeds.

Lewis DF. (1979). Meat products. In Food microscopy, ed Vaughan JG, publ Academic Press London, New York and San Francisco, pp. 233–272.

Nicol JM. (1940). Sugar bloom on chocolates III. *Research Records* No. 75.

Smith PR. (1949). Principles of the manufacture of lemon curd and their application. *Leatherhead Food RA Research Report* No. 19.

Smith PR, Hallas CA. (1943). The jelly properties of starches. *Research Records* No. 91.

Acknowledgements

All photographs are reproduced by permission of the Leatherhead Food Research Association and I am grateful to the Director, Dr AW Holmes, for permission to present this paper. I am also grateful for the support of the Ministry of Agriculture, Fisheries and Food for much of the work on gels and meat products. Finally, I thank all those microscopists, past and present, who have established and maintained the role of microscopy at the Food RA and so made this paper possible.

Discussion with Reviewers

M. Kalab: Are reprints (copies) of all the Leatherhead Food RA reports listed in the section References available?
Author: Yes, although a charge may be made.

G.G. Jewell: Is there any evidence for change of polymorphic form of the fat phase during the production of replicas?
Author: Polymorphic changes would be most likely to occur during the freezing stage of specimen preparation. This freezing step is also used in most differential scanning calorimetry and from these studies the following conclusions may be drawn:

1) If a liquid tempered or untempered fat is frozen then some of the liquid fat will solidify as the lower melting point polymorphs.

2) In a solidified fat where there is relatively little liquid fat then the formation of the lower melting polymorphs is less marked and there is no evidence to suggest that the fat crystals present in the bar before freezing change their polymorphic form on freezing.

S.H. Cohen: Doesn't fast cooling give a better gloss to candy than slow cooling?
Author: The relationship between cooling rate and stability of the chocolate is not a simple one. The important point is to produce a chocolate with a large number of small crystals in a stable polymorphic form. Cooling rapidly to a low temperature may induce small crystals in an unstable form. The practical significance of this can be seen by comparing moulded bars and enrobed products. A moulded bar will normally cool more slowly than chocolate on an enrobed product and the chocolate will be tempered to differing extents to suit the likely cooling rate of the chocolate.

G.G. Jewell: How much of the fat phase survives the fixation and embedding procedures?
Author: A variable amount! The actual amount will depend upon the ratio of liquid to solid fat in the sample, the size of crystals, the size and extent of other particles in the system, the temperature of fixation in osmium tetroxide, concentration of osmium tetroxide, length of time of fixation in osmium tetroxide, conditions of dehydration and embedding, type of resin used and probably many more factors.

A. Bridges: In the preparation of gels for electron microscopy using a solvent drying technique, the inherent nature of the gel results in acute shrinkage at a critical solvent concentration. Have you observed this problem? Have you found a solution to it?
Author: We do encounter this problem although it is not so pronounced with starch gels as with some other gel systems. It is a problem encountered with dehydration of most biological specimens for SEM and TEM and micrographs have always to be interpreted with the specimen preparation procedures in mind. The main alternative to the use of solvents is to use freezing techniques and these are useful as a check for other techniques. Freezing techniques have their own problems with ice crystal damage.

A. Bridges: Have you been able to correlate your observations on corn (maize) starch gels in scanning electron micrographs with your results in freeze-fracture studies?
Author: Yes, micrographs prepared by freezing techniques can generally be correlated to solvent dehydrated systems. For example, freeze-etching of acid-thinned maize starch systems shows the fragmentation of the starch grains more clearly than any other technique (see Figure 20a). However, as can be seen, freezing in Freon or liquid nitrogen produces ice crystals which make interpretation difficult. This problem is particularly marked when trying to freeze larger specimens for SEM (see Figure 20b).

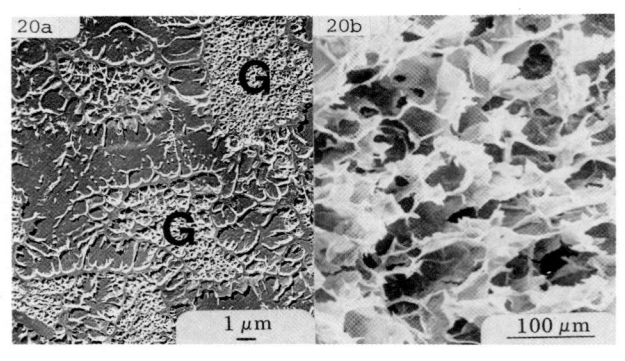

Figure 20a. Sample of acid-thinned maize starch cooked to peak viscosity prepared by freeze-etching. Note fragmented starch grains — G. Figure 20b. Sample of acid-thinned maize starch cooked to beyond peak viscosity SEM prepared by freeze-drying.

G.G. Jewell: In the case of pectin gel studies, was there any correlation between the thin-sectioning results and those obtained from freeze-etching or negative staining?
Author: Correlation was found in the final gels especially where clumping had occurred through too rapid gelling. Freeze-etching was not attempted on hot samples so that the build up of gel structure was not followed by freeze etching. The high sugar content made it impossible to etch the samples in the usual way and so untreated samples prepared by freeze-etching tended to show little structural detail. Freeze-etching revealed structures similar to those seen by thin-sectioning if the sugar was removed by similar methods to those used to prepare the gel for thin-sectioning, i.e. fixation with uranyl acetate. Negative staining was unrewarding.

C.A. Voyle: Did the use of shortened muscle with increased overlap of actin and myosin filaments result in a variation in the effect of salt/polyphosphate treatment?
Author: I did not deliberately look at shortened meat although there was a fair amount of variation within the samples examined. Variations in the patterns obtained after salt/polyphosphate treatment were found and the most likely explanation of this variation is connected with the degree of overlap in different cells.

S.H. Cohen: Has your laboratory had any experience using TPP on comminuted meats such as in the "Comitrol" process?
Author: We have looked at the performance of meats in comminuted products and at the effect of sodium tripolyphosphate on this performance but the results are currently only available to Member companies. We have not yet looked at the "flaked" products (such as "Comitrol") to any extent but have concentrated on products made in a mincer or bowl chopper.

Editor: Please provide details of Palade's fixative.
Author: For details of this fixative please see Palade, GE (1952): A Study of Fixation for Electron Microscopy, J.Exptl.Med., 95, 285. The details are also available in M.A.Hayat's Principles and Techniques of Electron Microscopy, Vol. 1, Van Nostrand, NY, p.335-336 (1970).

IDENTIFICATION OF FOREIGN MATTER IN FOODS

J. T. Stasny, F. R. Albright*, R. Graham*

Biological Division
Structure Probe, Inc.
West Chester, PA 19380

*Lancaster Laboratories, Inc.
Lancaster, PA 17601

Abstract

It is the purpose of this paper to review and illustrate recent, pertinent, advances in detection methods for those food particle contaminants that have potential as hazards to human health or that elevate the risk for product liability disputes.

Sources for the contamination of the human food supply include extraneous particulates associated with raw materials, food handling and packaging processes, as well as extraneous filth from insects and rodents. Likewise, environmental particulates, either airborne or in liquid suspension, can be considered as potential food contaminants.

Concern for the quality of human foods relates to the type and quantity of suspected contaminants. To reduce the risk of particle contaminations of all types in human food supplies and to insure high yields of safe, high quality products, industry safeguards include monitoring and identification of foreign matter.

Supportive methodology has been developed to detect, separate and identify particle contaminants of many types. Microscopic contaminants are characterized using light microscopy, electron microscopy and allied techniques. Specific applications of these techniques are described in brief detail.

KEY WORDS: Extraneous matter, particle contaminations, microscopic contaminants, analytical methods, quality control, product safety, product liability

Introduction

Extraneous material in foods, by definition, implies that it should not be present and is unwanted. However, most consumers would be surprised (and some horrified) to know that virtually no food is completely free of extraneous matter. In order to place some perspective on the issue of extraneous materials in foods it is often necessary to define the "foreign matter" in the context of a particular foodstuff and determine why and to whom it is important in our society.

The U.S. Food and Drug Administration is charged through federal legislation to monitor our food supply with the intent of insuring the delivery of wholesome and sanitary foods to consumers. The first attempt at regulating our food supply by examination of foods and drugs for adulterants dates back to the initial Federal Food and Drug Act passed by Congress in 1906. The first working guide aimed at training analysts in the identification of extraneous filth was Food and Drug Circular No. 1 entitled "Microanalysis of Food and Drug Products" and published in 1944. This has been superseded by FDA's Technical Bulletin No. 1, "Microscopic - Analytical Methods in Food and Drug Control," which is now out of print, and Technical Bulletin No. 2, "Training Manual for Analytical Entomology in the Food Industry," published in 1977. These documents provide useful background and scope not only to laboratory analysts, but also to inspectors, administrators and attorneys.

Specific methodology for the isolation and identification of extraneous materials usually appears as original contributions in the Journal of the Association of Official Analytical Chemists (AOAC) and subsequently in the Official Methods of Analysis of AOAC, assuming the method works in the hands of other volunteer collaborators. These collaborators are normally from both industry and government circles. Quite naturally most literature on the subject of extraneous matter in foods comes from government laboratories, because of their regulatory interest in the subject.

The definition of extraneous matter is dependent on our expectations. For example, we expect carrots freshly pulled from the earth to have soil particles clinging on the surface, but we don't expect such grit to be present in canned or frozen product. Likewise, we know that bees make honey and pollinate fruits and vegetables, but whole bees (or parts thereof) are considered extraneous matter in the honey or the dried fruits we eat. One way of differentiating extraneous matter is to categorize it as natural vs. unnatural. It never has been possible to grow crops, harvest, and process them totally free of natural contaminants, but it is possible to control this "natural" foreign matter to a reasonable and acceptable level. FDA has a long list of Defect Action Levels for various foods and ingredients which are considered upper limits of extraneous filth and above which legal action will be taken. It is also possible through Good Manufacturing Practices and proper sanitation to control unnatural contaminants like paint chips (from flaking ceilings or machinery) or rodent hairs.

The prime concern for all contaminants is whether or not they represent a hazard to human health. Extraneous matter can be classified as either a direct or indirect health hazard. Insect fragments are not a hazard per se, but the pathogenic or food spoilage organisms which insects can carry are a real concern. The same is true for rodent hairs. Hair identification is important in food analysis because hairs can be indicative of the presence of bacterial-laden feces (Vasquez, 1961). Rodents groom their fur by licking their coat and in the process they swallow some hair which survives the digestion process and ends up in pellets of excrement. Detection of rat or mice hair suggests defilement of stored product and improper storage conditions. The ability to identify the animal source of hairs by microscopic analysis is vital because hair from other fur bearing animals like rabbit are commonly used in gloves and coats. Finding such hairs in food, one would draw the conclusion that changes in handling the product are suspect rather than the more severe indictment of insanitary storage suggested by rat hairs. A very detailed description of hair analysis by light microscopy is reviewed by Vasquez (1961), but tutelege by an expert and comparisons with a battery of reference hairs offer the best chance of successful hair identification.

In recent years scanning electron microscopy (SEM) has supplemented conventional compound light microscopy as an identification tool. Unless a finished product has been invaded by insects after processing, the evidence for prior insect contamination is found in the form of fragments. A little detective work may be required to determine whether the fragments originated from the raw ingredients or came from the processing areas. A series of papers by Brickey and Gecan (1974) and Kvenberg (1977, '78, '81) examined the use of SEM to determine insect species from the morphology of elytral and mandible fragments. The three-dimensional depth of field afforded by SEM aids greatly in associating surface sculpture patterns, setae, and setal pits with insect family and even aids in species identification. Thus an assessment of the origin of the insect contamination can be made and appropriate processing changes implemented to correct the problem.

Apart from a health hazard posed by extraneous matter in foods and the potential for regulation and adverse publicity, there is also a potential for consumer lawsuits. Nothing has yet hit the food industry like the multitude of lawsuits filed against asbestos processing and manufacturing companies, but nuisance suits are a constant possibility. Claims of illness because of a piece of foreign matter, or breaking a tooth on a hard piece of debris in a food cannot be lightly dismissed; these usually end up in the hands of insurance companies and require the expertise of microscopists in identification of the extraneous material. Often third-party independent laboratories are engaged for this function.

Light and electron microscopy play an important diagnostic role in helping to assess the composition of a foreign substance, but frequently they also provide clues of how or at what stage in the processing the contamination occurred. Such laboratory detective work helps determine who is responsible or liable for the contamination--the provider of raw ingredients, the producer, the storage facility, the shipper, the retailer, or the supplier of the packaging materials. A simple illustration of assessing blame for insect infestation is provided by studies performed by Brickey, et. al. (1973), who showed that examination of various packaging materials can reveal whether the insects bored into the food item from the outside of the package (implicating poor storage conditions) or came from the inside (suggesting insanitary food production). A laboratory also needs to be alert to the possibility that a consumer has "planted" the rodent, insect, glass fragments, etc. in the food item. Knowledge about the food processing steps or site of manufacture are often helpful to the microscopist in deciding what to look for in his examination. An insect or rodent may be indigenous to the geographical area from which the complaint came and not from the area where the product was manufactured. Similarly, processing steps may alter the foreign matter in a way that would distinguish it from freshly added extraneous matter, or the identification and size classification of asbestos, for example, may prove that it couldn't have come from the processing plant.

Contaminants from the environment or the processing of food deserve further mention. Eisenberg (1974) discussed inorganic particulate matter which enters the food chain because of specific operational practices during production. These included sand and soil particles, asbestos, and glass particles. Sand and soil come primarily from ineffective cleaning of produce; asbestos may come from additives or filtration media; and glass particles arise from glass containers when chipping, rough handling, etc. occurs during filling or closure. Similarly, paint chips in food can reflect a poorly maintained physical plant and metal fragments might mean improper adjustment of or old machinery.

A more in-depth discussion of two inorganic contaminants posing potential health hazards, lead and asbestos, illustrates the increasing use of sophisticated instruments and methods in microanalytical techniques in the food industry.

Lead

In surveys conducted by the Food and Drug Administration (FDA) it was shown that about 30% of the lead in adult and infant diets is furnished by canned foods (FDA 1975). Likewise, Thomas et. al. (1973) demonstrated that levels of lead were higher in canned products than when the same products were fresh. It is generally believed that the higher concentrations of lead in the canned products are a result of contamination from the canning process or from the soldered seam and, at least partially, are due to the presence of particulate Pb dust produced during the can-seam soldering process. However, highly variable and conflicting results for lead analysis have appeared in the literature.

To ensure reliable reporting of Pb concentrations in foods, for possible future regulations governing allowable levels of Pb in foods, Jones and Boyer (1979) have developed procedures for the improved homogenization of canned food samples prior to Pb determination.

Based on model studies employing particulate Pb in dilute HNO_3, followed by blending in a Polytron homogenizer and subsequent monitoring of the particle size by SEM, commercially canned foods were similarly prepared for analysis of lead by atomic absorption spectroscopy. From this work it was determined that the precision of the analysis by this procedure was significantly improved, and reliable data can now be collected upon which to base regulations to limit exposure of consumers to lead intake.

Asbestos

Sources for asbestos contamination in air, water and foods are ubiquitous and result from the widespread employment of asbestos in many commodities where its presence is dictated by its desirable physical and chemical properties and its favorable economics.

It is well known that epidemiologic data clearly associate inhalation of asbestos with an increased incidence of cancer (Selikoff, et. al., 1970). Ongoing NIEHS oral asbestos studies in rats and hamsters represent a systematic attempt to assess the biological effects associated with primary ingestion of selected asbestos fibers (Moore, 1977), fed continuously in the diet at a 1% level.

The lifetime exposure phase of these experiments has now been completed, and a Status Report recently issued (Moore, 1981) related that there was no indication of major differences in the mortality rate between the hamsters receiving the asbestos diet or the control diet. Likewise, preliminary reports of the histopathologic examination of tissues from female hamsters indicate that no carcinogenic or cocarcinogenic effect was observed. A similar review of the male hamster is in progress. For rats, the lifetime exposure phase of the study has been completed, and preliminary evidence suggests that longevity was not affected by exposure to the various types of fibers.

Methods to isolate asbestos from a variety of foods and beverages have been developed (Stasny et. al., 1979; Albright et. al., 1979). This FDA-sponsored research details the separation and retrieval of deliberate additions of chrysotile asbestos fibers from alcoholic beverages and bulk food samples and their detection by SEM and energy dispersive spectroscopy (EDS) analysis. The results provide practical, workable protocols for potential diagnostic and/or regulatory application.

Finally, methodology to accommodate

the rapid, cost effective detection and identification of asbestos in a variety of environments and applications has been investigated using a fluorescent marker for light microscopy (Albright, et. al., 1980). The procedure can discriminate between chrysotile asbestos and amphiboles as well as non-asbestiform fibers.

Methods and Materials

Isolation, Detection and Characterization of Extraneous Fragments

Isolation of insect fragments, striated rodent hairs, synthetic fibers, paint chips, metal fragments, cellulose fibers, etc. from different food products seems like a routine laboratory procedure based on examination of the official AOAC methods (Chapter 44, 1980 edition). However, the ability to recover extraneous material reproducibly depends very greatly on experience and precise adherence to specified details of the method. Slight differences in manipulations between laboratories can result in significantly different findings. Moreover, final identification of isolated particulates requires training by experienced personnel and is considered something of an "art" by many in the field. The AOAC procedures generally employ techniques whose separatory features are based on specific gravity. The food sample may be dispersed in hot water, dilute acid or solvents, and after appropriate further steps (e.g. sieving), the particulates of interest are selectively segregated into organic solvents like mineral oil or heptane. Following filtration, the residues are examined by reflected light microscopy.

For the work reported here, reflected light optical microscopy (LM) was performed on isolated fragments from insects, paint and metal, as well as striated hairs and fecal matter. Sample material was transferred to the surface of either carbon SEM mounts or to silver membrane filters for recording of LM images. A Bausch & Lomb Stereo Zoom 7 microscope was used, fitted with a modified Polaroid SX-70^R Alpha Model II, featuring through-the-lens viewing. Specimens were originally photographed at a 37X enlargement using Polaroid SX-70 Time-Zero Supercolor film. The marker bar in each case represents an approximate distance of 0.5 millimeters.

Samples collected and transferred onto SEM specimen mounts were examined following sputter coating with a conductive layer of Au-Pd or by evaporating a carbon film over the mounted specimen. Secondary electron images were recorded onto Polaroid Land film Type 52 using a JEOL JSM-U3 SEM operated at 25 kV.

Magnifications of 300X were used, and the marker bar in each case represents a distance of 50 micrometers.

Silicates

Airborne silica or silica in liquid suspension is collected onto silver filter membranes, either directly or following filtration onto intermediate filters which are then ashed (Low Temperature Plasma Ash, LTA), resuspended and filtered onto final silver filters. All filtrations and subsequent preparations are performed in a class 100, Laminar Flow device. Air-jet milled quartz (AJMSiO$_2$) was filtered onto a silver (Ag) filter for subsequent primary analysis by XRD and confirmatory EDS analysis and observation by SEM.

Whole, remounted Ag filters containing SiO$_2$ are analyzed by x-ray diffractometry (XRD) in the continuous scan mode. The diagnostic reflections selected for SiO$_2$ were 26.66° (2 Theta, primary) and 20.85° (2 Theta, secondary). By comparison the expected diffraction peaks for Ag are 38.12° (2 Theta, primary) and 44.28 (2 Theta secondary).

Area EDS scans are recorded for samples of SiO$_2$, located on Ag filters fastened to carbon mounts, to establish a confirmatory analysis in support of XRD. EDS does not discriminate between crystalline and amorphous SiO$_2$. The EDS system is interfaced to a JEOL JSM-U3 SEM. For analysis, representative areas were photographed at a magnification of 100X followed by setting the beam for a scan of the same area. EDS was accomplished at 25 kV using x-ray counts of approximately 3000 cps.

To verify the results from XRD analysis, and to establish possible anomalous results due to particle distribution, select samples of SiO$_2$ on Ag filters (prepared for XRD) are examined by SEM following mounting on carbon supports and coating with an evaporated carbon film. The marker bar represents a distance of 10 micrometers.

Asbestos

Methods for the separation, isolation and detection of asbestos fibers from a variety of asbestos spiked beverages and foods have been reported (Albright et. al., 1979; Stasny and Albright 1977-79; Stasny et. al., 1979, Pattnaik, 1976). Likewise, the sample preparation methods reviewed by DeNee (1978) and by Chatfield and Dillon (1978) are useful and directly applicable to the isolation of asbestos contaminants from foods. For analyses reported here, the following methods were employed:

For SEM, mounted polycarbonate filters were sputter coated with a gold-palladium (Au-Pd) alloy. Filtered residues of spiked food samples were critically examined for the presence of chrysotile asbestos and non-asbestiform residues. Representative electron images were recorded at magnifications of 10,000X using a JEOL JSM U-3 SEM operated at 25 kV with 0° specimen tilt.

For EDS, residues on polycarbonate filters were mounted onto spectrographically pure carbon mounts and coated with evaporated carbon. The x-ray system interfaced to the SEM employed an Ortec detector and a multichannel analyzer. The area or fiber of interest was photographed, followed by setting the beam to scan a spot, for analysis of single fibers. X-ray count rates of approximately 2000 counts per second were used.

Although the use of a fluorescence light microscope affords no advantage in resolution of small fibers (< 1 micron) over other optical microscopy methods, the "tagged" chrysotile fibers > 1 micron in length are more easily discernible.

The NIOSH method of phase contrast microscopy involves a very subjective determination of what is actually an asbestos fiber by stipulating that any material with an aspect ratio of 3:1 or greater be counted as asbestos. Observation of only fluorescent fibers results in more definitive identification. An American Optical fluorescence microscope was used to detect chrysotile fibers. In this type of microscope, the excitation light is beamed onto the sample from above and a discrete band of emission light passes through barrier filters to the eye. Because of the wavelengths of the filter system for this microscope, morin hydrate was the only dye able to be observed.

Morin hydrate has an excitation maximum at 415 nm and fluoresces at 555 nm (emission). The specificity of binding of the morin hydrate was surprisingly narrow. Except for chrysotile asbestos, no other material examined displayed an affinity for this dye. Among the materials tested were: (1) natural minerals of similar chemical structure to chrysotile, (e.g. serpentine rock, fibrous silicates, talc); (2) man-made materials (e.g., glass, wool, rayon, ceiling tile); (3) common airborne contaminants (e.g. clays); and (4) other types of asbestos (i.e., amphiboles).

For the work reported here, a sample of asbestos dyed in bulk with morin hydrate was removed from the surface of the filter and spread on a glass slide and allowed to dry. A drop of mounting solution (used to match the refractive index of the immersion oil) was added, followed by a glass coverslip. The slide was observed with an oil immersion lens at 1000X magnification.

Results and Discussion

Useful approaches for detecting and identifying contaminants in human foods combine well established techniques for light microscopy (LM) with the advantages of SEM, EDS and allied techniques. More and detailed information is provided by the improved depth of focus and resolution of the SEM, in addition to supportive data obtained via EDS and XRD analyses.

In what follows, several examples of suspected food contaminants are shown that are amenable to further characterization and identification using one or more of the methods described.

Pest Fragments

Food processing and packing, through years of progressive development, has grown in scope and complexity. Present day food plants are departmentalized such that a food product may be handled by a number of departments during its preparation. Increased handling of foods also increases the likelihood for entry of pest fragment contaminants into the final product.

Representative examples of insect fragment and rodent hair contaminations, isolated from separate food items, are shown in Fig. 1 and 2 respectively. Fig. 1a shows an insect fragment as viewed with the light microscope at low magnification and reveals only the boundaries of the fragment and little or no surface detail. The corresponding SEM image of the same aspect shows distinct geometric structures characteristic of sculpturing pattern, setae and setal pits (Fig. 1b). These features can be used to differentiate among insect species when SEM is employed.

Another contaminant, indicative of extraneous filth in food is the detection of mammalian hair. The hair shaft shown in Fig. 2a is shown protruding through a particle of excrement, isolated from cereal grain and photographed via light microscopy. The corresponding SEM image shows greater detail and provides a more definitive identification of the hair (Fig. 2b).

Paint and Metal Fragments

The fragments shown in Fig. 3, were removed from two different food samples and photographed side by side. The original photomicrographs, recorded in color, show that isolated particle A is bright red and that particle B is white when viewed via reflected light microscopy at the surface of a spectroscopically pure carbon mount.

The enlarged secondary electron images of particle A and B, as well as their corresponding EDS spectra, are shown in Fig. 4 and 5 respectively. The SEM-EDS examination extends the information by providing additional morphological and identifying analytical details

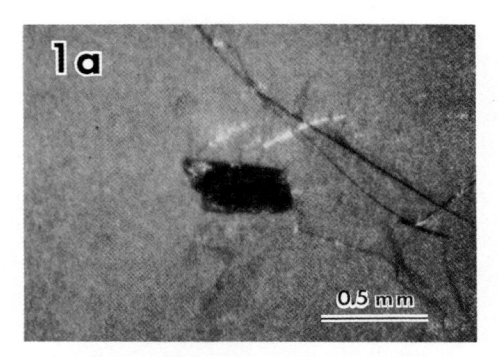

Fig. 1. Fragment. (a) Reflected light and (b) SEM images. The original light micrograph was in color, the SEM image reveals distinct characteristics of insect parts.

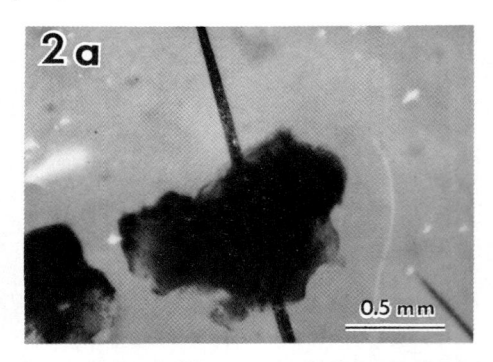

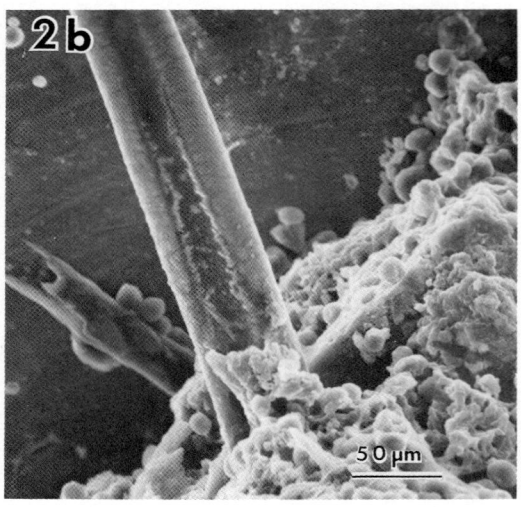

Fig 2. Hair shaft and attached rodent pellet found as a contaminant in cereal grains. (a) Light micrograph, original was in color; and (b) SEM image, showing more detail of the structure of the hair.

about the individual particles in question. Although, in this instance, the morphological information for these samples is of limited value, the analytical capability of the EDS mode provides spectral information which identifies the elemental composition of each particle. Primary peak values for the elements lead (Pb) and titanium (Ti) are shown for particle A (Fig. 4b) and B (Fig. 5b) respectively. Minor peaks, which are also present, are indicative of identifiable elemental components, present in lesser amounts. Such elemental analyses as these provide evidence toward the further classification of these particles as paint chips or fragments, traceable to a point of origin.

Using methods identical to those employed for separating, detecting and identifying paint particles, the fragment shown in Fig. 6a was also recovered as a contaminant from a food item. The SEM image reveals a slender, irregular fragment with nonremarkable surface features. The corresponding EDS result (Fig. 6b), however, reveals a singular prominent peak for aluminum (Al), thus identifying this contaminant as a metal fragment whose sole component is aluminum. Such contaminations are usually associated with filings from metal containers or machinery employed during the processing of food items.

Glass Fragments and Dusts

Contaminating glass fragments and dusts occur in a variety of foods, especially those packaged in glass containers. Glass particles may originate during manufacture of the glass containers and/or during handling and filling operations.

In an independent study (Eisenberg, 1974), it was determined that 93-96% of the glass particle contaminants originating in a manufacturing plant were in the 0.1 to 0.4mm size range. Likewise, glass dust particles of the size which could be airborne were also detected. Although injury to the digestive tract, including the mouth and teeth, would depend on the individual circumstances encountered, both the FDA and industry representatives are committed to preventing or eliminating

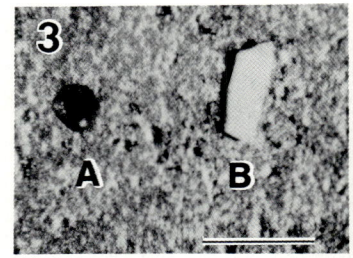

Fig. 3. Reflected light micrograph of paint fragments isolated as contaminants from food. In original color photograph particle A was red and particle B white.

Fig. 4. (a) SEM image and (b) EDS spectrum from paint particle A in Fig. 3. Major elements present are lead and titanium.

Fig 5. (a) SEM image and (b) EDS spectrum from paint particle B in Fig. 3. Major element present is titanium.
Bar lengths:
Fig. 3: 0.5 mm.
Fig. 4: 50 µm.
Fig. 5: 50 µm.

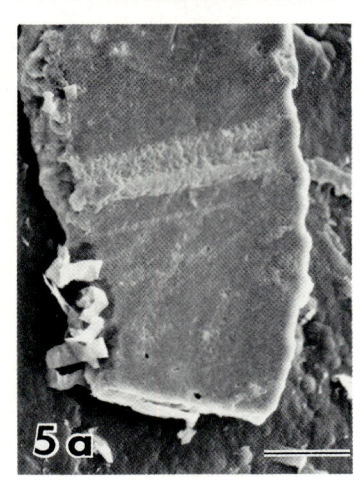

glass particle contaminations regardless of the size of the particle or the danger that the glass particle may or may not present to the consumer.

Similar to glass, sand and soil contaminants would be defined as water-insoluble silica based residues. Analytical methods have been developed for the detection and identification of silica or SiO_2, the dioxide of silicon, which in its pure crystalline form is known as quartz. Silica or quartz dust is an important environmental contaminant encountered during mining processes and is the cause of Silicosis which is considered to be one of the most important of the dust diseases of the lung.

It is for this reason, to evaluate worker's exposure to SiO_2, that environmental sampling and analysis for silica dust was first developed. The method for the determination of silica content in airborne dusts employs x-ray diffraction (XRD) as recommended in the NIOSH Manual of Analytical Methods (Method #259, 1977) and described in the original work of Bumstead (1973). The method is directly applicable to the detection of silica based contaminants. The results from a simulation of the detection and identification of such real world contaminants for demonstration purposes is shown in Fig. 7a-c.

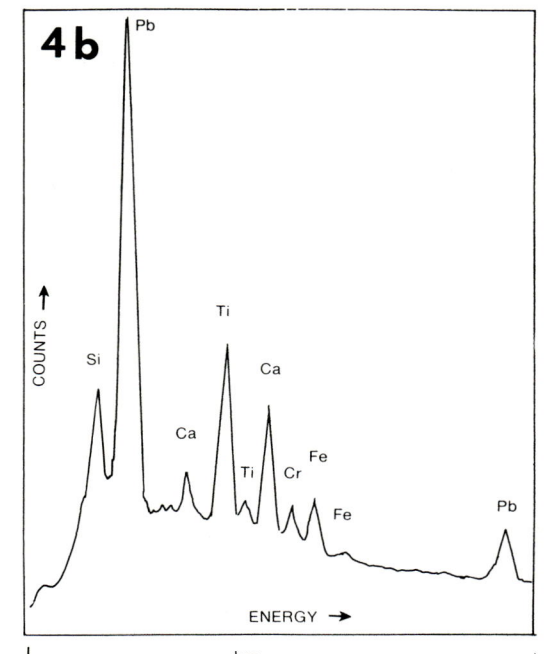

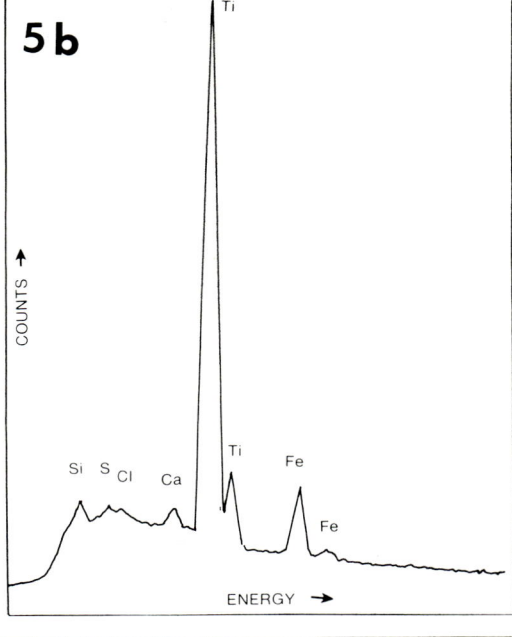

The resulting XRD pattern shown in Fig. 7b illustrates corresponding intensity peaks for Ag and SiO_2, obtained from the distribution of SiO_2 particles shown in Fig. 7a. The diagnostic peaks result from the SiO_2 "contaminants" at the surface of the silver filter. The peak heights have been proportionally reduced to accommodate the illustration. Only peaks for Ag can be detected when silver membranes without SiO_2 are analyzed by XRD. Low x-ray background is evident and the diffraction peaks for Ag do not interfere with the detection

45

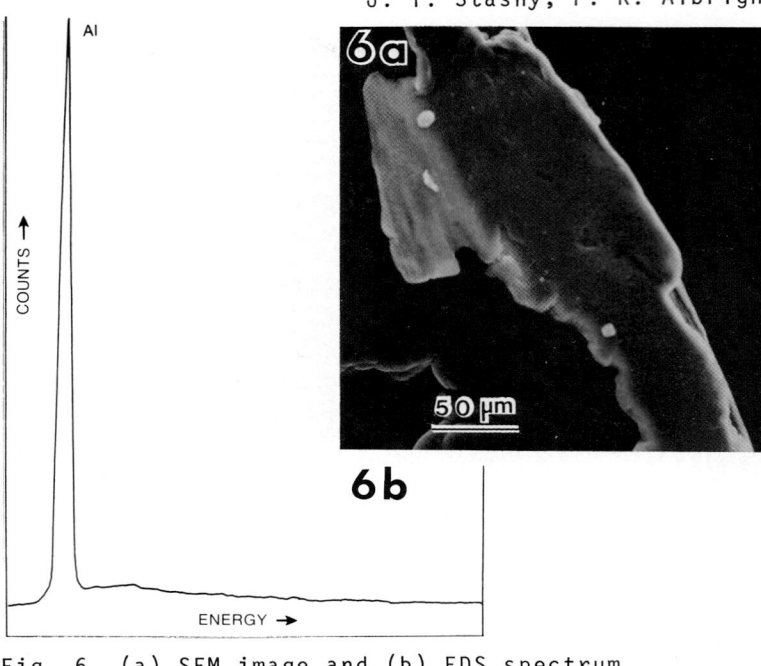

Fig. 6. (a) SEM image and (b) EDS spectrum from metal fragment identified as aluminum from the single peak in (b).

Fig. 7. (a) SEM image, (b) X-ray diffraction peaks (proportionally reduced for illustration) and (c) EDS spectrum (at full scale) of small, variously sized fragments (identified as particle contaminants of crystalline SiO_2) at the surface of a silver membrane collecting filter; (b) shows presence of SiO_2 and Ag, while (c) shows peaks for Si and Ag.

and measurement of SiO_2. The XRD method, however, discriminates between crystalline and non-crystalline SiO_2 and only crystalline SiO_2 can be characterized. EDS does not discriminate between crystalline and amorphous SiO_2 and the diagnostic elemental composition for crystalline Si and the Ag support filter is shown in the EDS area scan in Fig. 7c.

The analytical options reviewed here can be used to detect and characterize silica based contaminants found in foods and to classify them according to type, size, shape and chemical composition.

Asbestos

The separation of minute quantities of asbestos fibers from bulk food matrices without altering the physical or chemical nature of the mineral is a difficult problem both in concept and in practice. Moreover, any extraneous inorganic or organic particles from the food matrix which remain following isolation procedures can obscure the final detection of fibers. The final goal of these treatments was to retrieve asbestos fibers, unaltered and unobscured, at the surface of a membrane filter for

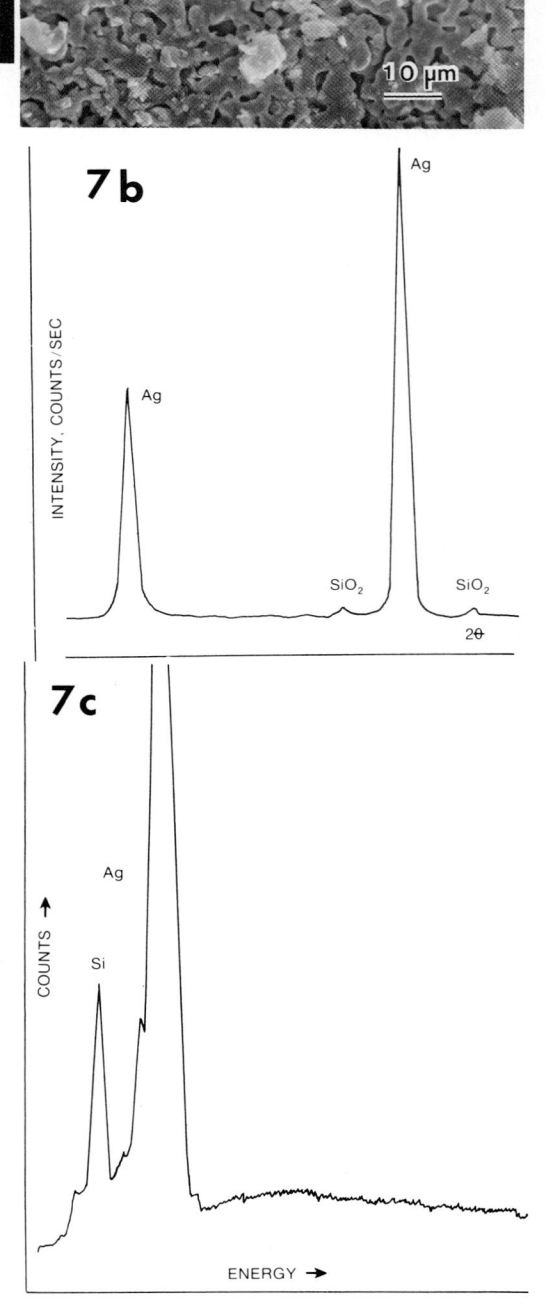

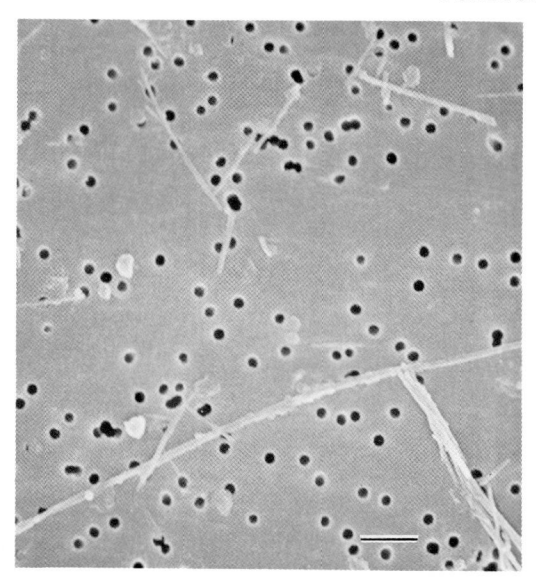

Fig. 8. SEM image of chrysotile asbestos retrieved from spiked ketchup and final filtered onto a 0.2μm pore dia. Nuclepore filter. Bar length = 1.0μm

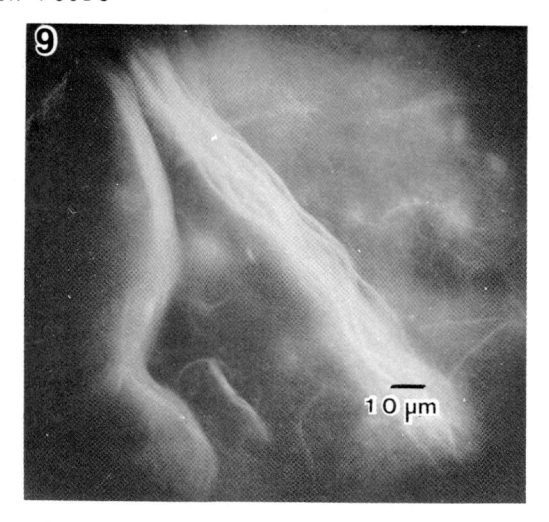

Fig. 9. Chrysotile asbestos fiber bundle tagged with morin hydrate and viewed using a fluorescence light microscope. Image shown in black and white. Bar length = 10.0μm

subsequent morphological identification by SEM and chemical confirmation using EDS.

Scanning electron microscopy (SEM) proved to be an excellent method for detecting asbestos fibers at magnifications of 5000X or 10,000X. The controlled treatments and experimental conditions employed were shown to have essentially no effect on the morphology of the spiked and retrieved fibers of chrysotile asbestos as demonstrated in Fig. 8 for fibers retrieved from spiked ketchup.

SEM provides for rapid observations and is capable of detecting fibers as small as 0.03μm in width. The use of a gold-palladium coating is critical to obtain good contrast for asbestos fiber detection in the secondary mode of operation. Thus the utility and versatility of the SEM provides a means to monitor the results of the preparative treatments during the normal progression of method development, as well as the final detection and characterization of the size, shape, numbers, and distribution of the isolated asbestos fibers from food, beverage and water samples.

In a study sponsored by the National Science Foundation (NSF) Albright et. al., (1980) developed a rapid screening method for identifying chrysotile asbestos by attaching fluorescent probes. The detectability of chrysotile asbestos by fluorescent light microscopy is greatly enhanced because fibers are more easily recognized among non-asbestiform fibers or extraneous debris at the filter surface (Fig. 9). The chrysotile fibers in the original color photograph exhibit a bright yellow color. The procedure for tagging chrysotile asbestos with a fluorescent dye appears to be an ideal screening technique for rapid detection of asbestos containing materials.

Summary and Conclusions

Processing and handling of foods from raw materials to finished food products is often a long and complicated procedure. During such time periods, accidental or inadvertent particulate contaminations from a variety of sources can present serious problems to manufacturers and consumers alike. Ongoing quality control programs emphasize product safety in all phases of the food industry to insure that the consumer has access to safe, high quality food products and to reduce industry's product liability exposure. In recent years there has been a marked trend toward the use of the analytical tools and allied methodology described in this paper to assist in the solving of difficult quality control problems presented by contaminants in foods.

References

Albright FR, Schumacher DV, Sweigart DS, Stasny JT, Husack C, Boyer K. 1979. Methods for Isolating and Identifying Asbestos Fibers in Spiked Beverages and Foods. Food Tech. pp. 69-76.

Albright FR, Schumacher DV, Mayer JA, Sweigart DS. 1980. Research on a Rapid and Simple Detection Method for Asbestos. R & D Report #209-80, supported by Grant No. PFR-7917183 A01 from the National Science Foundation, Washington, DC 20550

Brickey PM Jr, Gecan JS. 1974. Introductory Study of Scanning Electron Microscopy of Elytral Fragments of Stored Product Beetles, J. AOAC, 57, 1235-1247.

Brickey PM Jr, Gecan JS, Rothschild A. 1973. Method for Determining Direction of Insect Boring through Food Packaging Materials, J. AOAC, 56, 640.

Bumstead HE. 1973. Determination of Alpha-Quartz in the Respirable Portion of Airborne Particulates by X-Ray Diffraction. Amer. Ind. Hyg. Assoc. J. 34, 150, 158.

Chatfield EJ, Dillon ML. 1978. Some Aspects of Specimen Preparation and Limitations of Precision in Particulate Analysis by SEM and TEM. Scanning Electron Microsc. 1978; I: 487-496.

DeNee PB. 1978. Collecting, Handling and Mounting of Particles for SEM. Scanning Electron Micros. I, 1978; I:479-486.

Eisenberg WV. 1974. Inorganic Particle Content of Foods and Drugs, Environ. Health Persp., 9, 183-191.

Food and Drug Administration. 1980. The Food Defect Action Levels. pp. 1-12. FDA Industry Programs Branch, Bureau of Foods (HFF-326), 200 C Street, SW, Washington, DC 20204.

Food and Drug Administration. 1977. Technical Bulletin No. 2, "Training Manual for Analytical Entomology in the Food Industry." Gorham JR, editor.

Food and Drug Administration. 1975. Compliance Program Evaluation FY 1974 Heavy Metals in Foods Survey, 7320.13C, Bureau of Foods, Washington, DC.

Food and Drug Administration. 1960. Technical Bulletin No. 1, "Microscopic-Analytical Methods in Food and Drug Control." Harris KL, editor.

Jones JW, Boyer KW. 1979. Sample Homogenization Procedure for Determination of Lead in Canned Foods. J. Assoc. Off. Anal. Chem. 62, No. 1, pp. 122-128.

Kvenberg JE. 1981. Comparison of Larval Stored Product Beetle Mandibles by Light Microscope and Scanning Electron Microscope, J. AOAC, 64, 199.

Kvenberg JE. 1978, Scanning Electron Microscopic Study of Adult Stored Product Beetle Elytra, J. AOAC, 61, 76-87.

Kvenberg JE. 1977. Scanning Electron Microscopic Study of Adult Stored Product Beetle Mandibles, J. AOAC, 60, 1185-1209

Moore JA. 1981. Biological Effects of Ingested Asbestos (Status Report January 22, 1981), Department of Health, Education, and Welfare, Public Health Service, National Toxicology Program, P.O. Box 12233, Research Triangle Park, NC 27709.

Moore JA. 1977. NIEHS-ORAL ASBESTOS STUDIES In: Workshop on Asbestos: Definitions and measurement Methods. National Bureau of Standards, Gaithersburg, MD. Sponsored by NBS and OSHA. p. 42

NIOSH MANUAL OF ANALYTICAL METHODS 1977, Method No.: P&CAM 259. Appendix 1, Vol. 1, Supt. of Documents, U.S. Government Printing Office, Washington, DC 20402, GPO #017-033-00267-3, pp. 259-1 to 259-7.

Pattnaik P. 1976. Isolataion, identification and determination of asbestos in food, drugs and talc. Final report for contract 223-74-2217 to FDA, Rockville. MD. See Stasny, Albright ref. for availability.

Selikoff IJ, Hammond EC, Churg J. 1970. Mortality Experiences of Asbestos Insulation workers, 1943-1968. Pneumoconiosis: Proceedings of the International Conference (Johannesburg, 1969). pp. 180-186. HA Shapiro (ed) Oxford University Press NY.

Stasny JT, Husack C, Albright FR, Schumacher DV, Sweigart DS, Boyer K. 1979. Development of Methods to Isolate Asbestos from Spiked Beverages and Foods for SEM Characterization. Scanning Electron Microsc. 1979; I: 587-595.

Stasny JT, Albright FR. 1977-79. The Separation of Asbestos in Foods, Drugs and Talc for Identification and Determination. Quarterly report 1-6 for contract 223-77-2029 to DHEW/FDA, Rockville, MD from Structure Probe, Inc., West Chester, PA and Lancaster Laboratories, Inc., Lancaster, PA. Copy available from FDA, Freedom of Information Office, HFI-35, 5600 Fishers Lane, Room 12A-12, Rockville, MD 20857, USA.

Thomas B, Rougham JA, Watters ED. 1973. Lead and Cadmium Content of Some Canned Fruit and Vegetables. J. Sci. Food Agric. 24, 447-449.

Vasquez AW. 1961. Structure and Identification of Common Food-Containing Hairs J. AOAC, 44, 754-779.

Discussion with Reviewers

J.A. Brown: On Figure 2a, how was rodent excrement identified?
R.J. Carroll: How can you determine from Figure 2a that this is a rodent pellet?
E.J. Chatfield: How would one know from either Figures 2a or 2b that the material attached to the hair is excrement?
Authors: The presence of rodent hairs usually characterizes rodent pellets. A presumptive finding that certain particles originate from rodent droppings can be confirmed by a test on a gelatin plate to establish the presence of pro-

teolytic enzymes in the suspected particle. Rodents groom themselves with their mouths, swallow the hairs, and these hairs eventually appear in the feces. Although other tests have been used, the common diagnostic characteristic of rodent pellet fragments is the presence of rodent hairs.

E.J. Chatfield: When examining Nuclepore filters by SEM to search for asbestos fibers, is the instrument operated under slow scan conditions or at TV scan rate? The example of Figure 8 is a heavily-loaded sample, and I think that when searching for widely-spaced 300A fibers on the CRT image a large number of the small ones will be overlooked, if not on resolution grounds, certainly on the basis of poor contrast or image noise.
Authors: The SEM search for asbestos fibers includes both TV scan and slow scan conditions. To enhance the display of asbestos fibers on Nuclepore filters, one-half of the filter is coated with Au-Pd, which improves the electron emission from the sample surface and increases the topographic contrast of the fibers. In our experience we routinely detect fibrils of chrysotile asbestos in the 0.03 micrometer diameter range. The contrast and resolution obtained are adequate for the detection of such fibrils at a top magnification of only 10,000X. Finding or detecting only a few fibers of small diameter, within the image area is enhanced, and the clear advantage for using the SEM to detect asbestos fibers is the short time and relative ease with which sample preparations can be made and evaluated.

J.A. Brown: What was the specimen preparation technique used for separating the asbestos fibers from extraneous particles and how were these fibers identified as being asbestos?
R.G. Draftz: Could you describe new methods used by you and others to isolate particle contaminants from various food products?
Authors: A more extensive description of recent particle isolation techniques can be found in several of the references cited in the present paper. More specifically and in brief detail, the work of Albright et. al., 1979 and Stasny, et. al., 1979 describes the development of methods for the separation and retrieval of chrysotile asbestos fibers from alcoholic beverages and bulk food samples and their identification by SEM and EDS. Procedures were established for spiking and digesting various types of foods within the major categories, generally classified as liquids, semi-solid and solid food substances. Beer, wine, mayonnaise, animal tissue and animal feed were used for these studies. Alkaline hydrolysis and oxidation by hydrogen peroxide (H_2O_2), coupled with solvent extractions, were used to reduce or remove unwanted residues. Pressure filtration of the food digests and extracts allowed relatively large volumes of product to be analyzed. Allied chemical methods using brief acid treatment and low temperature ashing (LTA), also enhanced the detection of unobscured chrysotile asbestos spikes from the various food types examined. Other recent work, pertinent to the isolation of particle contaminants includes publications by Chatfield and Dillon (1978) and DeNee (1978) for asbestos, as well as Jones and Boyer (1979) for Pb determinations. In addition, improved methodology that is applicable to the collection and identification of silicates from food substances has been presented (NIOSH Method #259, 1977).

R.G. Draftz: Many of the particle contaminants described in this paper can be easily and rapidly identified by polarized light microscopy. What distinct advantages can be gained from the more time consuming SEM/EDS analysis?
Authors: In general, both light and electron microscopy are important, complementary, diagnostic aids. The decision to use either or both SEM and light microscopy is usually dictated by the nature of the problem(s) to be solved and the amount and type of information required. In our experience we prepare and evaluate particulate samples for SEM/EDS analyses with relative ease, on a routine basis. The value of SEM to the work describing the elytral (anterior wing) fragments of stored product beetles, was initially discussed by Brickey and Gecan (1974) and its expanded use for the taxonomy of insects and insect fragments was reported by Kvenberg (1977, 78, 81). The use of SEM is an advantage in illustrating insect elytral fragments because the SEM image reduces the complex patterns presented by light microscopes to black and white surface detail with good depth of field and clear images when using high magnifications. Light photomicrographs presented as illustrations are difficult to interpret because of the resulting narrow depth of field. However, observations made by light microscopy can be interpreted using characteristics such as setae, setal pits and striations, while disregarding complex subsurface patterns and color arrangements. When considering this question with respect to asbestos, the separation of minute quantities of asbestos fibers from food matrices without altering the physical or chemical nature of the mineral is a difficult problem, both in concept and in practice. Part per million concentrations of asbestos can mean the presence of literally millions of fine

fibers, yet a magnification of 5,000X to 10,000X or more is needed to detect these fibers, some of which can have a width of only 0.03 micrometers. Moreover, following secondary electron imaging, the direct chemical identification of a given asbestos fiber can be obtained by EDS analysis. Likewise, EDS spectra from suspect metal and/or paint fragments can expedite the definitive identification of these contaminants and may eliminate the need for further characterization.

R.J. Carroll: What is the advantage of using silver membranes over other types of membranes for collecting foreign matter and subsequent SEM and EDS or XRD analysis?
Authors: Within the context of collecting silicate samples for analysis, early work indicated that dust (environmental silicates) would have to be removed from the original, organic collection filter if satisfactory XRD patterns were to be obtained. Organic filters give a broad diffuse band and obscure the reflections in the area of interest. Silver membranes proved to be a suitable support for redeposited silicates, since the membranes produce a very low x-ray background and do not interfere with the pattern for SiO_2. In addition, the silver membranes offer a conductive surface which sheds no fibers and can withstand extreme temperature, pressure and solvent conditions not applicable to non-metallic filters. Because of these properties, silver membranes are most useful in the filtration of alcoholic beverages, high temperature filtrations of syrups and other viscous materials, high temperature sterilization procedures and the filtration of organic solvents. The structural characteristics of residues deposited at the surface of the silver membrane can then be examined by SEM. Provided that the spectrum for silver does not interfere with that of the suspected element, EDS analyses may also be employed for the chemical identification of the particle contaminant in question.

R.J. Carroll: What are the levels of detection of foreign matter in foods? Can an estimate be given?
Authors: The determination of detection levels for foreign matter, however they may be expressed, is a difficult problem, especially since such variables as type and size of contaminants, as well as preparative and analytical methods can and do influence the result. For example, when insect fragments and rodent hairs are suspect, the size of the sieve and the amount of product sampled become important. Particles smaller than the sieve will not be retained for recognition of the foreign matter using light optical microscopy. However, the detection of submicroscopic foreign matter, such as fibrous asbestos, may be limited by both, preparative methods and the detection limits of the instrumentation used. Reasonable estimations of the detection limits for food particle contaminants can only be made using these types of considerations.

R.J. Carroll: What other fluorescence dyes could be used to tag asbestos fibers?
Authors: Fluorescent compounds which exhibited a strong binding affinity for chrysotile asbestos were fluorescein at an acidic pH, and morin hydrate and 4,5-dihydroxy-naphthalene-2, 7-disulfonic acid, both at a basic pH of 11.4. These dyes were screened from a group of 27 fluorescent compounds, with differing functional groups, tested for binding affinity to chrysotile asbestos. Obviously, this by no means exhausts the list of possible candidates.

R.G. Draftz: The specification of mammalian hairs through scale patterns has not proved to be very successful. Wouldn't optical microscopy be more useful to distinguish the multi-serrated medullas common to rodents? What are the advantages that SEM analyses provide in hair analysis?
J.E. Kvenberg: Are external scale patterns considered diagnostic for the origin of a striated hair?
Authors: External scale patterns of striated hairs are made more visible by SEM but the patterns are not considered to be diagnostic. The visualization of internal structures of the striated hair (Guard or Fur hair) is necessary for a proper diagnosis.

D.N. Holcomb and M. Buri: Are food particle contaminants usually found in sufficient numbers or size as to pose a threat to health?
Authors: Aside from the physical size and nature of the contaminant(s), which in itself may be harmful, many contaminants in food products are also potential sources of disease organisms transmissible to man. The association of rodents, insects and their parts, with unsanitary conditions and the contamination of food, represents a danger to health although the specific agents of disease may be difficult to demonstrate.

For additional discussion see page 24.

SCANNING ELECTRON MICROSCOPY IN MEAT SCIENCE

C. A. Voyle

Agriculture Research Council
Meat Research Institute
Langford, Bristol, BS18 7DY
U.K.

Abstract

The textural quality of meat and meat products is closely related to structure. Morphological characteristics can be associated with toughness, tenderness, granularity or smoothness. Microscopic analysis can be carried out by light and electron microscopy to give information about these parameters. The handling of a carcass post mortem, the action of proteolytic enzymes in the 'aging' process, storage and cooking all bring about structural changes in muscle which affect its quality as meat. This paper presents and reviews information obtained by a number of workers on the structure of muscle fibres under various post-mortem conditions. The information was obtained by scanning electron microscopy and comparisons are made with the results of other methods of microscopical analysis.

KEY WORDS: Meat structure, muscle, aging, cooking, meat products.

Introduction

The scanning electron microscope was introduced commercially as a scientific instrument in 1965. It has been widely applied in both materials and biological research providing information in an area which overlaps the magnification range of both the light microscope and the transmission electron microscope.

In a review article on the application of scanning electron microscopy in food science and technology, Pomeranz (1976) quotes 343 references; of these, only five are concerned with meat. Twenty or so reports have been published since then, reflecting the increased availability of, and interest in, scanning electron microscopy in meat science. In the article referred to, meat is considered in the section entitled 'Miscellaneous Foods'. This is surprising in view of the importance attached to meat in the human diet.

The usefulness of the technique lies in the considerable depth of focus which is attainable, revealing details of structure at the surface of the material being examined and also allowing information to be obtained on the spatial arrangement of sub-cellular structures relative to each other. The variety of modes in which the scanning electron microscope can be used has yet to be exploited in meat science, and although the capital cost is high, the versatility of modern electron microscope systems will make it easier for those concerned in the study of the ultrastructure of meat to probe further into the problems which it presents. This paper summarizes the results reported by various workers and considers the prospects for further investigations.

Scanning electron microscopy (SEM) has been used to study the surface ultrastructure of muscle fibres subjected to a variety of treatments including the effect of restraint during the development of rigor, variation in storage time post-rigor, the effect of cooking conditions and of exposure to various enzymes in vitro. Processed meats such as hams and frankfurters have been examined and different methods of specimen preparation have been evaluated. In general, the morphological appearances observed with the scanning electron microscope have been considered in relation to the texture of meat. These appearances are complementary to those obtained by light microscopy (LM) and transmission electron microscopy (TEM).

Normal Structure

The earliest description of surface ultrastructure of skeletal muscle fibres was provided by Boyde and Williams (1968) who had examined single fibres or small bundles dissected from the M. semitendinosus and M.iliofibularis of the frog. The contours of the fibre surface were interpreted as being consistent with the location and composition of the sarcomeres which make up the myofibrillar structure of all skeletal muscle. Each sarcomere has a banded structure, the predominant features of which are the A-band consisting mainly of thick myosin filaments and the I-band occupied by thin actin filaments. The limit of each sarcomere is defined by the Z discs, the distance between them being a measure of the contractile state of the muscle. Thick and thin filaments overlap, sliding past each other during the process of contracton and relaxation, as described by Hanson and Huxley (1955), and widely studied by many workers during the quarter century that has since elapsed. Voyle (1979) has described the microscopy of meat, summarising the terms currently used for the identification of structural features.

Figs. 1 - 3 illustrate the micro-structure of muscle fibres as found in bovine M. psoas. Fig. 1 shows the array of the different structural features of the myofibril as observed by TEM. Fig. 2 illustrates the surface structure of fibres as seen by SEM, displaying the regular pattern of transverse bands. Fig.3 shows the perimysium - a sheath of connective tissue containing collagen fibres which surrounds bundles of muscle fibres. Boyde and Williams (1968) observed a similar structure in frog muscle in which the connective tissue fibres were either grouped in longitudinal bundles or as a random network. Both forms could be removed by treatment with bacterial collagenase prior to fixation.

Schaller and Powrie (1971) examined skeletal muscles from rainbow trout, turkey and beef animals. Their observations were concentrated on the location of elements of the sarcoplasmic reticulum, and they compared their findings with the reports of Franzini-Armstrong and Porter (1964), Porter and Palade (1957) and others who have described the location of sarcoplasmic reticulum in similar species using TEM. In fish muscle pronounced transverse elements were observed which were closely associated with the Z-disc (Bertaud et al.,1970). In avian and mammalian muscles a smaller transverse ridge was seen to lie on either side of the more prominent transverse protrusion. These are commensurate with the triad structures seen at the A-I junction by TEM. In Fig.4 which shows myofibrils from bovine M. psoas elements of sarcoplasmic reticulum in the vicinity of the A - I junction can be seen with, in some instances, remnants of elements of the T-system and also mitochondria.

Since Schaller and Powrie published their findings in 1971 new methods of specimen preparation have been introduced. If we allow the possibility that the prominence of the ridges they described was exaggerated by the collapse of surrounding structures, especially as a result of drying in vacuo, then the use of critical point drying may be found to avoid, or at least reduce, such an artefact. This method of specimen preparation is said by Carroll and Jones (1979) to give the best preservation of meat structure.

'Aged' meat

Changes in the chemistry and morphology of muscle are initiated at death with the process of glycolysis and the consequent fall in pH. These changes contribute to the process known as 'aging'. Proteins are denatured and become exposed to the proteolytic activity of endogenous enzymes. This has an effect on the texture of the tissue as meat, an effect which can be followed by microscopical observations of the change in the morphology of the muscle fibres. Changes may be seen in the sarcoplasmic reticulum (Schaller and Powrie, 1971), the sarcolemma (Varriano-Marston et al., 1976) and the myofibrils (Davey and Dickson, 1970, Sayre 1970, Stanley, 1974).

Schaller and Powrie (1971) commented on the apparent deterioration in the morphology of sarcoplasmic reticulum (SR) during storage. Whereas the location of transverse elements of SR was initially represented as ridges, changes occurring up to six days post mortem resulted in the appearance of troughs at these sites. This implies the collapse of the tubular system of the SR or, alternatively, the swelling of adjacent regions of the sarcomere. The scanning electron microscope facilitates the detection of such changes more readily than does the transmission mode because of the three-dimensional appearance of the image. Schaller and Powrie (1971) also reported breaks in myofibrils in the region of the transverse elements, an observation supported by the findings of Davey and Dickson (1970) and Sayre (1970), who described breaks near the Z-disc in aged bovine and chicken muscle examined by TEM. Preparations of bovine M. semitendinosus examined by TEM are illustrated in Figs. 5 and 6. The site of perturbation of the myofibrils, closely associated with the Z-discs is clearly seen in Fig. 5, while the more extensive disruption involving a number of adjacent myofibrils is illustrated in Fig.6.

Varriano-Marston et al.,(1976) reported changes in the fine structure of the sarcolemma of free and restrained muscle over a period of twelve days, during which the structure degenerated from a relatively smooth membrane around the fibre to a collection of randomly distributed aggregations of protein. Individual

SEM in Meat Science

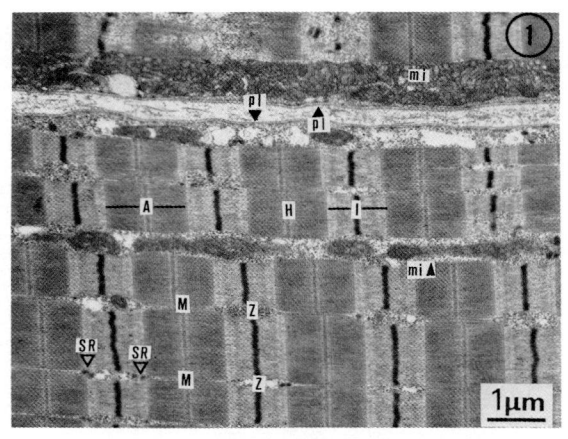

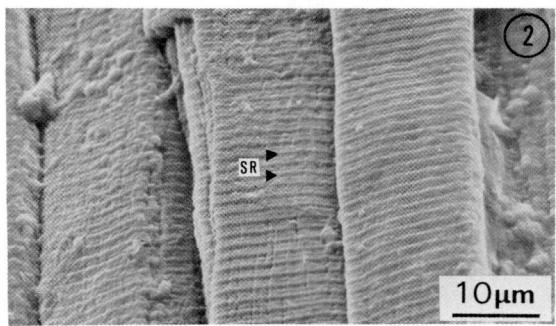

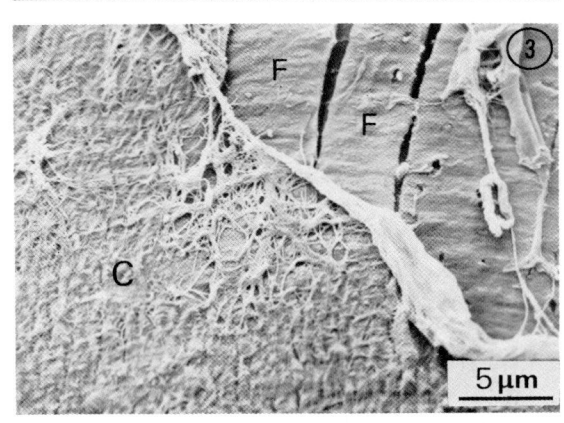

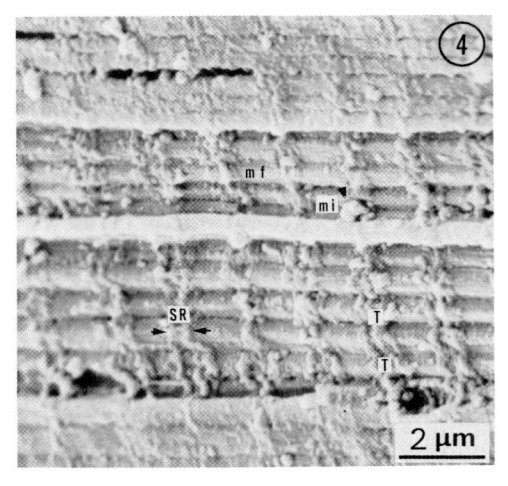

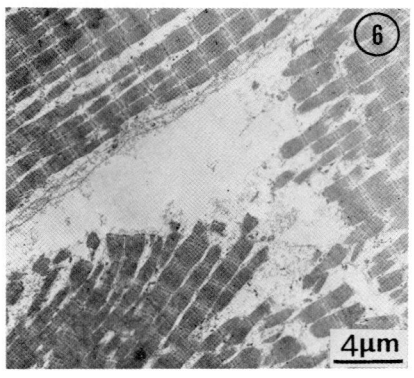

Fig.1 TEM micrograph of bovine M.sternomandibularis showing structure of sarcomeres in longitudinal section. Features illustrated include A-band (A), I-band (I), Z-disc (Z), H-zone (H), M-line (M), mitochondria (mi), sarcoplasmic reticulum (SR). The plasmalemma of adjacent fibres may be seen in the upper part of the micrograph (pl). Tissue fixed in 2% glutaraldehyde in 0.15% sodium chloride pH 7.0. Post fixation in 1% osmium tetroxide in 0.1M cacodylate buffer pH 7.0. Dehydrated in ethanol. Embedded in Epon 812. Sections stained with lead citrate and uranyl acetate.

Fig.2 SEM micrograph of bovine M. psoas showing surface structure of fibres in longitudinal array. Transverse elements of sarcoplasmic reticulum give rise to ridges (SR). The muscle is shortened. Tissue fixed in 2.5% glutaraldehyde. Dehydrated in ethanol followed by substitution in amyl acetate. Critical point-dried from liquid CO_2. Mounted specimens coated with gold-palladium.

Fig.3 SEM micrograph of bovine M. psoas showing perimysium - the connective tissue sheath which surrounds bundles of muscle fibres. Preparation as for Fig.2. Muscle fibres (F), connective tissue (C).

Fig. 4 SEM micrograph showing myofibrils (mf) in bovine M. psoas, exposed by freeze-fracturing after fixation. Preparation as for Fig. 2. Micrograph shows elements of sarcoplasmic reticulum (SR) with remnants of T-system (T) and mitochrondria (mi).

Fig. 5 TEM micrograph of bovine M. semitendinosus showing breaks in myofibrils adjacent to Z-discs (arrows). Tissue fixed in 2.5% glutaraldehyde in 0.1M cacodylate buffer followed by 1% osmium tetroxide in same buffer. Dehydration and embedding as for Fig. 1.

Fig. 6 TEM micrograph of bovine M. semitendinosus showing development of myofibrillar breaks leading to fragmentation of the fibre. Preparation as for Fig.5.

myofibrils were discernible beneath the aggregated layer. The implications of these changes with respect to meat texture are obvious and important, as are the changes in the structure of myofibrils. Stanley (1974) discusses the changes observed in bovine M. psoas and M. semitendinosus at 0, 6 and 12 days post mortem, using SEM, and relates his findings to the chemical and physical changes taking place in the tissue.

In a paper concerned mainly with methods of specimen preparation Jones et al.,(1976) used muscles from freshly slaughtered bovine animals and from aged carcasses, in particular M. semitendinosus (ST) and M. longissimus dorsi (LD). The results obtained with ST showed the importance of fixation prior to fracturing and dehydration. This muscle was obtained through a normal retail outlet and was, therefore, aged, but no comparison was made with ST from a freshly slaughtered animal. The LD, on the other hand, was exposed to conditions which readily induced cold-shortening (Locker and Hagyard, 1963). The appearances described closely resemble those observed by Voyle (1969) in M.sternomandibularis subject to similar conditions.

Cold-shortening, which is a phenomenon associated with the rapid chilling of pre-rigor muscle, is characterised by the presence of fibres which demonstrate severe shortening of sarcomeres - 'active' shortening - together with fibres which assume a zig-zag or wavy configuration - 'passive' shortening. These forms are easily recognised in stained sections viewed by LM as well as in preparations observed by SEM. Fig.7 shows the mixed reaction to cold-shortening seen in fibres of bovine M. sternomandibularis observed by LM. The severe contraction of muscle fibres which occurs in 'active' shortening is illustrated in Fig.8 using TEM. 'Passive' shortening, observed by SEM is seen in Fig.9.

Effect of heat

It is not surprising that the majority of reports on the use of SEM in meat science have been concerned with the effect of heat on the fine structure of muscle tissue. Of all the treatments that can influence texture, cooking is probably the most important because it acts on all the tissue components, irrespective of their identity; myofibrillar proteins are coagulated, collagen shrinks and is converted to gelatin, and water is released. Schaller and Powrie (1972) heated cubes of tissue from post-rigor bovine LD, 24 hour post-mortem ST from chicken, and similar dorsal muscle from rainbow trout to an internal temperature of 60° or 97°. At 60°, of the three species examined, trout muscle was subject to the most extensive damage. At 97° myofibrils from all species showed loss of organised structure of the thin filaments in the I band, although the main features of the sarcomere were still recognisable. Trout myofibrils heated to 97° showed a gap at the location of the H zone. This is the area in the mid-region of the A-band of sarcomeres in relaxed muscle, which is

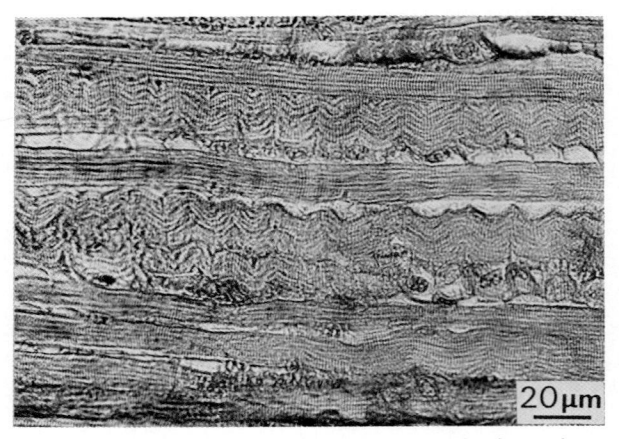

Fig. 7 LM micrograph of bovine M. sternomandibularis exposed to cold-shortening conditions. Straight fibres show 'active' shortening; wavy fibres show 'passive' shortening. Tissue fixed in formol-saline, dehydrated in ethanol and embedded in paraffin wax.

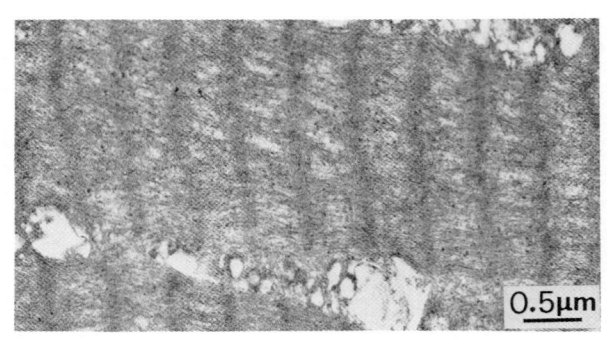

Fig. 8 TEM micrograph of severely contracted myofibrils from cold-shortened bovine M.sternomandibularis. Thick and thin filaments show total overlap. Dense lines are Z-discs (Z). Preparation as for Fig.5.

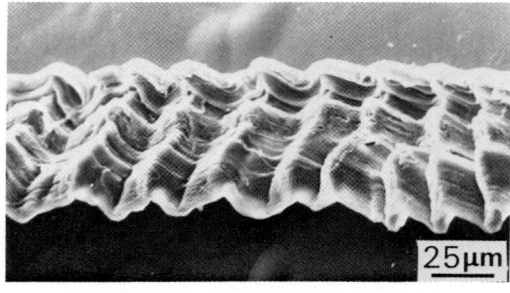

Fig. 9 SEM micrograph showing 'passive' contraction in fibres from cold-shortened bovine M.sternomandibularis. Tissue fixed in 5% glutaraldehyde in phosphate buffer pH 7.5. Teased fibres mounted on stub and air-dried prior to coating with gold-palladium.

occupied by thick filaments only (see Fig.1). The appearance of cooked bovine muscle reported by Paul (1963) using LM was confirmed by the observations of Schaller and Powrie (1972). Similarly, observations of granulation in heated fibres, reported by Paul (1963) and Doty and Pierce (1961) in chicken muscle, were confirmed. The granular material seen was thought to be formed by denaturation of sarcoplasmic protein. Fine collagen fibres in chicken muscle heated to 60° were no longer distinguishable, and larger fibres became swollen.

Cheng and Parrish (1976) compared bovine LD and psoas muscles which were aged and uncooked with samples of aged muscles heated to 60°, 70° and 80°. No report was made of tissue examined immediately after slaughter as uncooked fresh control. The treatment of the carcasses from which samples were removed suggests that some cold-shortening might have occurred in the LD. As the temperature was increased progressive changes occurred in the endomysial sheath which became swollen and crimped at 70° and disintegrated at 80°. Granular material - probably a mixture of heat-denatured collagen and coagulated sarcoplasmic protein - accumulated in the inter-fibre space. The endomysial sheath of fibres in the psoas muscle became swollen when heated but granular material did not accumulate. Coagulation and longitudinal splitting of myofibrils from LD occurred at 80° but not at the lower temperatures. Psoas muscle heated to 70° showed shortened sarcomeres and coagulated and shrunken myofibrils, in contrast to the unheated aged muscle. Similar observations which suggest that heat-shortening is more marked in initially long sarcomeres than in shorter sarcomeres have been widely reported using TEM. The posture of the carcass after slaughter and during the rigor period determined the sarcomere length. The influence of carcass position on sarcomere length in different bovine muscles and the effect on texture was described by Herring et al.,(1965). In long sarcomeres there is less overlap between thick and thin myofilaments with a consequent reduction in the protection afforded by crosslinking against thermal damage. Jones et al.,(1977) studied the effect of heating samples of bovine ST for forty five minutes at 50°, 60° or 90°. These experiments confirmed the view that the higher the temperature the more damage was sustained by the muscle fibres. Transverse fracturing was conspicuous at 90° but not at lower temperatures. At 60° fibres tended to cleave in the longitudinal plane, with only occasional transverse cleavage. Collagenous membranes showed no morphological change at 60°, but at 90° a granular structure was apparent. The sarcolemma assumed a banded appearance at 90°, the bands of coagulated material lying transversely across the fibre. Myofibrils showed an increasing tendency to fragment at 60° and above and, as has frequently been observed, the discontinuity occurred at the I - Z junction.

The texture of cooked meat is influenced by the water holding capacity of the tissue, although there are many other factors to be taken into account as well. Godsalve et al.,(1977) have studied the rate of water loss from bovine ST in varying oven conditions and used SEM to evaluate the structure of samples showing differences in this parameter. They reported that muscle of loose inter-myofibrillar packing, as in samples restrained during rigor, lost moisture more rapidly in the initial stages of cooking. Restrained muscle had longer sarcomeres and the myofibrils were loosely though regularly packed. Unrestrained muscle showed much tighter packing, together with some structural distortion which inhibits water loss during the initial stage of 'dry' cooking. Similar results have not been found under moist cooking conditions (Bouton et al., 1973; Locker and Daines, 1975).

Results compatible with those obtained by Jones et al.,(1977) were reported by Hearne and her co-workers (Hearne et al (1978).,in a series of experiments using bovine ST, in which the rate of heating to endpoint temperatures of 40°, 50°, 60° or 70° was designated 'slow' or 'fast'. SEM studies showed that higher temperatures resulted in more myofibrillar fragmentation than lower temperatures, but in addition a faster rate of heating tended to increase the degree of fragmentation. According to the authors this suggests that the rate of rise in temperature is as important as the temperature attained in determining the ultimate condition of meat when cooked. It is interesting to note that these workers did not find a change in shear values with the increase in fragmentation, suggesting that tenderness is influenced by other factors such as the coagulation of myofibrillar protein and shortening of sarcomeres at elevated temperatures. Solubilization of collagen is an additional factor to be taken into account.

Carroll et al.,(1978) have described a series of experiments in which the structure of muscle tissue subjected to tensile stress was observed. Bovine ST was used and heated samples were held at 90° for 45 minutes. Stress was applied in a parallel or perpendicular direction relative to the long axis of the muscle fibres in raw tissue. The change in orientation of connective tissue fibres was observed in material before and after heat treatment, and prepared for examination by SEM. Results obtained on heated tissue were generally similar to those obtained on raw tissue, the perimysial fibres becoming oriented in the direction of the applied stress. Stress applied parallel to the axis of the muscle fibres gave rise, initially, to rupture of the muscle fibres. A video tape recording of the LM image showed thin strands of perimysial connective tissue appearing at the point of rupture. The strands eventually broke as stress increased. Measurement of the force applied showed that rupture of connective tissue strands occurred at approximately double the force required to break muscle fibres. These findings have also been described by Carroll and Jones (1979).

A comparison of surface ultrastructure of bovine LD and ST muscles was described in a recent paper by Leander et al.,(1980) who cooked samples to internal temperatures of 63°, 68° and 73°. These authors state that detailed structure was not so readily observed in the LD as in the ST.

Effects of enzymes on structure

Several workers who have studied surface ultrastructure of muscle fibres have also been interested in the role of enzymes in the mechanism of tenderisation by the selective removal of connective tissue and changes in myofibrillar structure. In their examination of skeletal muscle with SEM, Boyde and Williams (1968) applied bacterial collagenase to frog sartorius muscle. This removed the connective tissue layer over the muscle fibres, exposing features of their structure in fine detail. These have been described earlier in this paper.

Eino and Stanley (1973 a and b) demonstrated that structural changes occurred concurrently with cathepsin activity. They also showed by <u>in vitro</u> experiments that cathepsins acting on muscle strips produce changes in tensile properties. Such changes occurred under the same conditions of pH and temperature as in-carcass aging. Collagenase was found to affect connective tissue elements and the sarcolemma, but the myofibrils were not greatly affected.

The action of bovine spleen cathepsin on bovine semimembranosus muscle (SM) was tested by Robbins and Cohen (1976) who found that the degree of structural change observed by SEM in the Z-disc region in myofibrils was determined by the concentration of the enzyme. This action is thought to be due to the presence of cathepsin D in the enzymes extractable from bovine spleen. Furthermore, the presence of cathepsin B1 probably accounts for the degrading action on the sarcolemma and other collagenous tissues.

More recently Robbins et al.,(1979) have studied the action of highly purified cathepsin D on bovine myofibrils incubated in the presence of the enzyme at 25° or 37° for up to 24 hours. At pH values of 5.2 - 5.3 it was observed that Z-disc degradation was apparently complete within 40-60 minutes. The myofibrils were prepared for SEM by a freeze-substitution method followed by critical point drying. This is a modification of the method described by Cohen (1976) for the fixation of myofibrils for observation by SEM.

Further experiments have been performed by Cho et al.,(1978) in which the effect of enzymes on muscle structure has been investigated. Lysosomal enzymes obtained from pig leucocytes were applied in <u>vitro</u> at pH 4.0 and 7.0, at either 37° or 4°. Strips of bovine psoas muscle were exposed to the enzyme preparation which caused changes in endomysial connective tissue, sarcolemma and the transverse ridges associated with the T-system. In samples held at 37° moderate to severe structural changes occurred between 12 and 24 hours. At 4° and the same pH there was partial degradation of the Z-disc and weakening of the A-I junction. At pH 4.0 less degradation was evident. Even after 7 days at 4° there was little difference between treated samples and controls, but some degradation of connective tissue membranes did occur.

Texture Studies

Many of the investigations into muscle structure using SEM indicate an association between texture and structure, even though this may not have been the primary purpose of the investigation. Such a relationship has been well established in meat research over many years, e.g. Locker (1960). Specifically, Stanley and Geissinger (1972) related the structure of contracted psoas muscle of the pig to texture, and described the surface ultrastructure in different degrees of contraction. Sarcomere lengths were measured on scanning electron micrographs and the values of other physical properties determined by objective and subjective measurement. Increased toughness was clearly associated with shorter sarcomere lengths. In another paper devoted to the consideration of meat texture, Stanley and Swatland (1976) compared scanning electron micrographs of unrestrained and restrained muscles. Morphological differences were apparent, with significant correlation between the experimental treatments and most of the parameters measured, the only exception being total water content.

In view of the importance of texture in assessing meat quality, and the complexity of the factors involved, there is bound to be an expansion of interest in the application of SEM to the study of this area of meat science.

Meat Products

The structure of certain processed meats has been examined by scanning electron microscopy by Theno et al.,(1978), Theno & Schmidt (1978) and Siegel et al.,(1979). The effect of massage on excised muscles of cured hams with varying levels of added salt and phosphate for periods of time between 0 and 24 hours was revealed in changes in the ultrastructure of the tissue (Theno et al., 1976, 1978). After 4 hours significant fibre disruption occurred, as shown by sarcolemma breakdown, with longitudinal disruption of muscle fibres resulting from further massage. The effects of massage were more pronounced in the presence of salt and phosphate.

In a comparison of commercial frankfurters, using LM and SEM, Theno and Schmidt (1978) demonstrated a wide variation in the microstructure of acceptable commercial products. In one brand only out of the three which they examined was a true meat emulsion observed, in which small fat droplets of uniform size were evenly distributed in a finely structured protein matrix. The scanning electron microscope made it possible to study detail within the recesses of the product, revealing the small size of the fat particles bound throughout the matrix. Other products were seen to have large fat globules in a very coarse protein matrix.

The identification of fat and protein components in meat emulsions has also been described by Ray et al.,(1979) who examined preparations of liver sausage by LM and SEM. Using serial sections of fixed material, the coagulated protein matrix, the fat component and structure bears upon the assessment of texture. Morphological parameters, such as muscle fibre diameter or cross-sectional area, and sarcomere length have been measured with varying degrees of correlation to texture. The connective tissue content of muscles is another variable factor, assessed chemically by the measurement of hydroxyproline. A histological approach has been described by Schmitt et al (1979). The identification of chemically and genetically distinct collagens at different sites has been enhanced by the use of immunofluorescence microscopy with specific antibodies (Duance et al, 1977). The use of stereological methods would provide data on the proportional distribution of the tissue components which are found in meat, and this could be aided by the application of image analysis techniques. It is only as measurements of several parameters are combined that we can get anything approaching a full assessment of quality. This topic is discussed by Asghar and Pearson in their recent review of muscle composition and meat quality (Asghar and Pearson, 1980).

H.D. Geissinger: The author implies that the fine transverse structures overlying the myofibrils are T-tubules. Are the more coarse structures also thought to be T-tubules, or could they be interpreted as transversely oriented mitochondria?
Author: Yes, mitochondria may be seen in transverse and longitudinal orientations in the SEM micrograph (Fig.4). They are somewhat larger than elements of SR as may be seen in micrographs prepared by TEM. It is probable that the regularly spaced pairs of transverse structures in Fig. 4 represent terminal cisternae of the sarcoplasmic reticulum.

R.J. Carroll: Please explain in more detail the differences in 'active' and 'passive' contractions associated with cold-shortening.
Author: When a muscle in the pre-rigor state receives a stimulus to contract, the response of the population of fibres is variable. Some fibres shorten by the sliding action of the myofilaments as is demonstrated by the increased over-lap of myosin and actin filaments. This may be measured in terms of a change in sarcomere length and is described as 'active' contraction. Such a length change effects a reduction in the overall length of the muscle, unless it is restrained. The fibres which have not responded to the contractile stimulus as described above must, however, conform to the change in muscle length. This is accommodated by these fibres adopting a wavy configuration, described as 'passive' contraction.

C.L. Davey: The paper does not refer to the electron microscopic observations that gaps open up in stretched muscle during aging, suggesting that distinct structural changes are occurring in post-rigor meat at sites other than those mentioned.
Author: This is an important observation described by Davey and Graafhuis (1976) and other workers at the Meat Industry Research Institute of New Zealand (MIRINZ). Since this feature can only be observed in excessively stretched muscle, involving experimental manipulation of the muscle while in the pre-rigor state, it has generally been obscured from view in microscopic studies on meat. Scanning electron microscopy has still to be applied to this area of investigation.

Discussion References

Asghar, A. and Pearson, A.M. Influence of ante- and postmortem treatments upon muscle composition and meat quality. Adv. Food Res., 26, 1980, 53-213.
Davey, C.L. and Graafhuis, A.E. Structural changes in beef muscle during aging. J. Sci. Fd Agric. 27, 1976, 301-306.
Duance, V.C., Restall, D.J., Beard H., Bourne, F.J. and Bailey, A.J. The location of three collagen types in skeletal muscle. FEBS Letters 79(2), 1977, 248-252.
Schmitt, O., Degas, T., Perot, P., Langlois, M. and Dumont, B.L. Etude morpho-anatomique du périmysium (Méthodes de description et d'évaluation). Ann. Biol. anim. Bioch. Biophys., 19(1A), 1979, 1-30.

Editor's Note: The reader is also referred to the paper "Preparation of Muscle Samples for Electron Microscopy" by H.D. Geissinger and D.W. Stanley in this issue.

H. D. Geissinger and D. W. Stanley

Discussion with Reviewers of the paper "Preparation of Muscle Samples for Electron Microscopy" from page 72.

C.A. Voyle: Are there risks of spurious contraction effects in fixation at temperatures around 310 K or 273 K in freshly dissected (pre-rigor) muscle? How would these be avoided?
Authors: In freshly dissected muscle there certainly are risks since parts of the muscle are frequently seen to contract as soon as the muscle is grasped with forceps. Thus, the muscle may not be in a true state of relaxation even though it may be mechanically stretched by immobilizing it later on. The only possible way to avoid this, that we are aware of, is to use the "non-touch-technique" for removing the sample advocated by Bethlem (Text reference #2), i.e. careful avoidance of touching the sample with forceps or fingers, and only taking the untouched portion of the specimen for further processing after it has been immobilized by pinning the ends, or placing sutures or a biopsy clamp at each end of the muscle and immersion of it into the fixative solution.

R.J. Carrol: Restraint of muscle or meat tissue by clamping or tying the tissue to prevent shortening of the sarcomeres is overlooked by many researchers. Can you discuss this aspect in more detail when comparing lengths of sarcomeres from different muscles, animals, species, or treatments?
Authors: In the ensuing discussion it is presumed that muscle samples are prepared for electron microscopy optimally and in such a fashion that the sarcomeres of individual myofibrils are well enough demarcated from one another that their length can be measured from the micrographs. Workers in the field will have to be aware of the fact that sarcomere length in different muscles may vary, e.g. it has been found (Dutson T R, Hostetler R L, Carpenter Z L. Effects of collagen levels and sarcomere shortening on muscle tenderness. J. Food Sci. 41 (1976) 863-866) that in two restrained muscles, sternomandibularis (S) and psoas major (PM) muscles, respectively, a variation in shortening due to different treatments could be produced in sarcomere length of 1.7-3.25 μm for the PM and 1.55 to 2.6 μm for the S muscles. This variation in sarcomere length also coincided with the amount of collagen in the muscles (less for PM) and, therefore, the ultimate tenderness of the meat.

Sarcomere length in muscles from different animal species may vary widely, e.g. the average sarcomere length of most amphibians or mammals is ~ 2 μm, whereas the sarcomere length of slow-moving muscles of the horseshoe crab (Limulus) may be six times greater and reach 12 μm (Szent-Györgyi A. Chemistry of Muscular Contraction, 2nd edn. Academic Press Inc., New York (1951) p. 12). However, although different types of muscle may show different sarcomere lengths (see above), we are unaware of a variation in sarcomere length in restrained "untreated" muscle in the different species of most meat-producing (domestic) animals.

It is well appreciated by meat researchers that the mean sarcomere length (SL) decreases as the muscle passes from the pre-rigor state (SL = 1.86 μm) to the in-rigor state (SL = 1.54 μm) and lengthens considerably in the post-rigor period (SL = 2.29 μm). (Ruddick JF, Richards JF. Comparison of sarcomere length measurement of cooked chicken pectoralis muscle by laser diffraction and oil immersion microscopy. J. Food Sci. 40 (1975) 500-501). Cooking of veal will appreciably shorten sarcomere length in pre-rigor Achilles tendon-hung or pelvic-hung muscle, and the latter shows twice the sarcomere length (3.20 μm) than the former (1.53 μm) (Bouton P E, Harris P V, Shorthose W R, Ratcliff D. Changes in the mechanical properties of veal muscles produced by myofibrillar contraction state, cooking temperature and cooking time J. Food Sci. 39 (1974) 869-875).

Lastly, we like to stress that the resulting electron micrographs of muscle are the results of a fresh muscle specimen which has been subjected to all sorts of preparation procedures to make it suitable for electron bombardment in vacuo. Therefore, any measurements made should not be considered as absolute measurements of sarcomere length but as a relative parameter which is valid as long as all samples in an experiment have been prepared the same way.

C.A. Voyle: A recently described freezing method using nitrogen slush is being increasingly used. (Sevéus, L. Preparation of biological material for x-ray microanalysis of diffusible elements. J. Microsc. 112 (1978) 269-279). Have the authors any experience of this, especially in relation to freezing damage?
Authors: We have not yet attempted the method, because the preservation of ultrastructural detail, even in the 5 to 8 μm ice crystal-free surface layer of the tissue, would not appear to be sufficient for our purposes (resolving details of sarcolemma and smooth endoplasmic reticulum in muscle fibers). Therefore we cannot comment on the freezing damage which may be caused by it. This method should be a promising starting point, once we get to the microanalysis in our specimens and therefore have to circumvent chemical fixation procedures.

M. Ashraf: Can the authors recommend a method for preparation of muscle tissue (previously paraffin embedded/epon embedded or unembedded) for backscatter electron imaging?
Authors: One method which recently gave promising results (i.e. we could see lighter staining T-tubules at the A-I junctions of sarcomeres of skeletal muscle against a darker background of the myofibrils and their associated structures) is as follows: Pinned skeletal muscle segments were placed for two hours in 3% glutaraldehyde (GA) - 8% tannic acid (TA) fixative, then in 1% OsO4 for two hours, and finally into the GA-TA fixative followed by OsO4 for an additional four hours. The pieces of tissue were infiltrated with a cryoprotective agent, placed on 255 K cryostat stubs and sectioned (~10μm). The cryostat sections were critical point dried in CO2 and were ready for examination in the SEM in the backscattered mode.

PREPARATION OF MUSCLE SAMPLES FOR ELECTRON MICROSCOPY

H. D. Geissinger and D. W. Stanley

Department of Biomedical Sciences and Department of Food Sciences, University of Guelph, Guelph, Ontario, Canada, N1G 2W1.

Abstract

The relevant literature on EM preparation techniques for muscle is reviewed. The currently available methodologies are presented and critically discussed from the point of view of the biologist and the food scientist. Those SEM and TEM procedures which have proved to be most useful in the present state of the art of specimen preparation for EM are summarized. Some of the more recently available staining and fracturing procedures for muscle, as well as specimen preparation procedures for intermicroscopic (LM, SEM, TEM) correlation are stressed.

It is proposed that the normal ultrastructural appearance of muscle should serve as a baseline for the ultrastructural examination of meat.

KEY WORDS: Histological preparation procedures, meat texture, skeletal muscle, cardiac muscle, smooth muscle, scanning electron microscopy, transmission electron microscopy, correlative microscopy

Introduction

Muscle (also called contractile tissue) comprises one of the four basic tissues in animals; it is categorized as striated and nonstriated (or smooth) muscle. Striated muscle may be again subdivided into skeletal (voluntary) and cardiac (non-voluntary) muscle[1]. Since muscle represents the major bulk of tissue in man and animals, it is not surprising that a vast amount of literature regarding its interpretation at the ultrastructural level is available. It is not within the scope of this paper to discuss this work critically since many of these papers would be of historical interest only.

The histological structure of muscle at the light microscopic (LM)[2] and transmission electron microscopic (TEM)[3] levels of examination is well known. Obviously, great strides have been made in understanding the structure of striated muscle using these two techniques. One only has to look at micrographs of skeletal muscle which appeared in older histology tests[1] and compare these to more recent TEM micrographs in newer tests[4], in order to appreciate the improvements in specimen preparation and instrumentation that have taken place in the last three decades.

While specimen preparation procedures and instrumentation for striated muscle have improved considerably for the scanning electron microscope (SEM) over the past 15 years as well (compare the first published micrographs of frog skeletal muscle[5] to more recent micrographs of cardiac muscle[6]), the structure of muscle as revealed by SEM has not received nearly the attention as the ultrastructural findings derived from the TEM. There is also a better understanding of the structure of smooth muscle at the TEM level. Very few investigations concerning smooth muscle appear to have been undertaken by the SEM to date[7,8,9].

The food scientist views muscle tissue as the precursor of meat, a food that has always held a rather special place in man's diet. It is the conversion of muscle to meat, with the concomitant biochemical and structural alterations, that has proved enigmatic to researchers in the field. Meat has the most complex structural organization of any foodstuff and the influence of structural factors on meat quality is only now being fully appreciated.

Meat, for the most part being post-mortem

striated muscle with associated connective and vascular tissue is an aggregate of many structural elements. Of particular interest because of their relation to quality are connective tissue, myofibrils and their degree of contraction and integrity, and muscle fiber arrangement. It is now recognized that a study of the microstructure of food such as meat can reveal a great deal about quality characteristics such as texture and about the influence upon it of processing operations such as heating, aging and cooling or freezing. A skilled microscopist can learn much from the microstructure of muscle which is of use in maintaining and improving meat quality.

Although it is not within the realm of this paper to delineate the advances made in this area, the reader is directed to papers that deal with this area[10-16] for an overview of an important subject, since the purpose of this paper lies more with sample preparation of muscle for electron microscopy. Of the two major techniques available, SEM has proven invaluable and for this reason it is so popular for all biological research. It is apparent also that, if time and facilities allow, correlative microscopy is the best approach.

The purpose of this paper is to acquaint the reader with the commonly accepted preparation methods peculiar to muscle samples for SEM and TEM. It is recommended that optimal preparation methods for electron microscopy should serve as experimental controls for the examination of pathologically altered muscle, as well as for the examination of meat.

Specimen Preparation

Anaesthesia and Killing and Perfusion (smaller samples)

Several suitable anaesthetics are available for the sacrifice of animals. Smaller animals can be rendered unconscious with ether or halothane or their necks can be dislocated. Larger animals can be anaesthetized with pentobarbital (50 to 60 mg/kg of body weight) or be killed by a stunning blow to the head. Ashraf and Sybers[17] recommended the following procedure for the removal of hearts:
1. The heart is removed with a short segment of aorta attached.
2. A cannula attached to a syringe containing lactated Ringer's solution is inserted rapidly into the aorta which is then perfused retrogradally.
3. If the heart muscle should be in a relaxed state, Procaine or potassium chloride[18] are added to the perfusate.
4. After a brief flush to remove the blood cells from the coronary vessels the Ringer's solution is replaced with fixative and perfusion is allowed to continue for 10 minutes.
5. Small tissue blocks (0.5 cm x 0.05 cm x 0.05 cm) are removed from the desired regions in the heart and immersed into fresh fixative.

For skeletal muscle of small animals (rat, rabbits) perfusion with Locke's solution containing 0.1% $NaNO_2$ through the abdominal aorta is followed by a 12% solution of glutaraldehyde. The procedure is described in detail by Hayat[19]. It has been shown experimentally that perfusion fixation proved superior to immersion fixation in rat lumbrical muscles[20].

The practice of immersing the entire hind- or forelimb into fixative for the initial fixation of samples is followed by some workers[21]. These muscles are attached to their origins and insertions which is said to prevent shortening.

The above-described methods are obviously not suitable for skeletal muscles of most meat producing animals because the bones and the attached musculature of these animals are much too large.

Smooth muscle, depending on the site of the animal it is taken from, will likely have to be fixed by immersion. Although perfusion fixation is undoubtedly superior to fixation by immersion[19,22] and should be carried out wherever possible, practical considerations usually dictate that muscles are fixed by immersion.

Initial Handling of Larger Samples

Nothing is more frustrating to a histologist who works with muscle than to have fixed muscle samples submitted that are distorted and curled. Therefore, it is strongly recommended to affix strips of striated or smooth muscle in some fashion to prevent this. Human biopsies are obtained with a biopsy clamp which prevents the muscle from shortening; and careful handling of this tissue results in excellent ultrastructural preservation[23].

Firstly, the muscle sample should be removed from its surrounding tissue as speedily as possible and immersed in isotonic saline. Tying pieces of suture to each end of the biopsy and fastening these to a cork prevents the piece of tissue from contracting. Care has to be exercised not to overstretch the muscle unless this is intended and inherent in the design of the experiment, and to keep it wet at all times. The muscle may then be immersed into the fixative solution of choice as fast as possible (Figs. 1 and 2). Omitting this most important first step in pre-rigor samples of muscle may result in uneven contractions of some muscle bundles (Figs. 3 and 4). In post-rigor samples of meat, it may not be desirable to immobilize the muscle samples.

Fixation of Muscles

If only the surface texture of skeletal muscle is to be studied at low and medium magnifications, simple LM fixatives (eg. formalin) may suffice[12-15]. However, if ultrastructural features of muscle are to be studied, the use of LM fixatives will result in poor preservation of the tissue (Figs. 5 and 6).

In his recent book on specimen preparation for the SEM, Hayat[19] described the optimal properties of fixatives for SEM and TEM. Fixatives for SEM do not differ remarkably from those used for TEM, except in their respective osmolalities. Those used for TEM should be of higher osmolalities than those used for SEM, because the hypertonic fixatives used for TEM will actually become isotonic or hypotonic by the time the fixative has penetrated the deeper layers of the tissue block. Since for most examinations in the SEM one is primarily concerned with adequate fixation of the surface layers, the ideal fixative for SEM should be isotonic. Hayat[19] stresses the

Muscle preparation for Electron Microscopy

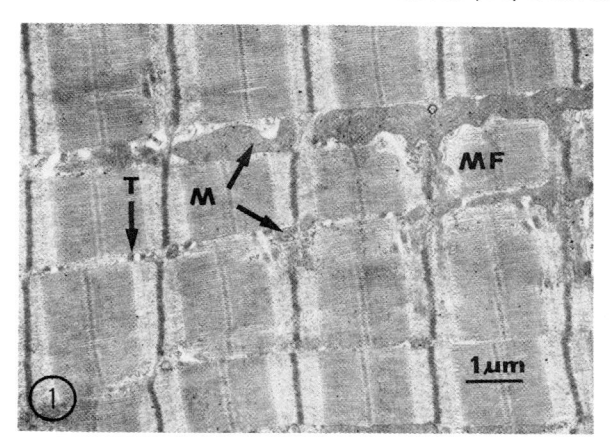

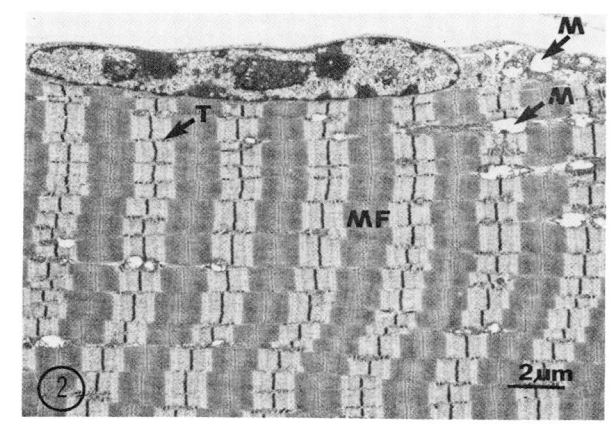

Fig. 1. Biopsy of porcine gracilis showing well preserved myofibrils (MF), mitochondria (M) and sarcoplasmic triads (T). (Courtesy of Sandra Frombach).
Fig. 2. Porcine gracilis collected post mortem showing adequate myofibrillar (MF) and triad (T) preservation, however very poor preservation of mitochondria (M). (Courtesy of Sandra Frombach).

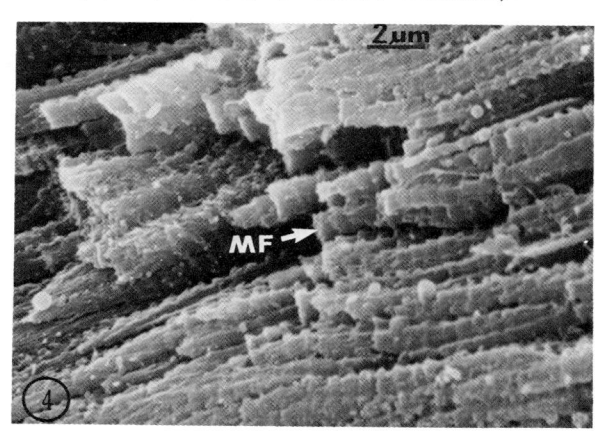

Fig. 3. Porcine semitendinosus which had been restrained prior to immersion into glutaraldehyde showing relaxed sarcomeres (S).
Fig. 4. Porcine semitendinosus which had been immersed into glutaraldehyde without prior restraint showing hypercontracted myofibrils (MF).

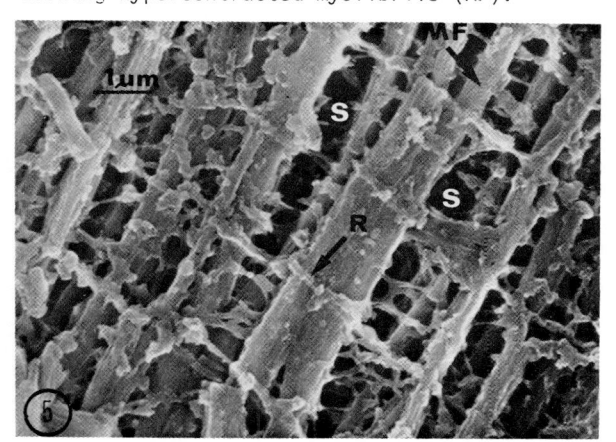

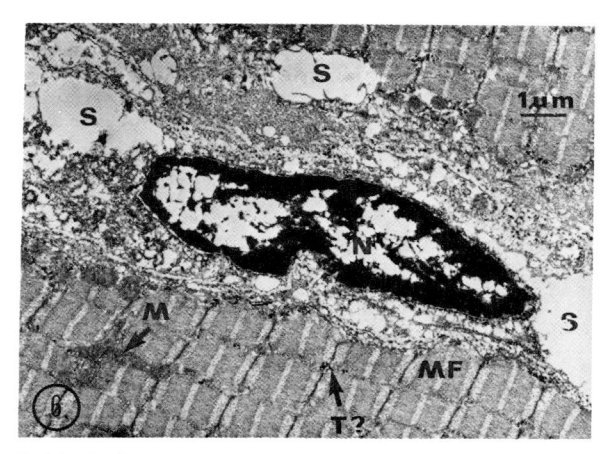

Fig. 5. Murine gastrocnemius which had been fixed with Heidenhain-Susa fixative showing artifactual spaces between the myofibrils (MF) which are crossed by transverse ridges (R).
Fig. 6. Murine gastrocnemius which had been fixed with Bouin's fluid and osmium tetroxide showing barely adequate myofibrillar (MF), triad (T?) and mitochondrial (M) preservation, as well as artefactual spaces (S) in the perimysium surrounding the nucleus (N) probably of a satellite cell.

experimentally proven fact[24] that the osmolality of the fixative solution depends to a large degree on the osmolality of the buffer.

The fixatives most commonly employed in the fixation of muscles for TEM and SEM are glutaraldehyde[25] and Karnovsky's[26] fluid (a combination of glutaraldehyde and paraformaldehyde). Glutaraldehyde, at a concentration of 2% of the aldehyde in 6.9% sucrose has an osmolality of 480 milliosmols, whereas Karnovsky's fixative (3% glutaraldehyde and 1.5% formaldehyde) has an osmolality exceeding 2000 milliosmols. Although Karnovsky's fluid was primarily intended for TEM, it has been used for SEM of muscle as well[21,27].

Concentrations of glutaraldehyde (GA) varying from 1.7 to 3.3% were found optimal for the ultrastructural presentation of striated muscle[28,29].

Phosphate or cacodylate buffers are usually employed with aldehyde solutions, and both have been given excellent results with heart muscle[6,17] although cacodylate contains arsenic, and therefore constitutes a health hazard.

The pH of the fixative solution should not exceed 7.5[19,22] and is best kept between 7.2 and 7.4 for fresh muscle. For meat, which may have as low a pH as 5.5 to 5.6, the fixative solution should be adjusted accordingly. Jones et al.[27] used fixative solution pH's of 7 for fresh muscle and 6 for aged muscle.

At higher temperatures (25 to 37°C) fixation with glutaraldehyde is more rapid, while at lower fixation temperatures (0 to 4°C), fine structural detail of microtubules in muscle may be lost. Since it is desirable to fix at refrigeration temperatures in order to prevent autolytic changes, Hayat[22] recommends a brief preliminary fixation of muscles at 4°C, while fixation should continue at room temperature. It has been demonstrated that delicate filamentous processes and filamentous projections are better preserved at higher temperatures[30]. As a general rule, Hayat[19] suggests that specimens designed for SEM should be fixed at higher temperatures.

Depending on the size of the tissue (less than 1 mm³ for TEM, and up to 5 mm³ for SEM), variable lengths of fixation (from 7 to 8 hours) in aldehydes are employed.

Post-fixation in osmic acid is routinely employed in most EM laboratories, unless it is omitted for histochemical reasons. A significant difference can be shown in skeletal muscle which has been fixed with Karnovsky's fluid alone, or with Karnovsky's fluid and OsO_4. The osmicated muscle showed superior ultrastructural preservation[21] (Figs. 7 and 8).

Storage of Fixed Muscle

The usual practice in most electron microscope laboratories is that the tissue, once it is adequately fixed, will be further processed to the embedding stage. Sometimes circumstances render this normal routine impossible.

Sabatini et al.[25] has stated that tissues can be stored in most aldehyde solutions up to periods of several months.

The practice recommended by Hayat[19] and which is followed in this laboratory, allows the aldehyde-fixed muscle to be stored for several weeks before it is post-osmicated, dehydrated and embedded in Epon for TEM, or fractured, critical-point-dried (CPD) in CO_2[31] and mounted for SEM.

TEM Preparation of Muscle

Muscle (striated or smooth) which has been fixed appropriately (glutaraldehyde and OsO_4) generally has well-preserved structure. Dehydration in alcohols or acetone is the same as for other tissues. Since striated muscle is a relatively tough tissue, embedding for sectioning is preferably carried out in plastics which, when cured, are of hard consistency[22,32]. For precise details of TEM preparation of muscle the reader is referred to many excellent texts which deal specifically with this procedure (e.g. Hayat[22]).

SEM Preparation of Muscle

Drying. If the intention of the SEM investigation is to show the texture of myofibers and myofibrils, drying the samples in air may be preferable to critical-point-drying[12-15]. The shrinkage of the sample which ensues is, of course, artifactual, but it undoubtedly exaggerates structures which may otherwise be masked. Thus it becomes obvious that the A-band is raised with respect to the I-band of the sarcomere, except where the I-band is crossed by the Z-line which is again raised[12]. Minute differences in texture of surfaces of meat from rainbow trout, turkey and beef have been reported[10].

However, since drying the samples in air from organic solvents (which had been used for dehydrating the samples) often introduces unpredictable and therefore uninterpretable artifacts, which are due to the recession of the air-solvent interface across the tissue as the dehydrating fluid evaporates, "air drying" of muscle is not a procedure to be recommended. The most commonly used methods of avoiding the injurious interface are CPD[33] and freeze-drying[34]. The latter, although it produces less volume change[35] than CPD, and therefore may be preferable to it, does not appear to be in common usage for muscle tissue, although its use has been reported on bovine muscle, in which it yielded variable results[27].

CPD can be carried out in CO_2 without using amyl acetate as the last intermediate fluid[31], and this procedure is followed in many laboratories which prepare muscle samples for SEM. The CPD procedure, which we use in our laboratory, is as follows:

1. Before starting the procedure the critical point drying bomb is chilled with cold water.
2. The small pieces of muscle (∿ 5 x 5 x 2 mm) or the microtome sections of muscle on standard microscope slides (25 x 75 mm) are transferred "dripping wet" from the last dehydrating solution[33] (absolute alcohol in our case, since we omit amyl acetate) to the bomb. Speed is essential in this step, since inadvertent air drying might result in the transfer of the specimen from the extremely volatile absolute alcohol.
3. The bomb is first flushed with liquid CO_2 for a period of 30 to 40 minutes and then it is filled with CO_2.
4. The water mixing valve is adjusted to "Hot" until the pressure gauge shows between 8270 and 10340 kPa. (The critical pressure of CO_2 is 7240 kPa).
5. The temperature gauge on the bomb is brought

Muscle preparation for Electron Microscopy

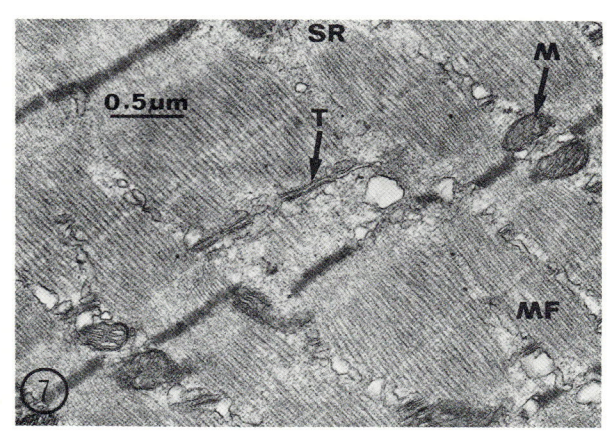

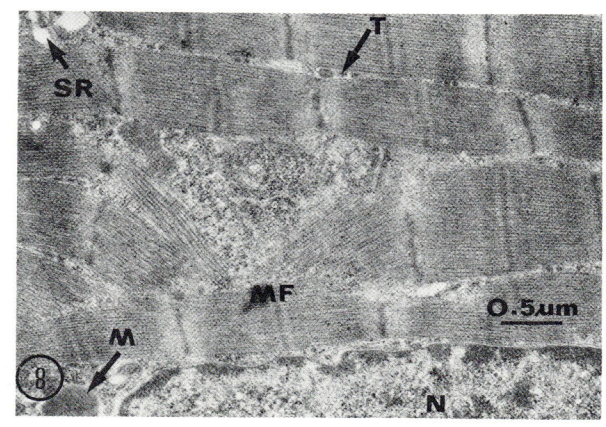

Fig. 7. Murine gastrocnemius which had been fixed in Karnovsky's fluid and osmium tetroxide showing excellent preservation of myofibrils (MF), T-tubules (T), mitochondria (M) and sarcoplasmic reticulum (SR).
Fig. 8. Murine gastrocnemius fixed with Karnovsky's fluid (no post-osmication) showing well-preserved myofibrils (MF) and T-tubules (T), however, poorly preserved mitochondria (M) and sarcoplasmic reticulum (SR).

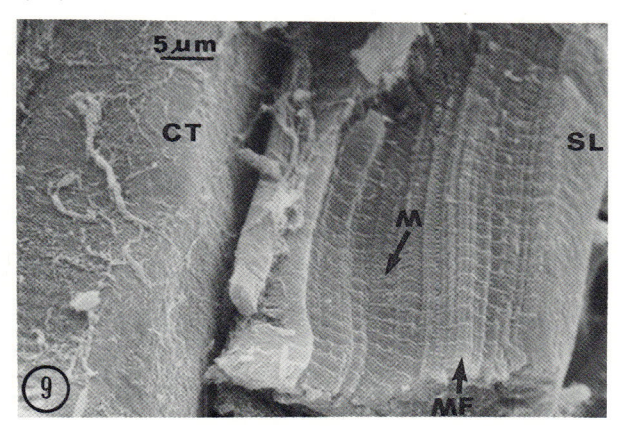

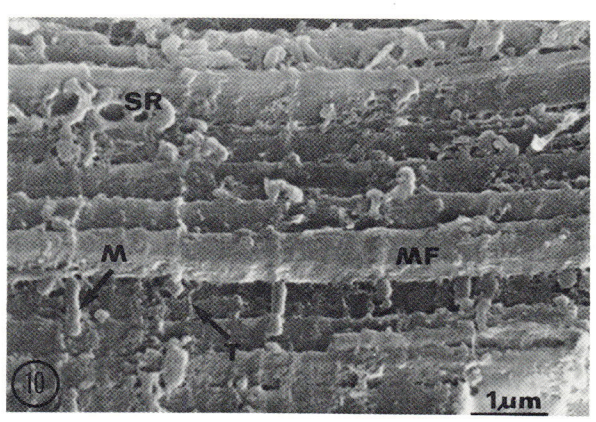

Fig. 9. Standard SEM (GA fixation) of porcine gracilis showing sarcolemma (SL), transversely oriented mitochondria (M) with respect to the myofibrils (MF), and some connective tissue (CT) of the endomysium.
Fig. 10. Freeze fracture (Karnovsky's fixation) of murine gastrocnemius showing well defined myofibrils (MF), transverse mitochondria (M), T-tubules (T) and sarcoplasmic reticulum (SR).

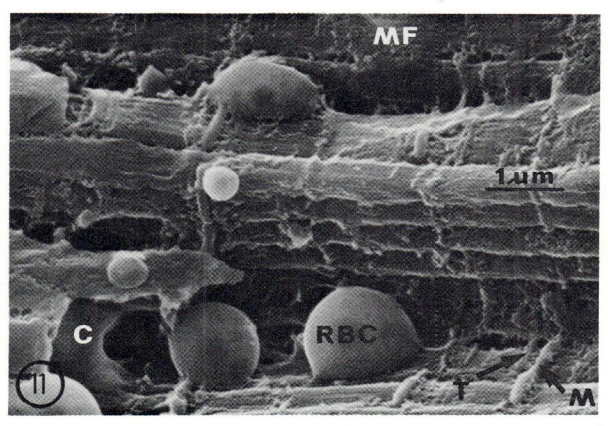

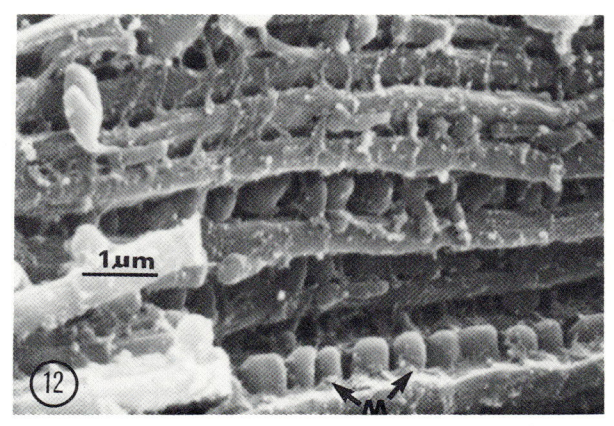

Fig. 11. Freeze fracture (Karnovsky's fixation) of murine gastrocnemius showing red blood cells (RBC) emanating from a broken capillary wall (C), well-defined myofibrils (MF), T-tubules (T) and mitochondria (M).
Fig. 12. Dry fracture (Karnovsky's fixative) of murine gastrocnemius showing intermyofibrillar mitochondria (M).

H.D. Geissinger and D.W. Stanley

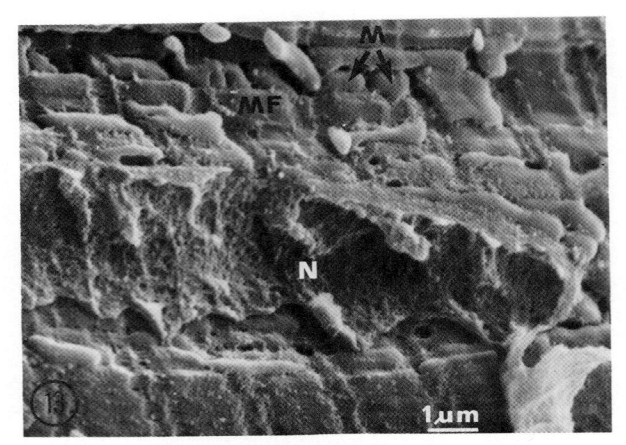

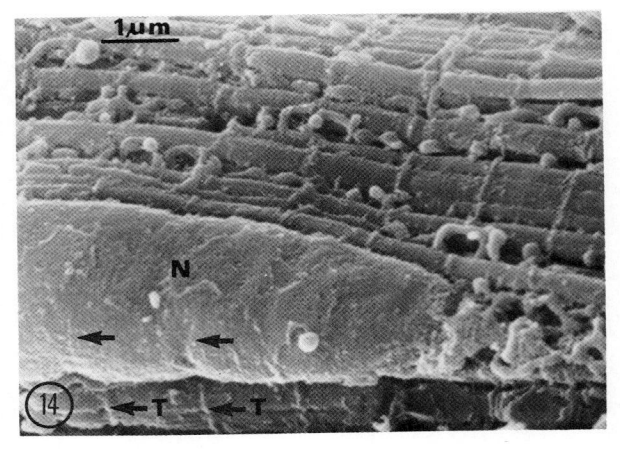

Fig. 13. Dry fracture (Karnovsky's fixative) of murine gastrocnemius showing the porous surface of the nucleolemma of a fractured myonucleus (N) and mitochondria (M) adjacent to myofibrils (MF).
Fig. 14. Dry fracture (Karnovsky's fluid) of murine gastrocnemius showing myonucleus (N) the surface of which (arrows) appears to be in register with T-tubules (T).

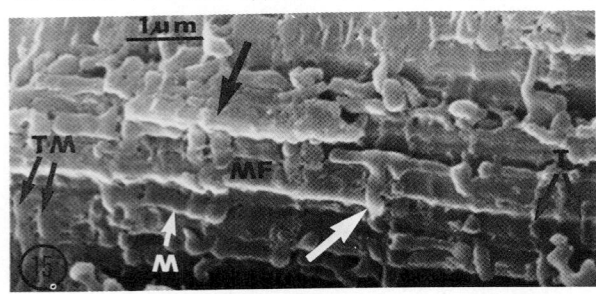

Fig. 15. Cryostat section (Karnovsky's fixative) of murine gastrocnemius showing transversely oriented mitochondria (TM), myofibrils (MF), intermyofibrillar mitochondria (M) and apparently branching mitochondria (white arrow), as well as T-tubules(T). Part of a myofibril consisting of several sarcomeres (black arrow) is also shown. The main advantage of this preparation over standard preparations is that the possibility of SEM→LM→TEM correlation exists, therefore doubtful structures in the SEM-image can be interpreted.

to 46°C (not above 49°C--critical temperature of CO_2 is 31°C[31]). The mixing valve and the rate of water flow are adjusted accordingly. The upper outlet valve is bled as required to keep the pressure within range.

6. As soon as the temperature and pressure are stabilized, the procedure is timed for ten minutes (critical point drying of the specimen). At the end of this step, the upper outlet valve may be opened slowly to reduce the pressure in the bomb to zero.

7. The tissue blocks or glass slides are removed from the CPD bomb and coated with metal. Then they are stored in a desiccator until SEM examination. The water mixing valve of the CPD apparatus is turned to "Cold" or "Off" depending on whether the method is to be repeated immediately. Freon has been used very successfully for CPD in the study of cardiac muscle[6,17,36].

Further SEM Preparation. In order to study the interior of myofibers, the sarcoplasmic organelles have to be exposed to the electron beam of the SEM in some fashion. This is commonly achieved using either "standard" or "fracture" methods of SEM preparation.

(a) Standard Method. In this method the myofibers are cut parallel to the long axis of the fibers with a sharp, clean razor blade. The tissue is then dehydrated and dried. Good results have been reported for this method using cardiac tissue[6,17], although it does not seem to be as reliable as the fracture methods of preparation since the surfaces obtained by the standard method are not smooth and myofibers are masked by connective tissue of the surrounding perimysium (Fig. 9). Also transverse and oblique sections of the muscle are often obtained, that are of limited use to the investigator, because they are difficult to interpret[17].

(b) Freeze Fracture Method. In the first attempt of this method, the fixed nondehydrated tissue was frozen in liquid nitrogen, cleaved with a single edge blade, thawed and dehydrated in alcohol before it was critical-point dried[37]. Using this method, the tissue is fractured along points of structural weakness. Haggis[37] used various murine tissues including muscle. Artifactual spaces between myofibrils occurred which were thought to be due to ice crystal formation. It was sought to overcome this problem by the use of fluids with low freezing points such as DMSO,* ethylene glycol or glycerol. Nemanic[38] using glycerol partially overcame the problem of ice crystals, but had to wash the tissue overnight in 20% alcohol to remove the glycerol. The method, which is now almost universally accepted and practiced in many laboratories, is that of preventing ice crystal formation in tissues by partial dehydration of the fixed tissues with alcohols or Freon 113 before the tissue is frozen in liquid nitrogen cooled Freon 113, and fractured with a precooled knife. Sybers and Ashraf[39], Ashraf and Sybers[6,17] and McCallister et al.[36] have reported excellent results using this method on normal and infarcted myocardial tissue. Skeletal muscle yields equally good results

*dimethyl sulfoxide

66

(Figs. 10 and 11). For further and precise details of the cryofracture method for muscle, the reader is referred to the appropriate chapter in Hayat's "Principles of SEM" authored by Ashraf and Sybers[17].

Cryofracture of ethanol-infiltrated specimens yields an uncontaminated crisp fracture face. This procedure has been used for the examination of meat by Jones et al.[27], who found this fracture preparation method preferable to any other. In view of the fact that the ethanol fracture method eliminates artifacts which may have been caused by freezing of the tissue, the method can also be used on specimens which are destined for TEM[40].

Recently a new preparation technique for scanning electron microscopic study of the internal structure of skeletal muscle cells was described by Jósza et al.[41] The specimens were fixed, dehydrated and embedded in Epon or Durcupan--without polymerization--and thereafter they were freeze-fractured.

(c) <u>Dry Fracture Method</u> (Flood[42]). In this method of preparation, the fixed and CPD tissue is fractured with the edge of a razor blade. The method followed in this laboratory is as follows:
1. The blade is rinsed in alcohol or acetone to remove any traces of grease or oil.
2. Muscle tissue is positioned on a stub and broken by applying moderate pressure to the blade which is held in a hemostat.
3. The two halves are pulled apart with fine forceps and positioned on the stub with the fracture face uppermost.

This method yielded comparable results to freeze fracture on skeletal muscle in our laboratory[43,44] (Figs. 12 to 14) and is much simpler and less time-consuming than the freeze fracture method. An alternate method for dry fracture of muscle fibers, also described by Flood[42], consists of applying the cut face of the dried tissue to double-sticky adhesive tape and applying moderate vertical pressure to the tissue. In this fashion some of the fibers stick to the tape, and can be examined in the SEM after suitable coating procedures. Essentially the same procedure, called "stripping" by Watson et al.[45], has been reported for the exposure of intracellular organelles.

<u>Staining Procedures for TEM</u>. Fixed and plastic-embedded muscle can either be stained with uranyl acetate and lead citrate[46] in the block before it is cut, or the stain can be applied to ultrathin sections on the grid. This procedure greatly enhances the intracellular appearance of skeletal muscle[22].

Peachey[47] used horseradish peroxidase for a better visualization of the longitudinal elements of the T-system in frog skeletal muscle. An even more elegant method for the staining of the sarcotubular system in murine skeletal muscle has been published by Forbes et al.[48]. Postfixation with osmium-ferrocyanide of tissue fixed with aldehydes containing Ca^{++} or other cations but lacking phosphate resulted in a very crisp demarcation of T-tubules and sarcoplasmic elements surrounding myofibrils.

When tannic acid is used together with OsO_4 on glutaraldehyde-fixed muscle, it stains the sarcolemma, sarcoplasmic reticulum and T-tubules of striated muscle[49].

By the use of colloidal iron (0.1 to 0.4% ferric chloride solution) which appears to be specific for acid mucopolysaccharides, certain organelles in smooth muscle have been stained[50,51].

<u>Staining Procedures for SEM</u>. Although heavy metals have been used in SEM[19], the authors are aware of only two published papers where lanthanum or silver particles were used to stain T-tubules in cardiac muscle[52,53].

<u>Correlative (LM, SEM, Scanning Transmission Electron Microscopic (STEM), TEM) Procedures</u>. Although correlation procedures are time consuming, they should be employed where doubts regarding the correct SEM interpretation of a certain structure persist.

(a) <u>LM→SEM→LM</u>. Correlative studies on microtome sections of cardiac or skeletal muscle have been performed[54-56].

In this procedure, suitably fixed and dried striated or smooth muscle which has been embedded in paraffin is cut on a rotary or sledge microtome with a steel knife, and stained with a suitable histological stain. The section on the slide is then covered with a coverslip as is done in histological procedures for LM preparations. Areas of interest in the preparation are photographed with the LM, the coverslip is removed by immersion of the slide in xylol, and after dehydration, critical-point-drying and metal-coating the specimen is ready for the examination of the same areas in any SEM, the stage of which has been fitted with a slide adapter[57-59]. For re-examination of the areas which had been photographed in the SEM with the LM, it is only necessary to remount the section in glycerol or any mounting medium. The metal coat does not interfere with the quality of the LM-image[55]. Since the tissue was primarily prepared for LM, SEM interpretation of the muscle was uncertain, however, the SEM-microscopist was at least certain what area in a preparation he was looking at, because he had an LM-micrograph to guide him. The SEM-image of ocular muscle, which had primarily been prepared for TEM, was also very difficult to interpret, after 1 µm sections embedded in Epon had been cut and the embedding medium had been partially removed by the application of an iodine solution[60]. However, since these sections were from the same block from which the ultrathin sections had been cut, the possibility of correlating the SEM-image with the TEM-image existed. The sections were examined with the SEM at low kV either on carbon-coated or uncoated coverslips. The preparation on coated glass showed less charging of the surface.

(b) <u>STEM→LM→SEM</u>. A pilot study of this procedure has been published[61]. Skeletal muscle, which had been fixed in Heidenhain-Susa[2] fixative, was embedded in paraplast, cut at ~ 1 µm and placed onto a formvar-covered indexed (Fullam) grid. The section was first examined in the STEM, then the area was relocated under an LM and photographed. Precisely the same area was then examined and micrographs were taken with the SEM. Since the section had originally been embedded in paraplast, the resolution obtainable with the STEM was rather poor, but it was possible to see the same structures crossing the myofibrils using the STEM and SEM.

(c) SEM→LM→TEM. Since it has been established that CPD tissue, that had been examined with the SEM, was still suitable for TEM-examination[62], the possibility of correlating the SEM to the TEM image became a feasibility. Accordingly, the following preparation methods were tried on murine muscle which had been fixed in Heidenhain-Susa or Karnovsky's fluid[21]. Paraplast sections were cut, the sections were immersed in xylol to remove the paraplast, stained with Hematoxylin and Eosin or were left unstained, and the slide was examined and micrographs were taken with the SEM. After the mounting of the preparation with a coverslip and its photography with the LM using polarized light, the coverslip was removed and the entire section was embedded in Epon in an aluminium weigh boat. The Epon used was of a harder consistency according to the recommendations by Luft[32]. After looking again at it with the LM, areas of interest were sawed out with a jeweller's saw, the piece of Epon-embedded muscle was glued onto an Epon dummy block, and serial sections were cut on an ultramicrotome. These preparations were examined in the TEM, and when the best correspondence to the image of the previously taken SEM-micrograph was reached, micrographs were taken in the TEM. Although the method is tedious and time-consuming, and could, therefore, not be recommended as a routine procedure, a positive correlation of LM, SEM and TEM images of myofibrils was possible. Refinements to this method were made[43]; the most striking one being the use of frozen cryostat sections instead of paraffin sections of skeletal muscle. Therefore, cryostat sections are used for muscle correlative microscopy work in our laboratory (Fig. 15); preliminary results on human myopathies[44] have shown that this procedure might be useful in the diagnosis of diseased muscle. It is suggested that this approach might also be useful in the examination of meat. After all, meat is muscle which has to a greater or lesser extent been altered from the in vivo condition.

Further Preparation for SEM

A detailed treatise on the subjects of mounting the specimens on stubs and coating them with metals is presented by Hayat[19].

Mounting. Specimens of muscle are usually mounted on a specimen stub with conducting silver paint. There are no special requirements for muscle which would make them more or less difficult to mount than the requirements needed for conventional SEM. Specimens can be mounted on standard microscope slides or glass coverslips for LM→SEM correlation. The precoating of glass coverslips with carbon may prevent the build up of charge during the SEM examination[60].

Coating. Muscle specimens are usually coated with a thin layer of precious metal (e.g. Au, Pt, etc.) to prevent the build-up of electrical charge on the specimen surface. Two common coating techniques, viz. vacuum evaporation and sputter coating, are in use. Both techniques have been discussed fully by Echlin and Hyde[63] and Echlin[64].

Uncoated Muscle Specimens. Although coating would appear to be necessary for the examination of most muscle specimens, it may not be necessary for the examination of individual myofibrils. Lin and Lamvick[65] have demonstrated that high resolution SEM can be achieved if individual myofibrils are placed on a carbon-coated grid. However, the surface of most muscle specimens has to be made conductive in some fashion. Malick and Wilson[66,67] have used a modification of the original Osmium Tetroxide-Thiocarbohydrazide-Osmium Tetroxide (O-T-O) method[68]. Glutaraldehyde fixed, post-osmicated muscle is rinsed six times in distilled water and then placed into a saturated solution of Thiocarbohydrazide (TCH) for 20-30 minutes. After rinsing the specimen in several changes of distilled water, it is placed into 1% OsO_4 for 2-3 hours. Then the specimen is rinsed as before in distilled water. It is placed again into TCH, rinsed in distilled water and placed into osmium for 2-3 hours. After a final thorough rinse in distilled water, the specimen can be dehydrated and critical point-dried. This appears to render the entire tissue block sufficiently conductive (TCH acts as a mordant for osmium) so that SEM of muscle is made possible, at least at low magnifications[67]. A fine granular precipitate was noted, however, in TEM specimens which had been cut from specimens previously examined with the SEM. A modification of the procedure described by Malick and Wilson[66] was necessary for specimens destined for correlative microscopy, i.e. for specimens which had been sectioned on a cryostat and mounted on glass slides[43]. This modified method produced a superior TEM image (no contrasting was necessary, and no precipitates were noted at high magnifications) however, some globular artifacts were noted on SEM examination which might have been due to some reaction of the osmium or TCH with the "Optimal Cutting Temperature" (O.C.T.) embedding compound, used for frozen sections.

Less Frequently Used Methods

The methods which are cited below probably hold the greatest promise to resolve or analyse molecular detail in muscle cells. While some of them require sophisticated equipment and a great deal of expertise, the superior results may compensate for the high cost of research equipment and the hardships of specimen preparation.

Frozen Sections. A most authoritative review of this subject was published recently (Sjöström and Squire)[69]. The recent advances in specimen preparation methods of quick-frozen material make cryoultramicrotomy of striated muscle a very desirable alternative to conventionally fixed and embedded muscle, especially if the A-band and myosin filaments are studied. Sjöström and Squire[69] first prefixed the very carefully dissected out muscle tissue with 2.5% glutaraldehyde in Ringer's or other buffer solutions for 5 to 15 minutes (or more) at 227 K.

Glycerol treatment (30% for 30 min. at room temperature) as a cryoprotective agent is deemed mandatory, otherwise ice-crystal growth will almost completely destroy any ultrastructural detail in the specimen. Rapid freezing of very small (0.2 to 0.5 mm^3) specimen blocks is accomplished by immersion of these in a liquid N₂/chilled Freon 12 mixture. The frozen sections are cut on a cryo-ultramicrotome with cooled glass knives equipped with a trough containing 50% DMSO or 60% glycerol. The sections are negatively stained using 0.5% uranyl acetate, pH 4.6, or 2% ammonium molybdate, air dried and examined in a TEM. The authors show very convincingly that the definition in the M-band region in cryosections is far superior to the definition of the same band in conventionally embedded plastic sections. Thus fiber typing into red (type I) and white (type II) fibers is much facilitated. On the other hand, the authors admit that glycogen is more readily seen in plastic sections unless histochemical glycogen-labelling procedures have been carried out. For such sections to show ultrastructural surface features, they could be critical point dried, coated with a very thin layer of metal and the grid could be examined in the SEM in the secondary electron emission mode. These sections of muscle could also be examined in a STEM. This would solve the problem of poor contrast in dried tissue sections, because the STEM is capable of inducing contrast instrumentally[70]. Since most STEM attachments have a secondary electron detector, correlation of fine surface (SEM) and interior (TEM) detail would seem to be a feasibility.

Replicas. Very small (0.1 mm³) specimens can be sandwiched between copper sheets, ultrarapidly frozen and shadow-cast[71] which appears to result in optimal replicas, *i.e.* the native structure appears to be preserved and there is no ice-crystal damage. Replica studies on skeletal muscle gave very significant information regarding the sarcoplasmic reticulum and T-tubules, and the cross-bridges in the myosin filaments[72].

X-Ray Microanalysis of Freeze-Dried Sections Rapid freezing of unfixed muscle tissue, drycutting it in the frozen state and freeze-drying the sections is the method of choice, if data on diffusible ions are desired[73]. This approach has been tried on striated skeletal[74] and cardiac muscle[75,76], and vascular smooth muscle[75]. Unfortunately the preservation of ultrastructural morphology in freeze-dried sections is poor, which is due to the fact that no prefixation with glutaraldehyde (which may result in visualization of superior ultrastructural detail in skeletal muscle--see Sjöström and Squire[69]) can be carried out.

X-Ray Analysis of Bulk Specimens. Readers are directed to the recent review by W. Fuchs and H. Fuchs[77] which deals with the main problems encountered during the SEM examination of frozen hydrated bulk specimens. Since specimen charging is the main problem, even lightly carbon-coated specimens will have to be examined at low operating voltages, which in turn results in poor peak-to-background ratios, thus the sensitivity of the method is reduced. Nevertheless, there are reports in the literature on the x-ray analysis on muscle[78,79] but at present reliable and good preparation methods for muscle appear to be lacking.

The requirements for the most commonly used specimen preparation procedures of muscle for TEM, SEM and correlative microscopy (SEM→LM→TEM) are summarized in Table 1.

Table 1

Summary of Specimen Preparation Procedures

Fixation	\multicolumn{3}{l}{1.7 to 3.3% phosphate or cacodylate buffered glutaraldehyde (GA) solution (pH 5.5 - 7.4) or buffered mixture of 2% GA and 1.5% formaldehyde solution}		
Dehydration	Graded series of alcohols or acetones		
	TEM	SEM	SEM→LM→TEM
Drying	Air drying	Critical-point-drying or freeze-drying	Critical-point-drying
Exposure of cell surface and/or interior	Embedding in Epon	Freeze fracture or dry fracture	1) Cryostat section (∼ 10 μm) for LM and SEM 2) Epon section (∼ 90 nm) for TEM
Specimen support	Grid	Stub	1) Glass slide for LM and SEM 2) Indexed grid for TEM
Specimen conductivity	No special preparation procedures needed	Au-Pd or OTOTO	Au-Pd or OTOTO

Concluding Remarks

Some of the EM specimen procedures for muscle (e.g. resin sections for TEM, fractured material for SEM) have stood the test of time and are so well worked out that improvements in them are not too likely. Other methods (e.g. freeze-dried sections for TEM, and bulk samples for x-ray-analysis with the SEM) are still in the state of infancy, and will require further investigative work to perfect them.

While some of the above-mentioned specimen preparation procedures are at least time-consuming, (e.g. correlative SEM→LM→TEM procedures) it is our opinion that they may be followed in certain cases of uncertainty for the preparation of samples of muscle.

In order to appreciate the morphological changes which occur in treated or untreated meat, the normal ultrastructural appearance of the structure of muscle should serve as a baseline. The need for this is especially prevalent in muscle samples which have been prepared for SEM, because, in many cases, the interpretation of the surface ultrastructure of muscle is ambiguous.

Acknowledgements

The research cited in this paper which has been conducted by the authors was supported by grants from the Ontario Ministry of Agriculture and Food, and grants from the Natural Sciences and Engineering Research Council. We thank Ms. Diana Markusic for photographic, and Ms. C.A. Thomson for secretarial assistance.

References

1. Ham A W. *Histology* 2nd edn. J B Lippincott, Philadelphia, Montreal (1953) pp. 318, 326.

2. Bethlem J. *Muscle Pathology. Introduction and Atlas* North-Holland Publishing Company, Amsterdam, London (1970) pp. 12, 62, 63.

3. Mair W G P, Tomé F M S. *Atlas of the Ultrastructure of Diseased Human Muscle* Churchill Livingstone, Edinburgh, London (1972) pp. 1-45.

4. Ham A W, Cormack D H. *Histology* 8th edn. J B Lippincott, Philadelphia, Toronto (1979) pp. 545, 548, 558.

5. Boyde A, Williams J C P. Surface morphology of frog striated muscle as prepared for and examined in the SEM. J. Physiol. 197 (1968) 10P-11P.

6. Ashraf M, Sybers H D. Scanning electron microscopy of the heart after coronary occlusion. Lab. Invest. 32 (1975) 157-162.

7. Uehara Y, Suyama K. Visualization of the adventitial aspects of the vascular smooth muscle cells under the scanning electron microscope. J. Electron. Microsc. (Tokyo) 27 (1978) 157-519.

8. Uehara Y, Desaki J, Nagato S. Scanning microscopic view of smooth muscle cells. Nippon Heikatsukin Gakkai Zasshi 14 (1978) 272-274.

9. Yoder M J, Baumann F G, Goodyear J I, Imparato A M. Endothelial alterations in the constricting rabbit ductus arteriosus: relationship to smooth muscle cell bleb formation. Scanning Electron Microsc. 1980; III: 271-276.

10. Schaller D R, Powrie W D. Scanning electron microscopy of skeletal muscle from rainbow trout, turkey and beef. J. Food Sci. 36 (1971) 552-559.

11. Schaller D R, Powrie W D. SEM of heated beef, chicken and rainbow trout muscles. Can. Inst. Food Sci. Technol. J. 5 (1972) 184-190.

12. Stanley D W, Geissinger H D. Structure of contracted porcine psoas muscle as related to texture. Can. Inst. Food Sci. Technol. J. 5 (1972) 214-216.

13. Stanley D W, McKnight L M, Hines W G S, Usborne W R, DeMan J M. Predicting meat tenderness from muscle tensil properties. J. Texture Studies 3 (1972) 51-68.

14. Eino M F, Stanley D W. Catheptic activity, textural properties and surface ultrastructure of post-mortem beef muscle. J. Food Sci. 38 (1973) 45-50.

15. Stanley D W, Swatland H J. The microstructure of muscle tissue--a basis for meat texture measurement. J. Texture Studies 7 (1976) 65-75.

16. Cohen S H, Trusal L R. The effect of catheptic enzymes on chilled bovine muscle. Scanning Electron Microsc. 1980; III: 595-600.

17. Ashraf M, Sybers H D. Preparation of myocardial tissue: normal and infarcted. In: *Principles and Techniques of Scanning Electron Microscopy. Biological Applications.* vol. 5 (M A Hayat ed.) Van Nostrand Reinhold Co., New York, Cincinnati, Atlanta, Dallas, San Francisco (1976) 1-20.

18. Roy P E, Gailis L, Norin P J, Cote G. A new beating heart preparation. Ultrastructural studies with special reference to vesiculation of intercalated discs. Cardiovas. Res. 5 (1971) 178-193.

19. Hayat M A. *Introduction to Biological Scanning Electron Microscopy* University Park Press, Baltimore, London, Tokyo (1978) 91-123, 207-211, 219-238, 249-254.

20. Merrillees N C R. The fine structure of muscle spindles in the lumbrical muscles of the cat. J. Biophys. Biochem. Cytol. 7 (1960) 725-742.

21. Geissinger H D, Yamashiro S, Ackerley C A. Preparation of skeletal muscle for intermicroscopic (LM, SEM, TEM) correlation. Scanning Electron Microsc. 1978; II: 267-274.

22. Hayat M A. *Principles and Techniques of Electron Microscopy. Biological Applications.* vol. 1. Van Nostrand Reinhold Company, New York, Cincinnati, Toronto, London, Melbourne (1970) 76, 77, 105-107, 147, 242, 295.

23. Price H M, Howes E L, Sheldon D B, Hutson O D, Fitzgerald R T, Blumberg J M, Pearson C M. An improved biopsy technique for light and electron

microscopic studies of human skeletal muscle. Lab. Invest. 14 (1965) 194-199.

24. Rasmussen K E. Fixation in aldehydes, a study on the influence of the fixative, buffer and osmolarity upon the fixation of the rat retina. J. Ultrastruct. Res. 46 (1974) 87-102.

25. Sabatini D D, Bensch K, Barnett R J. Cytochemistry and electron microscopy. The preservation of cellular structure and enzymatic activity by aldehyde fixation. J. Cell Biol. 17 (1963) 19-58.

26. Karnovsky M J. A formaldehyde-glutaraldehyde fixative of high osmolality for use in electron microscopy. J. Cell Biol. 27 (1965) 137A-138A.

27. Jones S B, Carroll R J, Cavanaugh J R. Muscle samples for scanning electron microscopy: preparative techniques and general morphology. J. Food Sci. 41 (1976) 867-873.

28. Fahimi H D, Drochmans P. Essais de standardisation de la fixation an glutaraldehyde I. Purification et determination de la concentration du glutaraldéhyde. J. Microscopie 4 (1965a) 725-736.

29. Fahimi H D, Drochmans P. Essais de standardisation de la fixation an glutaraldéhyde II. Influence des concentrations en aldehyde et de l'osmolalite. J. Microscopie 4 (1965b) 737-748.

30. Barber T, Burkholder P. Relation of surface and internal ultrastructure of thymus and bone marrow derived lymphocytes to specimen preparatory technique. Scanning Electron Microsc. 1975; 369-378.

31. de Bault L E. A critical point drying technique for scanning electron microscopy of tissue culture cells grown on plastic substratum. Scanning Electron Microsc. 1973; 317-324.

32. Luft J H. Improvements in resin embedding methods. J. Biophys. Biochem. Cytol. 9 (1961) 409-414.

33. Lewis E R, Nemanic M K. Critical point drying techniques. Scanning Electron Microsc. 1973; 767-774.

34. Boyde A, Echlin P. Freeze and freeze drying --a preparative technique for SEM. Scanning Electron Microsc. 1973; 759-766.

35. Boyde A. Review of basic preparation techniques for biological SEM. Electron Microscopy 1980 vol. 2, P Brederoo & W de Priester (eds.). VII. Eur. Congr. on EM Fdn., Leiden, 768-777.

36. McCallister L P, Mumaw V R, Munger B L. Stereoultrastructure of cardiac membrane systems in the rat heart. Scanning Electron Microsc. 1974; 713-720.

37. Haggis G H. Cryofracture of biological materials. Scanning Electron Microsc. 1970; 99-104.

38. Nemanic M K. Critical point drying, cryofracture, and serial sectioning. Scanning Electron Microsc. 1972; 297-304.

39. Sybers H D, Ashraf M. Scanning electron microscopy of cardiac muscle. Lab. Invest. 30 (1974) 441-450.

40. Humphreys W J, Spurlock B O, Johnson J S. Critical point drying of ethanol infiltrated cryofractured biological specimens for scanning electron microscopy. Scanning Electron Microsc. 1974; 275-282.

41. Józsa L, Järvinen M, Reffy A. A new preparation technique for scanning electron microscopy of skeletal muscle. Microscopica Acta 83 (1980) 45-47.

42. Flood P R. Dry-fracturing techniques for the study of soft internal biological tissues in the scanning electron microscope. Scanning Electron Microsc. 1975; 287-294.

43. Vriend R A, Geissinger H D. An improved direct intermicroscopic (LM→SEM→TEM) correlative procedure for the examination of mammalian skeletal muscle. J. Microsc. 120 (1980) 53-64.

44. Geissinger H D, Vriend R A, Ackerley C A, Yamashiro S. Correlative light optical, scanning electron and transmission electron microscopy of skeletal muscle from muscular dystrophy and muscular atrophy: a pilot study. Ultrastruct. Path. 1 (1980) 327-335.

45. Watson J H L, Page R H, Swedo J L. A technique for determining the interior topography of single cells. Scanning Electron Microsc. 1975; 417-424.

46. Reynolds E S. The use of lead citrate at high pH as an electron-opaque stain in electron microscopy. J. Cell Biol. 17 (1963) 208-212.

47. Peachey L D. Longitudinal elements of the T-system of vernier band displacements in frog skeletal muscle. J. Cell Biol. 67 (1975) 327a.

48. Forbes M S, Plantholt B A, Sperelakis N. Cytochemical staining procedures selective for sarcotubular systems of muscle: modifications and applications. J. Ultrastruct. Res. 60 (1977) 306-327.

49. Bonilla E. Staining of transverse tubular system of skeletal muscle by tannic acid-glutaraldehyde fixation. J. Ultrastruct. Res. 58 (1977) 162-165.

50. Curran R C, Clark A E, Lovell D. Acid mucopolysaccharides in electron microscopy. The use of the colloidal iron method. J. Anat. 99 (1965) 427-434.

51. Wetzel M G, Wetzel B K, Spicer S S. Ultrastructural localization of acid mucosubstances in the mouse colon with iron-containing stains. J. Cell Biol. 30 (1966) 299-315.

52. Ashraf M, Livingston L H, Bloor C M. SEM of T-tubules of myocardial cells. Scanning Electron Microsc. 1976; II: 179-186.

53. Ashraf M, Bloor C M. The nature of myocardial Z-line ridges seen with the scanning electron microscope. J. Molec. Cell. Card. 8 (1976) 489-495.

54. Geissinger H D, Bond E F. Nomarski differential interference contrast microscopy and scanning electron microscopy of tissue sections and fibroblast cell culture monolayers. Mikroscopie 27 (1971) 32-39.

55. Geissinger H D. Correlated light optical and scanning electron microscopy of Gram smears of bacteria and paraffin sections of cardiac muscle. J. Microsc. 93 (1971) 109-117.

56. Poh T, Altenhoff R L J, Abraham S, Hayes T. Scanning electron microscopy of myocardial sections originally prepared for the light microscopy. Exp. Molec. Path. 14 (1971) 404-407.

57. McDonald L W, Hayes T L. Correlation of SEM and light microscope images of individual cells in human blood and blood clots. Exp. Molec. Path. 10 (1969) 186-198.

58. Geissinger H D, Kamler H. Precise and fast correlation of light microscopic and scanning electron microscopic images. Can. Res. Development 5 (4) 1972, 13-16, 20.

59. Geissinger H D. A precise stage arrangement for correlative microscopy for specimens mounted on glass slides, stubs or EM grids. J. Microsc. 100 (1974) 113-117.

60. Pachter B R, Penha D, Davidowitz J, Breinin G M. Technique for examining uncoated specimens in the SEM with light microscope and TEM correlation. Scanning Electron Microsc. 1973: 387-394.

61. Geissinger H D, Grinyer I. Correlated scanning electron microscopy in transmission (STEM) and reflection (SEM) on sections of skeletal muscle. Mikroskopie 32 (1976) 329-333.

62. Wickham M G, Worthen D M. Correlation of scanning and transmission electron microscopy on the same tissue sample. Stain Technol. 48 (1973) 63-68.

63. Echlin P, Hyde P J W. The rationale and mode of application of thin films to non-conducting materials. Scanning Electron Microsc. 1972: 137-146.

64. Echlin P. Sputter coating techniques for scanning electron microscopy. Scanning Electron Microsc. 1975: 217-224, 332.

65. Lin P S D, Lamvick M K. High resolution scanning electron microscopy at the cellular level. J. Microsc. 103 (1975) 249-257.

66. Malick L E, Wilson R B. Modified thiocarbohydrazide procedure for scanning electron microscopy: routine use for normal, pathological, or experimental tissue. Stain Technol. 50 (1975) 265-269.

67. Malick L E, Wilson R B. Evaluation of a modified technique for SEM examination of vertebrate specimens without evaporated metal layers. Scanning Electron Microsc. 1975; 259-266.

68. Kelley R O, Dekker R A F, Bluemink J G. Ligand-mediated osmium binding: its application in coating biological specimens for scanning electron microscopy. J. Ultrastruct. Res. 45 (1973) 254-258.

69. Sjöström M, Squire J M. Cryoultramicrotomy and myofibrillar fine structure: a review. J. Microsc. 111 (1977) 239-278.

70. Jones A V, Leonard K R. Scanning transmission electron microscopy of unstained biological sections. Nature 271 (1978) 659-660.

71. Costello M J. Ultra-rapid freezing of thin biological samples. Scanning Electron Microsc. 1980; II: 361-370.

72. Rayns D G. Myofilaments and cross bridges as demonstrated by freeze-fracturing and etching. J. Ultrastruct. Res. 40 (1972) 103-121.

73. Sjöström M, Thornell L-E. Preparing sections of skeletal muscle for transmission electron analytical microscopy (TEAM) of diffusible elements. J. Microsc. 103 (1975) 101-112.

74. Somlyo A P, Somlyo A V, Shuman H, Stewart M. Electron probe analysis of muscle and x-ray mapping of biological specimens with a field emission gun. Scanning Electron Microsc. 1979; II: 711-722.

75. Hagler H K, Burton K P, Greico C A, Lopez L E, Buja L M. Techniques for cryosectioning and x-ray microanalysis in the study of normal and injured myocardium. Scanning Electron Microsc. 1980; II: 493-498, 510.

76. Wendt-Gallitelli M F, Stöhr P, Wolburg H, Schlote W. Cryoultramicrotomy, electron probe microanalysis and STEM of myocardial tissue. Scanning Electron Microsc. 1980; II: 499-509.

77. Fuchs W, Fuchs H. The use of frozen hydrated bulk specimens for x-ray analysis. Scanning Electron Microsc. 1980; II: 371-382, 296.

78. Zierold K, Schafer D. Quantitative x-ray microanalysis of diffusible ions in the skeletal muscle bulk specimen. J. Microscopy 112 (1978) 89-93.

79. Zierold K, Schäfer D, Gullasch J. Application of x-ray microanalysis to studies on diffusible ions in skeletal muscle tissue. In: Microscopica Acta Suppl. 2. Microprobe analysis in biology and medicine P Echlin & R Kaufman (eds.) Hirzel Verlag, Stuttgart (1978) 92-101.

For discussion with reviewers see page 60.

The Effect of Catheptic Enzymes on Chilled Bovine Muscle

S. H. Cohen and L. R. Trusal*

Biochemistry and Nutrition Group
Food Sciences Laboratory
US Army Natick Research and Development Command
Natick, MA 01760

*Experimental Pathology Division
US Army Research Institute of Environmental Medicine
Natick, MA 01760

Abstract

The toughening of meat which has been caused by cold shortening prior to the onset of rigor is of significant commercial importance. Various studies have shown that catheptic enzymes produce degradative changes to meat which are very similar to those which occur during the natural aging process and which lead to a more tender meat product. Because of the tenderizing action of cathepsins, this study was undertaken to determine what effect these enzymes had on the cold shortening process of bovine sternomandibulatis muscle. Samples which were cold shortened for 24 or 72 hrs were soaked in either a control solution or one containing catheptic enzymes. Microstructural observations and measurement of sarcomere length by laser diffraction, transmission and scanning electron microscopy, indicated that enzyme treatment hastened the change from rigor to aged muscle.

KEY WORDS: Cathepsin, cold shortening, sarcomere, Z-band, laser diffraction, rigor, muscle, electron microscopy

Introduction

The chilling of bovine skeletal muscle prior to the onset of rigor mortis produces a reversible cold shortening effect (Locker and Hagyard, 1963) which causes toughening in cooked meat when there is a shortening of between 20 to 50% (Marsh and Leet, 1966). However, when shortening exceeds 50%, the severe contraction of the myofibrillar ultrastructure causes fiber fracture leading to a decline in toughness (Marsh, Leet and Dixon, 1974).

As a result of postmortem aging meat becomes tender, but this tenderness is caused by factors unlike the supraphysiological shortening mentioned above. Aged meat is tender because of the degradation of Z discs (Davey and Gilbert, 1967) and disruption of the sarcolemma (Varriano-Marston et al, 1976) as well as certain other factors including loss of Ca^{++} accumulating ability (West et al, 1974) and modification of collagen (Marsh, 1977). Eino and Stanley (1973 a,b) and Robbins and Cohen (1976) found that catheptic enzymes produce degradative changes to the bovine myofibrillar ultrastructure which were quite similar to those which occur during postmortem aging.

Since there are similarities between morphological changes within the myofibrils caused by physiological contraction and cold shortening (Davey and Gilbert, 1974), the addition of catheptic enzymes to cold shortened muscle might act to reverse some of the effects induced by cold shortening. The purpose of the present paper is to investigate this possibility.

Material and Methods

Within 15 min of slaughter, a 5 cm^2 strip of sternomandibularis muscle was dissected from a cow of undetermined age. The muscle was divided into 3 smaller strips, one to be used for laser diffraction (LD) one for transmission electron microscopy (TEM) and one for scanning electron microscopy (SEM). Sarcomere length measurements were determined by all three methods (LD, SEM, TEM).

Sampling periods were as follows: at death control (C); 24 hr cold shortened control (C1); 24 hr cold shortened enzyme treated (E1); 72 hr cold shortened control (C2); and 72 hr cold shortened enzyme treated (E2).

The strips were refrigerated at 2°C in an unrestrained condition within freezer wrapping paper for 24 or 72 hrs. At these times the samples were soaked overnight (16 hrs) in a 1% KCl control solution or in a catheptic enzyme solution (1.2 activity units/ml) prepared from bovine spleen (Robbins and Cohen, 1976).

Laser Diffraction-Sarcomere Length Measurements

Muscle fiber bundles from each sampling period were teased from the larger portion of muscle and fixed for 1 hr in 2.5% glutaraldehyde in KCl-borate buffer (pH 7.1), and then washed twice in the buffer solution (10 min each).

Next, individual fibers were placed into a drop of the buffer on a glass slide and a coverslip placed over the drop. The slide was mounted on a modified microscope stage and positioned so that the muscle fiber was in the path of a laser beam (Spectra Physics HeNe 632.8 nm) so that a diffraction pattern was formed on a ground glass screen 10 cm from the sample. The sarcomere length was determined by measuring the distance between the 0th and 1st order diffraction bands and using the formula $d \sin \theta = n\lambda$ where d = sarcomere length, θ = angular separation between the 0th and 1st order bands, λ = the wavelength of laser light (632.8 nm) and n = the order of particular diffraction band (Cleworth and Edman, 1969). A total of 50 sarcomere length measurements were made for each sampling period.

TEM Processing

Small muscle strips taken at each sampling period were macerated and then fixed in 2.5% glutaraldehyde in 0.10 M cacodylate - 0.11 M sucrose buffer (pH 7.3) for 24 hrs. This was followed by three (30 min ea) rinses in cacodylate sucrose buffer, post fixation in 1% OsO_4 in 0.04 M cacodylate - 0.14 M sucrose buffer for 1 hr at 4°C, three buffer rinses (10 min ea) and dehydration in a graded series of ETOH (70, 95, 100 and 100%). After dehydration the strips were transferred to propylene oxide for 30 min and then into a mixture of propylene oxide and Epon-Araldite (1:1) {Epon 812 (25 ml), Araldite 6005 (15 ml), DDSA (55 ml), DMP-30 (2 ml), DBP (3 ml)} for 1 hr. This was followed by Epon-Araldite (1:3) for an additional 1 hr before final infiltration in Epon-Araldite. The resin mixture was then polymerized for 48 hrs at 60°C.

A trapezoid shaped block face was hand trimmed, smoothed by a glass knife with final ultrathin sectioning by diamond knife. Sections were picked up with 300 mesh copper grids and stained with 5% uranyl acetate in 50% methanol for 15 min followed by 2% lead citrate for 7 min, and then viewed using a JEOL Model 100B (JEOL, Medford, MA) at an accelerating voltage of 60-80 kV. At least 5 sections from each sample were examined and measurements of 40 sarcomeres were made.

SEM Processing

Following overnight washing in the KCl-borate buffer solution, mentioned in the laser diffraction preparation, samples were dehydrated in 70, 90, 95, and 100% ETOH for 1 hr each. Then the fibers were critical point dried from liquid CO_2. Following this step the fibers were mounted on SEM stubs with double sided sticky tape and sputter coated (Commonwealth Sci., Alexandria, VA) for 10 min with gold-palladium.

After sputtering the samples were placed in a Coates and Welter Model 100-2 field emission SEM (Sunnyvale, CA) for observation. Photographs were taken using an emission current of 10 μA and an accelerating voltage of 15kV. Twenty-five sarcomere length measurements were made for each treatment.

Statistical Analysis

Sarcomere length measurements obtained by laser diffraction, transmission and scanning EM were subjected to a two way analysis of variance with repetitions followed by the Tukey test for multiple comparison of means. Significance was computed at the 95% confidence level ($p < 0.05$) for all mean comparisons.

Results and Discussion

The SEM micrograph (Fig. 1) of a pre-rigor control (C) shows the even register of sarcomeres with transverse ridges representing the Z-bands. In Fig. 2, the A, I and Z-bands as well as the mitochondria can be clearly distinguished in the corresponding TEM micrograph. The average sarcomere lengths of pre-rigor, pre-cold shortened muscle, as seen in Table 1, ranged from 2.05 μm (LD), to 2.15 μm (SEM) to 2.29 μm (TEM) which is consistent with the at death sarcomere lengths of 2.4 μm (TEM) determined by Henderson, Goll and Stromer (1970).

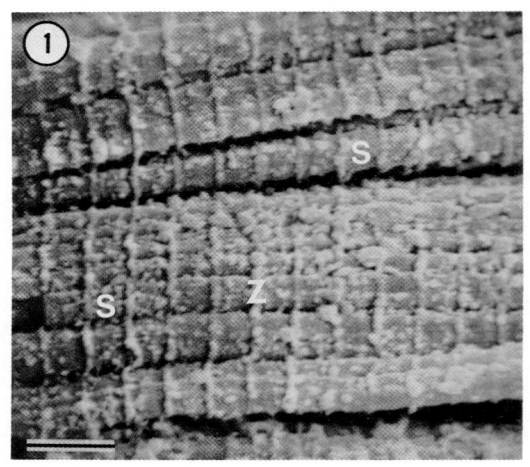

Fig. 1. SEM micrograph of pre-rigor muscle showing the register of sarcomeres (S) and Z bands (Z). Bar = 5 μm

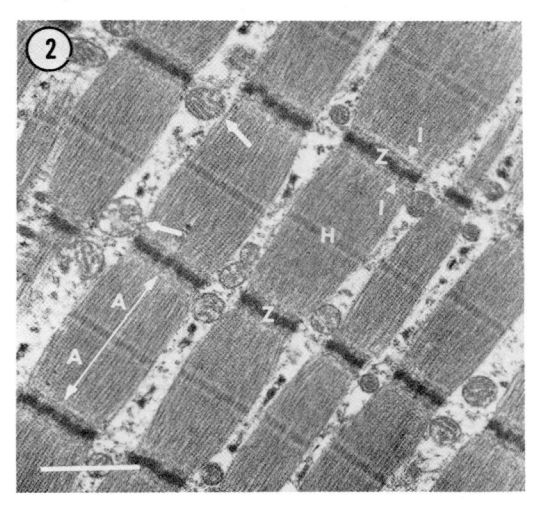

Fig. 2. TEM micrograph of pre-rigor muscle. Typical A, H, I and Z bands as well as mitochondria (long arrows) are easily distinguished. Bar = 1 μm

After cold shortening for 24 hrs (C1), the values decreased to 1.03 μm (LD), 1.21 μm (SEM) and 1.28 μm (TEM). Once again, the mean sarcomere length as determined by TEM (1.28 μm) was almost identical to the value published by Henderson, Goll and Stromer (1970) who obtained 1.3 μm (TEM) after cold shortening for 24 hrs.

According to Locker and Hagyard (1963), the degree of muscle shortening was approximately the same (47.7%) at 0°C and 2°C. Our results show that sarcomere lengths of muscle cold shortened at 2°C for 24 hrs and then soaked in a control solution (2°C) overnight, shortened between 43.7 and 48.8% (Table 1).

Chilling the samples for 24 hrs resulted in severe contraction of the myofibrils and altered the surface morphology considerably (Fig. 3). The degree of contraction was such that it was difficult to distinguish the untreated (C1) sample from the treated sample (E1) (Fig. 4).

As seen in Fig. 5, chilling induced sarcomere contraction producing an overlapping of myofibrillar filaments resulting in the disappearance of the I-band causing the sarcomere to take on a concave appearance. Separation and distortion of the individual myofilaments also occurred. The overlapping of actin and myosin filaments by sliding across one another substantiates the sliding-filament hypothesis of Marsh and Carse (1974) who explained that filament overlapping is a significant stage in the cold-shortening process.

When the 24 hr cold shortened muscle (C1) was soaked in a catheptic enzyme solution (E1), there was a significant increase ($p < 0.05$) in sarcomere length (Table 1) in the LD and SEM samples but not the TEM sample. The degree of chilling induced contraction made it difficult to differentiate the untreated sample (Fig. 5) from the enzyme treated one (Fig. 6) in the TEM micrographs.

After 72 hrs, the sarcomere lengths of the cold shortened muscles (C2) were significantly longer ($p < 0.05$) than the 24 hr group (C1) when measured by all three methods. The sarcomere lengths of the control samples (C2) were 1.72 μm (LD), 1.69 μm (SEM) and 1.46 μm (TEM). However, the enzyme treated samples (E2) had sarcomere lengths of 2.08 μm (LD), 2.06 μm (SEM) and 1.95 μm (TEM) which ranged from 21 to 22 to 34% longer than the controls, respectively. The significant differences were probably caused by the synergistic effect of soaking in the catheptic enzyme solution.

While Eino and Stanley (1973a) found that the resolution of rigor usually occurred at five days postmortem, we have found that the rigor process seems to be largely resolved by 72 hrs (plus overnight soaking). The surface morphology of the 72 hr control sample (C2) has changed so there is no longer any difficulty in differentiating the myofibrillar features (Fig. 7). The enzyme treated sample (E2), although similar in appearance to the untreated sample (C2) has areas, especially in the Z-I band region, where a certain amount of degradation appears to have taken place (Fig. 8). Eino and Stanley (1973a) and Variano-Marston et al. (1976) found similar changes in naturally aged muscle as did Eino and Stanley (1973b) and Robbins and Cohen (1976) in muscle treated with catheptic enzyme.

The greatest differences between untreated (C2) and treated (E2) samples are seen in the TEM micrographs (Figs. 9, 10) of the 72 hr samples. The untreated sample (Fig. 9) is similar in appearance to Fig. 5, but the sarcomere has increased significantly ($p < 0.05$) in width, some degradation in the Z-band has occurred and the sarcolemma is devoid of normal appearing mitochondria.

Table 1

SARCOMERE LENGTH MEASUREMENTS* (μm) OF SAMPLES PREPARED FOR LASER DIFFRACTION (LD) SCANNING (SEM) AND TRANSMISSION (TEM) ELECTRON MICROSCOPY

		C	C1	E1	C2	E2
LD (n=50)	MEAN ± S.D.	2.01±.02	1.03±.03	1.23±.02	1.72±.03	2.08±.01
SEM (n=25)	MEAN ± S.D.	2.15±.21	1.21±.09	1.39±.19	1.69±.14	2.06±.12
TEM (n=40)	MEAN ± S.D.	2.29±.06	1.28±.12	1.32±.15	1.46±.04	1.95±.15

C- PRE-RIGOR CONTROL
C1- CHILLED AT 2°C FOR 24 HRS., SOAKED IN KCl SOLUTION OVERNIGHT
E1- CHILLED AT 2°C FOR 24 HRS., SOAKED IN ENZYME SOLUTION OVERNIGHT
C2- CHILLED AT 2°C FOR 72 HRS., SOAKED IN KCl SOLUTION OVERNIGHT
E2- CHILLED AT 2°C FOR 72 HRS., SOAKED IN ENZYME SOLUTION OVERNIGHT

*THE FOLLOWING PAIR COMPARISONS WERE SIGNIFICANT AT $p < 0.05$

LD SEM } C1-E1   LD SEM TEM } C1-C2   C2-E2   C-C1   C-C2   C2-E1

Fig. 3 and 4. SEM micrographs of untreated (C1) (Fig. 3) and enzyme treated (E1) (Fig. 4) 24 hr chilled muscle. Degree of contraction makes differentiation of surface features of both samples difficult. Bar = 5 μm

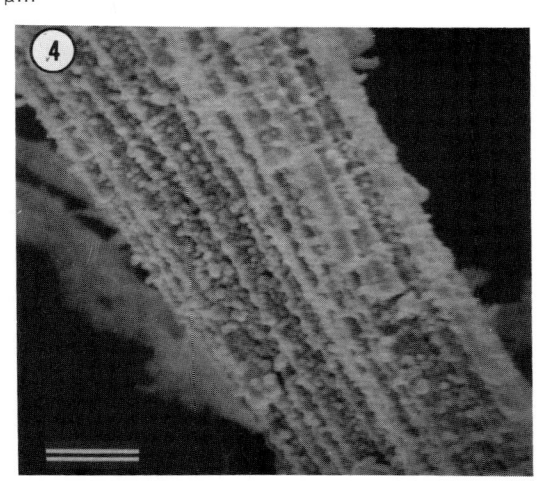

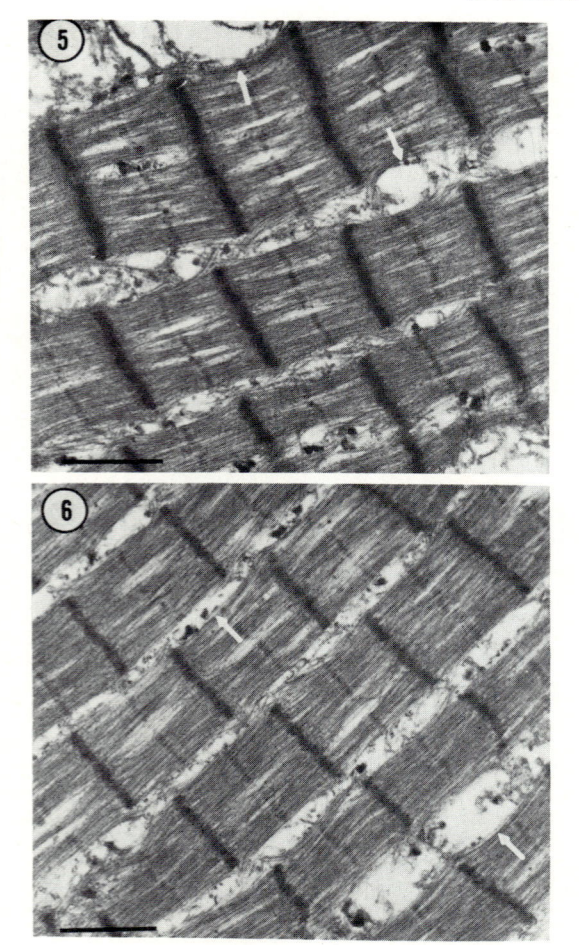

Figs. 5 and 6. TEM micrographs of untreated (C1) (Fig. 5) and enzyme treated (E1) (Fig. 6) 24 hr chilled muscle. Note disappearance of I band, concavity in A band region (arrows) and apparent separation of myofilaments in both figures. Bar = 1 μm

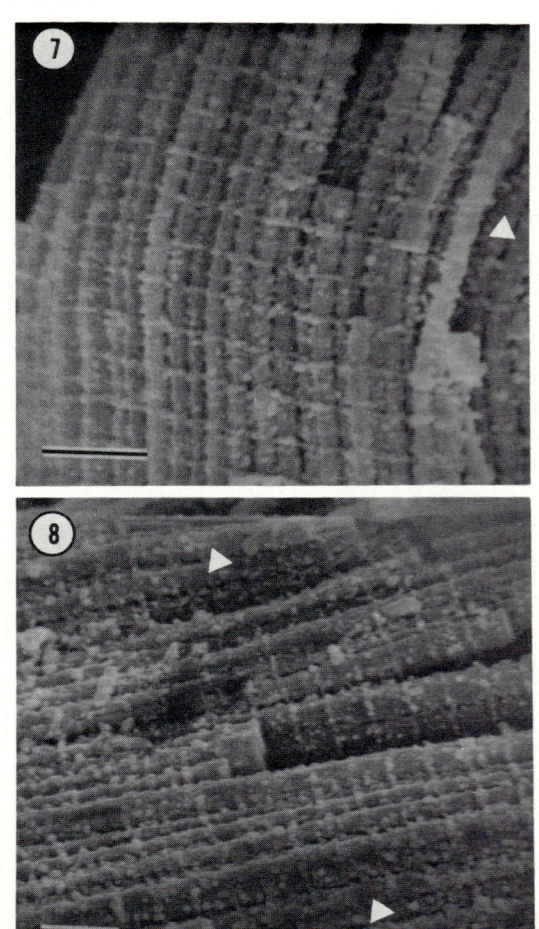

Fig. 7 and 8. SEM micrographs of 72 hr muscle. Chilling-induced contraction is reduced in untreated sample (C2) (Fig. 7) and surface features can be better differentiated. With enzyme-treatment (E2) some degradation (arrows) has occurred. (Fig. 8) Bar = 5 μm

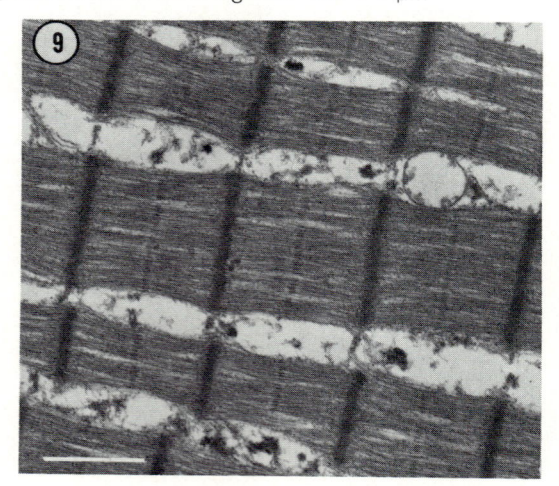

Fig. 9. TEM micrograph of 72 hr muscle. Untreated myofibrils (C2) have undergone some aging-related changes including increased sarcomere length. Bar = 1 μm

Fig. 10. TEM micrograph of 72 hr enzyme treated myofibrils (E2) shows extensive degradation of the Z-I band regions (I-Z-I). Bar = 1 μm.

The contraction of the untreated 72 hr samples (C2) has been largely reversed. Because of the higher concentration of catheptic enzymes, the treated (E2) myofibrils exhibit the typical ultrastructural morphology found in aged muscle (Davey and Dickson, 1970). Furthermore, the sarcomere lengths of the 72 hr enzyme treated (E2) sample more closely approach those of the original pre-rigor, pre-cold shortened control (C).

Although statistical analysis clearly showed significant differences between untreated and enzyme treated samples no matter which preparative technique was used (LD, SEM, TEM), the value differences of the three methods was probably due to the techniques themselves (Varriano-Marston, 1978). If one is solely interested in measuring sarcomere lengths, the laser diffraction method is the simplest, fastest and probably most accurate method. It also involves the fewest manipulative procedures. On the other hand, if information is desired concerning surface structure or ultrastructure, then SEM and TEM examination would be necessary.

Thus, there is little doubt that cathepsins both aid and speed the processes responsible for significantly increasing the sarcomere length in rigor meat by degrading the Z-I band region and sarcolemma microstructure. Our results suggest that the addition of cathepic enzymes to cold shortened muscle hastens the change from rigor to aged muscle.

Acknowledgment

We are grateful to Dr. Frederick M. Robbins and John Walker for providing us with the catheptic enzymes and for very helpful technical advice and guidance throughout this investigation. We also thank Ella Monro, Leonora Kundla and Dr. Edward Ross for help with the statistical analysis and Albert Guzman for his excellent technical assistance.

References

Cleworth, D. and K. A. P. Edman. Laser diffraction studies on single skeletal muscle fibers. Science 163:1969, 296-298.
Davey, C. L. and M. R. Dickson. Studies in meat tenderness and ultrastructural changes in meat during aging. J. Food Sci. 35:1970, 56-60.
Davey, C. L. and K. V. Gilbert. Structural changes in meat during aging. J. Food Technol. 2:1967, 57-59.
Davey, C. L. and K. V. Gilbert. The mechanism of cold-induced shortening in beef muscle. J. Food Technol. 9:1974, 51-58.
Eino, M. F. and D. W. Stanley. Catheptic activity, textural properties and surface ultrastructure of post-mortem beef muscle. J. Food Sci. 38:1973a, 45-50.
Eino, M. F. and D. W. Stanley. Surface ultrastructure and tensile properties of cathepsin and collaganase treated muscle fibers. J. Food Sci. 38:1973b, 51-55.
Henderson, D. W., D. E. Goll and M. H. Stromer. A comparison of shortening and Z line degradation in post-mortem bovine, procine and rabbit muscle. Am. J. Anat. 128:1970, 117-136.
Locker, R. H. and C. J. Hagyard. A cold shortening effect in beef muscles. J. Sci. Food Agric. 14:1963, 787-793.
Marsh, B. B. Symposium. The basis of quality in muscle foods. The basis of tenderness in muscle foods. J. Food Sci. 42:1977, 295-297.
Marsh, B. B. and W. A. Carse. Meat tenderness and the sliding filament hypothesis. J. Food Technol. 9:1974, 129-139.
Marsh, B. B. and N. G. Leet. Studies in meat tenderness. III. The effects of cold shortening on tenderness. J. Food Sci. 31:1966, 450-459.
Marsh, B. B., N. G. Leet and M. R. Dickson. The ultrastructure and tenderness of highly cold-shortened muscle. J. Food Technol. 9:1974, 141-147.
Robbins, F. M. and S. H. Cohen. Effects of catheptic enzymes from spleen on the microstructure of bovine semimembranosus muscle. J. Texture Studies 7:1976, 137-142.
Varriano-Marston, E., G. A. Davis, T. E. Hutchinson and J. Gordon. Scanning electron microscopy of aged free and restrained bovine muscle. J. Food Sci. 41:1976, 601-605.
Varriano-Marston, E., E. A. Davis, T. E. Hutchinson and J. Gordon. Postmortem aging of bovine muscle: A comparison of two preparation techniques for electron microscopy. J. Food Sci. 43:1978, 680-683.
West, R. L., P. W. Moeller, B. A. Link and W. A. Landmann. Loss of calcium accumulating ability in the sarcoplasmic reticulum following degradation by cathepsins. J. Food Sci. 39:1974, 29-31.

Discussion with Reviewers

J. D. Fairing: It is interesting to note that except for samples C2 and E2 the sarcomere lengths measured by TEM and SEM are larger than those from laser diffraction. The shrinkage process commonly encountered in sample preparation would have led one to expect just the opposite results. Have you any explanation for these findings?
Authors: There is an average sarcomere length difference of 13% between LD-TEM, 8% between LD-SEM and 7% between SEM-TEM. While this is somewhat higher than we would normally expect, there are several possible explanations. These include various degrees of contraction of individual myofibrils when beef is subjected to temperatures just above the freezing point (Voyle, C.A., Some observations on the histology of cold-shortened muscle. J. Food Technol. 4: 1969, 275-281), variations in the measurement accuracy by the three methods used and differences in the tonicity and ionic constituency of the KCl-borate buffer used for the LD and SEM samples compared to the cacodylate-sucrose buffer used for the TEM.

J. D. Fairing: At what temperature was the 16 hr. KCl or enzyme treatment carried out?
Authors: All treatments were carried out at 2°C.

E. Varriano-Marston: Would injection of catheptic enzymes into the animal prior to death produce the same effects as soaking the muscle in catheptic enzyme solutions? Or is this an impossibility?
Authors: There have been attempts by some investigators to inject proteolytic enzymes into animals prior to slaughter. Although we have not as yet tried this technique, we feel it may have some merit. Endogenous cathepsins are present at a lower level than added cathepsins. Also, they have to diffuse to sites of attack after release from lysosomes. The addition of cathepsins forces more active enzyme into the system increasing the number of sites attacked by the enzyme and improving the tenderizing efficiency.

C. L. Davey: The sampling is quite small and the number of measurements per mean is low. Also, most published work from about 1965-1975 relates to mean sarcomere lengths from over 100 measurements using more elaborate experimental design.

Authors: Laser diffraction offers a simple and accurate method of sarcomere length measurement. Since between 500 and 1000 sarcomeres are "averaged" into a diffraction pattern this method is superior statistically to those using photographs. For laser diffraction 5 fibers from each of 50 samples per group were measured (250 total measurements); 10 measurements were made on each of 40 samples per group used for TEM (400 total measurements); and 5 measurements were made on each of 25 samples per group used for SEM (125 total measurements).

C. L. Davey: It is of interest that fully stretched muscle (twice equilibrium length) tenderized due to aging, thus the tenderness of aged meat is due to a loss of tensile strength of the muscle fibers and is not due to sarcomere lengthening, as could be implied from this paper.

Authors: Cathepsin D degrades myofibrils under post-mortem conditions (at pH 5.1-5.3) altering Z disc structure and breaking down myosin heavy and light chains as well changing the troponin-tropomyosin complex (Robbins, F.M., et. al., Action of proteolytic enzymes on bovine myofibrils, J. Food Sci. 44: 1979, 1672-1677). Tenderness of aged meat is due, therefore, to enzymatic action on molecular structure leading to a loss of tensile strength and change in sarcomere length.

MICROSCOPICAL OBSERVATIONS ON ELECTRICALLY STIMULATED BOVINE MUSCLE

C. A. Voyle

Agriculture Research Council
Meat Research Institute
Langford, Bristol, BS18 7DY
U.K.

Abstract

The application of electrical stimulation to beef carcasses is known to limit the adverse effects of rapid chilling during the pre-rigor period. Structural changes have been observed in the fibres of longissimus dorsi following electrical stimulation and storage at 15° for about 24 h post mortem. The changes were not observed in the samples of semitendinosus muscle which were examined, although this may indicate a limitation in the sampling procedure.

Iron haematoxylin densely stains bands of protein which can be observed by light microscopy. In the electron microscope these bands were seen to be zones of severe contraction. Fragmentation of the myofibrils also occurs.

Key words - Bovine muscle, meat, muscle structure, muscle contraction, electrical stimulation.

Introduction

It was thirty years ago that Harsham and Deatherage (1951) reported that electrical stimulation of a freshly slaughtered beef carcass gave rise to an accelerated fall in pH, with early onset of rigor in the musculature. It was also found that meat from the carcass was generally more tender than that from non-stimulated carcasses. The earliest report of this effect was apparently made by Benjamin Franklin in 1749, as quoted by Lopez and Herbert (1975).

More recently electrical stimulation has been applied as a means of reducing the adverse effects of rapid chilling of pre-rigor carcasses (Carse, 1973; Bendall et al.,1976). The phenomenon of cold-shortening described by Locker and Hagyard (1963) causes an increase in toughness due to the severe contraction of the muscle fibres when cooled to temperatures below 10° while still in an excitable state (Marsh and Leet, 1966). The biochemical changes which follow the death of an animal, such as breakdown of adenosine triphosphate (ATP) and the fall in pH, are accelerated in electrically stimulated muscle (Bendall, 1976). As a result of the accelerated changes electrically stimulated muscle will no longer react to lowering of temperature which would normally evoke a contractile response. This is a reason for the considerable interest in the beneficial effects of electrical stimulation on meat quality. However, it also appears that electrical stimulation has a tenderising effect which is unrelated to the prevention of cold-shortening (Savell et al.,1977; George et al.,1980; Bouton et al.,1980).

Electrical stimulation induces severe contraction of the musculature producing dramatic, though transient, distortion of the carcass. Savell et al.,(1978) examined bovine longissimus dorsi muscle (LD) after electrical stimulation, using light and electron microscopy. They observed densely stained bands more or less transversely arrayed in muscle fibres viewed in longitudinal section by light microscopy. Control samples from non-stimulated muscles did not display these bands. Electron microscopy of thin sections of electrically stimulated muscle showed adjacent sarcomeres in neighbouring myofibrils with disturbed structure. I-bands and Z-discs were poorly defined and the sarcomere length noticeably shorter than in the flanking areas where the sarcomeres were stretched and sometimes ruptured.

Bendall and his co-workers (George et al., 1980) carried out experiments in which beef half-carcasses were electrically stimulated, the other half serving as a non-stimulated control. They also found densely stained bands in histological preparations of LD. Some years earlier Bendall and Wismer-Pedersen (1962) had described bands of similar appearance in histological preparations of pig muscle in which a very rapid pH fall occurred while the muscle temperature was still high. Muscle affected in this way is characterised by its pale, soft, exudative (watery) appearance - a condition referred to as PSE. Bendall and Wismer-Pedersen (1962) thought that the densely stained bands consisted of denatured sarcoplasmic protein deposited on the myofibrils. It is not surprising, therefore, that George et al.,(1980) considered the densely stained bands they observed in electrically stimulated muscle to be of similar origin to those observed in PSE porcine muscle. Cassens et al.,(1963) observed similar densely stained bands in porcine LD which was permitted to undergo thaw rigor. From a study of the fine structure of these bands Cassens and his co-workers concluded that they were probably the result of violent contraction, rather than the deposition of denatured sarcoplasmic protein.

The fast-growing interest in electrical stimulation as an advance in meat science and technology highlighted the need for a careful and detailed study of its effect on the structure of muscle fibres. Experimental material was available from a series of investigations being carried out by other workers in the Meat Research Institute. The overall design of these experiments imposed some limitations on the histological study, however.

Briefly, examination of electrically stimulated bovine muscles showed densely staining bands in LD, but under the same experimental conditions they were not observed in semitendinosus muscle (ST). Light and electron microscopy was used to determine the morphology and fine structure of the bands.

Materials and Methods

The method of electrical stimulation used was that described by Bendall (1976) and George et al.,(1980), in which clip electrodes were attached to the severed neck region and the Achilles tendon. Twenty five pulses per second were delivered to the entire carcass for two minutes at 700 peak volts. Polarity of the electrodes was reversed at 30 second intervals. The carcasses were stimulated approximately one hour after slaughter. Non-stimulated carcasses were used as control.

Randomly selected samples, measuring about 1 cm x 1 cm x 0.5 cm were removed from the LD and ST immediately after electrical stimulation.

Similar samples were removed from the non-stimulated muscles approximately 1 h after slaughter. Three electrically stimulated LD and six ST muscles were examined, together with three LD and two ST controls. These samples were placed into fixative, as described below, and used for the measurement of sarcomere length, light and electron microscopy. A portion of each muscle was stored at 15°*and similar samples were removed after 24 h post mortem and placed into fixative. As was demonstrated by Locker and Hagyard (1963) shortening in pre-rigor bovine muscle is minimal between 14° and 19°.

A slice of each muscle sample was fixed for about 30 min in 2.5% glutaraldehyde in sucrose buffered with 0.1 M sodium cacodylate (pH 7.2). Bundles of 2-3 fibres were carefully teased from the fixed slice and mounted for sarcomere length measurement using the optical diffraction method described by Rome (1967) and modified by Voyle (1971).

Material for light microscopy (LM) was fixed in 10% formol-saline, dehydrated in ethanol and embedded in paraffin wax. Sections of about 5 µm thickness were stained with iron haematoxylin. For transmission electron microscopy (TEM) portions of tissue measuring 2 x 1 x 1 mm were fixed in 2.5% glutaraldehyde buffered in 0.1 M sodium cacodylate at pH 7.2 for 18-24 h, post-fixed in 1% osmium tetroxide in the same buffer for 1 h, dehydrated in ethanol and embedded in epoxy resin. Thin sections were stained with lead citrate and uranyl acetate for examination in an AEI EM 6B electron microscope.

Samples fixed in glutaraldehyde were also prepared for scanning electron microscopy (SEM). Fixation was followed by washing in sodium cacodylate buffer. After washing, the samples were individually wrapped in aluminium foil and rapidly cooled by immersion in nitrogen slush. The frozen samples were fractured with a razor blade on a cold plate which had been immersed in liquid nitrogen, and the fragments of tissue transferred to an Edwards-Pearse tissue freeze-drier. Dehydration was complete in 3 to 4 hours at a plate temperature of -60°. The dried tissue fragments were mounted on bulk specimen carriers using double-sided adhesive tape. After coating with platinum or gold-palladium the prepared samples were stored in a desiccator over silica gel. The samples were examined in a JEOL CX-100 Temscan, by courtesy of JEOL (UK) Limited, Colindale, London.

Results

Table 1 lists the mean sarcomere lengths in samples of LD and ST in stimulated and non-stimulated muscles before and after storage at 15° for 24 h. Initial values of sarcomere length were obtained from samples immediately after electrical stimulation and, in the case of the controls, after an equivalent interval (about 1 h) after slaughter. The level of significance (P) of the difference in mean sarcomere length was also determined, using Student's t-test (Fisher, 1970).

---

*All temperatures are in degrees Centigrade.

These data show that at 15° there was no significant change in sarcomere length in electrically stimulated or non-stimulated ST during the period in which rigor develops, nor in non-stimulated LD. Electrically stimulated LD, however, did show a slightly significant (P=10%) reduction in mean sarcomere length, an observation which may be correlated with the incidence of densely stained bands in histological sections of this muscle, described below.

Longitudinal sections of paraffin-embedded samples of electrically stimulated LD and ST were prepared. The samples were taken immediately after stimulation ('initial') or 24 h post mortem. The sections were stained with iron haematoxylin and examined by light microscopy. Sections from non-stimulated control muscles were stained in the same way. By this method A-bands and Z-discs could be clearly observed, as the structural proteins in these components exhibit a high affinity for the stain. I-bands may be seen as lightly stained areas in relaxed muscle fibres. When the space between A-bands and Z-discs is reduced in shortened sarcomeres I-bands disappear.

In stained sections of stimulated and non-stimulated ST a generally uniform distribution of sarcomeres was observed in both initial and 24 h samples. Initial samples of electrically stimulated LD showed a similar appearance, but in stored samples an array of irregular densely stained transverse bands was observed (Fig.1). The bands which stain densely with iron haematoxylin could also be clearly observed by phase contrast microscopy, as zones of increased optical density. The bands were of variable width, from about 2.0 μm to about 20 μm. They sometimes appeared to be bifurcated and were separated by up to thirty sarcomeres. Sarcomeres between the densely stained bands, measured on the micrographs, were about 1.6 μm in length. Fibres were seen to bulge in the location of dense bands.

In their examination of samples of electrically stimulated LD, removed from the carcass 20-24 h post mortem, Savell et al.,(1978) observed similar, though less densely stained, bands using toluidine blue. It is not clear what storage conditions were used following electrical stimulation in this experiment.

Examination by TEM of thin sections of initial samples of electrically stimulated ST, in which no densely stained bands had been detected by light microscopy, showed a uniform distribution of sarcomeres (Fig. 2), although some reduction of the sarcomere length from that associated with the 'resting state' had occurred because the sample had been cut from the entire muscle while still in the pre-rigor condition. The sarcomere length in Fig. 2 is about 1.6 μm, measured from the micrograph.

Thin sections of electrically stimulated LD which had been stored at 15° up to 24 h post mortem showed that the areas which appeared as densely stained bands in iron haematoxylin stained sections were, in fact, zones of severe contraction, (Fig. 3). The structure of these zones was similar to that briefly described by

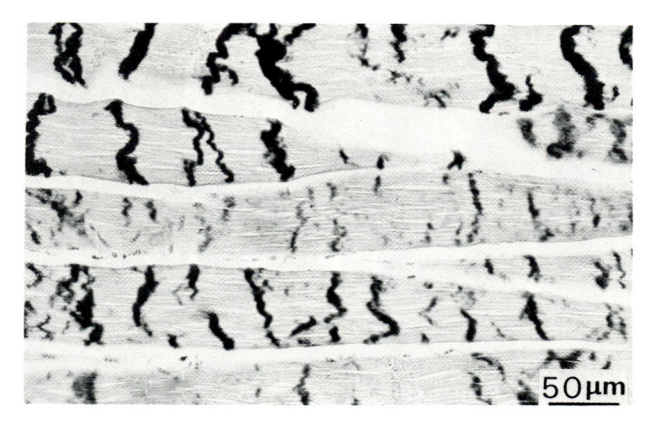

Fig.1 Longitudinal section of electrically stimulated longissimus dorsi muscle stained with iron haematoxylin. Densely stained bands are clearly seen. Transverse striations corresponding to sarcomere repeat can be seen between the bands.

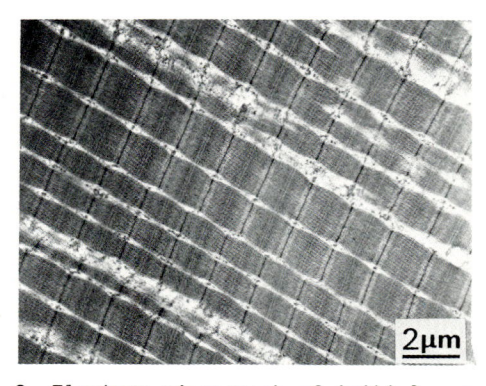

Fig. 2 Electron micrograph of initial sample of electrically stimulated ST showing uniform sarcomere length of myofibrils. Structure is well preserved.

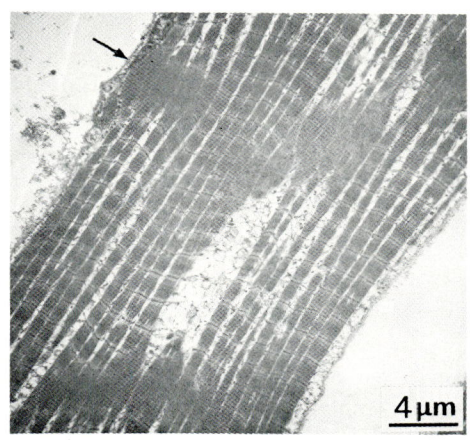

Fig. 3 Electron micrograph of a zone of severe contraction (arrow) involving several adjacent myofibrils in electrically stimulated LD.

Table 1. Sarcomere lengths following electrical stimulation and after storage for 24 h at 15°.

Muscle	Sarcomere length (μm) Mean $\pm$ SE Initial	24 hour	% change	P%
ST(C)	1.73 $\pm$ .029 (12)	1.79 $\pm$ .037 (12)	+3.5	NS
ST(S)	1.89 $\pm$ .014 (36)	1.82 $\pm$ .065 (36)	-3.7	NS
LD(C)	1.87 $\pm$ .016 (36)	1.88 $\pm$ .024 (24)	+0.5	NS
LS(S)	1.87 $\pm$ .020 (42)	1.68 $\pm$ .034 (24)	-10	10

C = control
S = electrically stimulated
NS = not significant
Numbers in parenthesis = number of observations made by optical diffraction method.

% change and significance values relate to comparison of mean sarcomere lengths in initial and 24 h samples.

Savell et al.,(1978). The A-bands and I-bands of individual sarcomeres within such a zone could no longer be discerned. The Z-discs had lost their contrast, due, presumably, to excessive overlap of thick and thin filaments. The Z-discs in the nearest recognisable sarcomeres were bow-shaped, appearing to be drawn in towards the contraction zone as may be seen in Fig. 3. Loss of Z-disc contrast has been observed in cold-shortened muscle (Voyle, 1969), where it appeared that the very severe contraction occurring in parts of the muscle caused the thick filaments of one sarcomere to penetrate through the Z-disc into the neighbouring sarcomere. In some preparations it has been possible to measure sarcomeres in the zone of severe contraction and these were found to be as small as 0.5 μm. In electrically stimulated muscle we have also observed that, in the region of zones of severe contraction, the sarcolemma was frequently corrugated in appearance compared with the relatively smooth profile of the remainder of the muscle fibre membrane.

In addition, as observed by Savell et al., (1978), the zones of contraction were often flanked by sarcomeres displaying partial or complete rupture in the I-band. Fig. 4 shows an early stage in the fragmentation process in which the widening gap between some sarcomeres is bridged by thin filaments. In Fig.5 rupture is complete in several myofibrils. The site of rupture appears to be at the junction of the thin filaments and the Z-disc, separation occurring on one side only of the latter. Structural breakdown at the same sites during the normal aging process has been reported by Davey and Dickson (1970), and the possible involvement of gap filaments in this process has been discussed by Davey and Graafhuis (1976). In a separate investigation on the effect of electrical stimulation on ST stored under different time/temperature conditions it has been found that fragmentation occurred earlier at a given temperature in stimulated muscle than in those which were not stimulated (Voyle - unpublished results).

An examination by SEM of preparations from electrically stimulated LD, stored at 15° for 24 h, showed transversely arrayed bands involving many myofibrils (Fig.6). The latter may be seen running approximately from left to right in the micrograph. The bands, marked 'b', were similar in shape and distribution to those observed by LM and TEM. They involved several neighbouring myofibrils which had a thickened appearance so that the inter-myofibrillar space was considerably reduced.

Fig. 6 also shows an array of regularly spaced inter-myofibrillar connections between the transverse bands. These have a periodicity of about 1.5 μm but close examination at a higher magnification (Fig.7) shows them to be complex structures making multiple connections between adjacent myofibrils. These structures are more closely packed in the transverse bands than in other parts of the myofibrils.

Discussion

The results we have obtained from electrically stimulated bovine LD are similar to those reported by Savell et al.,(1978) in that electrical stimulation appears to give rise to densely staining bands detectable by LM and zones of severe contraction as seen by TEM and SEM. The time elapsing between the application of electrical stimulation and the selection of samples which show this feature was probably similar in the experiments described by Savell and his co-workers and in those described in this paper, being about 24 h post mortem. However, in addition we examined samples of LD immediately after electrical stimulation and found no densely stained bands or zones of contraction. It is not clear why a contraction effect was not detected in these early stages after electrical stimulation, especially since George et al.,(1980) reported densely stained bands in LD 1.5 h after stimulation. It is intended to look more closely at the pH/time/temperature conditions associated with the development of the zones of contraction.

The report by Savell et al.,(1978) in which they describe 'contracture bands' does not indicate how soon after electrical stimulation the sides (treated and control) were chilled. The absence of contraction bands in the control, assuming identical chilling conditions for both sides, suggests that the incidence of such bands in the treated muscle was a consequence of electrical stimulation. However in some further preliminary experiments to be described more fully elsewhere, we found zones of contraction

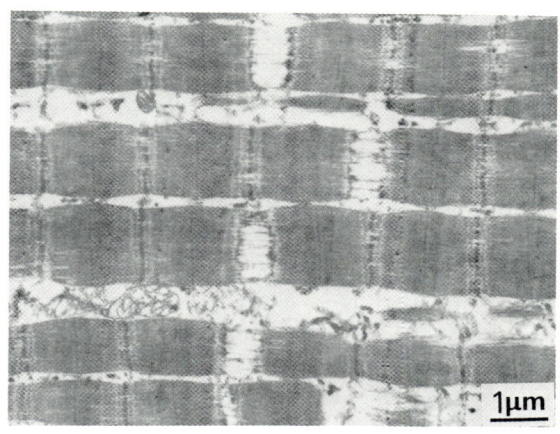

Fig. 4. Electron micrograph showing partial rupture of I-band in electrically stimulated LD.

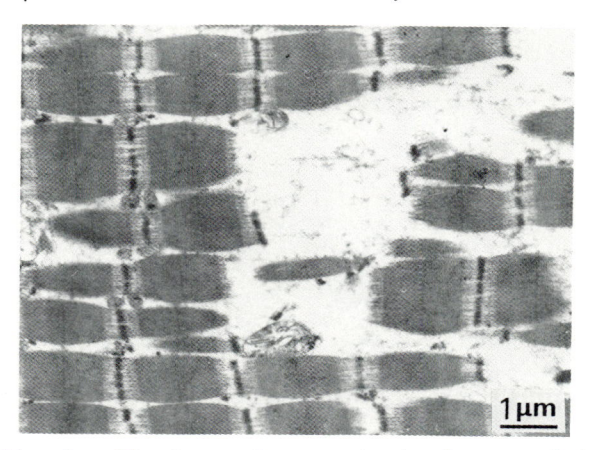

Fig. 5. Electron micrograph showing complete breaks in myofibrils in electrically stimulated LD.

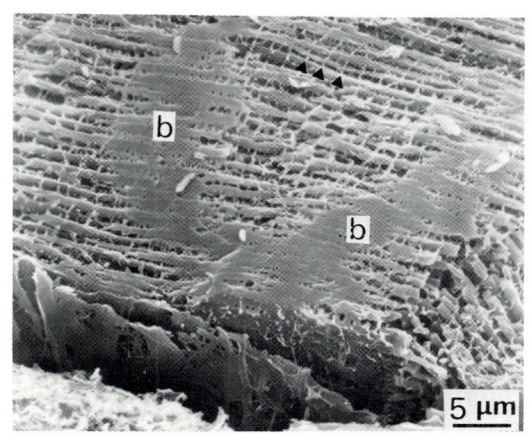

Fig. 6. Scanning electron micrograph of electrically stimulated LD (24 h post mortem at 15°) showing zones of severe contraction ('b') involving adjacent myofibrils. Note the regular spacing of the inter-myofibrillar connections (arrows).

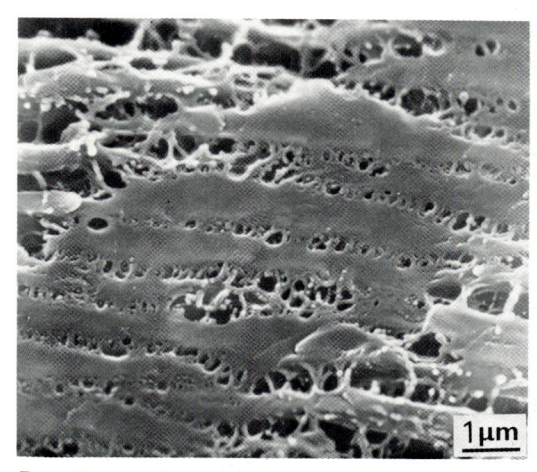

Fig 7. Part of sample seen in Fig. 6 showing close packing of inter-myofibrillar connections in zone of severe contraction.

in LD when cooled to 0°, whether previously exposed to electrical stimulation or not. ST, on the other hand, did not show localised contraction when stored at 0° following electrical stimulation, but non-stimulated controls were markedly affected. Electrical stimulation, therefore, appeared to be effective in inhibiting the effects of rapid chilling in ST.

The reason for the different response to electrical stimulation in LD and ST under similar conditions of storage is not clear. In pigs the ST is characterised by well-defined zones, being red or white in colour. These zones are comprised of a predominance of red fibres and white fibres respectively. So far as is known, such differentiation has not been made in bovine ST but it is possible that some areas have a predominance of one fibre type over the other. A sample of tissue in which muscle fibres were predominantly white would be more reactive ('fast') than a sample from a mainly red population ('slow'). Therefore very careful control over the sampling procedure is required in order to make valid comparisons between different muscles. The classification of muscle fibres as 'red', 'white' or 'intermediate' has been described by a number of workers using histochemical methods or ultrastructural morphology (Padykula and Gauthier, 1967, Dutson et al., 1974).

The different postural aspects of LD and ST on the carcass during electrical stimulation may also contribute to the different response found in these muscles.

Since only electrically stimulated LD displayed a significant change in mean sarcomere length as a result of storage at 15° for about 24 h post mortem it is interesting that only samples subjected to these conditions displayed zones of severe contraction made visible by dense staining and detectable by LM. It is reasonable to suppose that the method of optical diffraction used to measure sarcomere length was measuring the periodicity in the ordered parts of the muscle fibre between the zones of severe contraction and the mean values quoted in Table 1 probably represent these sarcomeres in particular.

It seems, therefore, that contraction has occurred throughout the muscle fibre as a result of electrical stimulation but locally it was much more severe.

It is interesting to speculate on why contraction occurred in such a localised manner. During the application of electrical stimulation contraction was general and the entire carcass was affected. A consequence of this treatment may be damage to the membrane of the sarcoplasmic reticulum. When the muscles relax on removal of the stimulus local leaks of calcium ions into the sarcoplasm may occur giving rise to contractile activity. It would be interesting to know if the sarcoplasmic reticulum from electrically stimulated muscle suffers an impairment in its ability to function as a calcium pump. Indeed this has been suggested recently by Joseph et al., (1980).

Acknowledgements

The author wishes to thank A. Cousins for technical assistance, Dr D.B. MacDougall for making available samples of electrically stimulated bovine muscle and Dr G.W. Offer for his interest and constructive suggestions in the compilation of this Paper.

References

Bendall, J.R. Electrical stimulation of rabbit and lamb carcasses. J. Sci. Fd Agric. 27, 1976, 819-826.
Bendall, J.R. and Wismer-Pedersen, J. Some properties of the fibrillar proteins of normal and watery pork muscle. J. Fd Sci. 27, 1962, 144-159.
Bendall, J.R., Ketteridge, C.C. and George, A.R. The electrical stimulation of beef carcasses. J. Sci. Fd Agric. 27, 1976, 1123-1131.
Bouton, PE., Ford, A.L., Harris, P.V. and Shaw, F.D. Electrical stimulation of beef sides. Meat Sci. 4, 1980, 144-155.
Carse, W.A. Meat quality and the acceleration of post-mortem glycolysis by electrical stimulation. J. Fd Technol. 8, 1973, 163-166.
Cassens, R.G., Briskey, E.J. and Hoekstra, W.C. Similarity in the contracture bands occurring in thaw-rigor and in other violent treatments of muscle. Biodynamica 9, 1963, 165-175.
Davey, C.L. and Dickson, M.R. Studies in Meat Tenderness. 8. Ultrastructural changes in meat during aging. J. Fd Sci. 35, 1970, 56-60.
Davey, C.L. and Graafhuis, A.E. Structural changes in beef muscle during aging. J. Sci. Fd Agric. 27, 1976, 301-306.
Dutson, T.R., Pearson, A.M. and Merkel, R.A. Ultrastructural postmortem changes in normal and low quality porcine muscle fibres. J.Fd Sci. 39, 1974, 32-37.
Fisher, R.A. Statistical methods for research workers 14th Edn., Oliver and Boyd, Edinburgh, Scotland, 1970, chapter 5.
George, A.R., Bendall, J.R. and Jones, R.C.D. The tenderising effect of electrical stimulation of beef carcasses. Meat Sci. 4, 1980, 51-68.
Harsham, A. and Deatherage, F.E. Tenderization of meat. U.S. Patent 2544681, 1951.
Joseph, A.L., Dutson, T.R. and Carpenter, Z.L. Morphology and calcium uptake of bovine sarcoplasmic reticulum as affected by electrical stimulation and time post mortem. Proc. 26th European Meeting of Meat Research Workers, Colorado Springs, Colorado, U.S.A. 1980, 77-80.
Locker, R.H. and Hagyard, C.J. A cold-shortening effect in beef muscles. J. Sci. Fd Agric. 14, 1963, 787-793.
Lopez, C.A. and Herbert, E.W. The private Franklin, the Man and his Family, W.W. Norton & Co. New York, 1975, 44-45.
Marsh, B.B. and Leet, N.G. Studies in Meat Tenderness. III. The effects of cold-shortening on tenderness. J. Fd Sci. 31, 1966, 450-459.
Padykula, H.A. and Gauthier, G.F. Morphological and cyto-chemical characteristics of fibre types in normal mammalian skeletal muscle. In: Exploratory concepts in muscular dystrophy and related disorders, A.T. Milhorat (ed.), Excerpta Medica Foundation, New York, U.S.A., 1967, 117-131.
Rome, E. Light and x-ray diffraction studies of the filament lattice of glycerol-extracted rabbit psoas muscle. J. Mol. Biol. 27, 1967, 591-602.
Savell, J.W., Smith, G.C., Dutson, T.R., Carpenter, Z.L. and Suter, D.A. Effect of electrical stimulation on palatability of beef, lamb and goat meat. J. Fd Sci. 42, 1977, 702-706.
Savell, J.W., Dutson, T.R., Smith, G.C. and Carpenter, Z.L. Structural changes in electrically stimulated beef muscle. J. Fd Sci. 43, 1978, 1606-1607, 1609.
Voyle, C.A. Some observations on the histology of cold-shortened muscle. J. Fd Technol. 4, 1969, 275-281.
Voyle, C.A. Sarcomere length and meat quality. Proc. 17th European Meeting of Meat Research Workers, Bristol, U.K. 1971, 95-97.

Discussion with Reviewers

R.G. Cassens: I understand that the technique of electrical stimulation of carcasses is not at all standardised and various researchers use quite different conditions. Would you comment on this especially concerning how morphology could be affected?
Author: Voltages between 15 and 1200V have been applied, with variation in the pulse width and frequency. Low voltage stimulation is less of a hazard to the operator, although the carcass is still subject to severe limb movement. Bouton et al.,(1980 - see list of References) reported some difference in sarcomere length changes between low and high voltages, depending also on stimulation time and the time elapsing post-slaughter to the collection of samples.

T.R. Dutson: Were any steps taken to prevent pre-rigor shortening?
R.J. Carroll: Would the results of sarcomere length, SEM and TEM observations differ if the 24 h samples were taken from a hanging carcass?

Electrical stimulation and muscle structure

S.H. Cohen: Have you any values for at-death non-stimulated muscle sarcomere lengths?

H.D. Geissinger: Why were no muscle samples fixed immediately after slaughter?

Author: Muscles were not removed from the carcass until after electrical stimulation or, in the case of controls, 1 h after slaughter. For a number of reasons it was not possible to obtain at-death samples or to take steps to prevent pre-rigor shortening when the muscles were divided for the various analyses to be performed.

The main effect of leaving the muscles on the carcass for 24 h prior to sampling would be one of restraint, which would be reflected in longer sarcomeres. However, both LD and ST are somewhat slack in the hanging carcass, so I do not expect any great difference in the results which would be obtained.

R.J. Carroll: Since the pH drops significantly after slaughter (George et al 1980) both in the control and the stimulated samples, would not better preservation of muscle fine structure be obtained if the fixative pH be adjusted to the corresponding pH of the specific sample?

Author: This may well be true. It is generally recommended that the pH of a fixative should be in the physiological range 6.5 - 8.0, no doubt on the assumption that the tissue is in a physiologically normal condition. I would suggest, however, that in post mortem muscle the damage to the fine structure occurs during the treatment phase e.g. electrical stimulation, prior to sampling and fixation. Adequate controls are essential.

S.H. Cohen: Is there a difference between the effect of electrical stimulation on the mitochondria of the LD compared with the ST muscles?

Author: There are no reports to suggest that this is so. Swollen and ruptured mitochondria have been observed in both muscles (Will et al.,1980) following electrical stimulation, and also in controls examined 24 h post mortem.

T.R. Dutson: Since, at 15° you find a varied response in shortening when comparing stimulated and non-stimulated ST and LD muscles, (LD was shortened and ST was lengthened), do you feel that electrical stimulation increases tenderness by prevention of shortening when muscles are subjected to colder temperatures?

Author: It is widely recognised that a major benefit of electrical stimulation is that meat can be chilled soon after slaughter without the detriment of increased toughness due to the phenomenon of cold-shortening. It is also becoming more evident that increased tenderness, as opposed to prevention of increased toughness, results from the enhanced activity of autolytic enzymes following electrical stimulation. Improved texture, is therefore, due to a combination of these factors.

T.R. Dutson: In the LD muscles (which are mixed fibre type muscles) it appears that all fibres show contraction bands when electrically stimulated (Fig.1). Were any differences in response found in different fibres which could be related to your hypothesis that red and white fibres might be affected differently, relative to contractive band production, by electrical stimulation?

Author: It was not infrequently observed by TEM that adjacent fibres displayed contrasting contraction patterns. One fibre showed a widespread array of shortened sarcomeres, its neighbour showing a much more localized array. It is necessary for us to examine these fibres more carefully and in greater numbers to see if they can be identified in terms of fibre type.

H.D. Geissinger: It is very surprising to note the effects of electrical stimulation are not histologically visible immediately after stimulation, but became visible in the meat after a period of several hours of storage at 15°. How do you view the pathogenetic development of the lesion (contraction bands and ruptured sarcomeres)? Also, do you have any more results on the histology of the transverse inter-myofibrillar connections shown in Fig. 7?

Author: It is true that in these experiments we did not observe zones of severe contraction in samples of LD taken soon after electrical stimulation. This caused us some surprise since other workers (Will et al., 1980; George et al., 1980) did observe such bands as soon as 1 or 1.5 h post mortem. We cannot say, therefore, that they do not occur, only that we did not observe them in our experiment.

There is still much to be done in the investigation concerning the fine structure of the zones of contraction, their pathogenesis, and the inter-myofibrillar connections. This will form the basis of a further report.

S.H. Cohen: Since normal muscular contraction occurs in waves, could the formation of zones of severe contraction be caused by two zones coming together?

Author: Maybe, but I think it is clear that the zones are the result of abnormal conditions, in this case electrical stimulation. Similar appearances can result from other conditons, e.g., the application of heat to pre-rigor muscle.

R.J. Carroll: Is fragmentation in ST muscle due to electrical stimulation or some other factor, since fragmentation is observed in non-stimulated ST?

Author: The point is made in the text that fragmentation occurs sooner in electrically stimulated muscle than in non-stimulated muscle. The response to stimulation is not to cause fragmentation so much as to accelerate it. In addition, Dutson et al.,(1980) have shown that enzyme activity is increased following electrical stimulation due to damage to lysosomal membranes.

The released enzymes are known to be capable of degradation of myofibrillar proteins, contributing to the fragmentation of the myofibrils.

R.J. Carroll: Are the complex structures in Fig. 7 triads?

Author: This is a possibility since the mean sarcomere length of the electrically stimulated LD is about the same as the spacing between these structures. This distance also approximates to the myosin filament (A-band) length. In mammalian muscle the triads are in register with the A-I junction but this fact is obscured in such a shortened sample as the one illustrated.

R.A. Lawrie: It has been suggested that the initial absence of marked exudation from the musculature of electrically stimulted bovine and ovine muscles, despite temperature/pH combinations which approach those in pale, soft, exudative pork, reflect a more robust sarcolemma in the former species, which opposes egress of intracellular fluid. The present paper clearly indicates that substantial structural damage can occur in electrically stimulated muscles. Can the author suggest why massive exudation has not been a feature of the hitherto published work on electrical stimulation?

Author: In our observations the most conspicuous structural damage has been at the myofibrillar level. We have also noticed considerable sub-sarcolemmal accumulations of amorphous granular material in thin sections observed by TEM, especially in ST after prolonged storage. There has not been any evidence of an extracellular presence of this material. Will et al.,(1980) reported some sarcolemmal damage in several muscles, but this seemed to occur under both stimulated and non-stimulated conditions. Discussion with colleagues in the UK Meat Research Institute supports the idea that there is a species difference in the properties of the sarcolemma which could explain the ready release of fluid in porcine muscle, contrasting with bovine muscle which does not give rise to fluid loss to the same extent.

Discussion with reviewers references

Dutson, T.R., Smith, G.C. and Carpenter, Z.L. Lysosomal enzyme distribution in electrically stimulated ovine muscle. J. Food Sci. 45, 1980, 1097-1098.
Will, P.A., Ownby, C.L. and Henrickson, R.L. Ultrastructural post mortem changes in electrically stimulated bovine muscle. J. Food Sci. 45, 1980, 21-25, 34.

Note added in Proof

A number of Papers on the general topic of electrical stimulation were presented at the 26th European Meeting of Meat Research Workers, 1980, Colorado Springs, Colorado, USA. The reader is referred to the Proceedings of that Meeting for further information.

SCANNING AND TRANSMISSION ELECTRON MICROSCOPY OF NORMAL AND PSE PORCINE MUSCLE

J. D. Cloke, E. A. Davis, J. Gordon, S-I Hsieh, J. Grider, P. B. Addis and C. J. McGrath*

Department of Food Science and Nutrition, University of Minnesota, 1334 Eckles Avenue, St. Paul, MN 55108

and

*Department of Veterinary Clinical Science, University of Minnesota, 1352 Boyd Avenue, St. Paul, MN 55108

Abstract

Structural differences between normal and pale, soft, exudative (PSE) longissimus dorsi (LD) muscle from purebred and crossbred pigs were observed using scanning and transmission electron microscopy. The effect of severe freeze-thaw contraction on the muscle samples was observed. The membranes of many different organelles (sarcoplasmic reticulum, mitochondria, sarcolemma and connective tissue) appeared markedly disrupted after cryofracture of PSE muscle with frequent breaks, especially near the Z-lines. In addition, there was little definition of the M-line and H-zones in either the PSE or intermediate animals.

KEY WORDS: Porcine muscle ultrastructure; muscle ultrastructure; TEM muscle; SEM muscle; PSE porcine muscle.

Introduction

Stress susceptibility in pigs and the resulting pale, soft, exudative (PSE) carcasses are of concern to food scientists interested in overcoming the poor quality characteristics of the meat, and to animal scientists wishing to alleviate the problem by better breeding and improved management. Stress susceptibility in pigs is also of interest to medical scientists, since the disease malignant hyperthermia in humans has many similarities (Jones et al., 1972; Nelson et al., 1974). Many studies on the fundamental differences between stress susceptible and stress resistant animals have been conducted (Briskey, 1964; Cassens et al., 1975). Fundamental postmortem studies have revealed that PSE muscle is caused by an abnormally rapid fall in pH while the carcass temperature is still high (Briskey, 1964; Briskey and Wismer-Pedersen, 1961; Bendall and Wismer-Pedersen, 1962).

Differences between stress susceptible and stress resistant pigs include changes in the membranes of several micro- or ultrastructural components of muscle, such as the mitochondria, sarcoplasmic reticulum, and connective tissue as well as the fiber. It has been suggested that because membranes are either immature or defective, permeability may be altered (Cassens et al., 1975; Schmidt et al., 1972; Briskey, 1964; Field et al., 1970; McClain et al., 1968 and 1969; Wipf et al., 1970; Greaser, 1967).

Light and electron microscope studies have indicated differences in PSE and normal muscle postmortem (Cassens et al., 1963a; Greaser et al., 1969a; Muir, 1970; Dutson et al., 1974; Bergman, 1975; Wesemeier, 1974). There has been much discussion as to the possible relationship between PSE and fiber type or fiber size (Linke, 1972; Andersen et al., 1975; Cassens et al., 1975; Sair et al., 1972).

Light microscopy and transmission electron microscopy (TEM) studies have shown that PSE muscles exhibit more disruption of components after death than the corresponding normal muscle (Cassens et al., 1963a; Greaser et al., 1969a; Muir, 1970; Dutson et al., 1974; Bergmann, 1975). PSE muscles showed changes in the mitochondria, including swelling and decreased matrix density (Greaser et al., 1969a; Dutson et al., 1974) and changes in the sarcoplasmic reticulum (Greaser et al., 1969c; Dutson et al., 1974; Cassens et al., 1963a). Sometimes myofibrillar breakdown with loss of material at the Z-lines was observed, particularly in studies of postmortem times up to 24 hr (Greaser et al., 1969a; Cassens et al., 1963a; Dutson et al., 1974; Muir, 1970). White fibers were found to be more labile to autolysis than red fibers (Abbott et al., 1977).

Scanning electron microscopy (SEM) has not been used in the study of PSE porcine muscles, although it has been used in studies of the texture of normal porcine psoas muscle (Stanley and Geissinger, 1972) and in some studies of bovine muscle (Varriano-Marston et al., 1976; Schaller and Powrie, 1971).

Although the above studies have shown some differences between normal porcine muscle and muscle exhibiting PSE characteristics, conventional methods of sample preparation were employed in such a way as to minimize the effects of muscle contraction. The assumption being that this is more representative of in vivo muscle. However, differences in degree and type of protein bonding in vivo may be quite different from that which results after glutaraldehyde and osmium fixation. Also, there are no definitive differences in contraction state or ultrastructure development that can be concluded. In fact, the literature results are often contradictory from one study to the next. In the past, conventional fixation methods have been deliberately chosen to minimize the effects of contraction. The assumption is made that this minimizes the effects of contraction.

At the present time, no one has reported whether normal versus PSE muscles respond differently than normal and PSE longissimus dorsi (LD) muscles by SEM and TEM following severe freeze-thaw contraction.

Materials and Methods

Animals

A total of 16 pigs were used, comprised of 4 each from purebred Pietrain (PxP), purebred Minnesota No. 1 (MxM) and 2 groups crossbred (Px(PxM)) pigs. Px(PxM) pigs resulted from mating Pietrain boars to F_1 (PxM) crossbred females. Pigs were classified as stress-susceptible or stress-resistant based upon their reaction to halothane anesthesia at 12 weeks of age by a slightly modified method of Eikelenboom and Minkema (1974). Pigs were classified as stress-susceptible (PSS) or malignant hyperthermia positive (MH+) if muscle rigidity occurred within 5 min using 3% halothane in oxygen (Hwang et al., 1978). In this experiment, all PxP were MH+ and all MxM were MH-. Four MH+ and four MH- crosses were also selected. The Px(PxM)MH- pigs were retested with the same halothane treatment described earlier plus simultaneous treatment with succinyl choline (0.22 mg/kg body weight). The lack of muscle rigidity and other symptoms of PSS

Fig. 1. SEM of LD porcine fiber and membrane structure. Arrows; membranous complex. a) MxM; b) Px(PxM)MH-; c) Px(PxM)MH+; d) PxP.

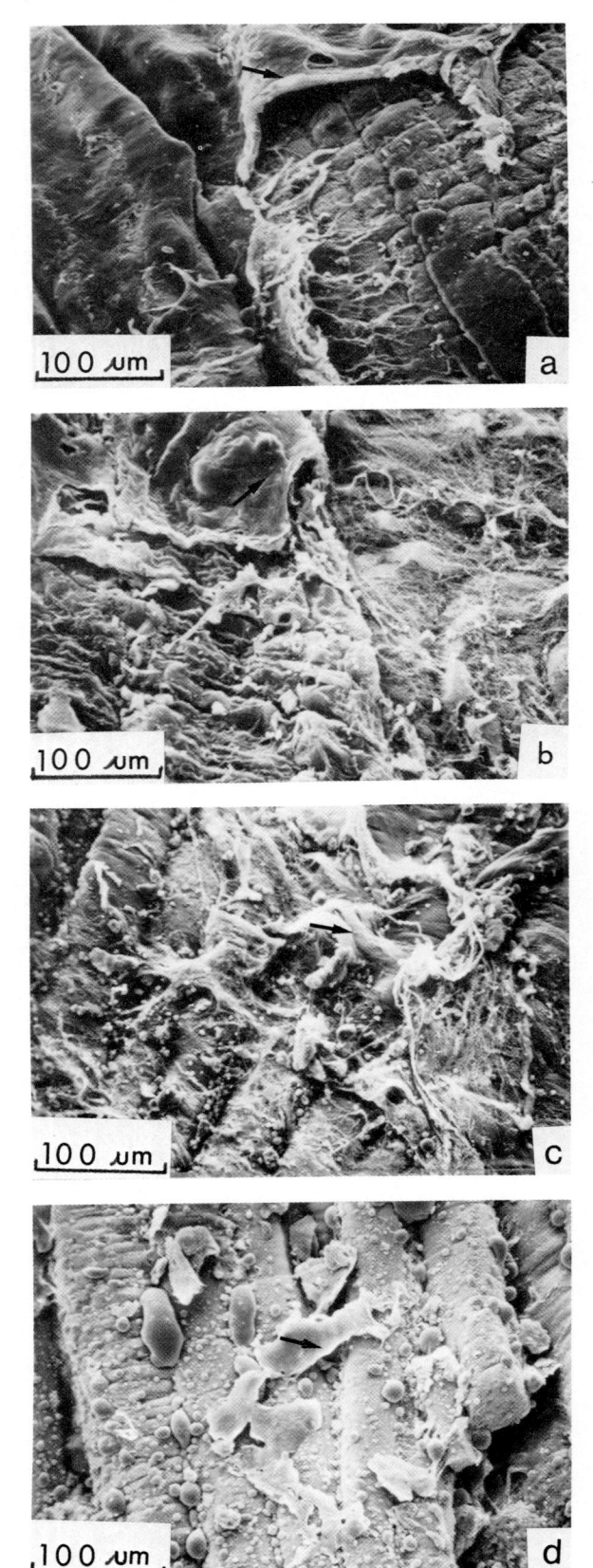

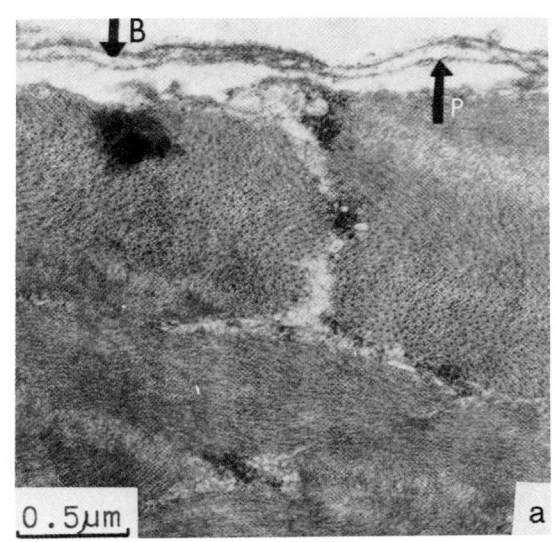

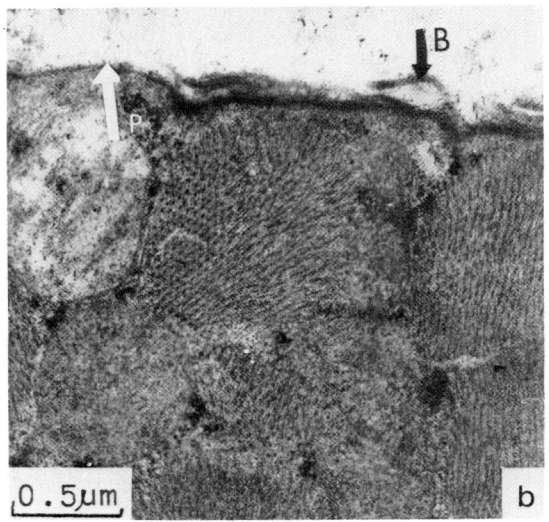

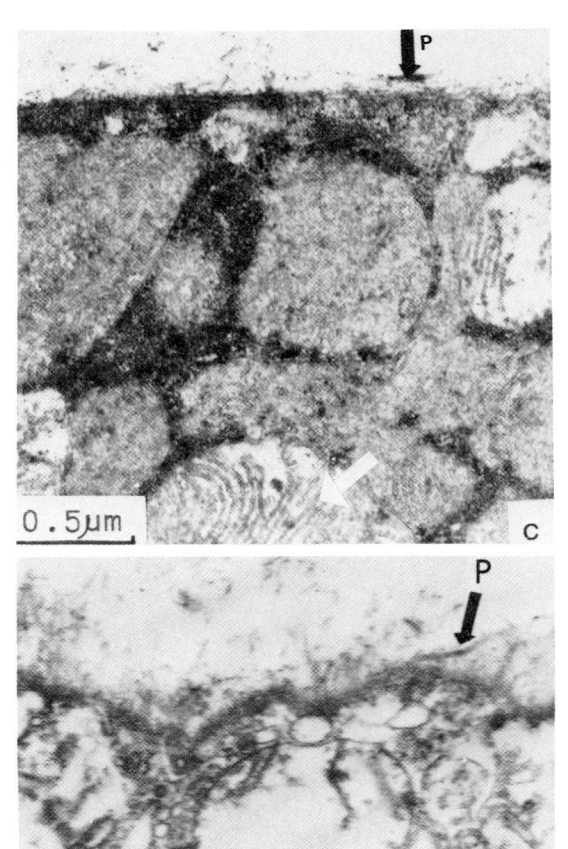

Fig. 2. TEM of LD porcine muscle, cross-sections. Basement membrane (B); Plasmalemma (P); mitochondria (white arrow)   a) MxM; b) Px(PxM)MH-; c) Px(PxM)MH+; d) PxP.

and MH during succinyl choline treatment established beyond doubt that these pigs were stress-resistant.

Pigs were dispatched at 21 weeks of age (82±5 kg live weight). Samples of the LD (5.0 mm x 20 mm x 2 mm) were taken from the left side of the carcass and frozen in isopentane cooled by liquid nitrogen within 5 min of death.

SEM procedure

The methods of Varriano-Marston et al. (1976) were used. Samples were cryofractured in liquid nitrogen, fixed in cacodylate-buffered glutaraldehyde, post-fixed in buffered osmium-tetroxide, dehydrated in acetone, critical point dried, mounted on aluminum stubs and coated with gold-palladium. The samples were viewed and electron micrographs were taken on the Cambridge Stereoscan 600 operated at 15 kV.

TEM procedure

The modified method of Varriano-Marston et al. (1977) was followed which involved cryofracturing in liquid nitrogen, cutting and fixation in buffered glutaraldehyde, post-fixation in buffered osmium-tetroxide, serial acetone dehydrations and embedding in Epon 812. Buffers used in the SEM and TEM procedures were maintained at pH 6.5 or pH 7.0 which was the pH of the muscle. (Differences were not noted, therefore samples were processed at pH 7.0 to compare with earlier work.) The embedded material was sectioned on a Sorvall microtome to sections having a gold to silver color and stained with uranyl acetate and lead citrate. Micrographs were taken on the Hitachi HU 11 C and Philips 300 transmission electron microscopes operated at 75 kV and 60 kV, respectively.

Results

Scanning electron micrographs representative of the general features of the perimysium of normal and PSE LD muscle samples are shown in Fig. 1. Although the muscle samples examined are in the supercontracted state, differences are noted in

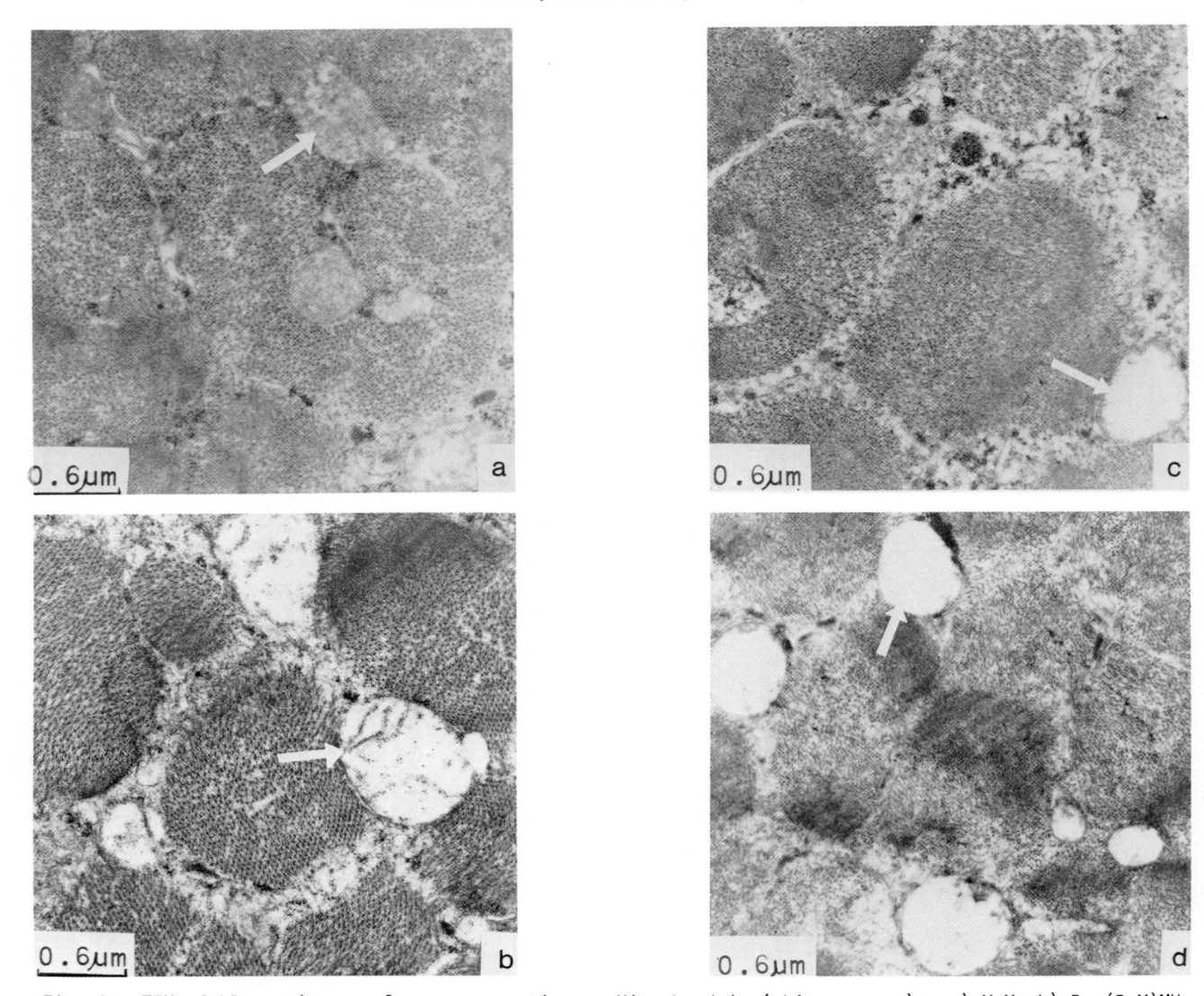

Fig. 3. TEM of LD porcine muscle, cross-sections. Mitochondria (white arrows). a) MxM; b) Px (PxM)MH-; c) Px(PxM)MH+; d) PxP.

the structures relative to each muscle genotype after similar treatment. The perimysium covers most of the muscle bundles, however, there are few areas in MxM muscles in which the perimysium was pulled back to reveal the underlying muscle fibers (Fig. 1a).

The surface of the connective tissue of MxM is smooth and falls in folds over the muscle bundles. It appears loosely fitting, as if the fibers, but not the connective tissue, have contracted. This may be similar to the "wrinkled sock" appearance reported by Locker and Leet (1975). Several layers of fibrils can be seen in the areas where the connective tissue complex has been broken.

The perimysium of the Px(PxM)MH- muscle (Fig. 1b) does not have the smooth external appearance of the MxM muscle although the connective tissue covers most of the muscle bundles. Parts of the connective tissue appear to be aggregated together (Fig. 1b, arrow), forming clumps rather than being spread in a uniform layer over the muscle fibers, suggesting some structural weaknesses or degradation. The perimysium of the Px(PxM)MH+ (Fig. 1c, arrow) is mostly absent indicating that structural weaknesses or degradation is more pronounced in the MH+ group of muscles. The perimysium in the PxP muscle (Fig. 1d, arrow) is essentially absent.

Transmission electron micrographs are shown in Figs. 2-4. The nomenclature used is by Franzini-Armstrong (1973) which divides the sarcolemma about the muscle fibers into three components: the plasmalemma, the basement membrane, and the collagen fibril network. In Fig. 2a, MxM, the plasmalemma (P) can be observed above the surface of the myofibrils. The fixation procedure has probably lifted the sarcolemma above the myofibrils, because in other micrographs of the MxM muscle, it was more tightly associated with the surface of the myofibril. However, the micrograph was chosen to show the membrane structure

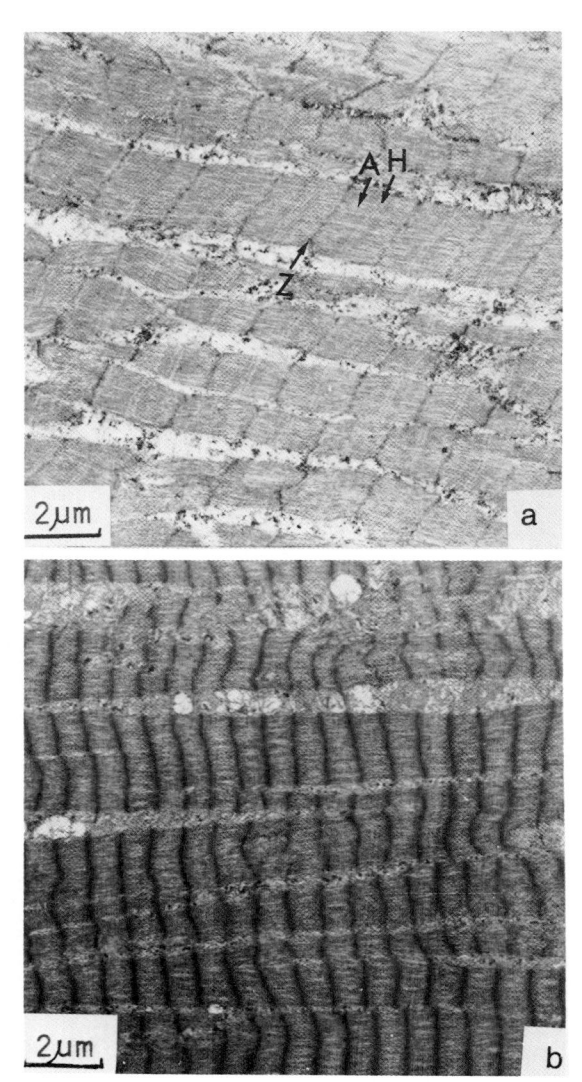

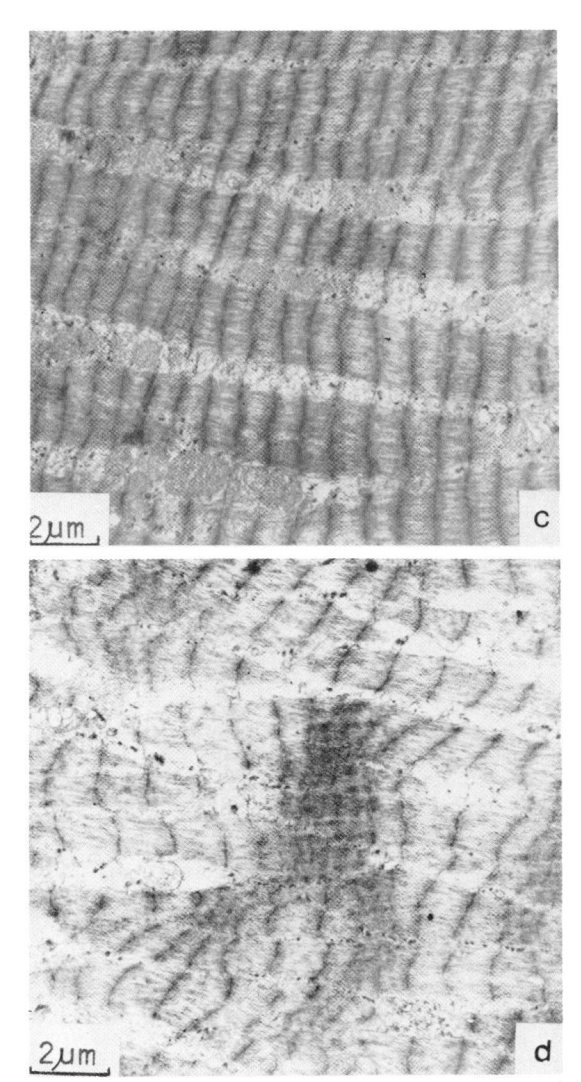

Fig. 4. TEM of LD porcine muscle, longitudinal sections. a) MxM. Z line (Z); A band (A); H band (H); b) Px(PxM)MH-; c) Px(PxM)MH+; d) PxP.

more clearly. The basement membrane (B) which coats the plasmalemma (Franzini-Armstrong, 1973) is intact in the MxM samples and is of the same order of thickness (100nm) reported by Franzini-Armstrong (1973). The loose network of collagen fibrils is generally associated with the fiber surface, but in this micrograph they are not visible.

In the Px(PxM)MH- animal the basement membrane is still visible but the arrow (Fig. 2b) shows the membrane partially pulled off. In other areas only the plasmalemma remains. Possibly some of the granular material above the basement membrane is part of the loose collagen network. The loss of the membranous components is progressive, and in the PxP PSE animal remnants of the plasmalemma are seen flaking off the surface. The micrograph for the PxP PSE muscle (Fig. 2d) is from a longitudinal section, whereas the pictures for the other groups are transverse sections. The difficulty of obtaining a true cross-section

increased with increasing severity of PSE because of the more extreme bands of contracture. The severe contraction increases the ease with which the membrane tears or breaks about fiber bundles. However, in looking over a large portion of the muscle fiber surface, the membrane is not simply broken but it is clearly not visible anywhere on the sample surface.

Mitochondria (Fig. 3) also display a gradual progression from densely staining organelles in MxM muscles (Fig. 3a) to the (PxP) mitochondria (Fig. 3d) which do not show any cristae. The mitochondria in the MxM animals were often difficult to locate because the contrast was poor between the organelles filled with many cristae membranes and the myofibrils. The PxP group, on the other hand, appeared to have many mitochondria because the low density matrix material made them readily visible. The crosses sometimes had both degenerate and normal-appearing mitochondria. For example, the MH+ sample shown in Fig. 3c has a mitochondrion with no cristae, whereas another MH+ muscle sample seen in Fig. 2c, from another animal, had mitochondria with many cristae.

Transverse sections are shown in Fig. 4. The MxM muscle sample (Fig. 4a) has Z-lines which are usually in register and with some A band, H band and M line resolution even though greatly contracted. This type of appearance was evident in muscle samples from all four normal animals studied. There was occasional stretching and tearing of myofibrils. The Px(PxM)MH- (Fig. 4b) muscle samples exhibited very regular sarcomere patterns that were contracted to the point where little banding pattern appeared between Z lines. The Px(PxM)MH+ (Fig. 4c) muscle samples exhibit similar sarcomere patterns but more irregularity in the pattern and loss of filaments of the sarcomere. The PxP muscles (Fig. 4d) show considerable overall disruption and the sarcomeres are not in register and the sarcomere banding pattern is extremely irregular. Bands of severe contracture (arrow) are readily visible and the sarcomeres are shortened such an extent that only the Z-lines are apparent, and in particularly bad areas, it is difficult to differentiate between one Z-line and the next. The width of a band contracture is approximately 8 µm and could run either partially or completely across the fiber. Therefore, PSE muscle seems far more sensitive to severe contraction than the normal muscles. This is supportive of the findings of Cassens et al. (1963b) who have shown that the bands of contracture that form in PSE muscle pre-rigor will still be present 24 hr postmortem and believed the phenomenon was primarily a mechanical disturbance of the fibrillar system caused by violent contraction. In our case it also may be caused by the freeze-thaw procedure.

The degree of contraction is reflected in the sarcomere length. Based on measurements of at least 50 sarcomeres from several micrographs of each group of animals, the sarcomere lengths for MxM, Px(PxM)MH-, Px(PxM)MH+ and PxP, respectively were: 1.8 ± 0.3 µm (mean $\pm$ standard deviation), 0.9 ± 0.3 µm, 0.8 ± 0.2 µm and 0.7 ± 0.3 µm. Sarcomere length for the MxM animals was greater than the distance found by Stanley and Geissinger (1972). They reported 1.5 µm (range 1.43-1.82 µm) for normal psoas muscle and 1.22 µm (range 0.99-1.3 µm) for contracted muscles. Muir (1970) also reported a value of 1.5 µm for porcine LD. It would appear, therefore, that the MxM sarcomere length found in our study is slightly longer than in those other two studies, but also that the value is within the expected experimental variation. This would suggest that the freeze-thaw procedure had minimal effect on sarcomere lengths and that the muscle genotype was the major determinant of degree of contraction.

Discussion

The differences seen in MxM, Px(PxM)MH-, Px(PxM)MH+ and PxP muscle samples which had undergone severe freeze-thaw contraction although not indicative of the true state of the muscle in the live animals are indicative of the relative fragility of the tissue components. In fact, the structural characteristics are systematic and even predictive of the type of muscle.

Connective tissue of porcine muscles and PSE muscles, in particular, has not been studied extensively by EM. Conclusions about the state of connective tissue have been based primarily on chemical analysis. McClain et al. (1969) found that connective tissue had a lowered ground substance content and concluded that the collagenous proteins were less mature, containing fewer intra- and probably also fewer intermolecular cross-links. Briskey (1964) also reported that PSE muscles can lose their connective tissue attachments. The results of our examination with EM would be consistent with those findings.

Throughout studies of PSE muscles, the question has been raised as to whether the observations are an indication of the condition within the animal in vivo or are the result of rapid postmortem changes. McClain et al. (1969) were unsure whether the altered connective tissue was a causative factor in PSE or the result of postmortem changes.

In our study, samples for EM studies were excised and frozen within 5 min of death. Elizondo et al. (1976) found that MxM LD had a pH of 6.5 at 5 min postmortem and PxP LD, 6.2. It was in part for this reason as well as for the reasons stated earlier, that the muscles were frozen within 5 min after death to help stop enzymatic induced postmortem changes. In this way, the biochemical results (Elizondo et al., 1976) which were determined immediately after death could be compared for the same genetic type muscle samples. Briskey (1964) did not find pH's below 6.5 at 5 min postmortem even in extremely PSE muscles. Bendall and Wismer-Pedersen (1962) attributed a granulated surface to a layer of denatured sarcoplasmic protein firmly bound to the surface of the myofilaments. Thus the observations in our SEM study suggests that underlying differences in the muscle fibers existed even before extensive glycolysis occurred postmortem.

The disruption of the sarcolemma as seen in Fig. 2 has important implications, regardless of whether it occurs in vivo, postmortem, or is an artifact of the preparation that reflects underlying fragility of the structure. If the sarcolemma is disrupted, no barrier to diffusion exists. The plasmalemma helps maintain the osmotic equilibrium of the muscle fiber and also serves the important function of conveying the action potential along the fiber from its nervous origin at the motor end-plate (Bendall, 1970).

As with the other structures, because of the difficulties in obtaining an experimentally unambiguous answer, it is not clear as to the cause of the degeneration of the mitochondria and whether it is an in vivo condition or part of the complex postmortem change and subsequent sample preparation methods.

Eickelenboom and Van den Bergh (1973) concluded that the yields of mitochondria from different breeds of pigs may not be different, but other workers (Wesemeier, 1974, Bergmann, 1974, Bergmann and Wesemeier, 1970) have observed degeneration of mitochondria. Bergmann (1974) observed transitional and degenerative forms suggestive of latent and permanent insufficiency of aerobic energy exchange in pre-slaughter biopsy samples. Greaser et al. (1969a), in an examination of the mitochondrial fraction found at death that most mitochondria had a conformational pattern similar to that of normal muscle, but some were markedly swollen, had a decreased density, and were more easily disrupted during homogenization. Dutson et al. (1974), in examination of fibers at 15 min postmortem found that mitochondria from βR fibers of PSE and normal fibers were similar, but those from PSE αW fibers had an open-structured appearance and showed considerable disruption. They also concluded that disruption of the mitochondria, sarcoplasmic reticulum and glycogen occurs simultaneously in the same fibers.

It would seem, therefore, on the basis of our studies and previous studies, that some degree of membrane disruption characterizes the stress-susceptible animal, and concomitant changes in membrane-associated biochemical events occur. These changes then can induce further deleterious effects. For example, the enzymes of the aerobic oxidation system are membrane-bound. Elizondo et al., (1976), for example, suggested the use of creatine kinase levels as an indicator of stress-susceptibility in swine. For this reason, they froze the serum immediately after death as we did muscle, in order to quench the production of creative kinase levels prior to analysis and quench the enzyme activity to the serum. In another study, Cassens et al. (1975) suggested that the permeability of membranes is affected and leakage of proteins, water, salts, and ions into the blood stream can occur. It is also possible that inability to bind materials, especially ions such as calcium, may be another characteristic of PSE muscles that have undergone extensive changes in membrane. This was suggested by Greaser et al. (1969b, c).

Regardless of the sequence of biochemical events, the practical consequences of the degradative reactions are considerable. The overall effects of free water, lack of diffusion barriers, and alteration of control of enzyme systems combine to change the response to processing and preparation. Thus, the severe contraction and structural differences that develop during fast freezing and thawing into fixatives prior to EM sample viewing appear to be systematic and reproducible for each type.

The more definitive experiment will come when we take biopsies and/or at-death muscle and look at the frozen-hydrated unfixed sample by low temperature SEM and TEM, with subsequent X-ray microanalysis to localize the calcium in order to determine its role in the muscle chemico-physical differences.

Acknowledgements

This study was supported in part by the University of Minnesota Agricultural Experiment Station Projects No. 18-27 and No. 18-63, and Contract No. ENG 76-09808, National Science Foundation, Washington, D. C.

The authors thank Dr. William E. Rempel of the University of Minnesota Animal Science Department who made possible the animals used in this study and for helpful technical advice.

References

Abbott M T, Pearson A M, Price J F, Hooper G R. Ultrastructural changes during autolysis of red and white porcine muscle. J. Food Sci. 42: 1977, 1185-1188.

Andersen L D, Parrish F C Jr, Topel D G. Histochemical and palatability properties of M. longissimus from stress-resistant and stress-susceptible porcine animals. J. Anim. Sci., 41: 1975, 1600-1610.

Bendall J R. Muscles, Molecules and Movement. Heinemann Educational Books, Ltd., London, England, 1970, p. 75.

Bendall J R, Wismer-Pedersen J. Some properties of the fibrillar proteins of normal and watery pork muscle. J. Food Sci. 27: 1962, 144-157.

Bergmann, V. Zur Ultrastruktur der Mitochondrien in der Skelettmuskulatur des Schweines. Archiv für Experimentelle Veterinärmedizin. 28: 1974, 225-241.

Bergmann V. Die Ultrastruktur von PSE-Muskulature vor and nach der Schlachtung. Monatshefte für Veterinärmedizin. 30: 1975, 285-288.

Bergmann V and Wesemeier H. Elektronenmikroskopische Befunde an der Skellerttmuskulatur bei Fleischschweinen. Archiv für Experimentelle Veterinärmedizin. 24: 1970, 1241-1255.

Briskey E J. Etiological status and associated studies of pale, soft, exudative porcine musculature. Adv. Food Res. 13: 1964, 89-178.

Briskey E J, Wismer-Pedersen J. Biochemistry of pork muscle structure. I. Rate of anaerobic glycolysis and temperature change versus the apparent structure of muscle tissue. J. Food Sci. 26: 1961, 297-305.

Cassens R G, Briskey E J, Hoekstra W G. Electron microscopy of postmortem changes in porcine muscle. J. Food Sci. 28: 1963a, 680-684.

Cassens R G, Briskey E J, Hoekstra W B. Similarity in the contracture bands occurring in thaw-rigor and in other violent treatments of muscle. Biodynamica. 9: 1963b, 165-175.

Cassens R G, Marple D N, and Eikelenboom G. Animal physiology and meat quality. Adv. Food Res. 21: 1975, 71-155.

Dutson T R, Pearson A M, Merkel R A, Spink G C. Ultrastructural postmortem changes in normal and low quality porcine muscle fibers. J. Food Sci. 39: 1974, 32-37.

Eikelenboom G, Van den Bergh S. G. Mitochondrial metabolism in stress-susceptible pigs. J. Anim. Sci. 37: 1973, 692-696.

Eikelenboom G, Minkema D. Prediction of pale, soft, exudative muscle with a non-lethal test for the halothane-induced porcine malignant hyperthermia syndrome. Tijdschr. Diergeneesk. 99: 1974, 421-426.

Elizondo G, Addis P B, Rempel W E, Madero C, Martin F B, Anderson D B, Marple D N. Stress response and muscle properties in Pietrain (P), Minnesota No. 1 (M) and PxM pigs. J. Anim. Sci. 43: 1976, 1004-1014

Field R A, Pearson A M, Koch D E, Merkel R A. Thermal behavior of porcine collagen as related to postmortem time. J. Food Sci. 35: 1970, 113-116.

Franzini-Armstrong C. Membranous systems in muscle fibers. In: Structure and Function of Muscle. Vol. 2, 2nd ed., G H Bourne (ed.), Academic Press, New York, U.S.A., 1973, p. 531.

Greaser M L; Cassens R G, Hoekstra W G. Changes in oxalate-stimulated calcium accumulation in particulate fractions from postmortem muscle. J. Ag. Food Chem. 15: 1967, 1112-1117.

Greaser M L, Cassens R G, Briskey E J, Hoekstra W G. Postmortem changes in subcellular fractions from normal and pale, soft, exudative porcine muscle. 2. Electron microscopy. J. Food Sci. 34: 1969a, 125-132.

Greaser M L, Cassens R G, Briskey E J, Hoekstra W G. Postmortem changes of subcellular fractions from normal and pale, soft, exudative porcine muscle. 1. Calcium accumulation and adenosine triphosphate activities. J. Food Sci. 34: 1969b, 120-124.

Greaser M L, Cassens R G, Hoekstra W G, Briskey E J, Schmidt G R, Carr S D, Galoway D E. Calcium accumulating ability and compositional differences between sarcoplasmic reticulum fractions from normal and pale, soft, exudative porcine muscle. J. Anim. Sci. 28: 1969c, 589-592

Hwang P T, McGrath C J, Addis P B, Rempel W E, Thompson E W, Antonik A. Blood creatine kinase as a predictor of the porcine stress syndrome. J. Anim. Sci. 47: 1978, 630-633

Jones E W, Nelson T E, Anderson I L, Kerr D D, Burnap T K. Malignant hyperthermia of swine. Anesthesiology. 36: 1972, 42-51.

Linke H. Histologische Untersuchungen bei wässerigem blassem Schweinefleisch. Fleischwirtschaft. 52: 1972, 493-496.

Locker R H, Leet N G. Histology of highly stretched beef muscle. I. The fine structure of grossly stretched single fibers. J. Ultrastruc. Res. 52: 1975, 64-75.

McClain P E, Pearson A M, Brunner J R, Crevasse G A. Connective tissue from normal and PSE porcine muscle. 1. Chemical characterization. J. Food Sci. 34: 1969, 115-119.

McClain P E, Pearson A M, Fenell R A, Merkel R A. Metachomasia of epimysial connective tissue from normal and from pale, soft and exudative porcine muscle. Proc. Soc. Exp. Biol. Med., 128: 1968, 624-627.

Muir A R. Normal and regenerating skeletal muscle fibers in Pietrain pigs. J. Comp. Path. 80: 1970, 137-142.

Nelson T E, Jones E W, Henrickson R L, Falk S N, Kerry D D. Porcine malignant hyperthermia: Observations on the occurrence of pale, soft, exudative musculature among susceptible pigs. Am. J. Vet. Res. 35: 1974, 347-350.

Sair R A, Kastenschmidt L L, Cassens R G, Briskey E J. Metabolism and histochemistry of skeletal muscle from stress-susceptible pigs. J. Food Sci. 37: 1972, 659-663.

Schaller D R, Powrie W D. Scanning electron microscopy of skeletal muscle from rainbow trout, turkey and beef. J. Food Sci. 36: 1971, 552-559.

Schmidt G R, Goldspink G, Roberts T, Kastenschmidt L L, Cassens R G, Briskey E J. Electromyography and resting membrane potential in longissimus muscle of stress-susceptible and stress-resistant pigs. J. Anim. Sci. 34: 1972, 379-383

Stanley D W, Geissinger H D. Structure of contracted porcine psoas muscle as related to texture. J. Inst. Can. Sci. Technol. Aliment. 5: 1972, 214-216.

Varriano-Marston E, Davis E A, Hutchinson T E, Gordon, J. Scanning electron microscopy of aged free and restrained bovine muscle. J. Food Sci. 41: 1976, 601-605.

Varriano-Marston E, Gordon J, Davis E A, Hutchinson T E. Cryomicrotomy applied to the preparation of frozen hydrated muscle tissue for transmission electron microscopy. J. Microsc. 109: 1977, 193-202.

Wesemeier H. Elektronenmikroskopische Untersuchungen an der Skelettmuskulatur von unbelasteten sowie experimentell belasteten Fleischschweinen. Ein Beitrag zur Ätiologie und Pathogenese des blassen, wässrigen Schweinefleisches (PSE-Fleisch). Archiv für Experimetelle Veterinärmedizin. 28: 1974, 329-383. Abstract only. Food Sci. and Technol. Abstracts 7: 1975, p. 1538.

Wipf V K, Mullins A M, Passbach F L Jr. Acid mucopolysaccharides and porcine muscle quality. J. Anim. Sci. 30: 1970, 355-359

Discussion with Reviewers

S. H. Cohen: You mentioned cryofracturing prior to fixation. How does that method compare with post-fixation cryofracture?

R. J. Carroll: In TEM preparation, what is the purpose of cryofracturing prior to fixation, since cutting of the tissue to proper size precedes fixation?

Authors: This has now been answered, in part, in the text where reference to enzyme studies by Elizondo et al. (1976) was made. In that study they froze serum samples to stop creatine kinase activity for enzyme level determination. Higher levels of creatine kinase were found in PSE porcine muscle than normal porcine muscle.

Comparisons of different specimen preparation techniques on these types of muscles could not be rigorously completed at this time due to the non-availability of animals with known genetics. It is intended to do such a study as the animals come of age and are once again available to us. However, we did repeat the study on muscle that was determined to be intermediately PSE. In one set of samples the cryofracture step was omitted altogether. Figure 5 shows a comparison of TEM observed structures for cross and longitudinal muscle sections prepared both ways. We found that the non-cryofractured cross sectional muscle samples (Fig. 5a) did not look that different from the cryofractured samples (Fig. 5b). The basement

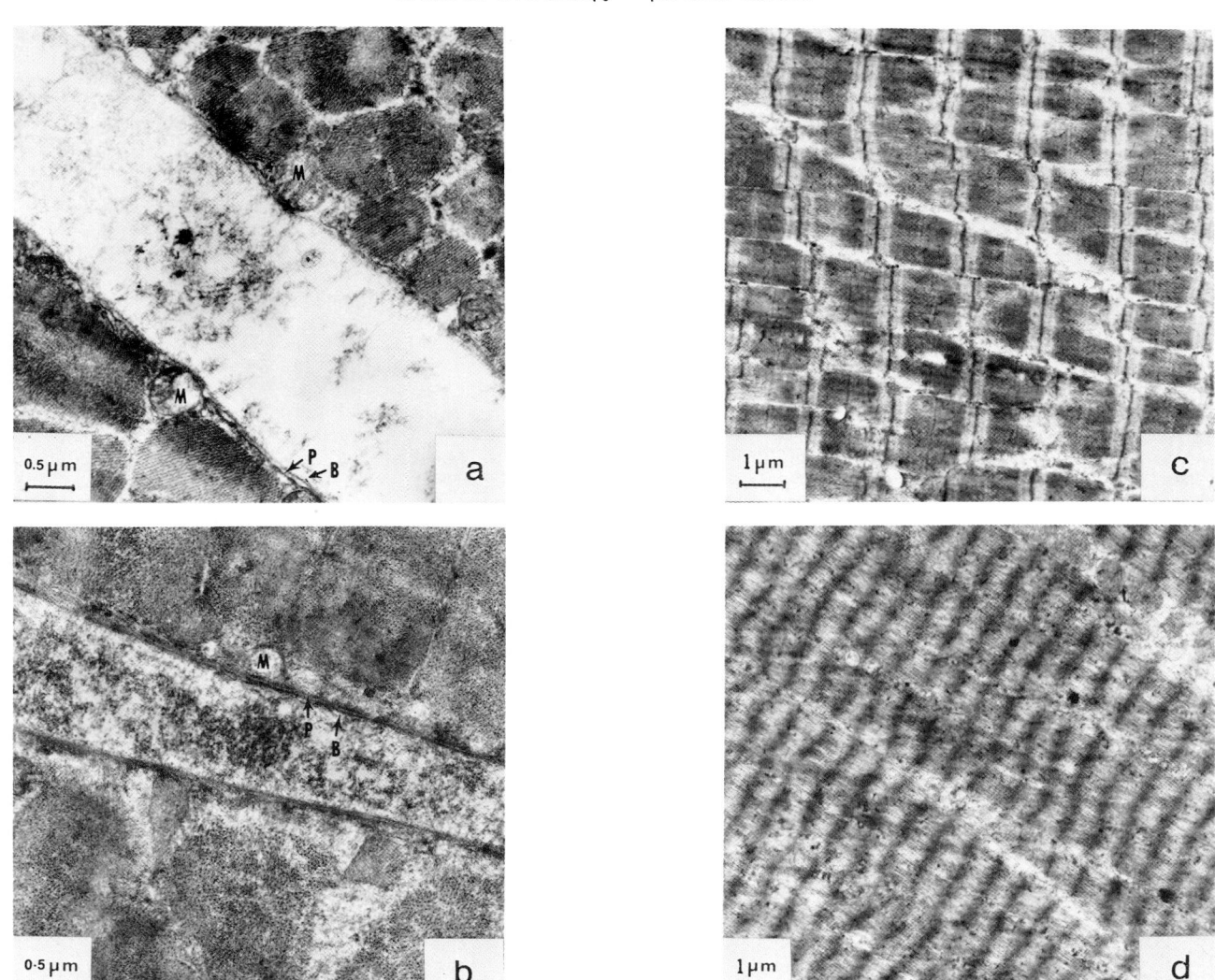

Fig. 5. TEM of LD porcine muscle. Basement membrane (B); plasmalemma (P); mitochondria (M); a) Non-cryofractured, cross-section; b) Cryofractured, cross-section; c) Non-cryofractured, longitudinal section; d) Cryofractured, longitudinal section.

(B) and plasmalemma (P) membranes are visible in both, with their appearance being similar to Fig. 2b of the text. Also, the mitochondria (M) are seen in both, and they are not always normal in appearance. Occasionally, breaks and some disintegration of the basement membrane are visible with both preparation methods. The biggest difference could be seen between the non-cryofractured longitudinal muscle samples (Fig. 5c) and that which was cryofractured (Fig. 5d). It was found that the non-cryofractured samples had sarcomere banding patterns that look very much like normal muscle and was approximately 1.8 μm in sarcomere length. However, the cryofractured samples had severe contraction patterns typical of those seen in Fig. 4b and 4c of the main text, with sarcomere length approximately 0.8 μm in length. Therefore, some of these features are present due to the freezing procedure, but they are quite predictable as a function of muscle type. We are not saying the structures exist in the animal to such a damaged or altered degree, but that those features in PSE muscle that do appear altered from normal porcine muscle, after ultra-low temperature freezing and subsequent thawing into fixatives, are predictably different.

C. L. Davey: Have the authors evidence to indicate at what post-mortem stage the shortening had occurred?
Authors: It was observed to occur maximally during thawing in glutaraldehyde. To what extent prior to that, and at what stage, cannot be reported from this study.

C. L. Davey: Since there has been shown there is considerable rupturing of the perimysial connective tissue, was there lowered tensile strength?
Authors: This has been the case from other experiments on these PSE genotypes. (Personal communication from P. Addis laboratory.)

S. H. Cohen: Could the appearance of the membrane being "pulled off" be caused by the angle at which the knife cuts the sample?
R. J. Carroll: Could the severe contractions upon freezing and/or thawing also cause disruption of the sarcolemma?
Authors: Disruption of the sarcolemma is possible due to freeze-thaw or some sectioning problems. However, the observation of a progression through the genotypes indicates regularity rather than randomness.

S. H. Cohen: It seems to me that calcium ions might play an important role in producing the characteristics of PSE muscle studied here. First, the involvement of calcium in nerve impulses is quite well known and, therefore, might affect the stress of the animal. Secondly, the "calcium activated factor" might be involved in altering membrane permeability, since its peak activity is at approximately pH 6.3-6.5. What are your thoughts on this matter?
Authors: This comment parallels our point that once the membrane is changed, a whole series of permeability-related changes can occur. Of course, we cannot tell which precedes the other in studies such as these. Is membrane disruption a result or a cause? Or is a feedback situation involved? In fact, a similar view is presented by Lucke, Halle and Lister (1979) where they conclude in this recent review that "the central factor that establishes a threshold of malignant hyperthermia sensitivity is the level of free calcium in the myoplasm, even though there is still no direct evidence to support this view. The threshold may be influenced by many factors that affect cell membranes and subcellular organelles, some of which may be inherited." (Reference: Lucke J N, Hall G M, Lister D. Malignant hyperthermia in the pig in the role of stress. In: Muscular Dystrophy and Other Inherited Diseases of Skeletal Muscle in Animals. J. B. Harris. Ann. N. Y. Academy of Sci. 317: 1979, 326-337. Quote found on page 332.)

S. H. Cohen: What is the role, if any, of catheptic enzymes in terms of PSE muscle?
Authors: The location and type varies. It would be attractive to hypothesize that if membrane disruption is generalized then membranes of lysozomes are also included in this disruption. One could then hypothesize that the deterioration seen in the PxP porcine muscle, partially reflects this fact. It is interesting that Franzini-Armstrong (1973, page 567) points out that "lysozomes are occasionally visible but that they are in significant quantities only in conditions of degeneration". Muir (1970) interpreted his experimental evidence as degeneration of muscle in vivo for PSE muscle. In addition to cathepsins, the inhibitors of the cathepsins might play a role. This is especially intriguing in that Barrett (1980) reports that the high molecular inhibitor found in bovine cardiac muscle interacts with calcium-dependent thiol proteinases. This, then, is another example of a calcium-dependent mechanism. (Barrett A J. The many forms and functions of cellular proteinase. Federation Proceedings, 39: 1980, 9-14.)

R. J. Carroll: Define βR and αW fibers.
Authors: R and W is red and white muscle fiber, respectively. Alpha and β will require somewhat longer definition. As a result, the following discussion on βR and αW definition follows: Historically, three main types of muscle fiber are recognized; red, white and intermediate (Bloom W, and Fawcett D W. A Textbook of Histology, 10th ed. W. B. Saunders Co., Philadelphia, U.S.A., 1975, p. 296.) Red fibers may be either physiologically slow or fast, they contain a higher oxidative enzyme activity through the fast-acting live fibers. Ashmore (Ashmore C R, Addis P B, and Doerr L. Development of muscle fibers in the fetal pig. J. Anim. Sci. 36: 1973, 1088-1093.) has suggested that the nomenclature of this βR, αR or αW system is more appropriate to the fiber type in porcine muscles. The βR is a slow twitch aerobic fiber, the αR is a fast twitch aerobic, while the αW is the fast twitch anaerobic fiber. Several studies have been performed to determine the distribution of fiber type in PSE porcine muscle. The result of histochemical studies, however, presents a confused picture because of the lack of uniform terminology (Cassens, et al. 1975), since some workers included intermediate fibers with red fibers into a single group labeled dark fibers. PSE characteristics are more apparent in the predominantly white muscles such as the L. dorsi than in the red muscle of the same carcass. The red fiber content of the L. dorsi in the Chester White breed is approximately 30% compared to 60% red fibers found in the trapezius (Moody W G, and Cassens R G. Histochemical differentiation of red and white muscle fibers. J. Anim. Sci. 27: 1968, 961-968.)

R. J. Carroll: Since isopentane is a solvent for lipids, could the amorphous material observed in Fig. 1d result from lipid-solvent interactions? This material to a lesser degree is also observed in Figs. 1a and 1c.
Authors: Since a very low temperature was used in that part of the study with the isopentane solvent, we have made the assumption that solubility of the lipid was minimal in this case. Furthermore, postmortem changes were observed progessively without isopentane in the SEM study of aged, free and restrained muscles by Varriano-Marston, et al. (1976).

H. D. Geissinger: Why are your results different from those in previous studies?
R. J. Carroll: Dutson et al. (1974) observed TEM of normal and stressed L. dorsi porsine muscle at 15 min. and 24 hrs. after slaughter. How do you explain the severe contraction observed in your tissues as opposed to Dutson's results which show no excessive shortening of the sarcomeres?
Authors: When we compare our results with studies using similar, but not same methods, the conclusions do present a unified picture of the morphologies or ultrastructures. The results of the major studies are summarized in the text, but we should point out that what has been assumed to be facts from studies using different methods of preparation in previously published results may, in fact, be artifact. For example, Greaser and Cassens (1969a) used homogenization and separation

of the organelles prior to fixation and found similar patterns. Muir (1970) in his studies defined antemortem animals as those that had been killed by exposure to etorphine and chloroform anesthesia. The postmortem samples were obtained from animals which had been electrically stunned followed by bleeding and scalding. Then muscle samples were obtained within 15 min, restrained and conventionally fixed. Therefore, in Muir's results, he called these postmortem samples. In Dutson's studies, the pietrain pigs did show a pattern similar to what we were reporting. Bergmann and Wesemeier (1970) show patterns of organization which are more extensive than we see in Fig. 4d. However, those muscle samples were recorded to have low pH values. A major difference in published literature citing is the pH at which muscle was starting from prior to fixation. Since Muir's work often had values as low as 5.85, the fiber coagulation may be more extensive and of the type discussed by Bendall and Wismer-Pedersen (1962). Antemortem muscle is different from postmortem muscle. We also know that there are differences in the same PSS animal having different degrees of PSE. On the other hand, Jönsson L, and Johansson G. (1979) (Cardiac muscle cell damage of the porcine stress syndrome: In: Muscle function in porcine meat quality. Wegger I, Hyldgaard-Jensen, J, and Moustgaard J. (ed). Acta Agriculturae Scandinavica, Suppl. 21, 1979, p. 330) have reported similar ultrastructural changes as we have reported for the LD muscle for crossbred Yorkshire and Swedish Landrace. One thing we do wish to point out is that creatine kinase used as a diagnostic tool is simply that - obviously more complex enzyme chemistry is going on in a single enzyme system, as we have indicated in our response to Drs. Cohen and Carroll.

H. D. Geissinger: Are the differences noted in Fig. 5a and 5b intended to point out that the differences are due to the severe contraction state?
Authors: In Fig. 5a and 5b the authors were asked whether it was intended to point out that the differences are due to the severe contraction state. The authors feel that it is difficult to obtain true cross-sectional images in comparisons on fiber diameters, etc. Therefore, it may or may not be due to severe contraction state. It is difficult to make a comparison between the two methods other than the membrane structures that we have pointed out. We are looking at different elements in Fig. 5a and 5b versus 5c and 5d. Therefore, membranes are not that different but sarcomeres are different.

H. D. Geissinger: Is the coagulated material in Figs. 1b and 1c different? Also, are the components marked as basement membrane and plasmalemma, in fact, collagen?
Authors: The coagulated fiber differences in Figs. 1b and 1c, the authors feel, are actually different materials. It was after careful examination and with reference to the article by Franzini-Armstrong (1973), that a measurement of the structures in terms of the thickness of those membranes were made and found to be about 100 nm, it was decided that these structures are, in fact, basement membrane and plasmalemma rather than collagen. In fact, there is material above those membranes that we see in other micrographs which we have not pointed out because they are badly broken (or as Dr. Geissinger refers to as coagulated) but we believe is the collagenous material Franzini-Armstrong describes.

H. D. Geissinger: Why are there mixtures of muscle contraction patterns for PSE muscles?
Authors: In the Swedish study by Jönsson and Johansson (1979), and also the study by Greaser, et al (1969a), similar patterns of muscle structure were found although in the former article, the patterns were for heart muscle. In the article by Greaser et al., it is pointed out that one does see mixtures of muscle contraction although their muscle is not as pietrain as the muscle that we have been looking at. Also, the study by Cassens, et al. (1963 a, b) does not contradict our results. It is concluded then that in muscles that are not as pietrain as ours, or as PSS as ours, one does see a mixed state of contraction state. The point that we have made in the study we have done is that in muscle that is more pure in its PSE characteristics, the differences are predictable. In discussing the PxP strains with the geneticists at the University of Minnesota, it has been pointed out to us that they believe that the group of animals we have been studying are the most pietrain available when compared to those studied to-date in the literature. We believe that the impossibility of obtaining a truly PSS animal is one of the reasons that the literature is not as clear cut as we would like it to be when cross-comparing studies from different laboratories. Geneticists do not completely understand the genetics of the pietrain animals in terms of whether the PSE characteristic is recessive. The fact that an animal is halothane sensitive and then termed PSS does not mean that those animals are the same in terms of their PSE characteristics in the muscle. In fact, it has been determined very conclusively that the differences in PSS are considerable from animal-to-animal, even though the halothane sensitivity may be the same. If there are differences in mitochondria that cause degeneration of the membrane of the mitochondria, this might be induced by free radical breakdown. This could be an uncoupling of oxidative phosphorylation. Thus, the glycolytic cycle could be speeded up. This would be an exothermic reaction which would disturb the ATP production. It has been determined that there are higher values for potassium and calcium, and it is believed that these higher values for them reflect cell membrane damage. When this happens in cases where it is applied to cardiac muscles, it might result in cardiac arrhythmia.

E. Varriano-Marston: Your data on sarcomere length clearly shows that the values were not different from data reported by other authors where no freeze-thaw procedure was employed. That fact should indicate that your method was satisfactory. Furthermore, one does not have to freeze-thaw to obtain supercontracted muscle; I have observed violent contraction of muscle due to fixation only. In fact, within the same TEM block of muscle tissue, sections can be found which represent extreme contraction as well as relaxed states. One way to check your method is to confirm sarcomere lengths and general structure with light microscopy (polarizing) of fresh (unfixed) muscle. One must also remember, as Dr. Luyet has pointed out in his many articles, that the rate of freezing and the rate of thawing dramatically determines the state of the muscle on freeze-thawing. To obtain more rapid penetration of the glutaraldehyde fixative into the muscle during thawing, the O_2 level in the fixative should be increased as described by Johnson and Rash (J. Cell Biol. 87: 231a).

<u>Authors:</u> These suggestions are worthy of further investigations.

IDENTIFICATION OF FAT AND PROTEIN COMPONENTS IN MEAT EMULSIONS
USING SEM AND LIGHT MICROSCOPY

F. K. Ray, B. G. Miller[1], D. C. Van Sickle[2], E. D. Aberle[3], J. C. Forrest[3] and
M. D. Judge[3]

Animal Science Dept.
004 Animal Husbandry
Oklahoma State Univ.
Stillwater, OK 74074

[1]Dept. of Anatomy
Indiana Univ.
School of Medicine
1100 West Michigan St.
Indianapolis, IN 46202

[2]Dept. Veterinary Anatomy
[3]Dept. Animal Science
Purdue Univ.,
W. Lafayette, IN 47907

Abstract

A new technique for identifying lipid (fat) and non-lipid (protein) components in a cooked meat emulsion product by SEM was developed. Serial sections from a sample block of product were prepared in the following sequence: Section 1) Zenker's solution was used to selectively fix the protein component. The fat component was extracted with ether, and the section was then evaluated by SEM. Section 2) Osmium tetroxide was used to fix both fat and protein components and that section was also evaluated by SEM. Section 3) This section was stained by oil red O (lipid stain) and counterstained with Delafield's hematoxylin (protein stain), and then evaluated by light microscopy to provide a reference for identifying the fat and protein components of the preceding sections. The sequence of section preparation was repeated at random. This technique identified the coagulated protein matrix, fat component, and air pockets, and allowed characterizing the surface texture of the protein matrix and the fat in sections evaluated by SEM. This characterization can be useful in confirming the protein matrix and the fat demonstrated in meat emulsion products by SEM micrographs of freeze-fractured surfaces.

KEY WORDS: Meat Emulsions, Correlative Microscopy, Histological Stains, Liver Sausage, Fat, Protein

Introduction

Hansen[1] was the first to show histological views of the protein matrix surrounding the fat globule in model emulsions and in commerical frankfurters. A frozen sectioning technique was used to prepare thin sections which were stained with Heidenhain's stain for light microscope observation. Froning et al.[2] modified this technique to study poultry meat emulsion by use of other stains or paraffin embedded samples. Cassens et al.[3] evaluated the effect of textured soy protein on the morphology of frankfurters using the frozen sectioning technique. Sections were stained with oil red O and Harris hematoxylin and observed with the light microscope. More recently, Cassens et al.[4] published on the microscopic structure of commercial sausage products using light microscope histological techniques. Histochemical methods revealed the lipids and connective tissues, and a description of the structure of some common sausages.

In recent years, electron microscopy has been shown to be a promising tool for evaluating meat emulsion structures. Borchert et al.[5] used both light and transmission electron microscopy to evaluate meat emulsion structures. Electron photomicrographs indicated fat globules as small as 0.1 μm in diameter are surrounded by distinct protein membranes. Scanning electron microscopy (SEM) also has been employed in evaluation of meat products.[6,7] Theno and Schmidt[8] published on evaluation of emulsion structures of three commercial frankfurters using SEM and light microscopy. However, the primary drawback to using SEM has been how to clearly distinguish fat from protein components.

The purpose of this investigation was to correlate specific emulsion components as illustrated by light microscopy (LM) and scanning electron microscopy (SEM).

Materials and Methods

Sectioning of Samples

Sampled meat emulsions were taken from laboratory prepared liver sausage, commercial liver sausage and commercial wieners, all of which were cooked meat emulsions. A 3-4 mm slice was removed from the sampled meat emulsion, from which two 4 mm x 4 mm pieces were cut with a razor blade and mounted for sectioning. These representative sample blocks (3 x 4 x 4 mm) were frozen for 10 minutes on the specimen holder plate using Lipshaw M-1 embedding matrix and the quick freeze bar in the cryostat chamber. The frozen blocks were shaved and squared before initiating collection of the samples which were sectioned 10 μm thick using the Model CTI International microtome-cryostat. The temperature of the cryostat chamber was maintained at -20° C. The frozen sections were cut and mounted on a warm gelatin - coated slide cover slip (18 x 18 mm). the cover slips were pre-coated with a thin layer of gelatin to enhance the adhesion of the section to the cover slip. The sections were air-dried for an hour before being treated with specific fixative or stain. Table 1 shows the sequence by which the sections were marked for stain or fixative and respective microscope.

Table 1. Sequence of serial sections taken from the frozen samples.

Samples	Stain	Microscope
1	Zenker's	SEM
2	Osmium	SEM
3	Oil red O	LM
4	Zenker's	SEM
5	Osmium	SEM
6	Zenker's	SEM
7	Oil red O	LM
8	Osmium	SEM
9	Zenker's	SEM
10	Osmium	SEM
11	Zenker's	SEM
12	Oil red O	LM
13	Osmium	SEM
14	Zenker's	SEM
15	Oil red O	LM
16	Osmium	SEM
17	Zenker's	SEM
18	Osmium	SEM
19	Oil red O	LM
20	Zenker's	SEM

Preparative Technique for Light Microscopy (LM)

Sections for LM were immersed in absolute propylene glycol for 2 minutes, and then in a solution of 0.5% oil red O and propylene glycol solution for staining. Sections were stained for 8 hours in enclosed containers and differentiated in 85% propylene glycol solution for 1 minute followed by two distilled water rinses. Next, the sections were stained in Delafield's hematoxylin solution for 5-7 minutes followed by 3 minutes in Scott's bluing solution.[9] Sections were washed for 5 minutes in tap water prior to mounting on glass slides using glycerin jelly. LM was performed with an American Optical series 20 microscope and AO Spencer 35 mm photomicrographic camera.

Preparative Techniques for Scanning Electron Microscopy (SEM)

After the selected sections had been dried, they were fixed either in 1% osmium tetroxide in phosphate buffer (pH 7.4) or in Zenker's solution.[10] Osmium tetroxide treated sections were fixed one hour, rinsed in phosphate buffer at pH 7.4 and then dehydrated with serial washings of 40, 70, 80, 95, 100 and 100% ethanol for at least 4 minutes per washing. The sections fixed in Zenker's solutions were dipped for 45 minutes and then dehydrated in 40, 70, 80, and 100% ethanol followed by petroleum ether utilizing the same washing times as osmium tetroxide. Both the osmium tetroxide treated sections and the Zenker's solution treated sections were mounted on 2 cm aluminum specimen holders using double stick adhesive tape. Conductive silver paint was used to ground the cover slips to the aluminum specimen mounts. The sections were coated with gold and observed using a low accelerating voltage of 5kV to prevent beam damage to the samples.

Results

Light microscopy (LM) techniques, using oil red O stain and counterstained with Delafield's hematoxylin, excellently differentiated fat (lipid) and protein (non-lipid) components. The fat components stained orange-red or red and the protein components stained blue. The objective was to compare the histological techniques to SEM procedures and identify fat and protein in SEM micrographs.

With SEM techniques and following the procedures outlined under the Materials and Methods section, it has been shown that fat and protein can be identified by selectively fixing. By comparing the osmium-tetroxide treated to the Zenker's solution treated section, fat components could be easily identified.

With this idea, the technique utilizes serial sections of a sample for alternate staining for LM or fixing for SEM. By using oil red O for LM and the selective fixatives for SEM, the serial sections can be compared to identify the fat and protein components.

Figure 1 shows this interrelationship of LM and SEM with a series of sections photographed from the same area. Concentrated areas of emulsified liver protein bordered by fat globules were identified with this comparative technique as illustrated in figure 1. These protein areas could not be distinguished from the fat using only osmium tetroxide for SEM evaluation. However, using serial sections and specific fixatives, both fat and protein components can be identified in the osmium tetroxide sections by comparison. The shape of the fat and protein components is not identical since each section was photographed 10 μm from the next section.

Figure 1a, a photo of liver sausage fixed in Zenker's solution shows the protein component left following the ethanol dehydration and pet ether rinse which removed the water and fat. A comparable section for SEM was fixed with buffered 1% osmium tetroxide (fig. 1c) in which all components are retained by the osmium fixative. The center section (fig. 1b) is a LM slide stained with oil red O and counterstained with Delafield's hematoxylin. The LM section is used as the standard for positive identification of the fat and protein components in the SEM sections.

Figure 2 is a representative series of sections taken from commercial liver sausage. Large protein areas can be observed in these sections which have been confirmed with the LM, 2b. Within this heavily concentrated area, fat components are also visible as illustrated in figure 2b and 2c. These fat areas appear to

Fat and Protein Components in Meat Emulsions

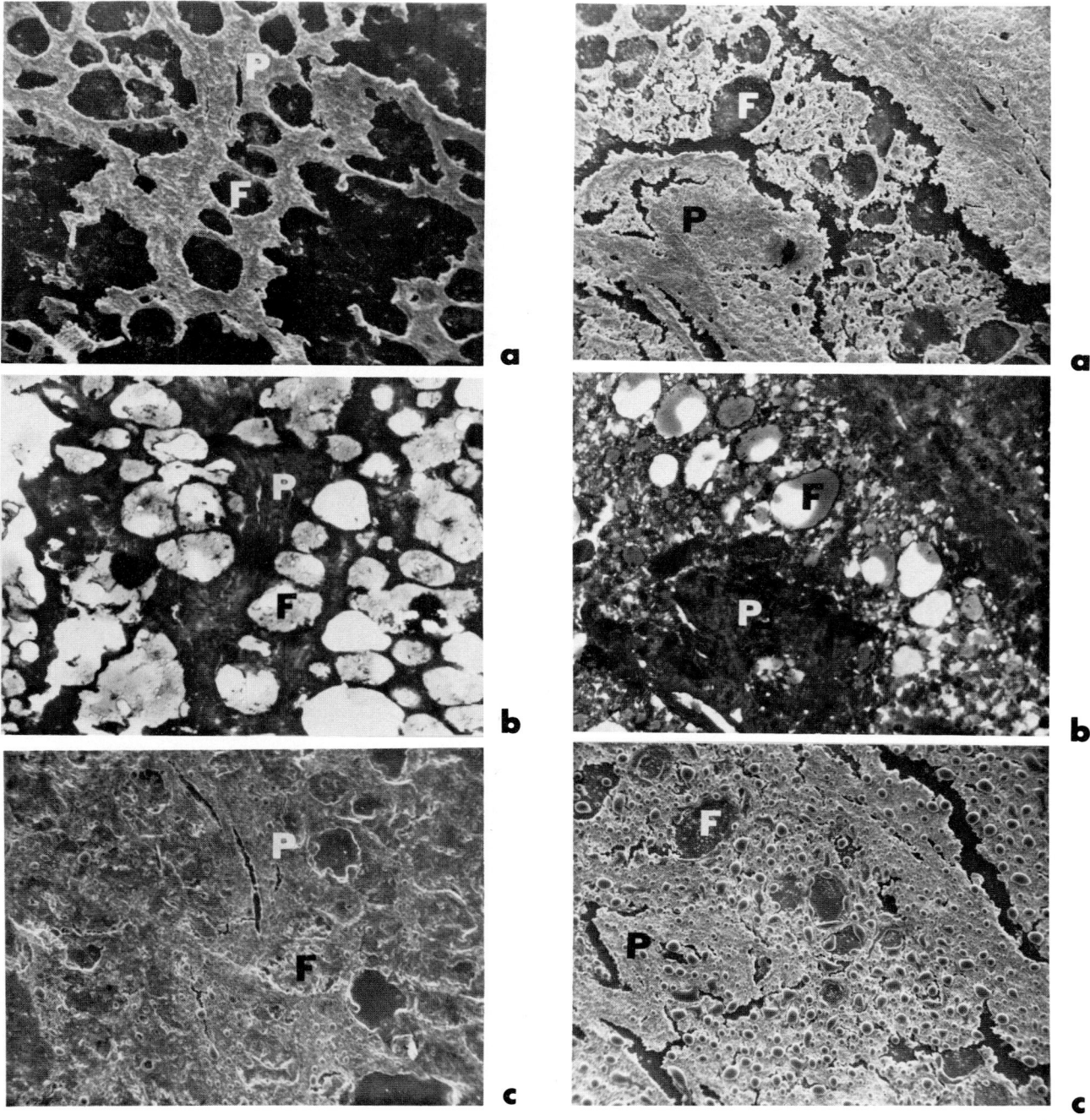

Fig. 1. Photomicrographs of serial sections of laboratory prepared liver sausage. (a) SEM micrograph, Zenker's fixed, F identifies where fat was present, PW=904 μm; (b) LM micrograph, oil red O stained, PW=786 μm; (c) SEM micrograph, Osmium tetroxide fixed, PW=912 μm. F and P identify common fat and protein components, respectively.

Fig. 2. Photomicrographs of serial sections of commercial liver sausage; (a) SEM micrograph, Zenker's fixed, F identifies where fat was present, PW=991 μm; (b) LM micrograph, oil red O stained, PW=950 μm; (c) SEM micrograph, Osmium tetroxide fixed, PW=991 μm. P and F identify common protein and fat components, respectively.

PW=Photo Width.

have a membrane or an absorbed layer of protein particles around them, similar to the micrographs shown by Borchert et al. [5]

In figure 2c, fat appears globular in shape on the surface of the osmium tetroxide fixed section. The fat globules are also visible in the oil red O and Delafield's hematoxylin sections, 3b which confirms these globular structures are fat. This idea is further confirmed in the Zenker's section by the absence of the globular structure, fig. 3a.

In addition to being globular, another characteristic of fat is that is appears shiny, fig. 3c. This shiny characteristic of fat could be of value in observing fat particles in fractured surfaces of freeze-fractured meat emulsion samples.

This comparative technique was also used on commercial wieners to identify fat and protein components in these products. Figure 4, a serial section of a wiener, shows comparable areas depicting the fat and protein arrangements. The fat and protein components showed similar identifying characteristics as previously mentioned in other emulsion products. The fat and protein arrangements did not change significantly over the 6-9 sequential sections. The reproducibility of the technique and the consecutive sections may not always be needed to compare areas.

Discussion

This procedure demonstrates that additional information can be obtained by similar areas through LM and SEM evaluations. The protein and fat components in these areas were distinguishable by comparing serial sections and using specific stains for either LM or SEM. Previous experience with SEM on freeze-fractured samples did not show components in meat emulsion samples. This comparative technique confirmed what was fat and what was protein in the osmium tetroxide fixed samples prepared for SEM. The common method for identifying emulsion components has been light histological procedures. However, more information was needed to identify fat and protein components in SEM micrographs. With this concept, serial sections of a sample were taken and fixed with specific fixatives for SEM and a third section was prepared for LM, which was used to confirm what was being observed in SEM micrograph.

Several similarities in structures were observed while comparing the three serial sections but there were some differences in exact shape of the components. That can be attributed to the sectioning of the samples due to a compression during sectioning which will change the shape of the particles present. The size of the particles also varied between serial sections due to size of the particle and how many times it was sectioned. However, with this comparative technique we have demonstrated that fat and protein components can be identified in osmium tetroxide sections prepared for SEM when compared to oil red O and Delafield's hematoxylin counterstained LM sections.

In addition, this technique helped characterize the surface texture of fat and protein

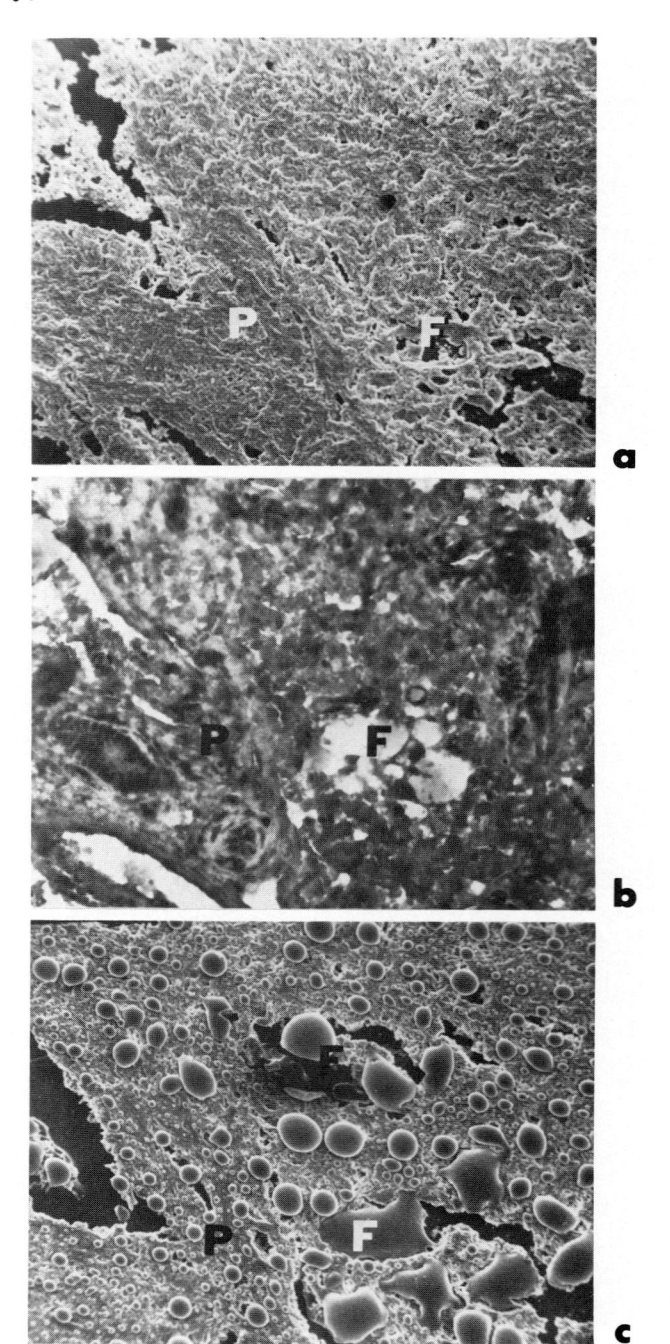

Fig. 3. Photomicrographs of serial sections of commercial liver sausage; (a) SEM micrograph, Zenker's fixed, F identifies where fat was present, PW=283 µm; (b) LM micrograph, oil red O, PW=356 µm; (c) SEM micrograph, Osmium tetroxide fixed, PW=283 µm. P and F identify common protein and fat components, respectively.

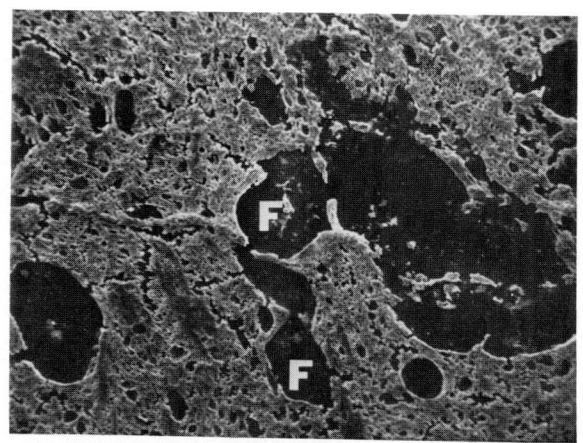

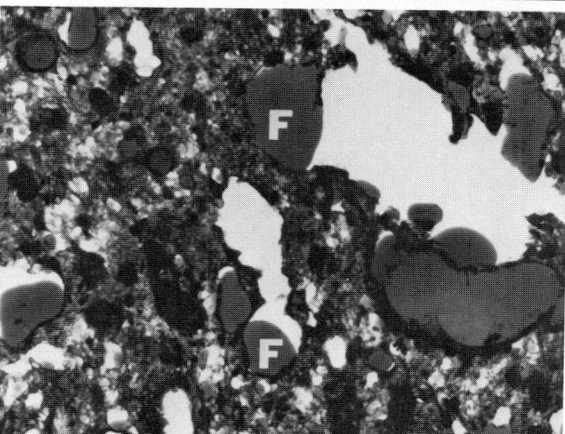

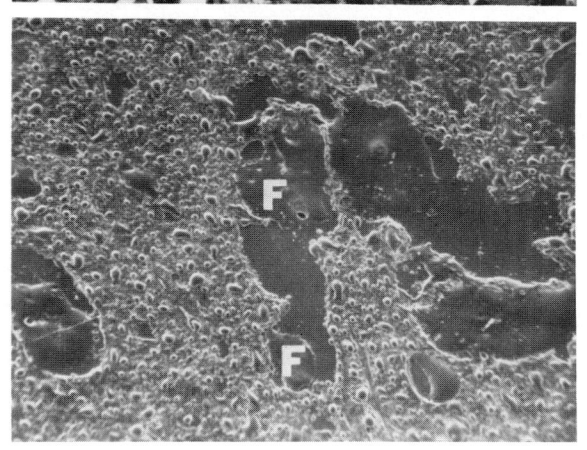

Fig. 4. Photomicrographs of serial sections of commercial wiener; (a) SEM micrograph, Zenker's Fixed, F identifies where fat was present, PW= 981 μm; (b) LM micrograph, oil red O stained, PW= 876 μm; (c) SEM micrograph, Osmium tetroxide fixed, PW=981 μm. In b and c, F identifies fat components.

components. This surface texture may be used in determining those components in a freeze-fractured surface. Furthermore, this technique may be used as a means of evaluating emulsion structures if it is further refined. Incorporation with other techniques such as transmission electron microscopy (TEM) could be beneficial while studying the ultrastructure of emulsions.

Scanning electron microscopy could be a good tool for examining emulsion structures, but at this point it cannot be used alone, since there is not enough information available to verify what is actually observed with SEM. Light microscopy must be used in conjunction with SEM to substantiate observations in the SEM micrographs. Further research is needed in SEM preparation methods and their effect on meat emulsion structures.

References

1. Hansen, L.J. Emulsion formation in finely comminuted sausage. Food Technol. 14, 1960. 565-569.
2. Froning, G.W., Andersen, J. and Mebus, C.A. Histological characteristics of turkey meat emulsions. Poul. Sci. 49, 1970. 497-503.
3. Cassens, R.G., Terrell, R.N. and Couch, C. The effect of textured soy flour particles on the microscopic morphology of frankfurters. A Research Note. J. Food Sci. 40, 1975, 1097-1098.
4. Cassens, R.G., Schmidt, R., Terrell, R. and Borchert, L.L. Microscopic structure of commercial sausage. Research Report R. 2878, Research Division, College of Agriculture, University of Wisconsin, Madison, WI. 1977.
5. Borchert, L.L., Greaser, M.L., Bard, J.C. Cassens, R.G. and Briskey, E.J. Electron Microscopy of a meat emulsion. A Research Note. J. Food Sci. 32, 1967, 419-421.
6. Theno, D.M., Siegel, D.G. and Schmidt, G.R. Meat massaging techniques. Proceedings of the Meat Industry Research Conf. Am. Meat Sci. Assoc. and Am. Meat Inst. Foundation, Chicago, IL., 1977, pp. 53-68.
7. Theno, D.M., Siegel, D.G. and Schmidt, G.R., Meat Massaging: Effects of salt and phosphate on the ultrastructure of cured procine muscle. J. Food Sci. 43, 1978, 488-492.
8. Theno, D.M. and Schmidt, R.G. Microstructural comparisons of three commercial frankfurters. J. Food Sci. 43, 1978, 845-848.
9. Humason, G.L. Animal Tissue Techniques, W.H. Freeman and Company, San Francisco, CA. 1962, pp. 253-254.
10. McManus, J.F.A. and Mowry, R.W. Staining Methods: Histologic and Histochemical, Hoeber Medical Divison, Harper and Row Publishing Co., New York, N.Y. 1963. pp. 10-19.

Discussion With Reviewers

T. R. Dutson: It appears that most of the information gained from use of these combined techniques could be obtained from the combined oil red O and Delafield's hematoxylin staining. Could the authors elaborate further on the advantages of using the three different techniques?
Authors: The advantage of using the three different techniques was to further substantiate what was being seen in SEM micrographs. Research in SEM evaluation of meat emulsion has been rather limited and to apply information from other meat research areas would be pure assumption. So the purpose of this project was to confirm what has

previously been seen.

T. R. Dutson: Would the authors please comment on what they feel the clear areas not containing fat or protein in the micrographs are and what techniques might be employed to determine their exact nature?
Authors: We feel the clear areas are air pockets that are induced by mixing of the product. We found these areas to be consistent from section to section which makes us believe it is due to the product and not from loss of material during fixative preparation. This was the reason for using oil red O and hematoxylin counterstain to see if we could not identify those areas better in SEM. Better vacuum mixing would be one way to find out if the void areas would be reduced.

R. D. Sullins: Do the authors have an opinion whether the open space or voids described are real or artifact caused during sectioning of unfixed samples?
Authors: We are of the opinion these areas are real and they are more than likely air pockets. We grant there may be artifacts induced from sectioning, but we do not believe they present any great limitations on the technique. Further refinement of technique and other investigative procedures need to be further developed to control possible artifacts.

R. G. Cassens: The authors indicate the technique they developed could be used as a means of evaluating emulsion structures. I ask that they explain not only how the technique could be used to evaluate emulsion structure but also what practical use such on evaluation would be to industry or consumer?
Authors: The limitation of the present technique is the magnification, but with further refinement of the technique and reducing chance of artifacts this magnification could be increased possibly to the level of TEM areas for further comparison. If structures can be verified in SEM micrographs at this level then SEM and TEM procedures could be used and alleviate light microscopy. As far as industry, we believe it has more value for basic research than practical use at this point. I have had researchers from the industry show an interest in this area. For the consumer this is of little importance, however, in the future SEM may help in understanding emulsion technology that would provide better processed meat products for the consumer.

R. D. Sullins: SEM responds to surface topography. How does sectioning and producing a smooth surface effect the structure and interpretation of your samples?
Authors: We realize there may be artifacts on the surface due to sectioning, but there were no facilities to get good repeatable freeze-fracturing. From the freeze-fractured surfaces observed, we could not positively identify what was present. Those problems are what brought about this technique for trying to confirm what was being observed. LM sections were used to interpret the results from SEM micrographs.

R. D. Sullins: What are the samples you are attempting to compare in this study and how do they differ in composition and physical characteristic?
Authors: The samples were cooked meat emulsions which differed mainly in meat ingredient. The products were liver sausage and wieners which differ in consistency, texture and stability properties.

R. D. Sullins: Would you describe the Zenker's solution and method for fixation of your samples with the Zenker's fluid?
Authors: Zenker's solution is a good general protein fixative. The fluid contains potassium dichromate, mercuric chloride, water and glacial acetic acid. The mercuric chloride coagulates the protein and preserves the connective tissue elements; the potassium dichromate hardens the tissue. The sections were mounted on slide cover slips and put in an enclosed container of Zenker's solution.

MEAT EMULSIONS - FINE STRUCTURE RELATIONSHIPS AND STABILITY

R. J. Carroll[*] and C. M. Lee[†]

[*] Eastern Regional Research Center, Agricultural Research, Science and Education Administration, USDA, 600 East Mermaid Lane, Philadelphia, PA 19118
[†] Department of Food Science and Nutrition, University of Rhode Island, Kingston, RI 02881

Abstract

Fine structure relationships to the thermal stability of meat emulsions were examined by light, scanning and transmission electron microscopy. Modifications of the procedures for fixation, dehydration, and embedding improved the preservation of lipid structure. Structural changes, as a result of increasing chopping temperatures, were evaluated by examining emulsions prepared at 16, 21, and 26°C and after being cooked. These changes were related to fat and water retention, shear and compression values. The lipid organization distinctly changed from oval globules to irregularly shaped structures accompanied with the formation of channels within the protein matrix.

KEY WORDS: Meat, emulsion, frankfurter, protein matrix, fat, stability, temperature, microscopy, structure, sample preparation

Introduction

Emulsion type meat products make up a significant share of processed meats for the consumer. Approximately one third of all sausage products are produced as frankfurters and in 1976 amounted to approximately 1.5 billion pounds (Agricultural Statistics 1977). Thus additional knowledge of the factors contributing to the stability and shelf life of these products would be advantageous for both the processor and the consumer.

The stability of meat emulsions has been evaluated by determining the extent of fat and water released when heated at 75°C (Swift et al. 1961). Factors known to affect the fat stabilization in the meat emulsion products include: protein solubility (Bard 1965), fat dispersion particle size (Borchet et al. 1967, Ackerman et al. 1971), chopping temperature (Ackerman et al. 1971, Schut 1976, Townsend et al. 1968), chopping temperature-time relationships (Townsend et al. 1968), melting properties of fat (Ackerman et al. 1971), consistency of protein gel (Townsend et al. 1971 and Bard 1965), and water binding (Hamm and Grabowski 1978).

The structure of emulsions in relation to stability was examined first by Hanson (1960) using light microscopy (LM) and later by Borchet et al.,(1967), employing transmission electron microscopy (TEM). They observed a dense membrane-like layer surrounding the fat globule, but it did not resemble a true bilayer membrane.

Recently, scanning electron microscopy (SEM) has been used to elucidate emulsion structure. Theno and Schmidt (1978) found differences in the protein matrix and fat globule distribution in commercial frankfurters. Ray et al.,(1979) also combined light and scanning electron microscopy to study emulsion type products.

We investigated the structure of all-beef emulsions using LM, SEM, and TEM observations correlated with measurements of physical properties of compression, shear, and water retention as a function of three different processing temperatures (16, 21, and 26°C). In addition, we improved preservation of the lipids by modifying existing techniques of fixation, dehydration and by use of a low viscosity embedding medium.

The application of electron microscopy to elucidate emulsion microstructure has probably been held back by lack of suitable sample preparation procedures. Hopefully, the techniques presented in this paper will aid in the use of electron microscopy to elucidate lipid-protein structures in food products.

Materials and Methods

Beef emulsions obtained either commercially or prepared in our laboratory were used in this study. Laboratory emulsions were prepared in a Fleetwood* food chopper. No vacuum was used. Ground lean beef (600 g) was blended with 262 ml of prechilled 10% brine and 35 ml of 50% corn syrup for 15 min to solubilize the protein. Each batter was chopped for 15 min after the addition of ground beef trimming fat (300 g). Fat content of the batter was 26% on the average. Temperatures of the batter were varied to 16, 21, and 26°C by altering either the ingredient or ambient temperature. The ingredients for the batter were not frozen. The prepared batter was stuffed into cellulose casings (23 mm diameter) and cooked in a smokehouse so that the internal temperature rose to 66°C within 30 min and held there for another 30 min.

Light Microscopy

For light microscopy, specimens were frozen in liquid nitrogen, sectioned at 16 µm using a microtome-cryostat held at -30°C, fixed in vapors of 25% glutaraldehyde, and stained with oil Red-O for the lipids followed with hematoxylin for the protein matrix, and examined in a Leitz microscope.

Electron Microscopy

Conventional sample preparation procedures for electron microscopy usually employ a primary fixation with glutaraldehyde, post fixation in OsO_4, dehydration through increasing ethyl alcohol series, and embedding in epon 812 resin. Results obtained from emulsions prepared by these procedures were poor; the fat globules gave a nonuniform appearance and many holes in the sections were found. An improved method of sample preparation was required. The method which gives better uniform preservation of the fat globule is as follows.

Samples (1 x 1½ mm) from the frankfurters (pH = ~5.9) were fixed in 1% OsO_4 in 0.07 M phosphate buffer at pH 5.9 for 4 hr. The samples were washed four times in the phosphate buffer at pH 5.9. The secondary fixation was carried out overnight in 3% glutaraldehyde - 1% paraformaldehyde in the same buffer. The samples were washed four times in water and rapidly dehydrated (5 min each) in increasing concentrations of acetone - H_2O mixture [50, 70, 80, 95, and 100% (3 times)]. The samples were prepared both for the SEM and the TEM.

For TEM, the samples were placed in 50% acetone - Spurr resin (Spurr 1969), followed by two changes of fresh resin, and placed in capsules to be cured overnight at 65°C. Sections were cut on an LKB-8800 ultramicrotome and stained 5 min each with uranyl acetate and lead citrate. A Zeiss 10-B TEM operating at 60 kV was used to examine the sections.

The samples for the SEM were cryofractured (Humphreys et al. 1974) in acetone at liquid nitrogen temperature and critical-point dried from carbon dioxide in a Denton DCP-1 unit. The samples were sputter-coated with gold in a Denton DSM-5 unit and examined with a JEOL 50-A SEM operating at 15 kV.

The amount of fat and water released from heated emulsions in a water bath at 60°C for 30 min was used as a stability index. This test was a modification of the method reported by Townsend et al., 1968. The internal temperature of batter at which such release commenced was measured by monitoring internal temperature changes of the batter which had been placed in a water jacketed funnel attached to a graduated cylinder. The temperature of the water jacketed funnel was maintained at 60°C by circulating water.

The mechanical strength of the cooked emulsion products was determined by compression and shear test using a straight blade (1 mm thick) with an Instron Universal testing machine (Model 1122). Cylindrical specimens (five replications) of uniform geometry (20 mm height and 20 mm diameter) were used for both tests at a deformation rate of 50 mm/min and at 1:2 ratio of chart speed to crosshead speed. The data were analyzed by an analysis of variance including emulsion treatment contrast.

Results

Frankfurters made from the emulsion prepared at 26°C showed visual evidence of liquids on the surface of the casings. In addition, these frankfurters had a mushy texture and were difficult to cut precisely with a razor blade.

LM photomicrographs of meat emulsions processed at 16, 21, and 26°C are shown in Fig. 1. At 16°C, Fig. 1A, a homogeneous mixture of lipid droplets surrounded by the protein matrix was observed. The lipid droplets ranged in size from > 100 µm down to the limits of resolution of the light microscope. As processing temperature was increased to 21°C, Fig. 1B, the fat droplets tended to be irregularly shaped. Some voids or air pockets were seen and the protein matrix was less homogeneous. At 26°C, Fig. 1C, the air pockets increased in size, some forming channels through the protein matrix; the lipid droplets tended to coalesce, indicating melting of the fat. The channels which have developed may have allowed egress of the lipids to the outside casing with release to the exterior of the frankfurter.

SEM photomicrographs of the same emulsions, fixed in osmium tetroxide and glutaraldehyde-formaldehyde are presented in Fig. 2. Emulsion at 16°C, Fig. 2A, showed the fractured interior surface of the protein matrix surrounding various size fat droplets, some of which have been lost during sample preparation. At 21°C, Fig. 2B, evidence for the beginning of lipid coalescence was observed. Irregular shaped fat droplets of various sizes were found. The protein matrix in this emulsion was less homogeneous than that in

---
*Reference to brand or firm name does not constitute endorsement by the U. S. Department of Agriculture over others of a similar nature.

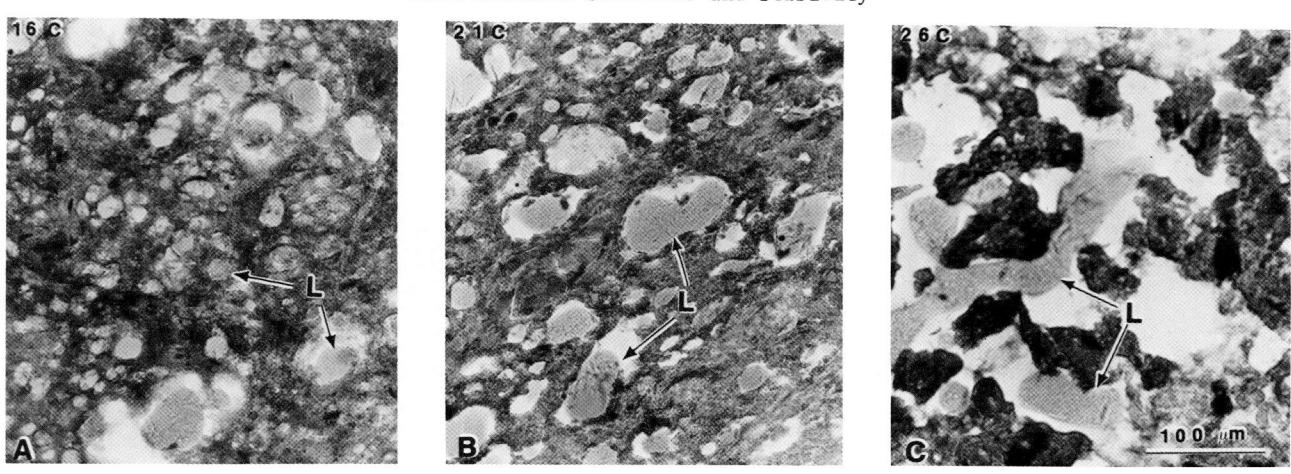

Fig. 1 LM micrographs of emulsions prepared at (A) 16°C, (B) 21°C, and (C) 26°C. L = lipid.
Note: Increased open space in (C) where channels have formed.

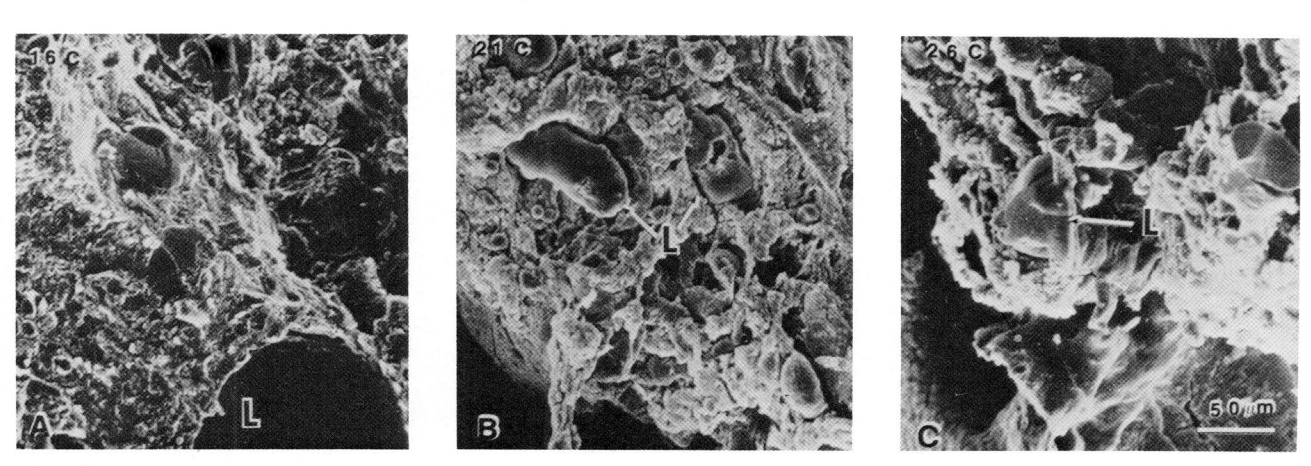

Fig. 2 SEM micrographs of emulsions prepared at (A) 16°C, (B) 21°C, and (C) 26°C. L = lipid.
Note: In (A) lower right a void results from lipid removal during fracture.

the emulsion processed at 16°C. The emulsion processed at 26°C, in Fig. 2C, showed a change in the protein matrix as well as larger fat droplets. Little or nothing can be discerned in the SEM regarding the protein-lipid interface even at higher magnifications.

Alternate embedding procedures explored to improve retention of the fat globule structures involved water miscible resins, bovine serum albumin, and gelatin. All of these procedures resulted in either poor penetration of embedding mediums or the inability to obtain good uniform sections on the ultramicrotome. Conventional fixation, dehydration and epon embedding of commercial frankfurters resulted in the extraction of the lipid droplets and development of large holes as seen in Fig. 3A. Fig. 3B shows the improved lipid stabilization obtained with the revised procedure; the lipid is in close contact with the electron dense peripheral layer indicating minimal lipid extraction. In order to preserve the lipid moiety, the initial fixation was performed in osmium tetroxide, followed by aldehyde fixation, in phosphate buffers at the pH of the specific emulsion sample. Rapid acetone dehydration at 0°C was followed by embedding in Spurr low viscosity resin.

TEM micrographs of thin sections of the frankfurter emulsions prepared at 16, 21, and 26°C are shown in Fig. 4. The protein matrix, Fig. 4A, consisted of disrupted muscle tissue components around the lipid droplets. The lipid droplets were limited by an electron dense membrane similar to that reported by Borchet et al. 1967. This membrane surrounded the lipids and was in partial contact with the protein matrix of the emulsion. Fat globule sizes varied with some being as small as 0.1 - 0.3 μm.

When the processing temperature was raised to 21°C, the fat globules tended to change to irregular shapes (Fig. 4B). The protein matrix was similar to that of the 16°C processed emulsion. In contrast, emulsions prepared at 26°C showed a drastic change in fat globule size and shape as seen in Fig. 4C. The fat globules increased in size, formed irregular shapes and

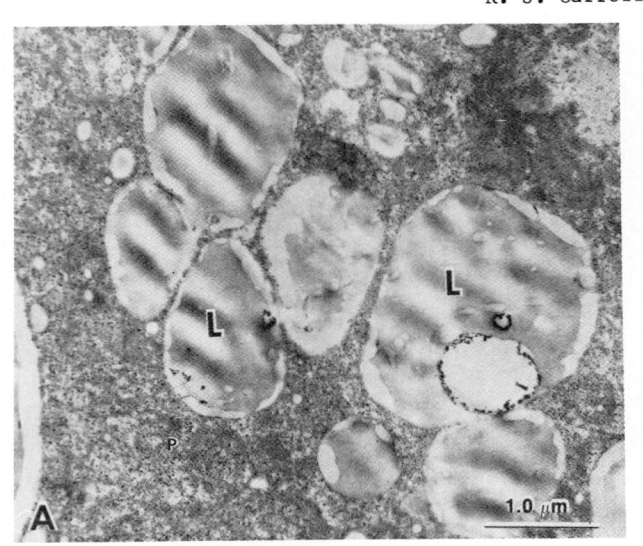

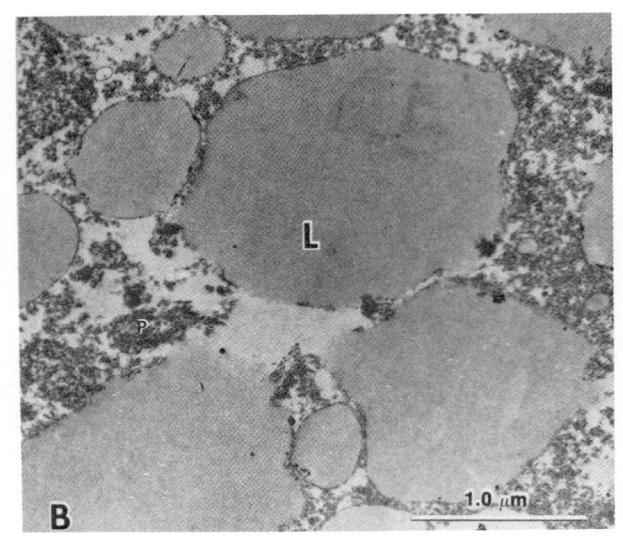

Fig. 3 TEM micrographs of commercial frankfurters: (A) prepared in buffered glutaraldehyde-paraformaldehyde, post fixed in OsO4, dehydrated in ethyl alcohol and embedded in epon 812; and (B) fixed in buffered OsO4 - post fixed in glutaraldehyde-paraformaldehyde, dehydrated in cold acetone, embedded in Spurr resin. L = lipid. P = protein.
Note: Much better preservation of lipid structure in (B) than in (A).

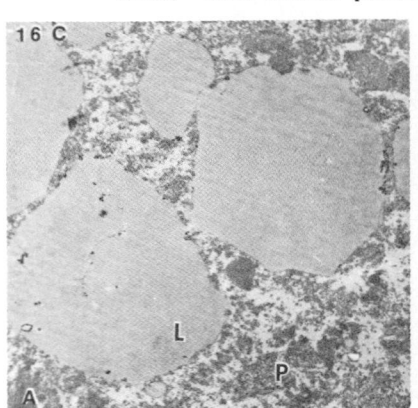

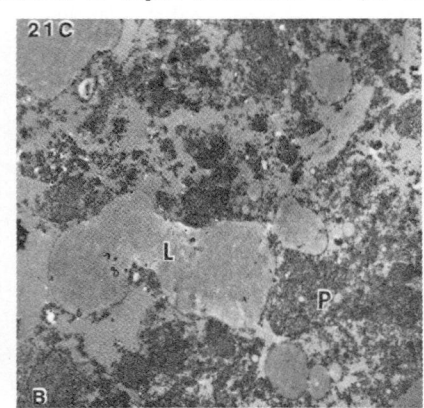

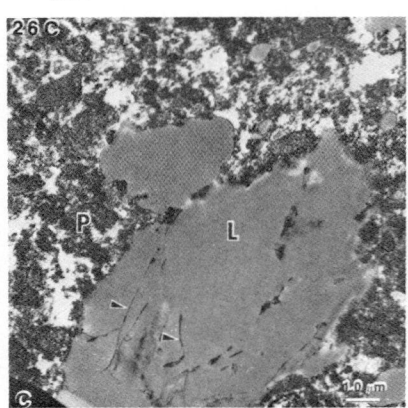

Fig. 4 TEM micrographs of emulsion prepared at (A) 16°C, (B) 21°C, and (C) 26°C. Coalescence of lipids is evident in (B) and (C). L = lipid. P = protein. Arrowheads = electron dense material.

incorporated electron dense membrane-like material in the interior of the fat globule (arrowheads). Not all the fat globules were affected, since some small globules were still observed.

These observations correlated well with results of physical tests of shear force, compression and water retention of the emulsions prepared at increasing processing temperatures. Shear force, Table 1, significantly (p <0.05) decreased with increasing processing temperature. The shear force required to shear emulsions prepared at 26°C was 65% of that required for those prepared at 16°C. Compression force required to compress emulsions was 76% less (significant, p <0.05) for those prepared at 26°C than for those prepared at 16°C, indicating a soft and mushy texture.

Emulsions prepared at increasing temperatures decreased in the ability to hold water and fat in suspension at 60°C, Table 2. Emulsions prepared at 26°C retained 22% less water and 37% less fat than emulsions prepared at 16°C (p <0.05).

These physical measurements reflect a decrease in emulsion stability with increasing processing temperature, and correlate with the formation of channels observed in the LM and the changes in the appearance of the fat globules found by TEM. Formation of the channels observed in the emulsion prepared at 26°C suggests a possible mechanism for release of water and fat from the protein matrix.

Summary

Observations with light microscopy and transmission electron microscopy showed definite changes in emulsion structure with increasing processing temperature. The formation of fat channels, accompanied by changes in size and shape of the fat globules, is observed. Decreases in shear strength, compression, and water and fat retention ability are correlated with structure changes in emulsions prepared at increasing temperatures. This study reflects definitive changes occurring in emulsion morphology with decreases in emulsion stability as a function of processing temperature.

An improved procedure for preparing emulsion samples for TEM observations is presented. This procedure preserves the fat globules in the emulsion with minimum extraction of the lipids.

Acknowledgment

The authors gratefully acknowledge the statistical assistance furnished by Dr. John G. Phillips. This work was performed at Drexel University, Philadelphia, PA.

Table 1
Mechanical Properties of Emulsions

Emulsion temp °(C)	Shear force (kg)	Compression force (kg)
16	0.60 ± 0.07	4.6 ± 0.25
21	0.57 ± 0.04	2.3 ± 0.10*
26	0.39 ± 0.04*	1.1 ± 0.21*

*Significantly different from emulsion prepared at 16°C ($p < 0.05$).

Table 2
Retention by Emulsion at 60°C

Emulsion temp °(C)	Water %	Fat %
16	97.63 ± 3.76	98.75 ± 2.65
21	86.04 ± 5.27*	72.50 ± 6.21*
26	75.93 ± 6.11*	62.50 ± 8.17*

*Significantly different from emulsion prepared at 16°C ($p < 0.05$).

References

Ackerman SA, Swift CE, Carroll RJ and Townsend WE. Effects of types of fat and rates and temperatures of comminution on dispersion of lipids in frankfurters. J Food Sci 36 1971 262-266.

Agricultural Statistics USDA 1977. US Government Printing Office Washington DC Catalog No A1.47: 977.

Bard JC. Some factors influencing extractability of salt soluble proteins. In "Proc Meat Ind Conf" Am Meat Inst Foundation, Chicago, IL 1965 96-98.

Borchet LL, Greaser ML, Bard JC, Cassen RG and Briskey EJ. Electron microscopy of a meat emulsion. J Food Sci 32 1967 419-421.

Hamm R and Grabowski J. Protein solubility and water binding under conditions present in brühwurst products, Part II: Effect of heating on the dissolved proteins. Fleischwirtschaft 58 1978 1345-1347.

Hanson LJ. Emulsion formation in finely comminuted sausage. Food Technol 14 1960 565-569.

Humphreys WJ, Spurlock BO and Johnson JS. Critical point drying of ethanol-infiltrated cryofractured biological specimens for scanning electron microscopy. Scanning Electron Microsc 1974 275-280.

Ray FK, Miller BG, Van Sickle DC, Aberle ED, Forrest JC and Judge MD. Identification of fat and protein components in meat emulsions. Scanning Electron Microsc 1979 III 473-478.

Schut J. Meat Emulsions. S. Friberg (ed) Marcel Dekker New York NY 1976 Chapter 8 Food Emulsions.

Spurr AR. A low viscosity epoxy resin embedding medium for electron microscopy. J. Ultrastruct Res 26 1969 31-43.

Swift CE, Lockett C and Fryar AJ. Comminuted meat emulsions - The capacity of meats for emulsifying fat. Food Technol 15 1961 468-473.

Theno DM and Schmidt GR. Microstructural comparisons of three commercial frankfurters. J Food Sci 43 1978 845-848.

Townsend WE, Witnauer LP, Riloff JA and Swift CE. Comminuted meat emulsions: Differential thermal analysis of fat transitions. Food Technol 22 1968 71-75.

Townsend WE, Ackerman SA, Witnauer LP, Palm WE and Swift CE. Effects of types and levels of fat and rates and temperatures of comminution on the processing and characteristics of frankfurters. J Food Sci 36 1971 216-265.

Discussion with Reviewers

F.K. Ray: Do the authors have an opinion what the open space or clear areas not containing fat or proteins are in the micrograph, Figure 1?

Authors: The irregular clear areas represent voids in the emulsion resulting from trapped air or water. Some oval areas result from loss of fat during sectioning or staining. When the stained sections are viewed under the light microscope, it is much easier to discern the empty fat globules from the water or air voids in the emulsion.

D.N.Holcomb: Did you find uniform osmium staining on the meat emulsions?
F.K.Ray: Did you find any penetration problems with OsO4?
Authors: Yes, problems with osmium tetroxide penetration of the specimen blocks were encountered initially. This was overcome by trimming the blocks as small as possible, no more than 1-1.5 mm on a side. Usually four hours fixation time resulted in complete penetration of the osmium tetroxide.

D.N.Holcomb: In many liquid to semi-solid oil-in-water emulsions, the oil droplets assume spherical shapes. Are the fats spherical in the batter, then elongate after stuffing and working?
Authors: The oil droplets in oil-in-water type emulsions are bound by absorbed proteins forming an interface which stabilize the spherical shapes. In contrast, the fat globules in a meat emulsion are confined within a continuous protein matrix. When uncooked emulsions prepared at increasing temperatures were observed in the microscope, the fat globules tended to change from spherical to irregular and coalesced shapes prior to stuffing the emulsion in the casing. The change in shape of the fat globule results from a weakening of the protein matrix with increasing processing temperature.

M.A.Christman: What causes the striations across the lipid droplets in Figures 3A and 4A? Is this just the difference in density and/or hardness such that the fat is softer and the knife chatters when it contacts the fat?
Authors: The striations are not the result of knife chatter. The knife passes through the resin embedded tissue and meets the much softer lipid which tends to flow upon impact, causing a build-up of thick and thin areas within the fat globule. Chatter would extend across the entire section perpendicular to the direction of the block travel.

SCANNING ELECTRON MICROSCOPY OF DAIRY PRODUCTS: AN OVERVIEW

M. Kalab

Food Research Institute
Research Branch
Agriculture Canada
Ottawa
Ontario, Canada K1A OC6

Abstract

The use of scanning electron microscopy in dairy research is reviewed. Attention has been paid to dairy products based on protein, such as skim milk and whole milk powders, instant skim milk powder, dried whey protein, Cottage cheese, and cheeses. Current unpublished research results obtained with instant skim milk powder, defective yoghurt, and a hypothetical skin on Cottage cheese granules are also presented.

KEY WORDS: Dairy Products Microstructure, Milk Powder, Whey Protein, Yoghurt, Cottage Cheese, Cheese

Introduction

In general, dairy products may be divided into two groups based on their composition, one consisting mainly of milk proteins, that is, casein and whey proteins, and the other consisting mainly of fat. Only the former group (dried milk, yoghurt, Cottage cheese etc.) may be examined by regular scanning electron microscopy (SEM), whereas a special cold stage attachment would be required in SEM studies of the latter group (cream, ice cream, dairy spreads, butter etc.). Most cheeses are composed of both protein and fat, and hence require special preparative procedures depending on the subject of interest. Other electron microscopical methods such as thin-sectioning, negative staining, and freeze-fracturing are invaluable in supporting SEM findings, yet they will be not discussed in this overview as they are related to transmission electron microscopy and were reviewed elsewhere[1,2].

In this overview attention will be paid to the microstructure of dried milk, dried whey protein, yoghurt, Cottage cheese, and cheese as visualized by SEM and to the implications of these findings on the properties of the products.

Dried Milk

Most industrial milk and skim milk destined for drying is subjected to spray-drying. Other methods such as drum-drying and freeze-drying are being used on a considerably smaller scale and are almost non-existent in some countries.

SEM was first used to examine spray-dried milk by J.H. Prentice in 1970 in England[3] but the first micrographs were published by T.J. Buma and S. Henstra[3] in the Netherlands. The milk powder particles were attached to SEM stubs coated with an adhesive; loose particles were removed by a gentle stream of air, and the firmly adhering particles were coated with carbon and gold by vacuum evaporation. This treatment made it possible to eliminate charging effects almost completely. However, even properly mounted and coated dry milk particles are susceptible to producing artifacts as they crack easily during electron microscopical examination. T.J. Buma and S. Henstra[3] preferred to use low magnifications and to keep the observation time as short as possible.

In his series of papers, T.J. Buma studied[4]

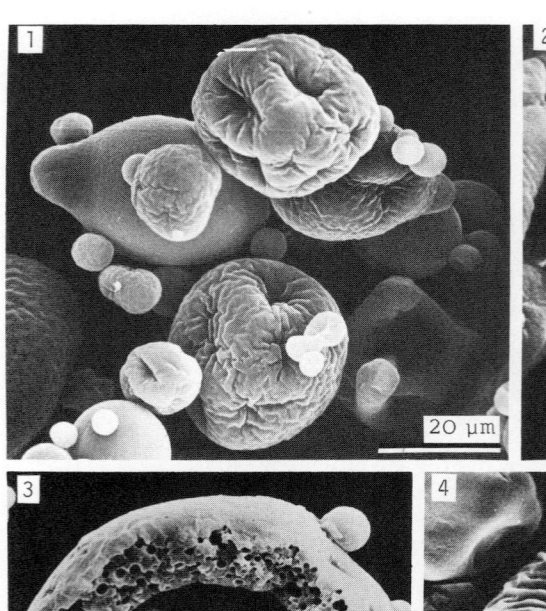

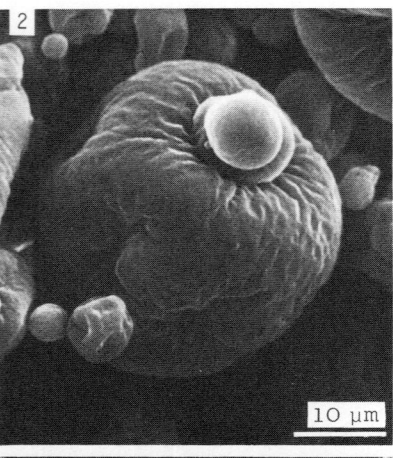

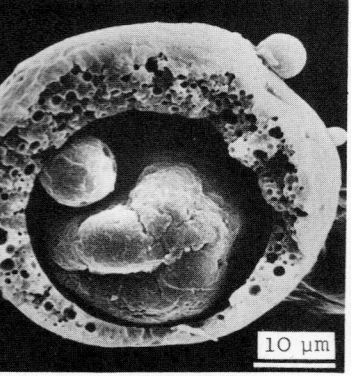

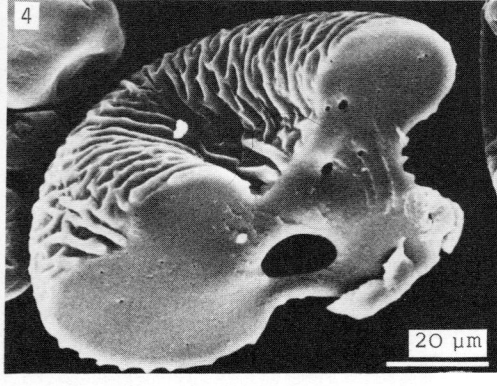

Fig. 1.
Spray-dried skim milk powder composed of smooth and wrinkled particles.

Fig. 2.
"Apple-like structure" of a globular particle in spray-dried skim milk.

Fig. 3.
Pores in a spray-dried whole milk particle; the central vacuole in the particle contains two trapped milk globules. (Reproduced by permission of Dr.T.J. Buma and the Netherlands Milk and Dairy Journal).

Fig. 4.
Compact interior of a skim milk particle displaying deep surface folds. (Reproduced by permission of Dr.T.J. Buma and the Netherlands Milk and Dairy Journal).

both the external appearance and the internal microstructure of whole milk and skim milk powder particles. Skim milk powder particles were round with different surface structures. Some particles were smooth, but most of them were severely wrinkled with deep surface folds (Fig. 1). There was a high proportion of particles with an "apple-like structure" (Fig. 2) which was caused, T.J. Buma believed, by an implosion during the last stage of the drying process, or during the cooling of particles which contained a relatively large vacuole.

To study the reason for the formation of the deep surface folds in spray-dried milk particles, T.J. Buma and S. Henstra[5] spray-dried calcium caseinate and lactose solutions under the same conditions under which they spray-dried whole milk and skim milk. They also examined spray-dried whey[3]. The authors found the deep surface folds only when casein was present in the spray-dried material.

In these studies fragmentation with a razor blade was used to reveal the internal structure of spray-dried milk particles. Almost every one of them contained a large vacuole, in which smaller globular particles were trapped. In whole spray-dried milk, the body of the particle was porous (Fig. 3), whereas in skim milk most particles were compact (Fig. 4). Particles prepared from unhomogenized concentrated whole milk were more porous than particles made from homogenized concentrated milk. In powders of the same origin, small particles were more porous than large particles. High porosity was usually associated with the occurrence of cracks and capillaries in the particles. SEM and extraction with organic solvents, also involving model experiments with butterfat dispersed in concentrated lactose solutions, led to the design of a model of fat distribution in spray-dried milk (Fig. 5). T.J. Buma concluded[4] that extractable fat consists, in principle, of four components:

f_s = Surface fat present in the form of pools or patches on the powder particle, particularly in surface folds and contact points between particles;

f_l = Outer layer fat consisting of fat globules in the surface layer of the powder particles which can be reached directly by fat solvents;

f_c = Capillary fat consisting of fat globules inside the powder particles which can be reached by fat solvents through capillary pores or cracks;

f_d = Dissolution or "second echelon" fat consisting of fat globules inside the powder particles which can be reached by fat solvents through the holes left by dissolved fat globules in the outer particle layer, or close to wide capillaries in the powder particles.

T.J. Buma also studied porosity[6] in relation to the penetration of gases under pressure into spray-dried milk particles. Using SEM and pycnometry the author concluded that there were two mechanisms causing particle porosity:
(a) rapid drying due to large temperature gradients and

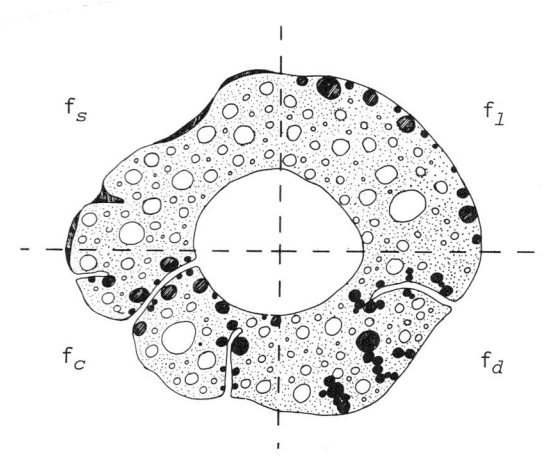

Fig. 5. Schematic model of four forms of extractable fat in a spray-dried whole milk particle with a central vacuole. Dark areas represent extractable fat. (Reproduced by permission of Dr. T.J. Buma and the Netherlands Milk and Dairy Journal).

(b) mechanical damage of the particles, for example during separation in cyclones or pneumatic conveying.

The milk powder is difficult to dissolve in the form in which it leaves the spray-dryer because of frothing and lumping. To avoid these problems, so-called "instant" milk powder is being manufactured. The principle of "instantizing" is the agglomeration of individual globular milk particles into clusters and the conversion of lactose from the glass form into a microcrystalline form, which makes the powder more wettable and less hygroscopic. This is achieved usually by wetting air-suspended powder with steam, atomized water, or a mixture of both, agglomeration of the sticky particles, and re-drying with hot air[7].

Two commercial samples of instant milk powders were examined by SEM in the author's laboratory and the micrographs are presented in Fig. 6 to 9.

Particles of one brand were smooth and consisted of fused primary globular particles (Fig. 6). At higher magnifications very small lactose crystals were arranged in the forms of rosettes on the particle surface (Fig. 7).

Instant milk particles of the other brand appeared at low magnifications to have rough surfaces (Fig. 8). Higher magnifications showed the surfaces to be covered with relatively large lactose crystals (Fig. 9). This study, aimed at correlating the appearance of the instant milk particles with the exact manufacturing processes, has not yet been concluded.

Similar to the initial skim milk powder they had been made from, the instant skim milk powder particles were hollow and the crust was compact.

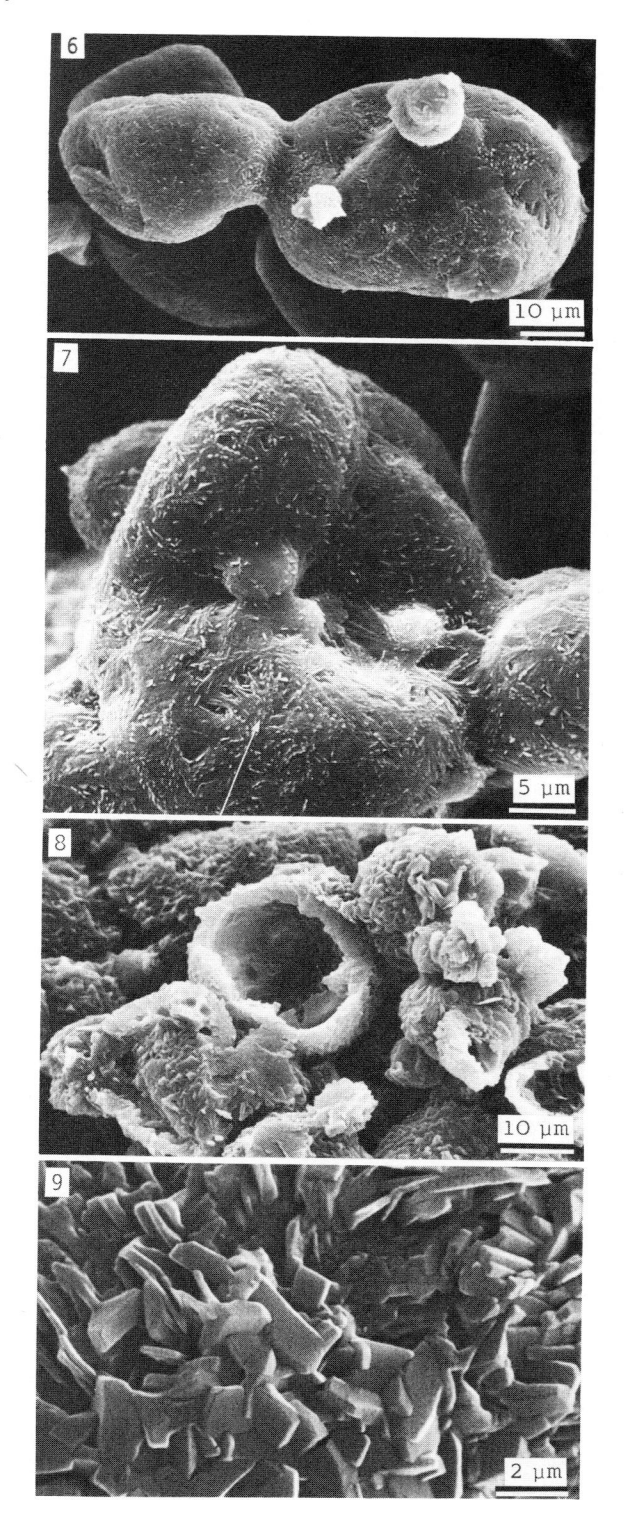

Fig. 6. Instant skim milk powder. Two particles are fused to each other by the instantizing process.
Fig. 7. Detail of a smooth particle cluster in instant skim milk powder. Lactose crystals are arranged in the forms of rosettes (arrow) on the particle surface.
Fig. 8. Instant skim milk powder of another brand. Particle surfaces are covered with relatively large lactose crystals.
Fig. 9. Detail of lactose crystals on the surface of an instant skim milk particle.

The surface of the instant milk particles is very fragile and T.J. Buma's remark on the sensitivity of spray-dried milk to electron beam damage also applies to instant milk powders.

Moisture relations in spray-dried skim milk were studied by Sylvia Warburton and S.W. Pixton[8]. As there is over 50% of lactose in dried skim milk, lactose has a significant effect on the moisture relations in the milk powder. When a lactose solution is dried very rapidly, the viscosity increases too rapidly for crystallization to take place, and an amorphous glass mixture of α- and β-lactose is produced. Slow drying leads to the formation of a crystalline α-hydrate at low temperature, whereas β-anhydride crystals are formed above 93.5°C. The shapes of the α-hydrate crystals depend on the conditions of drying. Prism-shaped crystals are formed at a high precipitation pressure and at rapid crystallization. With the decreasing pressure the crystal forms change to diamonds, pyramids, tomahawks, and 13-sided crystals[9].

Prisms and tomahawks are the most common lactose forms found in dried milk. SEM examination of dried skim milk less than a week old by S. Warburton and S.W. Pixton[8] revealed the presence of small prism-shaped crystals ranging from 5 to 30 μm in length. These crystals were inside the milk particles and the authors explained that this was probably the reason why their presence had not been noticed by earlier researchers.

In skim milk powder stored for 1 to 3 years, tomahawk crystals ranging from 60 to 170 μm in length were observed. The tomahawk crystals have thus been viewed as a possible sign of age of the milk powder although they may also be found in high-moisture content samples.

In dried milk temporarily exposed to high equilibrium humidity, no prism-shaped lactose crystals were seen. S. Warburton and S.W. Pixton concluded that freshly dried skim milk contained lactose glass and small prism-shaped α-hydrate lactose crystals inside the smooth-surface particles. This structure has not changed if the glass has not crystallized, although tomahawk-shaped α-hydrate crystals developed after 1 to 3 years and the number of prism-shaped crystals was substantially reduced. When the glass crystallizes, a coating of crystals forms on the outside of the particles and the prism-shaped crystals disappear. Tomahawk-shaped crystals appear soon after the crystallization of the glass, particularly at a very high moisture content, and they persist even when the moisture content is reduced.

Microstructure of skim milk powders in relation to the drying procedure was studied by M. Kalab and D.B. Emmons[10]. Spray-dried, drum-dried, and freeze-dried skim milk was examined by both SEM and thin-section electron microscopy (TEM). Only spray-drying preserved the casein micelles in a way which made it possible to reconstitute the skim milk without severely changing its properties compared with fluid skim milk. Heat-induced milk gels made from the reconstituted skim milk were composed of globular casein micelles[10]. Drum-drying produced flakes (Fig. 10), in which casein micelles were aggregated in clusters as was shown by TEM. This skim milk powder produced a sediment after dispersion in water. Freezing of skim milk for freeze-drying was accomplished either rapidly in liquid nitrogen (-195.8°C), or slowly at -23°C. Flakes obtained by freeze-drying indicated the development of ice crystals during freezing, particularly at -23°C (Fig. 11). The development of ice crystals was also evident by TEM, which demonstrated the presence of displaced and distorted casein micelles. However, the freeze-dried skim milk powder dissolved in water without leaving any sediment and produced firm heat-induced milk gels, which consisted of fused casein micelles[10]. When unheated skim milk, reconstituted from freeze-dried powder, was used to make yoghurt, extremely large casein micelle clusters (1.1 μm in diameter) were found[11] in contrast to considerably smaller micelle dimensions in yoghurt made from fresh skim milk (see Yoghurt).

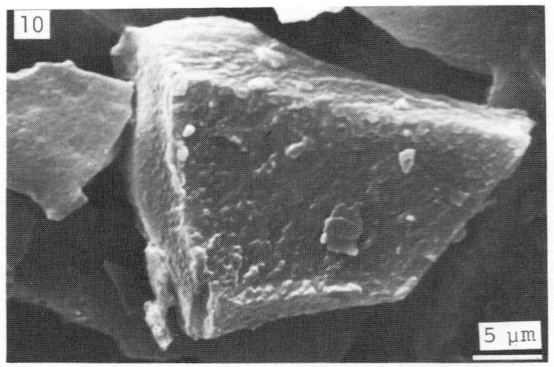

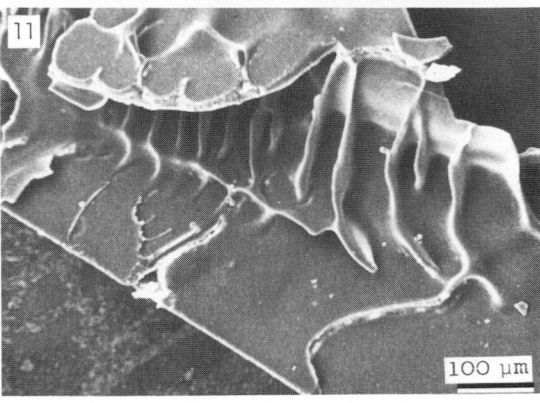

Fig. 10. Drum-dried skim milk powder. Fragment of a flake displays compacted casein micelles.

Fig. 11. Freeze-dried skim milk powder. Parallel partitions were formed as the result of ice crystal development during slow freezing at -23°C.

Dried Whey Protein

Dried whey protein is another dairy product, the microstructure of which has been studied by electron microscopy[12,13]. Considerable effort has recently been focused on the recovery of protein from whey. Heat denaturation, whereby the protein is coagulated, is one of several possibilities of such a recovery. The whey protein curd has a high water-holding capacity (WHC) for which it is valued in the food industry. However, because of the high water content, the curd is heavy and has

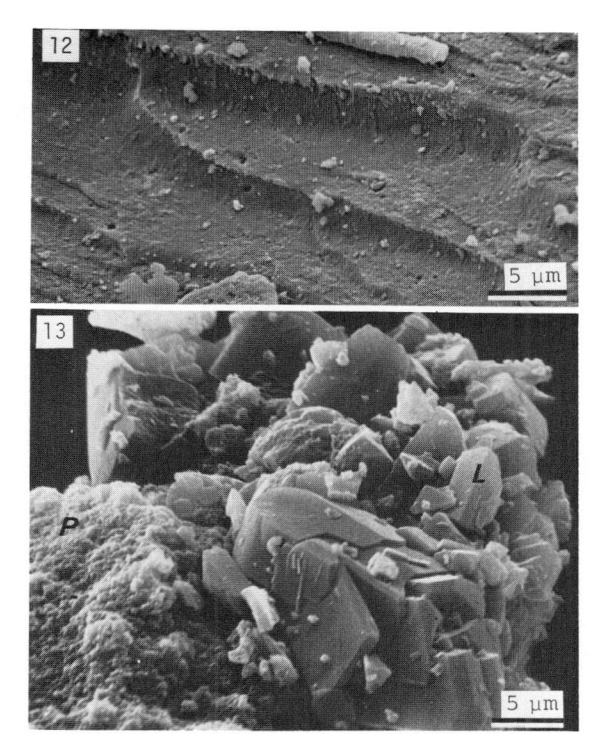

Fig. 12. Compact surface of an air-dried particle of heat-coagulated whey curd.

Fig. 13. Drum-dried heat-coagulated whey curd. P = Whey protein crust; L = Lactose crystals.

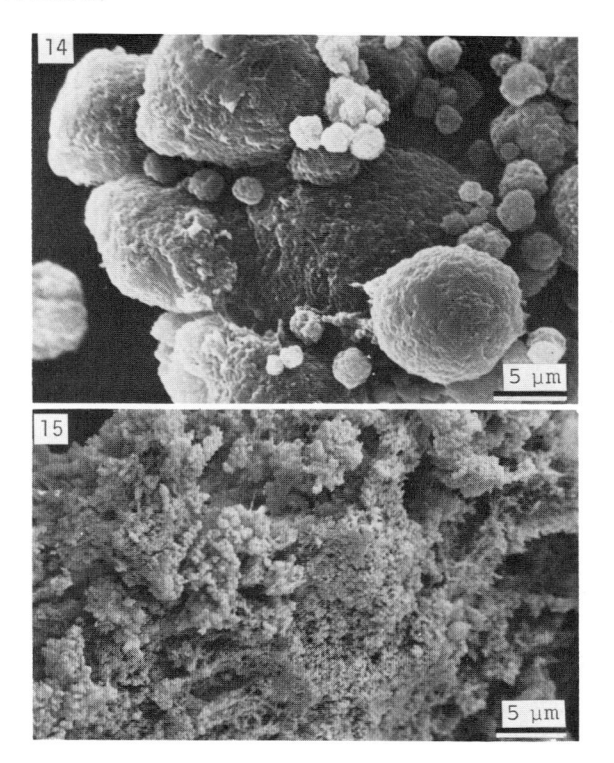

Fig. 14. Spray-dried heat-coagulated whey curd. TEM reveals that the surface of the larger globules in this micrograph is formed by a crust approximately 5 μm thick; smaller globules are compact.

Fig. 15. Freeze-dried heat-coagulated whey curd. Particles are porous as the result of the protein matrix immobilization and ice crystal formation during freezing.

a limited shelf life. It has to be dried to become marketable. J.L. Short et al.[12] showed by SEM the differences in particle geometry of dried lactalbumin as related to spray-, roller, and fluid-bed-drying. P. Jelen et al.[13] studied the WHC of dried heat-coagulated whey protein in relation to various drying processes. In contrast to whole or skimmed milk, whey protein curd was used either in the form of a slurry for spray- and drum-drying, or was pressed to remove excessive water for freeze- and air-drying. SEM revealed a compact and smooth surface of air-dried particles (Fig. 12), a ragged surface consisting of lactose crystals and protein clusters on drum-dried particles (Fig. 13), a globular appearance of spray-dried whey curd (Fig. 14), and an absence of any superficial crust on freeze-dried curd particles (Fig. 15). SEM was complemented by TEM which demonstrated the presence of a compact crust on all dried whey curd specimens except the freeze-dried one. WHC was significantly correlated with the microstructure and was the highest with the fluffy freeze-dried curd which had also the lowest bulk density, and

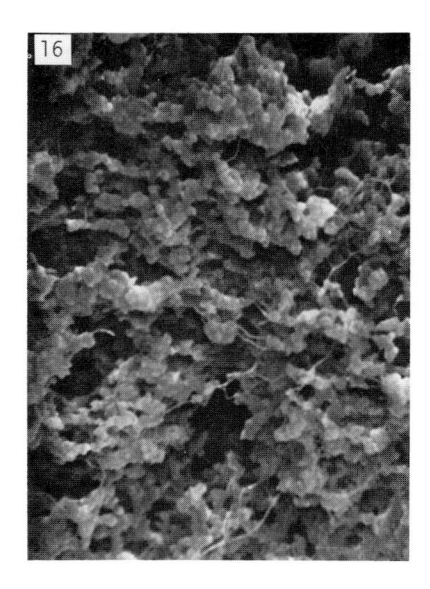

Fig. 16.
A pair of stereo micrographs of stirred yoghurt.
(Reproduced by permission of the Journal of Dairy Research).

was the lowest with the compact air-dried curd of the highest bulk density. In the whey curd the development of ice crystals during freezing and their sublimation during freeze-drying led to a porous microstructure which allowed penetration of water more readily than the encrusted curd particles obtained by the other drying processes.

Yoghurt

After its introduction to the local market several years ago, yoghurt became a popular dairy snack in North America. In general, it is a fermented dairy product, made from partially skimmed milk fortified with skim milk powder to increase the total solids content. A combined culture of *Lactobacillus bulgaricus* and *Streptococcus thermophilus* is used to gel the milk by bringing its pH value below 4.3. The product is marketed in the gel form as set-style natural yoghurt, or the gel is broken down and offered as stirred natural yoghurt. A greater part of the production is supplemented with natural fruit jams or artificial fruit flavours. Stabilizers, such as pre-gelatinized starch, alginates, and gelatin, or special viscous bacterial cultures are used to modify the physical properties of viscosity and mouthfeel, and to stabilize the protein matrix of the yoghurt against syneresis, thereby increasing its shelf life.

The protein matrix of yoghurt is composed of casein micelles which, in contrast to fluid milk, are fused together and form chains and clusters, thus immobilizing the liquid phase. The protein matrix is continuous in the set-style yoghurt, but is disrupted by stirring and pumping in the stirred yoghurt. Individual submicroscopical clusters of casein micelles, formed by the stirring, only touch each other in the stirred yoghurt, and make it difficult to examine them under SEM. The difficulties start with fixation in glutaraldehyde, in which the yoghurt droplets have a tendency to disperse. Following freeze-drying or critical-point-drying, the droplets that had remained intact need to be fractured to have their internal structures exposed. Mounting of the fragments on SEM stubs using a conductive silver cement is particularly cumbersome as the fragments easily decompose when touched with a brush. Because of the loose composition of stirred yoghurt, so-called "island charging"[14] has been a common defect of the micrographs. Nevertheless, pairs of stereo micrographs were published of both the set-style and stirred yoghurt[15]. The stereo technique contributed significantly to the distinction of the fluffiness or discontinuity of stirred yoghurt (Fig. 16).

The presence of lactic bacteria in yoghurt is best documented by SEM. Bacterial colonies frequently form "pockets" in the protein matrix (Fig. 17). These pockets are best seen in smooth surfaces produced by fracturing of freeze-dried yoghurt particles[16]. This phenomenon is not considered to be an artifact caused by the freezing of yoghurt as the incidence of pockets is supported by TEM. It is rather probable that the pockets are formed as a result of bacterial action.

Some bacterial cultures produce extracellular polysaccharides more readily than others. They are

Fig. 17. "Pockets" in yoghurt occupied by lactic bacteria.

Fig. 18. Filaments anchoring lactic streptococci in the protein matrix of yoghurt.

Fig. 19. Yoghurt containing pre-gelatinized starch (2%) as a stabilizer.

Fig. 20. Yoghurt containing carrageenan (0.4%) as a stabilizer.

frequently referred to as so-called "ropy" cultures. Filaments, by which the bacteria are attached to the protein matrix, are presented in Fig. 18. B.E. Brooker[17] studied them in cheese by TEM. In Cottage cheese these filaments are believed to keep the bacteria from being drained away with whey by anchoring them to the curd. Yoghurt produced by ropy cultures usually resists syneresis better than yoghurt produced by regular cultures. To prevent syneresis in yoghurt, also various viscosity-increasing additives are used. They frequently change the microstructure. Gelatin was the only additive of several additives studied[15], the presence of which could be observed by neither SEM nor TEM. Pre-gelatinized corn starch (2%) appeared in the form of fibres either linking casein clusters or having free terminations (Fig. 19). Carrageenan (0.4%) caused the casein micelles to aggregate into large clusters in agreement with the findings of F.L. Hood and J.E. Allen, who reported[18] aggregation of casein micelles to various degrees depending on the variety of the carrageenan used. T.H.M. Snoeren et al.[19] found that high-molecular weight κ-carrageenan led to the formation of threadlike structures in the presence of κ-casein, whereas with low-molecular weight carrageenan aggregates were formed. Long thin fibres permeating the protein matrix were clearly visible in the yoghurt (Fig. 20) even when the carrageenan concentration was lowered[15] to 0.13%. Since the fibres were approximately 400 to 500 nm thick whereas T.H.M. Snoeren et al.[19] reported the thickness of carrageenan fibres to be only 5 to 25 nm, it is highly probable that, in accordance with the observation of the above

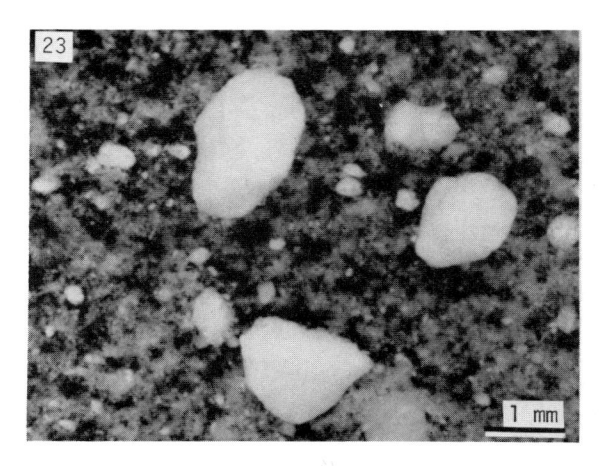

Fig. 23. Lumps in stirred natural yoghurt photographed in a 1.4% glutaraldehyde solution, in which the body of the yoghurt rapidly dispersed.

authors, κ-casein particles became adsorbed on the carrageenan chains and made them appear thicker.

SEM along with TEM was also used to study the microstructure of yoghurt as related to the heating of milk. It has been known that heating to 90°C has been the prerequisite for the production of firm yoghurt[20]. Heating denatures whey proteins, that is α-lactalbumin and β-lactoglobulin, and leads to the formation of a complex between β-lactoglobulin and κ-casein on the casein micelle surface. The latter complex, observed by TEM by F.L. Davies et al.[21], had been earlier postulated to inhibit casein micelle fusion[22,23]. The postulate received support by SEM and TEM observations of large micellar clusters in yoghurt made from unheated skim milk (Fig. 21); the absence of this complex apparently made it possible for the micelles to fuse into large clumps compared to regular yoghurt made from heated skim milk (Fig. 22).

Recently SEM and TEM have been used in the author's laboratory to examine a seasonal defect in stirred yoghurt, characterized by the incidence of small lumps (Fig. 23). The lumps were particularly noticeable in the blueberry-flavoured yoghurt as they were not penetrated by the dark blueberry colouring and remained white. The lumps were found to consist of a protein matrix only slightly denser than the surrounding medium. No excessive accumulation of bacteria could be observed in those lumps.

Cottage Cheese

In contrast to yoghurt, which is made from partially skimmed heated milk, Cottage cheese is made from pasteurized skim milk[24]. This is because syneresis is desirable in Cottage cheese curd, the manufacture of which requires expulsion of whey during cooking.

Of all the dairy products, Cottage cheese has received the least attention by electron microscopists. Although two SEM micrographs have been published to show freeze-dried and critical-point-dried Cottage cheese[16], most of the work is still at the experimental stage. J. Glaser et al.[25] in California and M. Kalab et al. in Ontario have

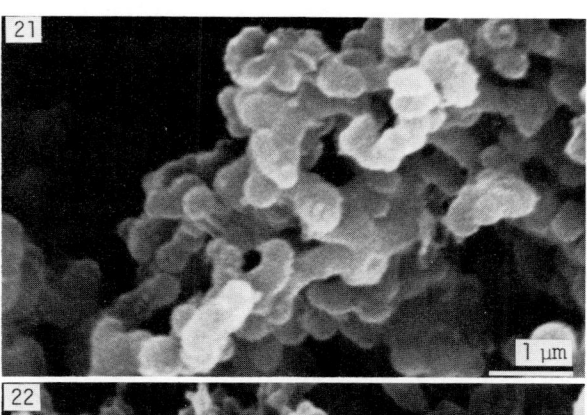

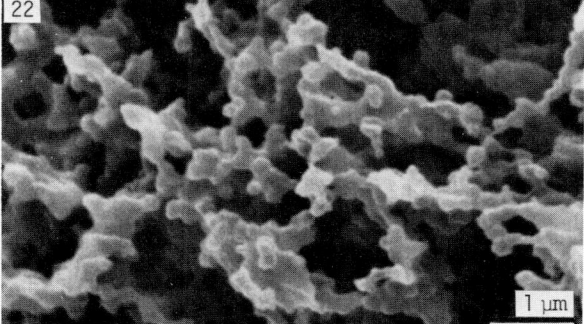

Fig. 21. Yoghurt made from unheated skim milk.

Fig. 22. Regular yoghurt made from skim milk which had been heated to 90°C.

been studying the question as to whether a hypothetical "skin" develops on the surface of Cottage cheese granules. This skin was visualized by S.L. Tuckey[26] to be a very active and sensitive surface which would act as a semipermeable membrane; if formed excessively, it would inhibit the expulsion of whey and the absorption of the Cottage cheese dressing. However, SEM and TEM examinations failed to find such a hypothetical membrane. In fact, J. Glaser et al.[25] found that the microstructure of the surface of laboratory-made and commercial Cottage cheese granules was similar to that of the interior portions of the granules (Fig. 24). Curd that was subjected to severe heating (70°C for one hour), not typical of commercial cooking treatments, showed a marked difference between the surface and interior microstructure (Fig. 25). The surface was covered with fibrous particles, smaller than casein micelles. The crosslinking of the fibres formed a network with varying degrees of porosity and some areas appeared impervious[25]. M. Kalab has been examining Cottage cheese during commercial manufacture and found the casein micelles in the superficial layer of the granules to be clustered more densely than in the interior portions of the granules[2] (Fig. 26). However, neither of the above research groups found any membrane-like structures that form during cooking treatments typical of those used in commercial manufacture of Cottage cheese. Experimental work aimed at relating the microstructure of the superficial layer of Cottage cheese granules with some defects is in progress.

Cheese

Most cheeses are made from whole milk, the coagulation of which is achieved by a combined action of starter lactic bacteria and a proteolytic coagulant such as rennet.

SEM is better suited for studies of more advanced stages of cheesemaking than the initial stages of milk coagulation, as it is considerably easier to examine the more compact structures produced after the removal of excessive water in the form of whey. Substantial changes take place in the ratios of the remaining substances forming the curd: most of lactose and water is removed with whey and the curd is composed mainly of casein ($\approx 24\%$), fat ($\approx 34\%$), and salts ($\approx 5.5\%$); there is approximately 35% of water in Cheddar cheese.

Casein micelles in milk aggregate during cheesemaking and fat globules are entrapped in the coagulum. To study the microstructure of curd by SEM it is necessary to remove the fat unless, as was mentioned in the Introduction, a cold stage is used. Artifacts caused by the failure to do so were shown earlier[16]. Cheese is fixed in a glutaraldehyde solution, post-fixed with OsO_4 to preserve fat globule membranes composed of lipoproteins, dehydrated in a graded series of alcohol, defatted in chloroform, and critical-point-dried. Another way is to freeze-dry the fixed cheese specimen at -80°C and to extract fat from it with chloroform. Experiments in the author's laboratory proved[27] that the robust protein matrix in cheese was not affected by the extraction; in fact, fat globule membranes seemed to be better preserved in the freeze-dried cheese consecutively extracted. However, the extracted cheese specimens were transferred into absolute alcohol and critical-point-dried in CO_2 to avoid the development of artifacts during drying.

Dehydration and defatting may be achieved in one step by using 2,2-dimethoxypropane[27,28].

It has been practice in the author's laboratory to fragment any dried specimens before mounting on SEM stubs to reveal their internal

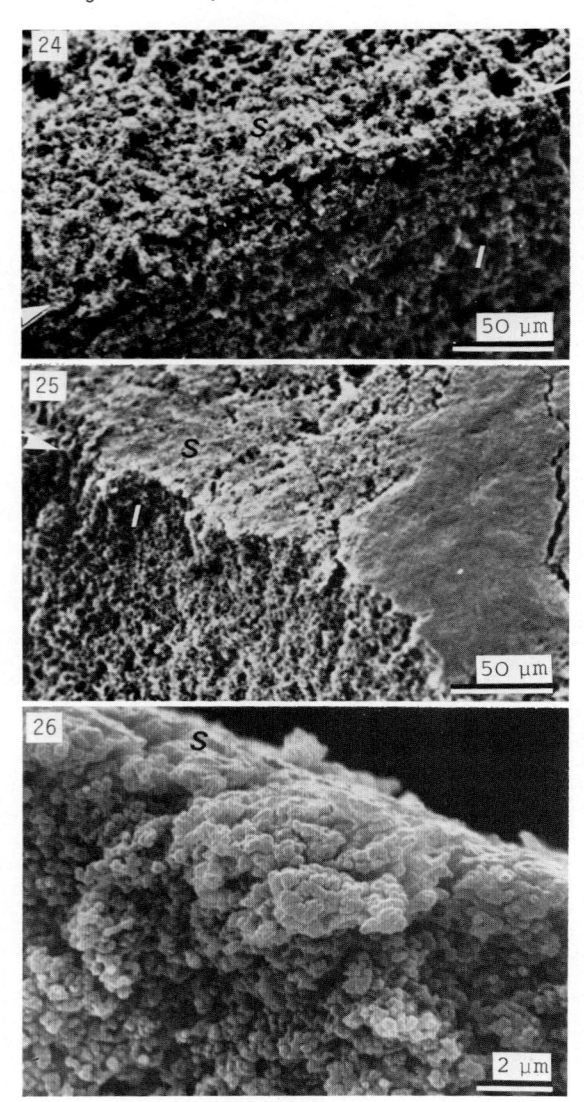

Fig. 24. Surface (S) and the interior (I) of a Cottage cheese granule. Edge is indicated by arrows. (Reproduced by permission of Mr. J. Glaser and the American Dairy Science Association).

Fig. 25. Surface (S) and the interior (I) of a Cottage cheese granule heated at 70°C for an hour. Edge is indicated by an arrow. The curd granule surface is composed of matted casein micelles. (Reproduced by permission of Mr. J. Glaser and the American Dairy Science Association).

Fig. 26. Fracture of a commercial Cottage cheese granule. Casein micelles are packed more densely at the surface (S).

microstructure. In this way the true microstructure is observed rather than the surfaces which may have become contaminated during the preceding preparative steps.

Most cheeses were found[27] to reveal two kinds of structure:
(a) The macrostructure arises from the way in which curd granules are fused, for example the plain fusion in cheeses such as Brick, Edam, Gouda etc., or from the stretching of fused curd granules by various processes in cheeses such as Cheddar[29], Mozzarella, or Provolone. The differences can be observed in fixed, dehydrated, and defatted cheese sections under a light microscope at a very low magnification (Fig. 27). Different patterns were observed with various cheeses[27].
(b) The microstructure of the protein matrix inside the curd granules is also affected by the manufacturing process though to a different extent.

In freshly cheddared curd, for example, the macrostructure of stretched granules is reflected at the microscale by the development of protein fibres which may be observed by both SEM (Fig. 28) and TEM (Fig. 29). Following the removal of the tensile force, however, the nature of the fibrous microstructure is rapidly changed as the protein fibres in the matrix contract, whereas the parallel orientation of the curd granule junctions is preserved. On the basis of light and electron microscopy a model has been designed which explains

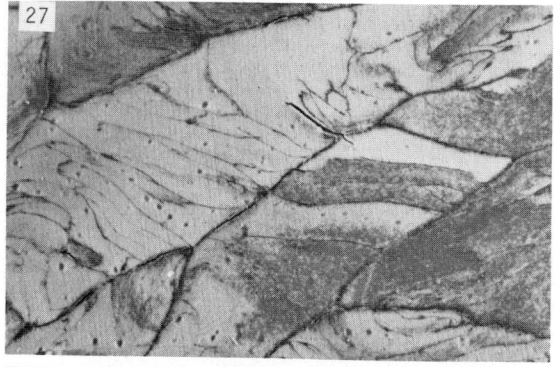

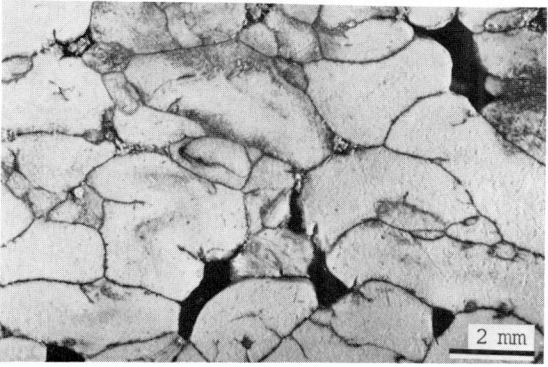

Fig. 27. Curd granule junctions (light microscopy) in Cheddar cheese (upper micrograph) and in Brick cheese (lower micrograph). In Cheddar cheese the thin dark lines represent curd granule junctions, originating during cheddaring, and the thick dark lines represent milled curd junctions, originating by pressing the curd after milling.

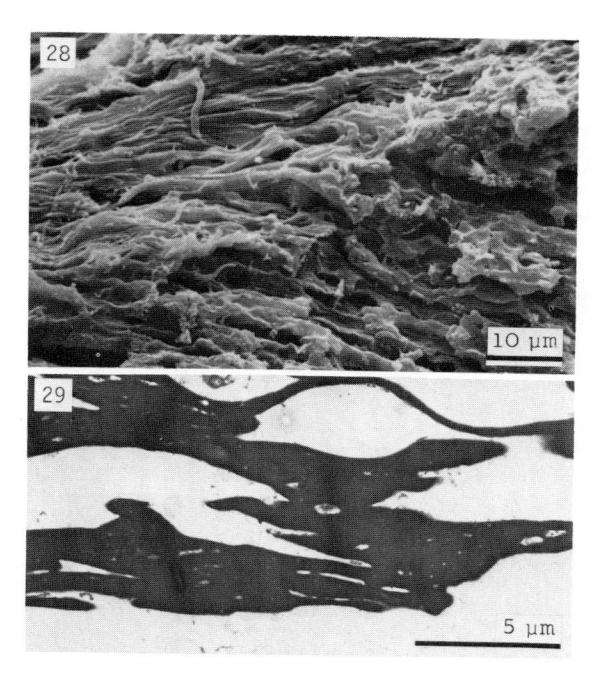

Fig. 28. SEM of cheddared curd following fat extraction.

Fig. 29. TEM of cheddared curd. Dark areas represent protein.

the development of the curd texture and microstructure during cheddaring and cheese ripening[30]. In addition to protein, fat as the other major component of cheese was found to play an important role in these processes. This indicates, however, that the exclusive use of SEM in the case of cheese is of limited value. Nevertheless, SEM was used alone in studying the effect of calf rennet and substitutes such as bovine pepsin[31,32] and porcine pepsin[32] on the microstructure of Cheddar cheese: the traditional calf rennet cheese was softer but was found to be more compact during the first 6 months of ripening than the bovine pepsin cheese.

D.M. Hall and L.K. Creamer[33] used both SEM and freeze-fracturing to compare the microstructures of Cheddar, Cheshire, and Gouda cheeses. For SEM the cheese specimens were etched with trypsin in a phosphate buffer at pH 9 to remove the surface protein between fat globules, freeze-dried at -30°C, and vacuum-coated with a 20 nm layer of carbon followed by two layers of gold-palladium. On the basis of both techniques, Gouda cheese was found to consist of globular units 10 to 15 nm in diameter. The protein matrix in Cheshire cheese was less uniformly organized and strands and globules 3 to 4 nm in diameter were apparently observed.

The microstructure of curd granule junctions in cheese was studied by M. Kalab[27]. The junctions, which appeared as dark veins under a light microscope (Fig. 27), were found by SEM to contain considerably less fat and, hence, more protein than the interior portions of the curd granules. This was evident from the high number of empty cavities in the interior areas of the curd granules following the extraction of fat in contrast to very few cavities found in the curd granule junctions (Fig. 30). Consequently, the areas rich in cavities scattered light and appeared lighter than the compact junctions. The author cautioned that the differences in the microstructure of the junctions and the interior portions of the granules should be taken into consideration when cheeses are subjected to electron microscopical examination.

Following fat removal, residues of fat globule membranes, lactic bacteria, and microcrystalline inclusions of calcium phosphate, earlier observed by B.E. Brooker et al.[34] by TEM, became available for SEM examination[27] (Fig. 31 and 32). Correlations between SEM and TEM of such components as well as the protein matrices in Cheddar, Gouda, Mozzarella, and Provolone cheeses showed the way in which the two different electron microscopical techniques complement each other[27].

M. Rüegg et al.[35] used SEM in their studies of the microstructure of cheese curd and rind. The cheese specimens were fixed in a mixture of acrolein and glutaraldehyde buffered at the pH value of the particular cheese, dehydrated in a graded series of alcohol, defatted in chloroform, and only then, in the dry state, post-fixed with OsO_4 vapours. Lactic bacteria were freeze-dried and similarly fixed in the dry state with OsO_4 vapours. A great number of micrographs displayed the differences in the microstructures of Romadur, Emmental, and Tilsit cheeses. Micrographs of lactic bacteria in the above cheeses as well as in pure cultures and micrographs of the hyphae and sporangia of the mold used in the manufacture of Camembert cheese were also presented.

Conclusions

Casein micelles, whey proteins, fat globules, and lactose as the major constituents of milk are capable of undergoing a great variety of changes, particularly under the effects of proteolytic enzymes and/or lactic bacteria, leading to various dairy products. The microstructure of such dairy products is controlled, to a great extent, by the manufacturing processes. On the other hand, the microstructure determines some properties of the product, for example viscosity, syneresis, firmness, or mouthfeel. SEM has proved to be useful, particularly in conjunction with other electron microscopical techniques, in studying the microstructure of dairy products as it relates to the effects of manufacturing processes and the properties of the products.

Acknowledgment

Appreciation is expressed to Dr. D.N. Holcomb for an invitation to write this overview, to Dr. T.J. Buma for his permission to reproduce two micrographs and a diagram, to Mr. J. Glaser and Dr. W.L. Dunkley for information concerning their research on Cottage cheese and for their permission to reproduce two micrographs, to Dr. D.B. Emmons for useful comments and for reviewing the manuscript, and to Mr. Pavel Kalab for linguistic assistance.

The author thanks the Electron Microscope

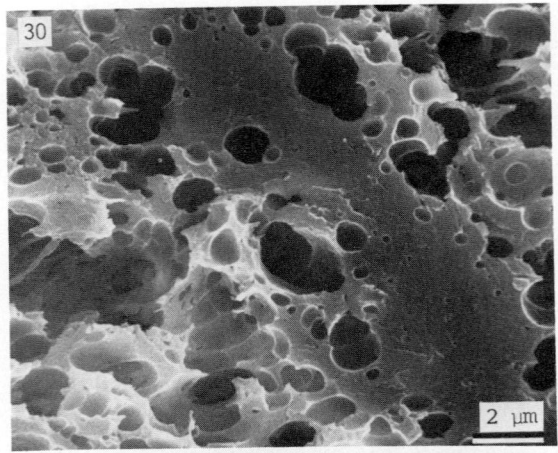

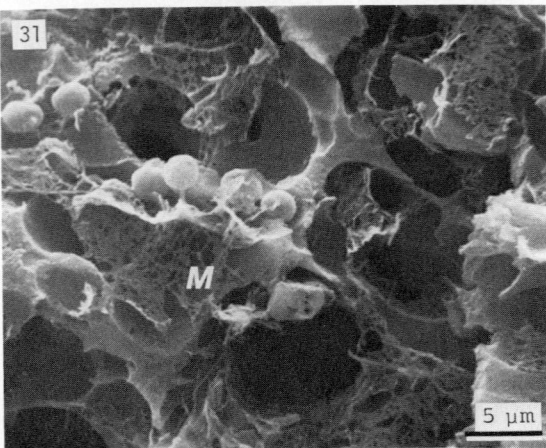

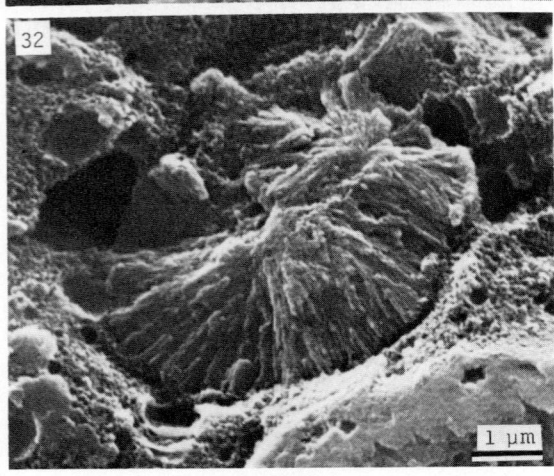

Fig. 30. SEM of a curd granule junction in Brick cheese. The junction is characterized by a fat content lower than the interior portions of the curd granule.

Fig. 31. Residues of fat globule membranes in Gouda cheese resemble a lace (M).

Fig. 32. A fragmented crystalline inclusion of calcium phosphate in cheese.

Centre, Research Branch, Agriculture Canada, Ottawa, for providing facilities.

This overview is Contribution 383 from Food Research Institute, Research Branch, Agriculture Canada, Ottawa.

References

1. M. Kalab: Electron microscopy in dairy research. Microscop. Soc. Canada Bull. 5(4), 1977, 4-10.
2. M. Kalab: Microstructure of dairy foods. 1. Milk products based on protein. J. Dairy Sci. (in press).
3. T.J. Buma and S. Henstra: Particle structure of spray-dried milk products as observed by a scanning electron microscope. Neth. Milk Dairy J. 25, 1971, 75-80.
4. T.J. Buma: Free fat and physical structure of spray-dried whole milk. Dissertation - a collection of 10 research papers published in Neth. Milk Dairy J. between 1968 and 1971, Wageningen, the Netherlands, 1971, 131 pp.
5. T.J. Buma and S. Henstra: Particle structure of spray-dried caseinate and spray-dried lactose as observed by scanning electron microscope. Neth. Milk Dairy J. 25, 1971, 278-281.
6. T.J. Buma: Teilchenporosität von sprühgetrockneter Milch. Milchwissenschaft 33(9), 1978, 538-540.
7. C.W. Hall and T.I. Hedrick: Drying Milk and Milk Products, The Avi Publishing Co., Inc., Westport, Connecticut, 1966, 152-155.
8. Sylvia Warburton and S.W. Pixton: The moisture relations of spray-dried skimmed milk. J. Stored Prod. Res. 14, 1978, 143-158.
9. T.A. Nickerson: Fundamentals of Dairy Chemistry, B.H. Webb and A.H. Johnson (ed.), The Avi Publishing Co., Inc., Westport, Connecticut, 1965, 224-260.
10. M. Kalab and D.B. Emmons: Milk gel structure. III. Microstructure of skim milk powder and gels as related to the drying procedure. Milchwissenschaft 29, 1974, 585-589.
11. M. Kalab, D.B. Emmons, and A.G. Sargant: Milk gel structure. V. Microstructure of yoghurt as related to the heating of milk. Milchwissenschaft 31(7), 1976, 402-408.
12. J.L. Short, H.R. Cooper, and R.K. Doughty: The effect of manufacturing variables on lactalbumin for use in high protein biscuits. New Zealand J. Dairy Sci. Technol. 13, 1978, 43-48.
13. P. Jelen, M. Kalab, and R.I.W. Greig: Water-holding capacity and microstructure of whey protein powders. Milchwissenschaft (in press).
14. A. Boyde: Do's and don'ts in biological specimen preparation for SEM. SEM/1976/I, IIT Research Institute, Chicago, IL, 60616, 683-690.
15. M. Kalab, D.B. Emmons, and A.G. Sargant: Milk gel structure. IV. Microstructure of yoghurt in relation to the presence of thickening agents. J. Dairy Res. 42, 1975, 453-458.
16. M. Kalab: Milk gel structure. VIII. Effect of drying on the scanning electron microscopy of some dairy products. Milchwissenschaft 33(6), 1978, 353-358.
17. B.E. Brooker: Cytochemical observations on the extracellular carbohydrate produced by *Streptococcus cremoris*. J. Dairy Res. 43, 1976, 283-290.
18. L.F. Hood and J.E. Allen: Ultrastructure of carrageenan-milk sols and gels. J. Food Sci. 42, 1977, 1062-1065.
19. T.H.M. Snoeren, P. Both, and D.G. Schmidt: An electron-microscopic study of carrageenan and its interaction with κ-casein. Neth. Milk Dairy J. 30, 1976, 132-141.
20. H. Grigorov: Effect of heat treatment of cow's milk on the hydrophilic properties of the protein in Bulgarian yoghurt. XVII International Dairy Congress, Section F, 5, 1966, 649.
21. F.L. Davies, P.A. Shankar, B.E. Brooker, and D.G. Hobbs: A heat-induced change in the ultrastructure of milk and its effect on gel formation in yoghurt. J. Dairy Res. 45, 1978, 53-58.
22. A.-M. Knoop and K.-H. Peters: Die Ausbildung der Gallertenstruktur bei der Labgerinnung und der Säuregerinnung der Milch. Kieler Milchwirt. Forschungsber. 27(3), 1975, 227-248.
23. A.-M. Knoop and K.-H. Peters: Die strukturellen Veränderungen der Labgallerten während der Alterung. Kieler Milchwirt. Forschungsber. 27(4), 1975, 315-330.
24. D.B. Emmons: Cottage Cheese and Other Cultured Milk Products. Chas Pfizer & Co., Inc., New York, N.Y., 1967, 24.
25. J. Glaser, P.A. Carroad, and W.L. Dunkley: Surface structure of Cottage cheese curd by electron microscopy. J. Dairy Sci. (in press).
26. S.L. Tuckey: Properties of casein important in making Cottage cheese. J. Dairy Sci. 47(3), 1964, 324-326.
27. M. Kalab: Milk gel structure. VI. Cheese texture and microstructure. Milchwissenschaft 32(7), 1977, 449-458.
28. L.L. Muller and T.J. Jacks: Rapid chemical dehydration of samples for electron microscopical examination. J. Histochem. Cytochem. 23, 1975, 107-110.
29. C.G. Rammell: The distribution of bacteria in New Zealand Cheddar cheese. J. Dairy Res. 27, 1960, 341-351.
30. M. Kalab and D.B. Emmons: Milk gel structure. IX. Cheddared curd. Milchwissenschaft 33(11), 1978, 670-673.
31. D.W. Stanley and D.B. Emmons: Cheddar cheese made with bovine pepsin. II. Texture-microstructure-composition relationship. Can. Inst. Food Sci. Technol. J. 10, 1977, 78-84.
32. M.F. Eino, D.A. Biggs, D.M. Irvine, and D.W. Stanley: A comparison of microstructure of Cheddar cheese curd manufactured with calf rennet, bovine pepsin, and porcine pepsin. J. Dairy Res. 43, 1976, 113-115.
33. D.M. Hall and L.K. Creamer: A study of the submicroscopic structure of Cheddar, Cheshire and Gouda cheese by electron microscopy. New Zealand J. Dairy Sci. Technol. 7, 1972, 95-102.
34. B.E. Brooker, D.G. Hobbs, and A. Turvey: Observations on the microscopic crystalline inclusions in Cheddar cheese. J. Dairy Res. 42, 1975, 341-348.
35. M. Rüegg, R. Sieber, and B. Blanc: Untersuchungen der Feinstruktur von Käseteigen und Käserinden mit Hilfe der Raster-Elektronen-Mikroskopie. Schweiz. Milchwirt. Forchung 3, 1974, 1-5.

Discussion with Reviewers

<u>D.N. Holcomb</u>: The structure of the spray-dried whole milk particles (Fig. 3) with smaller entrapped globules is analogous to the structures of fly ash reported by G.L. Fisher, D.P.Y. Chang, and M. Brummer[1] (Science 192, 1976, 553). Do these structural similarities suggest similarities of the formation mechanisms?

<u>Author</u>: G.L. Fisher *et al.*[1] found by SEM that fly ash agglomerates, composed of spherical particles, were typical of fly ash derived from coal combustion. They hypothesized that the observed sphere-within-sphere structures may have resulted from a bubbling process occurring in the high-temperature combustion zone; there the surface of the particle melts whereas the particle interior is still solid or highly viscous because of the temperature gradient. Thermal decomposition of some minerals produces gases such as CO_2 and H_2O which may form a bubble around the core. Additional gas formation and the high temperature eventually cause the core to boil away from the interior surface and to form small microspheres in the process before the particle is carried out of the high-temperature combustion region and freezes. A term "plerosphere" has been suggested by the above authors for the sphere-within-sphere structure.

In spray-drying, spherical milk droplets enter a high-temperature zone where water is rapidly evaporated from the particle surface; the loss of water in the superficial layer increases the viscosity of this layer and leads to its solidification. Because of the high heat-transfer rates in the droplets, the liquid at the centre vaporizes causing the outer shell to form a hollow sphere. In general, hollow particles are produced by spray-drying of any aqueous solution which tends to form a tenuous skin on drying[2]. Central vacuoles may also be formed by the deposition of solids on the inner surface of the shell which leaves the centre of the particle hollow. The formation of plerospheres in dried milk is difficult to explain as not all spray-dried milk particles exist in that form. Normally the short contact time causes the liquid droplets to remain relatively cool while evaporation is taking place in the hot zone of the dryer, whereas at the point when they are no longer wet, the surrounding gases have been cooled by the evaporation process thereby eliminating overheating. Thus it seems that there is little similarity between the mechanisms of the formation of the microstructures of the fly ash and spray-dried milk particles.

References

1. G.L. Fisher, D.P.Y. Chang, and Margaret Brummer: Fly ash collected from electrostatic precipitators: Microcrystalline structures and the mystery of the spheres. Science 192, 1976, 553-555.
2. W.R. Marshall, Jr., and S.J. Friedman: Drying in Chemical Engineers' Handbook, J.H. Perry (ed.), 3rd edition, McGraw-Hill Book Co., Inc., New York, Toronto, London, 1950, 840.

ELECTRON MICROSCOPY OF MILK PRODUCTS: A REVIEW OF TECHNIQUES

M. Kaláb

Food Research Institute
Research Branch
Agriculture Canada
Ottawa
Ontario, Canada K1A 0C6

Abstract

Electron microscopical techniques are reviewed as they apply to various milk products. The following techniques are discussed and illustrated with representative micrographs: Scanning electron microscopy (SEM): Conventional SEM at ambient temperature; cold-stage SEM; replication of freeze-fractured specimens with gold for SEM.
Transmission electron microscopy (TEM): Negative staining; metal-shadowing; thin-sectioning; freeze-fracturing and freeze-etching; replication of dried specimens with platinum.
Personal experience of the author with some more recent electron microscopical techniques is also presented.

KEY WORDS: Electron microscopical techniques; Milk products; Review; SEM, TEM.

Introduction

Milk products form a large group of foods based on milk proteins and fat. They range from products with a high water content (fluid milk) to dry products (milk powders), and from products of a low fat content (skim milk, yoghurt) to products high in fat (butter). Cheeses are high in protein and may be low in fat (skim milk cheese) or high in fat (most cheeses). Some products are marketed and consumed in the frozen state (ice cream, frozen yoghurt dessert).

When microstructure of milk products is studied by electron microscopy, the composition of the product is taken into consideration before the proper technique is selected. There is a wide variety of electron microscopical techniques available for different specimens. They are generally divided into scanning (SEM) and transmission electron microscopy (TEM). The principles on which these techniques are based may be found in any textbook on electron microscopy. Table 1 summarizes the various techniques along with their suitability for studies of certain milk products. It is advisable to select and use more than one technique and compare the results to avoid reporting artifacts.

Electron microscopy has been used to study the microstructure of the individual components in milk products such as casein micelles and fat globules, and changes which these components undergo either alone or by interactions with each other or with additional ingredients such as stabilizers, thickeners, emulsifiers, lactic bacterial cultures *etc.* during manufacturing processes. Being basically of biological origin, milk products are prepared for electron microscopy in ways similar to other biological specimens[1-3].

The objective of this review is to summarize electron microscopical techniques most widely used in studies of milk products and to present examples of their use in practice. For details and step-by-step procedures, however, the reader is referred to the respective bibliographic sources.

1. Scanning electron microscopy

Specimens examined by SEM are either dry (conventional SEM) or frozen (cold-stage SEM). Replication of freeze-fractured specimens with gold and examination of the gold replica by

Table 1. A review of electron microscopical techniques used in studies of milk products

Technique:	Nature of specimen:	Example of milk product for which suitable:	References:
SCANNING ELECTRON MICROSCOPY			
Conventional	Dry	Powdered milk, whey, buttermilk etc.	5-14, 28
	Dried	Low-fat milk products (yoghurt, Cottage cheese, some cheeses)	15-17, 28, 42, 47
	Dried	Cheese (fat extracted)	18, 19, 21-23, 25, 43, 53, 54, 111
	Dried	Cheese (trypsin-etched)	20, 24
	Freeze-fractured and replicated with gold	Viscous and high-fat milk products (cream, butter, cheese, etc.)	4
Cold stage	Freeze-fractured and coated with carbon and/or gold	Viscous, whipped, high-fat products (ice cream, whipped cream, butter etc.)	57-59, 61
TRANSMISSION ELECTRON MICROSCOPY			
Negative staining	Suspension	Fluid milk, cream	64, 66-74
Metal shadowing	Suspension	Fluid milk, cream	29, 68, 75-77, 82
Thin-sectioning	Suspension	Liquid products (microcapsulation)	84-88
	Solid	Products solid by nature (cheese) or solidified by mixing with agar (cream)	13, 15-17, 21, 24, 28, 54, 70, 89-92, 95-103, 106-110
Freeze-fracturing (Freeze-etching)	All products	All products	61, 70, 114-123, 128-142, 145
Replication of dried specimen	Solid	Milk products based on protein	46, 145

conventional SEM at ambient temperature has been suggested recently[4].

1.1. SEM of dry specimens

This technique is suitable for products which are marketed in the dry state such as milk[5-9], whey[10-12], buttermilk[13], and other powders[14], and for products which contain water but may be dried for SEM (yoghurt[15,16], Cottage cheese[17], cheese[18-25] etc.).

Preparation of dry powders for SEM is simple and consists of mounting the powder particles on metal SEM stubs and coating them with gold. Johari and DeNee[26] and DeNee[27] dealt with various preparatory steps in great detail.

Double sticky tape is easier to use[5,28] than a liquid cement[11], especially with particles as porous as dried milk products. Excess (loose) particles are removed from the tape by a stream of air or nitrogen or are shaken off[5] to prevent overlapping of the particles and charging artifacts. Painting the tape edge with a conductive silver cement provides an uninterrupted conductive surface between the specimen and the metal stub after the powder particles are coated with a thin ($\simeq$20nm) layer of gold by sputter coating or vacuum evaporation.

SEM shows the external appearance of the dry powder particles as well as their internal microstructure (porosity) if crushed particles[5-10,13] are viewed. Fig. 1 shows both intact and crushed particles of a fresh spray-dried buttermilk powder.

Milk powders are susceptible to electron beam damage[5,11,13,28] during electron microscopical examination. This susceptibility, particularly evident in the presence of lactose, may be caused by the removal of water from α-hydrate crystals *in vacuo* inside the electron microscope leaving them in an unstable fluffy state as the anhydride retains the original crystalline shape of the hydrate. Effects of air humidity on the microstructure of milk powders were studied by several authors[7,9,11,13]. In a humid atmosphere the initially amorphous glassy mixture of α- and β-lactose in milk is rapidly turned into the crystalline α-hydrate; the latter may eventually cover the entire powder particle and obscure all the characteristic details on the surface. Differences between nonfat dry milk and buttermilk powders, clearly seen in the freshly spray-dried products or in products stored in a dry atmosphere, were rapidly eliminated after the powders were exposed to air humidity[13]. Crystallization of lactose was observed at a relative humidity as low as 40%[11]. Thus, SEM reveals the true microstructure of spray-dried milk products only if exposure of the powders to humidity is prevented. For this reason fixation of milk powders, destined for SEM, with OsO_4 vapours is not recommended as the fixation is carried out in the presence of water vapour[13].

Preparation of water-containing milk products for SEM requires more steps than do milk powders because the specimen must be dried. As drying may alter the original microstructure of the product, the protein network is first fixed.

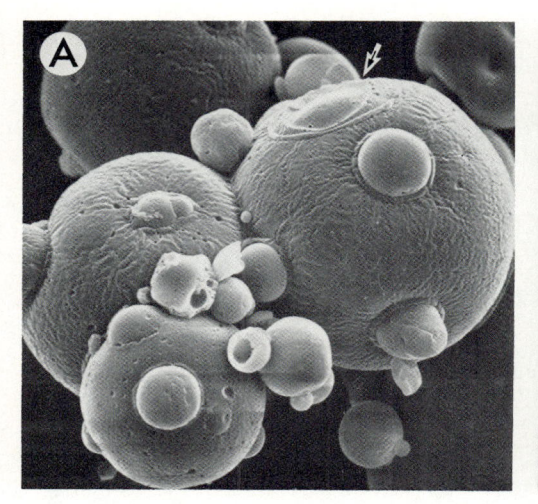

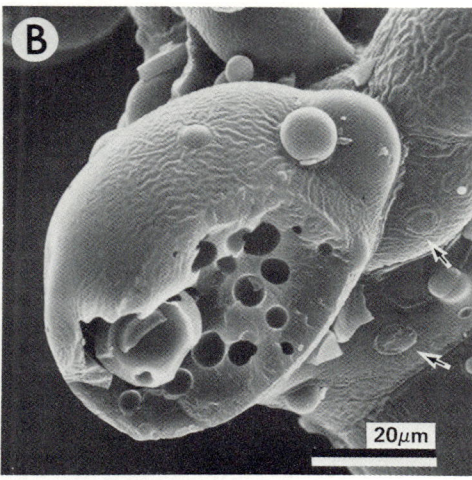

Fig. 1. SEM of intact (A) and crushed (B) particles of spray-dried buttermilk.

Shallow wrinkles on large particles, rims around small globules emerging from the large particles, and crater-like scars (arrows) characterize spray-dried buttermilk.

Glutaraldehyde is the most commonly used fixative[29]. The concentrations used by various authors vary within a wide range and so does the duration of fixation. Hermansson and Buchheim[30] cautioned that in the case of very fine protein sols there was an interaction between glutaraldehyde and the sol leading to gelation of the latter. However, milk gels are considered to be coarse gels[30] and no artifacts were reported to be caused by this fixative. Carroll et al.[29] studied its effects and recommended it for the fixation of casein micelles.

Although air drying of fixed, dehydrated, and defatted cheese specimens was reported[23], drying is most frequently accomplished by freeze-drying or critical-point drying. Formation of ice crystals and the accompanying alteration of microstructure is reduced in specimens frozen rapidly. The specimen (≃1 mm in diameter) is frozen in liquid freon or isopentane cooled with liquid nitrogel (LN_2). LN_2 is not suitable for rapid freezing when used at or near its boiling point; after the specimen is immersed in LN_2, the latter rapidly vaporizes and forms an insulating layer around the specimen thus reducing the rate of freezing. Melting nitrogen slush, which does not boil around the specimen[31], is a considerably better cryofixative. The slush is prepared by exposing LN_2 in an insulated cup to a reduced pressure of approx. 94 torr for ≃1 min; LN_2 vigorously evaporates and its temperature decreases to the freezing point of N_2. No samples destined for electron microscopy should be frozen at temperatures above 198 K (-75°C), as the risk of ice crystal formation increases with temperature[32]. Ice crystals disrupt the protein network[10] and aggregate individual casein micelles[28] leaving irregular void spaces in the specimen after the ice has sublimed off during freeze-drying.

Critical-point drying (CPD) is another technique[33-35] suitable for drying most milk product specimens which had been previously fixed and dehydrated in a graded series of alcohol or acetone, or in acidified 2,2-dimethoxypropane[21,36,37]. The technique is based on the conversion of liquid CO_2 (as one of several suitable media[33]), in which the dehydrated specimen is immersed in a closed container and with which it is gradually impregnated, into the gaseous form by increasing the temperature above the critical point[35,38] (304.3 K; 31.3°C). Above this temperature, CO_2 exists only as a gas irrespective of the pressure applied. The specimen thus does not pass through any phase boundary, as is common with all other drying techniques[39] (Fig. 2).

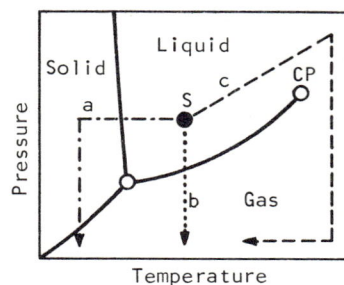

Fig. 2. The course of different drying processes indicated in a phase diagram after Moor[39]. S = specimen. CP = critical point. a = freeze-drying. b = air drying. c = critical-point drying.

In spite of its advantage, CPD is not absolutely free of producing artifacts[40]. Although it is suitable for most milk products based on protein, it should not be applied to products which contain gelatinized starch[41], used, for example, as a stabilizer in some yoghurts[15]. Gelatinized starch, which cannot be fixed, is affected by the organic solvents used to dehydrate the specimen[41,42]. Fat may also be affected during CPD because the organic solvents and liquid CO_2 have lipophilic properties. However, high-fat products such as cheese, destined for SEM in the dry form, are usually defatted, because fat left in the specimen may cause various difficulties. It may, for example, obscure the protein matrix and cause charging artifacts[42]. Fat may be extracted with acetone[25] during dehydration of the cheese specimen, with a mixture of petroleum ether and diethyl ether[43], or with chloroform[21] from specimens impregnated with absolute alcohol prior to CPD. Chloroform is replaced with alcohol, acetone, or amyl acetate, which are miscible[44] with liquid CO_2, as no information has yet been published on the miscibility of chloroform with liquid CO_2, in which CPD is carried out.

Alternatively, fat may be extracted from freeze-dried specimens[21]. Hall and Creamer[20], Taranto et al.[24], and Eino et al.[43] retained fat in the cheese and exposed the fat globules to view by trypsin-etching the protein matrix. Effects of drying on the SEM images of some milk products were studied earlier[42].

Dry specimens are fractured[15,16,21,24] before being mounted on SEM stubs. Fracturing is done by hand under a low-magnification microscope. The tip of a scalpel (blade) is briefly pressed into the edge of the particle and the particle is broken. Details on dry-fracturing were presented by Flood[45]. An alternative procedure consists of freeze-fracturing the specimen and drying and mounting the fragments[46]. In this case the fractures are smoother (Fig. 3 A and B) than those obtained by fracturing a dry specimen (Fig. 3 C) and, consequently,

with the silver cement as closely to the fracture plane as possible (Fig. 4), and by coating the

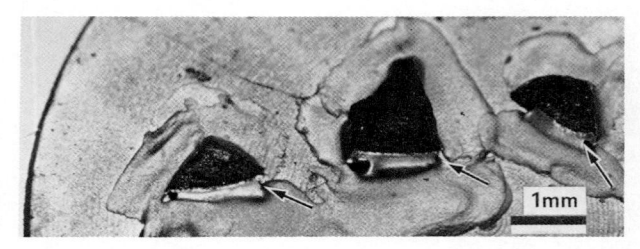

Fig. 4. Dry fractured cheese particles mounted on an SEM stub using a conductive silver cement painted with a brush as close to the fracture planes as possible (arrows). Particles are shown coated with gold.

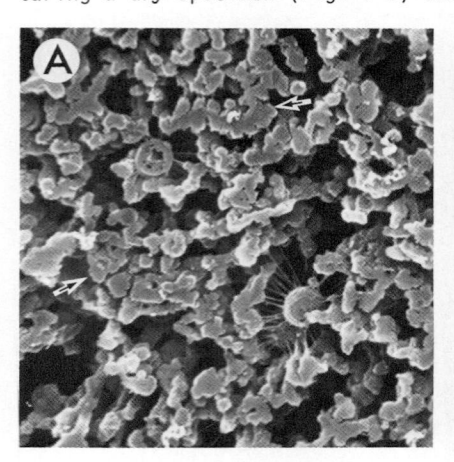

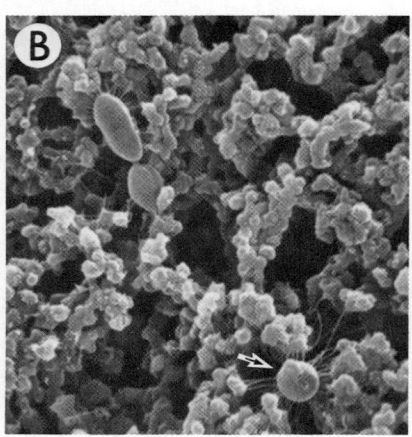

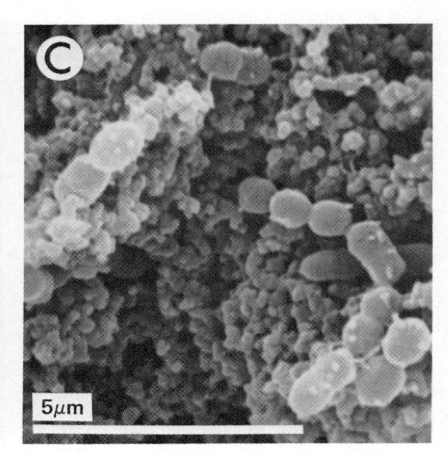

Fig. 3. SEM of Cottage cheese specimens fractured by different procedures.
A = The specimen was impregnated with a 30% glycerol solution, freeze-fractured, fixed in a glutaraldehyde solution, and critical-point dried. Arrows point to fractured casein micelle clusters; fractured bacteria are attached by filaments to the protein matrix.
B = The same specimen was fixed in a glutaraldehyde solution, impregnated with absolute alcohol, freeze-fractured, melted in alcohol, and critical-point dried. An arrow points to a fractured bacterium.
C = Another Cottage cheese specimen was fixed in a glutaraldehyde solution, freeze-dried, and fractured when dry. Neither casein micelle clusters nor bacteria were fractured.

better images of the protein matrix, distribution of lactic bacteria etc. are produced. Fig. 3 shows the microstructure of a Cottage cheese specimen fractured by the above procedures. Failure to fracture the specimen may result in viewing surfaces which have originated from cutting or breaking the crude specimen and which have been contaminated by smeared material rather than viewing the true internal microstructure of the product[19,43].

Specimen fragments are mounted on SEM stubs using a conductive silver cement; the use of other mountands, for example a nail polish[47], was also reported. Almost any viscous cement is suitable provided that it does not generate bubbles in vacuo when dry[48,49]. The importance of using the silver cement at a proper consistency was emphasized earlier[47]: a cement too thin may penetrate the dry fragment and collapse it whereas a cement too thick does not attach the fragment to the stub securely enough to provide an uninterrupted conductive surface. Charging artifacts can be reduced by carefully painting the walls of the fragments

fragments with gold at higher pressures (13 to 133 mPa, i.e. 10^{-4} to 10^{-3} torr), which allows the gold atoms to collide with nitrogen molecules and to be deposited on the specimen from various directions.

Charging artifacts in extremely porous specimens such as gelled milk were avoided[50] by using a low gun potential (5 kV) and a low beam current ($\simeq 40$ μA).

The use of non-coating techniques to render specimens conductive[51,52] has not yet been reported with milk products although such techniques have been used successfully with a variety of biological materials.

The use of conventional SEM in studies of the microstructure of milk products was reviewed earlier[37]. Recently, lactic microorganisms used in the manufacture of Gruyere cheese were examined by SEM[53]. SEM was also used to examine the microstructure of Cheddar cheese made with chicken pepsin[22] and to compare it with Cheddar cheese made by the traditional calf rennet process. In a study of the differences in the microstructure of Cheddar cheese made from whole milk and from homogenized low-fat milk[54], SEM was used in conjunction with thin-sectioning; a model was proposed which explains the development of curd granule junctions[21,37] as areas depleted of fat during the cutting of curd.

1.2. SEM of frozen specimens

Milk products which cannot be dried without artifactual changes, such as cream, butter etc., may be examined with SEM while frozen. No chemical fixation of the specimen is required because the specimen is cryofixed, i.e. fixed by rapid freezing. A cold stage mounted inside the electron microscope, which keeps the specimen at a low temperature (173-203 K; -100 to -70°C) is the prerequisite for this technique[55,56] although preliminary viewing may be carried out even without the cold stage provided that the specimen is cooled with liquid nitrogen before insertion into the electron microscope, and is quickly examined before its temperature rises too high (personal communication by Mr. D.J. Saunders of the Cambridge Instruments Company of Canada Ltd.). However, there is a risk that the specimen will be contaminated with condensed frost inside the electron microscope because any cold object forms a focus for contamination[55]. In a rigorous approach, the specimen is freeze-fractured inside the electron microscope and examined either directly or after coating with gold.

Requirements for low-temperature SEM of biological specimens were discussed by Robarts and Crosby[56], who also reviewed the commercial equipment used. Because of the high price of equipment, which makes it possible to prepare the specimen for SEM inside the microscope, the individual operations such as freeze-fracturing and metal-coating are less expensively performed outside the microscope, while the same precautions to prevent ice crystal formation are observed as mentioned earlier with freeze-drying. The specimen is then transferred into the microscope equipped with a cold stage. A protective carrier is available for this transfer[56].

A technique for cold-stage SEM of cheese was developed by Schmidt et al.[57,58]. Using this technique, Schmidt and van Hooydonk[59] studied the whipping of cream: a narrow aluminum cylinder coated on the inside with a conductive carbon layer was filled with whipped cream and the cream was frozen and freeze-fractured. The fragments were coated with a 20 nm layer of carbon at 173 K (-100°C) in a cooled vacuum evaporator and examined by SEM using a cold stage.

In a technique suggested by Katoh[4], which will be discussed in the following section, the freeze-fractured specimen is coated with gold in a conventional uncooled vacuum evaporator. Low temperature of the specimen in the evaporator is maintained during gold-coating by placing the freeze-fractured specimen in an aluminum dish filled with LN_2. As soon as vacuum is applied, LN_2 vigorously boils and freezes solid within several seconds thus further cooling the specimen. Solid nitrogen is allowed to sublime off as is the frozen water vapour condensed on the dish. The dish is then rotated and the fractured surface of the specimen is coated with gold evaporated at 2 different angles. Care is taken not to allow the wall of the dish to mask the specimen during the gold-coating procedure, after which the specimen is immediately returned into LN_2 for transport to the cold stage in the electron microscope or for short-time storage.

The above technique[4] was subjected to several modifications in this laboratory. Dimensions of the tubular specimen holder were considerably reduced: the outer diameter was reduced from 15 to 5 mm, the inner diameter was decreased from 7 to 3 mm, and the length of the holder was shortened from 10 to 6 mm. The aluminum dish was made to accommodate 4 holders, as this is considered to be the optimum number of specimens for processing within a half-day session at the electron microscope in this laboratory. One objective of these modifications was to reduce ice crystal formation during freezing by using considerably smaller specimens. The other objective was to process more than a single specimen at a time and utilize gold for coating more economically. Alternatively, hollow grid holders for SEM, 3.0 mm in diameter (designed by E.F. Bond and mentioned but not shown by Soni et al.[60]), were also used (Fig. 5).

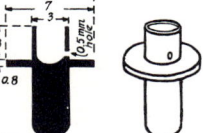

Fig. 5. Hollow grid holder designed by E.F. Bond (see[60]).

In spite of the precautions, severe ice crystal formation occasionally occurred in the centre of some high-water specimens as demonstrated at a low magnification with Cream cheese in Fig. 6. However, provided that ice crystal formation is reduced to a minimum by plunging the holder into melting freon with the specimen going in first, and that areas near the perimeter of the specimen are examined, there is a good correlation[61] with micrographs obtained by other techniques, as demonstrated on a Cream cheese specimen in Fig. 7 (Fig. 18 and 22 may be used for a comparison). In some cases it is possible to preserve the gold coating on the specimen in the form of a replica for an additional SEM examination at ambient temperature as will be discussed in the following section.

SEM of biological specimens at low temperature in general was reviewed earlier[55,56,62,63]. Several disadvantages of low-temperature SEM were mentioned, one of them being the lack of control over the removal of the frozen water vapour contaminating the fractured surface and over the extent of freeze-etching, that is the sublimation of ice forming a part of the specimen. The only controls are the pressure inside the bell jar (the pressure decreases after the frost condensed on the dish vanishes) and the time of sublimation of the frost from the polished surface of the dish[4]. If coating is done prematurely while some of the frost is still covering the surfaces, the conductive elements (carbon or gold) are deposited on the frost particles and peel off after the specimen is returned into LN_2.

This technique, however, has also several advantages. Low temperature makes it possible to examine milk products without fixation within a relatively short time. It takes less than 90 min between freezing the specimen and taking the micrographs, of which approximately 60 min are required

Fig. 6. Cold-stage SEM of a Cream cheese specimen, frozen by the standard[125] procedure in a hollow SEM holder[60], freeze-fractured under LN₂, and coated with gold by the Katoh's procedure[4].
A = An overall view of the entire fracture shows ice crystal formation (lines converging toward the centre of the specimen).

B = Detail of the borderline between areas affected (lower half) and unaffected (upper half) by the ice crystal formation.

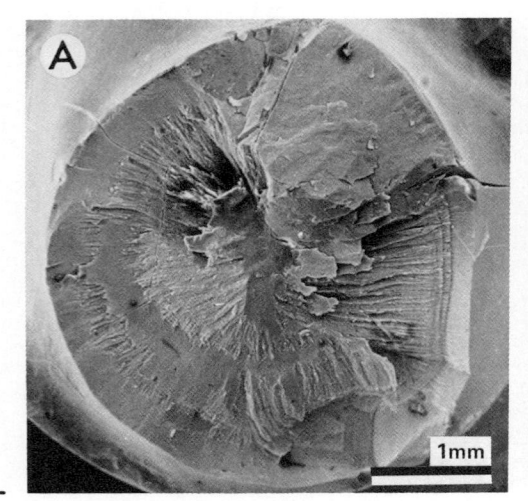

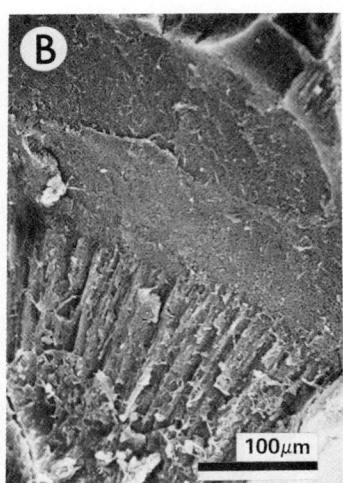

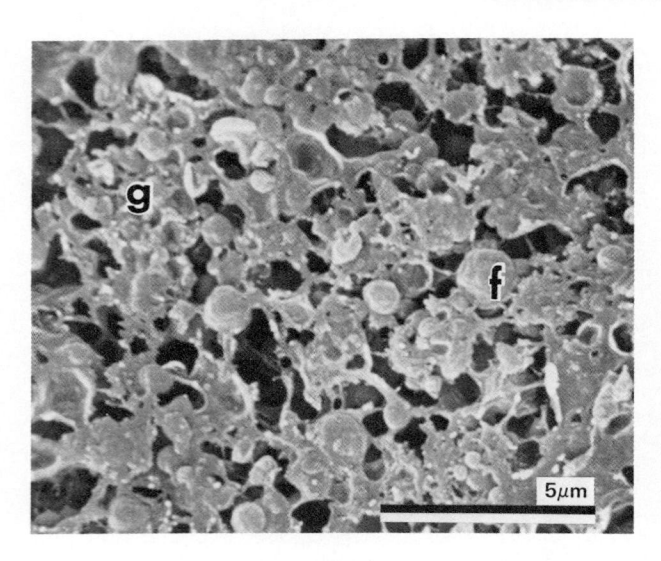

Fig. 7. Cold-stage SEM (173 K; -100°C) of a Cream cheese specimen coated with gold. Individual fat globules (f) and fat globule clusters surrounded by protein (g) are seen in this deeply freeze-etched specimen. [Compare with Fig. 18 and 22].

for the sublimation of solid nitrogen and the condensed frost from the specimens in the vacuum evaporator before coating the specimens with gold. This time may be longer if the bell jar is not cleaned inside after every run and if it contains a gold coating on its inner wall, because the specimens are shielded from heat radiation in the room and their temperature rises more slowly than in a clean bell jar.

SEM of uncoated biological specimens is based on the presence of ions in the frozen aqueous phase which may provide sufficient conductivity to operate the microscope at a low accelerating voltage of 3 to 5 kV. However, because of the fine microstructure of milk products, accelerating voltage of 5 to 20 kV is required to obtain the necessary resolution. No results have yet been reported on cold-stage SEM of uncoated milk products. If fracturing is done outside the electron microscope and the specimen cannot be transferred in a closed container into the microscope, water vapour condenses on the fractured surface and contaminates it. If SEM is carried out before the frost sublimes off, the frost particles may, in the case of some milk products, simulate fat globules (Fig. 8) and lead to artifacts.

Fig. 8. Contamination of the surface of an uncoated specimen with condensed frost in the form of globules and minute crystals (arrows).

The frost vanishes at -100°C and 0.3x10⁻⁴ torr within 30-40 min, after which freeze-etching of the specimen is started. Exposure of fat globules and recession of the aqueous phase in fractured milk products reduce conductivity and may cause charging artifacts. Fig. 9 shows an uncoated Cream cheese specimen examined at 173 K (-100°C)(40 μA at 10 kV). Uncoated specimens gradually change their appearance under the microscope. The quality of micrographs is worse than with metal-coated specimens (Fig. 7).

1.3. SEM of replicated specimens

The impossibility of drying high-fat milk products for conventional SEM without introducing artifacts in their structure and the need to use a cold-stage attachment in the electron microscope to examine freeze-fractured specimens as a low temperature were by-passed by Katoh[4], who devised a technique, in which a freeze-fractured specimen is coated with gold, the gold coating is separated from the specimen in the form of a replica, and the replica is examined by conventional SEM at ambient temperature.

Further modifications, in addition to those mentioned in the preceding section, involved the design of aluminum dishes holding 3 to 7 specimens (Fig. 10):

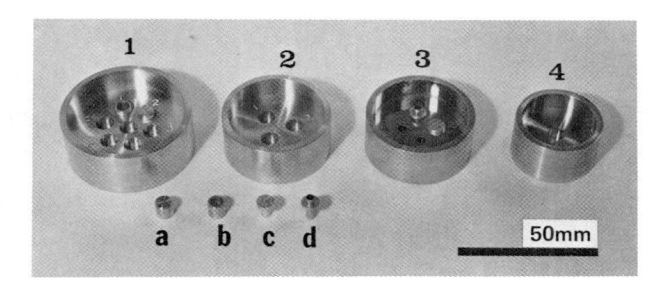

Fig. 10. Aluminum dishes for gold-coating of freeze-fractured specimens by modified Katoh's procedure[4]. Cavities in the bottoms of dishes 1, 2, and 4 accommodate modified tubular holders (b) or aluminum plugs (a). Dish 3 accommodates hollow SEM holders[60] (d; Fig. 5) or special T-shaped holders (c), which can be also used as plugs to fill unused cavities.

If less than 3 or 7 specimens are processed at the same time, the cavities drilled in the bottom of each dish for positioning of the specimen holders are filled with aluminum plugs to eliminate void spaces in which solid nitrogen and frost might persist. A circular groove in the outer side of the bottom of each dish has been provided to fit 3 pins in the plate of the rotary coater to facilitate positioning of the LN₂-filled dish in the vacuum evaporator (Fig. 11). Dimensions of the dishes allow a sufficient volume of LN₂ to be accommodated to protect the specimens, yet the wall of each dish is low to permit the specimens to be coated with gold at an angle acute with respect to the fracture plane. The other portion of gold is evaporated at an angle of ≈80°.

Fig. 11. Arrangement illustrating Katoh's procedure[4] for rotary coating of fractured specimens in an aluminum dish. Arrow points to gold, which will be evaporated at an acute angle (the other source of gold is not shown).

Gold-coated specimens are brought to ambient temperature and the coating on each specimen is separated in the form of a replica. The replicas are floated on water, cleaned by techniques common to TEM replicas (discussed in the section on freeze-fracturing and freeze-etching), and lifted with a platinum loop and placed on single-slot grids (3.05 mm in diameter) coated with formvar and carbon films (Fig. 12). Cleaning of replicas of high-fat milk products is easier if the replicas are lifted on formvar film-coated 150-mesh (hexagonal) grids and gently bathed in chloroform; both the residual fat and the formvar film are

Fig. 9. (facing page) Cold-stage SEM (173 K; -100°C) of an uncoated Cream cheese specimen frozen by the standard procedure[125] and freeze-fractured under liquid nitrogen. Deep freeze-etching resulted in decreased conductivity and charging artifacts, yet fat globules (f) and fat globule clusters covered with protein (g) are clearly evident.

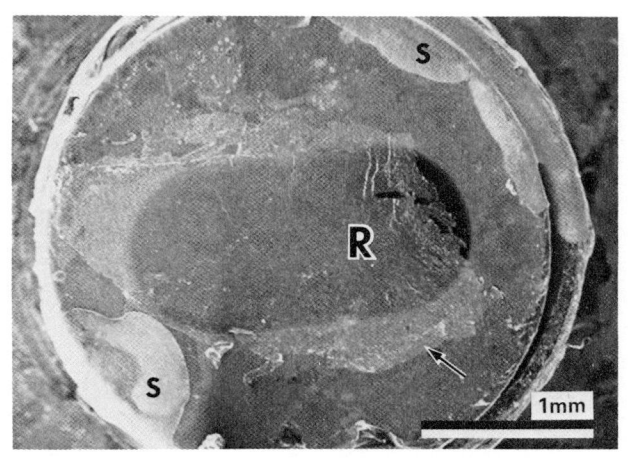

Fig. 12. A gold replica (R) positioned on a single-slot grid, 3.05 mm in diameter, coated with formvar film and carbon. The grid is attached to a hollow SEM holder[60] using a conductive silver cement (s). Interference by backscattered electrons from the metal support appears in the form of markedly lighter areas (arrow) of the replica lying on the metal part of the grid.

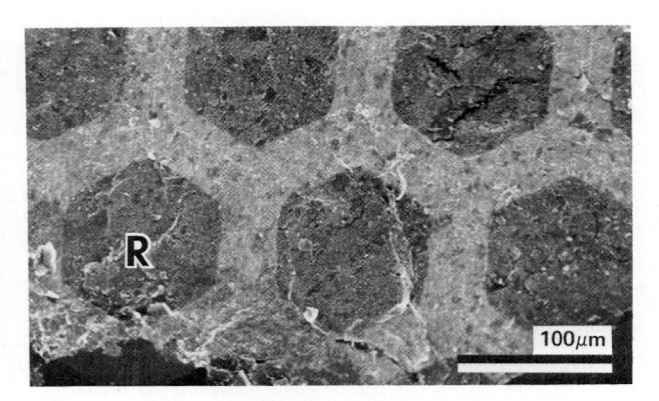

Fig. 13. Part of a gold replica (R) positioned on a bare 150-mesh grid, 3.05 mm in diameter. The grid is attached to a hollow SEM holder[60]. Interference by backscattered electrons from the metal support makes parts of the replica lying on grid bars lighter.

dissolved leaving the gold replica clean. The grids are mounted on hollow SEM holders[60] (Fig. 5) to provide free space under the replica. Should the replica be positioned directly on a metal support as suggested by Katoh[4], backscattered electrons from the metal support might interfere with electrons coming from the replica (G.H. Haggis - personal communication), particularly at a higher electron gun potential[3]. This interference is demonstrated in portions of the replicas lying on the solid grid (Fig. 12) or on the grid bars (Fig. 13) as markedly lighter areas.

Micrographs of the gold replicas of cheese and butter obtained by Katoh[4] closely resemble those obtained by cold-stage SEM in this laboratory. A good agreement was also obtained with gold replicas of Cream cheese (Fig. 14) as compared to cold-stage SEM of gold-coated (Fig. 7) and uncoated specimens (Fig. 9).

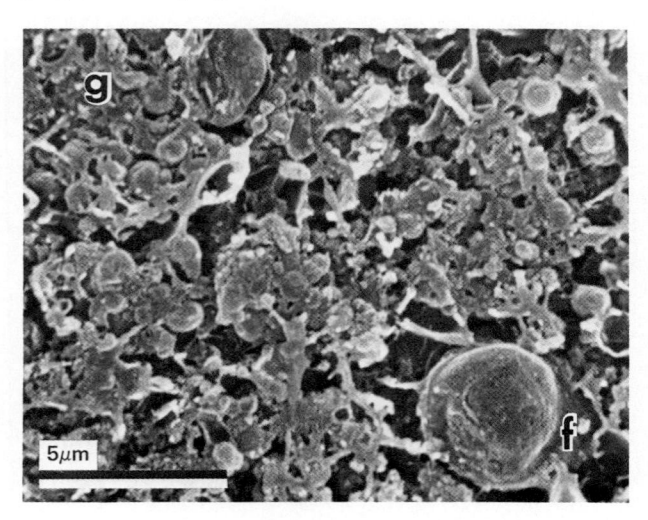

Fig. 14. Gold replica of a Cream cheese specimen examined by SEM.
f = Individual fat globules. g = Clusters of fat globules cemented with protein. [Compare with Fig. 7 and 9].

2. Transmission electron microscopy

TEM comprises all techniques in which the specimen is placed in the electron beam and the enlarged shadow is examined (Table 1). There are various methods of preparing the specimen for this kind of study.

2.1. Negative staining

Negative staining is probably the easiest TEM technique. The specimen is in the form of a suspension of submicroscopical particles semitransparent to the electron beam. Addition of phosphotungstic acid (PTA), sodium phosphotungstate, or ammonium molybdate solutions[64] to the suspension makes the medium, but not the particles, electron-dense. After a thin layer of the suspension is dried, the electron beam passes only through the semitransparent particles under study and is absorbed by the surrounding stain. The particles appear light against a dark background in the micrographs (Fig. 15).

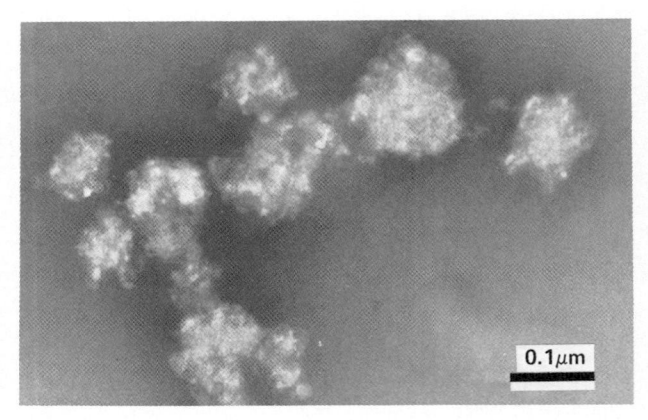

Fig. 15. A negatively stained preparation of fixed, washed, and resuspended casein micelles.

Irradiation of formvar- and carbon-coated grids with ultraviolet light or exposure to glow discharge[65] make the grids wettable for aqueous suspensions. Most suspensions are diluted prior to mixing with the PTA solution to prevent overlapping of the particles. A proper pH of the mixture is also an important factor. A preliminary fixation with formaldehyde[66,67] or glutaraldehyde[29] is advisable particularly with casein micelles. Interfering whey proteins are separated from casein micelles by ultracentrifugation and washing of the sedimented micelles prior to their negative staining[68]. A suspension under study is usually mixed with an equal volume of a 2% PTA solution and a small droplet of the mixture is placed on prepared grids. Excessive liquid is removed after several min by touching the droplet with a piece of a filter paper and a thin layer of the mixture left on the grid is allowed to dry for the electron microscopical examination.

This technique is limited to dilute suspensions and, hence, has been used mostly to study the ultrastructure of casein micelles[66-68], lipoprotein membrane fragments[69], and bacteriophage[70] in lactic cultures. As PTA penetrates porous particles such as the casein micelles, it shows their corpuscular ultrastructure in great detail.

Calapaj[66] used negative staining to study the ultrastructure of bovine and human casein micelles in fresh milk and in milk acidified with lactic acid. Uusi-Rauva et al.[67] examined the effects of various pasteurization and storage temperatures on the disintegration of casein micelles into their subunits. Recently, Creamer and Matheson[71] showed by negative staining that casein micelles increased in size with heat treatment, the effect being greater at higher pH and higher temperature. Surfaces of the casein micelles were found to be more diffuse after heating to 403 K (130°C) than to 373 K (100°C). Snoeren[72] and Snoeren et al.[73] used negative staining to study interactions of milk proteins with κ-carrageenan. Keenan et al.[74] found distinct morphological differences between plasma membrane and milk fat globule membrane.

2.2. Metal-shadowing

Metal-shadowing is another technique most useful for studying suspensions. A dilute suspension of particles such as casein micelles is fixed and dried on a formvar film-coated grid or on freshly cleaved mica sheet or on a glass slide coated with a carbon film. Rose and Colvin[75,76] substituted a dilute $CaCl_2$ solution for another fixative to prevent disintegration of casein micelles during the dilution of milk. The dried particles are then shadowed with platinum[77] or a platinum and palladium alloy[68].

The dimensions of the suspended particles under study determine the optimal angle for shadowing; the smaller the particles, the more acute the angle (2 to 30°). Sharp shadow outlines are obtained at high vacuum (1.3 to 0.13 mPa, i.e. 10^{-5} to 10^{-6} torr) on cold surfaces when the scattering of platinum atoms and their deposition on areas shielded by the particles is greatly reduced. If shadowing is done on a mica sheet or on a microscope glass slide, the shadowed particles are coated with carbon at 90° and the carbon film is floated on water, cleaned, and lifted on a grid for electron microscopical examination[77].

Because the electron beam passes through the shadowed area and exposes photographic material, the shadow appears dark on the negative, whereas areas with a platinum deposit produce lighter images. Human eyes perceive the negative as an object lit with light at a low angle. To obtain a print of this more familiar version, an intermediate negative is made by copying or by photographing[78] the initial negative on another photographic film or plate, by copying the print[79], or by developing the exposed film using an image-reversing process[80] similar to slide processing[81]. The advantage of the last process is the elimination of contamination of the film with dust particles, which may take place during film copying; film exposure differs from the exposure used with the conventional film development and is established experimentally. The above techniques are also useful for TEM of replicas.

Fig. 16 shows casein micelles shadowed with platinum at a 30° angle; prints made from an intermediate negative (Fig. 16 A) as well as from the original negative (Fig. 16 B) are presented. The negative version of the images of shadowed specimens is also quite frequently encountered in the literature.

Using several techniques, Shimmin and Hill[68]

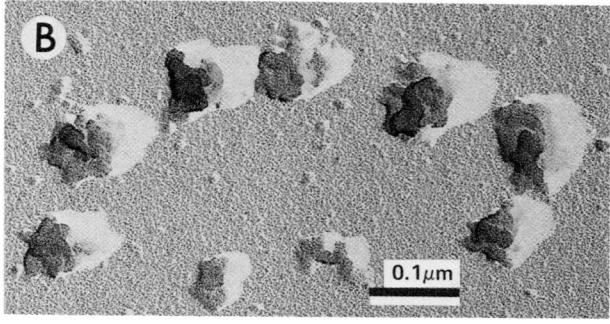

Fig. 16. Fixed, washed, and resuspended casein micelles shadowed with platinum at 30°.
A = Print made from an intermediate negative.
B = Print made from the original negative.

studied the ultrastructure of casein micelles in great detail; fine structure made up of spherical units was best seen in micrographs of shadowed micelles which had been previously fixed with OsO_4, centrifuged, and resuspended in water. The risk of reporting artifacts which could occur, for example, by performing shadowing with relatively coarse metal particles, was eliminated by a comparison with negative staining and thin-sectioning. Parry and Carroll[82] used antibodies to κ-casein to study the distribution of κ-casein in milk micelles.

2.3. Thin-sectioning

Techniques reviewed in the preceding sections show primarily the external features rather than the internal microstructure of the specimen. The standard technique used to study the internal microstructure is TEM of thin sections. It is suitable for suspensions as well as for solid specimens. The specimen is fixed, dehydrated in a graded series of alcohol or in acidified 2,2-dimethoxy-propane[21,39,41], impregnated with a resin monomer[83], and the monomer is polymerized. The embedded specimen is sectioned and the sections (≤100 nm thick) are stained and examined[1].

To handle suspensions and emulsions, Salyaev[84] devised the so-called microcapsulation technique. A glass rod, 0.5 mm in diameter, is dipped in a warm 4% agar sol and the agar gel cylinder thus formed is filled with the liquid specimen by aspirating it with the glass rod functioning as a piston[85]. The ends of the agar gel tube are sealed with agar gel (Fig. 17).(A simpler modification of this technique is shown by G.G. Jewell in this volume). The agar gel tube encasing the specimen is placed in a fixative and is processed for

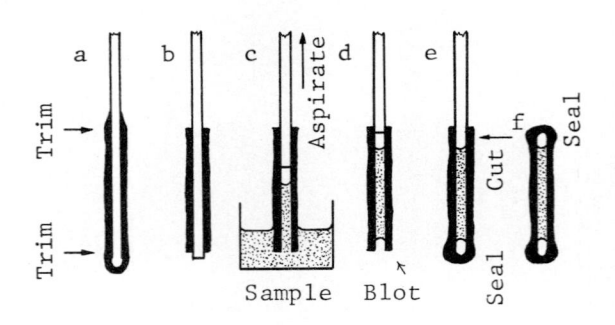

Fig. 17. Salyaev's[84] microcapsulation technique.
a = A glass rod is coated with agar gel (solid black line) and the gel is trimmed.
b = An agar gel tube is formed around the glass rod.
c = The glass rod is used as a piston to aspirate a liquid specimen.
d = The lower end of the agar gel tube is blotted with filter paper.
e = The clean lower end of the agar gel tube is sealed and the upper end is cut off the tube.
f = Also the upper end of the tube is sealed with agar and the encapsulated specimen is immediately placed in a fixative.

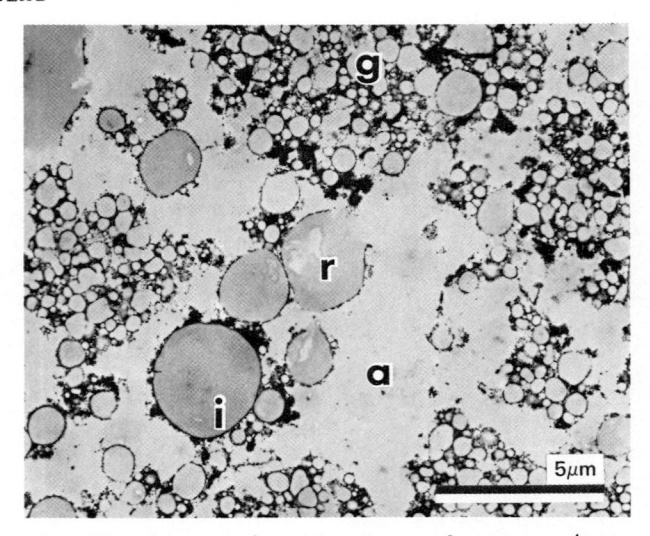

Fig. 18. Thin section of a Cream cheese specimen stained with uranyl acetate and lead citrate. The Cream cheese is composed of large intact (i) and ruptured (r) fat globules and of fat globule clusters (g) in which small fat globules are cemented together by milk proteins (dark bodies). a = Aqueous phase. [Compare with Fig. 7, 9, and 14].

embedding in the same way as a solid specimen. Corpuscular components of the suspension are retained in the gel tube which is permeable to fixing, dehydrating, and embedding agents. The technique has been extensively used[85-88]. Henstra and Schmidt[87] used it to study the effects of homogenization on the association of milk proteins with fragmented fat globules. Harwalkar and Vreeman[88] studied changes in casein micelles in UHTST-sterilized concentrated skim milk also using the microcapsulation technique. The advantage of the microcapsulation technique is the preservation of the initial distribution of the components in the specimen, for example in concentrated skim milk or in cream.

Highly viscous emulsions are difficult to aspirate in the agar gel tube. In such cases, a small piece of the specimen (0.15 to 0.40 mm^3) on the tip of a stainless steel needle is briefly dipped in a lukewarm agar sol. If necessary, the dip-coating is repeated and the needle is withdrawn from the specimen; the hole left by the needle in the specimen is sealed with a droplet of the agar sol and the beaded specimen is processed for electron microscopy in the same way as a solid specimen[61]. Fig. 18 shows Cream cheese examined in this way.

A technique also using agar gel for the preparation of fat globules for electron microscopy was developed by Hobbs[89]. Milk is expressed into an equal volume of a buffered glutaraldehyde solution (pH 7.2) and the mixture is centrifuged. The resulting cream layer is dispersed in a cacodylate-HCl buffer, retrieved by centrifugation, and postfixed in a 1% OsO$_4$ solution. The fat globules, washed and resuspended in distilled water, are collected on a Nucleopore filter (200 nm pore size) and covered with a thin layer of 2% agar. The agar gel is block-stained with a uranyl acetate solution, cut into small blocks, and embedded in Araldite for sectioning.

Another method of preparing suspensions and emulsions for embedding is mixing them with an agar sol[90-92], although agar fibres are visible under the electron microscope. Some authors separate the suspended particles such as casein micelles and fat globules by ultracentrifugation and process the pellet[13, 29, 69] or the floating compacted cream layer (as mentioned by Hobbs[89]) in the same manner as a solid specimen. However, all the techniques except the one by Salyaev[84] concentrate, dilute, or contaminate the specimen and disrupt the initial particle distribution.

A high fat content or extremely large fat globules in the milk product under study may cause difficulties during sectioning. Fat (lipids) is partially fixed during postfixation of the specimen in OsO$_4$[93, 94]; this treatment prevents the fat from being extracted with lipophilic solvents during the embedding procedure. Once embedded, fat particles form a heterogeneity in the resin, because they are softer than the resin medium and become compressed during sectioning. This may result in a striated image of the fat globules (Fig. 19). The use of a softer resin somewhat remedies the problem.

Although the sections are most frequently stained with uranyl acetate and lead citrate, there have been exceptions to this rule. Knoop et al.[95] studied the distribution of calcium phosphate in casein micelles and resorted to the absorption of the electron beam by calcium alone. Alternatively, the authors replaced calcium with barium, lead, and uranium.

Specific cytochemical methods at the ultrastructural level have been used by several authors, particularly by those studying the distribution of κ-casein in casein micelles. Kudo et al.[96] used periodic acid-silver methenamine staining; compared to uranyl acetate or lead citrate, silver

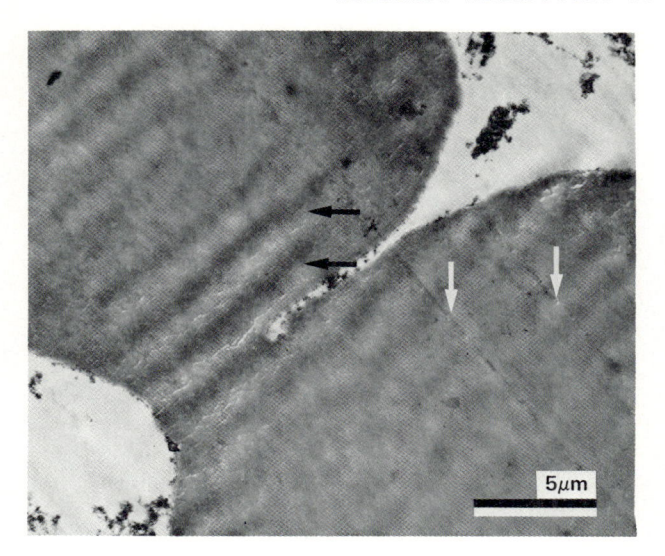

Fig. 19. Striation of large fat globules due to a defect (periodic compression) during sectioning. The striations (dark arrows) are perpendicular to the direction of sectioning indicated in this micrograph by scratch marks (light arrows) caused by a defective knife.

staining was less uniform as the silver granules were quite large. The hypothesis by the authors that κ-casein is distributed evenly throughout casein micelles smaller than 100 nm in diameter but is localized mainly in the outer shell of larger micelles has not been confirmed by Horisberger and Vonlanthen[97], who worked with plant lectin-labelled gold granules. It is possible that a slow penetration of the periodic acid-silver methenamine stain into larger micelles was responsible for the difference. Using a similar technique, Horisberger et al.[98] studied the location of glycoproteins on the fat globule membrane.

Electron microscopy of thin sections was used by several authors in studies of intermicellar relationships in rennet-treated milk[99-100] and in studies of the development of microstructure in yoghurt[15,16,91], Cottage cheese[17], and cheese[21,24,54,101-103]; results of such studies were reviewed earlier[104,105]. Recently, soft (melting) and hard (nonmelting) process cheeses were found by Kimura et al.[106] and Taneya et al.[107] to differ in the microstructure of the protein matrix at high magnifications: the soft cheese consisted of single protein particles, 20-25 nm in diameter, whereas the hard process cheese was composed of a network-like structure of longer protein strands. A comparison of process cheese and Gouda cheese microstructure was made by Kimura and Taneya[108]. Emulsification of fat and gradual degradation of the emulsifying salt crystals in process cheese were followed by Rayan et al.[109] using both thin-sectioning and SEM. It is probable that the water-soluble crystals in the process cheese were dissolved and left cavities in the protein matrix while the protein had been fixed; the shapes of the crystals were thus preserved as shown in Fig. 20. Probably the same phenomenon took place in specimens examined by SEM. Microcrystalline inclusions of calcium salts in cheese were studied by Brooker et al.[110] Thin-sectioning was used in conjunction with freeze fracturing in a study of crystalline inclusions in Emmental (Swiss) cheese[102]; similar inclusions in Emmental cheese were studied by SEM and analyzed by energy dispersive spectrometry[111] and were considered to consist of tyrosine.

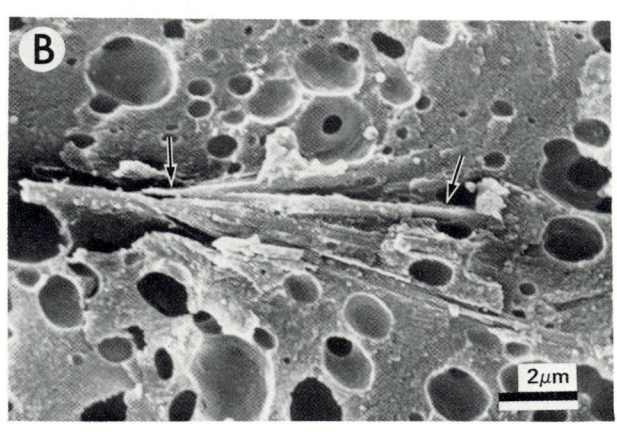

Fig. 20. Presence of tetrasodium pyrophosphate crystals in process cheese is demonstrated by TEM of a thin section (A) and by SEM of a defatted specimen (B). It is probable that the original crystals were washed out from both specimens during preparatory steps; consequently, cavities (light crystal-shaped areas in A) were left in the protein matrix. Unless the presence of Na and P in the crystal-like structures in B (arrows) is proven by energy dispersive spectrometry, such structures may be considered to consist of protein compacted between the real crystals in the original curd (arrows in A).

To calculate dimensions of globular particles from their sections, a mathematical procedure was developed[112] and later amended[113]. The former procedure[112] was used[114,115] to calculate the size distribution of casein micelles in cow's milk.

2.4. Freeze-fracturing and freeze-etching

Any milk product may be examined by freeze-fracturing and freeze-etching, but these techniques are particularly suited for studies of liquid specimens (milk and cream[114-119]) and high-fat milk products (butter[120, 121]). They are the only electron microscopical techniques which make it possible to examine the genuine microstructure of ice cream[122, 123].

The specimen is rapidly frozen, fractured at 143 to 173 K (-130 to -100°C) and the fractured surface is replicated with platinum either immediately or after a certain period of freeze-etching, during which a thin layer of ice sublimes from the specimen and exposes structures initially covered with ice. The deposited platinum is reinforced with carbon.

Rapid freezing is a crucial step in the entire procedure and most artifacts develop at this stage, as was mentioned earlier. Solid specimens such as yoghurt, Cottage cheese etc. are frozen either in their natural state without fixation or are fixed and consecutively impregnated with a cryoprotective agent such as glycerol[116, 119]. Only limited freeze-etching is possible if cryoprotective agents are used[124] as they do not sublime. The natural state of emulsions and of high-fat products is examined without the use of cryoprotective agents. In such cases, so-called spray-freezing may be applied[125]. This technique made it possible to obtain quantitative data such as molecular weights, dimensions, and shapes of molecules of individual milk proteins[126, 127]. Buchheim[124] developed a method of forming a fine emulsion of milk or a fine suspension of a protein gel in paraffin oil; standard freezing of these emulsions and suspensions did not lead to the ice crystal formation in the specimens under study ("standard freezing" is the term[125] for dipping a small metal specimen holder with approximately 1 mm^3 of the sample into liquid freon at 123 K (-150°C). The use of organic solvents was extended to milk powder particles. Milk powders were suspended in paraffin oil, glycerol, or polyethylene glycol[128-130], rapidly frozen, freeze-fractured, and replicated. A different fracturing behaviour obtained with dioxane[131] was demonstrated with casein micelles. Whereas in aqueous media the casein micelles as well as the submicelles were fractured in the same plane as the aqueous phase, casein micelles suspended in dioxane were fractured along their surfaces and the surface structures were revealed by appropriate freeze-etching.

Fracturing is accomplished with the specimen frozen in one of a variety of holders. One type of holder is a thin-walled metal (silver) tube consisting of 2 parts: the lower part is firmly attached to the freeze-fracturing table cooled to the desired temperature and the upper part is knocked off by a cooled mechanical arm after the desired temperature and vacuum are achieved. Incorporation of air bubbles in the tubular holders is difficult to avoid. As soon as the specimen coated with platinum and carbon is melted for the separation of the replica, contraction of the air bubbles leads to breakage of the replica. To eliminate this defect, open-end holders have been designed. In addition, the freeze-fracturing tables have been modified in this laboratory to accommodate 3 specimens at the same time.

In another mode of operation the specimen is placed on a metal disc and fractured with a cooled blade which shaves off thin layers of the specimen. The fractured surface obtained by the latter technique is smoother[119, 122] than that obtained by the former[61]. Photographs of 3 different freeze-fracturing tables are presented in Fig. 21:

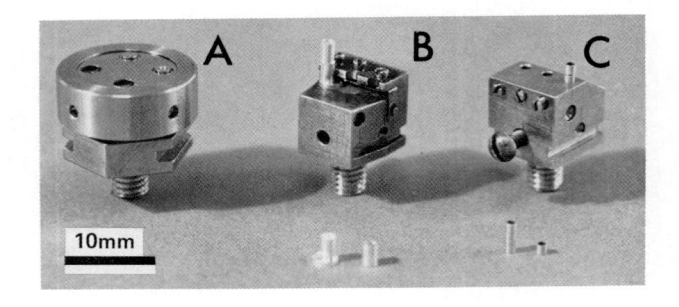

Fig. 21. Freeze-fracturing tables modified (A) or designed (B and C) to fit the Polaron Freeze-Fracturing Module.
A = Balzers' table accommodating 4 dish-like specimen holders.
B and C = Tables developed in the author's laboratory to accommodate 3 footed tubular Polaron holders (B) or 3 open-end tubular holders (C). The holders are shown in the foreground.

Cleaning of replicas obtained with milk products is usually simple unless additives, which might be difficult to remove, are present. Sodium hypochlorite[119, 120], concentrated ammonia[132], sulfuric acid[133], and chromic acid[46] solutions remove all the proteins. Large amounts of fat are removed most easily with organic solvents such as acetone, chloroform etc. The surface tension of such solvents is lower than that of water and, hence, the replicas sink into such solvents. An easy way to retrieve them is to transfer the clean replica into a solvent immiscible with water and lighter than water such as petroleum ether[132, 134]; water is pipetted under this solvent and after the layer of the organic solvent is removed and evaporated, the replica floats on the water surface. This procedure also serves as a criterion of the replica cleanliness: a greasy film on the water surface indicates that additional cleaning is necessary[134]. If the replicas are lifted on bare grids, gentle bathing in chloroform removes all the residual fat as was mentioned earlier.

Freeze-fracturing has been extensively used in various studies, for example on casein micelle subunits[117], on the size distribution of casein micelles[115, 118, 135], on the distribution of fat and casein in partly and fully homogenized milk[136] and in evaporated milk[137], on the composition of the fat globule membrane[138-140], on the composition and crystallization of fat in fat globules in milk[132, 141] cream[119], and butter[120, 121], on the development of microstructure in cheese[102, 142], in ice cream[122, 123] etc.

Fig. 22 shows a freeze-fractured and replicated Cream cheese specimen. The higher resolution of TEM reveals considerably more detail than SEM (Fig. 7, 9, and 14) and is in a good agreement with thin-sectioning (Fig. 18).

2.5. Replication of dried specimens

The presence of water in freeze-fractured milk products causes the replica of the fracture to appear two-dimensional rather than

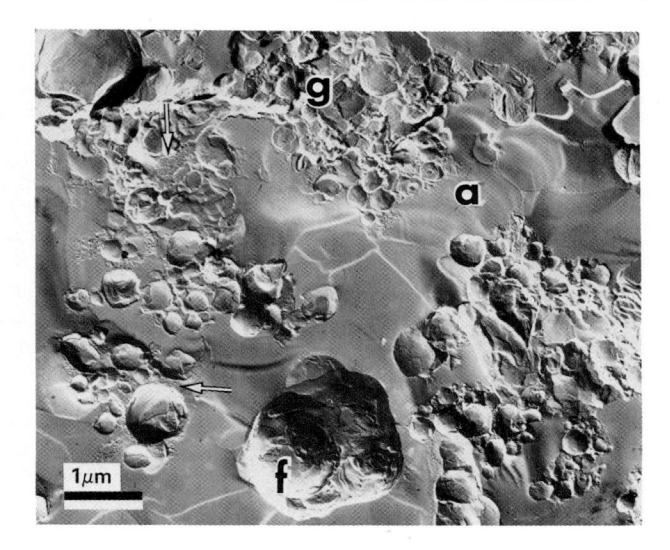

Fig. 22. Platinum replica of a Cream cheese specimen. The Cream cheese is composed of large fat globules (f) and of fat globule clusters (g) in which small fat globules are cemented together by milk protein (arrows). a = Aqueous phase.

three-dimensional although not as flat as a thin section. Freeze-etching removes some ice from the fractured plane, but the amount of ice removed is limited to a certain extent, otherwise the structure may collapse[143]. Recently Haggis and Bond[144] devised a technique for replication of dried specimens. The technique has been adapted for some milk products[46]. The specimen is sectioned by hand ($\simeq$0.5 mm) and the sections are left in the original state or are fixed and impregnated with a 30% glycerol solution as a cryoprotective agent, sandwiched in an aluminum foil, frozen by the standard freezing procedure[125], and freeze-fractured under LN_2. The fragments are transferred into a warm (293 to 298 K; 20 to 25°C) glutaraldehyde solution where they are melted and simultaneously fixed (if unfixed specimens had been freeze-fractured). The specimens are then dehydrated and CPD. In this way, it is possible to use cryoprotective agents in order to eliminate ice crystal formation during freezing. Alternatively, the specimen is fixed first, dehydrated, impregnated with absolute alcohol, and frozen in LN_2, as there is no risk of crystallization of alcohol by freezing; the specimen is fractured and the fragments are melted in absolute alcohol and CPD. Dry fragments are wrapped in an aluminum foil leaving only the fractured surface exposed, and coated with platinum at 45° while the specimen is rotated. The platinum coating is reinforced with carbon and the replicated portion of the specimen is cut off from the fragment, placed on a glass slide, and covered with a perspex film held on a rubber ring. The film is made from a 0.6% (w/v) perspex solution in chloroform[46,144] using the same procedure as that for making formvar film[2]. The glass slide with the replica is placed in a Petri dish saturated with chloroform vapours; there the film stretches, adheres firmly to the replica, and becomes attached to the glass slide. The film is then floated on water and the material adhering to the replica is digested away with a chromic acid solution. The replica is cleaned with water, lifted on a bare grid or on a grid covered with a carbon film, and is bathed in chloroform to dissolve the perspex film. The dry replica is examined by TEM.

This technique makes it possible to obtain electron microscopic images similar to SEM at a considerably better resolution. For example, the annular space around casein micelles in glucono-δ-lactone-induced skim milk gels[145], 50 to 80 nm wide, was apparently obscured by gold used as a coating for SEM, but was observable by replication of the dried specimen[46] (Fig. 23). The corpuscular ultrastructure of casein micelles was also revealed by this technique. The technique is suitable for studies of the ultrastructure of the individual components of milk products (casein micelles, protein aggregates, corpuscular additives, bacterial surfaces and filaments etc.) rather than for studies of the overall microstructure (porosity, distribution of fat globules, casein micelles, bacteria etc.). Because of the nature of the replica, pairs of stereo micrographs provide three-dimensional images of great depth[46]. However, platinum deposited on vertical walls of casein micelles, bacteria etc. lead to very high contrast of the micrographs (Fig. 23). Specimens which are too soft and would not withstand handling, may be compacted by centrifugation[46] or be embedded in a gel[144]. The initial stages of this technique may

Fig. 23. Platinum replica of a dried specimen (A) and a comparison micrograph of a thin-sectioned specimen (B) of a glucono-δ-lactone-induced skim milk gel[50,145] compacted by centrifugation. Free annular space between the casein micelle core and the outer lining is indicated with arrows.

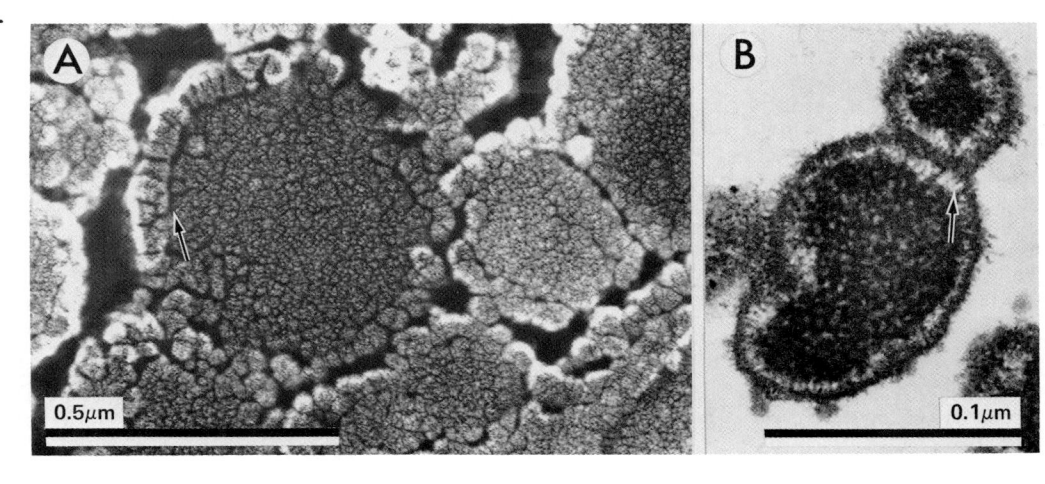

135

be used for SEM, *i.e.* the dried fragments may be coated with gold rather than with platinum, as was shown in Fig. 3.

Impregnation of unfixed specimens with glycerol followed by fixation of the fragments has been designed[146] to prevent retention of soluble proteins in the network which might obscure insoluble structures. Interestingly, the fracture plane runs through the submicroscopical particles in the specimen impregnated with glycerol and around such particles in specimens impregnated with ethanol (Fig. 3). This finding has been in agreement with that by Murphy (J.A. Murphy - personal communication).

Conclusion

Electron microscopy of milk products is a rapidly expanding discipline. The initial period, characterized by collecting data as in every other scientific discipline, has gradually developed into the research of mutual relationships. It is recognized today that changes at the ultrastructural level (structure proper of the individual components such as casein micelles and fat globules) and the microstructural level (structure of complexes, aggregates, networks *etc.*) reflect the manufacturing conditions and the presence of additional ingredients (lactic microorganisms, emulsifiers, stabilizers *etc.*) and affect the physical and sensory properties of the product. Many phenomena have yet to be explained, for example the differences in the microscopical appearance of various spray-dried milk powders, the core-and-lining ultrastructure developing in casein micelles in milk heated to 363 K (90°C) at pH 5.5, or the effects of various emulsifying salts on the ultrastructure of the protein matrix in process cheese to name a few. Electron microscopy is providing useful clues to the above and to other problems as new techniques are being developed to meet specific needs. The objective of this review was to bring the techniques and reports on them to the attention of the concerned researchers.

Acknowledgment

All micrographs and photographs presented in this review were obtained in the author's laboratory. Mr. A.F. Yang is the author of Fig. 15 and 16. Technical assistance provided by Mr. J.A.G. Larose is acknowledged. The author thanks Dr. V.R. Harwalkar for useful suggestions, Mr. E.F. Bond for technical demonstrations, discussion, and reviewing the manuscript, and Mr. R.P. Hocking of the Engineering and Statistical Research Institute, Research Branch, Agriculture Canada in Ottawa, for design and manufacture of aluminum dishes, freeze-fracturing tables, specimen holders, and other aids. Electron Microscope Centre, Research Branch, Agriculture Canada in Ottawa provided facilities. This review is Contribution 453 from Food Research Institute, Agriculture Canada, Ottawa.

References

1. Hayat M.A. Principles and Techniques of Electron Microscopy: Biological Applications. Vol. 1. Van Nostrand Reinhold Co., Toronto, Ontario, Canada, 1970, 412 pp.

2. Mercer E.H. and Birbeck M.S.C. Electron Microscopy. A Handbook for Biologists. 3rd ed. Blackwell Scientific Publications, Oxford, England, 1972, 55.

3. Hayat M.A. Principles and Techniques of Scanning Electron Microscopy. Biological Applications. Vol. 1. Van Nostrand Reinhold Co., New York, N.Y., 1974.

4. Katoh M. SEM replica technique for butter and cheese. J. Electron Microscopy 28, 1979, 199-200.

5. Buma T.J. and Henstra S. Particle structure of spray-dried milk products as observed by a scanning electron microscope. Neth. Milk Dairy J. 25, 1971, 75-80.

6. Buma T.J. Teilchenporosität von sprühgetrockneter Milch. Milchwissenschaft 33, 1978, 538-540.

7. Warburton S. and Pixton S.W. The moisture relations of spray-dried skimmed milk. J. Stored Prod. Res. 14, 1978, 143-158.

8. Warburton S. and Pixton S.W. The significance of moisture in dried milk. Dairy Ind. Internatl. 43(4), 1978, 23-27.

9. Roetman K. Crystalline lactose and the structure of spray-dried milk products as observed by scanning electron microscopy. Neth. Milk Dairy J. 33, 1979, 1-11.

10. Jelen P., Kalab M. and Greig R.I.W. Water-holding capacity and microstructure of heat-coagulated whey protein powders. Milchwissenschaft 34, 1979, 351-356.

11. Saltmarch M. and Labuza T.P. SEM investigation of the effect of lactose crystallization on the storage properties of spray-dried whey. Scanning Electron Microsc. 1980; III: 659-665.

12. Short J.L. The water absorption capacity of heat precipitated whey proteins. New Zealand J. Dairy Sci. Technol. 15, 1980, 167-176.

13. Kalab M. Possibilities of an electron microscopical detection of buttermilk made from sweet cream in adulterated skim milk. Scanning Electron Microsc. 1980; III: 645-652.

14. Buma T.J. and Henstra S. Particle structure of spray-dried caseinate and spray-dried lactose as observed by scanning electron microscope. Neth. Milk Dairy J. 25, 1971, 278-281.

15. Kalab M., Emmons D.B. and Sargant A.G. Milk gel structure. IV. Microstructure of yoghurt in relation to the presence of thickening agents. J. Dairy Res. 42, 1975, 453-458.

16. Kalab M., Emmons D.B. and Sargant A.G. Milk gel structure. V. Microstructure of yoghurt as related to the heating of milk. Milchwissenschaft 31, 1976, 402-408.

17. Glaser J., Carroad P.A. and Dunkley W.L. Surface structure of Cottage cheese curd by electron microscopy. J. Dairy Sci. 62, 1979, 1058-1068.

18. Eino M.F., Biggs D.A., Irvine D.M. and Stanley D.W. A comparison of microstructure of Cheddar cheese curd manufactured with calf rennet, bovine pepsin, and porcine pepsin. J. Dairy Res. 43, 1976, 113-115.

19. Eino M.F., Biggs D.A., Irvine D.M. and Stanley D.W. Microstructural changes during ripening of Cheddar cheese produced with calf rennet, bovine pepsin, and porcine pepsin. Can. Inst. Food Sci. Technol. J. 12, 1979, 149-153.

20. Hall D.M. and Creamer L.K. A study of the submicroscopic structure of Cheddar, Cheshire and Gouda cheese by electron microscopy. New Zealand J. Dairy Sci. Technol. 7, 1972, 95-102.

21. Kalab M. Milk gel structure. VI. Cheese texture and microstructure. Milchwissenschaft 32, 1977, 449-458.

22. Stanley D.W., Emmons D.B., Modler H.W. and Irvine D.M. Cheddar cheese made with chicken pepsin. Can Inst. Food Sci. Technol. J. 13, 1980, 97-102.

23. Rüegg M., Sieber R. and Blanc B. Untersuchungen der Feinstruktur von Käseteigen und Käserinden mit Hilfe der Raster-Elektronen-Mikroskopie. Schweiz. Milchwirt. Forschung 3, 1974, 1-5.

24. Taranto M.V., Wan P.J., Chen S.L. and Rhee K.C. Morphological, ultrastructural and rheological characterization of Cheddar and Mozzarella cheese. Scanning Electron Microsc.1979; III: 273-278.

25. Stanley D.W. and Emmons D.B. Cheddar cheese made with bovine pepsin. II. Texture-microstructure-composition relationship. Can. Inst. Food Sci. Technol. J. 10, 1977, 78-84.

26. Johari O. and DeNee P.B. Handling, mounting and examination of particles for scanning electron microscopy. Scanning Electron Microsc. 1972: 249-256.

27. DeNee P.B. Collecting, handling and mounting of particles for SEM. Scanning Electron Microsc. 1978; I: 479-486.

28. Kalab M. and Emmons D.B. Milk gel structure. III. Microstructure of skim milk powder and gels as related to the drying procedure. Milchwissenschaft 29, 1974, 585-589.

29. Carroll R.J., Thompson M.P. and Nutting G.C. Glutaraldehyde fixation of casein micelles for electron microscopy. J. Dairy Sci. 51, 1968, 1903-1908.

30. Hermansson A.M. and Buchheim W. Methods for characterization of protein gel structures by scanning and transmission electron microscopy. J. Colloid Interface Sci. 81, 1981, 519-529.

31. Umrath W. Cooling bath for rapid freezing in electron microscopy. J. Microsc. 101(1), 1974, 103-105.

32. Angold R.E. Cereals and bakery products. In: Food Microscopy, Vaughan J.G. (ed.), Acad. Press, London, England, 1979, 75-138.

33. Cohen A.L. Critical point drying. In: Principles and Techniques of Scanning Electron Microscopy. Biological Applications. Hayat M.A. (ed.), Vol. 1. Van Nostrand Reinhold Co., Toronto, Ontario, Canada, 1974, 44-112.

34. Cohen A.L. A critical look at CPD. Scanning Electron Microsc. 1977; I: 525-536.

35. Cohen A.L. Critical point drying - Principles and procedures. Scanning Electron Microsc. 1979; II: 303-323.

36. Muller L.L. and Jacks T.J. Rapid chemical dehydration of samples for electron microscopic examination. J. Histochem. Cytochem. 23, 1975, 107-110.

37. Kalab M. Scanning electron microscopy of dairy products: An overview. Scanning Electron Microsc. 1979; III:261-272.

38. Bartlett A.A. and Burstyn H.P. A review of the physics of critical point drying. Scanning Electron Microsc. 1975; I: 305-316, 368.

39. Moor H. Cryotechnology for the structural analysis of biological material. In: Freeze-Etching Techniques and Applications. Benedetti E.L. and Favard P. (eds.), Soc. Franç. Microsc. Electr., Paris, France, 1973, 11-20.

40. Boyde A. Pros and cons of critical point drying and freeze drying for SEM. Scanning Electron Microsc. 1978; II: 303-314.

41. Chabot J.F. Preparation of food science samples for SEM. Scanning Electron Microsc. 1979; III: 279-286.

42. Kalab M. Milk gel structure. VIII. Effect of drying on the scanning electron microscopy of some dairy products. Milchwissenschaft 33, 1978, 353-358.

43. Eino M.F., Biggs D.A., Irvine D.M. and Stanley D.W. Microstructure of Cheddar cheese: sample preparation and scanning electron microscopy. J. Dairy Res. 43, 1976, 109-111.

44. Lewis E.R., Jackson L. and Scott T. Comparison of miscibilities and critical-point drying properties of various intermediate and transitional fluids. Scanning Electron Microsc. 1975; I: 317-324.

45. Flood P.R. Dry-fracturing techniques for the study of soft internal biological tissues in the scanning electron microscope. Scanning Electron Microsc. 1975; I: 287-294.

46. Kalab M. Milk gel structure. XII. Replication of freeze-fractured and dried specimens for electron microscopy. Milchwissenschaft 35, 1980, 657-662.

47. Washam C.J., Kerr T.J. and Todd R.L. Scanning electron microscopy of Blue cheese: Mold growth during maturation. J. Dairy Sci. 62, 1979, 1384-1389.

48. Wells O.C. Scanning Electron Microscopy. McGraw-Hill Book Co., Montreal, Quebec, Canada, 1974, 333.

49. Nickerson A.W., Bulla L.A. and Kurtzman C.P. Spores. In: Principles and Techniques of Scanning Electron Microscopy. Vol. 1. Hayat M.A. (ed.), Van Nostrand Reinhold Co., Toronto, Ontario, Canada, 1974, 159-180.

50. Harwalkar V.R. and Kalab M. Effect of acidulants and temperature on microstructure, firmness, and susceptibility to syneresis of skim milk gels. Scanning Electron Microsc. 1981; III: 503-513.

51. Murphy J.A. Non-coating techniques to render biological specimens conductive. Scanning Electron Microsc. 1978; II: 175-194.

52. Murphy J.A. Non-coating techniques to render biological specimens conductive. Scanning Electron Microsc. 1980; I: 209-220.

53. Rüegg M., Moor U. and Blanc B. Veränderungen

der Feinstruktur von Greyerzerkase im Verlauf der Reifung. Eine Studie mit dem Raster-Elektronen-Mikroskop. Milchwissenschaft 35, 1980, 329-335.

54. Emmons D.B., Kalab M., Larmond E. and Lowrie R.J. Milk gel structure. X. Texture and microstructure in Cheddar cheese made from whole milk and from homogenized low-fat milk. J. Texture Stud. 11, 1980, 15-34.

55. Echlin P. Scanning electron microscopy at low temperature. In: Freeze-Etching Techniques and Applications. Benedetti E.L. and Favard P. (eds.), Soc. Franç. Microsc. Electr., Paris, France, 1973, 211-222.

56. Robards A.W. and Crosby P. A comprehensive freezing, fracturing and coating system for low temperature scanning electron microscopy. Scanning Electron Microsc. 1979; II: 325-344, 324.

57. Schmidt D.G., Henstra S. and Thiel F. Eine einfache Methode für die raster-elektronenmikroskopische Präparation von Käse bei tiefer Temperatur. Beitr. Elektronenmikroskop. Direktabb. Oberfl. 10, 1977, 415-418.

58. Schmidt D.G., Henstra S. and Thiel F. A simple low-temperature technique for scanning electron microscopy of cheese. Mikroskopie (Wien) 35, 1979, 50-55.

59. Schmidt D.G. and van Hooydonk A.C.M. A scanning electron microscopical investigation of the whipping of cream. Scanning Electron Microsc. 1980; III: 653-658, 644.

60. Soni S.L., Kalnins V.I. and Haggis G.H. Localization of caps on mouse β-lymphocytes by scanning electron microscopy. Nature 255(5511), 1975, 717-719.

61. Kalab M., Sargant A.G. and Froehlich D.A. Electron microscopical study of commercial Cream cheese. Scanning Electron Microsc. 1981;III:473-482.

62. Echlin P. and Moreton R. Low temperature techniques for scanning electron microscopy. Scanning Electron Microsc. 1976; I: 753-762.

63. Echlin P. Low temperature electron microscopy: A review. J. Microsc. 112(1), 1978, 47-61.

64. Creamer L.K., Berry G.P. and Matheson A.R. The effect of pH on protein aggregation in heated skim milk. New Zealand J. Dairy Sci. Technol. 13, 1978, 9-15.

65. Milne R.G. and Luisoni E. Rapid immune electron microscopy of virus preparations. In: Methods in Virology. Maramorosch K. and Koprowski H. (eds.), Vol. VI, Acad. Press, New York, N.Y., 1977, 265-281.

66. Calapaj G.G. An electron microscope study of the ultrastructure of bovine and human casein micelles in fresh and acidified milk. J. Dairy Res. 35, 1968, 1-6.

67. Uusi-Rauva E., Rautavaara J.-A. and Antila M. Über die Einwirkung von verschiedenen Temperatur-Behandlungen auf die Caseinmicellen. Eine Elektronen mikroskopische Untersuchung unter Verwendung von Negativfärbung. Meijeritieteellinen Aikakauskirja (Helsinki) 31, 1972, 15-25.

68. Shimmin P.D. and Hill R.D. Further studies on the internal structure of the casein micelles of milk. Austral. J. Dairy Technol. 20, 1965, 119-122.

69. Stewart P.S., Puppione D.L. and Patton S. The presence of microvilli and other membrane fragments in the non-fat phase of bovine milk. Z. Zellforsch. 123, 1972, 161-167.

70. Knoop A.-M. Milchforschung mit dem Elektronenmikroskop. Z. Lebensm. Unters. Forsch. 168, 1979, 305-313.

71. Creamer L.K. and Matheson A.R. Effect of heat treatment on the proteins of pasteurized skim milk. New Zealand J. Dairy Sci. Technol. 15, 1980, 37-49.

72. Snoeren T.H.M. Kappa-Carrageenan. A Study on Its Physico-chemical Properties, Sol-Gel Transition and Interaction with Milk Proteins. NIZO-Verslagen, Ede, the Netherlands, 1976, 64-91.

73. Snoeren T.H.M., Both P. and Schmidt D.G. An electron microscopic study of carrageenan and its interaction with κ-casein. Neth. Milk Dairy J. 30, 1976, 132-141.

74. Keenan T.W., Morré D.J., Olson D.E., Yunghans W.N. and Patton S. Biochemical and morphological comparison of plasma membrane and milk fat globule membrane from bovine mammary gland. J. Cell. Biol. 44, 1970, 80-93.

75. Rose D. and Colvin J.R. Internal structure of casein micelles from bovine milk. J. Dairy Sci. 49, 1966, 351-355.

76. Rose D. and Colvin J.R. Appearance and size of micelles from bovine milk. J. Dairy Sci. 49, 1966, 1091-1097.

77. Carroll R.J., Thompson M.P. and Melnychyn P. Gelation of concentrated skimmilk: Electron microscopic study. J. Dairy Sci. 54, 1971, 1245-1252.

78. Boye H. and Rønne M. Low cost internegatives from electron micrographs of metal shadowed objects. J. Microsc. 112(3), 1978, 353-358.

79. Towe K.M. Electron micrographs of metal shadowed materials: A simple technique for making negative prints. J. Microsc. 116(2), 1979, 281-283.

80. Eastman-Kodak Co., Rochester, N.Y.: How to use the Kodak Direct Positive Film Developing Outfit. 1979, 4 pp.

81. Schulz W.W. and Reynolds R.C. Enhancement of three-dimensional appearance of freeze-fracture images by reversal processing of electron microscopy sheet film. J. Microsc. 112(2), 1978, 249-252.

82. Parry, Jr., R.M. and Carroll R.J. Location of κ-casein in milk micelles. Biochim. Biophys. Acta 194, 1969, 138-150.

83. Luft J.H. Embedding media - old and new. In: Advanced Techniques in Biological Electron Microscopy. Koehler J.K. (ed.), Springer-Verlag, New York, N.Y., 1973, 1-34.

84. Salyaev R.K. A method of fixation and embedding of liquid and fragile materials in agar microcapsulae. Proc. 4th Europ. Regional Conf. Electron Microsc. Rome, II, 1968, 37-38.

85. Henstra S. and Schmidt D.G. Ultradünnschnitte aus Milch mit Hilfe der Mikrokapselmethode.

Naturwissenschaften 57(5), 1970, 247.

86. Henstra S. and Schmidt D.G. The microcapsule technique. An embedding procedure for the study of suspensions and emulsions. LKB Application Note No. 150, 1974.

87. Henstra S. and Schmidt D.G. On the structure of the fat-protein complex in homogenized cow's milk. Neth. Milk Dairy J. 24, 1970, 45-51.

88. Harwalkar V.R. and Vreeman H.J. Effect of added phosphates and storage on changes in ultra-high temperature short-time sterilized concentrated skim-milk. 2. Micelle structure. Neth. Milk Dairy J. 32, 1978, 204-216.

89. Hobbs D.G. An improved method for preparing bovine milk fat globules for electron microscopy. Milchwissenschaft 34, 1979, 201-202.

90. Kalab M. Milk gel structure. VII. Fixation of gels composed of low-methoxyl pectin and milk. Milchwissenschaft 32, 1977, 719-723.

91. Davies F.L., Shankar P.A., Brooker B.E. and Hobbs D.G. A heat-induced change in the ultrastructure of milk and its effect on gel formation in yoghurt. J. Dairy Res. 45, 1978, 53-58.

92. Andrews A.T., Brooker B.E. and Hobbs D.G. Properties of aseptically packed ultra-heat-treated milk. Electron microscopic examination of changes occurring during storage. J. Dairy Res. 44, 1977, 283-292.

93. Crozet N. and Guilbot A. A note on the influence of osmium fixation on wheat flour lipids observation by transmission and scanning electron microscopy. Cereal Chem. 51, 1974, 300-304.

94. Geyer G. Lipid fixation. Acta Histochem., Suppl. XIX, 1977, 209-222.

95. Knoop A.-M., Knoop E. and Wiechen A. Electron microscopical investigations on the structure of the casein micelles. Neth. Milk Dairy J. 27, 1973, 121-127.

96. Kudo S., Iwata S. and Mada M. An electron microscopic study of the location of κ-casein in casein micelles by periodic acid-silver methenamine staining. J. Dairy Sci. 62, 1979, 916-920.

97. Horisberger M. and Vonlanthen M. Localization of glycosylated κ-casein in bovine casein micelles by lectin-labelled gold granules. J. Dairy Res. 47, 1980, 185-191.

*98. Horisberger M., Rosset J. and Vonlanthen M. Location of glycoproteins on milk fat globule membrane by scanning and transmission electron microscopy using lectin-labelled gold granules. Nestlé Research News 1978/1979, 67-72.

99. Green M.L., Hobbs D.G. and Morant S.V. Intermicellar relationships in rennet-treated separated milk. 1. Preparation of representative electron micrographs. J. Dairy Res. 45, 1978, 405-411.

100. Green M.L., Hobbs D.G., Morant S.V. and Hill V.A. Intermicellar relationships in rennet-treated separated milk. II. Process of gel assembly. J. Dairy Res. 45, 1978, 413-422.

101. Kimber A.M., Brooker B.E., Hobbs D.G. and Prentice J.H. Electron microscope studies of the development of structure in Cheddar cheese.

J. Dairy Res. 41, 1974, 389-396.

102. Rüegg M. and Blanc B. Beiträge zur elektronenmikroskopischen Struktur der Labgallerte und des Käseteiges. Schweiz. Milchwirt. Forschung 1, 1972, 1-8.

103. Knoop A.-M. and Peters K.-H. Die submikroskopische Struktur der Labgallerte und des jungen Camembert-Käseteiges in Abhängigkeit von den Herstellungsbedingungen. Milchwissenschaft 27, 1973, 153-159.

104. Kalab M. Microstructure of dairy foods. 1. Milk products based on protein. J. Dairy Sci. 62, 1979, 1352-1364.

105. Brooker B.E. Milk and its products. In: Food Microscopy. Vaughan J.G. (ed.), Acad. Press, London, England, 1979, 273-311.

106. Kimura T., Taneya S. and Furuichi E. Electron microscopic observation of casein particles in process cheese. Brief Communs. 20th Internatl. Dairy Congress, Paris, France, 1978, 239-240.

107. Taneya S., Kimura T., Izutsu T. and Buchheim W. The submicroscopic structure of processed cheese with different melting properties. Milchwissenschaft 35, 1980, 479-481.

108. Kimura T. and Taneya S. Electron microscopic observation of casein particles in cheese. J. Electron Microsc. 24, 1975, 115-117.

109. Rayan A.A., Kalab M. and Ernstrom C.A. Microstructure and rheology of process cheese. Scanning Electron Microsc. 1980; III: 635-643.

110. Brooker B.E., Hobbs D.G. and Turvey A. Observations on the microscopic crystalline inclusions in Cheddar cheese. J. Dairy Res. 42, 1975, 341-348.

111. Blanc B., Rüegg M., Baer A., Casey M. and Lukesch A. Essais comparatifs dans le fromage d'Emmental avec et sans fermentation secondaire. Schweiz. Milchwirt. Forschung 8, 1979, 27-36.

112. Bach G. Bestimmung der Häufigkeitsverteilung der Radien kugelförmiger Partikel aus den Häufigkeiten ihrer Schnittkreise in zufälligen Schnitten der Dicke δ. Z. Wissenschaft. Mikroskopie 66, 1964, 193-200.

113. Rose P.E. Improved tables for the evaluation of sphere size distributions including the effect of section thickness. J. Microsc. 118(2), 1980, 135-141.

114. Schmidt D.G., Walstra P. and Buchheim W. The size distribution of casein micelles in cow's milk. Neth Milk Dairy J. 27, 1973, 128-142.

115. Schmidt D.G. and Buchheim W. Particle size distribution in casein solutions. Neth. Milk Dairy J. 30, 1976, 17-28.

116. Eggmann H. Elektronenmikroskopische Untersuchungen an Milch und Milchprodukten. 2. Anwendung der Gefrierätztechnik. Milchwissenschaft 24, 1969, 479-483.

117. Schmidt D.G. and Buchheim W. Elektronenmikroskopische Untersuchung der Feinstruktur von Caseinmicellen in Kuhmilch. Milchwissenschaft 25, 1970, 596-600.

118. Rüegg M. and Blanc B. Influence of

*See editor's note at the end of Discussion with reviewers.

pasteurization on UHT processing upon the size distribution of casein micelles in milk. Milchwissenschaft 33, 1978, 364-366.

119. Precht D. and Buchheim W. Elektronenmikroskopische Untersuchung der Kristallisationsvorgänge in den Fettkügelchen während der Rahmreifung. Milchwissenschaft 34, 1979, 657-662.

120. Precht D. and Buchheim W. Elektronenmikroskopische Untersuchungen über die physikalische Struktur von Streichfetten. I. Die Mikrostruktur der Fettkügelchen in Butter. Milchwissenschaft 34, 1979, 745-749.

121. Precht D. and Buchheim W. Elektronenmikroskopische Untersuchungen über die physikalische Struktur von Streichfetten. II. Die Mikrostruktur der zwischenglobulären Fettphase in Butter. Milchwissenschaft 35, 1980, 393-398.

122. Buchheim W. Elektronenmikroskopische Darstellung der Struktur von Speiseeis. Süsswaren 16, 1970, 763-767.

123. Berger K.G. and White G.W. An electron microscopical investigation of fat destabilization in ice cream. J. Food Technol. 6, 1971, 285-298.

124. Buchheim W. A new technique for improved cryofixation of aqueous solutions. Proc. Fifth Europ. Congr. Electron Microsc. 1972, 246-247.

125. Bachman L. and Schmitt-Fumian W.W. Spray-freeze-etching of dissolved macromolecules, emulsions and subcellular components. In: Freeze-Etching Techniques and Applications. Benedetti E.L. and Favard P. (eds.), Soc. Franc. Microsc. Electr., Paris, France, 1973, 63-72.

126. Buchheim W. and Schmidt D.G. On the size of monomers and polymers of β-casein. J. Dairy Res. 46, 277-280.

127. Schmidt D.G. and Buchheim W. On the size of α-lactalbumin and β-lactoglobulin molecules as determined by electron microscopy using the spray-freeze-etching technique. Milchwissenschaft 35, 1980, 209-211.

128. Buchheim W. Elektronenmikroskopische Präparationsmethode zur Darstellung von Oberflächen- und Innerstruktur wasserlöslicher Pulverteilchen. Kieler Milchwirt. Forsch. Ber. 24, 1972, 97-107.

129. Buchheim W. The applicability of electron microscopy for studying the structure of liquid and solid foods. Proc. IV. Internal. Congress Food Sci. Technol. II, 1974, 5-12.

130. Buchheim W. A comparison of the microstructure of various dried milk products by application of the freeze-fracturing technique. Scanning Electron Microsc. 1981; III: 493-502.

131. Buchheim W. Freeze-etching of dehydrated biological material. Proc. 4th Europ. Congr. Electron Microsc., Jerusalem, Israel, 1976, 122-124.

132. Buchheim W. Der Verlauf der Fettkristallisation in den Fettkügelchen der Milch. Elektronenmikroskopische Untersuchungen mit Hilfe der Gefrierätztechnik. Milchwissenschaft 25, 1970, 65-70.

133. Buchheim W. and Prokopek D. Elektronenmikroskopische Untersuchungen an Ultrafiltrationskonzetraten aus Magermilch und daraus hergestelltem Käse. 1. Verhalten der Caseinmicellen während der Ultrafiltration. Milchwissenschaft 31, 1976, 462-465.

134. Kalab M. Cleaning replicas of freeze-fractured oil-water emulsions. Microsc. Soc. Canada Bull. 6(4) 1978, 24-25.

135. McGann T.C.A., Donnelly W.J., Kearney R.D. and Buchheim W. Composition and size distribution of bovine casein micelles. Biochim. Biophys. Acta 630, 1980, 261-270.

136. Buchheim W. and Knoop E. Zur Verteilung von Fett und Casein in vollhomogenisierter und teilhomogenisierter Milch. Kieler Milchwirt. Forsch. Ber. 22, 1970, 323-327.

137. Schmidt D.G., Buchheim W. and Koops J. An electron-microscopical study of the fat-protein complexes in evaporated milk, using the freeze-etching technique. Neth. Milk Dairy J. 25, 1971, 200-216.

138. Buchheim W. Zur Struktur der Hülle von Milchfettkügelchen. Naturwissenschaften 57, 1970, 672-673.

139. Pinto da Silva P., Peixoto de Menezes A. and Mather I.H. Structure and dynamics of the bovine milk fat globule membrane viewed by freeze fracture. Exptl. Cell Res. 125, 1980, 127-139.

140. Oortwijn H., Walstra P. and Mulder H. The membranes of recombined fat globules. I. Electron microscopy. Neth. Milk Dairy J. 31, 1977, 134-147.

141. Buchheim W. Die molekulare Ordnung in doppelbrechenden Fettkügelchen. Milchwissenschaft 25, 1970, 223-227.

142. Resmini P. Struttura e microstruttura dei prodotti lattiero caseari. Ind. Latte 15, 1979, 33-60.

143. Bohler S. Artefacts and Specimen Preparation Faults in Freeze Etch Technology. Balzers Aktiengesellschaft, Balzers, Liechtenstein, 1976, Fig. 19.

144. Haggis G.H. and Bond E.F. Three-dimensional view of the chromatin in freeze-fractured chicken erythrocyte nuclei. J. Microsc. 115(3), 1979, 225-234.

145. Harwalkar V.R. and Kalab M. Milk gel structure. XI. Electron microscopy of glucono-δ-lactone-induced skim milk gels. J. Texture Stud. 11, 1980, 35-49.

146. Haggis G.H., Bond E.F. and Phipps B. Visualization of mitochondrial cristae and nuclear chromatin by SEM. Scanning Electron Microsc. 1976; I: 281-286.

Discussion with Reviewers

M.V. Taranto: From your description of the procedure for SEM of frozen specimens, it is not clear how the fractured specimens are fixed to the SEM stubs. Please clarify.
Author: Depending on which stubs are used, the specimen is either cemented with glycerol to the stub which has a knurled face (Fig. 10 c), or is placed in the hollow SEM holder (Fig. 5 and 10 d). In either case, the holder with the specimen is

plunged into Freon 12 (-150°C) with the sample going in first. Knurled stub surface was designed to reduce the incidence of specimen separation from the stub during freezing and fracturing, which was frequent with smooth stubs. The work is considerably easier with the hollow SEM stubs designed by E.F. Bond (Fig. 5), because the specimen is placed in a 3-mm tubular holder and protrudes approximately 1 mm out of it. After the specimen is frozen in Freon and transferred in LN_2, it is easy to fracture the protrusion with a scalpel. However, if an uncoated specimen is examined by SEM, a clearance develops between the specimen and the holder, which results in charging artifacts; the knurled holder is better for this latter type of work.

M.V. Taranto: The microcapsulation technique, which you describe for fixing suspensions and emulsions, is intriguing. You are, in a way, forming a dialysis bag. Have you ever tried using conventional cellulose dialysis tubing to hold the sample during fixation?
Author: No, I have not tried to use cellulose dialysis tubing. The Salyaev's procedure is quite simple and has the advantage that the diameter of the tubing can be controlled by the thickness of the glass rod used. Thus, it may be even thinner than 0.5 mm. Using this technique, we succeeded in embedding undiluted mayonnaise and obtaining good micrographs of thin sections in spite of the high fat content in mayonnaise. For another modification of the microcapsulation technique please see the paper by G.G. Lewis in this volume.

M.V. Taranto: You describe several cytochemical methods for the study of fine structure in ultrathin sections. One technique used by some researchers is the selective digestion of structural components in the sections with specific enzymes. Valuable information concerning the localization and structural organization of components can be deduced using this technique. Have you used or do you know of any reported work that used this selective digestion technique on ultrathin sections of dairy products?
Author: I have not used this technique and have not found information on its use with milk products.

M.V. Taranto: One technique used by microscopists to observe the same structures in biological samples with TEM and SEM is to prepare the sample by fixing, dehydrating and embedding in epoxy resin. After thin-sectioning, the sample block is etched with the appropriate solvent to remove the epoxy resin, and is coated with gold and viewed with SEM. Have you ever used or do you know of any reported work that used this technique on dairy products?
Author: No. Maybe the availability of specimens has not necessitated to use this technique with milk products. However, I wish to thank you for mentioning these techniques which may stimulate work in this direction.

J.F. Chabot: During the procedure for replication of dried specimens, the materials are subjected to the worst parts of several techniques. They are frozen, melted, treated with dehydrating agents, exposed to glycerol, and critical-point dried. Is this going to be a technique in which the results will be trusted? In particular, isn't there a problem in wetting a dry specimen during fixation and causing immediate and important changes?
Author: I agree that the procedure may appear to break several rules. Using it for the first time, I felt as if I was committing blasphemy. However, the results obtained with Cottage cheese and their comparison with results obtained by other techniques and, in particular, the results obtained with glucono-δ-lactone-induced skim milk gels proved that this method can be trusted provided that appropriate precautions are observed. The specimen must be small to freeze instantly in Freon 12 (-150°C) and also melt instantly in absolute alcohol or a glutaraldehyde solution, depending which version of the technique is selected. From my own experience, the most critical steps are those of wrapping the dried specimen in an aluminum foil, cutting off the platinum-carbon-coated fractured surface with as little specimen as possible, and cleaning of the replica. The reason for wrapping the specimen is to prevent coating of any other part of the specimen than the fractured surface; such a side coating would collapse after the specimen is digested and washed off, and would contaminate the replica of the fractured surface. The perspex film must adhere to the replica firmly, otherwise if would peel off during cleaning. Care must be taken during cleaning that the specimen would not swell, shrink, or warp. For this reason, alkaline cleaning agents were not used with milk products. Finally, the perspex film must be completely removed by bathing the replica in chloroform after it had been lifted on a bare grid.
In this technique, the specimen is not dried prior to fixation and, hence, no wetting of the dry specimen takes place. The specimen is frozen, freeze-fractured in LN_2, and the fragments are melted in a warm (20-25°C) fixative. Ideally, soluble proteins are removed from the fractured surface and the insoluble protein matrix is fixed. Alternatively, the sample is fixed first, dehydrated, and impregnated with absolute alcohol. Freezing of a sample thus prepared poses no risk of ice crystal formation. The fragments are melted in warm (20-25°C) absolute alcohol and CPD. It is possible to examine the fractures by SEM as shown in Fig. 3.

O. Johari: Have you had any experience or problems with the decomposition of particles under the electron beam, contamination of the SEM column, distortion of image *etc.*? (See ref. 26).
Author: Yes. In some spray-dried milk powders, the particles were so susceptible to electron-beam damage that focussing was achieved with great difficulty. Depending on their origin, dried milk particles ruptured in different ways and the ruptures appeared as if occurring either in an amorphous (glassy) or a crystalline material; the micrographs were published[28]. A nonfat dried milk particle, ruptured accidentally during focussing at a higher magnification, was also shown (M. Kalab: Electron microscopy in dairy research. Microsc. Soc. Canada Bull. 5(4), 1977, 4-10). Contamination of the SEM column or distortions of the images were not encountered.

O. Johari: Please comment on the specimen preparation of dairy products for x-ray microanalysis.
Author: I found only one report on x-ray microanalysis of cheese[111], in which the authors

concluded that the crystalline inclusions found were most probably composed of tyrosine. To test my hypothesis postulated in this review that soluble salt crystals are removed from process cheese samples during preparatory steps for electron microscopy (see Fig. 20), crystals of soluble and insoluble salts are being embedded in cheese and x-ray microanalysis is being used to determine the composition of the inclusions. However, my personal experience with this technique is insufficient to allow me to comment.

D.N. Holcomb: Would you be willing to forecast future developments in this area? Obviously, continued application of the techniques covered here will lead to future advances in understanding and elucidating food microstructure. But in addition, do you see areas where microscopic analysis would be valuable, but not now possible because adequate techniques are not yet available?

Author: Before making an attempt to forecast any future developments in electron microscopy of foods, I wish to point to an interesting fact that in spite of its importance to every man, woman, and child, food has not received the same attention of science as did other less essential needs. It is probably because we have accepted food for granted. However, with the growing concern for possible effects of various food ingredients on our well-being and with the introduction of unconventional foods, scientific research has been gradually developing even in this discipline. Electron microscopical studies belong to one of the most recent branches of this endeavour. Such studies are being carried out along several lines. Electron microscopy is being used as an analytical tool to detect corpuscular ingredients alien to foods (for a review on contaminants such as asbestos fibres in beverages see J.T. Stasny et al., Scanning Electron Microsc. 1979; I: 587-596). As an aid to processing, electron microscopy is used to follow interactions of food components, for example between proteins, fat, and emulsifying salts in process cheese. Finally, electron microscopy may explain the behaviour of the individual food ingredients on the basis of their ultrastructure, for example the distribution of caseins and calcium phosphate in the casein micelle etc.

In my opinion, we are currently experiencing a very important development in the use of microscopical techniques in food science and that is the organization of the meetings under the auspices of SEM, Inc. I sincerely hope that the next development will be an active participation of scientists working in large food companies. It is well known that these colleagues have access to modern research facilities and are producing excellent results. I believe that in the future, they will be able to share some of their results with others, as these meetings cannot be based on a one-way flow of scientific information. Hopefully, we will experience a greater exchange of ideas in the future.

I assume that the trend in using electron microscopy as an analytical tool in the control of raw materials for alien components will be expanded. Energy dispersive spectrometry associated with the use of computerized image analyzers will be capable of providing both qualitative and quantitative data. Image analyzers have been little used in electron microscopy of foods although numerical processing can make the micrographs of emulsions, suspensions etc. considerably more meaningful and comparable. In fact, there are few other disciplines where the image analyzers could be as important as in food science.

Cold-stage SEM will also probably be used to a greater extent although not necessarily in the form reported in this review. Bio-chambers attached to the microscopes, in which the specimen can be fractured and coated and, without any exposure to the atmosphere, inserted in the microscope for an examination eventually allowing manipulation during the examination, will provide better results more quickly than the procedures in which the individual steps are carried out outside the microscope.

A further development of ultrarapid freezing methods could prove useful in the study by x-ray microanalysis of the distribution of soluble ions in foods. These ions are lost during fixation and redistributed during slow freezing.

Relationships between microstructure and sensory attributes of foods will be studied to a greater extent, for example, the relationships between food particle composition and dimensions on the one hand and their sensory perception (smoothness, grittiness, stickiness etc.) on the other hand.

I assume that this section on food microstructure will rapidly expand and continue to attract the attention of food scientists on an international scale.

I find it difficult to answer your question concerning techniques yet to be developed. I believe that there is a wide variety of techniques available and that progress could be achieved if these were used appropriately even without further development of new techniques. Too often we see some authors relying on a single technique. The other problem is that many food scientists still depend on someone else for sample preparation and operation of electron microscope. Thus, in my opinion, a greater attention paid to the nature of the specimen, understanding of the electron microscopical technique used, and personal involvement would greatly contribute to progress in this area.

M. Rüegg: Fixation with OsO_4 can also be accomplished in a dry atmosphere.
Author: To my knowledge this has not been proven. H. R. Müller (Milchwissenschaft 19, 1964, 345-356) used this technique and recommended that a droplet of water be present in the closed container in which dry milk powder and OsO_4 crystals were held. My own experience (Scanning Electron Microscopy 1980/III, 645-652) has confirmed that: dry buttermilk powder remained white for 24 h in the presence of OsO_4 crystals in a dry atmosphere, i.e. in the complete absence of any water vapour. The powder particles turned dark after the admission of a droplet of water in the closed container indicating that water vapour was necessary to serve as a vehicle for OsO_4 to fix specimens not immersed in the OsO_4 solution.

*Editor's Note: A comprehensive paper on "Colloidal Gold: A Cytochemical Marker...." is also being included in Scanning Electron Microsc.1981:II.

MICROSTRUCTURE AND RHEOLOGY OF PROCESS CHEESE

A. A. Rayan, M. Kaláb[*], and C. A. Ernstrom

Department of Nutrition and Food Sciences
Utah State University, Logan Utah 84321
*Food Research Institute, Research Branch Agriculture
Ottawa, Ontario, Canada K1A 0C6

Abstract

Four batches of pasteurized process cheese were prepared from the same Cheddar cheese by cooking to 82°C in the presence of sodium citrate, disodium phosphate, tetrasodium pyrophosphate or sodium aluminum phosphate. Each batch contained the same moisture (40.6%) and emulsifying salt concentration (2.5%). The process cheese was sampled for microstructural and rheological examination after 0, 5, 10, 20 and 40 min in the cooker at 82°C.

Even though each emulsifying salt affected the physical properties of the process cheese differently, the cheese generally became firmer, more elastic and less meltable as cooking time increased from 0 to 40 min. These changes were accompanied by a decrease in the dimension of fat masses and an increase in the degree of emulsification as evidenced by scanning electron microscopy and transmission electron microscopy. Sodium citrate and tetrasodium pyrophosphate crystals remained undissolved in the cheese after 40 min in the cooker while sodium aluminum phosphate crystals were still undissolved after 10 min.

KEY WORDS: Process cheese, microstructure, rheology, emulsion, emulsifying salts, cheese, salt crystals, cheese protein, meltability, firmness.

Introduction

Pasteurized process cheese is made by heating comminuted natural cheese to at least 65.5°C, but more commonly to about 85°C, in the presence of an approved emulsifying salt (1,2). The cheese is mechanically mixed while heating and its moisture and fat content may be adjusted by adding water and cream (2). The emulsifying salts are calcium sequestering agents which also increase the pH of the cheese (3), and together with heat, convert the relatively insoluble natural cheese curd into a smooth more soluble and homogeneous mass in which the fat is emulsified (4). Other products such as pasteurized process cheese food and pasteurized process cheese spread are made similarly, but they contain more moisture and less fat than process cheese (2).

In natural Cheddar cheese the native milk fat globules are forced together into aggregates within a protein matrix, and even though distorted, may still maintain their globular integrity (5,6). When made into process cheese, free fat is separated during initial heating of the cheese and is then reemulsified. The key to reemulsification is the emulsifying salt which under the influence of heat increases the emulsifying properties of the cheese proteins (7).

The commercially important physical properties of process cheese such as meltability and firmness are affected significantly by pH and age of the natural cheese from which it is made (8). Also, they are affected by the kind of emulsifying salt used, the time the cheese is held at the cooking temperature, and the inclusion of previously processed cheese in the batch (4,9,10). Little is known, however, about the relationship between these processing variables and the microstructure of process cheese, or whether there is a correlation between microstructure and physical properties.

Taranto, et al. (5) reported that subtle differences in the microstructure of Cheddar and mozzarella cheese may be partially responsible for their rheological differences.

The purpose of this study was to examine the effect of different emulsifying salts and cooking times on the microstructure and rheological properties of pasteurized process cheese.

Materials and Methods

Cheese

Thirty-five day old Cheddar cheese at pH 5.1 was processed in a Damrow (Damrow Co., Fond Du Lac, Wisconsin, USA) pilot size horizontal cheese cooker with indirect heating. Four 18-kg. batches of the same cheese were prepared to contain approximately 40% moisture and 2.5% emulsifying salt. Each batch was prepared with a different emulsifying salt viz. sodium citrate (CIT), disodium phosphate (DSP), tetrasodium pyrophosphate (TSPP) and sodium aluminum phosphate (SALP). The natural cheese for each batch was comminuted in a Hobart vertical cutter model VCM 25 (Hobart Mfg. Co., Troy, Ohio) and added to the cooker with dry emulsifying salts and an appropriate amount of water. The cheese mixture was heated to $82^\circ C$ within 6 min at which time a 0 time sample was removed and vacuum treated to remove air bubbles. Subsequent samples were removed after 5, 10, 20 and 40 min at $82^\circ C$. Process cheese samples were kept under refrigeration in sealed 0.45 kg containers until examined.

SEM

Process cheese specimens were fixed in a 1.4% glutaraldehyde solution, dehydrated in a graded alcohol series, defatted in chloroform, and critical-point dried from CO_2. Dry specimens were fractured and the fragments mounted on SEM stubs, coated with carbon (50) and gold (200A) by vacuum evaporation and examined under a Cambridge Stereoscan electron microscope operated at 20 kv, as reported earlier (11).

TEM

The same procedures were observed as described earlier (11).

Physical Properties of the Cheese

Meltability was measured in duplicate as described by Olson and Price (12) except that the heating time at $110^\circ C$ was extended from 6 to 30 min.

Firmness and degree of elasticity were measured on a universal testing machine (MTS Tensile Testing Machine, type T5002, J.J. Lloyd Instruments Ltd., Southampton, England). Cheese samples were equilibrated in their original containers for 48 h at $15^\circ C$, then cut into cylindrical plugs 19 mm in diameter and 20 mm high immediately before testing. Firmness was measured by the force required for a 0.4 mm stainless steel wire to shear through the cylinder of cheese (13) at a rate of 30 mm per min. A slotted stainless steel cylinder as described by Emmons and Price (14) was modified in size to hold the cheese plug during the test. Degree of elasticity was measured as described by Weaver and Kroger (15)

Chemical Analysis of Cheese

Cheese moistures were determined as described by Price, et al. (16) Fat was determined by a modified Babcock procedure for cheese (17).

Results

All four batches of process cheese contained 40.6% moisture and 31.0% fat when removed from the cooker at 0 time. After 40 min in the cooker, the moisture had decreased to 40.2% in the CIT batch and 40.3% in the other three batches. The pH of the process cheese was 5.4 in the CIT samples and 5.8 in the DSP, TSPP and SALP samples.

Physical Properties

At 0 time, the process cheese containing CIT or SALP had excellent melting qualities while that emulsified with DSP or TSPP had poor meltability (Figure 1). The melting quality of all samples decreased as cooking time increased, and the most rapid decrease occurred in samples with the best initial meltability. There was little change in meltability of the samples after 20 min at $82^\circ C$.

The firmness of all process cheese samples tended to increase with increased cooking time (Figure 2), but cheese emulsified with TSPP was the firmest at all stages of cooking and that made with SALP remained softest. DSP and CIT cheese also were soft initially but increased in firmness during cooking more rapidly than SALP cheese.

Degree of elasticity or ability to recover after partial deformation by compression was greatest in cheese emulsified with TSPP and least in SALP cheese (Figure 3). CIT and DSP cheese samples were intermediate. In general, degree of elasticity tended to increase in all samples with increased exposure to heat.

There were marked differences in the effects produced by the four emulsifying salts used in this study, but there was a general tendency toward greater firmness, poorer meltability and a greater degree of elasticity with increasing cooking time at $82^\circ C$.

Microstructure

CIT Cheese (Figure 4). SEM of fractured process cheese heated to $82^\circ C$ for 0 and 40 min (A and B) shows a marked reduction in the dimension of fat masses during cooking. At 0 time the fat particles (seen as empty cavities) vary greatly in size from very large ones (left lower corner) to small ones the size of bacteria. This image is confirmed by TEM at 0 time (C) where large fat masses may be seen along with fat particles smaller than bacteria. During cooking the large fat particles disintegrated; the process has been captured in TEM micrographs D (10 min) and E (40 min). The constrictions associated with the large fat particles are sites where they will disintegrate into smaller particles.

Added CIT is in the form of needle-like crystals at 0 time (C) where dimensions gradually decreased during cooking D (10 min) and E (40 min). After 40 min in the cooker some small CIT crystals still remained undissolved. A crystal of calcium phosphate (18) that may have originated in the natural cheese is seen in micrograph D.

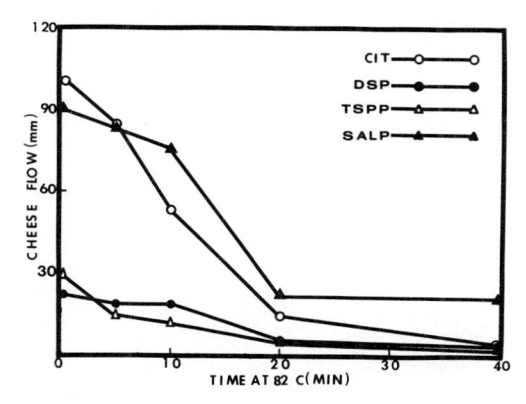

Fig. 1. Effect of cooking time at 82°C on the meltability of pasteurized process cheese emulsified with sodium citrate (CIT), disodium phosphate (DSP), tetrasodium pyrophosphate (TSPP) and sodium aluminum phosphate (SALP).

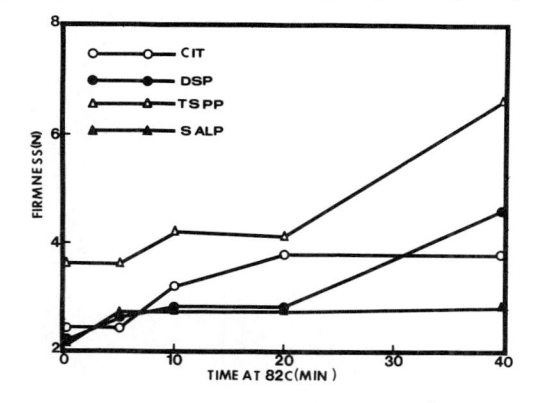

Fig. 2. Effect of cooking time at 82°C on the firmness of pasteurized process cheese emulsified with sodium citrate (CIT), disodium phosphate (DSP), tetrasodium pyrophosphate (TSPP) and sodium aluminum phosphate (SALP).

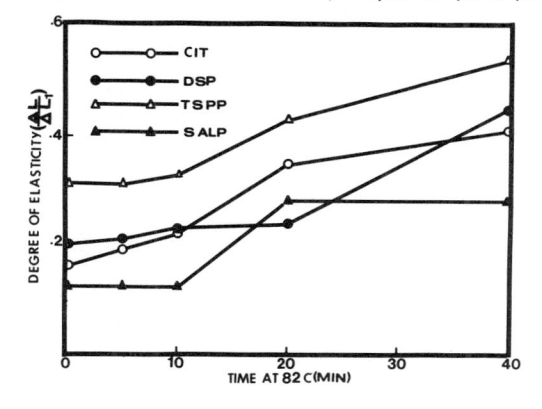

Fig. 3. Effect of cooking time at 82°C on the degree of elasticity of pasteurized process cheese emulsified with sodium citrate (CIT), disodium phosphate (DSP), tetrasodium pyrophosphate (TSPP) and sodium aluminum phosphate (SALP).

DSP Cheese (Figure 5). As with the CIT sample there was a reduction in the size of the fat particles with increasing time in the cooker. Compare SEM micrographs A (0 time) and B (40 min). The same disintegration of irregularly sized fat masses as seen with other salts can be found in TEM micrographs C (0 time) and E (40 min). A crystal of calcium phosphate with a hairy structure as reported by Brooker et al. (18) is shown under high magnification in TEM micrograph D. A similar crystal under lower magnification is in E. Cheese processed with DSP generally contained larger pockets of fat than CIT cheese at 0 time and the emulsion was not as fine after 40 min in the cooker (compare Figure 4B with 5B).

TSPP Cheese (Figure 6). Reduction of the fat from a very coarse to very fine emulsion during cooking characterized cheese processed with TSPP. Compare SEM micrographs A (0 time) with B (40 min), then TEM micrographs F (0 time) and G (40 min). "Torchlike" TSPP crystals are evident in TEM micrographs D (N) and in SEM micrograph C which suggest that the crystals are growing. Another rather common but unidentified crystalline form found in TSPP cheese is illustrated in micrograph E.

SALP Cheese (Figure 7). The emulsification process with SALP proceeded more slowly than with other salts as judged by both SEM and TEM (Figure 7). The SEM emulsion shown in B (10 min) has not progressed much beyond that shown in A (0 time), and in C and E (40 min) emulsification is still progressing. Very large fat masses at 0 time are shown in TEM micrograph D. Undissolved SALP crystals were found in abundance during the initial 10 min in the cooker (F).

Discussion

The degree of emulsification (fineness of fat particles) seems directly related in some way to firmness, poor meltability and elasticity of process cheese. This does not suggest a direct cause and effect relationship because the emulsifying salts produced their own specific effects on the protein matrix which in turn could have affected the physical properties as well as the state of the fat emulsion in the cheese. Emulsification with TSPP produced the firmest, least meltable and most elastic product at most stages of cooking. TSPP also produced the most rapid emulsification and finest emulsion after 40 min in the cooker. On the other hand, emulsification with SALP seemed slower and less complete than with the other salts, and the cheese was generally softer, more meltable and less elastic than the other cheese samples at most stages during cooking.

Some emulsifying salt crystals remained undissolved in process cheese after 10 min (Figure 7F) and 40 min (Figures 4E and 6B and G) in the cooker. If the salt must be in solution to properly function, the rate of solution may account for the time required to go from a coarse (0 time) to a fine (40 min)

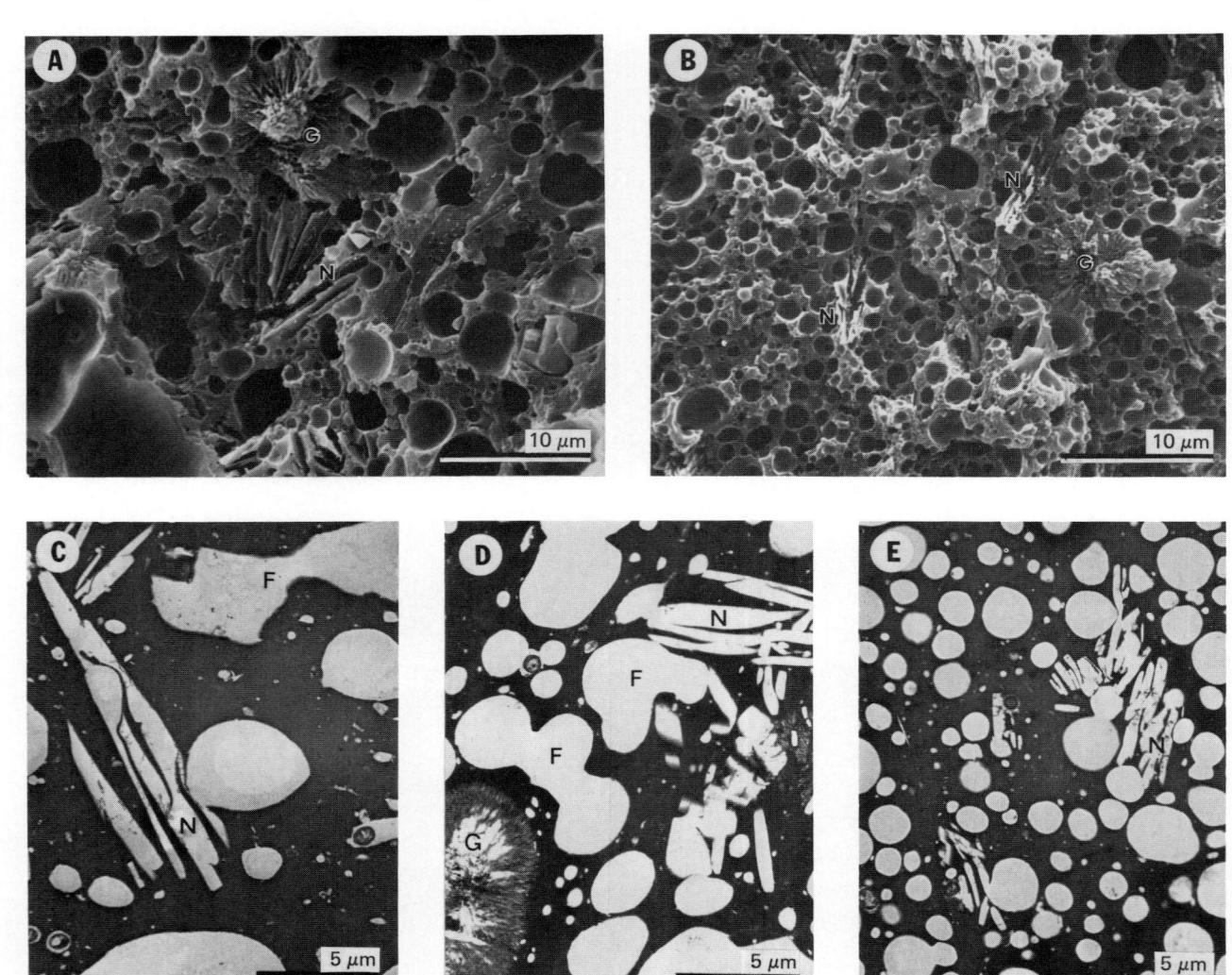

Fig. 4. Development of microstructure in process cheese in the presence of sodium citrate (CIT). A = SEM at 0 time. G = Initially present calcium phosphate crystal. N = Crystals of added sodium citrate. B = SEM after 40 min in the cooker. G and N same as in A. C = TEM at 0 time. Dark areas are the protein matrix. F = Fat particles undergoing emulsification. N = Crystals of added sodium citrate. D = TEM after 10 min in the cooker. F = Fat particles being divided into smaller particles at the constrictions. G = Initially present crystal of calcium phosphate. N = Needle-like crystals of added sodium citrate. E = TEM after 40 min in the cooker. Dimensions of the fat particles (round light areas) and crystals of added sodium citrate (N) have been reduced during the 40 min process.

emulsion. It also could explain the claim by Bosy et al. (19) that the concentration of emulsifying salts can be cut in half if put into solution before being added to the cooker. Studies are now underway to test this possibility by comparing the microstructures and physical properties of process cheese made by the method of Bosy et al. (19) with those of similar cheese processed with dry emulsifying salts.

As the emulsifying salts dissolved, they interacted under the influence of heat and mechanical work with the cheese proteins. This appeared to modify the proteins to enhance their emulsifying properties and enable them to form a finer and more uniform network as cooking progressed. The more uniform protein network probably reduced the number of "weak links" and allowed a more uniform distribution of stress, thus exhibiting a greater firmness and less meltability.

Water molecules could become oriented between protein strands in such a network to act as a "lubricant" which permits the strands to slide and interdigitate more freely and thus contribute to the elasticity of the cheese (M.V. Taranto, Personal Communication, 1980).

Excessive firmness that develops when previously processed cheese is added to a new batch is known in the industry as the

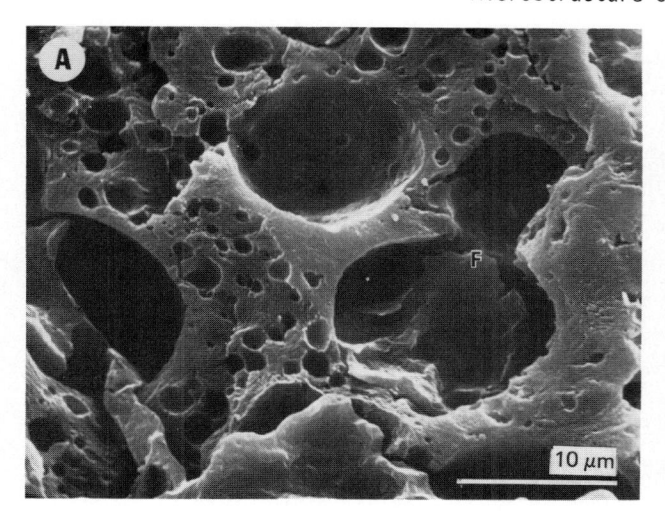

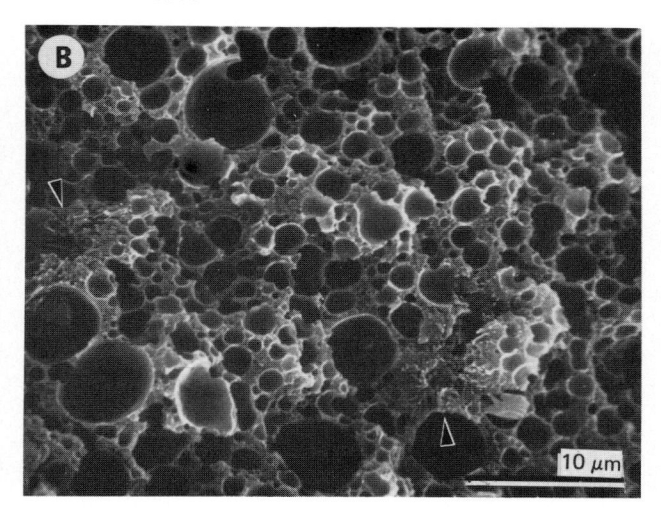

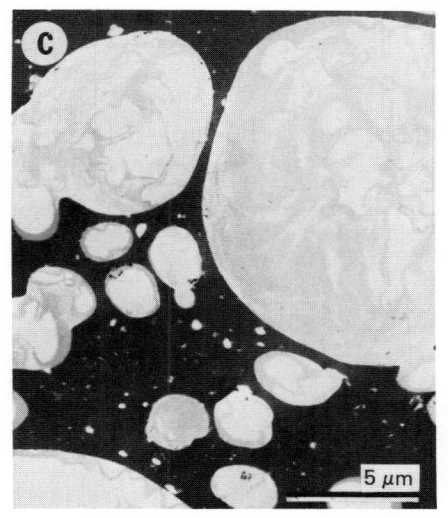

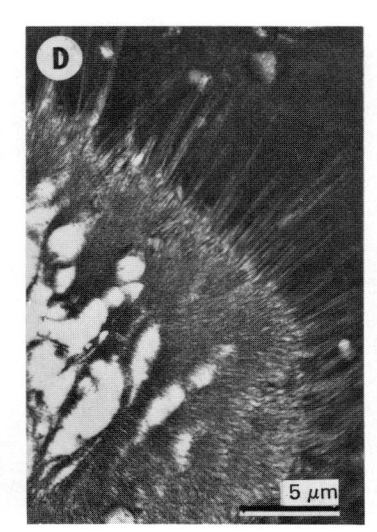

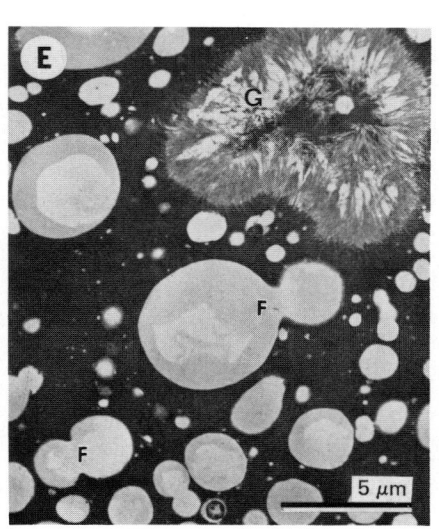

Fig. 5. Development of microstructure in process cheese in the presence of disodium phosphate (DSP). A = SEM at 0 time. Large fat particles (F) have started to be emulsified into smaller particles. B = SEM after 40 min in the cooker. Fat has undergone emulsification and is present in the form of small globular particles. Arrows show calcium phosphate crystals. C = TEM at 0 time. Dark areas are protein, light areas were occupied by fat. Grey areas indicate residual fat that had not been removed during preparation of the specimen for TEM. D = TEM detail of calcium phosphate crystal showing additional growth of fine spikes at the perimeter. E = TEM after 40 min in the cooker. G = Calcium phosphate crystal showing signs of additional growth. F = Fat particles undergoing emulsification.

"rework defect". This might be explained by the previously processed cheese protein being already in the modified form so it can hasten the emulsifying process and induce rapid over emulsification during cooking.

Acknowledgements

Appreciation is expressed to Dr. D. N. Holcomb for an invitation to submit this paper.

This paper is Utah Agricultural Experiment Station Journal Article No. 2537, and contribution 418 from Food Research Institute, Research Branch, Agriculture Canada, Ottawa.

References

1. Kosikowski, F.V. 1978. Cheese and Fermented Milk Foods. 2nd ed. F.V. Kosikowski and Associates, Brooktondale, N.Y., USA. Ch. 21.
2. U.S. Food and Drug Administration. 1979. Code of Federal Regulations. 21: 133.169, 133.173, 133.179. U.S. Govt. Printing Office, Washington, D.C.
3. Holtorff, F., V. Mularz and E. Traisman. 1951. A study of process cheese emulsifiers. J. Dairy Sci. 34: 486.
4. Price, W.V. and M.G. Bush. 1974. The process cheese industry in the United States.

A review. II. Research and development. J. Milk and Food Technol. 37: 179-198.
5. Taranto, M.V., P.J. Wan, S.L. Chen, et al. 1979. Morphological, ultrastructural and rheological characteristics of Cheddar and mozzarella cheese. SEM/1979/III, SEM Inc., AMF O'Hare, IL 60666. 273-278.
6. Kimber, M., B.E. Brooker, D.G. Hobbs, et al. 1974. Electron microscope studies of the development of structure in Cheddar cheese. J. Dairy Res. 41: 389-396.
7. Ellinger, R.H. 1972. Phosphates as Food Ingredients. The Chemical Rubber Co., Cleveland, OH, USA. p. 73.
8. Olson, N.F., D.G. Vakaleris and W.V. Price. 1958. Acidity and age of natural cheese as factors affecting the body of pasteurized process cheese spread. J. Dairy Sci. 41: 1005-1016.
9. Templeton, H.L. and H.H. Sommer. 1936. Studies on the emulsifying salts used in process cheese. J. Dairy Sci. 19: 561-572.
10. Lauk, R.M. 1972. Using salvage cheese in preparing pasteurized process cheese. U.S. Patent, 3,697,292.
11. Kalab, M. 1977. Milk gel structure. VI. Cheese texture and microstructure. Milchwissenschaft 32:(7), 449-458.
12. Olson, N.F. and W.V. Price. 1958. A melting test for pasteurized process cheese spreads. J. Dairy Sci. 41: 999-1000.
13. Voisey, P.W. 1971. Modernization of texture instrumentation. J. Texture Stud. 2: 129-195.
14. Emmons, D.B. and W.V. Price. 1959. A curd firmness test for cottage cheese. J. Dairy Sci. 42: 553-556.
15. Weaver, J.C. and M. Kroger. 1978. Free amino acid and rheological measurements on hydrolyzed lactose Cheddar cheese during ripening. J. Food Sci. 43: 579-582.
16. Price, W.V., W.C. Winder, A.M. Swanson et al. 1953. The sampling of Cheddar cheese for routine analysis. J. Assoc. Off. Agric. Chem. 36: 524-538.
17. Van Slyke, L.L. and W.V. Price. 1953. Cheese. 2nd ed. Orange Judd Publishing Co., New York, NY, USA. Ch. 28.
18. Brooker, B.E., D.G. Hobbs and A. Turvey. 1975. Observations on the microscopic crystalline inclusions in Cheddar cheese. J. Dairy Res. 42: 341-348.
19. Bosy, G., E.N. Edwards, W.M. Hoffbeck et al. 1978. Method for the manufacture of process cheese. U.S. Patent, 4,112,131.
20. Katoh, M. 1979. SEM replica technique for butter and cheese. J. Electron Microsc. 28:(3), 199-200.
21. Schmidt, D.G., S. Henstra, and F. Thiel. 1979. A simple low-temperature technique for SEM of cheese. Mikroskopie 35: 50-55.
22. Kalab, M. 1978. Milk gel structure. VIII. Effect of drying on the scanning electron microscopy of some dairy products. Milchwissenschaft 33:(6), 363-358.
23. Kalab, M. 1980. Decayed lactic bacteria as a possible source of crystallization nuclei in cheese. J. Dairy Sci. 63: 301-304.

Discussion with Reviewers

D.N. Holcomb: Will changes in the cheese properties continue if the cooking time is extended beyond 40 minutes, and are there changes such as de-emulsification and crystal growth that occur on storage of process cheese?
Authors: I would expect the observed changes in cheese properties to continue, but at a slower rate if the cooking time were extended beyond 40 min. Forty minutes in the cooker is well beyond any treatment that would be likely in commercial cheese processing where a heating time of 0 to 5 minutes is customary. Growth of salt crystals can and does take place during the storage of process cheese if the salt concentration is too high. Deposits of calcium phosphate on the surface of process cheese constitute a serious defect that gives the appearance of surface mold growth (7). I am unaware of problems with de-emulsification of process cheese during storage.

M.V. Taranto: Hot water processing (heating and mechanical working) is used to develop stretching and stringing characteristics in acid ripened curd after milling. What happens to cheese structure when it is heated in the absence of emulsifying salts? Are similar structural and rheological changes observed?
Authors: The stringy characteristics you described are characteristic of cheese curd heated in the absence of emulsifying salts. The fat tends to separate from the curd and the protein remains nondispersable in hot water. Cheese proteins that have been modified by emulsifying salts and heat become very dispersable in hot water. The properties of the proteins are quite different, but additional work is needed before the nature of the transformation can be understood.

M.V. Taranto: You mention that sodium citrate and tetrasodium pyrophosphate crystals remained undissolved in the cheese after 40 min and sodium aluminum phosphate remained undissolved after 10 min. How long did it take for the various emulsifying salts to dissolve? Is there a correlation between salt solubility and

Fig. 6. Development of microstructure in process cheese in the presence of tetrasodium pyrophosphate (TSPP). A = SEM at 0 time, F = Two cavities initially occupied by fat. B = SEM after 40 min in the cooker. Arrows show crystals of added TSPP. C = An SEM detail of a TSPP crystal in the cheese protein matrix. D = TEM detail of TSPP crystals. N = Long thin spikes extending from the compact bodies of the crystals give them a torch-like appearance, which is a sign of recrystallization. B = Bacterium. E = TEM detail of abundant yet unidentified crystal-like structures. F = TEM of the cheese at 0 time. Lace-like structure (arrow) is a tip of an unidentified crystal. G = TEM after 40 min in the cooker. Fat still undergoing emulsification (F). N = Spikes of torch-like TSPP crystals.

Microstructure of Process Cheese

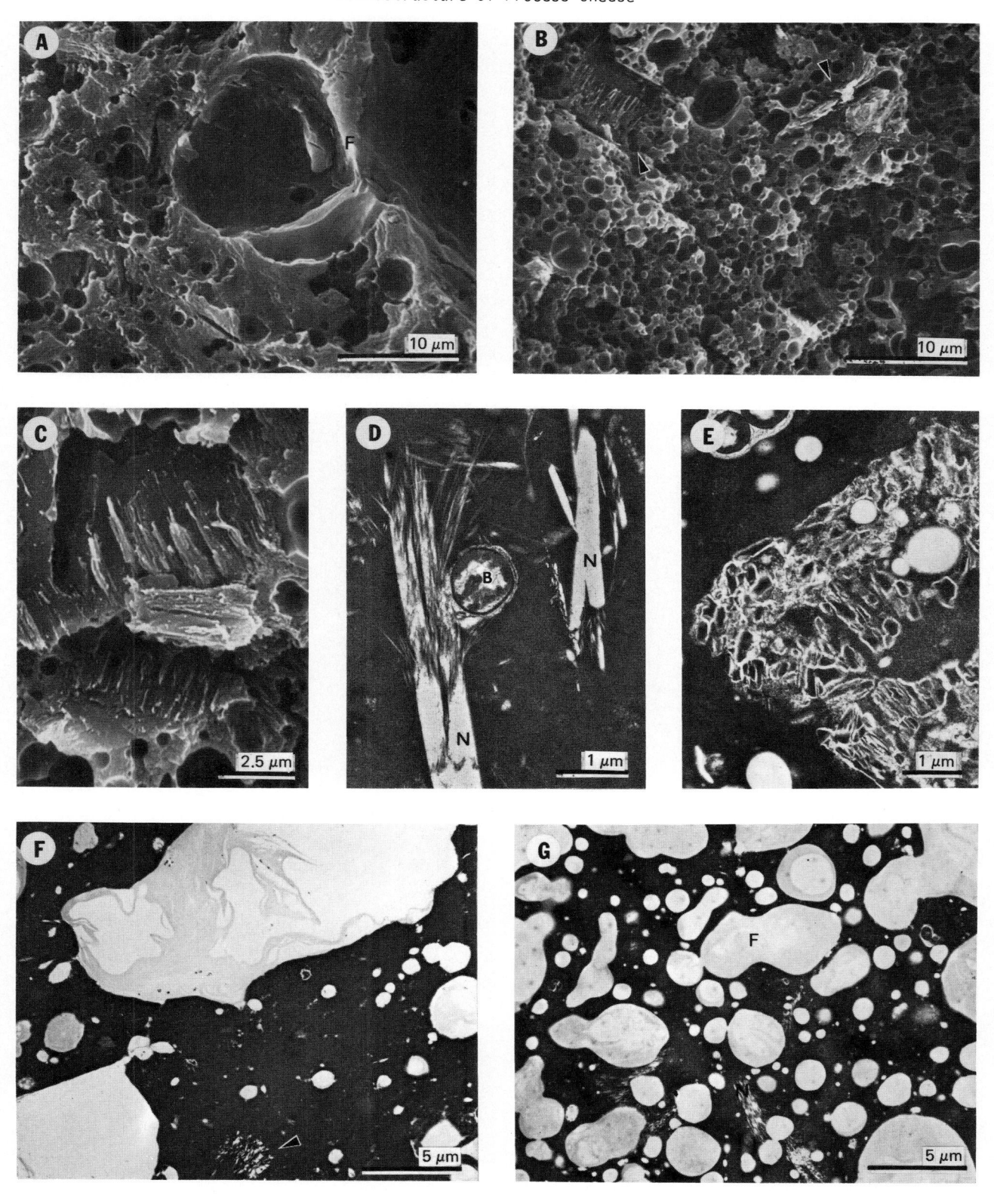

Fig. 7. Development of microstructure in process cheese in the presence of sodium aluminum phosphate (SALP). A = SEM at 0 time. Large fat particles (F) started to be degraded into smaller particles. G = Fragment of a calcium phosphate crystal. B = SEM after 10 min in the cooker. Fat is still in the form of large particles, many of which are being emulsified (F) into smaller particles. C = SEM after 40 min in the cooker. Some fat particles (F) are still undergoing emulsification. D = TEM at 0 time. Dark areas are the cheese protein matrix, light areas indicate fat. E = TEM after 40 min in the cooker. The emulsification process has not been completed and fat particles (F) are still undergoing emulsification. F = TEM detail of one of the added SALP crystals found in abundance during the initial 10 min in the cooker.

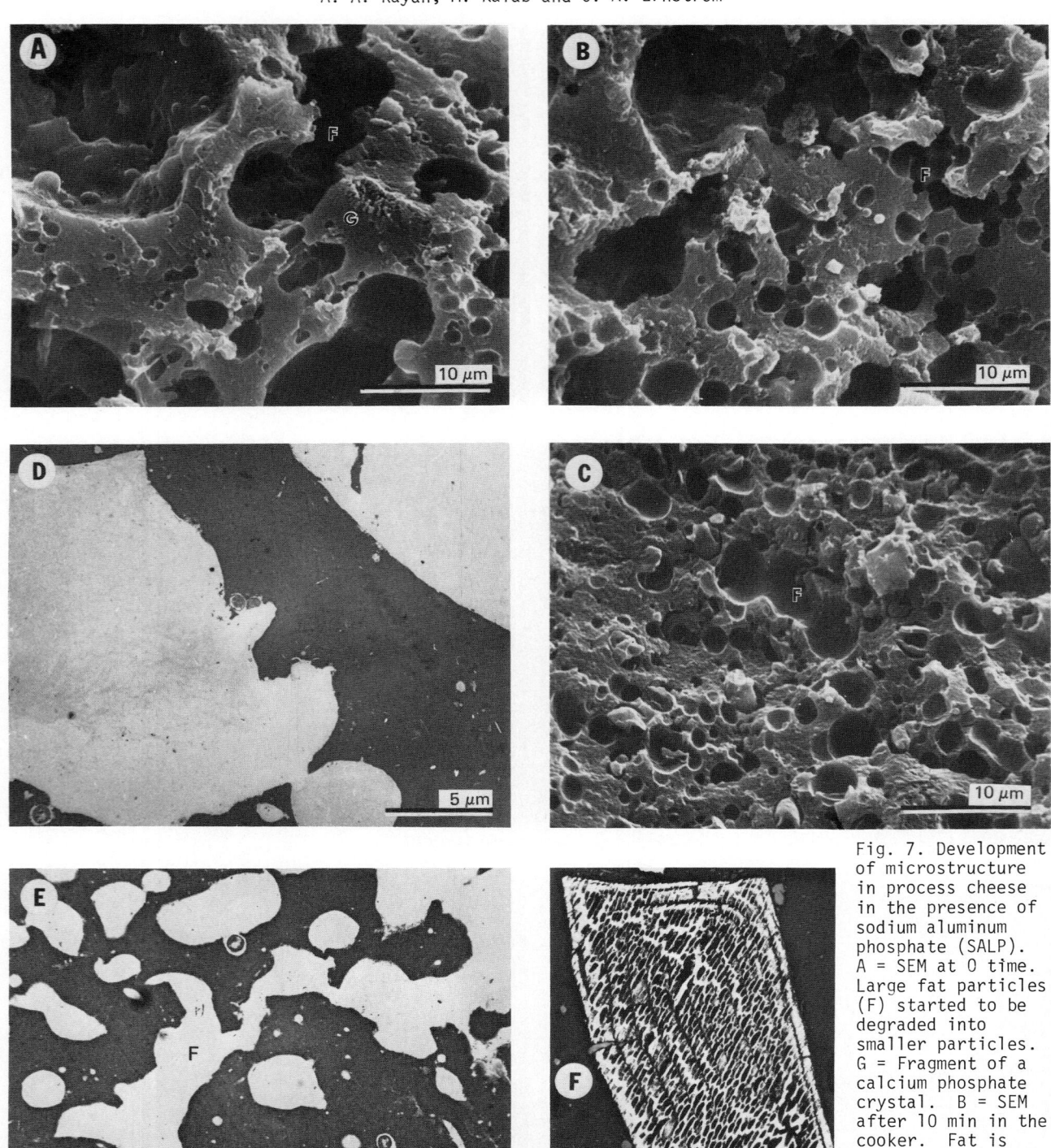

the changes observed in cheese structure and rheology?
Authors: All of these salts should easily dissolve in the water portion of cheese if their solubility in water is the only consideration. Obviously there are other considerations and additional experiments will be required to answer these questions.

M.V. Taranto: What were the dimensions of the fat droplets before and after heat processing with the various emulsifying salts?
Authors: As the cheese mixture was being heated in the cooker and before the temperature was high enough to cause the salt-protein interaction, the fat in the natural cheese was completely de-emulsified. The cheese mass was literally swimming in a sea of free fat. Emulsification was very rapid after reaching 65°C. After heat processing fat droplets varied in size but some were less than 1 µm in diameter.

M.V. Taranto: Were any studies conducted on model systems to determine the nature and extent of the interaction between the emulsifying salts and the cheese proteins? This interaction may have altered the functional properties of the proteins.
Authors: We do not know the extent or nature of the interaction between emulsifying salts and the cheese proteins, and I am unaware of studies on model systems that would make that determination. The interaction certainly does alter the functional properties of the proteins. All emulsifying salts increase the pH of the cheese and sequester calcium. Both of these actions will increase the solubility of caseinates.

M.V. Taranto: Were any experiments run to determine the degree of hydration of the protein matrix as a function of the interaction with the emulsifying salts and the cooking process?
Authors: Such studies were not run. We are aware that interaction of cheese proteins with emulsifying salts increases the solubility and presumably the hydration of the proteins.

M.V. Taranto and D.G. Schmidt: How was the identity of the various salt crystals observed in SEM and TEM micrographs verified?
Authors: On the basis of studies by Brooker et al (18) and our own findings (11,23) we became familiar with the typical appearance of calcium phosphate microcrystals in Cheddar cheese. As the addition of a single melting salt to cheese led to findings of crystals different from the initially present calcium phosphate, we assumed that these new crystals were those of the salt added, particularly as each different melting salt produced different images. We were unable to identify crystal-like structures such as the one in Fig. 6E. In other cases, particularly with sodium citrate, we suspect that the crystals had been washed out of the thin sections during sectioning and/or staining and that we have presented cavities in the fixed protein matrix initially occupied by the crystals.

J.H. Prentice: I am a little concerned that the authors found it necessary to fix and dehydrate the specimens for SEM, as that appears to detract from one of the main advantages of the method in that only minimal sample preparation is necessary. I would like to ask the authors whether they have compared SEM photographs of fixed and unfixed specimens, and are satisfied that glutaraldehyde and alcohol have no effect on the microstructure observed.
Authors: Sample preparation for SEM is minimal only with dry samples. Otherwise the preparation of biological samples for SEM is as critical as for TEM. Unless a replica of a freeze-fractured sample is examined by SEM (20) or the sample itself is examined in the frozen state (21), the sample has to be dehydrated because it is exposed to a high vacuum in the microscope. However, with cheese, dehydration alone is not sufficient to reveal the microstructure because of the high fat content in the sample; this fat obscures the protein matrix and must be removed. Taranto et al. (5), on the other hand, removed protein by a proteolytic enzyme and exposed fat globules. Collapse of the cheese protein matrix during dehydration and fat extraction is prevented by fixation, for example in glutaraldehyde. Artifacts caused by improper sample preparation in cheese were described earlier (22).

D.G. Schmidt: I wonder whether the crystalline materials are artifacts resulting from the specimen preparation procedure (in the dehydrating agent, alcohol, these salts are almost insoluble) or that they do occur as such in the cheese. In my own EM investigation of cheese I found such crystals only in ultrathin sections in TEM after alcohol dehydration and embedding, but not in freeze-fractured specimens, whereas in SEM specimens after cryofixation (D.G. Schmidt, S. Henstra and F. Thiel, Mikroskopie 35: 50. 1979) such crystals could not be found either. What is the author's opinion about this?
Authors: The composition of microscopic crystalline inclusions in cheese was studied in great detail by Brooker et al (18). The presence of the inclusions was confirmed by SEM in a great variety of Cheddar cheese specimens, some of which were prepared on laboratory and pilot plant scales. Calcium phosphate crystals were found irrespective of the dehydration technique used (freeze-drying or critical-point drying) and their incidence was highest in cheese made from milk to which calcium chloride had been added.

Additional discussion with reviewers of "Electron Microscopy...Cream Cheese" contined from page 162.

D.N. Holcomb: In plotting data from Table III it appears that the small lack of correlation is due primarily to samples #7 and #8 which are both immitation Cream cheeses. Please comment.

Authors: Data in Table 3 are the mean values produced in triplicate by 10 judges. Instead of presenting the standard deviation, these data were processed statistically to indicate whether the differences between them are significant at the 5% level. If the same letter is present at two different values (means) in the same column, this indicates that the values (means) are not significantly different and that they are within the limits of experimental error. I would find it difficult to comment on differences, which statistically do not exist. Probably more imitation Cream cheeses could be examined to see whether they would differ as a group from the two groups of traditional-style and newly formulated Cream cheeses, but no conclusions can be made concerning data in Table 3. Figure 9 shows the plots in which the data for imitation Cream cheeses are identified. There is, of course, no causal relationship between instrumental compressibility and properties other than sensory firmness, yet the correlation coefficients are high for all the sensory attributes plotted.

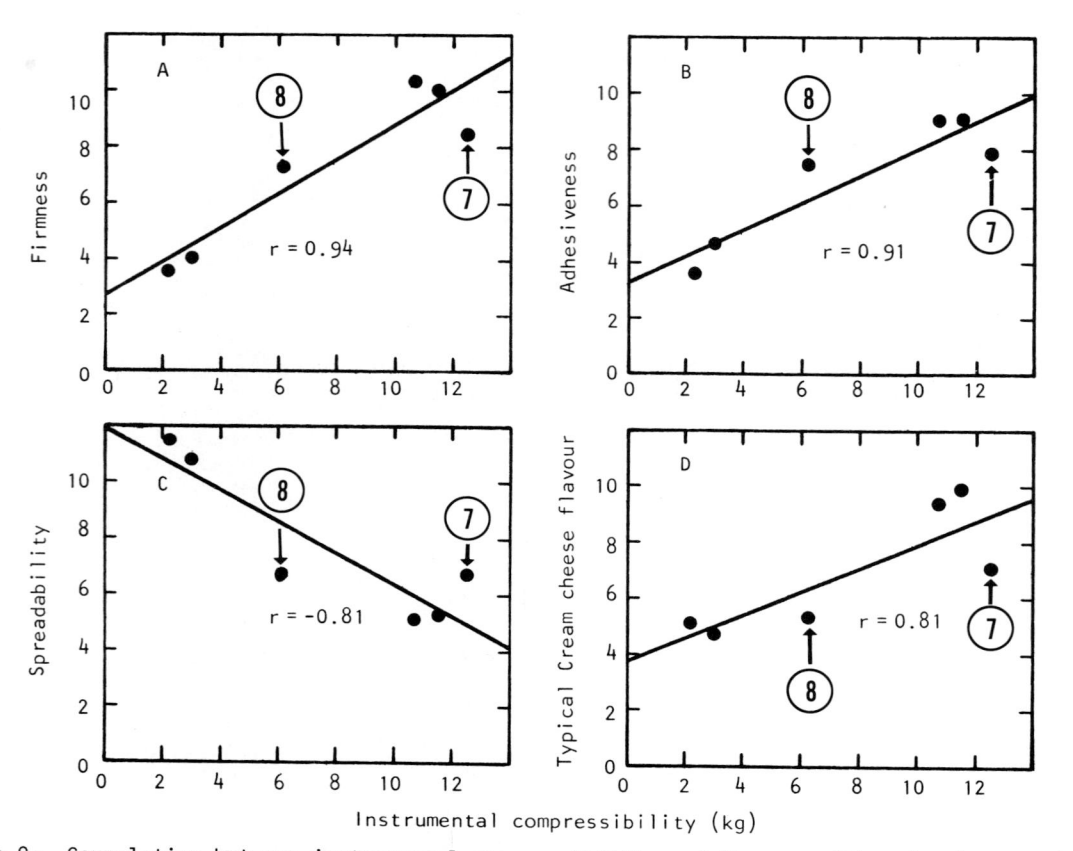

Figure 9: Correlation between instrumental compressibility and firmness (A), adhesiveness (B), spreadability (C), and typical Cream cheese flavor (D). Plot of data from Table 3. Points 7 & 8 represent immitation Cream cheeses.

ELECTRON MICROSCOPY AND SENSORY EVALUATION OF COMMERCIAL CREAM CHEESE

M. Kaláb, A. G. Sargant* and D. A. Froehlich

Food Research Institute
Research Branch
Agriculture Canada
Ottawa
Ontario, Canada K1A OC6

*Silverwood Dairies Limited
London
Ontario, Canada N6A 4E5

Abstract

Significant differences in microstructure, sensory attributes, and instrumental compressibility were found in commercial Cream cheeses; five specimens of traditional and new formulated Cream cheeses, one specimen of Neufchâtel cheese, and two specimens of imitation Cream cheese were studied.

Traditional Cream cheese was composed of fat globule clusters with fat globule membranes mostly preserved. Protein was aggregated at the surface of the fat globule clusters or in their vicinity. The new formulated Cream cheeses consisted of large fat particles formed probably by the coalescence of fat globules ruptured during the treatment of the curd; fat globule membranes were mostly absent, particularly on the large fat globules. Protein was fairly evenly distributed in the form of small clusters or chains. The corpuscular microstructure of all the Cream cheeses was probably due to stirring (in the traditional products) or homogenization (in the new formulated products) of the coagulated and cooked Cream cheese curd prior to packaging.

The traditional-style Cream cheese rated high in firmness, adhesiveness, typical Cream cheese flavour, and instrumental compressibility. The new formulated products were superior in spreadability; the differences in these attributes were statistically significant at the 5% level. Creaminess did not relate to the manufacturing processes used.

KEY WORDS: Cream cheese. Imitation Cream cheese. Neufchâtel cheese. Electron microscopy. Sensory evaluation. Microstructure. Spreadability. Firmness. Adhesiveness. Cream cheese flavour. Instrumental compressibility.

Introduction

Cream cheese is a soft unripened cheese with a rich, mildly acid flavour and a smooth buttery consistency, made from cream or from mixtures of cream and milk or skim milk[1]. Cream cheeses of various manufacturing origin possess various properties as may be evaluated by sensory and instrumental examination.

Canadian Cream cheese contains a minimum of 30% fat[2] whereas the minimum fat content in the American Cream cheese is 33%[3,4]. Neufchâtel cheese is the same type of cheese as Cream cheese but contains less fat (20-33%) and may contain more moisture than Cream cheese[3,4]. Traditional manufacturing processes for both cheeses have been described in detail[1,4].

In the traditional system of manufacturing, the cream mixture is pasteurized, homogenized, and ripened with a lactic culture until it attains pH of approximately 4.6. The curd is heated to 52-63°C either by steam or by the addition of hot water, after which it is drained. The curd is either cold-packed or hot-packed.

In the newly formulated method of manufacturing, a mixture of cream and milk solids with the final desired composition of the cheese is pasteurized, homogenized, and cooled to the incubation temperature of approximately 30°C. The mixture is then inoculated with a lactic culture and incubated until the desired acidity (pH≈4.6) is obtained, after which the mixture is homogenized and packaged without cooling. This procedure does not require the draining of whey.

The development of microstructure in other cheeses was studied by various workers and reviewed earlier by Brooker[5] and Kalab[6], but reports on the microstructure of Cream cheese have not been found in the literature. The objective of this research work was to examine the microstructure of various commercial Cream cheeses using electron microscopy in relationship to sensory attributes and instrumental compressibility.

Materials and Methods

All the Cream cheeses (8 specimens) were obtained from commercial sources.
Fat, moisture, and pH analyses. The fat content was

determined by consecutive extractions with ammonia, ethanol, diethyl ether, and petroleum ether using a Mojonnier Milk Tester[7]. Total solids and the moisture contents were determined by the method of Emmons et al.[8] The samples, 10 g, were frozen at -18°C, which led to the formation of large ice crystals. The resulting coarse texture of the samples permitted easy removal of the water vapour during consecutive freeze-drying at a pressure of 0.2 to 0.5 torr for 16 h and, subsequently, by drying in an evacuated oven for 5 h (98-100°C at a pressure of <100 torr). A Beckman miniature glass electrode and a Radiometer pH-Meter 26 were used to measure pH in the original undiluted specimens.
Electron microscopy.

Samples were taken from approximately 1 to 2 cm under the surface of the Cream cheese blocks. A small particle of the specimen, <1 mm in diameter, was picked up on the tip of a stainless steel needle and briefly dipped in a 35°C warm 3% agar sol. After the sol covering the Cream cheese particle had gelled within several seconds, the needle was withdrawn and the hole left in the particle was filled with a droplet of agar[9]. The sample thus encased in agar gel was then fixed in a 1.4% glutaraldehyde solution and postfixed in a 2% OsO_4 solution in 0.05M veronal-acetate buffer, pH 6.75, as reported earlier[10], dehydrated in a graded alcohol series, and embedded in a Spurr's low-viscosity medium[11] for thin-sectioning. Thin sections (≈90 nm) were stained with uranyl acetate and lead citrate solutions[12]. The encapsulation in the agar gel protected the soft Cream cheese specimens from disintegration during fixation[9].

Some Cream cheese specimens were freeze-fractured at -130°C: samples in metal tubular holders (Polaron) were frozen in liquid Freon 12 cooled to -150°C with liquid nitrogen and fractured by knocking off the upper holder section with a cooled mechanical arm[9]. The fractures were immediately shadowed with platinum evaporated at 45° and the coating was reinforced with carbon evaporated at 90°. The replicas were cleaned with a sodium hypochlorite solution followed by a bath in acetone to remove fat, and were retrieved from a petroleum ether-water phase boundary[13]. The clean replicas were placed on 150-mesh copper grids, 3 mm in diameter, coated with a formvar film and carbon.

A Philips EM 300 electron microscope was operated at 60 kV and micrographs were taken on 35-mm film. Intermediate negatives were used to obtain the final prints[9].

Some Cream cheese specimens were also examined by cold-stage scanning electron microscopy (SEM) at -100°C using samples freeze-fractured and gold-coated according to a modified procedure by Katoh[14]. A Cambridge Stereoscan electron microscope was operated at 20 kV and micrographs were taken on 35-mm film.

All Cream cheeses were examined in duplicate.
Sensory evaluation.

All the Cream cheeses except #2 and 6 were evaluated for sensory properties listed in Table 1 by 10 judges using the descriptive analysis with scaling[15]. Each judge recorded the perceived intensity of each attribute on unstructured 15-cm lines with anchor points 1.5 cm from each end. The descriptive terms assigned to the anchor points are shown in Table 1.

Table 1
Evaluation of sensory attributes on 15-cm unstructured lines with anchor points 1.5 cm from the line ends.

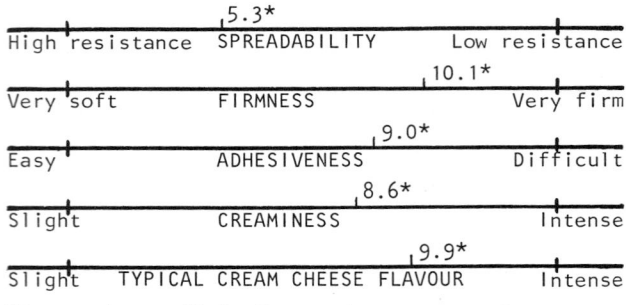

High resistance	SPREADABILITY 5.3*	Low resistance
Very soft	FIRMNESS 10.1*	Very firm
Easy	ADHESIVENESS 9.0*	Difficult
Slight	CREAMINESS 8.6*	Intense
Slight	TYPICAL CREAM CHEESE FLAVOUR 9.9*	Intense

*Cream cheese #1 is featured as an example (see Table 3).

The Cream cheeses at +5°C were cut into 5-g samples and evaluated for spreadability within an hour; unsalted crackers and knives were provided for this test. Other sensory attributes were evaluated by the judges during the same sessions using 10-g samples. Evaluation was replicated 3 times.
Instrumental test for compressibility.

Compressibility was measured with 50-g samples of all the Cream cheeses using the Universal Food Rheometer[16] equipped with a back extrusion cell of the Ottawa Texture Measuring System[17].

A plunger was forced at a rate of 5 cm/min into the sample extruding it through the 1-mm annular clearance between the container wall and the plunger and the force applied was plotted vs. time to obtain a force-deformation curve. The initial phase was approximately linear as the sample was compressed. The end of the compression phase was indicated by an abrupt change in the slope of the force-deformation curve. The point of inflexion was interpreted to indicate the force required to initiate spreading of the Cream cheese sample which would relate most closely to the sensory firmness evaluation. All measurements were replicated 3 times with each Cream cheese specimen.
Statistical analysis.

Mean values for each Cream cheese were subjected to analysis of variance and Tukey's test[18] was used to determine differences significant at the 5% level.

Results and Discussion

The Cream cheese specimens under study are listed in Table 2, which also indicates the method of manufacture, fat and moisture contents declared on the labels of the products, declared presence of emulsifiers and/or stabilizers, fat and moisture contents determined by chemical analyses[7,8], and pH.

There were considerable differences between the declared and actual fat content, which was probably due to the declaration of the minimum fat content on the labels. The relative difference was greatest in Neufchâtel cheese (20% vs. 34.3%) and smallest in new formulated Cream cheese #4 (36% vs. 39.0%) and in the dairy imitation Cream cheese #7 (12.5% vs. 13.9%). Individual ingredients in the nondairy imitation Cream cheese were listed in a descending order; this product was claimed to contain 1/3 less calories than regular Cream cheese.

Table 2

Cheeses examined in this study.

Cheese number:	Manufacture*	Declared: Fat (%)	Declared: Water (%)	Determined by analysis: Fat (%)	Determined by analysis: Water (%)	pH:	Declared emulsifiers and/or stabilizers:
1	T	31	54	35.1±1.0	53.60±0.05	4.6	Carob bean gum
2	T	30	55	40.2±2.1	50.61±0.01	5.2	Carob bean gum
3	T	30	55	37.1±1.5	54.50±0.07	5.0	Carob bean gum
4	F	36	55	39.0±1.5	51.63±0.07	4.8	Propylene glycol alginate; mono- and diglycerides
5	F	30	--	40.3±0.8	53.41±0.01	--	Propylene glycol alginate; carrageenan; locust bean gum
6	N	20	65	34.3±1.2	47.44±0.04	4.8	---
7	DI	12.5	--	13.9±0.9	59.36±0.39	--	Sodium citrate; carob bean gum
8	NI	--	--	26.7±0.3	63.35±0.01	5.1	---

* T = Traditional; F = New formulated; N = Neufchâtel cheese; DI = Dairy imitation Cream cheese; NI = Nondairy imitation Cream cheese.

Results of electron microscopical examination.

Three electron microscopical techniques were used to examine the Cream cheese specimens under study: cold-stage SEM of a freeze-fractured and gold-coated sample, freeze-fracturing and replication with platinum and carbon, and embedding in a resin followed by thin-sectioning. The objective was to compare the results and to select the most convenient technique.

To prevent the Cream cheese samples from disintegration in aqueous fixatives during fixation, encapsulation in agar gel was used as outlined in Materials and Methods. Embedding and thin-sectioning (Fig. 1 A and B) correlated well with both former techniques (Fig. 1 D and E) and was selected as the only technique for the examination of the microstructure in all the Cream cheese specimens. The resolution was better than that obtained by cold-stage SEM and the technique was easier to perform and evaluate than replication with platinum and carbon.

Cream cheese #1. Electron microscopy reveals (Fig. 1 A and B) that this cheese consists of small (<2 μm in diameter) fat globules with fairly uniformly dispersed protein and a smaller number of large (>15 μm in diameter) intact or ruptured individual fat globules. A large proportion of the small fat globules forms spherical clusters of various dimensions; the fat globules in the clusters are cemented to each other by protein which is also accumulated in even larger quantities at the surface of the clusters. At a higher magnification the cementing protein appears to be either in a diffuse or micellar form (Fig. 1 C). A bacterium attached to fat globule clusters by extracellular filaments[19] is also shown in Fig. 1 C; the stained electron-dense particles composing the bacterial filaments permeate the entire Cream cheese matrix. This phenomenon is also encountered in other Cream cheeses.

As was mentioned earlier, a platinum-carbon replica (Fig. 1 D) shows the same components and a similar microstructure in a cheese sample which had not been encapsulated in agar gel and chemically fixed but was fixed by rapid freezing and was freeze-fractured. A comparison with thin-sectioning indicates a close correlation between both techniques. Thus, thin-sectioning is assumed to produce images of the true microstructure of the Cream cheese specimens under study; the clustering of the fat globules is real and is not the result of the preparatory steps.

Cold-stage SEM of a freeze-fractured and gold-coated Cream cheese sample shows a similar microstructure (Fig. 1 E) at a considerably worse resolution; however, individual fat globules and fat globule clusters encased in protein envelopes are clearly visible.

Cream cheese #2. Microstructure of this specimen is to a great extent similar to that of the preceding Cream cheese; small fat globules are also aggregated in spherical clusters. Micellar protein is accumulated at the perimeter of the clusters (Fig. 2 A) forming an interface between the aqueous and the lipid phases in the Cream cheese (Fig. 2 B). Unlike in the preceding Cream cheese, micellar protein is packed around clusters consisting of only several fat globules and also around larger individual fat globules. The amount of protein ranges within wide limits: some fat globules are encased only in membranes with a thin layer of electron-dense particles probably of bacterial origin adhering to them with no protein present, but some other fat globules are surrounded by dense protein envelopes (Fig. 2 C). A closer examination reveals that most fat globules have their membranes preserved.

Cream cheese #3. This product is also composed of fat globule clusters cemented by protein. The clusters, however, are not spherical as in specimens #1 and 2, and the micellar protein is not accumulated at the surface of the clusters (Fig. 3 A). At a higher magnification (Fig. 3 B) this cheese resembles specimens #1 and 2.

In addition to fat globules and protein particles, there are short fibres present in this cheese connecting casein micelles with each other or with fat globules (Fig. 3 B). Long strands of extraneous material are found in this cheese winding through the matrix for several tens of micrometers. Fig. 3 C shows a short section of such a strand.

Cream cheese #4. Cream cheese #4 is substantially different from the preceding three cheeses as far as microstructure is concerned. The greater dimensions of the fat globules (>2 μm in diameter) (Fig. 4 A) and the absence of fat globule membranes (Fig. 4 B) are the typical features of this cheese. Protein is usually aggregated in relatively thin layers around the membrane-less fat globules. If the fat globule membranes are present, most of them are ruptured, as evident in Fig. 4 A and B. Casein micelles present in this cheese in the form of aggregates are considerably smaller

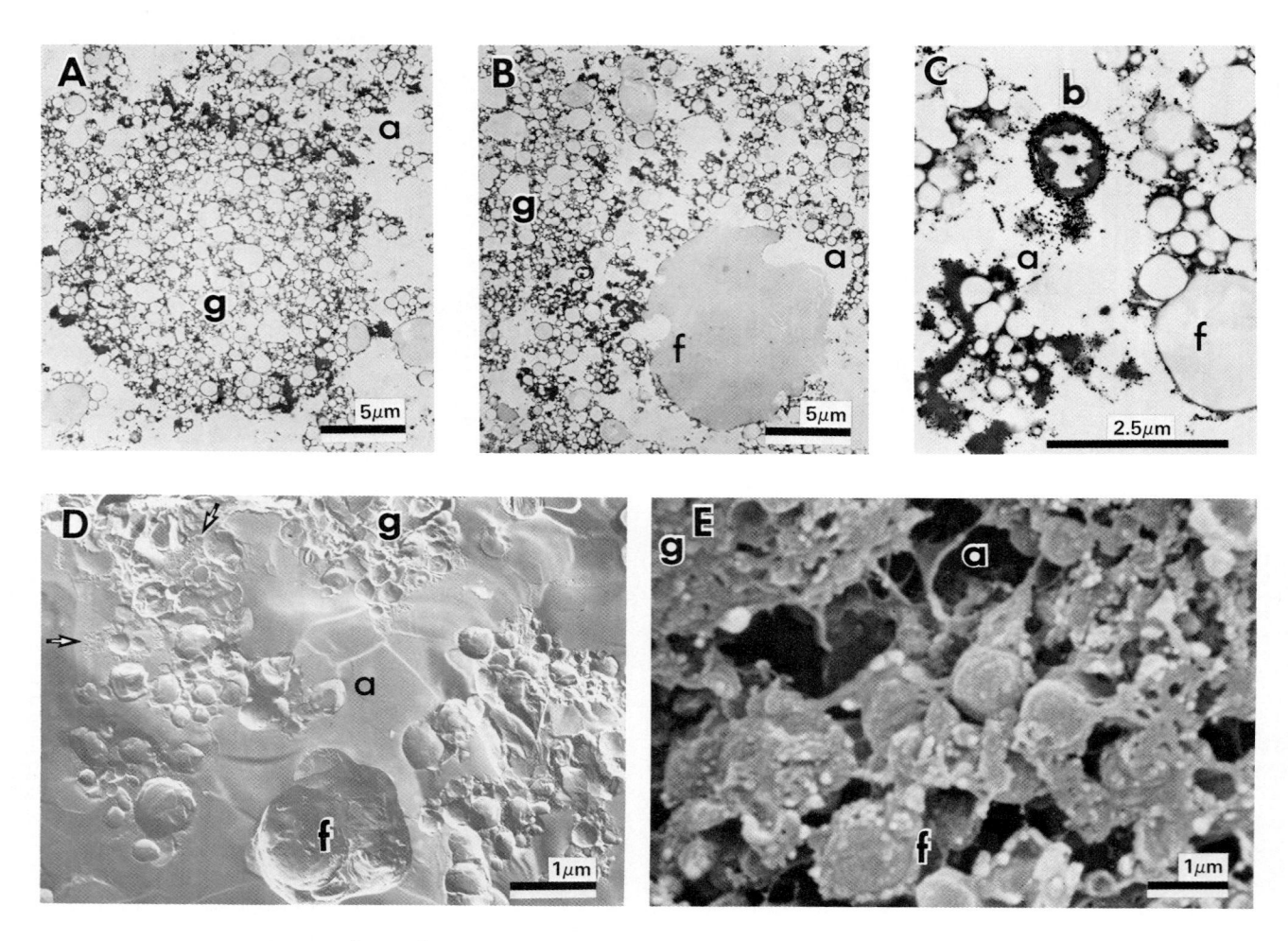

Fig. 1. (above) For caption see top of facing page.

Fig. 2. (below) For caption see bottom (left column) of facing page.

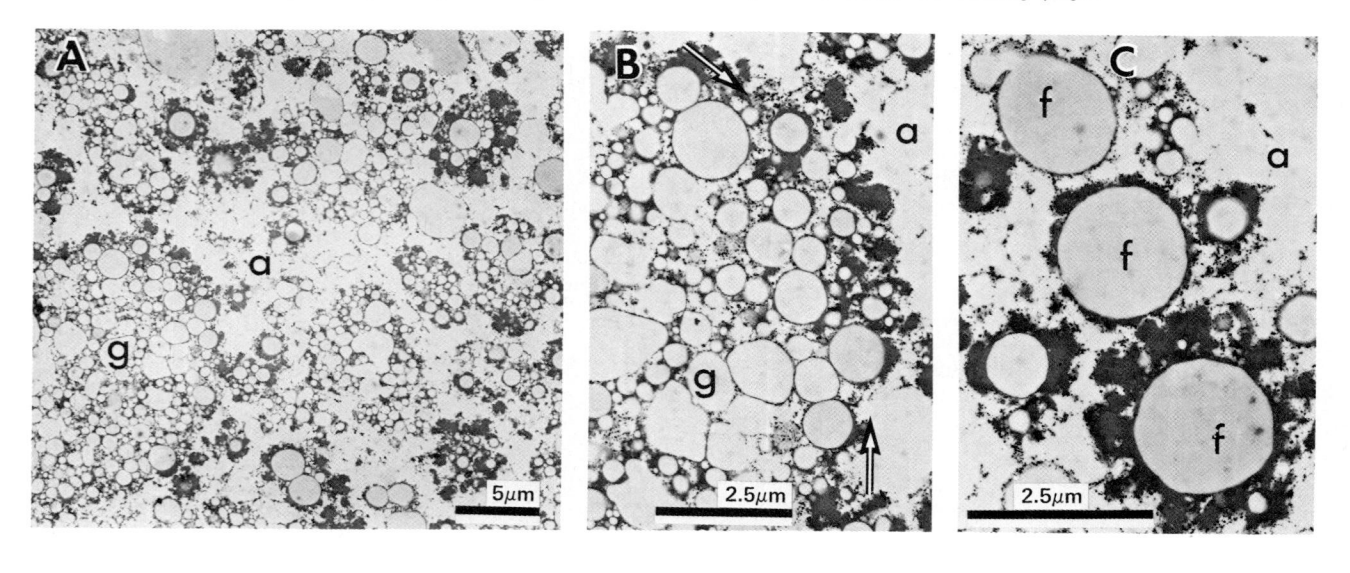

Fig. 1. (facing page - top) Microstructure of Cream cheese #1 (traditional system of manufacture).
A = A large spherical fat globule cluster (g) characteristic of this Cream cheese with protein (dark bodies) concentrated at the surface of the cluster. a = Aqueous phase.
B = Another area with a large, partly ruptured fat globule (f) and a part of a large spherical fat globule cluster (g); other fat globule and protein clusters are not spherical.
C = A detail of a bacterium (b) attached by filaments to fat globules; the filaments consist of minute dark particles permeating the entire Cream cheese matrix. a = Aqueous phase. f = Fat.
D = A platinum-carbon replica of a freeze-fractured specimen showing discrete fat globule clusters (g), large fat globules (f), and the aqueous phase (a). Arrows point to protein.
E = Cold-stage SEM of a freeze-fractured and gold-coated specimen. g = Part of a large fat globule cluster. f = Individual fat globules covered with protein. a = Space initially occupied by the aqueous phase.

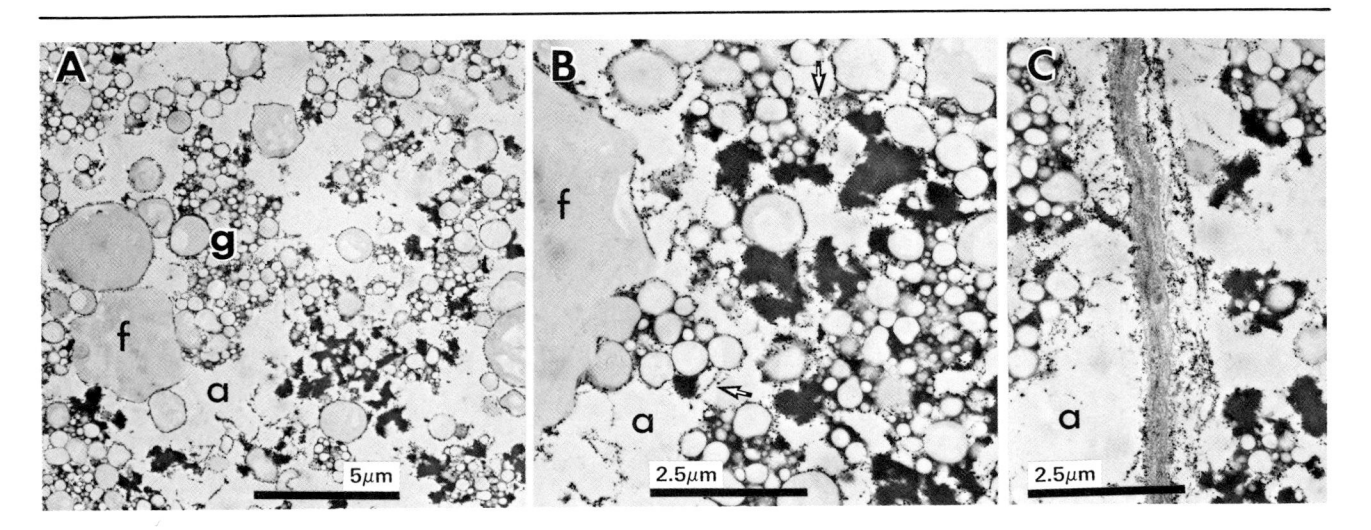

Fig. 3. (above) Microstructure of Cream cheese #3 (traditional system of manufacture).

A = Irregular fat globule clusters (g), individual fat globules (f), protein aggregates (dark bodies), and the aqueous phase (a).
B = A detail of the cheese matrix. f = A large fat globule with a partly ruptured membrane. Additives (possibly a vegetable gum) are visible (arrows) in the form of fine fibres in the aqueous phase (a).
C = An example of coarse filaments present in this specimen. a = Aqueous phase.

Fig. 2 (facing page - bottom) Microstructure of Cream cheese #2 (traditional system of manufacture).

A = A large spherical fat globule cluster (g) with protein (dark bodies) concentrated at its surface. Some smaller clusters and individual fat globules are surrounded with large amounts of protein producing a microstructure characteristic of this Cream cheese.
B = A detail of protein (dark bodies) concentrated at the surface of a fat globule cluster (g) forming an interface between the lipid (g) and the aqueous (a) phases (arrows).
C = Three fat globules (f) surrounded with small (upper globule) to large amounts of protein (lower globule). Minute dark particles permeating the Cream cheese matrix are probably of bacterial origin. a = Aqueous phase.

than in any other Cream cheese under study (Fig. 4 B); occasionally very compact protein clusters are present (Fig. 4 D and E), some of which display the typical micellar microstructure (Fig. 4 D) whereas others are very dense. Unlike in the three preceding cheeses, where individual bacterial cells are dispersed fairly evenly, the bacteria in Cream cheese #4 are present in large groups of up to 50 cells (as counted in the two-dimensional sections, Fig. 4 F).

Cream cheese #5. Also in this specimen a great proportion of fat is present in the form of large irregular particles either completely lacking membranes or having the fat globule membranes preserved to some extent (Fig. 5 A and B). The electron microscopical proof of the presence of fat in membrane-less particles is based on the phenomenon that fat fixed with OsO_4 appears darker[20] in the micrographs than the aqueous phase because of the presence of osmium in the fat as the result of the fixation. The ragged surface of the fat particles is more clearly visible at a higher magnification (Fig. 5 C); casein micelles are shown to be in the form of small clusters or chains, the surfaces of which are laced with electron-dense particles probably of bacterial origin, although the bacterium in Fig. 5 C shows no evidence of filamentous material.

Neufchâtel cheese #6. In this specimen of Neufchâtel cheese the fat is aggregated in particles so large that it is impossible to

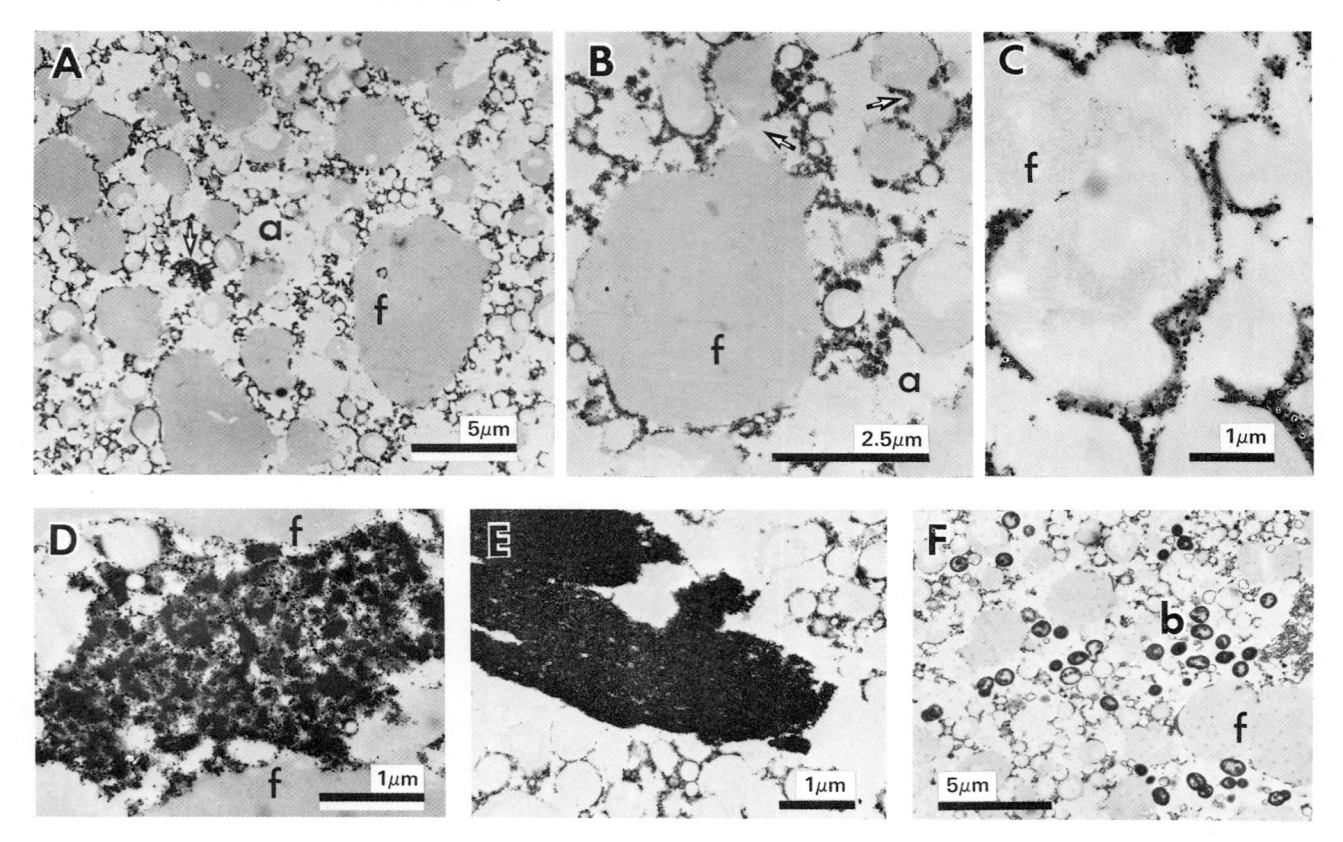

Fig. 4. Microstructure of Cream cheese #4 (new formulated method of manufacture).
A = Most fat globules (f) have their membranes ruptured. Protein is aggregated around small fat globules and in the form of compact clusters (arrow). a = Aqueous phase.
B = A detail of protein aggregates (dark bodies) and coalescing (arrows) fat particles (f).
C = A detail of ruptured and coalescing fat globules (f) in another Cream cheese sample.
D = A compact protein cluster (dark body) between two fat globules (f) is evidently of micellar nature.
E = A more compact protein cluster (dark body) in the same sample as presented in C.
F = Accumulation of bacteria (dark round bodies, b). f = Fat globules with ruptured membranes.

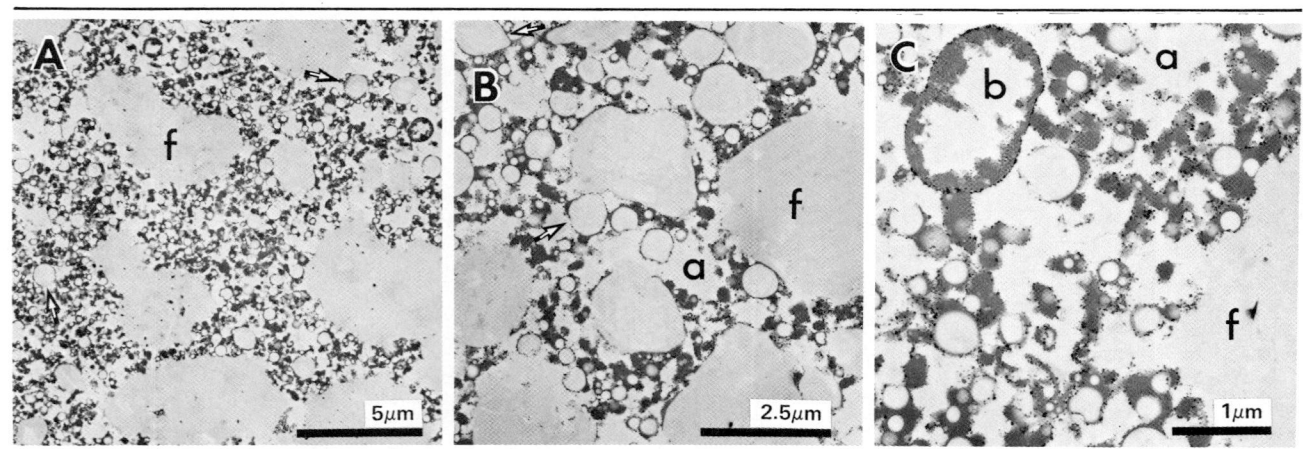

Fig. 5. Microstructure of Cream cheese #5 (new formulated method of manufacture).
A = The protein matrix is composed of micellar clusters (dark bodies). Fat globules greatly vary in dimensions from the small ones surrounded with membranes and/or protein (arrows) to the large ones (f) lacking membranes.
B = A detail of protein (dark bodies), small fat globules encased in membranes (arrows), and larger ruptured fat globules (f). a = Aqueous phase.
C = A detail at a higher magnification. f = Fat particle. a = Aqueous phase. b = Bacterium.

COMMERCIAL CREAM CHEESE

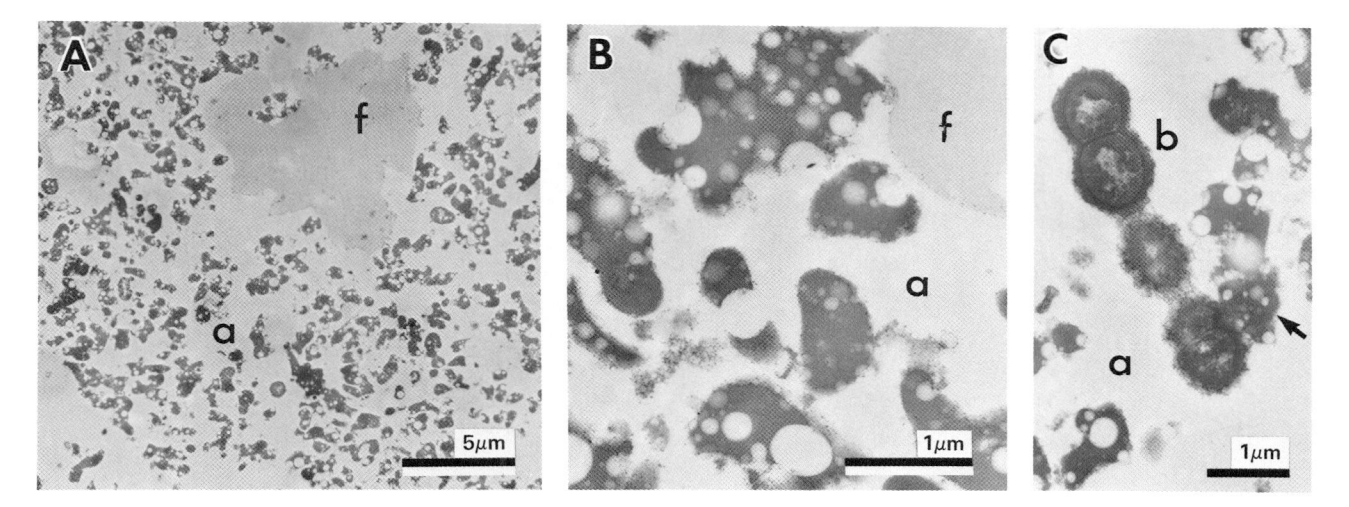

Fig. 6. Microstructure of Neufchâtel cheese #6.
A = The protein matrix is composed of protein particles which encase several minute fat globules. Most fat particles are several tens of micrometers in diameter (particle "f" is one of the smaller ones). a = Aqueous phase.

B = A detail of protein particles (dark areas) encasing minute fat globules (light circles).
f = Part of a fat particle. a = Aqueous phase.
C = Detail of a chain of bacteria (b) attached to a protein particle (arrow) which contains minute fat globules (light circles).

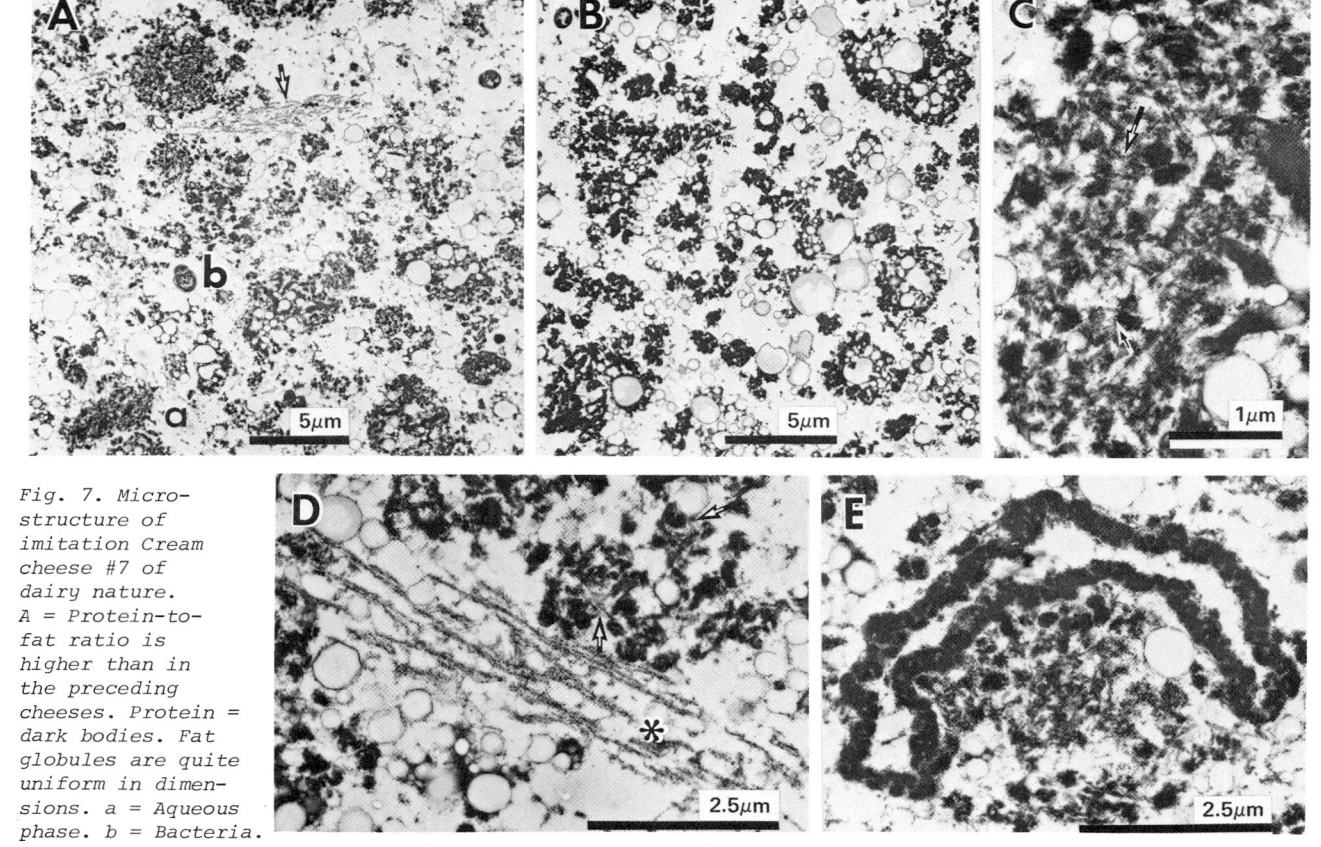

Fig. 7. Microstructure of imitation Cream cheese #7 of dairy nature.
A = Protein-to-fat ratio is higher than in the preceding cheeses. Protein = dark bodies. Fat globules are quite uniform in dimensions. a = Aqueous phase. b = Bacteria. Arrow points to long filaments of an additive (possibly a vegetable gum) shown at a higher magnification in D.
B = Another specimen shows a coarser microstructure than A.

C = Detail of a protein cluster with an additive in the form of short needles (arrows).
D = Two types of additive: free long filaments (*) and short needle-like structures (arrows).
E = A structure of unknown nature (probably casein).

159

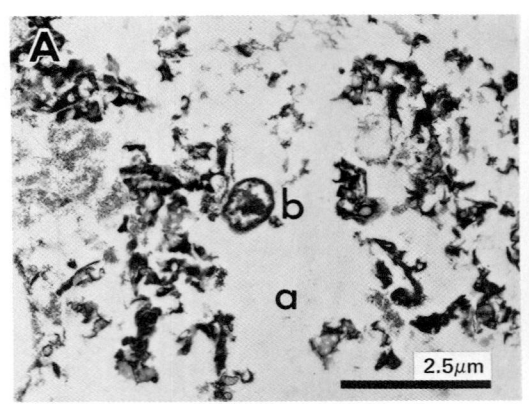

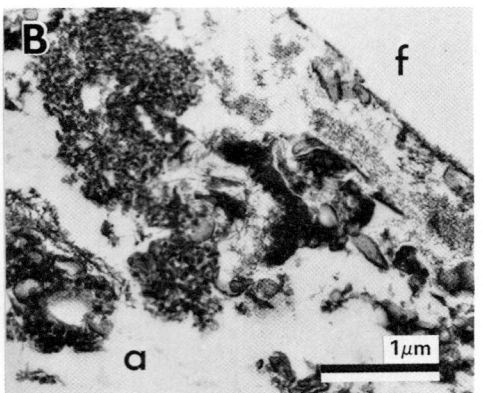

Fig. 8. Microstructure of imitation Cream cheese #8 of nondairy nature.
The nondairy origin is evident from the absence of casein micelles and milk fat globules.
A = Protein bodies (dark particles) are dispersed in the aqueous phase (a).
b = Bacterium.
B = A detail of the protein matrix. f = Fat globule. a = Aqueous phase.

accommodate them in the field of vision even at the lowest magnification used (2570 x). Such fat particles contain neither membranes nor aggregated protein at their surfaces (Fig. 6 A). In addition to the individual large fat particles, fat is present in the form of minute globules encased in small protein clusters (Fig. 6 B). Bacteria occasionally form longer chains (Fig. 6 C). The microstructure of the protein matrix in this cheese resembles the matrix in cheese #5 although by definition[3,4] it is supposed to contain considerably less fat; the specimen shown in Fig. 6, however, contained 34.3% of fat (Table 2), which is very close to the fat content in Cream cheese #1 (35.1%, Table 2).

Imitation Cream cheese #7. Two different specimens examined are of a slightly different microstructure (Fig. 7 A and B) but, in general, the microstructure of this cheese to a great extent resembles the microstructure of Cream cheeses #1 to 3 taking into consideration the markedly lower fat content in the imitation Cream cheese (13.9%, Table 2) as compared to the fat content in cheeses #1 to 3 (35.1-40.2%). The micrographs reflect the higher protein-to-fat ratio in this product declared to consist of a mixture of Cream and Cottage cheeses. Some of the protein is in the form of relatively unchanged casein micelles (Fig. 7 B to D) and some is in the form of larger conglomerates (Fig. 7 A to C). Structures of unknown nature resembling collapsed hollow shells probably composed of casein micelles (Fig. 7 C) are also present. Small fat globules are incorporated in the protein clusters. The presence of declared additives such as vegetable gums is well documented in Fig. 7 C to D; the additives are either in the form of short needles or in the form of long filaments (Fig. 7 D).

Imitation Cream cheese #8. The nondairy nature of this specimen is clearly visible under the electron microscope (Fig. 8). There is no evidence of casein micelles, as sodium caseinate is declared to be the proteinaceous ingredient. Although individual bacteria are found in this product (Fig. 8 A), the presence of bacterial cultures is not declared on the label. Fat globules of hydrogenated vegetable oil such as coconut, soybean, or corn oil are so large that electron microscopy is not suitable to study them even at the lowest magnification used. The microstructure of this product differs from the other Cream cheeses under study (Fig. 8 B). This is in agreement with the declared composition, as even tapioca flour is one of the ingredients in the imitation Cream cheese.

Discussion of microstructure.

Cream cheeses #1 to 3 represent the traditional manufacturing process. In spite of some differences in the microstructure within the group (such as the aggregation of micellar protein around individual fat globules in cheese #2 but not in #1 and 3, the smaller dimensions of fat globule clusters in cheese #3 than in #1 and 2, and the accumulation of protein at the surface of fat globule clusters in cheeses #1 and 2 but not in #3), the microstructure of the Cream cheeses in this group is generally similar. Fat globule membranes are mostly intact and the smaller fat globules are in the form of clusters. Large individual fat globules are quite rare. Protein is unevenly distributed either at the surface of the fat globule clusters (cheeses #1 and 2) or is aggregated separately (cheese #3). Such a microstructure is indicative of manufacturing processes which have not totally destroyed the initial microstructure of the cream used.

The microstructure of cheeses #4 and 5, which represent the newly formulated manufacturing process, is essentially different. Fat is present in the form of large particles, formed probably by the rupture and coalescence of fat globules during the manufacturing process. Protein is in the form of small clusters or chains or, as in cheese #4, in the form of particles smaller than casein micelles or very compact clusters.

Electron microscopical comparison with other cheeses[10,21-23] reveals that the Cream cheeses and Neufchâtel cheese do not have a solid uninterrupted protein matrix. It is assumed that the discrete fat globule clusters and protein clusters in the traditional Cream cheeses are the results of stirring the coagulated and cooked curd until a smooth and homogeneous consistency is obtained. Fragmentation of the protein matrix is particularly well evident in the #6 Neufchâtel cheese.

In the new formulated Cream cheeses, homogenization also leads to the disruption of the protein matrix and to the development of a corpuscular microstructure.

The #7 imitation Cream cheese resembles the #1 Cream cheese as far as both microstructure and sensory attributes are concerned. This similarity may arise from similar manufacturing processes as both these products are made by the same manufacturer.

Microstructure of the #8 specimen is completely different from the microstructure of any dairy Cream cheese and is presented here as an example of a product which significantly differs from the dairy products in microstructure and chemical composition but is comparable with them as far as sensory attributes are concerned.

Results and discussion of sensory evaluation and instrumental compressibility tests.

Analysis of variance indicates significant differences between Cream cheeses #1 to 3 on the one hand and cheeses #4 and 5 on the other hand in the sensory attributes except creaminess and in instrumental compressibility as shown in Table 3. Cheeses #4 and 5 spread more easily than cheeses #1 to 3 and are not as firm. The traditional-style group is more adhesive than the new formulated group. Creaminess is similar in all the groups. As far as the typical Cream cheese flavour is concerned, the traditional-style group rates significantly higher than any other Cream cheese under study.

Table 3
Sensory attributes and instrumental compressibility in commercial Cream cheeses.

Specimen number:	Spreadability:	Firmness:	Adhesiveness:	Creaminess:	Typical Cream cheese flavour:	Instrumental compressibility (kg)*:
1	5.3a	10.1ab	9.0a	8.6a	9.9a	11.5ab
2	--	--	--	--	--	9.1abc
3	5.1a	10.4a	9.1a	8.2a	9.4a	10.7ab
4	11.5b	3.7c	3.6c	8.6a	5.2b	2.2d
5	10.8b	4.1c	4.6c	8.9a	4.7b	3.0d
6	--	--	--	--	--	7.4bc
7	6.7a	8.6ab	7.9ab	8.1a	7.0b	12.5a
8	6.6a	7.4b	7.4bc	8.4a	5.3b	6.2cd

* Any two means followed by the same letter in the "sensory attribute" and "instrumental compressibility" columns are not significantly different at the 5% level.

Instrumental compressibility measurements indicate that the traditional-style Cream cheeses are significantly firmer (Table 3). The calculated correlation coefficient of 0.94 indicates a close correlation between the sensory firmness and the instrumental compressibility values expressing firmness.

Conclusion

In this study, 5 specimens of commercial Cream cheese, 1 specimen of Neufchâtel cheese, and 2 specimens of imitation Cream cheese were examined by electron microscopy, sensory evaluation, and instrumental compressibility test. Results indicate that there are substantial differences between Cream cheeses made by the traditional system of manufacturing and the new formulated Cream cheeses. The examination of Cream cheeses and imitation Cream cheeses yields an overall picture of the differences in microstructure and sensory attributes in a variety of related products. This study was facilitated by preserving the initial microstructure in the specimens destined for electron microscopy. Stirring and/or homogenization, which are the essential steps in the Cream cheese manufacture, disrupt the curd and produce a corpuscular microstructure in the finished product. Such products are susceptible to disintegration in aqueous fixatives used in preparing the samples for embedding in a resin and for sectioning. Encapsulation of minute samples in agar gel prevented them from disintegration. The large amounts of fat particles in the samples were fixed with OsO_4 and the small dimensions of the samples facilitated embedding in the resin. Other electron microscopical techniques such as cold-stage SEM and replication with platinum-carbon of samples which were fixed by rapid freezing and not by chemical fixatives, confirmed the results obtained with thin sections.

Each of the two systems of manufacturing was found to be associated with a distinct type of microstructure and sensory attributes. However, the mechanisms responsible for all the mutual relationships have yet to be explained. The next stage in this research includes an experimental preparation of Cream cheese specimens on a pilot plant scale under controlled conditions and examination of the curd at the gradual stages of development.

Acknowledgment

The authors thank Kraft Inc., Glenview, Illinois, who kindly donated some of their products for this study, and Mr. D.C. Beckett for the procurement of other specimens and for chemical analyses. Skillful assistance provided by Mr. J.A. G. Larose and Mr. D.P. Raymond is acknowledged. Electron Microscope Centre, Research Branch, Agriculture Canada in Ottawa provided facilities. Appreciation is expressed to Ms. Elizabeth Larmond and Dr. D.M. Irvine for reviewing the manuscript. This study is Contribution 454 from Food Research Institute, Research Branch, Agriculture Canada in Ottawa.

References

1. Van Slyke L.L. and Price W.V. Cheese. Orange Judd Publ. Co., Inc., New York, N.Y., 1952, 406-430.

2. Canada Food and Drug Regulations. B-08.035 Cream Cheese. Dept. of Natl. Health and Welfare, Ottawa, Ontario, Canada, 1979, 43b.

3. Cheese Varieties and Descriptions. U.S. Dept.

of Agriculture, Agriculture Handbook No. 54, Washington, D.C., 1953, 36.

4. Kosikowski F. Cheese and Fermented Milk Foods. 1966, 119-134.

5. Brooker B.E. Milk and its products. In: Food Microscopy. J.G. Vaughan (ed.), Acad. Press, London, England, 1979, 273-311.

6. Kalab M. Microstructure of dairy foods. 2. Milk products based on protein. J. Dairy Sci. 62, 1979, 1352-1364.

7. Instruction Manual for Setting up and Operating the Mojonnier Milk Tester. Mojonnier Bros. Co., Chicago, Illinois, 1925, 26-52.

8. Emmons D.B., Larmond E. and Beckett D.C. Determination of total solids in heterogeneous heat-sensitive foods. J. Am. Offic. Anal. Chem. 54, 1971, 1403-1405.

9. Kalab M. Electron microscopy of milk products: A review of techniques. Scanning Electron Microsc. 1981; 3: 453-472.

10. Kalab M. Milk gel structure. VI. Cheese texture and microstructure. Milchwissenschaft 32, 1977, 449-458.

11. Spurr A.R. A low-viscosity epoxy resin embedding medium for electron microscopy. J. Ultrastruct. Res. 26, 1969, 31-43.

12. Kalab M. and Harwalkar V.R. Milk gel structure. II. Relation between firmness and ultrastructure of heat-induced skim-milk gels containing 40-60% total solids. J. Dairy Res. 41, 1974, 131-135.

13. Kalab M. Cleaning replicas of freeze-fractured oil-water emulsions. Microsc. Soc. Canada Bull. 6(4), 1978, 24-25.

14. Katoh M. SEM replica technique for butter and cheese. J. Electron Microsc. 28, 1979, 199-200.

15. Larmond E. Laboratory Methods for Sensory Evaluation of Food. Canada Dept. of Agriculture, Publ. 1637. Ottawa, Ontario, Canada, 1977, 47-48.

16. Voisey P.W. and Randall C.J. A versatile food rheometer. J. Texture Stud. 8, 1977, 339-358.

17. Voisey P.W. Ottawa Texture Measuring System. Canad. Inst. Food Sci. Technol. J. 4, 1971, 91-103.

18. Steel R.G.D. and Torrie J.H. Principles and Procedures of Statistics. McGraw-Hill Book Co., Inc., New York, N.Y., 1960, 99-160.

19. Brooker B.E. Cytochemical observations on the extracellular carbohydrate produced by *Streptococcus cremoris*. J. Dairy Res. 43, 1976, 283-290.

20. Crozet N. and Guilbot A. A note on the influence of osmium fixation on wheat flour lipids observation by transmission and scanning electron microscopy. Cereal Chem. 51, 1974, 300-304.

21. Emmons D.B., Kalab M., Larmond E. and Lowrie R.J. Milk gel structure. X. Texture and microstructure in Cheddar cheese made from whole milk and from homogenized low-fat milk. J. Texture Stud. 11, 1980, 15-34.

22. Resmini P. Struttura e microstruttura dei prodotti lattiero caseari. Ind. Latte 15, 1979, 32-60.

23. Taranto M.V., Wan P.J., Chen S.L. and Rhee K.C. Morphological, ultrastructural and rheological characterization of Cheddar and Mozzarella cheese. Scanning Electron Microsc. 1979; III: 273-278.

Discussion with Reviewers

M.A. Amer: Please explain how adhesiveness of the Cream cheeses was evaluated.
Authors: After cleansing the mouth with a plain soda biscuit followed by a water rinse, the judge placed a piece of the Cream cheese in the mouth, pressed it up to the palate with the tongue, pushed the cheese forward to the teeth, and evaluated the force needed to remove the tongue from the cheese.

D.N. Holcomb: There does not seem to be *good* correlation between sensory attributes and instrumental compressibility, except, possibly, in the case of firmness. Could you explain the lack of correlation?
Authors: A correlation coefficient of 0.94 indicates that the correlation between instrumental compressibility and firmness is very close. Instrumental compressibility is also highly correlated with adhesiveness ($r=0.91$) and is inversely correlated with spreadability ($r=-0.88$).

M. Rüegg: If a series of interdependent tests is carried out, the confidence levels should be raised (error probabilities lowered). Therefore, the 5% level is somewhat low. Would you please comment?
Authors: The 5% level has been traditionally accepted in sensory evaluation of foods to be significant.

M. Rüegg: Please comment on the fat globule membrane. Do you mean the original membrane surrounding fat globules after secretion of milk or have the products been homogenized sometimes during manufacture? Have you observed single or double layers?
Authors: The cream mixes are usually homogenized. Fat globule membranes undergo rapid changes in expressed milk (*e.g.* F.B.P. Wooding: J. Ultrastruct. Res. *37*, 1971, 388-400) and the manufacturing conditions presumably induce additional changes. Single layers were observed.

M. Rüegg: The observation of milk fat in the form of membrane-less globules in fresh cheese is very interesting. Do you exclude the possibility of a membrane consisting of whey protein or casein monomers which had not been stained or cannot be observed at the magnifications used?
Authors: Membranes formed at the surface of homogenized fat globules and consisting of whey proteins and/or casein submicelles are usually visible under the electron microscope at the magnifications used. The question is whether there is a sufficient amount of protein in the cream mix to cover all the newly formed surfaces on the fat globules resulting from homogenization.

M. Rüegg: Have you speculated on the nature of the strands in Cream cheese #3?
Authors: The strands are probably the carob bean gum used as an additive, but this would have to be confirmed experimentally. For this reason and to solve some problems mentioned above, we are now studying the development of microstructure in Cream cheese under controlled conditions.

For additional discussion see page 152.

MORPHOLOGICAL, ULTRASTRUCTURAL AND RHEOLOGICAL CHARACTERIZATION OF CHEDDAR AND MOZZARELLA CHEESE

M. V. Taranto*, P. J. Wan, S. L. Chen and K. C. Rhee

*Dept. of Food Science
Univ. of Illinois
Urbana, IL 61801

Food Protein Research
and Development Center
Texas A&M Univ.
College Station, TX 77843

Present address: ITT Continental Baking Co.
P.O. Box 731,
Rye, NY 10580.

Abstract

Commercial cheddar and mozzarella cheeses were characterized with scanning electron microscopy (SEM), transmission electron microscopy (TEM) and transmitted light microscopy (TLM). The rheological properties of the cheese were evaluated by objective testing methods.

Trypsin etching of the cheese sample surface followed by freeze-drying clearly revealed the form and distribution of the milk-fat globules when viewed with SEM. The fat globules in cheddar cheese were non-uniformly distributed and tended to aggregate as a direct consequence of the cheddaring process. The mozzarella cheese exhibited a scattered distribution with little aggregation of the fat globules.

The mozzarella cheese exhibited a compact protein matrix with no specific orientation. This indicates that this particular mozzarella was manufactured without stretching the curd during processing. The cheddar cheese exhibited an open, fibrous protein matrix. The aggregation of the milk fat globules in the cheddar tends to disrupt the protein matrix.

Compression tests indicated that the cheddar cheese required a larger compression force for a given deformation than the mozzarella cheese. The mozzarella cheese exhibited a larger work ratio of the first two consecutive compressions (cohesiveness), elastic recovery after the first bite (springiness) and adhesiveness than the cheddar cheese.

The higher cohesiveness, springiness and adhesiveness and lower hardness of the mozzarella cheese is probably due to its higher moisture content and compact protein matrix. The increased moisture results in a greater hydration of the mozzarella protein matrix. This increased hydration makes the protein matrix more elastic, compact and less firm.

KEY WORDS: Cheddar Cheese, Mozzarella Cheese, Fat Globules, Protein Matrix

Introduction

Baron and Scott Blair (1953)[1] stated that "rheological work (on cheese) has been, hitherto, of an essentially empirical kind and this situation is only likely to be improved when other physical methods, capable of elucidating something of the very complex molecular structures, have been applied to cheese." Since that time, numerous studies on cheese microstructure have been published. King and Czulak[2] using light microscopy and Peters and Hansen[3], using electron microscopy showed that during the manufacturing process milk casein changed progressively from spherical micelles to filaments with a granular structure during rennetting and finally to a fibrous network during ripening. Reed[4], using freeze-etch techniques, showed that milk fat globules could be easily distinguished from other materials present in the cheese and that in cheddar cheese many of the fat globules were ruptured. Hall and Creamer[5] demonstrated that the fat globule distribution could be observed with scanning electron microscopy (SEM) using a simple trypsin etching technique. Eino et al.,[6] showed with SEM that the curd made with bovine and porcine pepsin were similar in structure and in orientation of the coagulated protein, whereas the curd produced with rennet possessed a compact and organized structure. Kalab[7,8] refined the methods of preparation of cheese samples for SEM and using this technique in conjunction with transmission electron microscopy (TEM), he demonstrated the differences in microstructure between several types of commercial cheeses and the effect of specific processing steps (e.g., stretching of mozzarella curd) on the cheese microstructure.

The important textural characteristics of cheese and objective methods for the measurement of the cheese texture have been summarized by Baron and Scott Blair[1] and more recently, by Prentice[9]. Most recently, Lee et al.,[10] evaluated the texture of cheese by a compression test using an Instron Universal Testing Machine Model 1122. Objective measurements were highly correlated with those made by a sensory evaluation system the authors called the Milestone method.

Cheese rheology and cheese microstructure each has been extensively studied. However, the correlation of cheese microstructure and cheese rheology has not been extensively examined. This basic information would be of value from an academic as well as practical point of view. Therefore, the objective of this study was to compare the microstructure of cheddar and mozzarella cheese and use this data to explain the differences in the rheological properties between these two types of cheese.

Materials and Methods

Cheese samples

Commercial cheese samples representing two distinctly different textures were selected for this study: sharp Cheddar and Mozzarella. Both types of cheese were manufactured by the same company and were purchased at a local supermarket in Urbana, Illinois.

Evaluation of cheese texture

The mechanical properties of each cheese were determined according to the procedure outlined by Lee et al.,[10] using an Instron Universal Testing Machine Model 1122. All samples were equilibrated at 25°C for 60 min prior to testing.

Chemical analysis of cheese

A proximate analysis of each cheese was made according to the AOAC methods[11].

Microscopy of cheese

Transmitted light microscopy (TLM). Three mm^3 pieces of cheese (randomly cut) were prefixed and postfixed according to the triple aldehyde procedure of Mollenhauer and Totten[12]. Fixed samples were embedded in glycol methacrylate according to the procedure of Feder and O'Brien[13]. One μm section was mounted on glass slides and stained with either Oil Red-O[14] or Amido Black 10 B[15].

SEM. Five mm^3 pieces of cheese (randomly cut) were prepared for SEM according to the trypsin-etching technique of Hall and Creamer[5]. Five mm^3 samples were also prepared for SEM according to Kalab[8]. All samples were examined with a JEOL JSM-U3 at 12 kv.

TEM. Two mm^3 pieces (randomly cut) were prefixed and postfixed according to Mollenhauer and Totten[14] and embedded in Spurr's resin[16]. Thin sections were first stained with lead[17] and then with uranyl acetate[18] and examined with a Hitachi HU-11E at 50 kv.

Results and Discussion

Mozzarella cheese, being made from low-fat milk, contains more protein and less fat than Cheddar cheese (Table 1). These differences in chemical composition are reflected in the differences observed in the cheese microstructure (Fig. 1,2).

Compression tests indicated that the cheddar cheese required a larger compression force for a given deformation (hardness) than the mozzarella cheese (Table 2). The mozzarella cheese exhibited a larger work ratio of the first two consecutive compressions (cohesiveness), elastic recovery or recovered height in cm (springiness) and adhesiveness (gumminess) than the cheddar cheese (Table 2). The results reported here agree with

Table 1. Proximate analysis of cheese samples[a].

Sample	N (%)	Dry wt. Basis Protein (Nx6.38) (%)	Fat (%)	Moisture (%)
Mozzarella	8.6	54.9	13.7	45.2
Sharp Cheddar	6.3	40.2	44.4	35.2

[a]Averages of duplicate analyses are reported.

data reported by Lee et al.,[10] for these types of cheese.

Both the trypsin-etching SEM and TLM techniques clearly revealed the form and distribution of the milk fat globules in the mozzarella and cheddar cheese (Fig. 1). The mozzarella cheese, containing less fat, exhibits larger fat globules than the cheddar cheese and the fat globules are uniformly scattered throughout the protein matrix with little aggregation (Fig. 1-a, b, c). The cheddar cheese fat globules were nonuniformly distributed and tended to aggregate (Fig. 1-d, e, f). Similar results were reported by Kalab[7,8] and Hall and Creamer[5].

The mozzarella cheese exhibits a compact protein matrix (Fig. 2-a, b, c). No specific orientation of the protein matrix, as described by Kalab[7], was detected in the mozzarella cheese used in this study. Therefore, this particular mozzarella cheese was manufactured without stretching the curd during processing. The cheddar cheese exhibited an open, fibrous protein matrix (Fig. 2-d, e, f). The aggregation of the fat globules tends to disrupt the protein matrix (Fig. 2-d, e). Hall and Creamer[5] reported that fat globule aggregation was a characteristic of cheddar cheese and these authors hypothesized that the fat aggregation was associated with the extensive flow induced in the curd during the cheddaring stage of processing. Hall and Creamer[5] also reported that higher cooking temperatures induced a greater amount of milkfat globule aggregation.

The differences in the rheological properties of these two types of cheese might be explained by the differences in the chemical composition and microstructure. The higher cohesiveness, adhesiveness and springiness of the mozzarella cheese could be due to its higher water content and compact protein matrix. A high water content is the cause of a cheese being less firm. The water in cheese is associated with the protein since the only other major component, fat, is hydrophobic. Fukushima et al.,[19] reported that the casein network is modified by the amount of water present in the cheese and that the modulus of the elastic component decreased as the water content increased. It is possible that the protein matrix of the mozzarella cheese is hydrated to a greater extent than that of the cheddar cheese. This would explain the greater compactness of the mozzarella protein matrix compared to that of the cheddar.

Characterization of Cheddar and Mozzarella Cheese

Figure 1. Milkfat globule distribution.
a and b - Scanning electron micrographs of trypsin-etched mozzarella cheese; a - no fixation, b - fixed in glutaraldehyde. c - Transmitted light micrograph of mozzarella cheese stained with Oil Red-O. d and e - Scanning electron micrographs of trypsin-etched cheddar cheese; d - no fixation, e - fixed in glutaraldehyde. f - Transmitted light micrograph of cheddar cheese stained with Oil Red-O. FG - fat globule; arrows in f indicate aggregated fat globules.

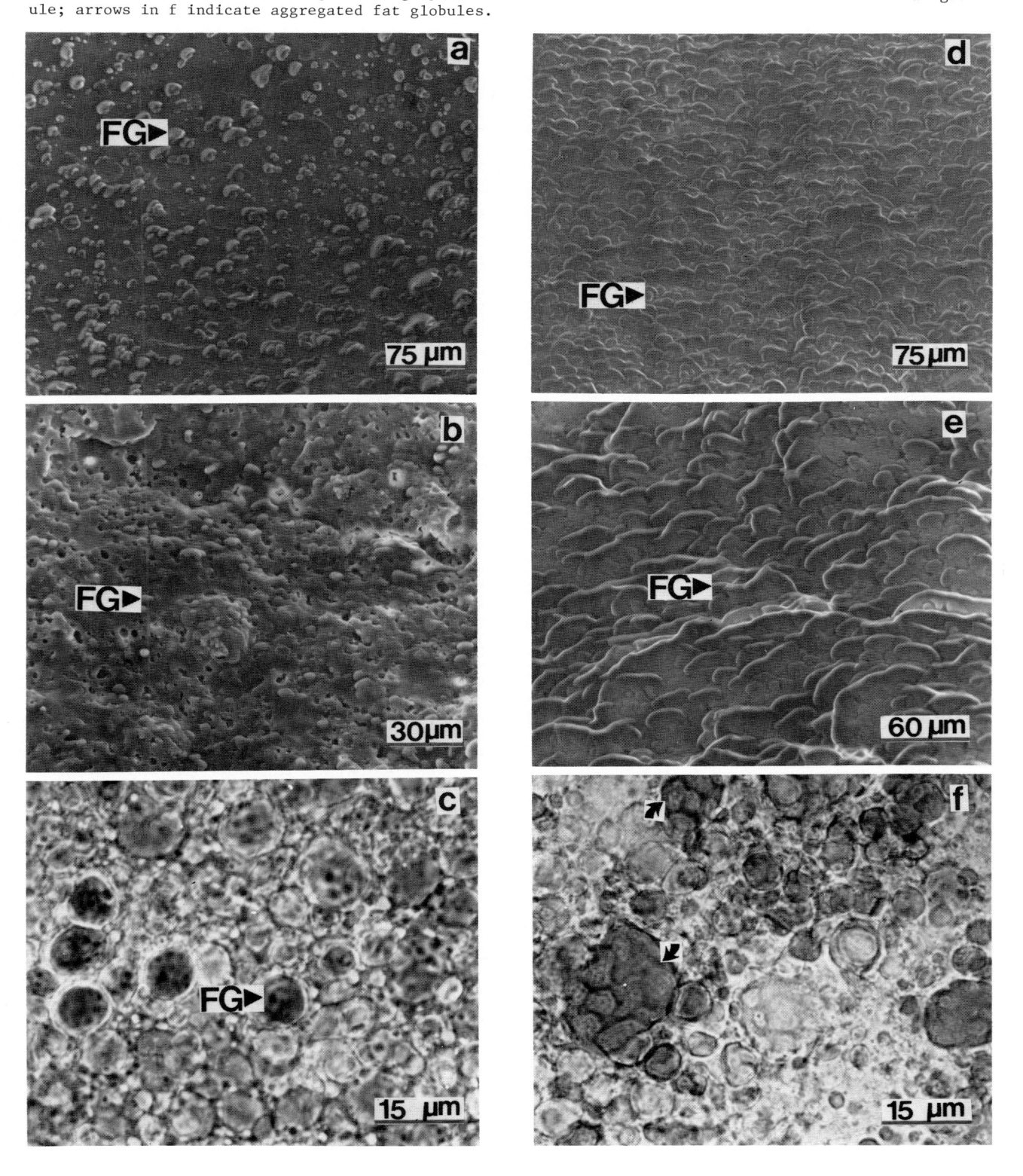

Table 2. Rheological properties of cheese samples.[a,b]

Sample	Compression Force Hardness (Newtons)	Work Ratio Cohesiveness (%)	Elastic Recovery Springiness (cm)	Adhesiveness Gumminess (Joule x 10^{-4})
Mozzarella[c]	65.3	73.2	0.31	2.3
Sharp Cheddar[c]	106.7	43.9	0.25	2.0

[a] Cycle: 0.625 cm compression, 2 cycles
Chart speed: 10 cm/min
Crosshead speed: 2 cm/min

[b] Averages of triplicate analyses are reported

[c] Force supplied: For Cheddar: 200 Newtons full scale
For Mozzarella: 50 Newtons full scale

It would also explain the lower hardness and greater cohesiveness, elastic recovery and adhesiveness of the mozzarella cheese. The design of the present experiment does not permit any assessment of the effect of fat content or distribution on the cheese rheological properties.

The present exploratory study indicates that there are very subtle differences in the microstructure of cheddar and mozzarella cheese. These structural differences are probably partially responsible for the rheological differences observed. However, much more experimentation which takes into account the effect of processing parameters and chemical composition of the final product is needed to accurately assess the correlation between cheese microstructure and rheological properties.

Acknowledgements

This work was supported in part by the University of Illinois Agricultural Experiment Station Hatch Project 30-15-50-304, Texas Peanut Producers' Board, and the Natural Fibers and Food Protein Commission of Texas.

References

1. Baron, M. and Scott Blair, G. W. 1953. Rheology of cheese and curd. In Foodstuffs: Their plasticity, fluidity and consistency, Scott Blair, G. W., ed. Interscience Pub. Inc., New York, NY U.S.A., Ch. 5 pp. 125-147.
2. King, N. and Czulak, J. 1958. Fibrous structure in cheese curd. Nature 181:113-114.
3. Peters, I. I. and Hansen, P. G. 1958. Electron microscopic observations on the structure of cheese. J. Dairy Sci. 41:57-60.
4. Reed, R. 1969. Green cheese scrutinized. New Scientist 43:377-385.
5. Hall, D. M. and Creamer, L. K. 1972. A study of the sub-microscopic structure of cheddar, cheshire and gouda cheese by electron microscopy. N. Z. J. Dairy Sci. Technol. 7:95-102.
6. Eino, M. F., Biggs, D. A., Irvine, D. M. and Stanley, D. W. 1976. A comparison of microstructures of cheddar cheese curd manufactured with calf rennet, bovine pepsin and porcine pepsin. J. Dairy Res. 43:113-115.
7. Kalab, M. 1977. Milk gel structure. VI. Cheese texture and microstructure. Milchwissenschaft 32(8):449-457.
8. Kalab, M. 1978. Milk gel structure. VIII. Effect of drying on the scanning electron microscopy of some dairy products. Milchwissenschaft 33(6):353-358.
9. Prentice, J. H. 1972. Rheology and texture of dairy products. J. Text. Studies 3:415-458.
10. Lee, C. H., Imoto, E. M. and Rha, C. 1978. Evaluation of cheese texture. J. Food Sci. 43(5):1600-1605.
11. AOAC. 1970. Official Methods of Analysis, 11th ed. Association of Official Agricultural Chemists, Washington, D. C.
12. Mollenhauer, H. H. and Totten, C. 1971. Studies on seeds I. Fixation of seeds. J. Cell Biol. 48:387-394.
13. Feder, N. and O'Brien, T. P. 1968. Plant microtechnique: Some principles and new methods. Amer. J. Bot. 55(1):123-142.
14. Pearse, A. G. E. 1968. Histochemistry-Theoretical and Applied, 3rd ed., Vol. I. Little, Brown and Company, Boston, MA U. S. A., p. 697.
15. Fisher, D. B. 1968. Protein staining of ribboned epon sections for light microscopy. Histochemie 16:92-98.
16. Spurr, A. R. 1969. A low viscosity epoxy resin embedding medium for electron microscopy. Ultrastruc. Res. 26:31-43.
17. Sato, T. 1967. A modified method for lead staining of thin sections. J. Electronmicroscopy 16:133-138.
18. Watson, M. L. 1958. Staining of tissue sections for electron microscopy with heavy metals. J. Biophys. Biochem. Cytol. 4(4):475-478.
19. Fukushima. M., Sone, T., and Fukada, E. 1965. The effect of moisture content on the viscoelasticity of cheese. J. Soc. Mater. Sci. Japan 14:270-279.

Discussion with Reviewers

M. Kalab and V. R. Harwalker: It is evident from Table 1 that the protein-to-moisture ratios were not much different in both cheeses (1.14 in Cheddar and 1.21 in Mozzarella), whereas the protein-to-fat ratios differed considerably (4.01 and

Characterization of Cheddar and Mozzarella Cheese

Figure 2. Structure of protein matrix.
a - Scanning electron micrograph of mozzarella cheese defatted with chloroform according to Kalab[7]. b - Transmission electron micrograph of mozzarella cheese. c - Transmitted light micrograph of mozzarella cheese stained with Amido Black 10 B. d - Scanning electron micrograph of cheddar cheese defatted with chloroform according to Kalab[7]. e - Transmission electron micrograph of cheddar cheese. f - Transmitted light micrograph of cheddar cheese stained with Amido Black 10 B. CJ - curd junction; FG - fat globule; FGM - fat globule membrane; IFG - impression left by fat globule after fracturing; PM - protein matrix.

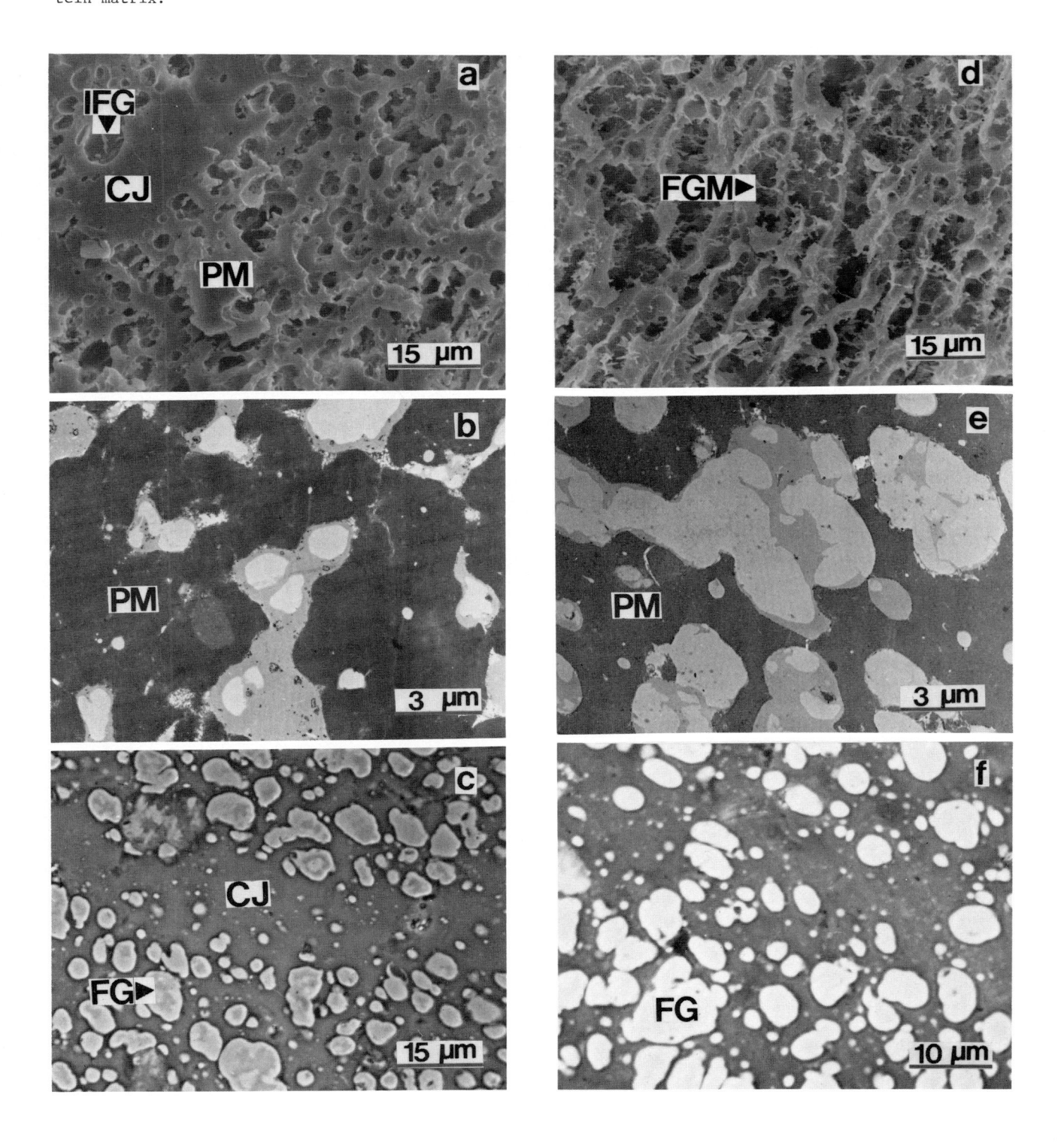

0.90, respectively. In view of these differences what makes you believe that the rheological properties were affected by the water content of the cheeses?
Authors: F. Kosikowski in his book, "Cheese and Fermented Milk Foods," (2nd printing, Edward Brothers, Inc., Ann Arbor, Michigan, 1966) indicates that the fat content of the finished cheese affects the texture, in particular the hardness of the cheese. For both mozzarella and cheddar cheese, finished products with lower fat contents have harder textures. Kosikowski states on p.202 of his book that the body texture of cheddar cheese made from skim milk is "usually hard as a rock." In view of the statements and data reported by Kosikowski, one would be tempted to predict that mozzarella cheese should exhibit rheological characteristics which reflect a hard texture compared to cheddar cheese due to its lower fat content. However, as seen in the data presented in Table 2, the opposite effects were indicated. Therefore, some other physico-chemical interaction must be influencing the rheological properties. The work of Fukushima et al.,[19] indicates that the interaction of the casein network and water in the finished cheese affects the modulus of the elastic component. We are proposing this casein-water interaction as a possible explanation for the results we obtained, but we realize that the "true" answer must take into account protein-fat as well as protein-water interactions.

M. Kalab and V. R. Harwalker: You have stated that the design of the present experiment has not permitted you to assess the affect of the fat content or distribution on the rheological properties. Do you visualize, with respect to the important role of fat in cheese, any design that would allow you to study such effects?
Authors: In the near future, we plan to produce mozzarella and cheddar type cheeses under standard conditions with varying fat contents and a constant moisture content. Also by the appropriate manipulation of the curd cooking, curd knitting and salting operations, we hope to be able to make mozzarella and cheddar cheese with varying moisture contents at a constant fat content. Rheological and microscopic analyses on these experimental products should provide data to more accurately assess the correlation between microstructure, chemical composition and rheological properties.

J. H. Prentice: The authors refer to the effect of "hydration" of the protein. The analyses show that over 40% by volume of the Cheddar cheese and over 50% by volume of the mozzarella is water. Have the authors' studies led them to any views on the residence sites of the water? How much is genuine hydration, i.e. bound to the protein, how much entrapped in the mesh? and how much is free to move between the fat and the protein? Each site would be expected to affect the rheological properties differently.
Authors: At the present time, we have not conducted any experiments to determine "bound" vs "free" water in our cheese samples. According to the water activity work conducted by M. P. Steinberg and co-workers in the Dept. of Food Science at the University of Illinois, "free" water in a food system would migrate between components. Therefore, the "free" water in a cheese product should be free to move between protein and the fat. We plan to conduct "bound" vs "free" water measurements by adapting the tests used by the meat scientist ("The Science of Meat and Meat Products," J. F. Price and B. S. Schweigert, eds., W. H. Freeman and Company, San Francisco, 1971, pp. 177-191).

J. H. Prentice: Fixation of specimens always presents a major challenge with dairy products as most commonly used fixatives attack one or other component with consequent damage to the overall structure. This is borne out by Figs. 1-a and 1-b, which are difficult to relate to each other. The globules in 1-b appear to be much smaller than in 1-a. Does this warrant a comment?
Authors: M. Kalab[7,8] has addressed this problem. Kalab concluded that the fixation methods employed for cheese samples did not introduce structural artifacts as borne out by the excellent correlation of TLM, SEM and TEM. In our study, we also observed good correlation between TLM, SEM and TEM observations of cheese samples. However, in reference to Figs. 1-a and 1-b, the difference in the size of the fat globules could be an artifact of sample preparation. Fat globules in Figs. 1-b, c and 2-a vary from 4-12 μm. The largest fat globule in Fig. 1-a measures ∼30 μm and could have been formed by aggregation of smaller ones during the sample preparation. Future experiments are planned in which a mixture of glutaraldehyde and osmium tetroxide will be used to simultaneously fix protein and fat.

M. Kalab and V. R. Harwalker: How does temperature affect the results of rheological tests with cheese?
Authors: As the test temperature is increased, the force required to achieve a given percent compression is decreased. A temperature is finally reached where the cheese becomes very soft and fat droplets will appear on the surface (i.e. the cheese begins to melt). This softening alters the nature of the contact between the sample and the Instron plates. (see J. Culioli and P. Sherman 1976. Evaluation of Gouda Cheese Firmness by Compression Tests. J. Texture Studies 7:353).

M. Kalab and V. H. Harwalker: Have you undertaken any microscopical examinations of fat globules released from the cheeses by trypsin treatment?
Authors: At this time we have made no study of the released fat globules.

MORPHOLOGICAL AND TEXTURAL CHARACTERIZATION OF SOYBEAN MOZZARELLA CHEESE ANALOGS

M. V. Taranto* and C. S. Tom Yang

Rm. 104 Dairy Manufactures Building
1302 - W. Pennsylvania Avenue
Department of Food Science
University of Illinois
Urbana, IL 61801

*Present Address

ITT Continental Baking Company
Research Laboratories
P.O. Box 731
Rye, New York 10580

Abstract

The morphology and texture of mozzarella cheese analogs prepared from soy proteins were compared to that of natural, low moisture-part skim mozzarella cheese and caseinate substitutes. The soybean mozzarella cheese analogs exhibited morphological features resembling natural mozzarella cheese, i.e., a protein matrix in which fat and other ingredients are embedded and dispersed. A Texture Profile Analysis and Weissenberg Test indicated that the soy protein mozzarella cheese analogs exhibited textural characteristics in the solid or gelled state and stretching properties in the melted state that are comparable to natural mozzarella cheese. The soybean analog prepared from soy protein concentrate exhibited textural and structural properties more closely resembling natural mozzarella cheese than any of the other analogs. These findings are important in light of the high cost and short supply of Ca/Na caseinates and indicate that mozzarella cheese analogs with adequate stretching and stringing properties can be made from a less expensive soy protein base.

KEY WORDS: Soybean proteins, natural mozzarella cheese, cheese analogs, texture, morphology, electron microscopy

Introduction

Cheese production in the United States has soared during the last twenty years. It continues to grow at a rate of about 10% per year (Siapantas 1980). The rising demand for cheese has been met in three ways: 1) increased imports; 2) development of new process cheese products containing less natural cheese and 3) development of cheese substitutes. In the area of cheese substitutes, mozzarella cheese analogs have become the "margarine" of the cheese industry (Siapantas, 1980).

The per capita consumption of Italian style cheese (in particular, mozzarella cheese) has increased approximately ten-fold between 1960 and 1975 and continues to increase today (Kasik and Peterson, 1977). The unusually rapid increase in the consumption of mozzarella cheese is due to the popularity of pizza. Mozzarella cheese is the most dominant cheese used in pizza manufacture. The substantial increase in the demand for mozzarella cheese has not only caused difficulties in obtaining predictable supplies, but has significantly increased the cost of the product (Kasik and Peterson, 1977; Vernon, 1972). The positive factors for the utilization of mozzarella cheese substitutes include price, shelf-life and availability. The price differential is substantial, approximately 50% (Siapantas, 1980; Peters, 1979).

At present, all of the mozzarella cheese substitutes are produced from caseinate (Kasik and Peterson, 1977; Bell et al., 1975). The supply of caseinates in 1980 were predicted to be approximately equal to that available in 1979 (Andres, 1980). The entire casein supply available to food processors is imported and no casein is currently produced in the U.S. (Andres, 1980). A U.S. International Trade Commission (ITC) report captioned "Casein and its Impact on the Domestic Dairy Industry" concluded that the cost of casein would have to rise to approximately $2.40/pound from the current $0.90/pound to induce domestic plants to divert fluid skim milk from the production of nonfat dry milk to casein (Andres, 1980). Based on the information in the ITC report, imports of casein are expected to stabilize in the range of 140-150 million pounds over the next five years. This leveling off of imports is expected to be a result of rising import prices and absolute world supply constraints (Andres, 1980). Therefore, the raw material presently used in cheese substitutes (casein) will be limited. This means that suitable alternatives to casein need to be developed to meet the rising demand for mozzarella cheese substitutes.

Availability, high nutritive value and protein content make soybeans and/or soy proteins the ideal raw material for mozzarella cheese analog production. The major objective of the present research is the development of mozzarella cheese analogs from soy proteins which have nutritional, flavor and textural parity with natural mozzarella cheese. This paper reports the initial results on the textural properties of soybean mozzarella cheese analogs.

Materials & Methods

Manufacture of Soybean Cheese Analogs

Cheese analogs were prepared with the following ingredients: soy proteins, gelatin, gum arabic and fat. Soy proteins were used in one of the following forms: soy milk, full-fat soy flour, defatted flour, soy protein concentrate or soy protein isolate. Soy milk was prepared in our pilot plant according to Nelson et al., (1976) (a flow diagram for soy milk production is presented in Fig. 1A). Full-fat and defatted soy flours (Nutrisoy) were purchased from Archer Daniels Midland Co., Decatur, IL. Soy protein concentrate (Promosoy 100) and soy protein isolate (Promine D) were purchased from Central Soya, Fort Wayne, IN. Gelatin (Type B, 128 bloom) was purchased from J.T. Baker Chemical Co., Philipsburg, NJ. Gum arabic was purchased from Fisher Scientific Co., Fair Lawn, NJ. Fat (coconut oil manufactured to exhibit a Wiley melting point of 38 °C) was purchased from Durkee, Cleveland, OH. The basic steps for cheese analog manufacture were as follows: 1) ingredient formulation, blending and emulsification; 2) gel formation and 3) packaging and storage. Blending and mixing were performed using a Kitchen Aide mixer with a flat beater. The mixing bowl was immersed in a water bath held at 80 °C. Ingredients were blended for a total of 10 min mixing time. After blending, the mixture was poured into preformed aluminum foil molds (10 cm in diameter by 8 cm in height). The molds were held at 4 °C for 24 hrs to allow gel formation. The composition of the various cheese analogs is presented in Table 1.

Textural Characteristics of Cheese Analogs

Texture Profile Analysis. An Instron Universal Testing Machine was used to perform a Texture Profile Analysis (TPA) (Friedman et al., 1963) on natural low moisture-part skim mozzarella cheese and the cheese analogs manufactured as described above. One hour before testing, four randomly located samples were taken from the cheese block using a cork borer (2.3 cm inside diameter) and cut with a sharp knife to a 2 cm height. The cylindrical samples were wrapped in air-tight plastic bags and allowed to equilibrate to 20 °C in the Instron testing room. A digital thermopenetrometer was used to measure the temperature of each sample prior to and after testing. The following parameters were used for the TPA of all cheese samples: Load range (full scale) - 2 kg force; Loading rate (crosshead speed) - 2 cm/min; Chart speed - 5 cm/min; Plunger diameter (flat head) - 0.64 cm; Penetration depth - 80% deformation (1.6 cm); Number of bites (cycles of loading) - 2 consecutive bites. A typical compression curve and data obtained from the curve for the TPA of the cheese samples is shown in Fig. 1B.

Weissenberg Test (Weissenberg, 1947; Garner and Nissan, 1946, 1947). Wiegand (1963) and Nelson (1980) have shown that the presence of viscoelastic flow in a concentrated solution may be demonstrated qualitatively and quantitatively

Fig. 1A - Flow diagram for the manufacture of soy milk

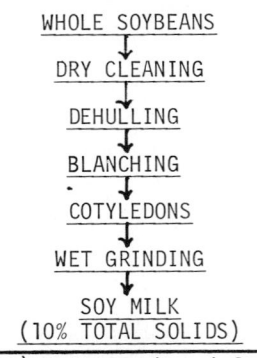

WHOLE SOYBEANS
↓
DRY CLEANING
↓
DEHULLING
↓
BLANCHING
↓
COTYLEDONS
↓
WET GRINDING
↓
SOY MILK
(10% TOTAL SOLIDS)

Table 1. Ingredient formulation for soybean cheese analogs[a]

SAMPLE NUMBER	FAT (g)	WATER (g)	GELATIN (g)	GUM ARABIC (g)	TYPE OF SOY PROTEIN AND AMOUNT (g)
1	10[b]	40	20	20	---
2	--	--	25	30	soymilk (10% TS) - 100
3	10[c]	60	24	24	full-fat soy flour - 10
4	10[c]	60	24	24	defatted soy flour - 10
5	10[c]	60	24	24	soy protein concentrate - 10
6	10[c]	60	24	24	soy protein isolate - 10
7	10[c]	60	24	24	soy protein isolate - 20

a) All samples contain 0.02 g of potassium sorbate to control mold growth.
b) Soybean oil purchased at the local supermarket; c) Hydro-100 coconut oil.

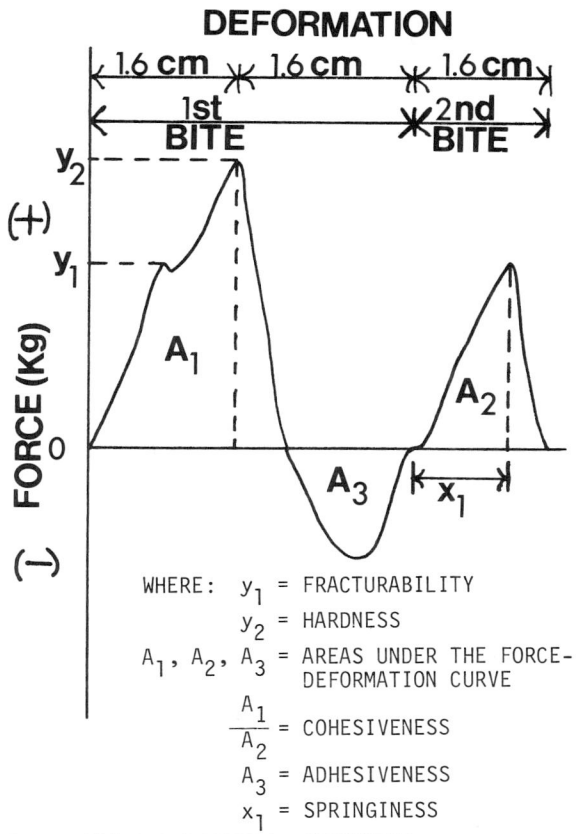

Fig. 1B - Typical TPA curve and data obtained from the curve.

WHERE:
y_1 = FRACTURABILITY
y_2 = HARDNESS
A_1, A_2, A_3 = AREAS UNDER THE FORCE-DEFORMATION CURVE
$\frac{A_1}{A_2}$ = COHESIVENESS
A_3 = ADHESIVENESS
x_1 = SPRINGINESS
(HARDNESS)(COHESIVENESS) = GUMMINESS
(GUMMINESS)(SPRINGINESS) = CHEWINESS

by the Weissenberg test. The experimental design is such that the liquid sample is sheared in a gap between an outer vessel rotating with a constant angular velocity and a rigidly fixed inner rod. As the vessel rotates, the sample undergoes a stationary laminar shearing movement because of the combined actions of the shear imposed at the boundaries and the forces of gravity and inertia (centrifugal forces).

By taking the initial height of the sample on the rod and the final height of the climbing sample after rotating at a fixed rpm for a certain period of time, a comparison of the stretchability among various samples can be obtained. For this experiment, a 50 ml beaker lined with a 4x4 wire mesh screen containing 30 g of sample is equilibrated to 63 ± 2 °C over a water bath. The initial submerging depth of the rod which is wrapped with a No. 2 filter paper is 1 cm. Both the screen and filter paper are used to reduce the lubricating effect of the exuded sample fat. The final climbing height is recorded after 1 min rotation at 60 rpm.

Preparation of samples for Scanning Electron Microscopy (SEM)

All samples were prepared for SEM according to Kalab (1978) in the following sequence: fixing; post-fixing; freezing in liquid nitrogen slush; freeze drying; fat extraction with chloroform; critical point drying; dry fracturing; mounting on SEM stubs and coating with gold. All samples were examined with a Cambridge Stereoscan microscope at a 30° tilt and 10 kV.

Results and Discussion

Texture Profile Analysis of Cheese Analogs

All of the soy mozzarella cheese analogs exhibited a fracturability and hardness greater than natural mozzarella cheese (Table 2). All soy analogs had adhesiveness values lower than natural mozzarella cheese, but of particular interest is the soy concentrate analog (sample no. 5) which exhibited an adhesiveness higher than all the other soy analogs and one-third the value of mozzarella cheese (Table 2). All soy analogs exhibited cohesiveness and springiness values comparable to natural mozzarella cheese (Table 2). Only the soy concentrate analog (sample no. 5), the full-fat soy flour analog (sample no. 3) and the soy milk analog (sample no. 2) had chewiness values comparable to that of natural mozzarella cheese (Table 2). All soy analogs exhibited gumminess values greater than natural mozzarella cheese (Table 2).

The TPA data provides an objective evaluation of the textural properties of the soybean analogs in the solid or gelled state. The data indicate that each cheese analog exhibits certain textural characteristics that are comparable to those of natural low moisture-part skim mozzarella cheese. There is only one cheese analog (sample no. 5 - made from soy protein concentrate) which approximates all the textural characteristics of natural mozzarella cheese.

The type of soy protein ingredient added to the formulation has subtle effects on the textural properties of the gel. Certain characteristics appear to be affected while others remain unchanged (Table 2). The chemical composition of natural cheese has been shown to be closely correlated with the textural properties of the cheese in the solid state (Chen et al., 1979). Therefore, it is not surprising to note the same effect with the soybean cheese analogs. At this time, it is not known whether the textural changes are caused by the selective removal of soluble and/or insoluble carbohydrates from the soy proteins, subtle changes in the manner in which the soy proteins interact with other system components or a combination of the two.

Weissenberg Test

All the samples made from the soy protein products, except the ones made from soy milk (sample no. 2) and defatted soy flour (sample no. 4) had Weissenberg test readings greater than the commercial mozzarella cheese (Table 3). Among the soy analogs, the soy isolate analog was found to exhibit the best stretchability. Since the amount of gum arabic and gelatin were fixed in each sample, the soy protein content must play an important role in the development of the stretching property of the analog.

Stretchability is a very important rheological parameter, especially for melted mozzarella cheeses. The Weissenberg test was found to be useful for stretchiness evaluation as long as the testing conditions are well controlled. The

Table 2. Data from the Texture Profile Analysis

SAMPLE	FRACTURABILITY (Kg)	HARDNESS (Kg)	ADHESIVENESS (Kg-cm)	COHESIVENESS	SPRINGINESS (cm)	GUMMINESS (Kg)	CHEWINESS (Kg-cm)
Low moisture-part skim mozzarella[a]	0.73	0.73	0.36	0.57	1.50	0.42	0.63
1[b]	0.86	0.86	0.02	0.46	1.47	0.87	0.77
2[b]	1.76	1.90	0.09	0.46	1.56	0.52	0.77
3[b]	1.09	1.12	0.08	0.49	1.48	0.54	0.79
4[c]	1.32	1.32	0.06	0.52	1.49	0.70	1.04
5[c]	1.23	1.29	0.13	0.42	1.56	0.53	0.83
6[c]	1.44	1.44	0.02	0.62	1.52	0.90	1.37
7[c]	1.30	1.30	0.08	0.57	1.50	0.74	1.12

a) Purchased at a local supermarket.
b) Data reported are averages of four analyses; c) Data reported are averages of two analyses.

Table 3. Weissenberg Test Data

SAMPLE	HEIGHT SAMPLE CLIMBED UP TEST ROD (cm)
Low moisture-part skim mozzarella	1.10
1	0.90
2	0.90
3	1.40
4	0.80
5	1.13
6	2.45
7	2.53

samples should be covered during tempering and the evaluation should proceed as rapidly as possible to avoid drastic temperature fluctuations during the test.

Morphology of Cheese Analogs

The typical structure of the gel matrix in natural mozzarella cheese is shown in Fig. 2A and 2B. The backbone of the matrix is the protein network in which fat globules are uniformly dispersed. The fat globules do not appear to aggregate into larger globules. These same structural features have been previously reported by other workers (Kalab, 1977, 1978; Taranto et al., 1979).

For comparison purposes, micrographs of a mozzarella cheese analog prepared from caseinate and the gel prepared from gum arabic, gelatin, fat and water (sample no. 1, Table 1) are presented in Fig. 3-6. The caseinate analog exhibits a protein network composed of thin protein strands. The fat globules are non-uniform in size and randomly dispersed throughout the protein network. The gum arabic-gelatin matrix is denser than the caseinate matrix. That is to say, the strands forming the network are thicker. At higher magnifications, the gum arabic-gelatin matrix (Fig. 6) can be seen to be composed of protein strands with fat globules (fairly uniform

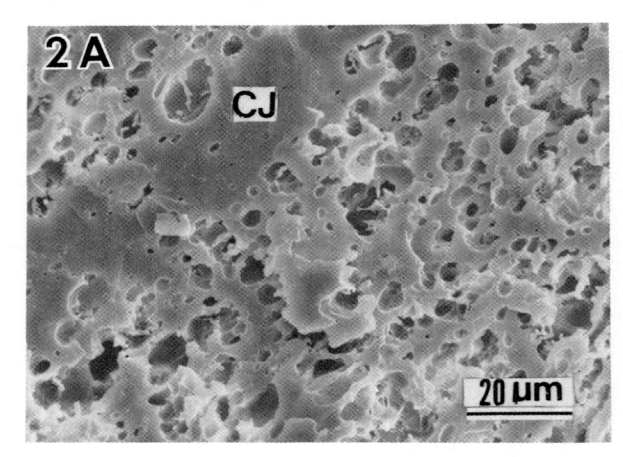

Fig. 2A,B - Low moisture-part skim mozzarella cheese. Note the dense protein area (curd junction zone, CJ) in the upper left corner of Fig.2A. Note the lactose and NaCl crystals (indicated by the arrow) embedded in the protein matrix in Fig. 2B. The holes in the micrographs are the impressions left by the fat globules after defatting with chloroform.

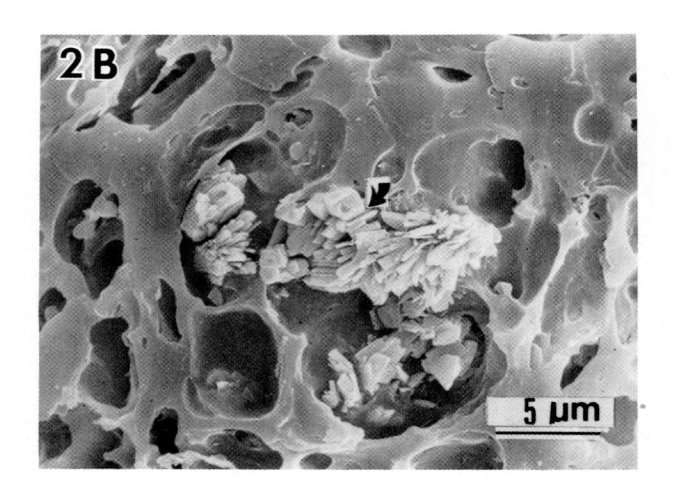

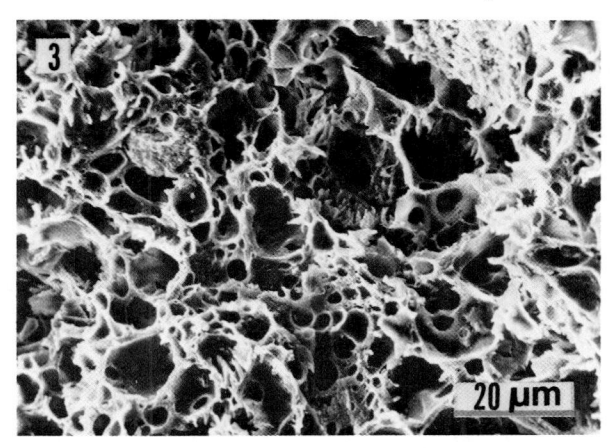

Fig. 3, 4 - Mozzarella cheese analog prepared from caseinate. This is a sample obtained from a local manufacturer of cheese substitutes.

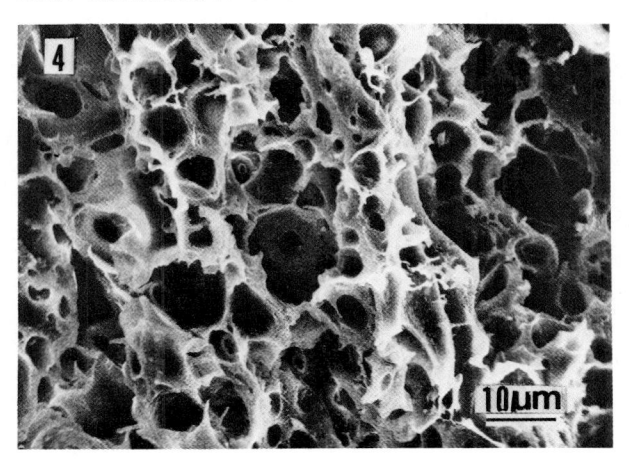

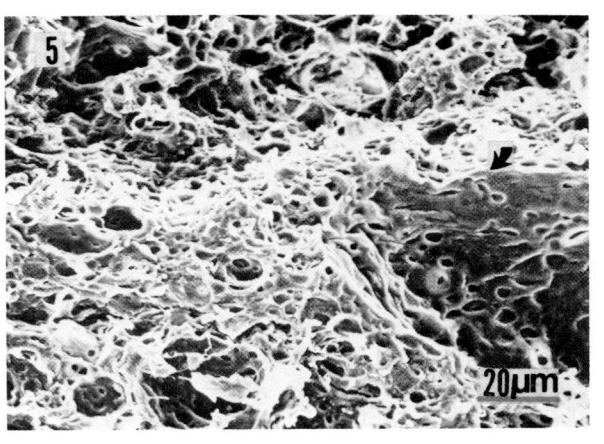

Fig. 5, 6 - Gel prepared from gum arabic, gelatin and fat. Note the impression left by an entrapped air bubble in the lower right corner of Fig. 5 (indicated by the arrow).

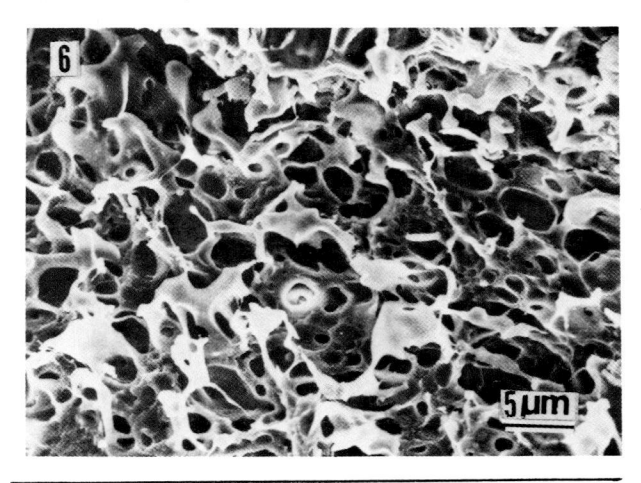

in size) evenly dispersed throughout the matrix. The actual location of gum arabic in the gelatin matrix cannot be determined from the present data. It is known that gelatin and gum arabic interact with each other to form coacervates (Glicksman and Schachat, 1959). Experiments are being conducted to determine the location of gum arabic in the gelatin gel and its role in modifying the gel's structure and will be reported at a later date.

The soy milk cheese analog (Fig. 7, 8) exhibits a gel matrix, at low magnification, that appears similar to that of the gum arabic-gelatin gel (Fig. 5). However, the soy milk gel matrix strands are much thicker with fewer and larger inclusions in the gel network (Fig. 8). The matrix strands appear to be interwoven in the gel network (Fig. 8). Both the full-fat (Fig. 9,10) and defatted (Fig. 11) soy flour analogs exhibit a gel matrix that is highly disrupted by cell fragments. These cell fragments are present in the soy flours and have been observed in these types of flours by other workers (Wolf and Baker, 1975; Taranto and Rhee, 1978). Although the analogs made from full-fat and defatted soy flours have some textural properties similar to natural mozzarella cheese, they exhibit gel structures quite different from those of either natural mozzarella (Fig. 2A,B) or the caseinate analog (Fig. 3, 4).

The gel matrices of the soy concentrate and soy isolate analogs is presented in Fig. 13-18. All three analogs exhibit gel structures similar yet distinct from the soy milk (Fig. 7, 8) and soy flour analogs (Fig. 9-12). The concentrate (Fig. 13) and isolate (Fig. 15, 17) gel networks are made up of interwoven strands with fat globules and other inclusions uniformly dispersed throughout the matrix. The matrix strands exhibit a substructure (Fig. 14, 16, 18) which is similar to the structure of pure (100%) soy protein gels reported by other workers (Furukawa et al., 1979; Saio, 1979). There are inclusions in the soy isolate gel matrix (Fig. 15, 17, 18) which are identical in morphology to the protein spheres described by Wolf and Baker (1975) in soy protein isolates. This indicates that the soy protein isolate was not completely dissolved during analog manufacture (the natural pH of the analogs is 5.3). The gel structure that is formed when the soy protein completely dissolves is seen in Fig. 14 and 16.

There is a striking similarity between the soy concentrate analog (Fig. 13, 14) and the caseinate analog (Fig. 3, 4). Both analogs have gel

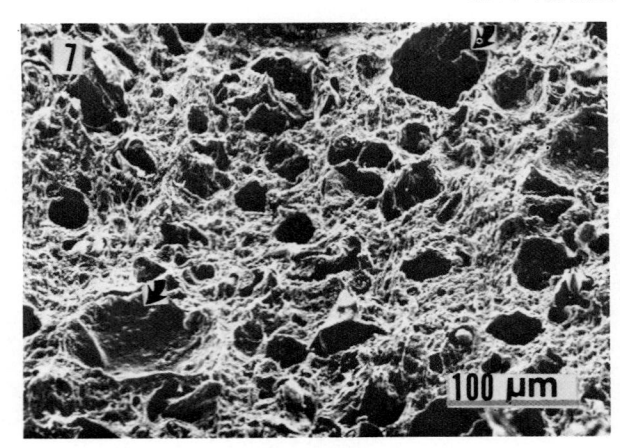

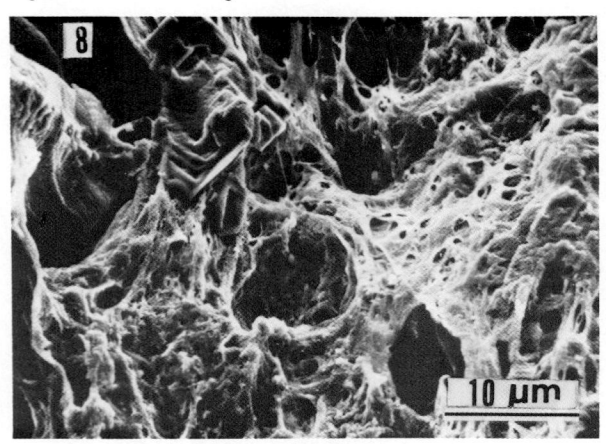

Fig. 7, 8 - Mozzarella cheese analog prepared from soy milk. Note the impressions left by entrapped air bubbles in the upper right and lower left corners of Fig. 7 (indicated by the arrows). Note the mesh-like character of the matrix in the right center of Fig. 8.

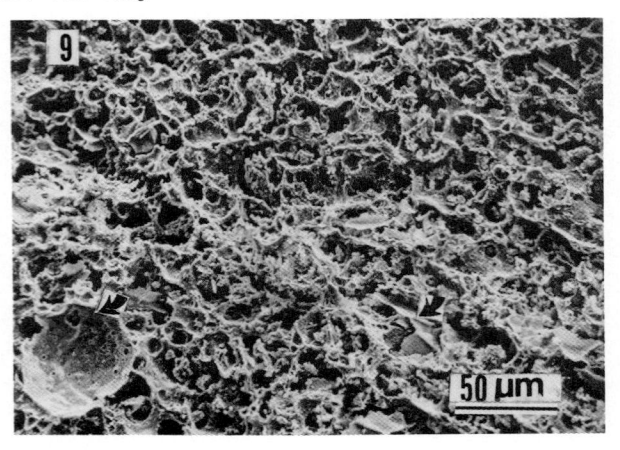

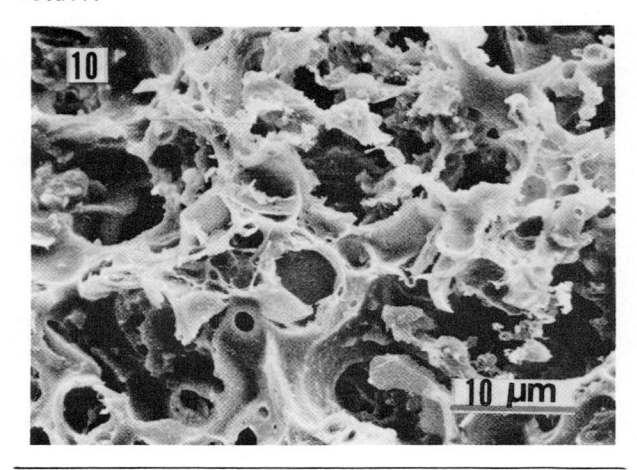

Fig. 9, 10 - Mozzarella cheese analog prepared from full-fat soy flour. Note the impression left by an entrapped air bubble in the lower left corner of Fig. 9 (indicated by the arrow). Cell wall fragments from the soybean cotyledons are seen in the lower right corner of Fig. 9 above and below the line scale (indicated by the arrow). These cell fragments tend to disrupt the gel matrix. Fig. 10 shows the gel matrix in greater detail.

matrices that exhibit similar coarse and fine structures. Except for the substructure of the gel matrix strands (Fig. 14), the soy concentrate analog is structurally similar to natural mozzarella cheese. This structural similarity could explain why only the soy concentrate analog has the best match of textural properties with natural mozzarella cheese.

Conclusions

Natural mozzarella cheese exhibits a pseudoplastic behavior in the melted state. Melted mozzarella will stretch and string and still retain a desirable degree of chewiness. These characteristics are critical and must be adequately duplicated for a cheese analog to be commercially acceptable. All of the soybean mozzarella cheese analogs, except the soy milk analog (sample no. 2), exhibit good melting and spreadability characteristics and fair stretching and stringing behavior. The stretching and stringing quality of the analogs is increased as the level of soluble carbohydrates present in the soy proteins is lowered. An objective evaluation of the rheological properties of the cheese analogs in the melted state is now being performed using a capillary rheometer.

Results will be reported at a later date. The findings presented in this report are important in light of the high cost and short supply of caseinates and indicate that mozzarella cheese analogs with adequate textural characteristics and stretching properties can be made from a less expensive soy protein base.

References

Andres, C. Barring government restrictions, casein supply should hold steady, Food Processing 41(4), 1980, 68-72.

Bell, R. J., Wynn, J. D., Denton, G. T., Sand, R. E., Cornelius, D. L. Preparation of simulated cheese, U.S. Patent No. 3,922,374, 1975.

Chen, A. H., Larkin, J. W., Clark, C. J., Irwin, W. E. Textural analysis of cheese, J. Dairy Sci. 62, 1979, 901-907.

Friedman, H. H., Whitney, J. E., Szczesniak, A. S. The texturometer - A new instrument for objective texture measurement, J. Food Sci. 28(4), 1963, 390-396.

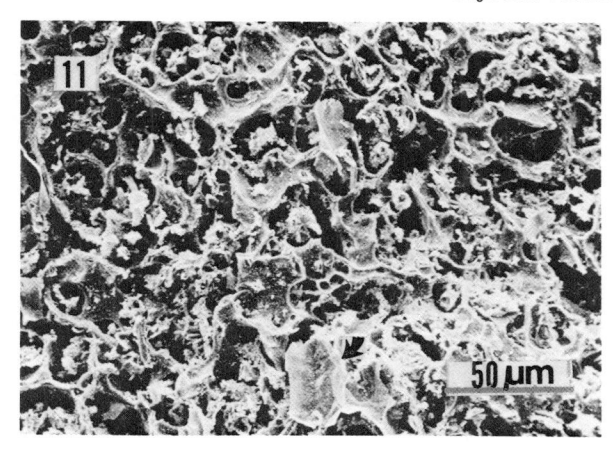

Fig. 11, 12 - Mozzarella cheese analog prepared from defatted soy flour. The gel matrix is severely disrupted by cell fragments seen in the lower right corner of Fig. 11 (indicated by the arrow). The details of the matrix are shown in Fig. 12.

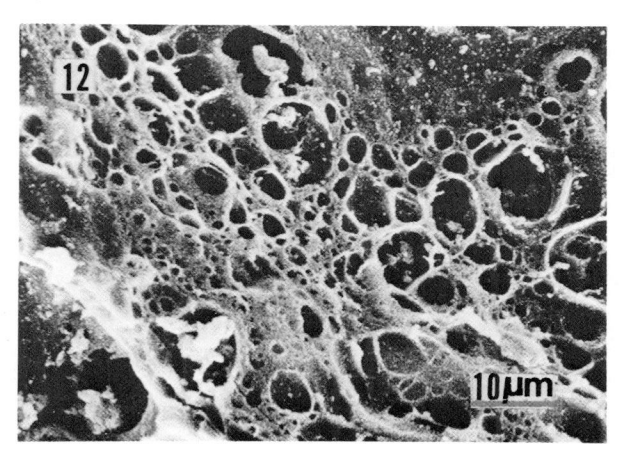

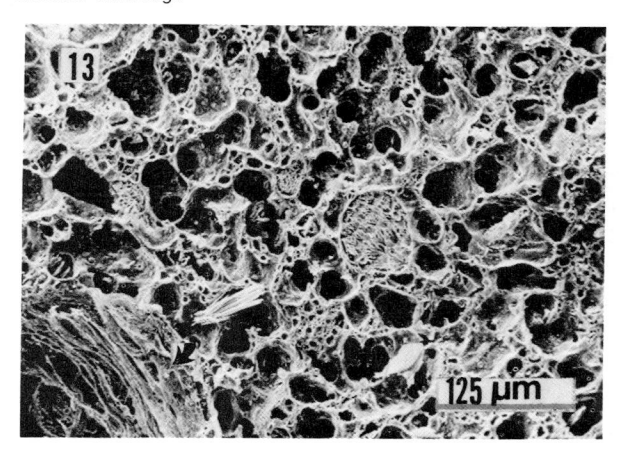

Fig. 13, 14 - Mozzarella cheese analog prepared from soy protein concentrate. Note the impression left by an entrapped air bubble and the fibrous detail of the air cell wall in the lower left corner of Fig. 13 (indicated by the arrow). The details of the gel matrix are shown in Fig. 14. Note the similarity of this gel matrix with that of natural mozzarella cheese (Fig. 2A) and the caseinate analog (Fig. 4).

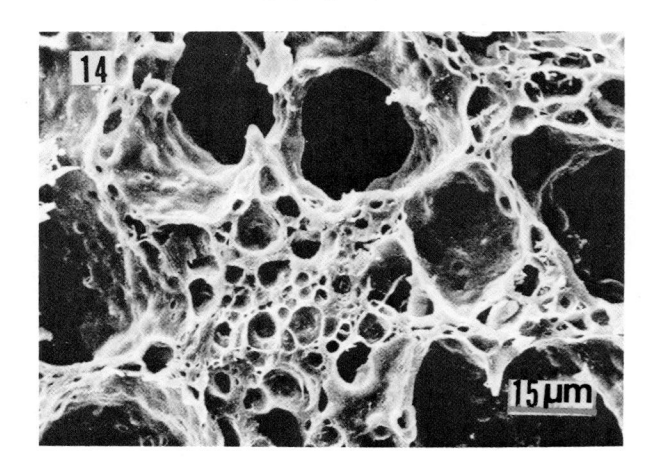

Furukawa, T., Ohta, S., Yamamoto, A. Texture-structure relationships in heat-induced soy protein gels, J. Texture Studies 10, 1979, 333-346.

Garner, F. H., Nissan, A. H. Rheological properties of high viscosity solutions of long molecules, Nature 158, 1946, 634-635.

Garner, F. H., Nissan, A. H. Rheological properties of high viscosity solutions of long molecules, Nature 164, 1949, 541-543.

Glicksman, M., Schachat, R. E. Gum Arabic, in Industrial Gums, R. L. Whistler, Ed., Academic Press, New York, U.S.A., 1959, 257-263.

Kalab, M. Milk gel structure VI. Cheese texture and microstructure, Milchwissenschaft 32(8), 1977, 449-457.

Kalab, M. Milk gel structure VIII. Effect of drying on the scanning electron microscopy of some dairy products, Milchwissenschaft 33(6), 1978, 353-358.

Kasik, R. L., Peterson, M. A. Cheese extender, U.S. Patent No. 4,016,298, 1977.

Nelson, A. I., Steinberg, M. P., Wei, L. S. Illinois process for preparation of soymilk, J. Food Sci. 41(1), 1976, 57-61.

Nelson, D. L. Measurement of the elasticity of mozzarella cheese by the Weissenberg test, Paper presented at the 40 th Annual Meeting of the Institute of Food Technologists, New Orleans, LA, June, 1980.

Peters, J. W. Pizza explosion necessitates changes in pricing, substitute ingredients, Food Product Devel. 13(12), 1979, 102.

Saio, K. Tofu - Relationships between texture and fine structure, Cereal Foods World 24(8), 1979, 342-354.

Siapantas, L. Engineered cheese products, Food Eng. 52(3), 1980, 20-24.

Taranto, M. V., Rhee, K. C. Ultrastructural changes in defatted soy flour induced by

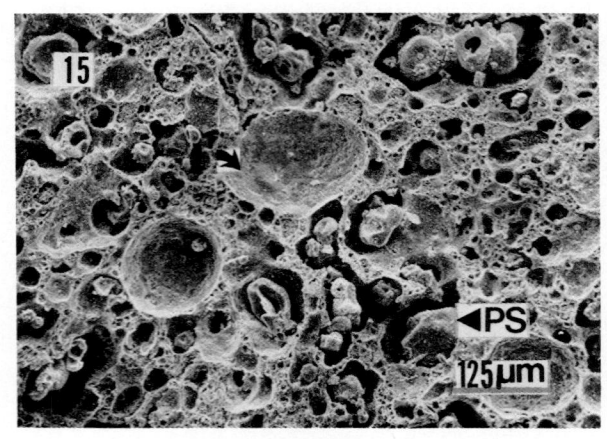

Fig. 15, 16 - Mozzarella cheese analog prepared from 10 g of soy protein isolate. Note the impressions left by air bubbles in Fig. 15 (indicated by the arrow). Large spherical inclusions (labeled - PS) which are identified as undissolved soy protein (see text) are embedded in the gel matrix causing interruptions in the gel network (Fig. 15). The details of the gel matrix are shown in Fig. 16.

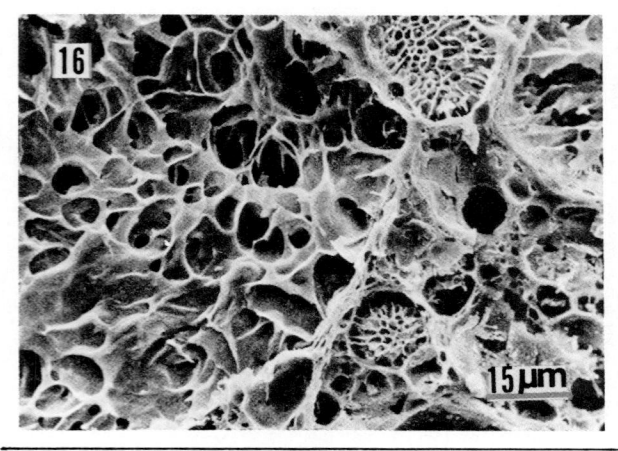

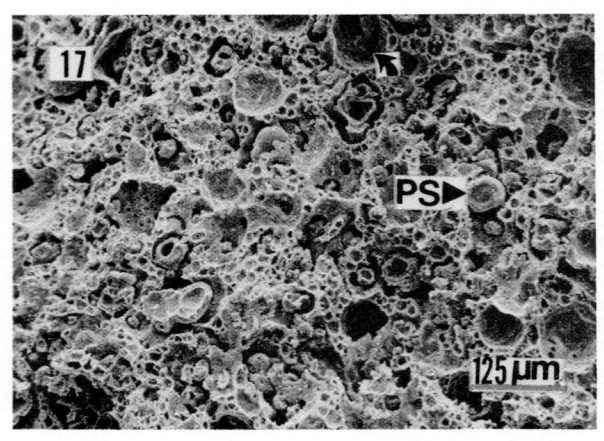

Fig. 17, 18 - Mozzarella cheese analog prepared from 20 g of soy protein isolate. Impressions left by air bubbles (indicated by the arrow) and globules of undissolved soy proteins (labeled - PS) are evident in the gel matrix (Fig. 17). The details of the gel matrix are shown in Fig. 18. Note the cross-section of a protein sphere in the lower portion of the micrograph (Fig. 18).

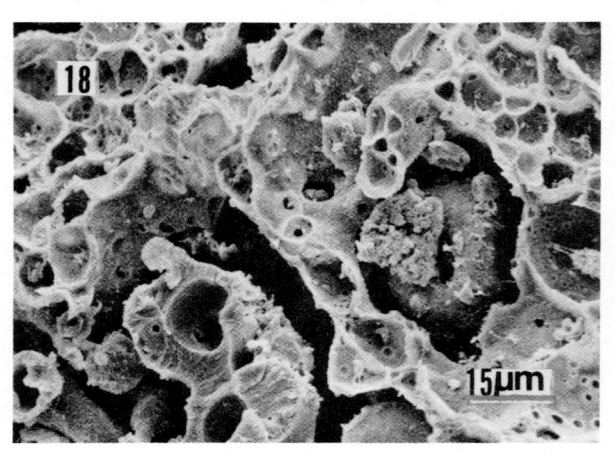

nonextrusion texturization, J. Food Sci. 43, 1978, 1274-1278.

Taranto, M. V., Wan, P. J., Chen, S. L., Rhee, K. C. Morphological, ultrastructural and rheological characterization of cheddar and mozzarella cheese, Scanning Electron Microsc. 1979, III: 273-278.

Vernon, H. R. Non-dairy cheese - A unique reality, Food Product Devel. 6(9), 1972, 22-26.

Weigand, J. H. Demonstrating the Weissenberg effect with gelatin, J. Chem. Educ. 40, 1963, 475-479.

Weissenberg, K. A continuum theory of rheological phenomena, Nature 159, 1947, 310-312.

Wolf, W. J., Baker, F. L. Scanning electron microscopy of soybeans, soy flours, protein concentrates and protein isolates, Cereal Chem. 52(3), 1975, 387-396.

Acknowledgements

This work was supported by the University of Illinois Agricultural Experiment Station Hatch Project # 1-30-15-50-304 entitled "Ultrastructure of Foods and Food Ingredients."

Discussion with Reviewers

R. D. Sullins: There are numerous manufacturers of natural and imitation mozzarella cheese in the U.S. What is the variability of product quality among these manufacturers?
Authors: We have run the TPA, as described in our paper, on several commercial natural mozzarella cheeses and found large variations in the TPA parameters which we attributed to compositional differences in the products. This conclusion is supported by data presented by Chen et al. All of these natural cheeses were not subjected to a microscopical analysis, therefore, we cannot comment on the possible correlation of their texture and microstructure. However,

Emmons et al.,(J. Texture Studies 11, 1980, 15-34) did study the effect of product composition of cheddar cheese on its textural and microstructural properties. These authors reported correlations between specific textural parameters and product microstructure. We have examined several different types of caseinate mozzarella analogs using scanning electron microscopy and found all to exhibit similar structural features (see Fig. 3, 4). However, not enough sample was available to run a TPA. All of the natural mozzarella cheeses and caseinate analogs we examined, exhibited similar stretching and stringing properties. However, those products with a higher moisture and fat content were more difficult to slice, cut or shred and tended to mat when slices were piled.

R.D. Sullins: Some mozzarella cheese substitutes made from rennet casein as a starting material. Was any cheese compared to these products?
Authors: No because we were not able to obtain any analogs prepared from rennet casein.

R.D. Sullins: Gelatin and gum arabic are expensive ingredients. Was any cost analysis made on cheese described in this paper versus any made with more conventional ingredients?
A. Bridges: Can suitable mozzarella cheese analogs be prepared from other types of relatively inexpensive protein, for example whey protein?
Authors: Numerous hydrocolloids were screened to find those which imparted the required stretching and stringing properties to the analog in the melted state. Gum arabic was the only hydrocolloid we tested that functioned as per our requirements. The gelatin is required to form the gel when the product is cooled. We found that a mixture of only soy proteins and gum arabic did not gel when cooled. At this time, we have not studied the use of other proteins (e.g. whey proteins) in our analog or performed a cost analysis. These are phases of our research plan which we will perform at a later date.

R. D. Sullins: Were any pizzas made using soybean mozzarella cheese analog?
Authors: At this time, no pizzas have been prepared using our analogs.

M. Kalab: Milk products have been the few remaining foods which have been providing human nutrition with more calcium than phosphorus. What is the situation with mozzarella cheese analogs?
Authors: We have prepared our analogs with the addition of various forms of calcium salts and found no adverse effects on the textural properties of the product. Therefore, we anticipate no problem in fortifying our analog to meet human nutrition requirements. In-depth study on this aspect of our research is planned and will be reported at a later date.

M. Kalab: Were attempts made to stretch melted mozzarella cheese analogs, fix the strings in this state and examine them by SEM?
A. Bridges: Does the fine structure of the mozzarella cheese and the mozzarella cheese analogs change when they are in a molten or melted state? Can the fine structural appearance be used to determine the physical state of the cheese?
Authors: At this time, no attemps have been made to fix the stretched analogs or natural mozzarella and examine them with SEM. Therefore, we do not know how the fine structure of natural mozzarella or the analogs changes when they are melted. Presumably, the fine structural appearance of the sample can be used to determine its physical state. This contention is supported by the work of M. Kalab (Milchwissenschaft 32(8), 1977, 449-457) who showed that mozzarella curd that was stretched during manufacture could be distinguished from unstretched curd due to the structural differences in the protein matrix observed with SEM. Taranto et al., (Scanning Electron Microsc. 1979, III: 273-278) made a similar observation.

C. A. Ernstrom: The pH of mozzarella curd made from milk has a very critical influence on the stretchability of the hot curd. Is the pH critical to the stretchiness of soy analogs? Also, how does the pH affect the microstructure and fat stability of the soy analogs?
Authors: The natural pH of our analogs is 5.3. Preliminary experiments have indicated that at pH values above 7 or below 4, the analogs stretching and stringing properties were negated. Acceptable stretchiness was obtained at pH values between 4 and 7. We are presently studying the effect of pH on the textural and structural properties of the analogs. A detailed report will be presented at a later date.

D. N. Holcomb: The adhesiveness (Table 2) for retail low moisture mozzarella is notably higher than for the analogs. Is this difference also apparent with melted samples?
Authors: Data for the Weissenberg test is presented in Table 3. These data indicate that the analogs exhibit stretching properties similar to commercial mozzarella. How the adhesiveness of the sample is related to the Weissenberg test data is not known at this time. We are presently planning to use a capillary rheometer to study the flow properties of the samples. The objective shear rate-shear stress data we obtain will enable us to more accurately compare our analogs to natural mozzarella cheese.

D. N. Holcomb: Can the authors make any correlation of TPA data and melting behavior with microstructure?
Authors: We feel it is premature to make any definitive statements on the correlation of microstructure and textural properties at this time since we are only now beginning to study in greater detail the effect of ingredient formulation on product performance. However, the data collected to date indicates that as the content of soy protein (free of insoluble and soluble carbohydrates) increases, the stretchiness and stringiness of the melted analog is increased. Examination of the respective micrographs indicates that the microstructure of the products change as the content of soy protein (free of carbohydrates) increases. The protein networks

are made up of finer strands. The strands exhibit a cross-linked substructure. Presumably, these structural changes are correlated with improved stretching properties. However, more work needs to be done to substantiate this contention.

K. Saio: I would like to ask how you distinguish the difference between the impressions left by the fat globules and entrapped air bubbles in the SEM images?
Authors: At the present time, we are making the distinction based on size and general appearance. Air bubble impressions are much larger than fat globule impressions and are usually devoid of structural detail. We have prepared a new set of samples for transmitted light, scanning electron and transmission electron microscopy to accurately locate and identify the components of the gel matrix via specific histological staining and enzyme digestion techniques. These results will be reported at a later date.

POSSIBILITIES OF AN ELECTRON-MICROSCOPIC DETECTION OF BUTTERMILK MADE FROM SWEET CREAM IN ADULTERATED SKIM MILK

M. Kaláb

Food Research Institute
Research Branch
Agriculture Canada
Ottawa
Ontario, Canada K1A 0C6

Abstract

Scanning electron microscopy showed that particles of pure spray-dried buttermilk had shallow wrinkles on their surfaces as opposed to deep wrinkles on skim milk particles. Minute globular particles emerging from the interior of large buttermilk spheres during the final stage of spray-drying were surrounded by rims; after breakage the minute particles left crater-like scars on the large spheres. The form of the buttermilk particles was quite typical, permitting detection of as little as 5% of spray-dried buttermilk in skim milk powder. In a humid atmosphere lactose in the buttermilk as well as in skim milk particles rapidly recrystallized; the crystals, which rapidly grew on the particle surfaces, significantly altered the particle appearance and obscured the differences between both products.

It was possible to distinguish precipitates obtained by acid coagulation of reconstituted buttermilk from those of skim milk using thin-section electron microscopy; the skim milk precipitates consisted almost exclusively of casein micelles whereas the buttermilk precipitates contained additional fat globule membrane fragments and cellular debris. The differences between both products were even more clearly evident by examining pellets obtained by ultracentrifugation of reconstituted skim milk and buttermilk. As little as 5% of reconstituted buttermilk, blended into two brands of the skim milk under study, was successfully detected. The sensitivity of the detection depended on the composition of the skim milk; in two other cases the sensitivity was as low as 10 and 25%.

This study has shown that electron microscopy might contribute to the detection of buttermilk made from sweet cream in adulterated skim milk.

KEY WORDS: Buttermilk, Skim Milk, Adulteration, Fat Globule Membrane, Cellular Debris, Lactose, Ultracentrifugation, Scanning Electron Microscopy, Thin-Sectioning.

Introduction

Sweet uncultured cream or cultured cream is used to produce butter. Churning separates the butterfat and leaves the serum called buttermilk. The appearance, flavour, and composition of the buttermilk depend to a great extent on the quality of the initial cream. This study is concerned with buttermilk made from sweet uncultured cream.

Fat is in a dispersed state in milk and, consequently, in cream separated from the milk. Fat globules, 0.5 - 10 μm in diameter[1], are encased in membranes composed of lipoproteins[2] which stabilize them in the milk and prevent them from aggregating. In addition to a high fat content, sweet cream also contains other milk components such as casein micelles, whey proteins, lactose, skim milk lipoprotein membranes, and cellular debris from the cow's mammary gland.

Churning disrupts most of the fat globule membranes, allowing the fat globules to aggregate into butter; most of the membrane fragments are released in the buttermilk and others are retained in the butter[3]. Hence, after the removal of the butterfat, the composition of buttermilk made from sweet cream is similar to that of skim milk, concerning protein and carbohydrates, but buttermilk contains, in addition, excessive membraneous material and a slightly higher lipid content. The higher the fat content of the cream, the higher the content of the unique substances in buttermilk[4].

Because of its high nutritional value, buttermilk is used in a variety of food products, for example as a protein-rich additive in bakery products etc. However, in North America buttermilk commands a lower price than skim milk. Hence there is a temptation to blend small amounts of buttermilk into skim milk; such a practice constitutes fraud and is illegal. Apart from the legal aspect, another reason for not blending both products is that the lipids in buttermilk oxidize easily to form off-flavours; presumably the blend would deteriorate more rapidly than pure skim milk powder. Chemical detection of buttermilk in skim milk may be difficult because of the similarity in the composition of both products. Detection based on the presence of fat globule membranes and their fragments in adulterated skim milk appears more promising. Buttermilk powder contains approximately 5% lipids. Skim milk powder can contain as little as 0.5% lipids. The limit for lipids in Canada First Grade powder is 1.2%. Thus, to be useful, the test for adulteration of skim milk solids with buttermilk solids must be able to detect the latter at ratios lower than 15%, more probably 5 to 10%.

This study examines two possibilities of detecting adulteration of skim milk with buttermilk: one using scanning electron microscopy (SEM) to detect differences in the morphology of buttermilk and skim milk particles in blends of both products, and the other using thin-section electron microscopy (TEM) to detect differences in the morphology of the reconstituted products.

Materials and Methods

Pure spray-dried skim milk (6 specimens) and pure spray-dried buttermilk (6 specimens) were provided by the Dairy Division, Food Production and Marketing Branch, Agriculture Canada in Ottawa. The products were obtained from various skim milk and buttermilk manufacturers in Canada.

Scanning electron microscopy (SEM)

The skim milk and buttermilk powders were mounted on a double-sticky tape attached to SEM stubs. The edges of the tape were painted with a silver cement for a better contact. The powders were used in the forms as they were obtained and, in separate experiments, were fixed in the dry state in an atmosphere saturated with OsO_4 vapours[5,6]. In addition, the powders were blended at various ratios and the mixtures were also examined. The mounted powders were coated with carbon (≃5nm) and gold (≃20 nm) by vacuum evaporation. A Cambridge Stereoscan electron microscope was operated at 20 kV and micrographs were taken on 35-mm films.

Thin-section electron microscopy (TEM)

The following specimens, fixed in aqueous 1.4% glutaraldehyde, post-fixed in a 2% OsO_4 solution in a 0.05 M veronal-acetate buffer, pH 6.75, dehydrated in a graded alcohol series, and embedded in a low-viscosity Spurr's medium[7] were examined in the form of thin sections stained with uranyl acetate and lead citrate[8]:

(a) precipitates obtained by coagulating reconstituted (20% total solids) buttermilk and skim milk with N HCl to pH 3.5 and compacting the precipitates by centrifugation (480 g for 30 min);

(b) pellets obtained by ultracentrifugation (8 x 10^4 g for 90 min) of reconstituted (20% total solids) buttermilk and skim milk (4 specimens each) and their mixtures (2.5, 5.0, 10.0, and 25.0% buttermilk in skim milk);

(c) the "fluff" and fat fractions obtained by the above ultracentrifugation were embedded in 2% agar for fixation and consecutive embedding in the Spurr's medium. The "fluff"[9] is an opalescent liquid fraction above the casein pellet and has no discrete boundary with the clear supernatant serum.

A Philips EM 300 electron microscope was operated at 60 kV and micrographs were taken on 35-mm films.

Results and Discussion

At the first glance, spray-dried skim milk and buttermilk look quite alike under SEM. They are both in the form of spheres or clusters of spheres widely ranging in dimensions. However, a closer examination at a higher magnification reveals some striking differences. In spray-dried skim milk there is a great number of spheres which are severely wrinkled and occasionally displaying the "apple-like" structure (Fig. 1) earlier described by Buma and Henstra[10]. The wrinkles on the spherical particles of spray-dried buttermilk made from sweet cream are not as deep (Fig. 2). Collapsed spheres, frequently found in spray-dried skim milk (Fig. 1) were never observed in any of the six different buttermilk specimens under study. Also the "apple-like" structure in buttermilk differed from that in skim milk; in buttermilk, minute globular particles were trapped inside the large "parental" spheres while they were still emerging from the interior through the surface (Fig. 2). Low rims around the emerging minute globular particles were formed by the material of the large particle as it was lifted during the process of penetration through the surface. After having been broken away from the large particle, the minute globular particles left crater-like scars.

Porosity of spray-dried buttermilk also differed from that of skim milk. Spherical vacuoles in the former were either empty or occupied by minute globular particles (Fig. 3 and 4). It is probable that the empty vacuoles had been occupied by smaller loose globular particles before the large particles were fractured (Fig. 4) but were lost during the preparative stages for SEM. The partitions between the vacuoles were almost compact. In general, the porosity of spray-dried buttermilk was lower than that of whole milk but higher than that of spray-dried skim milk as studied by Buma[11] and reviewed by Kalab[12]; this was in agreement with Buma's conclusion that porosity was related to the fat content.

These features were characteristic of spray-dried buttermilk and it was easy to detect typical buttermilk and skim milk particles in 1:1 mixtures. However, as the buttermilk content was decreased to 5% (w/w), a greater number of particles had to be examined under SEM to find reliable evidence of the buttermilk particles. Yet, examination of specimens, the composition of which was not known to the electron microscopist, proved that it was possible to detect a 5% admixture of spray-dried buttermilk in spray-dried skim milk. However, such a statement is difficult to document by micrographs for the following reason: At a 5% concentration of buttermilk in skim milk (w/w), there are approximately 5 buttermilk particles to 95 skim milk particles, assuming similar particle dimensions. Not all 5 buttermilk particles are presumably absolutely typical and may be distorted to varying extent. To show 3 to 4 typical buttermilk particles in the field of vision, the magnification used would have to be low to accommodate all the accompanying skim milk particles; such a magnification would be too low to identify the buttermilk particles clearly. By increasing the magnification to show typical details of the buttermilk particles, a great number of the skim milk particles would be eliminated from the field of vision. The resulting micrograph would resemble a micrograph obtained with a mixture containing a higher buttermilk concentration. This means that the area, in which the individual buttermilk

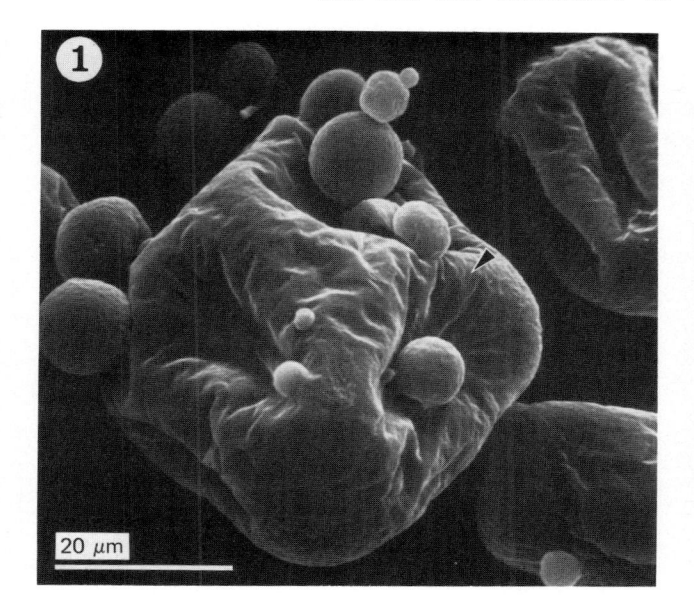

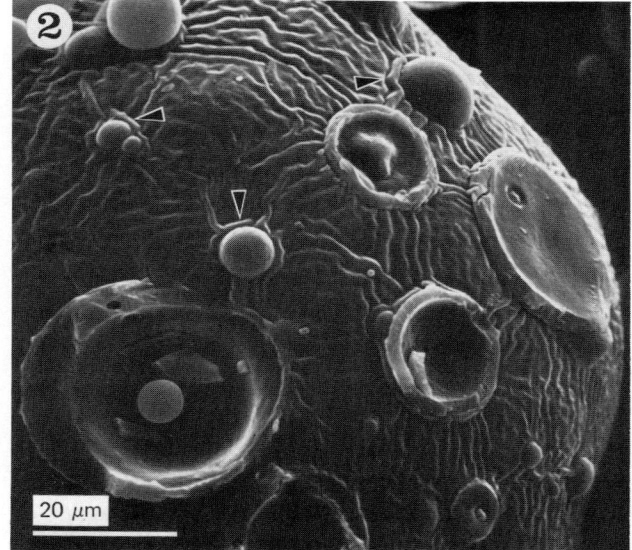

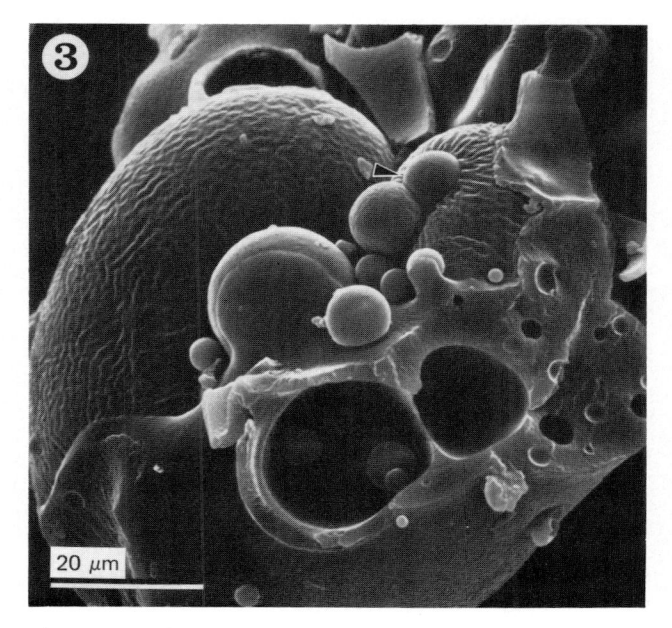

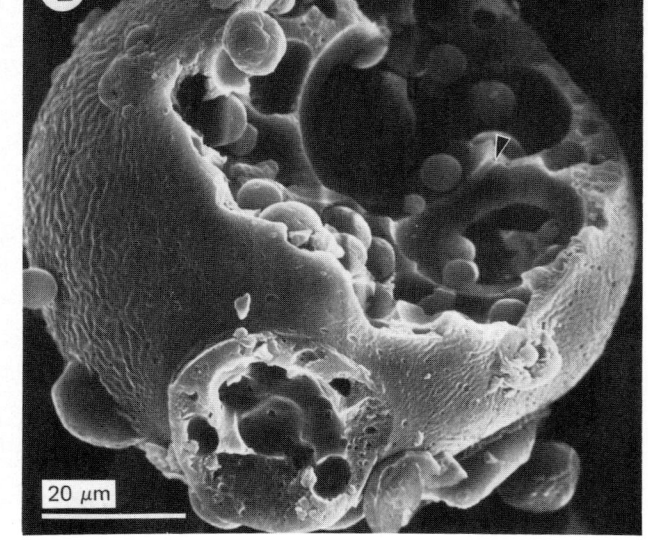

Fig. 1. Typical severely wrinkled and collapsed spherical particles of spray-dried skim milk. Particle in the centre displays the "apple-like" structure (arrow).

Fig. 2. An intact particle of spray-dried buttermilk with shallow wrinkles, minute globular particles emerging through the surface of the "parental" particle, surrounded with low rims (arrows), and crater-like structures originating from the breakaway of minute globular particles that had emerged through the surface of the large particle.

Fig. 3. Fractured particle of spray-dried buttermilk with vacuolized core, relatively compact walls, and minute globular particles emerging from the interior (arrow).

Fig. 4. Fragment of a vacuolized particle of spray-dried buttermilk. Particle surface is slightly wrinkled and the internal vacuoles are occupied with minute globular particles. Partitions between the vacuoles are compact (arrow).

particles are distributed, must also be taken into consideration.

Although it is probable that spray-dried buttermilk might be blended with spray-dried skim milk, the more likely possibility of adulteration is to blend fluid buttermilk with fluid skim milk and spray-dry the mixture. In this case the proof of the presence of buttermilk would have to be based on the presence of fat globule membrane fragments. This could be examined by TEM but not by SEM.

To embed buttermilk and skim milk powders, it is necessary to protect the particles from the effects of solvents by fixing them. Fixation of dry

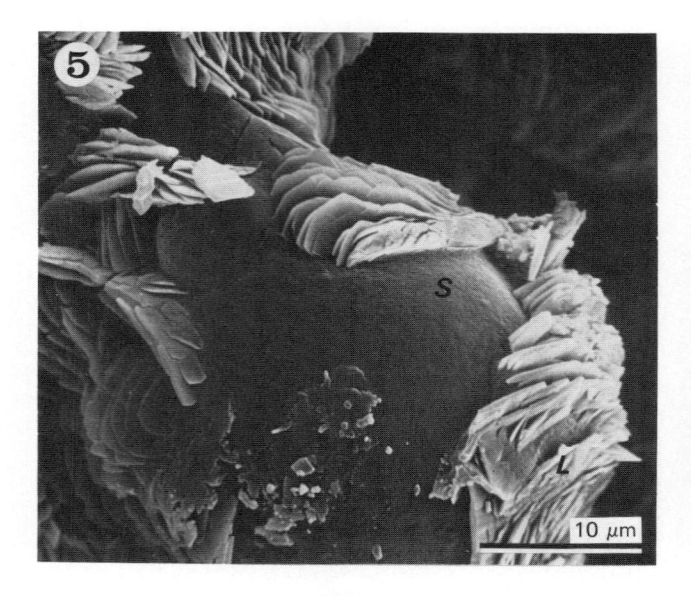

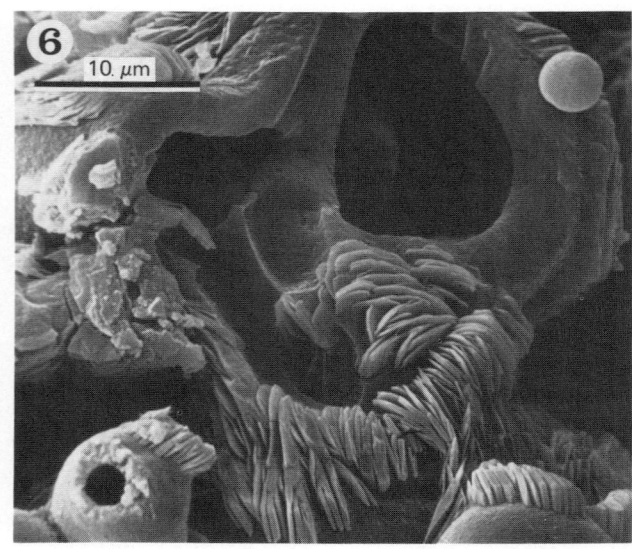

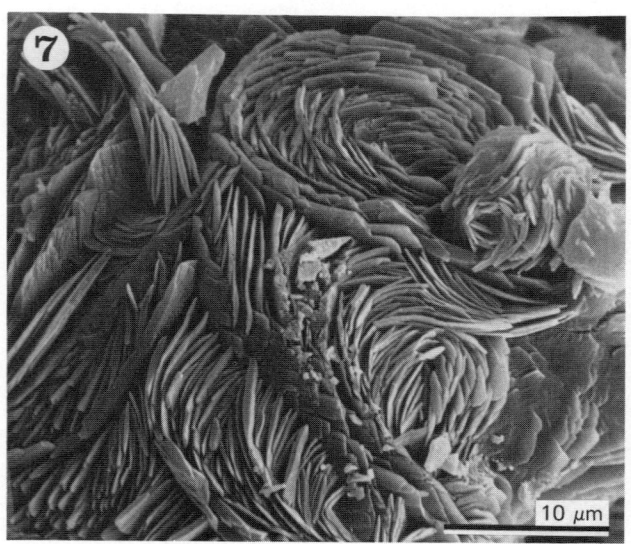

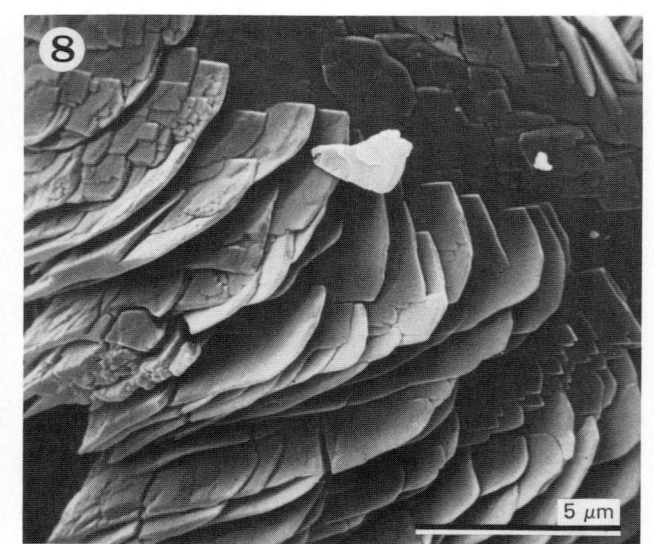

Fig. 5. Crystallization of lactose α-hydrate on the surface of a buttermilk particle before completion. L = Lactose crystals; S = Original particle surface; note the absence of wrinkles.

Fig. 6. Incomplete crystallization of lactose α-hydrate on a fractured particle of spray-dried buttermilk.

Fig. 7. Buttermilk particle completely covered with lactose α-hydrate crystals arranged in the form of rosettes. Typical features have been obscured by the lactose crystals.

Fig. 8. Detail of tomahawk-shaped α-hydrate lactose crystals covering the surface of a buttermilk particle that had been exposed to a humid atmosphere.

milk powders with OsO_4 vapours was described earlier by Müller[5]. After the powders under study were exposed to OsO_4 in a dry atmosphere, however, they remained light-coloured for more than 24 h. This indicated that no reaction took place between the particles and the OsO_4 molecules. To follow Müller's procedure and to facilitate the interaction with OsO_4, a droplet of water was placed inside the vial in which the fixation took place. Consequently the buttermilk and skim milk particles turned dark within an hour. This effect was examined under SEM. Micrographs in Fig. 5 - 8 show that the microstructure of spray-dried buttermilk particles underwent severe changes in the humid atmosphere. Lactose, present in the buttermilk particles in the form of an amorphous glass mixture of α- and β-lactose[13,14] turned into α-hydrate crystals and formed efflorescences on the surfaces of the particles. The crystallization was captured before completion in Fig. 5 (intact particle surface) and in Fig. 6 (fractured particle). A particle with its surface completely covered with lactose crystals is presented in Fig. 7. Details of the tomahawk-shaped lactose crystals,

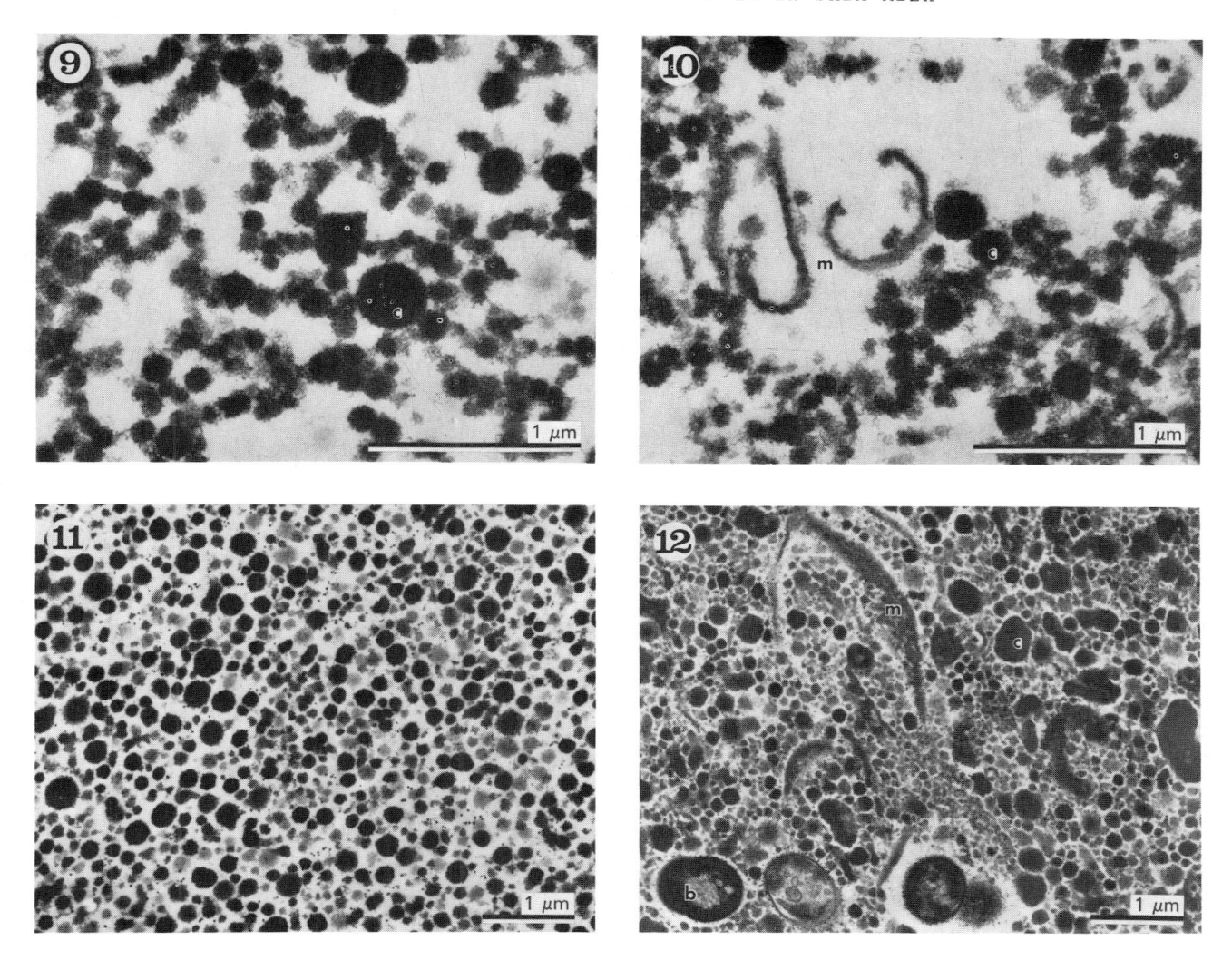

Fig. 9. Thin section of skim milk coagulated with N HCl (pH 3.5) and compacted by centrifugation (480 g for 30 min). The precipitate consists exclusively of casein micelles (c) forming a gel network.

Fig. 10. Thin section of buttermilk coagulated with N HCl (pH 3.5) and compacted by centrifugation (480 g for 30 min). The precipitate consists of casein micelles (c) similar to Fig. 9 and of membranous material (m) typical of buttermilk.

Fig. 11. Thin section of a pellet obtained by ultracentrifugation (8×10^4 g for 90 min) of reconstituted (20% total solids) skim milk. The pellet consists almost exclusively of casein micelles (dark discs).

Fig. 12. Thin section of a pellet obtained by ultracentrifugation of reconstituted buttermilk (same conditions as in Fig. 11). Casein micelles (c), membranous material (m), and bacteria (b) are evenly distributed throughout the section.

which are extremely susceptible to electron-beam damage, are presented in Fig. 8. It follows from these micrographs that crystallization of lactose in a humid atmosphere obscured all the typical features distinguishing spray-dried buttermilk from skim milk. This means that the SEM technique cannot be used alone to detect a probable adulteration of a product, which had been exposed to air humidity and in which recrystallization of lactose had occurred spontaneously. The other conclusion is that any TEM of the particles thus altered would not reveal the genuine microstructure of the dried powders; attempts to examine them failed because the powder particles were harder than the embedding medium and chipped off out of the resin during sectioning.

TEM examination of reconstituted and subsequently acid-coagulated buttermilk and skim milk showed differences in the composition of both products. The presence of fat globule membranes, their fragments, and cellular debris in buttermilk and the absence of these components in skim milk was thus confirmed (Fig. 9 and 10). Yet, the centrifugal force used was insufficient to properly compact the precipitates.

In another experiment ultracentrifugation was

used and reconstituted buttermilk and skim milk were fractionated into solid pellets at the bottom of the tubes, "fluffy" fractions above them, clear sera, and floating lipid layers at the surface as shown by Stewart et al.[9] Electron-microscopical examination of the pellets indicated that this was a prospective way to detect adulteration of skim milk. Casein micelles in skim milk and all the different components in buttermilk were concentrated in relatively small volumes of the pellets, yet retained their individual entities as shown in Fig. 11 and 12. The membranous material found in reconstituted buttermilk (Fig. 12) appeared to be somewhat different from the images obtained by other authors, who studied freshly secreted milk[15], 18% cream[16], or homogenized cream[17]. The manufacturing processes and storage most probably contributed to the changes, partially in agreement with Wooding's findings of a gradual decomposition of the fat globule membranes with time, probably due to enzymatic processes in milk[15] and, possibly, due to the heat treatment during pasteurization and spray-drying. Thus the membranous material in reconstituted ultracentrifuged buttermilk resembled a debris rather than well-defined organelles.

TEM of the "fluffy" and fat fractions provided no additional information; the most noticeable difference was the larger volumes of both fractions in buttermilk than in skim milk.

TEM of 4 different skim milk powder brands, to which different buttermilk brands were blended at 2.5 to 25% levels, did not produce uniform results. Two skim milk powders contained almost pure casein micelles and only traces of fat globule membranes in the casein pellets; it was possible to detect as little as 2.5 - 5.0% of added buttermilk. However, the other skim milk brands contained large quantities of membranous and cellular material; consequently, the added buttermilk was detected only after its content reached 25%. Evidently there is a need for an electron-microscopical survey of pure authentic skim milk before the findings of this presentation are used as a basis for an analytical procedure. However, such problems are beyond the scope of this study and will be discussed elsewhere.

Acknowledgment

Skillful assistance provided by Mr. J.A.G. Larose is acknowledged. The author thanks Dr. D.B. Emmons for useful comments and Mr. Peter Kalab for linguistic assistance. Electron Microscope Centre, Research Branch, Agriculture Canada in Ottawa provided facilities. This presentation is Contribution 417 from Food Research Institute, Agriculture Canada, Ottawa.

References

1. E. Knoop: Strukturaufklärung durch elektronenmikroskopische Untersuchungen an Eiweiss und Milchfett. Milchwissenschaft 27, 1972, 364-373.
2. J. R. Brunner: Milk lipoproteins. In: Structural and Functional Aspects of Lipoproteins in Living Systems. E. Tria and A.M. Scanu (eds.), Academic Press, London and New York, 1969, 545-578.
3. N. King: Milk Fat Globule Membrane and Some Associated Phenomena. Commonwealth Agricultural Bureaux, Farnham Royal, Bucks, England, 1955, 43-73.
4. H. Mulder and P. Walstra: The Milk Fat Globule. CAB, Farnham Royal, Bucks, England, and PUDOC, Wageningen, the Netherlands, 1974, 88-93.
5. H. R. Müller: Elektronenmikroskopische Untersuchungen an Milch- und Milchprodukten. 1. Strukturaufklärung in Milchpulvern. Milchwissenschaft 19, 1964, 345-356.
6. M. Kalab and D. B. Emmons: Milk gel structure. III. Microstructure of skim milk powder and gels as related to the drying procedure. Milchwissenschaft 29, 1974, 585-589.
7. A. R. Spurr: A low-viscosity epoxy resin embedding medium for electron microscopy. J. Ultrastruct. Res. 26, 1969, 31-43.
8. M. Kalab and V.R. Harwalkar: Milk gel structure. II. Relation between firmness and ultrastructure of heat-induced skim-milk gels containing 40-60% total solids. J. Dairy Res. 41, 1974, 131-135.
9. P. S. Stewart, D. L. Puppione, and S. Paton: The presence of microvilli and other membrane fragments in the non-fat phase of bovine milk. Z. Zellforsch. 123, 1972, 161-167.
10. T. J. Buma and S. Henstra: Particle structure of spray-dried milk products as observed by a scanning electron microscope. Neth. Milk Dairy J. 25, 1971, 75-80.
11. T. J. Buma: The relationship between free-fat content and particle porosity of spray-dried whole milk. Neth. Milk Dairy J. 25, 1971, 123-140.
12. M. Kalab: Scanning electron microscopy of dairy products: An overview. SEM/1979/III, SEM, Inc., AMF O'Hare, IL 60666, 261-272.
13. Sylvia Warburton and S. W. Pixton: The moisture relations of spray-dried skimmed milk. J. Stored Prod. Res. 14, 1978, 143-158.
14. Sylvia Warburton and S. W. Pixton: The significance of moisture in dried milk. Dairy Ind. Internatl. 43(4), 1978, 23-27.
15. F. B. P. Wooding: The structure of the milk fat globule membrane. J. Ultrastruct. Res. 37, 1971, 388-400.
16. M. Anderson, B. E. Brooker, T. E. Cawston, and G. C. Cheeseman: Changes during storage in stability of ultra-heat-treated aseptically-packed cream of 18% fat content. J. Dairy Res. 44, 1977, 111-123.
17. D. F. Darling and D. W. Butcher: Milk-fat globule membrane in homogenized cream. J. Dairy Res. 45, 1978, 197-208.

Discussion with Reviewers

R.J. Carroll: Is lactose the principal component of the amorphous wall of the buttermilk particles and of the skim milk particles?
Author: Yes, lactose is one of the principal components of the wall. It is probable that the particle surface consists of the glassy form of lactose β-anhydride as suggested by the high susceptibility of the surface to electron-beam damage (Fig. 13). However, detailed distribution of the other components such as whey proteins and casein micelles in the wall is not known. The conversion of lactose from the amorphous β-anhydride into

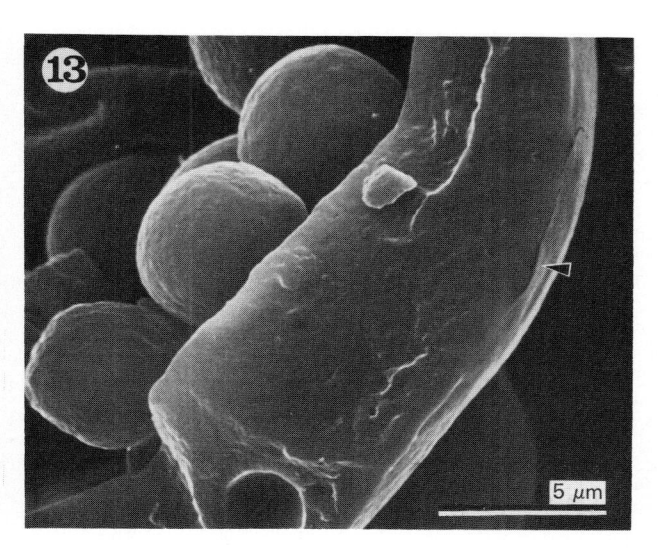

Fig. 13. Section of a skim milk particle wall. Arrow indicates electron-beam damage to the external surface during scanning.

crystalline α-hydrate in the humid atmosphere proceeded very rapidly. The crystals lost their water content again *in vacuo* during gold coating and SEM and this made them extremely sensitive to the electron-beam damage.

R.J. Carroll: Bacteria are observed in the reconstituted buttermilk sample (Fig. 12). Did you find bacteria in the skim milk samples?

D.N. Holcomb: One would expect bacteria counts in the order of 10^4 cells/mL in skim milk. Would the presence of remnants of these cells in the spray-dried product cause any problem in the method presented in this paper?

Author: Bacteria and, particularly, their remnants could cause serious problems as they would imitate adulteration of the skim milk with buttermilk following accumulation of the bacteria and their remnants in the pellet. However, bacteria sedimented first, and thus, examination of the central portion of the pellet, as practised in this study, revealed no bacteria. At this time we have already surveyed over a dozen skim milk brands and found isolated bacteria only in one of them.

R.J. Carroll: Were any membrane fragments observed in the reconstituted skim milk samples (lipid content 0.5-1.5%)?

Author: Some lipoprotein membrane fragments are always present in skim milk but are concentrated by ultracentrifugation in the fluffy fraction (Stewart et al.[9]). Consequently, pure casein micelles are accumulated in the pellet. This has probably happened with the two "pure" skim milk specimens. The presence of membranous material in the pellet of the two other skim milk specimens has caused great concern. We do not know the reasons for the remarkable differences in the morphology of the skim milk brands. Unless we are absolutely confident that the material, found in the "impure" skim milk, has originated from the addition of buttermilk, we cannot label such skim milk to be adulterated.

J.H. Prentice and M.V. Taranto: Could a represen-tative micrograph of a mixture of spray-dried skim milk and buttermilk be presented?

Author: The fact that not all spray-dried particles are typical makes the presentation of a representative micrograph difficult. When the microscopist looks for the presence of buttermilk particles in spray-dried skim milk, he/she does not care about what is in their immediate proximity. To publish a meaningful micrograph, however, it is necessary to find typical buttermilk particles surrounded by typical skim milk particles, all within a reasonably small area preferably free of charging artifacts. An example of such a setting is shown in Fig. 14; the proof that the particle marked with an arrow is really a buttermilk particle is presented in Fig. 15 at a higher magnification:

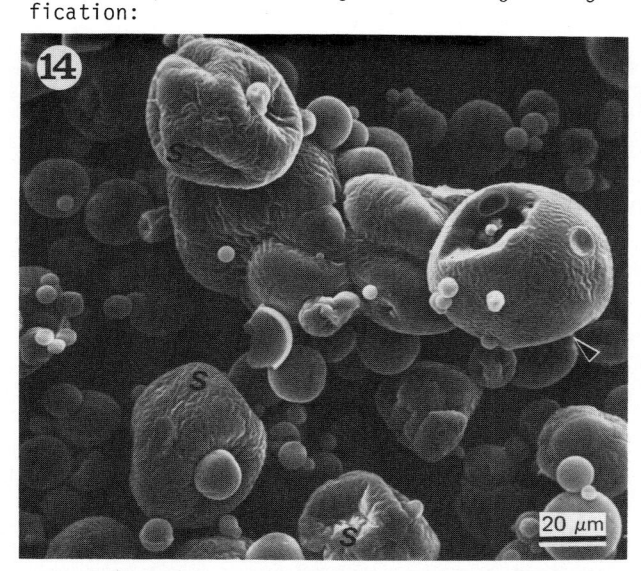

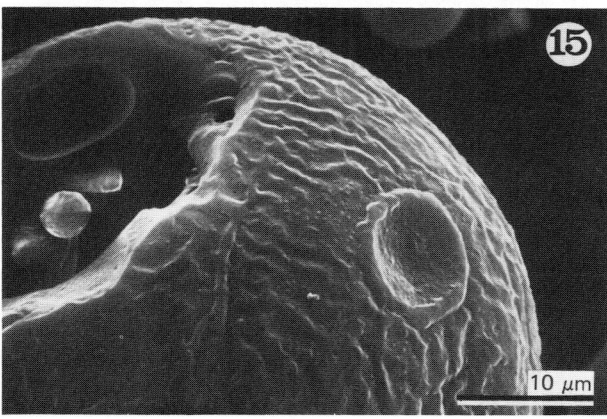

Fig. 14. A buttermilk particle (arrow) surrounded by typical collapsed skim milk particles (S) and small featureless globular particles. The spray-dried skim milk contained 10% (w/w) of buttermilk.

Fig. 15. Detail of the buttermilk particle shown in the preceding micrograph; the surface and the crater-like structure are typical of spray-dried buttermilk.

M.V. Taranto: J.F. Chabot (SEM/1979/III, pp. 279-286) advocates that "the best treatment is no treatment". Spray-dried soy proteins are commonly examined without fixation. Have you compared the structure of the spray-dried protein droplets of skim milk and buttermilk without OsO₄ fixation? If so, were there any structural differences between fixed and non-fixed samples?

Author: Morphology of unfixed spray-dried skim milk and buttermilk particles has been shown in Fig. 1-4. As indicated in Materials and Methods, the powders were mounted on a double sticky tape in the form as they were obtained. The reason for fixing the powders in a separate set of experiments was the intention to embed them in a resin for TEM. Because the powder particles turned dark only after a droplet of water had been placed in the atmosphere saturated with the OsO₄ vapours, a question arose whether the change in colour was accompanied by change in morphology. The answer has been presented in Fig. 5-8: the humid atmosphere, required for the interaction of OsO₄ with the specimens, led to the recrystallization of lactose and a complete alteration of the initial appearance of the particles. J.F. Chabot's advice should not be followed blindly, but here it has received additional support.

M.V. Taranto: Spray-dried soy sodium proteinate particles have a collapsed structure similar to the spray-dried skim milk particles. Both products contain less than 1% lipid. Do you think that the structural features of the spray-dried buttermilk particles can be attributed to its higher lipid content? If not, how would you explain the differences between the two types of particles?

Author: Wrinkles and collapsed structures in spray-dried milk and its derivatives such as sodium caseinate and lactose solutions were studied by T.J. Buma and S. Henstra ("Particle structure of spray-dried milk products as observed by a scanning electron microscope." Neth. Milk Dairy J. 25, 1971, 75-80 and "Particle structure of spray-dried caseinate and spray-dried lactose as observed by scanning electron microscope." *Ibid*. 278-281) and were associated with the presence of casein in the spray-dried products. Lipids presumably affected the porosity of such particles. I have no explanation for the differences in the morphology of skim milk and buttermilk particles but I assume that viscosity, in addition to the composition of the suspensions, and the spray-drying regimen contribute to the final appearance of the dry globular particles. Most particles are "typical" of the product, but there are also other "not-so-typical" particles; globules of diameters below a certain limit are usually smooth and featureless irrespective of the product (Fig. 13 and 14). This makes the distinction of the components in mixtures quite difficult.

D.G. Schmidt: In your paper you present a preliminary attempt to detect adulteration of skim milk powder with buttermilk by means of electron microscopy. Do you think that such a method might compete with, for instance, a lecithin determination?

Author: I am not aware of any chemical analytical method that would be capable of detecting as little as 5-10% (w/w) of buttermilk made from sweet cream added to skim milk. At present I do not dare to predict whether an analytical method will eventually be based on the findings of this study. This will depend on the differences in the composition and morphology of skim milk and buttermilk found by a survey of many specimens. If it is found that the pellets of most authentic pure skim milk brands contain no membrane fragments and that such fragments in the pellets are the signs of adulteration, an analytical procedure is feasible. If, however, our future research shows that even some pure skim milk brands contain membranous material in the pellets, we will investigate the reasons for such differences.

A SCANNING ELECTRON MICROSCOPICAL INVESTIGATION OF THE WHIPPING OF CREAM

D. G. Schmidt and A. C. M. van Hooydonk

Netherlands Institute for Dairy Research
P.O. Box 20, 6710 BA EDE
The Netherlands

Abstract

The whipping of normal and homogenized cream is investigated by SEM using a simple cryotechnique. During whipping of normal cream the entrapped air bubbles decrease in size until maximum foam strength is reached. In homogenized cream the size of the air bubbles passes through a minimum before maximum foam strength is reached. The entrapped air bubbles become surrounded by fat globules embedded in a layer of liquid fat, which has flowed out of the globules. In homogenized cream less globules remain intact than in non-homogenized cream.

KEY WORDS: Cream, Cryotechnique, Milk Fat Globules, Homogenized Cream, Whipping Of Cream

Introduction

The electron microscopical investigation of fat-rich foams, such as whipped cream, offers many difficulties. Conventional investigations with ultra-thin sections of embedded samples of whipped cream have been carried out by Graf & Muller (1). Such studies, however, are hampered by the solubility of the fat in most dehydrating agents and embedding media which leads to destabilisation of the fat globule and unpredictable extraction of fat. All this has for effect that conclusions drawn from such electron micrographs are dubious. Cryotechniques are expected to be much more reliable. Berger et al. (2) applied the freeze-fracturing technique to the study of ice cream, which product is actually a frozen fat-rich foam, and Buchheim (3) used the same method for the study of whipped cream. In this way both the fat phase and the air-serum interface could be studied.

For SEM-studies a cryotechnique in combination with a cold stage in the microscope seems to be appropriate. In previous publications (4, 5) a simple cryotechnique has been described for SEM-investigations on the structure of cheese, by which milk fat globules could be excellently detected. For the investigation described here, the method had to be modified slightly. Both normal and homogenized creams were studied.

Experimental

Cream with a fat content of 40% was prepared by centrifugation of thermisized (10s at 76°C) bulk milk at 40°C. Part of the cream was homogenized in two stages at 5 and 0.7 MPa. Both creams (homogenized and non-homogenized) were finally pasteurized in bottles for 5 min at 85°C and subsequently stored at 5 to 6 °C for three days. After storage whipping was carried out at the same temperature in a home-made apparatus equipped with two rods rotating in opposite directions and with a torsion wire to determine the strength of the foam. The time required to reach maximum foam strength was 158 and 328s for the normal and homogenized cream respectively.

Scanning electron microscopy was carried out by the cryotechnique that has been described earlier in regard to the investigation of cheese (4, 5). Because of the non-rigidity of the cream samples, the method had to be modified slightly. The specimen holder consisted of a copper disc, with a diameter of 1cm, provided with a central pin (see Fig. 1a). A piece of aluminium foil was wrapped around the specimen holder, thus forming a small cylinder (Fig. 1b) of which the inner walls were covered with conductive carbon cement according to Göcke (Neubauer Chemikalien, Münster). Next the cylinder was filled with the cream sample and immersed in liquid nitrogen. After freezing was completed, the aluminium foil was peeled off, and a film of conductive carbon cement was retained on the cylinder of frozen cream. This cylinder was fractured transversely with a razor blade cooled in liquid nitrogen (Fig. 1c). The sample was then transferred under liquid nitrogen to the preparation table of a Balzers BA301 freeze-etch unit, cooled to -100°C. For this purpose the original specimen holder of the preparation table had been replaced by a ring (Fig. 2a) just fitting the specimen holder with the cream cylinder (Fig. 2b). The layer of white frost, resulting from the transport of the sample, was removed by sublimation. To this end an aluminium plate (Fig. 2c) fixed to the microtome knife, which was cooled to -190°C, was placed just over the specimen. After the layer of frost had disappeared (within 15 min), the cleaned fractioned surface was coated by a carbon layer with a thickness of 20nm, which, in combination with the conducting carbon layer on the walls of the cream cylinder, accomplished a low degree of surface charging in the microscope. Next the specimen was transferred, under liquid nitrogen, to the cooled specimen stage of a JEOL JSM-U3 microscope. The layer of white frost resulting from this transport was removed by slightly heating the specimen surface with an infra red lamp through the lock window. The specimen was observed at -100°C at an acceleration voltage of 15kV.

The size distribution of the air bubbles in the whipped creams was determined on low-magnification (175 x) micrographs with the aid of a Zeiss TGZ3 analyser. The apparent distributions thus obtained were converted into true ones by the method of Goldsmith (6) for the transformation of histograms.

Results and Discussion

Figures 3 and 4 show the appearance of the milk fat globules in the non-whipped creams. The smaller size of the fat globules in the homogenized cream and the formation of so-called homogenization clusters are clearly visible. Further, the pictures do not show any noticeable segregation, which indicates that the applied freezing procedure did not cause large ice crystals that might damage large-scale structures. Figures 5 to 12 and 13 to 20 show the structure of the normal and homogenized creams at various stages of the whipping process.

The low-magnification pictures suggest that the air bubbles decrease in proportion as the time of whipping increases, and that they are much smaller in homogenized than in non-homogenized cream. This is confirmed by the results of the quantitative measurements, which are summarized in Figure 21. The number-average diameter of the air bubbles, $\bar{D}_n$, varies only slightly with the time of whipping in either type of cream. Their weight-average diameter, $\bar{D}_w$, however, strongly decreases in non-homogenized cream during whipping, and in homogenized cream it passes through a minimum, which indicates that coalescence of air bubbles has taken place as $\bar{D}_w$ started to increase after approximately 210s when it reached minimum. Maximum stability of the air bubbles in homogenized cream is apparently reached before maximum rigidity of the foam is obtained, whereas in non-homogenized cream the points of time at which maximum foam strength and overrun are reached almost coincide. Consequently the transition to the stage of overwhipping is less clear with homogenized than with non-homogenized cream, as is demonstrated by comparison of Figures 9 to 12 with Figures 17 to 20.

From the high-magnification pictures of normal cream (Figures 6, 8, 10 and 12) it might be concluded that at first fat globules are slowly adsorbed at the air-serum interface, so that after approximately 100s the interface is completely composed of tightly packed fat globules. Due to the high spreading pressure at the air-serum interface, the membranes surrounding the fat globules are ruptured, thus enabling the liquid fat to spread over the interface, with the result that the fat globules stick together, see Figure 10. In the lamellae between the air bubbles the vigorous agitation also leads to rupture of the fat globule membranes and to partial coalescence of the globules, so that a three-dimensional network is formed. Further whipping results in disappearance of the air bubbles and in formation of butter granules, similar to those found during churning, and the foam collapses (Figures 11 and 13). The electron micrographs thus confirm the current theory of the whipping process (7).

In homogenized cream the situation observed after 109s (Fig. 14) is comparable to that in normal cream after the same time of whipping (Fig. 8). However, with homogenized cream much more time is required to reach the point of maximum foam strength. During this period more fat globules appear to lose their liquid fat which, as a thick layer in which remaining fat globules are almost entirely buried, envelops the air bubbles (Figures 16 and 18). This suggests that in homogenized cream, in contrast to the situation found in normal cream, there is lack of balance between the stabilization of air

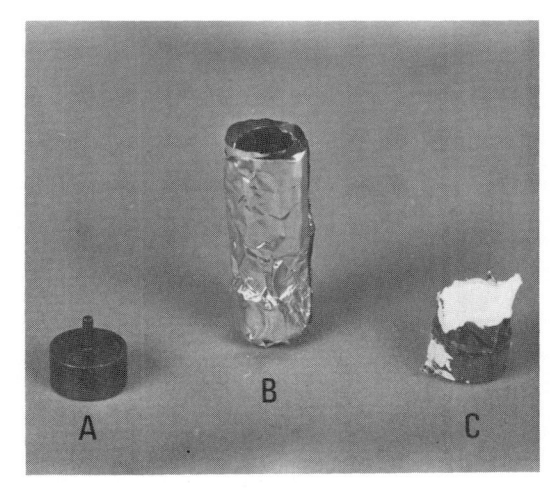

Fig. 1. First stages in the specimen preparation; a: the specimen holder; b: the specimen holder wrapped up with aluminium foil; c: the fractured cylinder of frozen cream, the side walls coated with carbon cement.

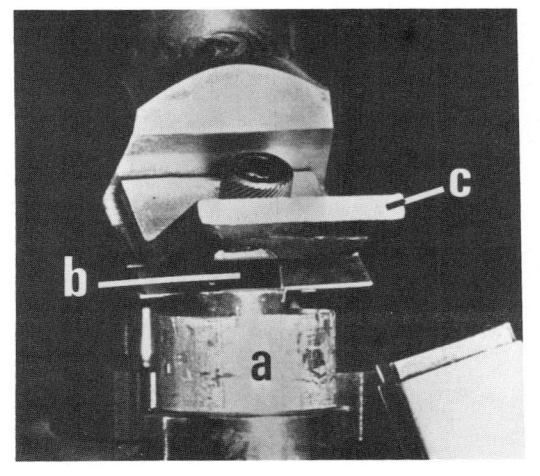

Fig. 2. The specimen in the freeze-etch unit; a: the ring on the cooled specimen table which fits the specimen holder; b: the specimen holder with the frozen fractured cylinder of cream; c: the cooled aluminium plate over the specimen.

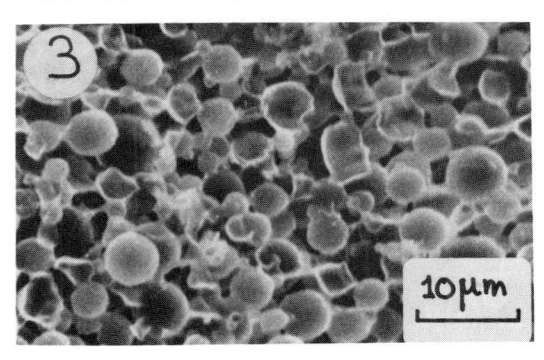

Fig. 3. Non-homogenized cream.

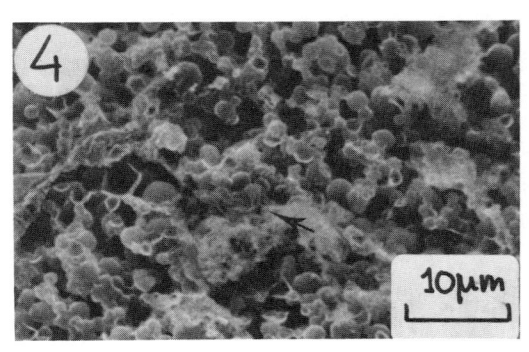

Fig. 4. Homogenized cream, arrow points to a homogenization cluster.

bubbles and the destabilization of fat globules which is responsible for the rigidity of the foam lamellae.

Concerning the accumulation of fat globules at the air-serum interface, Figure 22 suggests that some type of membrane, which envelops the air bubble, is penetrated by the fat globules. It is tempting to assume that this membrane is composed of protein adsorbed at the interface. This protein might be partly replaced by liquid fat in later stages of the whipping process. Such assumptions, however, require support from results of further electron microscopical studies at higher magnification.

Acknowledgment

The authors are much indebted to Mr. F. Thiel of the Technical and Physical Engineering Research Service, Wageningen, The Netherlands, for technical assistance and to Drs. A. Jongerius and D. Schoonderbeek, of the Netherlands Soil Survey Institute, Wageningen, The Netherlands, who made a Zeiss TGZ3 analyser available to us.

References

1. E. Graf & H.R. Müller: Fine structure and whippability of sterilized cream. Milchwissenschaft 20, 1965, 302-308.
2. K.G. Berger, B.K. Bullimore, G.W. White & W.B. Wright: The structure of ice cream. Dairy Industries 37, 1972, 419-425, 493-497.
3. W. Buchheim: Mikrostruktur von geschlagenem Rahm. Gordian 78 (6), 1978, 184-188.
4. D.G. Schmidt, S. Henstra & F. Thiel: Eine einfache Methode für die rasterelektronenmikroskopische Präparation von Käse bei tiefer Temperatur. Beitr. elektronenmikroskop. Direktabb. Oberfl. 10, 1977, 415-418.
5. D.G. Schmidt, S. Henstra & F. Thiel: A simple low-temperature technique for SEM of

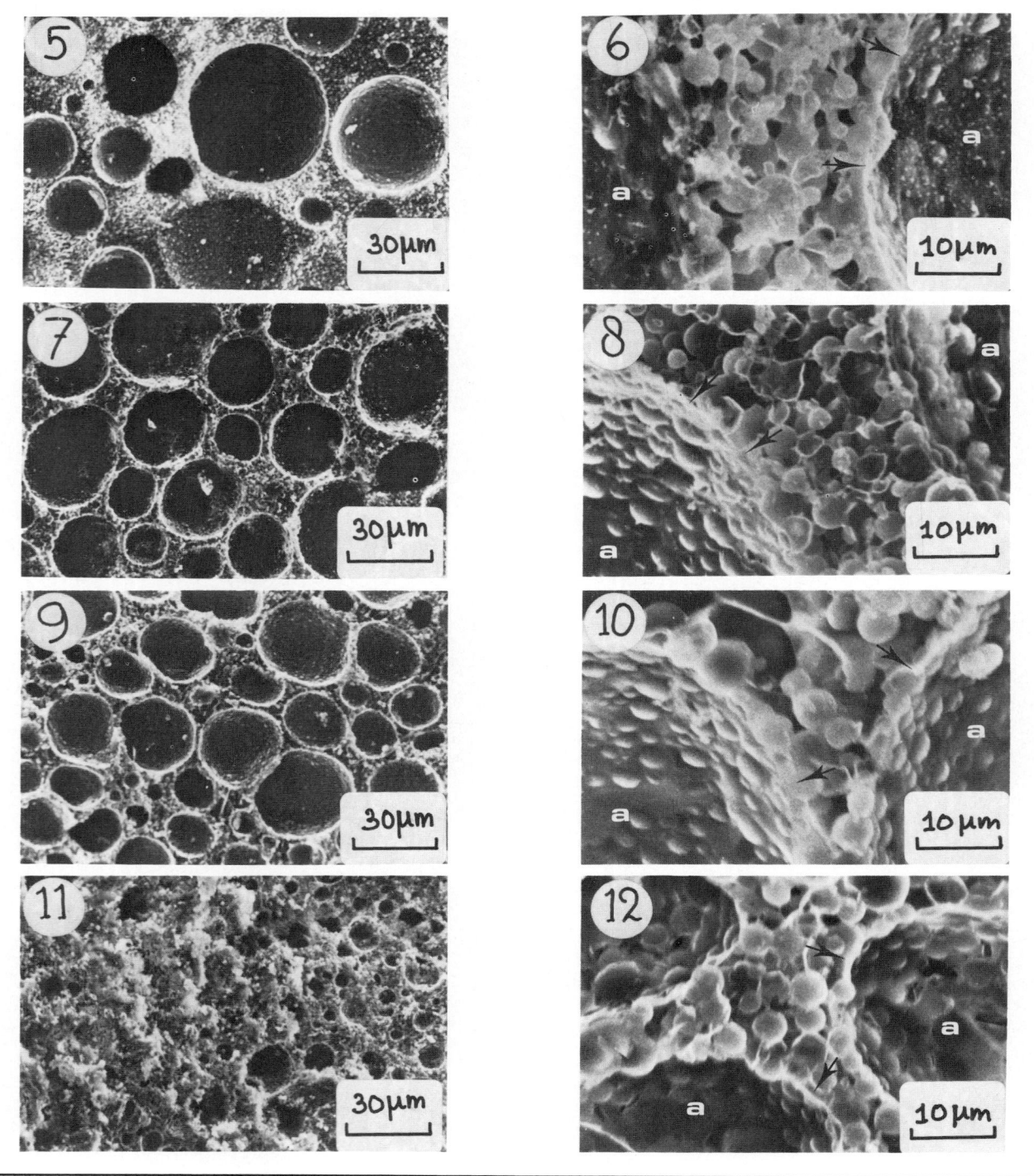

Figs. 5-12. Non-homogenized cream at various stages of whipping. Air bubbles are indicated by an a; the air-serum interface is indicated by an arrow. Figs. 5, 6: after 53s; Figs. 7, 8: after 105s; Figs. 9, 10: after 158s, at maximum rigidity of the foam; Figs. 11, 12: after 173s, overwhipped.

cheese. Mikroskopie 35, 1979, 50-55.
6. P.L. Goldsmith: The calculation of true particle size distributions from the sizes observed in a thin slice. Brit. J. Appl. Phys. 18, 1967, 813-830.
7. H. Mulder & P. Walstra: The Milk Fat Globule, Centre for Agricultural Publishing and Documentation, Wageningen, 1974, chapter 11.

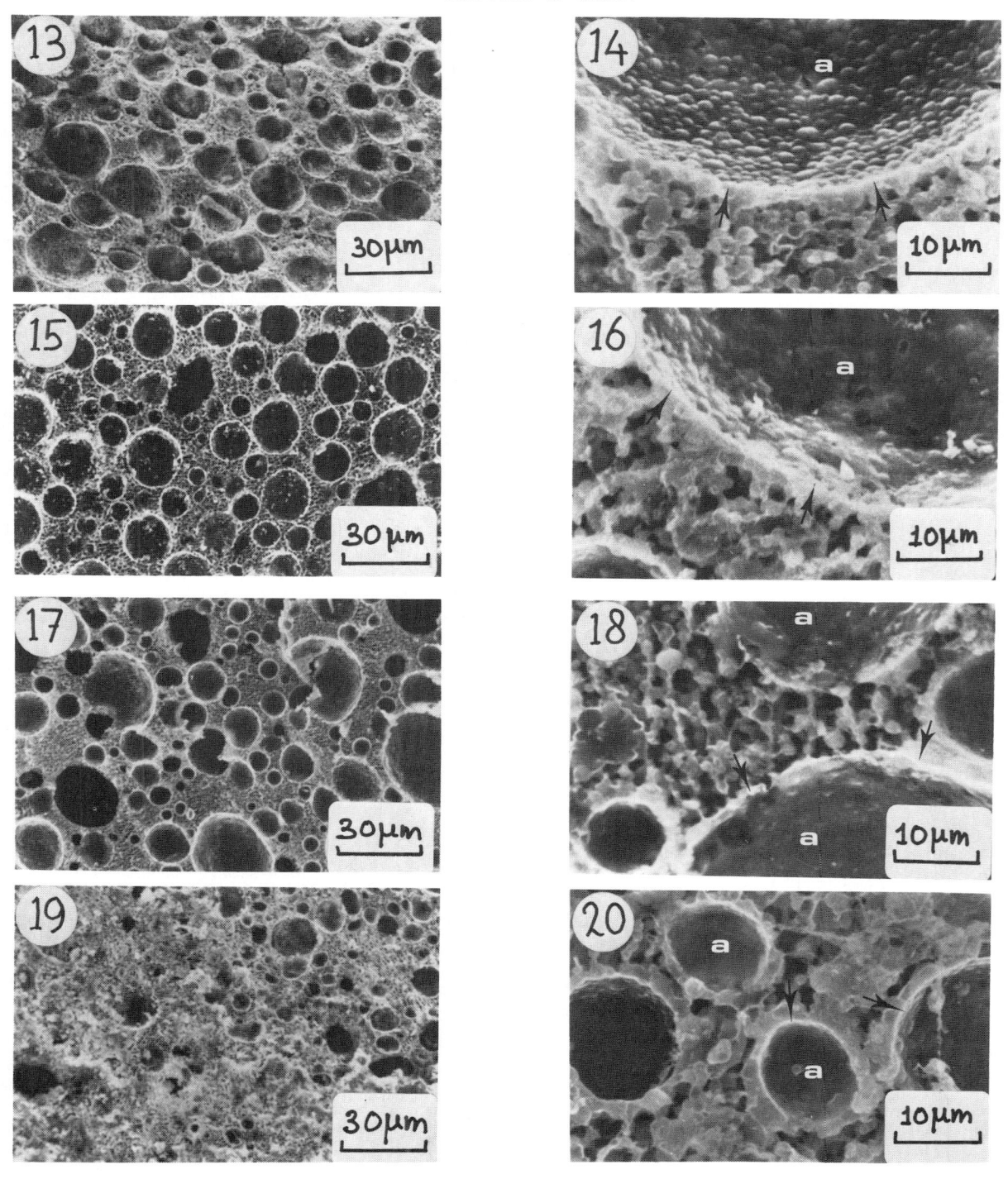

Figs. 13-20. Homogenized cream at various stages of whipping. Air bubbles are indicated by an a; the air-serum interface is indicated by an arrow. Figs. 13, 14: after 109s; Figs. 15, 16: after 219s; Figs. 17, 18: after 328s, at maximum rigidity of the foam; Figs. 19, 20: after 343s, overwhipped.

Discussion with Reviewers

D.N. Holcomb: Limitations of Graf and Müller's investigation on thin sectioned embedded samples were noted by the authors. Are there any specific examples where the information obtained by the present cryotechnique differs from the results of Graf and Müller?

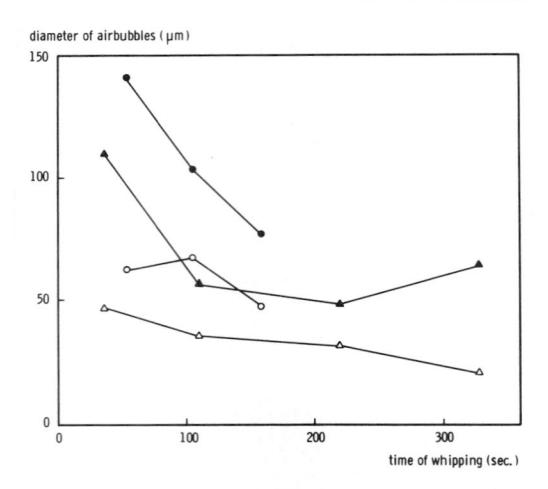

Fig. 21. Variation of number-average and weight-average diameters, $\bar{D}_n$ and $\bar{D}_w$, of the air bubbles in cream at various stages of whipping. Non-homogenized cream: $\bar{D}_n$ -o-o-, $\bar{D}_w$ -●-●-. Homogenized cream: $\bar{D}_n$ -△-△-, $\bar{D}_w$ -▲-▲-.

Fig. 22. The air-serum interface (arrows) of an air bubble (A) in non-homogenized cream whipped for 80s. Note the fat globules penetrating the membranous interface.

Authors: During the course of the embedding procedure fat is extracted, which may result in an unrealistic representation of the foam lamellae in the whipped product. Our technique shows that during whipping the fat globules are only partly damaged, resulting in bridge formation between individual globules (Fig. 22). Further, the thin-section technique fails to give any details concerning the structure of the air-serum interface, whereas our method shows that during whipping the fat globules penetrate through the air-serum interface (Fig. 22).

M. Ruegg: In the method of Goldsmith the "slice thickness" should be defined. The "slice thickness" corresponds to the thickness of the sample in which the particles to be counted are embedded. Usually, the calculated mean diameter increases with increasing "slice thickness". Which value has been used by the authors?
Authors: We took the "slice thickness" equal to the depth of etching. The latter, however, cannot be determined unambiguously, since removal of superficial frost and actual etching of the sample cannot be separated from each other. In Goldsmith's transformation the parameter v = slice thickness/width of size classes is the determining factor. We made the approximation v = 0. As will be mentioned in the answer to Kalab's first question, the maximum etching depth is 1.8μm, which in our case yields v = 0.144. This result is a size distribution of the air bubbles which is, within less than 1%, identical with that obtained with v = 0.

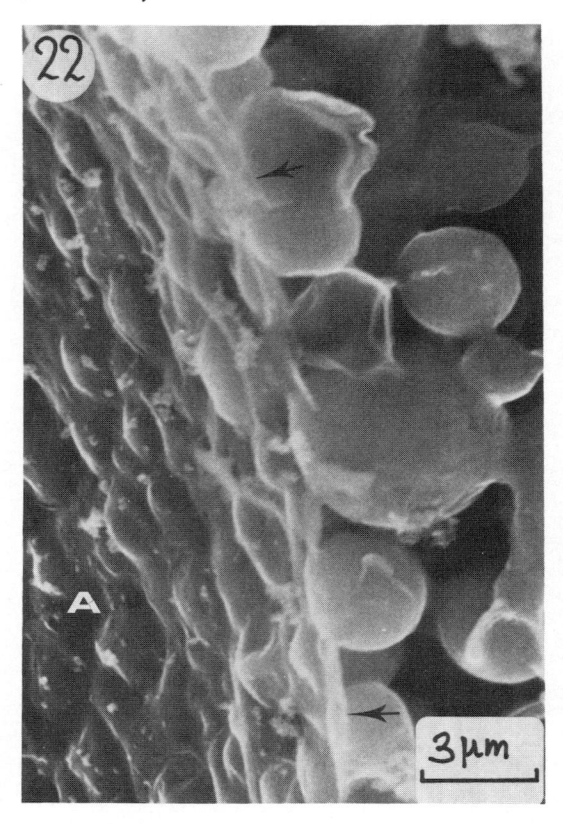

M. Kalab: Sublimation of frost from the fractured surface of the frozen cream specimen prior to carbon-coating removed an uncontrolled amount of ice from the specimen. Was it not possible to fracture the specimen in the freeze-etch unit and expose the fat globules in the cream by removing ice under controlled conditions? Was there no danger that the superficial layer of the fat globules might collapse if too much ice was removed from the specimen in spite of the support given to the cream column by the cylindrical wall of the conductive carbon cement?
Authors: By the technique we described it was not yet possible to fracture the large specimen required for the present investigation in the freeze-etch unit. We are aware of the hazards involved in our method and are therefore constructing an improved type of specimen holder with which large surfaces can be fractured inside the unit. We have not yet gained any experience with this device. We never observed that the superficial layer of fat globules collapsed during removal of the frost layer. Actually the sublimation time required never exceeded 15 min, which at -100°C results in a maximum etching depth of about 1.8μm, the equivalent of 1-2 fat globules. Any collapse of the structure seems therefore to be unlikely.

For additional discussion see page 210.

A COMPARISON OF THE MICROSTRUCTURE OF DRIED MILK PRODUCTS BY FREEZE-FRACTURING POWDER SUSPENSIONS IN NON-AQUEOUS MEDIA

W. Buchheim

Institut für Chemie und Physik
Bundesanstalt für Milchforschung
D-2300 Kiel (Germany)

Abstract

Freeze-fracturing of water-soluble powders for TEM is achieved by use of non-aqueous media such as polyethylene glycol for suspending the powder particles prior to cryofixation. Fine structures of the following dried milk products have been investigated: skim milk (low-and high-heat powders), whole milk, cream, sweet, acid, partly delactosed, and demineralized whey, whey protein concentrate, buttermilk from sweet and from cultured cream, rennet and acid casein, sodium caseinate, yoghurt and quarg. Distinct structural differences have been found between all samples except the low-and high-heat skim milk powders, the sweet and acid whey powders, the dried caseins and the caseinate. Protein in an aggregated or non-aggregated state is clearly detectable. Lipid particles have been clearly visible to sizes of less than 100 nm. It has been demonstrated that freeze-fracturing of dried milk products is well suited for a structural and compositional analysis of such powder particles and thus has to be considered as a complementary method of investigation to SEM.

KEY WORDS: Freeze-Fracture, Milk Powder Suspensions in Non-Aqueous Media, Cream powder, Whole Milk Powder, Skim Milk Powder, Whey Powder, Buttermilk Powder, Dried Casein, Yoghurt Powder, Quarg Powder.

Introduction

Drying of liquid milk products is nowadays a generally applied procedure in order to obtain powdered products which have good storage properties and which may easily be used in other foods. The microstructure of the powder particles of some dried milk products has been studied more frequently by scanning electron microscopy (1-12) than by transmission electron microscopy (13-15). Generally it can be stated that by SEM, studies of the surface morphology of powder particles can be performed in a unique manner thus allowing important conclusions to be drawn e.g. on the effects of composition and of processing parameters, especially those of drying and of storage conditions. By fragmenting single powder particles details on the internal morphology such as vacuoles, cracks and capillaries also become detectable by SEM. The identification of certain components of dried milk products by SEM seems more or less restricted to crystallized lactose (2, 8-11) and, to a very limited extend, also to lipid aggregates (1,5). An overview on the application of SEM to dried milk products has been given by M.Kalab (8).

In contrast to SEM, the application of TEM in principle offers better possibilities for visualizing certain components, particularly protein and lipid particles, either by studying specifically fixed and stained thin sections or freeze-fracture replicas of powder particles. As H.R.Müller(13)demonstrated, the casein micelles and fat globules become clearly detectable in thin sections of embedded powder particles which have been fixed and stained with osmium tetroxide vapour for 3 weeks.

For the application of the freeze-fracturing technique to powder particles of dried milk products it is necessary to first suspend the powder in a suitable water-free medium such as paraffin, glycerol or polyethylene glycol and then to cryofix this suspension (14,15). It has been demonstrated earlier that this preparatory approach results in a very distinct display of lipid and protein within cross-fractured powder particles (14,15). In addition, a surface-fractured powder particle may additionally reveal parts of its surface morphology.

As already stated by M.Kalab (12) compositional analysis of single powder particles (e.g. in powder mixtures of unknown composition) on the basis of electron micrographs may gain increasing importance since chemical detection is sometimes rather difficult.

In this paper a summary is given of the microstructure of 16 different milk products as it appears after freeze-fracturing. The main intention was to demonstrate how far this electron microscopical approach can be used for structural and compositional characterization of different products.

Materials and Methods

The following spray-dried milk products, all commercially available, have been studied: skim milk (low-heat and high-heat), whole milk, cream, whey (sweet and acid), partly delactosed whey, demineralized whey, whey protein concentrate, buttermilk (from sweet and from cultured cream), casein (rennet and acid caseins), sodium caseinate, yoghurt (from skim milk) and quarg (from skim milk). It should be mentioned that the exact processing conditions of all these samples could not be ascertained.

Polyethylene glycol (mol.weight 400) was used as the suspension medium for all powders studied. Individual powders were added to 1-2 ml of polyethylene glycol with stirring in amounts producing highly viscous yet flowing suspensions. Small amounts (1-2 μl) of this suspension were transferred onto normal freeze-fracture specimen holders (Balzers). The specimens were then cryofixed by immersion into melting Freon 22 (-160°C) and stored under liquid nitrogen. It should be mentioned that lower freezing rates may be applied without producing freezing artifacts within the powder particles.

A Balzers BA 360M unit equipped with a 4-specimen table was used for freeze-fracturing. The temperature of the specimen table was adjusted to -110°C. This temperature is far less critical than with water-containing samples since the powder suspensions represent water-free systems which do not allow any freeze-etching. It should perhaps not exceed -80°C in order to avoid melting artifacts of lipids during shadowing.

The specimens were fractured 5 - 10 times using only the coarse mechanical advance of the freeze-microtome. Thereafter the specimens were immediately shadowed with a 2 nm layer of platinum/carbon under an angle of 45 degrees using an electron gun. This layer was reinforced by pure carbon (15 - 20 nm). The thickness of these layers was controlled by a quartz crystal thin film monitor (Balzers). After opening the vacuum chamber the specimens were carefully immersed in pure polyethylene glycol to float off the replica film and then stepwise transferred onto pure water.

The cleaning procedure depended on the composition of the powders under study. Lipid-free systems were cleaned with a sodium hypochlorite solution; acetone was used to dissolve fat. The replicas were picked up on uncoated copper grids (200 mesh) which had been previously dipped into a 4% Bedacryl 122X (Imperial Chemical Industries, England) xylene/benzene (50:50) solution in order to produce better adhesiveness.

Micrographs were taken on a Siemens Elmiskop I electron microscope operated at 80 kV.

Results and Discussion

The final replicas exhibit areas corresponding to a cleave fracture of the specimen as well as areas where mainly marks from the microtome knife are visible. The freeze-fractured polyethylene glycol shows a more or less amorphous fine structure and therefore allows a clear differentiation from surface-fractured or cross-fractured powder particles. Only in cross-fractured particles becomes the fine structure of the powder matrix, i.e. the type and distribution of certain components detectable. Therefore the structural characteristics of the different dried milk products were based strictly on the 'internal' structure of the powder particles. Since cross-fractured powder particles often show marks from the microtome knife at their periphery it seems as if only those powder particles are cross-fractured which are directly cleaved by the knife during the last fracturing step.

A skim milk powder (Fig. 1 and 2) is easily distinguishable from other dried milk products by the typical fine structure and uniform distribution of the casein particles within the amorphous-looking continuous phase of lactose and salts. Although the packing density of the casein particles is quite high one can still recognize their typical micellar appearance and composition of subunits similar to the original micelles in milk (16). A low-heat skim milk powder (Fig.1) seems indistinguishable from a high-heat skim milk powder (Fig.2). This is not surprising since the only difference is that in the latter a higher portion of whey proteins is associated with casein as a result of heat denaturation. Only occasionally can fat globules be detected. The small cavities within fat globules (Fig.1) are discussed later.

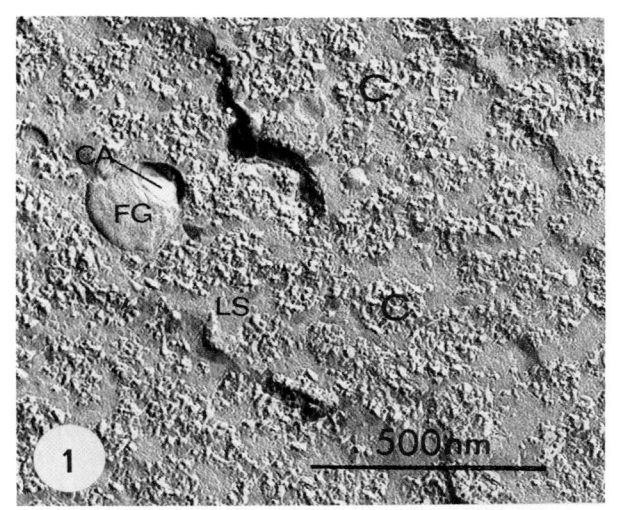

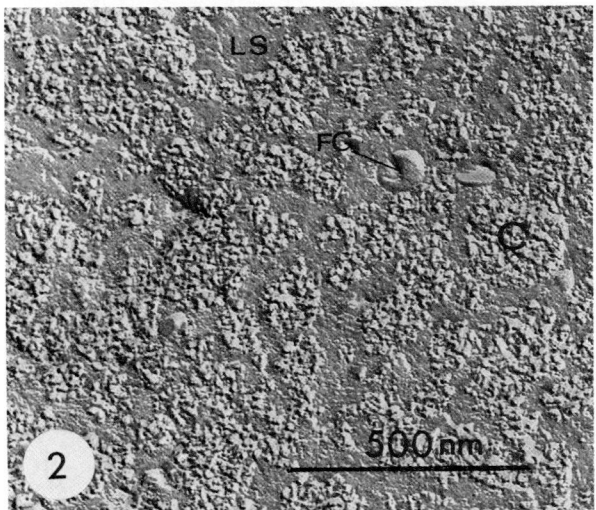

Fig. 1,2. Skim milk powders(1:low-heat; 2:high-heat). C: casein micelles; LS: lactose/salt matrix, FG: fat globule;CA: cavity in a fat globule. Please note:Fig. 5 - 18 are of the same final magnification as Fig.1.,2.

The microstructure of a whole milk powder is characterized by a high concentration of fat globules which mostly deviate slightly from the ideal globular shape and which when not centrally cross-fractured show the typical layering of crystallized fat (17). Figure 3 represents normal spray-dried whole milk powder with a homogenized fat phase. Most fat globules have a diameter of less than 0.5 μm. It would be possible to determine the degree of homogenization by the size distribution of the fat globules. As described earlier many fat globules show a central or peripheral cavity within the fat phase (14). Since such cavities are more or less absent in freeze-dried systems (14) and do never occur in freeze-fractured liquid milks, it is highly probable that this special structure is related to the drying step itself and not to the cryofixation or the freeze-fracturing.

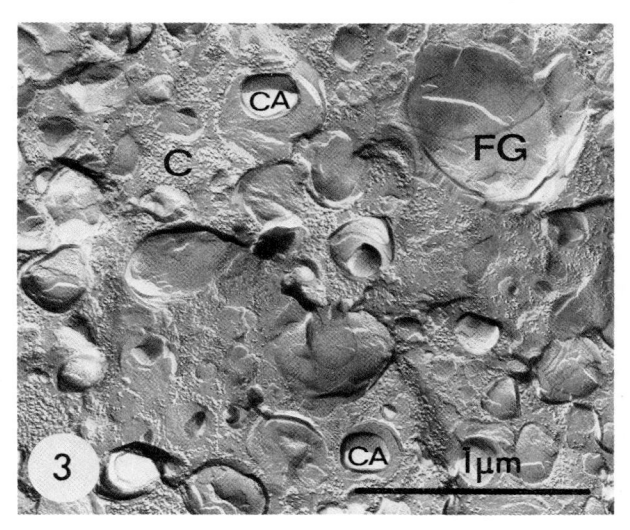

Fig. 3. Whole milk powder. FG: fat globule; CA: cavity in a fat globule; C: casein.

A cream powder has a similar appearance to whole milk powder except that the fat phase occupies a higher proportion of the micrographs. Figure 4 shows such a powder containing the fat phase of an unhomogenized or only slightly homogenized state. Many fat globules have the peculiar cavities described earlier.

The microstructure of whey powder particles, whether from sweet whey (Fig.5) or from acid whey (Fig.6) may be characterized by a non-uniform distribution of small protein particles, whose diameter is between 5 and 10 nm and are present in a predominantly non-aggregated or only loosely aggregated state. These loose protein aggregates have diameters up to 0.5 μm and are variable in shape. The overall appearance of these

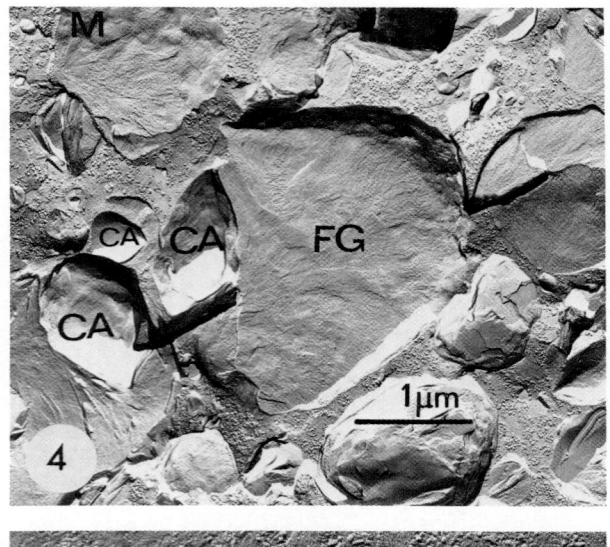

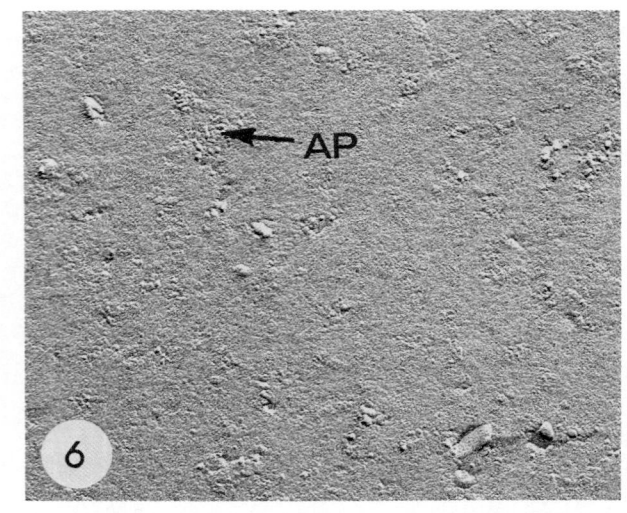

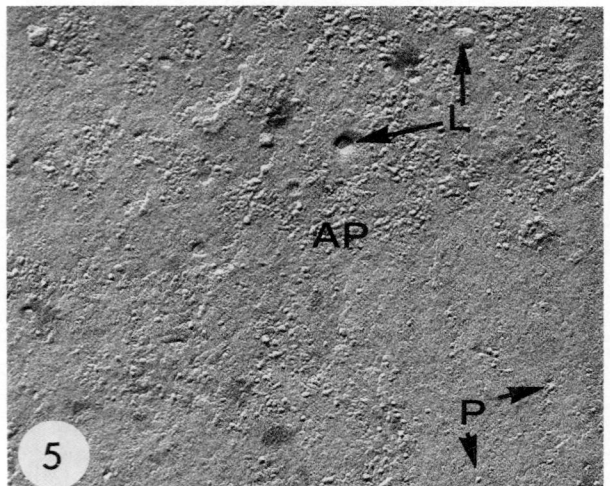

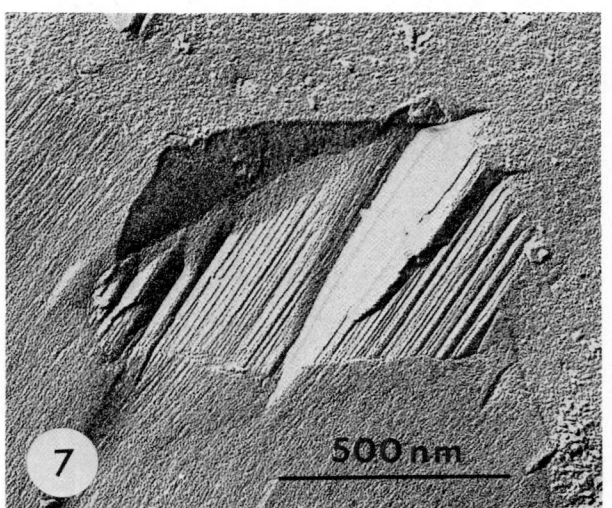

Fig. 4. Cream powder. FG: fat globule; CA: cavity in a fat globule; M: internal view of a fat globule membrane.

Fig. 5,6. Whey powders (5: sweet whey, 6: acid whey). AP: loosely aggregated protein; P: individual protein particles L: lipid particles.

Fig. 7. Crystallized lactose within a whey powder particle.

Fig. 5-7: Same magnification as Fig. 7.

whey protein particles or aggregates is distinctly finer than that of casein aggregates which permits differentiation between both types of proteins. If lactose is present in the amorphous state it shows a smooth fine structure (Fig.5, 6), but, if it is partly crystallized, the corresponding areas are easily detectable according to a regular crystalline layered structure (Fig.7). Small lipid particles, often only 100 nm or less in diameter, are found occasionally. Significant structural differences between a powder made from sweet whey (Fig.5) and from acid whey (Fig.6) could not be found.

In particles of partly delactosed whey powder (Fig.8) similar structural details prevail as in normal whey powder but it can be recognized that the concentration of the protein is correspondingly higher as is to be expected.

In contrast demineralized whey powder was found to differ markedly from the whey preparations described above. Figures 9 and 10 demonstrate that the protein is present in a very condensed state of either small or large and network-like aggregates, whilst the lactose phase appears almost void of small protein particles. This strongly aggregated state of the whey proteins resembles

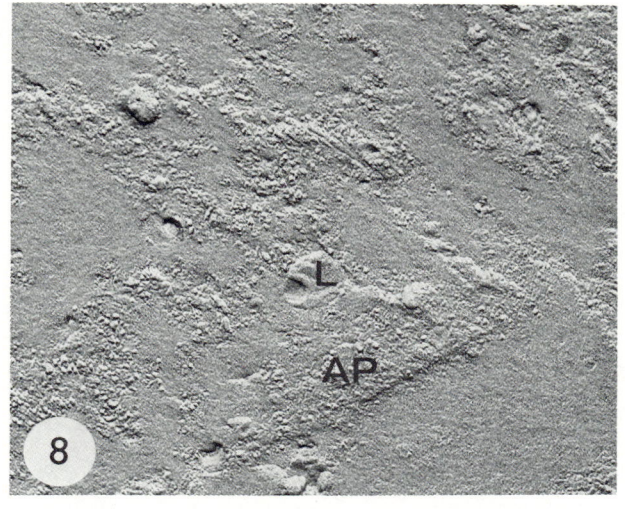

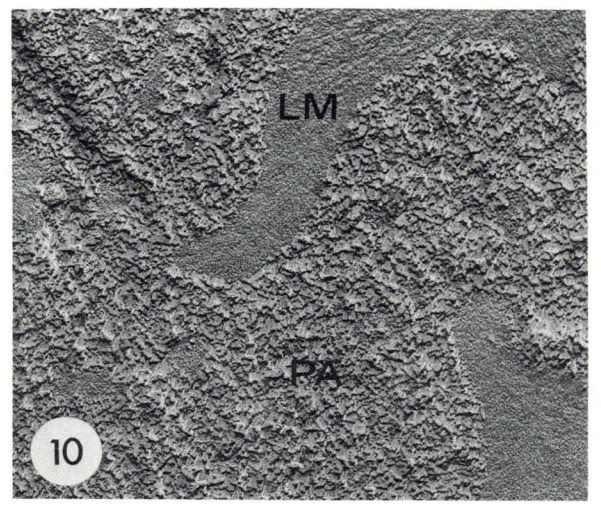

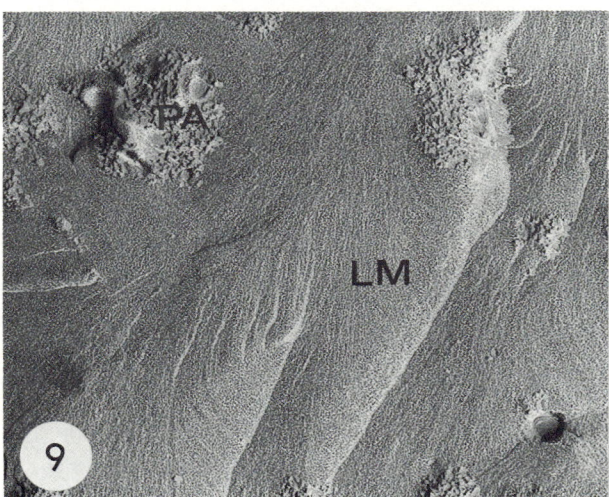

Fig. 8. Partly delactosed whey powder. AP: loosely aggregated protein; L: lipid particles.

Fig. 9,10. Demineralized whey powder. Both figures represent areas from the same powder particle and demonstrate the variation in protein aggregation. PA: protein aggregates; LM: lactose matrix.

Fig. 11. Whey protein concentrate. The visible fine structure corresponds to a dense packing of largely undenatured whey protein particles. Compare e.g. with Figures 5,6,8.

Figs. have same magnification as Fig. 11.

---

that of heat-denatured whey proteins which has been described elsewhere (18).

Figure 11 shows a whey protein concentrate which exhibits a dense packing of small protein particles of the type found in normal whey powders or in the partly delactosed whey powder. Since after dissolving of this powder in water only very small and loosely structured protein aggregates could be found the degree of heat denaturation of the whey proteins of this sample has been relatively low.

Particles of buttermilk powder also show a very characteristic fine structure, thus they can be differentiated from skim milk or whey powders. Figure 12 is representative of a freeze-fractured buttermilk powder sample made from sweet cream, whereas Figure 13 is representative of the structure of a buttermilk powder made from cultured cream. The common structural feature of both types of buttermilk powder is the high content of smaller (mostly less than 100 nm) lipid particles which are mainly globular in shape. This component of buttermilk powders may be the result of

Fig. 12,13. Buttermilk powders (12: from sweet cream; 13: from cultured cream). Note the relatively high concentration of small lipid particles (L) and the different state of aggregation of the casein (C).

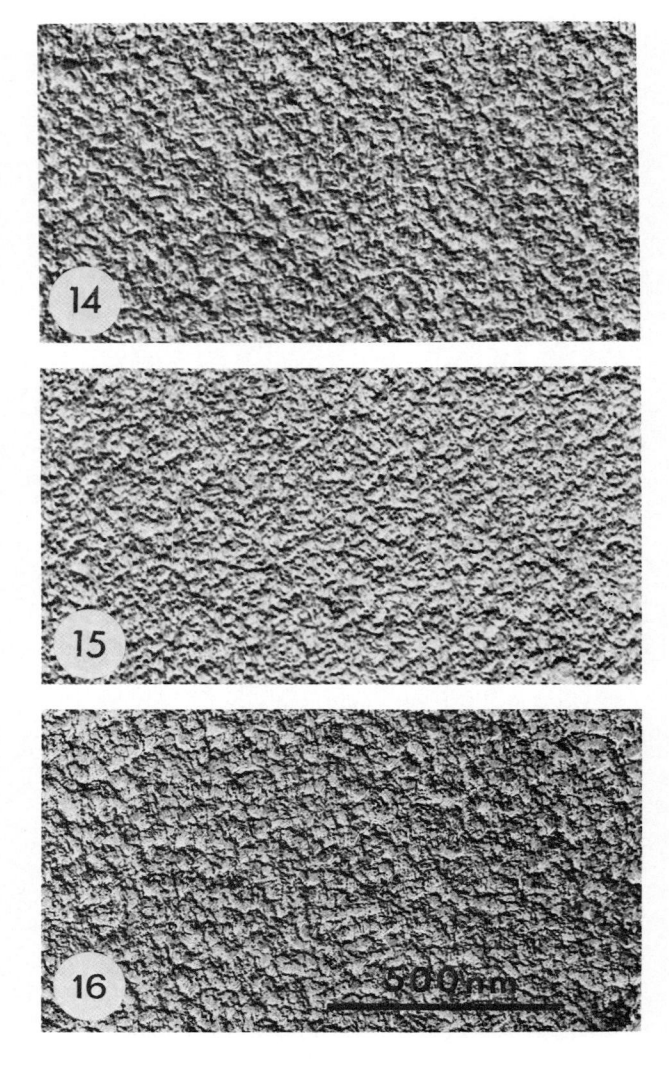

Fig. 14. Rennet casein.
Fig. 15. Acid casein.
Fig. 16. Sodium caseinate.
Note the coarser fine structure in comparison with whey proteins (Fig.11).
Figs. have same magnification as Fig. 16.

decomposed membranous material as suggested previously by M.Kalab (12). The aggregation state of casein in buttermilk made from sweet cream differs of course from that made from cultured cream. Whereas in the latter the casein is highly aggregated due to the lower pH-value, free casein micelles or only slightly aggregated casein can be seen in buttermilk from sweet cream.

Structural comparisons of dried rennet and acid caseins with sodium caseinate result in nearly identical structures for all 3 samples (Figs.14,15,16). Nevertheless these structures are clearly distinguishable from concentrated whey proteins (Fig.11) because of their distinctly coarser appearance.

Finally, the fine structure of 2 dried cultured milk products, a yoghurt powder (Fig.17) and a quarg powder (Fig.18) are compared. Both products were made from skim milk and thus contained only negligible amounts of fat. Figure 17 shows the typical network-like aggregation of the casein in a yoghurt gel (8) which is rather uniformly distributed within the powder particle. Within the lactose/salt

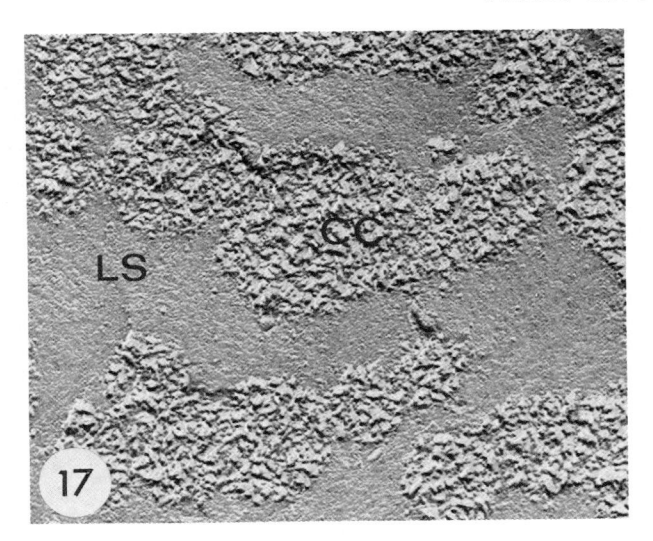

Fig. 17. Yoghurt powder (made from skim milk). CC: coagulated casein; LS: lactose/salt matrix.

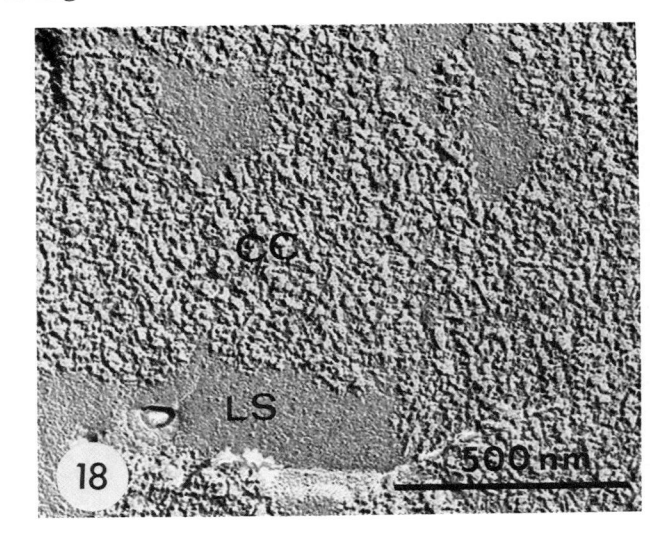

Fig. 18. Quarg powder (made from skim milk). CC,LS: see Fig.17.
Figs. have same magnification as Fig.18.

phase very small single protein particles can be seen which may represent whey proteins. In a dried skim milk quarg a coarser casein network which represents a higher proportion of the total area (volume) can be observed (Fig. 18) which is due to the higher casein content of this product.

Conclusions

The results described here demonstrate that freeze-fracturing of powders suspended in suitable non-aqueous media, which do not affect the integrity of the powder particles, represents a generally applicable method to study the microstructure of water-soluble powders by TEM at high resolution. For a complete structural characterization of powdered food systems a combined application of this method and of SEM methods appears most appropriate.

We have applied the freeze-fracture method to study certain milk powder samples which had been suspected to be illegal mixtures of different powders. In one case we were able to show that a so-called 'whole milk' powder consisted actually of a mixture of skim milk powder and cream powder. In another case a 'skim milk' powder was found to be a mixture of whey powder and caseinate. Thereby it could be demonstrated that electron microscopy may be useful in powder analysis.

The detailed display of different constituents of the powder particles will allow studies to be made on more general relationships between the structure of the powder, its properties and conditions during manufacturing, storage and further use in other foods.

Acknowledgment

I would like to acknowledge the numerous helpful comments of the reviewers of this paper. Furthermore I am grateful to Mr. E.F.Bond, Electron Microscope Centre, Agriculture Canada in Ottawa for linguistic assistance.

References

1. Buma,T.J.,Henstra,S.: Particle structure of spray-dried milk products as observed by a scanning electron microscope. Neth.Milk Dairy J. 25,1971,75-80.
2. Buma,T.J.,Henstra,S.: Particle structure of spray-dried caseinate and spray-dried lactose as observed by a scanning electron microscope. Neth.Milk Dairy J. 25,1971,278-281.
3. Verhey,J.G.P.: Vacuole formation in spray powder particles. 2. Location and prevention of air incorporation. Neth. Milk Dairy J. 26,1972,2o3-224.
4. Kalab,M.,Emmons,D.B.: Milk gel structure. III. Microstructure of skim milk powder and gels as related to the drying procedure. Milchwissenschaft 29,1974,585-589.
5. Gejl-Hansen,F.,Flink,J.M.: Application of microscopic techniques to the description of the structure of dehydrated food systems. J.Food Sci. 41, 1976,483-489.
6. Buma,T.J.: Teilchenporosität von sprühgetrockneter Milch.

Milchwissenschaft 33,1978,538-540.
7. Short,J.L.,Cooper,H.R.,Doughty,R.K.: The effect of manufacturing variables on lactalbumin for use in high protein biscuits. New Zealand J.Dairy Sci.Technol. 13,1978,43-48.
8. Kalab,M.: Scanning electron microscopy of dairy products. An overview. Scanning Electron Microsc.1979;III:261-272.
9. Jelen,P.,Kalab,M.,Greig,R.I.W.: Water-holding capacity and microstructure of heat-coagulated whey protein powders. Milchwissenschaft 34,1979,351-356.
10. Roetman,K.: Crystalline lactose and the structure of spray-dried milk products as observed by scanning electron microscopy. Neth.Milk Dairy J. 33,1979, 1-11.
11. Saltmarch,M.,Labuza,T.P.: SEM investigation of the effect of lactose crystallization on the storage properties of spray dried whey. Scanning Electron Microsc. 1980; III:659-665.
12. Kalab,M.: Possibilities of an electron-microscopic detection of buttermilk made from sweet cream in adulterated skim milk. Scanning Electron Microsc.1980; III: 645 -652.
13. Müller,H.R.: Elektronenmikroskopische Untersuchungen an Milch und Milchprodukten. 1. Strukturaufklärung in Milchpulvern. Milchwissenschaft 19,1964, 345-356.
14. Buchheim,W.: Elektronenmikroskopische Präparationsmethode zur Darstellung von Oberflächen-und Innenstruktur wasserlöslicher Pulverteilchen. Kieler Milchwirtschaftliche Forsch.Ber.24,1972, 97-107.
15. Buchheim,W.: The applicability of electron microscopy for studying the structure of liquid and solid foods. In: Proc. IV. Int.Congress Food Sci. Technol. E.P.Marco(ed.), Instituto Nacional de Ciencia y Tecnologia de Alimentos, Madrid,Spain, 1974,Vol.2, 5-12.
16. Schmidt,D.G., Buchheim,W.: Elektronenmikroskopische Untersuchung der Feinstruktur von Caseinmicellen in Kuhmilch. Milchwissenschaft 25,1970,596-600.
17. Buchheim,W.,Precht,D.: Elektronenmikroskopische Untersuchung der Kristallisationsvorgänge in den Fettkügelchen während der Rahmreifung. Milchwissenschaft 34,1979,657-662.
18. Buchheim,W.,Jelen,P.: Microstructure of heat-coagulated whey protein curd. Milchwissenschaft 31,1976,589-592.

Discussion with Reviewers

T.J. Buma: What are the advantages of using electron microscopy for detecting adulteration in milk powders over optical microscopy?

Author: Due to the high resolution of structural details on freeze-fracture replicas,when studied by TEM,just those components of the powders (e.g. protein aggregates and smaller lipid particles) can be seen which are well beyond the resolution limit of a light microscope. I would assume that in general electron microscopy will provide much more details on the composition of individual powders which will facilitate their identification.

D.Holcomb: The author notes that freezing rates used did not seem to be critical. What was the moisture content of the samples for which this was observed? Would the author expect to find freezing damages in higher moisture content powders?

Author: We studied the effect of low freezing rates (ca.10°C per minute) with a normal spray-dried whole milk powder, having a moisture content of 3%, and viscous paraffin as embedding medium. According to my experience I would not expect to find freezing damages in higher moisture content powders because this moisture should represent more or less bound water and thus should not be available for ice crystal formation.

D.Holcomb: It was noted that lipid melting can occur during shadowing above -80°C. Since most milk fat melts in the 0°C to 40°C range, does this imply that local heating effects causing temperature rises of 80°C to 120°C can occur during shadowing? Perhaps we should be more concerned about shadowing dried specimens at ambient temperature?

Author: The recommendation normally not to shadow at higher temperatures than ca. -80°C is based on the experience that the quality of the deposited platinum/carbon layer can deteriorate especially in areas of fractured fat aggregates. Of course, this effect will depend of the type of evaporation source used and its distance from the specimen as well as on the type of fat and the degree of final resolution required. Our experience is based on the use of an electron gun for shadowing freeze-fractured specimens.

M.Kalab: The non-aqueous media, in which milk powders are suspended for freeze-fracturing,evidently must not affect any of the powder components,i.e. lactose, protein and fat. What were the criteria for the selection of such media? Were liquids other than polyethylene glycol tested?

Author: The main criteria for the selection of non-aqueous media for milk powders have been (I) sufficiently high viscosity for easier handling of the suspensions,(II) good wetting properties for the powders under study,(III)favourable freezing and fracturing properties, i.e. amorphous solidification and largely structureless appearance on the replicas,(IV)favourable properties for the cleaning procedure. Apart from polyethylene glycol 400 we have used viscous paraffin and pure glycerol. Of course, paraffin will dissolve fat, especially readily accessible surface fat, but other negative effects could not be observed. It should be emphasized that glycerol and polyethylene glycol should be water-free, otherwise local dissolving of powder particles can occur. This effect would be recognizable e.g. if single protein or lipid particles are found in the phase of the embedding medium.

M.Kalab: The note on a successful detection of powders composed of artificial mixtures of skim milk powder + cream powder (=whole milk powder) and whey powder + caseinate (=skim milk powder) implies that a number of individual powder particles was examined in each case. Were such particles distributed in the suspensions uniformly or did you encounter clusters? Would it be possible to show micrographs of entire fractured particles at a lower magnification? Was the composition at the particle surface similar to that in the centre? Were the particle surfaces porous or compact? What were the images of the "more or less amorphous" freeze-fractured medium?

Author: Since the size of the powder particles is of the same order as the openings of the grids and moreover the replicas tend to disintegrate into smaller pieces during the cleaning procedure, it is difficult to obtain a reliable overview on the distribution of powder particles in the suspension. A pronounced tendency to form clusters has not been observed. Occasionally entire fractured particles can be found as shown in Figure 19. Already on this micrograph, but also at higher magnification (Fig.20) the smooth appearance of the freeze-fractured embedding medium is striking. In most products the composition within the powder particles showed no pronounced tendency to vary from the particle surface to the centre. Only in some products (e.g. dried wheys or buttermilks) a somewhat non-uniform distribution of powder components within one particle could be observed. It is assumed that such an uneven distribution preferentially occurs in systems which tend to form bigger aggregates (e.g. coagulated casein in buttermilk from cultured cream) either in the original liquid state or perhaps during the concentration process prior to drying. Surface-fractured powder particles generally had a rather compact and smooth structure except those of dried whole milk and cream where layers of surface fat are clearly visible. Small pores or fissures in the particle surface could not be observed. It appears possible that the embedding medium penetrates into such capillaries, if they really exist, thereby making them undetectable after fracturing.

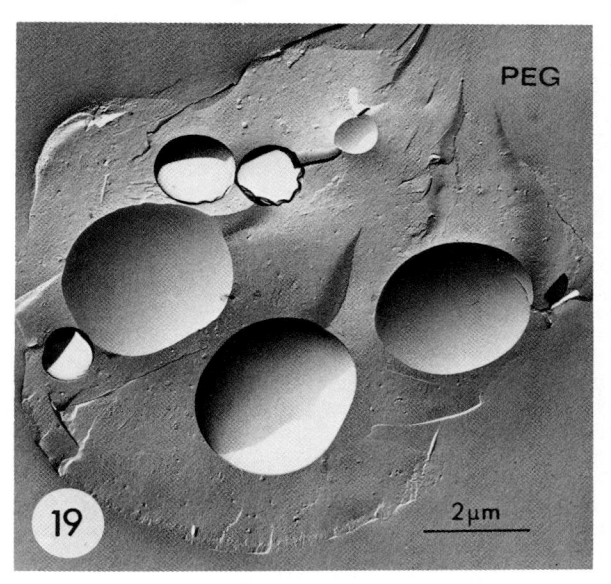

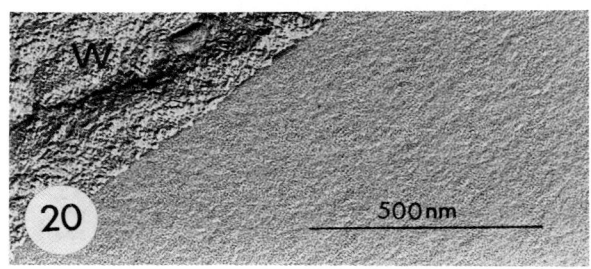

Fig. 19. Entire cross-fractured whey powder particle, showing 7 vacuoles. PEG: polyethylene glycol.

Fig. 20. This micrograph shows the amorphous structure of polyethylene glycol at higher magnification. W: part of a whey protein concentrate particle.

D.G.Schmidt: Does the author have an explanation for his observation that in demineralized whey powder much larger aggregates occur than in normal whey powder?

Author: Unfortunately an exact explanation cannot be given because we were not able to obtain the exact processing conditions of that powder from the manufacturer. I assume that the reason might be a strong reduction of the salt concentration in the whey, perhaps to less than 10% of its original value. It is known that under such conditions globular proteins tend to form bigger aggregates. Moreover it cannot be excluded that also some heat denaturation, which also leads to aggregates of the type observed, occurred during the manufacturing of this powder. In any case it will be necessary to consider all details of the manufacturing of an individual powder in relation to its final structure.

R.L.Steere: The author has stated that the cavities occurring in fat globules of spray-dried powders (Fig.3,4) are probably a result of the drying step itself and no result of the freeze-fracturing preparation. In freeze-etching, such cavities are generally the result of a portion of the fat globules fracturing away from the remainder. If there are any real cavities in his material, careful analysis of complementary replicas would be needed to demonstrate that both fracture faces have such cavities.
Author: The cavities observed in fat globules of dried milk products, especially of spray-dried samples, have such an appearance that it can be virtually excluded that they have been produced by the freeze-fracturing process. This interpretation is further supported by the observations that complementary protuberances on other fat globules are absent, that milk fat globules in freeze-dried products hardly show such cavities and that they never occur in freeze-fractured fat globules of liquid milk products. An explanation for this phenomenon might be that during storage of freshly formed powder particles at room temperature or below, the solidification (crystallization) of the different fat fractions within the small fat globules **takes place with considerable delay** due to the well-known supercooling properties whereas the lactose/salt matrix is already solid. Since there are differences between the density values and the dilation coefficients of crystallized and liquid fat, and of the different crystal modifications (α, β, β'), internal tensions will occur which could finally result in the cavities observed. Nevertheless it should be tried to study complementary freeze-fracture replicas in order to confirm the actual existence of those cavities.

R.L.Steere: How does the author explain the similarity between the 'protein aggregates'(PA) on Fig. 9 and 10 (demineralized whey powder) and the 'coagulated casein'(CC) on Figs.17 and 18?
Author: The fine structure of casein micelles and of coagulated casein, as it appears after freeze-fracturing, is more or less identical with that of whey protein aggregates produced by heat-denaturation. This has been described earlier when studying the original aqueous systems (see Ref.18). Since the demineralized whey powder consists mainly of whey protein and lactose, the big aggregates observed have to be considered as whey protein aggregates. As already mentioned in the answer to D.G.Schmidt's question, this strong aggregation of the whey proteins could be the result of a high degree of demineralization or possibly also some additional thermal denaturation.

SEM INVESTIGATION OF THE EFFECT OF LACTOSE CRYSTALLIZATION ON THE STORAGE PROPERTIES OF SPRAY DRIED WHEY

M. Saltmarch[*] and T.P. Labuza[**]

[*] Kelco, Div. of Merck & Co., Inc.
8355 Aero Drive
San Diego, CA 92123
[**] University of Minnesota, Dept. of Food Science and Nutrition
St. Paul, MN 55108

Abstract

The purpose of the present study was to evaluate by means of the scanning electron microscope the effects of water activity and temperature on the transition of lactose from amorphous to crystalline state. Results indicated that lactose crystallized at 0.40 water activity after one week storage at 25°C, at 0.33 water activity after one week at 35°C and at 0.33 water activity after one week at 45°C. The crystallization of lactose in the hygroscopic whey powder appeared to be related to the water activity related maximum for browning via the Maillard reaction in the hygroscopic whey powder.

KEY WORDS: Scanning Electron Microscopy, Whey Powder, Lactose Crystallization, Water Activity, Maillard Reaction, Non-Enzymatic Browning, Shelf Life, Kinetics, Amorphous-Crystalline Transfer

Introduction

Whey powders have received increased attention during the last ten years as inexpensive and nutritious alternative ingredients in many food products. With increased use, there has arisen a practical need to control and predict the shelf life of both whey powders and food products into which they are incorporated as a principle ingredient. Non-enzymatic browning via the Maillard reaction is one important mode of deterioration in dried milk and whey powders which limits shelf life. Relatively high concentrations of lactose and protein high in lysine are present in milk and whey powders. In the presence of moisture these components readily participate in the Maillard reaction. This results in decreases in protein quality which are accompanied or followed by undesirable color and functionality changes.

Both temperature and water content during storage have an effect on the Maillard reaction. In milk and whey powders, temperature and water content may also have an effect on the Maillard reaction by their influence on the physico-chemical state of lactose. A number of workers have evaluated the influence of water activity and temperature on the physico-chemical state of lactose in dried milk products by means of isotherms and various microscopic methods. Warburton and Pixton (1978) used the scanning electron microscope to relate the transistion of lactose from amorphous to crystalline form to the isotherm behavior of spray dried skim milk powder stored at 25°C. Roetman (1979) also utilized the scanning electron microscope to examine the changes in skim milk and whey powders brought about by the crystallization of lactose both before and after drying.

However, only Huss (1970, 1974 a,b) has associated the transition of lactose from the amorphous to the crystalline state with increased protein quality losses via the Maillard reaction in dried milk products.

In the present study, moisture sorption isotherms and scanning electron micrographs were utilized to evaluate the effect of temperature and water activity on the physico-chemical state of lactose in a hygroscopic spray dried sweet whey powder. This information was then related to the rate of brown pigment formation in the whey powder stored under similar conditions.

Materials & Methods

Whey Powder

The hygroscopic sweet whey powder utilized in the present study was a freshly spray dried fifty pound sample obtained from Mid-America Farms, Springfield, Missouri. The whey powder contained 66% lactose of which approximately 95% was in the amorphous state and 5% in the crystalline state (Mid-America Farms, personal communication).

Moisture Sorption Isotherms

Triplicate one to two gram samples of whey powder were placed into desiccators containing saturated salt solutions at water activities ranging from 0.11 to 0.85. The desiccators were then evacuated for the equilibration period. Adsorption isotherms were obtained at 25, 35 and 45°C. Total equilibration time for the 25°C isotherm samples was six weeks; for the 35°C isotherm, three weeks; and for the 45°C isotherm, one week. The equilibration periods were ended when brown pigments began to appear in the samples. Isotherm points were determined over the water activity range by weighing the samples initially and then periodically throughout the equilibration period.

Scanning Electron Micrographs

Samples of the whey powder were prepared for viewing in the scanning electron microscope by initially placing whey samples, at approximately 0.10 water activity, on 12 mm diameter aluminum specimen stubs coated with silver conductive paint (Tousimis No. 8010, Tousimis Research Corp.). The stubs with freshly applied whey powder were rapped at an anly on a table surface to remove excess powder leaving approximately 0.001 mg of powder on the stub surface. The stubs were then mounted in corks and placed in open screw cap jars.

The open jars were placed in dessiccators at one atmosphere pressure over salt solutions ranging from 0.11 to 0.85 water activities and equilibrated one week at 25, 35 or 45°C. The samples on the stubs were coated to a thickness of several hundred angstroms with a 60:40 mixture of gold and palladium in a Kinney vacuum evaporator Model KSE-2A-M. Duplicate samples were prepared for each treatment condition.

The coated samples were viewed in a Philips Model 50 Scanning Transmission Electron Microscope at 12 kV accelerating voltage in the secondary electron mode. The samples were systematically viewed at 80 to 1250X magnifications. Micrographs were selected from a total of five hundred representative micrographs.

Because of charging problems associated with the samples, great care had to be exercised in sample preparation. The thickness of the powder on the specimen stubs had to be kept at a minimum and the contrast in the micrographs was minimized. In addition, the samples could not be readily viewed at magnifications above 2000X as beam damage to the sample occurred almost immediately.

Brown Pigment Formation

Rates of brown pigment formation were measured by a three enzyme modification of the Choi et al. (1949) method for browning in milk powders. Samples of whey powder were stored at water activities of 0.33, 0.44 and 0.65 at 25, 35 and 45°C temperature conditions. In addition, the influence of water activity on the brown pigment formation maximum at 45°C was evaluated in whey powder samples stored at water activities ranging from 0.33 to 0.75.

Results & Discussion

Moisture Sorption Isotherms

In Figure 1, moisture sorption isotherms for the hygroscopic whey powder are shown. In the six week 25°C isotherm, a slight discontinuity occurred in the 0.33 to 0.44 water activity range, in the three week 35°C isotherm a discontinuity occurred in the 0.33 to 0.44 water activity range. The one week 45°C isotherm exhibits two discontinuities: the first in the 0.33 to 0.40 water activity range and the second in the 0.44 to 0.53 water activity range. According to Heldman et al. (1965) for hygroscopic milk powders at water activities below 0.40, adsorption of water occurred on polar sites of both amorphous lactose and proteins. At higher water activities moisture was adsorbed only by protein, since lactose was in the crystalline form. Based on this, the discontinuities exhibited by the 25, 35 and 45°C isotherms would be due to losses of moisture resulting from lactose crystallization which exceeds the moisture uptake by the protein within the powder. The hygroscopic whey powder isotherms demonstrated similar behavior to that noted by Berlin et al. (1970) for milk powder isotherms at different temperatures. They noted that as temperature increased, lower water contents were encountered at the same water activity.

The isotherm behavior of the whey powder appeared to indicate a physico-chemical change in lactose from the amorphous to crystalline state with increasing water activities but did not furnish specific information on initiation of lactose crystallization and crystal growth at different water activities and temperatures. To obtain this information, the scanning electron microscope was utilized.

Scanning Electron Micrographs

Figures 2 through 4 are scanning electron micrographs of the whey powder stored at 0.40, 0.44 and 0.53 water activities and 25°C for one week.

In Figure 2, at 0.40 water activity, the first evidence of lactose crystallization appears in the form of clumped prism-like structures on the surface of scattered whey particles. Warburton and Pixton (1978 a) noted smooth and wrinkled surfaces in scanning electron micrographs of spray dried skim milk powder stored at 25°C below a water activity of 0.42. At 0.47 water activity they noted feathery formations of crystals on the milk particle surfaces. Roetman(1979) and Buma and Henstra (1971) reported smooth or slightly wrinkled surfaces in scanning electron micrographs of freshly spray dried hygroscopic milk powders. Roetman (1979) noted that lactose crystallized into needle-like structures on the surface of milk powder particles when crystallization occurred during storage i.e. after processing.

By 0.44 water activity, shown in Figure 3, the frequency of crystal forms has increased.

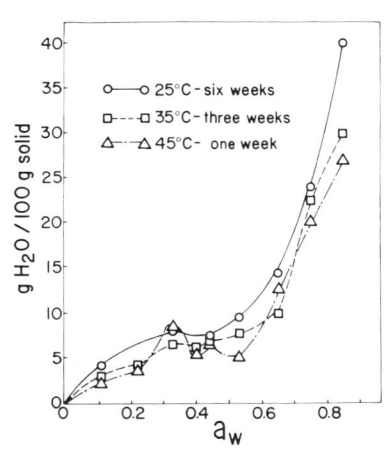

Figure 1. Hygroscopic whey powder moisture sorption isotherms at 25, 35 and 45°C

A variety of crystal shapes are encountered including prism, diamond and feathery shaped structures forming a bas-relief on the surface of the whey particles. This complex variety of irregularly shaped crystal forms probably results from interference by other whey powder components as noted by Nickerson (1962) for crystallizing lactose in milk powders. In Figure 3, some accretion or collapse between whey particles appears to be occurring as crystallization becomes more extensive.

Figure 4 depicts the 0.53 water activity condition. At this condition, elaborate enlarged bas-relief patterns are apparent on most whey particles. Although not shown here, the lactose crystals appeared to increase in size as the water activity increased above 0.53. Accretion and collapse between whey particles became extensive by 0.75 water activity.

Figures 5 to 7 show scanning electron micrographs of hygroscopic whey powder stored at 35°C and 0.33, 0.40 and 0.44 water activities. At 0.33 water activity, in Figure 5, prism shaped structures have appeared on some whey particles.

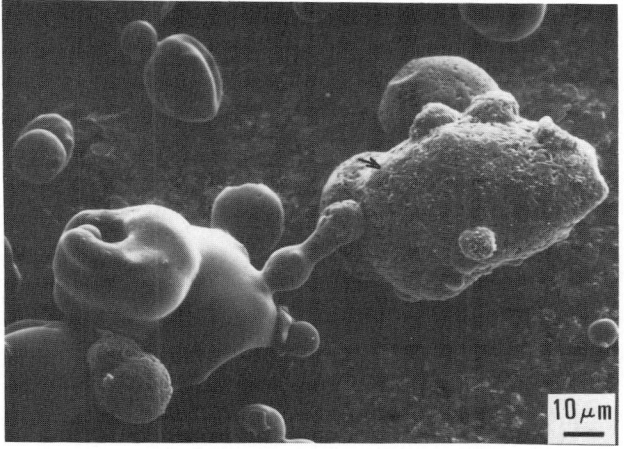

Figure 2. Hygroscopic whey powder stored one week at 0.40 water activity and 25°C

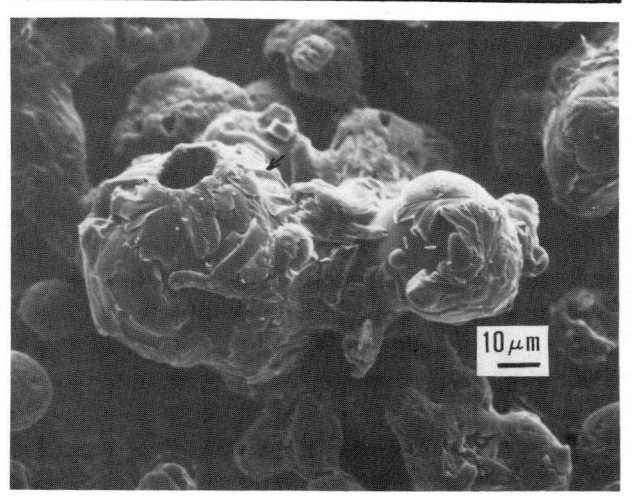

Figure 4. Hygroscopic whey powder stored one week at 0.53 water activity and 25°C

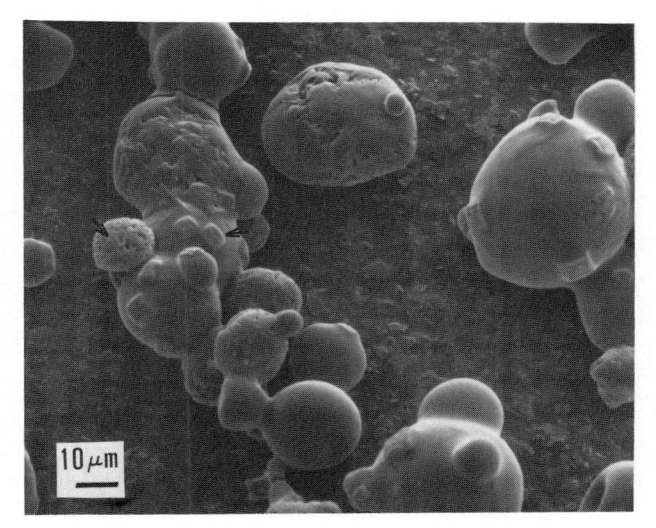

Figure 3. Hygroscopic whey powder stored one week at 0.44 water activity and 25°C

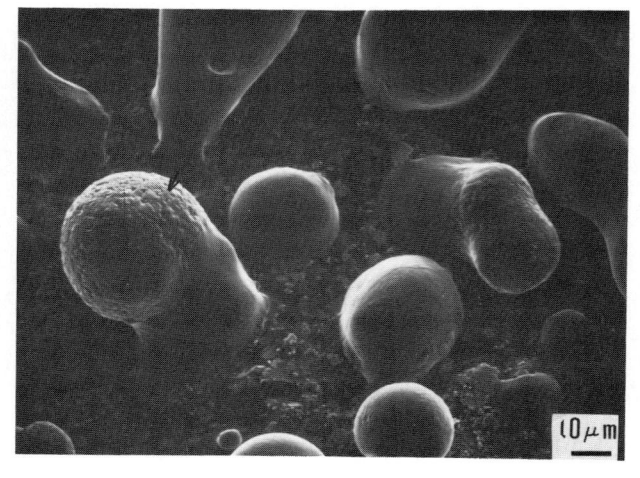

Figure 5. Hygroscopic whey powder stored one week at 0.33 water activity and 35°C

The whey particles have a relaxed slumped over appearance which is probably due to the increase in temperature.

By 0.40 water activity, in Figure 6, prism and irregularly shaped structures are present more frequently on the whey particle surfaces. Accretion is again apparent.

At 0.44 water activity, in Figure 7, lactose crystallization is apparent as a complex variety of irregular structures on the particle surfaces. Sintering of the whey particles due to lactose crystallization begins to become evident in addition to the accretion resulting from the increase in temperature.

Figures 8 to 10 show hygroscopic whey powder stored one week at 45°C and 0.33, 0.40 and 0.44 water activities. At 0.33 water activity, shown in Figure 8, prism and diamond shaped structures are present on the whey paricle surfaces. Accretion between whey particles is extensive. By 0.40 water activity, in Figure 9, lactose crystals are present as enlarged irregular structures which completely cover the whey particle surfaces. Collapse and sintering between the whey particles is extensive. Occasionally, whey particles with less well developed crystals were observed. At 0.44 water activity, in Figure 10, the whey powder surface structure is very similar to that seen in the 0.40 water activity sample shown in Figure 9. At 0.44 water activity, however, the occasional whey particle with less well developed crystals was no longer observed.

Lactose crystallization as observed by the scanning electron microscope appears to be initiated in hygroscopic whey powder at 0.40 water activity at the 25°C temperature condition and at 0.33 water activity at the 35 and 45°C temperature conditions. The shift in initiation of lactose crystallization to a lower water activity with increasing temperature was not as large as the shift in extensive crystallization with increasing temperature. At 25°C, extensive crystallization was evident by 0.65 water activity. At 45°C extensive crystallization was evident by 0.40 water activity. The initiation of lactose crystallization appeared to coincide with the isotherm discontinuities for the hygroscopic whey powder in the 25 to 45°C temperature range.

This is similar to the results of Warburton and Pixton (1978) who reported an isotherm discontinuity for spray dried skim milk powder at around 0.50 water activity. They reported lactose crystallization in scanning electron micrographs by 0.47 water activity. The initiation of crystallization in the hygroscopic whey powder occurred at a lower water activity than that noted by Warburton and Pixton (1978 a,b) for skim milk powder.

Brown Pigment Formation

Figure 11 shows the relative rates of brown pigment formation versus water activity for the whey powder stored 45°C. The browning maximum appears to occur in the 0.44 to 0.53 water activity range. This browning maximum is significantly lower than the 0.65 to 0.75 water activity maxima reported for browning via the Maillard reaction in a number of foods. Specifically, Loncin et al. (1968) reported browning and lysine loss maxima in the 0.60 to 0.70 water activity range for milk powder stored at 40°C. Huss (1974a) reported that lysine losses in skim milk powder stored at 37°C were greatest at 55% relative humidity, i.e. 0.55 water activity, and that they were related to the Maillard reaction and to the shift of lactose from the amorphous to crystalline state. However, Huss (1974a) did not measure lysine losses at water activities above a water activity of 0.55. The shift in the rate of brown pigment formation, to yield a maximum rate in the whey powders at a lower water activity than that noted for other dried food systmes, could be related to the transition of lactose within the whey powders from an amorphous to a crystalline state in the 0.33 to 0.44 water activity range.

As lactose shifts from the amorphous to a crystalline state, water is excluded which could then act as a mobilizing medium for reactive species. The result would be an increase in reaction rate (k). In addition, the amorphous lactose could undergo collapse and viscous flow as it took on liquid-like properties immediately prior to crystallization as proposed by To and Flink (1978) for freeze dried carbohydrates. The Maillard reaction most likely takes place within this collapsed structure.

Table 1 lists the brown pigment formation rates in the whey powder stored at 25, 35 and 45°C. The maximum rate of browning occurs at 0.44 water activity at each temperature condition, and again appears to be related to the shift of lactose from the amorphous to crystalline state.

TABLE I

Rate Constants × 10³ O.D./g solid/Day

a_w	25°C	35°C	45°C
0.33	0.4	1.8	13.7
0.44	0.8	3.5	23.6
0.65	0.6	2.1	15.3

Unfortunately, it was not possible to establish an exact water activity where brown pigment formation was at a maximum for each temperature condition from these storage studies. Based on the increase of the extent of lactose crystallization with increasing temperatures, as seen in the scanning electron micrographs of the whey powder, the water activity related brown pigment formation maximum could be expected to shift as the temperature increased.

Acknowledgements

We would like to acknowledge the help and cooperation of the Minnesota Agricultural Experiment Station Electron Microscopy Facility and in particular Gib Ahlstrand and Dr. R. Zeyen. This research was supported in part by projects 18-72

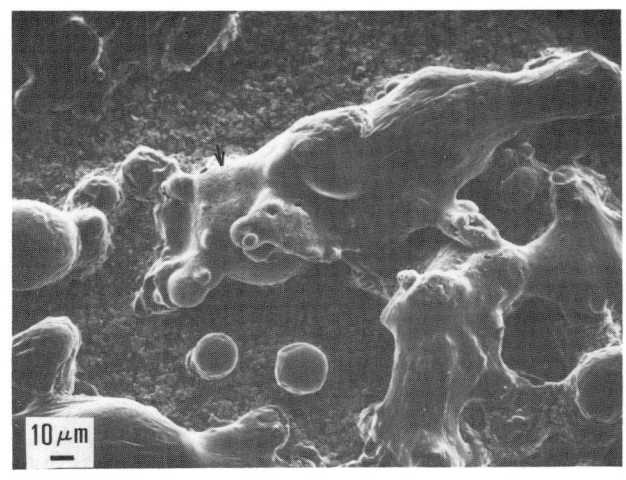

Figure 6. Hygroscopic whey powder stored one week at 0.40 water activity and 35°C.

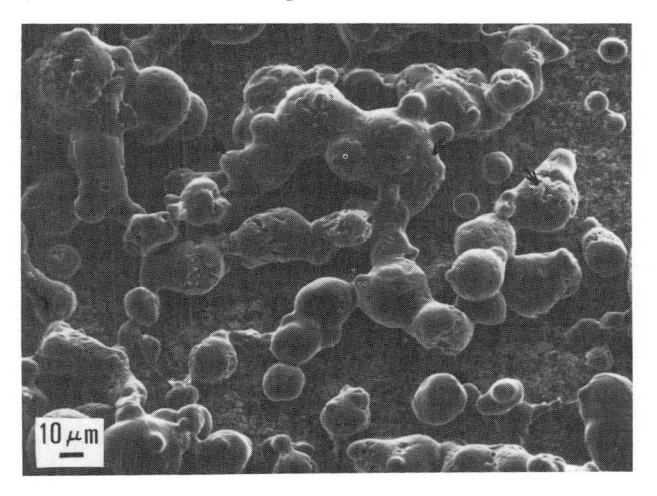

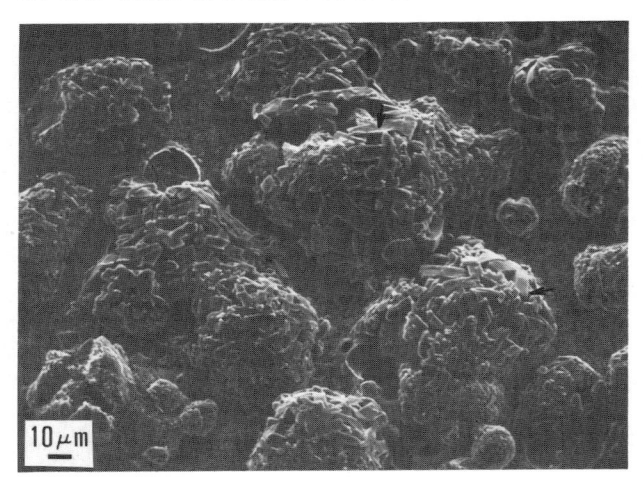

Figure 7. Hygroscopic whey powder stored one week at 0.44 water activity and 35°C.

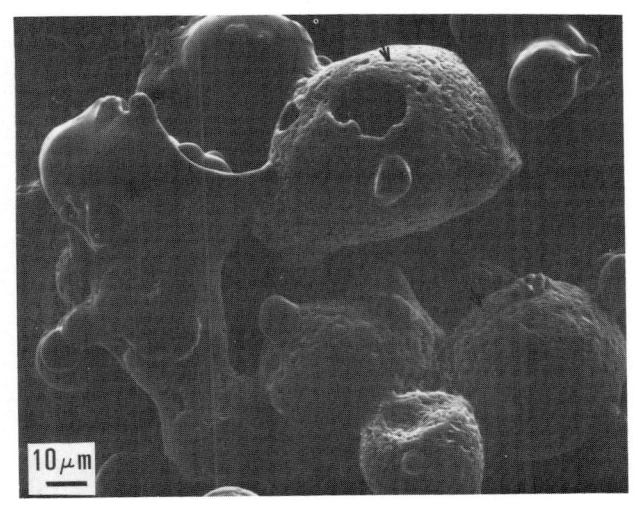

Figure 8. Hygroscopic whey powder stored one week at 0.33 water activity and 45°C.

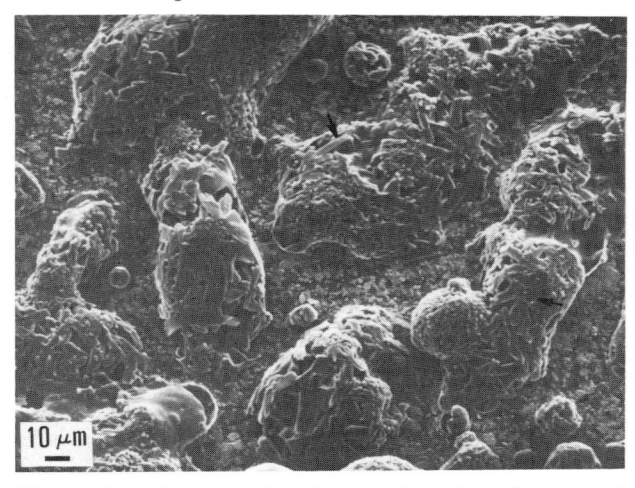

Figure 9. Hygroscopic whey powder stored one week at 0.40 water activity and 45°C.

Figure 10. Hygroscopic whey powder stored one week at 0.44 water activity and 45°C.

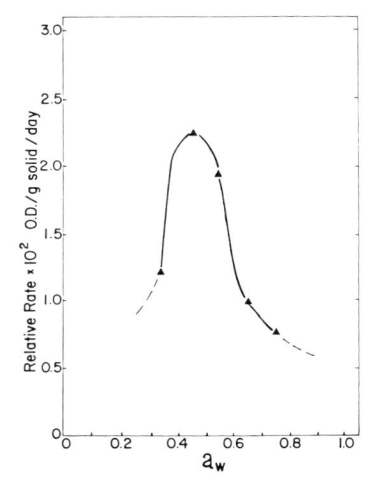

Figure 11. Relative rates of brown pigment formation versus water activity in hygroscopic whey powder stored at 45°C.

and 18-78 form the University of Minnesota Agricultural Experiment Station.

References

Berlin, E., Anderson, B.A., and Pallansch, M.J. 1970. Effect of temperature on water vapor sorption by dried milk powders. J. Dairy Sci. 53, 146-149.

Buma, T.J. and Henstra, S. 1971. Particle structure of spray-dried milk products as observed by a scanning electron microscope. Neth. Milk Dairy J. 25, 75-80.

Choi, R.P., Koncus, A.F., O'Malley, C.M., and Fairbanks, B.W. 1949. A proposed method for the determination of color of dry porducts of milk. J. Dairy Sci. 32, 580-587.

Heldman, D.R., Hall, C.W., and Hendrick, T.I. 1965. Vapor equilibrium relationships of dry milk. J. Dairy Sci. 58, 845-852.

Huss, V.W. 1970. Lactose Kristallisation und Lysin Verfugbarkeit nach Lagerung von Trocken Magermilchpulvern bei verschiedener Luftfeuchtigkeit. Landwirtsch. Forsch. 23, 275-88.

Huss, V.W. 1974a. Zeitlicher Verlauf der Lysinschadigung bei der Lagerung von Magermilchpulvern unter verschiedenen Bedingungen. Landwirtsch. Forsch. 27, 199-210.

Huss, V.W. 1974b. Untersuchungen zur Aminosaurenschadigung bei Verarbeitung und Lagerung von Molken und Molkenpulvern. Z. Tier physiol., Tierernahrg., u. Futtermittelkde. 34, 60-67.

Loncin, M., Bimbenet, J.J., and Lenges, J. 1968. Influence of the activity of water in the spoilage of foodstuffs. J. Fd. Tech. 3, 131-138.

Nickerson, T.A. 1962. Lactose crystallization in ice cream. IV. Factors responsible for reduced incidence of sandiness. J. Dairy Sci. 45, 354-360.

Roetman, K. 1979. Crystalline lactose and the structure of spray-dried milk products as observed by scanning electron microscopy. Neth. Milk and Dairy J. 33, 1-11.

To, E.C. and Flink J.M. 1978. "Collapse", a structural transition of freeze dried carbohydrates. II. Effects of solute concentration. J. Fd. Tech. 13, 567-581.

Warburton, S. and Pixton, S.W. 1978a. The moisture relations of spray dried skimmed milk. J. Stored Prod. Res. 14, 143-158.

Warburton, S. and Pixton, S.W. 1978b. The significance of moisture in dried milk. Dairy Industries International. April 1978, 25-27.

Discussion with Reviewers

D.N. Holcomb: Were higher water activity samples stable in the SEM? If samples were equilibrated at higher water activities (such as 0.7) would the structure be different from that at the water activity where the browning reaction is proceding at maximum rate?

Authors: Higher water activity samples i.e. equilibrated at 0.65, 0.75 and 0.85 water activities, appeared to be as stable as samples equilibrated to lower water activity levels. At 25°C, the samples equilibrated to 0.65, 0.75 and 0.85 water activities showed increased crystal sizes. In addition, crystallization became more extensive covering the entire specimen surface.

M. Karel: Over what time interval were the browning measurements followed? Were these the same as the crystallization measurements? The browning measurements were followed over a period of months, up to six months at 25°C, around three months at 35°C and about one month at 45°C.

Authors: The browning measurements were followed over a period of month: up to six months at 25°C, around three months at 35°C and about one month at 45°C. These periods were considerably longer than the one week period associated with the crystallization measurements. As amorphous lactose crystallizes, lactose molecules orient themselves into a crystal lattice. This rearrangement results in the exclusion of water from the lactose fraction of the whey powder. This water can then influence the Maillard reaction by mobilization and dilution effects depending upon the water activity and associated water content. Crystallization is a rapid, essentially irreversible process; the water released from which can participate in the Maillard reaction.

M. Kalab: Which salts were used to provide water activities of 0.11 to 0.85?

Authors: The salts used were as follows:

Salt	Water Activity
Lithium Chloride	0.11
Potassium Acetate	0.22
Magnesium Chloride	0.33
Zinc Nitrate	0.40
Potassium Carbonate	0.44
Magnesium Nitrate	0.53
Sodium Nitrite	0.65
Sodium Chloride	0.75
Potassium Chloride	0.85

M. Kalab: Why was a vacuum used in the determination of the moisture sorption isotherms whereas a normal pressure of 1 atm. was used to equilibrate the specimens destined for SEM?

Authors: A vacuum was used for the moisture sorption isotherm samples to speed up equilibration. Since the SEM specimens were so small i.e. since a very small amount of powder was placed on each specimen stub, it was judged that equilibration would be very rapid even at one atmosphere pressure. In addition, there was some concern that pulling and releasing a vacuum would disrupt the specimens on the stub surfaces.

M. Kalab: What was the initial appearance of the sweet whey powder before it was subjected to the humid atmosphere treatment?

K. Roetman: Could you supply a SEM of the whey powder before exposure to moisture?

Authors: The appearance was that of globular fairly smooth to wrinkled symmetrical particles as shown below in freshly spray dried whey at about 0.10 water activity. We also noted large, 60 to 100 μm diameter tomahawk shaped particles scattered among the smooth globular shapes. These particles (tomahawk shape) were considered to be lactose crystals present in the liquid whey prior to drying based on the very similar appearance of SEM's of pure alpha mono-hydrate crystalline lactose. This would constitute crystalline material in the initial sample without the whey particles themselves containing visible lactose crystals. Please see Fig. 12.

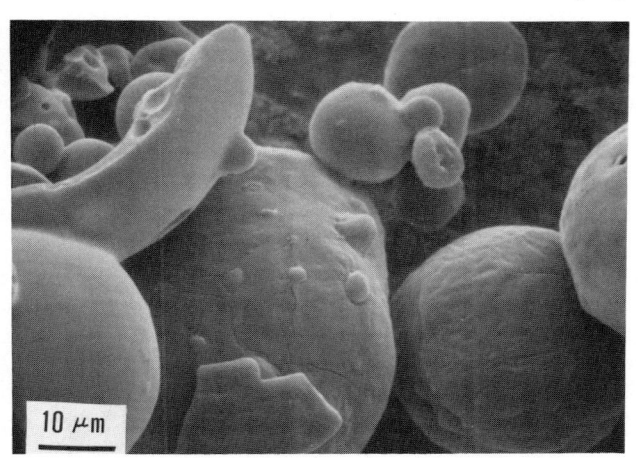

Figure 12. Hygroscopic whey powder at 0.10 water activity.

M. Kalab: What are the raised squares on the smooth globular particles in Figure 3?
Authors: We have concluded that these are also lactose crystals based on the fact that no such shapes were present on any of the whey particles viewed at water activities below that where crystallization was noted at any of the temperature conditions. According to Nickerson (1962), shapes of lactose crystals can be affected by other components in milk products. Diamond shapes are one of the shapes that lactose crystals can assume along with prism and the tomahawk shapes noted in the micrographs. It is possible that both prism and diamond shapes are noted due to different rates of crystallization due to these other components.

S. Patton: How much of the material that was viewed (number of fields) had how much crystalline lactose at the different water activity and temperature conditions?
Authors: Specimen stubs were viewed systematically i.e. twenty to thirty fields were viewed per stub in the 320 to 640X magnification range. If a percentage figure is used to estimate the amount of crystalline material present at each water activity and temperature conditions, the following estimates can be supplied:

Treatment Condition		% Lactose Crystallized
$25^\circ C$	0.40	10
	0.44	50
	0.53	90
$35^\circ C$	0.33	5
	0.40	15
	0.44	80
$45^\circ C$	0.33	50
	0.40	95
	0.44	95

It should be noted that information of this type is only approximate and gives no information on crystal shapes or sizes noted in the micrographs which were selected as representative of the whole sample systematically viewed.

D.N. Holcomb: Were any micrographs taken of samples equilibrated for six weeks at $25^\circ C$ as well as at one week?
Authors: Micrographs were only taken after one week of equilibration. The isotherm can only furnish information about a general physico-chemical change going on in a food material. Actual pictures of the material can furnish more precise information. It is anticipated that if six week micrographs had been taken, that crystallization would have been more advanced than in the one week samples. The scope of the present study unfortunately did not permit following the rate of crystallization of lactose in the whey powder stored at different water activities and temperatures. This additional information along with more complete information on the water related browning maxima at similar water activity and temperature conditions could lead to a better understanding of the relationship between lactose crystallization and the Maillard reaction in milk products.

Additional discussion with reviewers of the paper ""A SEM investigation of the whipping of cream" by D.G. Schmidt et al., continued from page 192.

M. Kalab: By suggesting that fat globules penetrate the membrane (Fig. 22), which envelops the air bubble, do you mean that they actually pierce the membrane?
Authors: Yes.

M. Ruegg: It is well known that cream cannot be homogenized at high pressures because the whippability is reduced. Could the authors comment on the relationship between the size distribution of fat globules and the whippability of the product?
Authors: The whippability of homogenized cream depends more largely on the size of the homogenization clusters than on the size of the fat globules proper. In two-stage homogenization a higher pressure in the first stage results in the formation of more clusters, the size of which is reduced in the second stage. In order to obtain good whippability, the clusters may neither be too small nor too large; a size of 15 to 20 μm, which is approximately the thickness of the foam lamellae, seems the optimum.

M. Kalab: Have you any experience with using conductive substances other than carbon to coat freeze-fractured surfaces for cold-stage SEM?
Authors: No.

M. Anderson: Have the authors any information on the effect of varying the processing temperatures on the structure of the air-serum interface?
Authors: We did not vary the processing temperatures during preparation of the cream, and we do not think that it will greatly affect the air-serum interface. However, some effect on the rate of the various processes taking place during whipping may be expected. Sterilization of cream, for example, results in a slightly increased surface tension at the air-serum interface, which theoretically reduces the foamability but increases the spreading pressure; this might result in a lower overrun and a shorter whipping time. In practice such an effect appears to be negligible. The centrifugation temperature of the cream influences the amount of phospholipids in the cream. If this temperature is lowered, the amount of phospholipids is increased and a shorter whipping time and a higher overrun are obtained.

EFFECT OF ACIDULANTS AND TEMPERATURE ON MICROSTRUCTURE, FIRMNESS AND
SUSCEPTIBILITY TO SYNERESIS OF SKIM MILK GELS

V.R. Harwalkar and M. Kaláb

Food Research Institute
Research Branch
Agriculture Canada
Ottawa
Ontario, Canada K1A OC6

Abstract

Skim milk was acidulated at 0°C with predetermined amounts of hydrochloric, citric, and oxalic acids to give final pH values of 4.6 or 5.5 at 22°C. The acidulated skim milk was heated under quiescent conditions to 40 or 90°C; gels were obtained with hydrochloric and citric acids, whereas oxalic acid led to flocculation of the milk proteins. Gels were also obtained by acidulating skim milk at 40 to 90°C with glucono-δ-lactone to a final pH of 4.6 or 5.5. As the temperature at which gelation occurred was increased, the gels became firmer; at 60°C it was not possible to obtain gels at pH 5.5. At pH 4.6, the firmness of gels made at 90°C was 3 times as high as the firmness of gels made at 70°C; at pH 5.5 the difference was even greater.

Gels made from skim milk previously heated to 90°C for 5 min were significantly firmer than gels made from skim milk which had not been previously heated. The latter gels were subject to severe syneresis. Gels made from fresh skim milk were firmer than gels from reconstituted nonfat dry milk.

Microstructure of the skim milk gels was markedly affected by the acidulant used, as was revealed by electron microscopy. Core-and-lining structures, initially observed in glucono-δ-lactone-induced skim milk gels at pH 5.5, were also observed with other acidulants. A probable involvement of a β-lactoglobulin-κ-casein complex in the formation of such structures is discussed.

KEY WORDS: Acidulated milk gels. Skim milk gels. Gel firmness. Syneresis. Microstructure of gels. Electron microscopy.

Introduction

Milk may be gelled in various ways. The formation of milk gels by acid precursors such as glucono-δ-lactone (GDL) was reported earlier[1]. The major feature of GDL-induced milk gels is the incorporation of the acid precursor at high temperature, enabling a slow generation of acid at the high temperature under quiescent conditions leading to the gelation of the milk proteins[1].

Gelation of milk by adding acids such as hydrochloric (HCl), citric etc. at high temperature can not be attained because an instantaneous precipitation of milk proteins takes place. However, it is possible to incorporate the above acids in milk at a very low temperature (0°C) and to obtain gels by heating under quiescent conditions to 20-25°C. Such gels can be used to produce Cottage cheese[2] or Cheddar cheese[3]. Microstructure and rheological properties of the acid-induced milk gels subjected to high temperature (i.e. 90°C) are unknown in contrast to the GDL-induced skim milk gels[4].

The objective of this study was to examine microstructure, firmness, and susceptibility to syneresis of milk gels made with HCl, citric, and oxalic acids under varying conditions of pH, temperature, preheat treatment of the milk etc. and compare them with the GDL-induced skim milk gels.

Materials and Methods

Gelation

Fresh skim milk used in this study was obtained by separating cream from whole pooled herd milk from the Central Experimental Farm of Agriculture Canada in Ottawa. Reconstituted skim milk was made from commercial low-heat nonfat dry milk powder (NDM).

The skim milk was cooled to 0°C and acidulated with predetermined amounts of HCl, citric, and oxalic acids to give a final pH value of 4.6 or 5.5 at 22°C. The acidulated skim milk was heated in beakers quiescently to 40 or 90°C to obtain gels. Skim milk gels were also obtained in the presence of varying amounts of glucurono-δ-lactone and glucono-δ-lactone to achieve final pH of 4.6 and 5.5. The lactone was added to skim milk cooled either to 0°C or at the temperature of gelation (40 to 90°C).

Firmness
Firmness was measured by a penetrometric method using a probe 12.4 mm in diameter at a speed of 27 mm/min and expressed as g/probe[1,5].

Syneresis
A measure of susceptibility to syneresis was obtained by a newly developed method based on varying centrifugal velocity at low g-force. Aliquots of 15 ml milk were allowed to gel in test tubes with acidulants and under different heating conditions. The gels were cooled to 22-23°C and then centrifuged in a Sorvall SS-34 head for 10 min each at 500, 1000, 1500, 2000, 2500, and 3000 rpm, i.e. in a range of 30 to 1000 g-force. The volume of the clear supernatant was measured and plotted against centrifugal force. A measure of syneresis was obtained from the initial slope of the curve; the steeper the slope, the greater the susceptibility to syneresis.

Electron microscopy
Microstructure of the gels was examined by electron microscopy. For scanning electron microscopy (SEM), gel samples (≃2 mm in diameter) were fixed in a 1.4% glutaraldehyde solution for 60 min at 22°C, dehydrated in a graded alcohol series, and critical-point dried in CO_2. The dry samples were fractured by hand. The fragments were mounted on SEM stubs with a conductive cement and coated with gold (≃20 nm) by vacuum evaporation. A Cambridge Stereoscan electron microscope was operated at 5-20 kV and micrographs were taken on 35-mm films[6].

For transmission electron microscopy (TEM), gel samples (≃1 mm in diameter) were fixed in a 1.4% glutaraldehyde solution for 60 min, postfixed in a 2% OsO_4 solution in a 0.05 M veronal-acetate buffer, pH 6.75, dehydrated in a graded alcohol series, and embedded in a low-viscosity Spurr's medium[7]. Thin sections (100 nm) were stained with uranyl acetate and lead citrate solutions and examined in a Philips EM 300 electron microscope operated at 60 kV. Micrographs were taken on 35-mm films[4,8,9].

Results and Discussion

The characteristics of gels made by acidulation of skim milk at 0°C varied depending upon the type of acidulant, pH, and temperature. A comparison of firmness (g/probe) of these gels is shown in Table 1.

At 40°C and pH 5.5, no gels were formed. This can be expected since the stability of the casein micelles at this temperature is considerably reduced only at pH close to the isoelectric point. At the isoelectric pH (4.6), the firmness of gels formed at 40°C with HCl and citric acid was not significantly different from the firmness of gels produced with lactones. In the presence of oxalic acid no gels were formed and the milk proteins flocculated.

When the temperature was raised to 90°C, a marked difference was observed between the gels made with GDL and gels made with HCl and citric acids. The differences in the firmness of the GDL-induced gels and gels made with HCl and citric acids were greater at pH 5.5 than at pH 4.6. There was no difference in firmness of gels made with either HCl or citric acid at either pH.

Table 1
Comparison of firmness of skim milk gels prepared with different acidulants.

Acidulant:	Firmness (g/probe) of gels made		
	at 40°C	at 90°C	
	pH 4.6	pH 5.5	pH 4.6
Glucono-δ-lactone	10.0±4.3	31.0±6.8	43.0±4.2
Glucurono-δ-lactone	11.0*	20.0*	33.0±5.7
Hydrochloric acid	6.3±1.6	4.7±0.5†	19.6±5.6
Citric acid	11.2±3.6†	4.8±0.9†	25.5±2.7
Oxalic acid	F l o c c u l a t i o n		

* One measurement.
† Firmness measured in gels showing spontaneous whey separation.

Skim milk gels made in the presence of GDL at higher temperatures (e.g. 90°C) were considerably firmer than those made at lower temperatures (e.g. 40°C), whereas in the presence of acids, there was only a slight difference between the firmness of gels produced at 90 and 40°C. As the temperature is raised, milk pre-acidulated with acids begins to gel at lower temperatures. Lactones added to milk hydrolyze slowly and require a higher temperature to produce acids; thus the gels are produced at a higher temperature.

Gels made with glucurono-δ-lactone were slightly softer than the gels made with GDL. The former lactone was hydrolyzed at a considerably slower rate than GDL and produced glucuronic acid which interacted with milk proteins to produce an off-flavour; for this reason, its effects were not investigated further.

The temperature at which GDL was added to milk influenced the gel characteristics. A comparison of changes in firmness of gels made from reconstituted NDM as depending on the temperature of gelation at pH 5.5 and 4.6 is presented in Fig. 1. At both pH values and at temperatures above 60°C the firmness increased with the temperature of gelation.

The firmness of gels made at pH 5.5 and 4.6 from fresh skim milk with or without preheating is shown in Table 2. The gels were formed at 70, 80, and 90°C. At each of these temperatures, the gels made from milk that had been preheated to 90°C for 5 min were considerably firmer than gels made without the preheat treatment of the milk. This is in agreement with previous observations of gels made at 90°C only[1]. The difference in firmness between gels made from milk with or without the preheat treatment was linked to the presence or absence, respectively, of a complex between β-lactoglobulin and κ-casein[4]. Gels made from fresh skim milk were generally firmer than those made from reconstituted NDM.

The different acid treatments that produced gels of variable firmness also produced gels varying in susceptibility to syneresis. Susceptibility to syneresis of the gels made with different acidulants at pH 4.6 is shown in Fig. 2. There was a greater tendency for syneresis in gels made with

Table 2
Effect of preheat treatment (90°C for 5 min), gelation temperature, and pH on firmness of gels made from fresh skim milk with glucono-δ-lactone†.

Temperature of gelation	Firmness (g/probe) of gels made at			
	pH 5.5		pH 4.6	
	not pre-heated	preheated	not pre-heated	preheated
70°C	1.0*	12.3±2.1	18.0*	19.7±3.1
80°C	9.3±2.1	27.0±1.0	41.0±5.2	53.0±2.1
90°C	29.0±2.8	36.0±7.8	50.0±4.1	68.3±2.6

† Glucono-δ-lactone was added at the temperature of gelation.
* One measurement.

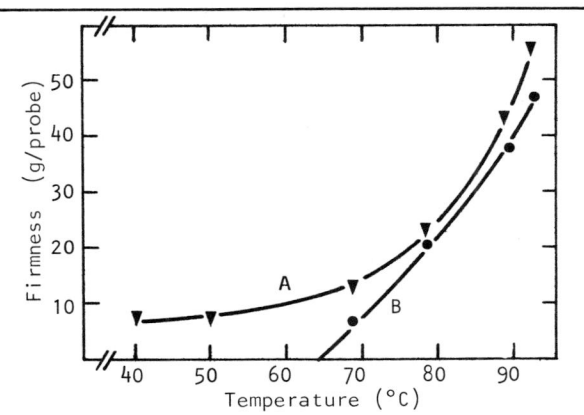

Fig. 1. Relationship between gel firmness (g/probe) and temperature of gelation (°C) of reconstituted nonfat dry milk at pH 4.6 and 5.5. Acidulant (GDL) was added at the temperature indicated. A: 1% GDL, pH 4.6; B: 0.5% GDL, pH 5.5.

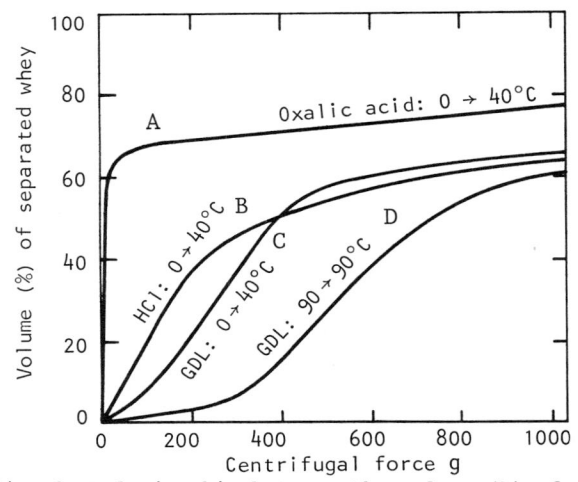

Fig. 2. Relationship between the volume (%) of whey, separated from skim milk floccules and gels, and centrifugal force (g). Oxalic acid (A), HCl (B), and GDL (C) were each added at 0°C and the mixtures were heated to 40°C. D: GDL was added at 90°C and gelation took place at this temperature.

acids than in GDL-induced milk gels. Oxalic acid flocculated casein in skim milk and instantly separated the whey. Other acids (HCl, citric and also lactic acids), added to skim milk at 0°C, produced gels after the acidulated milk was heated to 40°C. Separation of whey from gels subjected to centrifugation showed similar patterns when HCl, citric, and lactic acids were used. Because of this similarity, only HCl-induced gels are presented in Fig. 2. However, gels made in a similar way by adding GDL to skim milk at 0°C and warming to 40°C were less susceptible to syneresis. The susceptibility to syneresis was even further reduced when GDL was added to skim milk at 90°C and the milk was gelled at this temperature.

It is evident from the data[1,4] published earlier as well as from this presentation that firmness and syneresis of acid-induced skim milk gels are influenced by several factors such as pH, type of acidulant, preheat treatment of the milk, and temperature of gelation. To determine how these two rheological properties are correlated with microstructure, the gels were examined by electron microscopy. SEM micrographs of the gels, prepared under the same conditions as the gels used in the syneresis tests, are presented in Fig. 3 and 4. Corresponding gels are identified with the same letters as in Fig. 2. The lower magnification (Fig. 3) shows the overall microstructure of the gel matrix whereas the higher magnification (Fig. 4) shows details in the association of casein micelles. Micrographs A to D were obtained with samples progressively decreasing in syneresis and increasing in firmness. A good correlation is evident between the syneresis and firmness data on the one hand and the SEM images on the other hand. Floccules made with oxalic acid, which were subject to the highest syneresis (Fig. 2 A), showed a microstructure more corpuscular than gels B, C, and D; the casein micelles were aggregated in clusters. In contrast, gels made with GDL added at 90°C, which were least susceptible to syneresis, showed a fibrous microstructure of the network with most casein micelles associated in chains. Structures of the latter type were capable of immobilizing large volumes of water and showed a higher resistance to the penetrometer probe. Between these 2 extreme examples, there were gels made at 40°C with HCl (B) and GDL (D). The matrices of both gels consisted of casein micelle clusters and chains, but there was a somewhat higher incidence of the chains in the GDL-induced gel than in the gel made with HCl (compare micrographs B vs. C in Fig. 3 and 4). The micrographs correlate well with the lower firmness (Table 1) and the higher susceptibility to syneresis (Fig. 2) of the HCl-induced gel as compared to the GDL-induced gel. Casein micelles in the gel made with GDL at 40°C were slightly larger (Fig. 4C) than in the HCl-induced gel (Fig. 4 B); this is evident only at the higher magnification. Compared to the HCl- and GDL-induced gels produced at 40°C (Fig. 4 B and C, respectively), the GDL-induced gel made at 90°C (Fig. 4 D) shows considerably

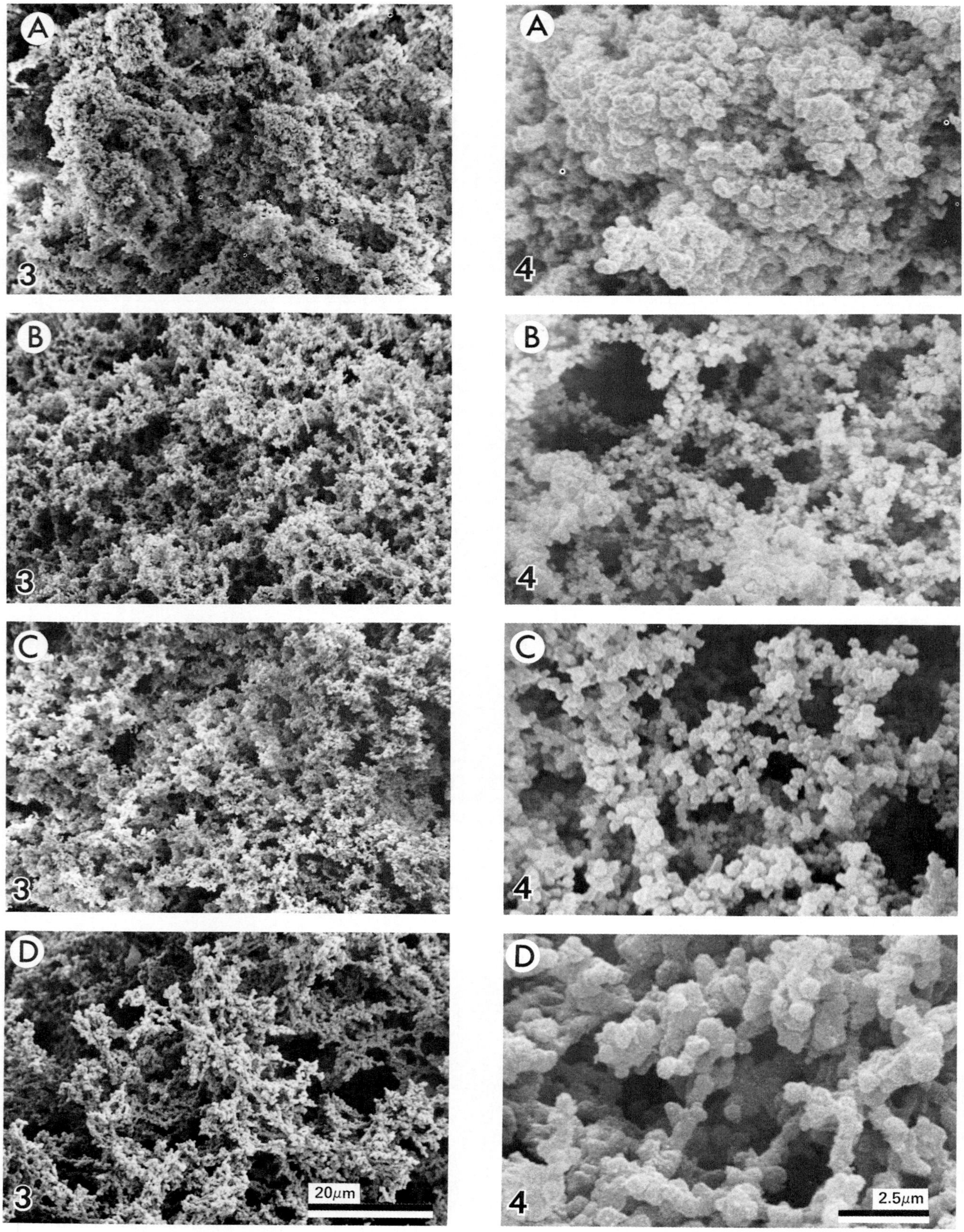
A
3
B
3
C
3
D
3
20μm
A
4
B
4
C
4
D
4
2.5μm

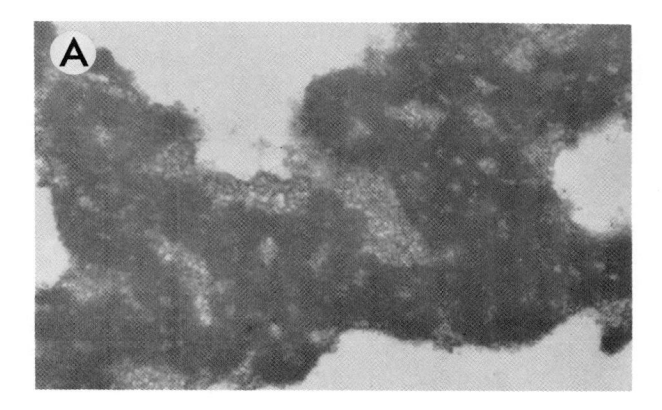

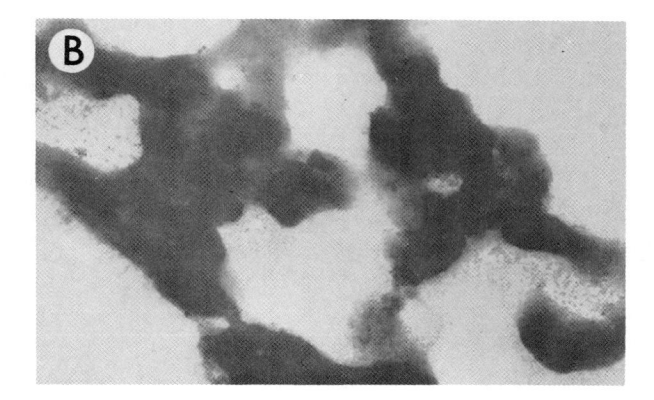

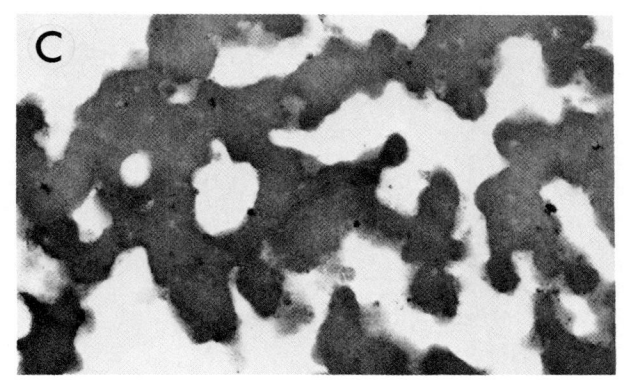

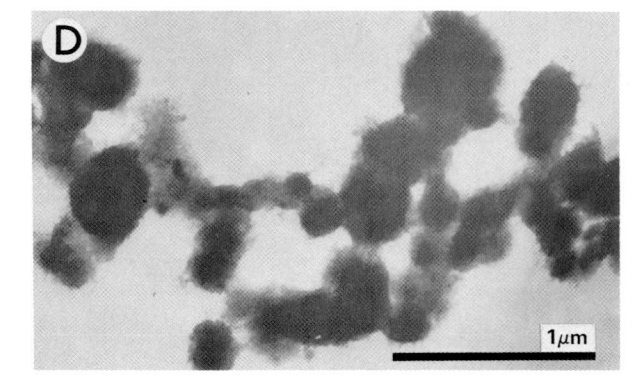

Fig. 5. Microstructure of floccules and gels obtained by heating acidulated milk (pH 4.6) to 90°C. The acidulants were: A: Hydrochloric acid; B: Citric acid; C: Oxalic acid; D: Glucono-δ-lactone.

larger micelles. The arrangement of the micelles in the network confirms the model presented earlier[4].

Thin sections of gels made by incorporating the above acidulants in skim milk at 0°C and by heating the mixtures to 90°C to a final pH of 4.6 were examined by TEM and are presented in Fig. 5. Whereas SEM was more suitable to study the gel matrix in general, TEM made it possible to examine the association of the casein micelles at the ultrastructural level. At pH 4.6, there was very little difference in the dimensions of the casein particles which were joined in the three-dimensional network. The extensive fusion of the casein micelles makes it difficult to evaluate their sizes. GDL-induced skim milk gels were almost twice as firm as the gels made with HCl or citric acid and heated to 90°C (Table 1), yet the particle dimensions were not significantly different in both kinds of gels (Fig. 5). Acid-induced gels apparently show more open spaces which presumably explains their greater susceptibility to syneresis. The microstructure of floccules produced with oxalic acid shows more compacted casein micelles.

The enlarged casein particles in all the gels result from the effect of the high temperature to which these gels were subjected. Gels made with the addition of different acids at 0°C and by heating the mixtures to 40°C show considerably smaller casein particles (Fig. 6). The dimensions of the micelles are similar in HCl- and citric acid-induced gels but were smaller than in the gel made with GDL at 40°C. The larger size of micelles in the GDL-induced gel is also evident in the SEM micrographs though not as clearly as in thin sections. The reason for this difference is not known. The density of casein particles in floccules produced by oxalic acid is higher than in gels made with HCl and citric acids.

Microstructures of gels made by incorporating various acidulants to milk at 0°C and by heating the mixtures to 90°C to obtain a final pH of 5.5 are shown in Fig. 7. The microstructures vary considerably with the type of acidulant used. The sizes of casein particles in gels made with citric acid and in floccules produced with oxalic acid are nearly the same although their internal structures are different. The GDL-induced gels also show an extensive fusion of the particles which are not significantly different from those in gels made with citric acid and in floccules produced by oxalic acid. The gels made with HCl, however, show markedly enlarged particles. There is a clear indication of a more extensive fusion

Fig. 3.(facing page, left column) SEM of skim milk floccules and gels at pH 4.6. A: Floccules produced with oxalic acid. B: Gel made with HCl. C: Gel made with GDL added to skim milk at 0°C and by heating the mixture to 40°C. D: GDL was added at 90°C and the mixture was gelled at the same temperature. Samples A to D are the same as those shown in Fig.2.

Fig. 4.(facing page, right column) SEM of the same samples as in Fig 3, viewed at a higher magnification.

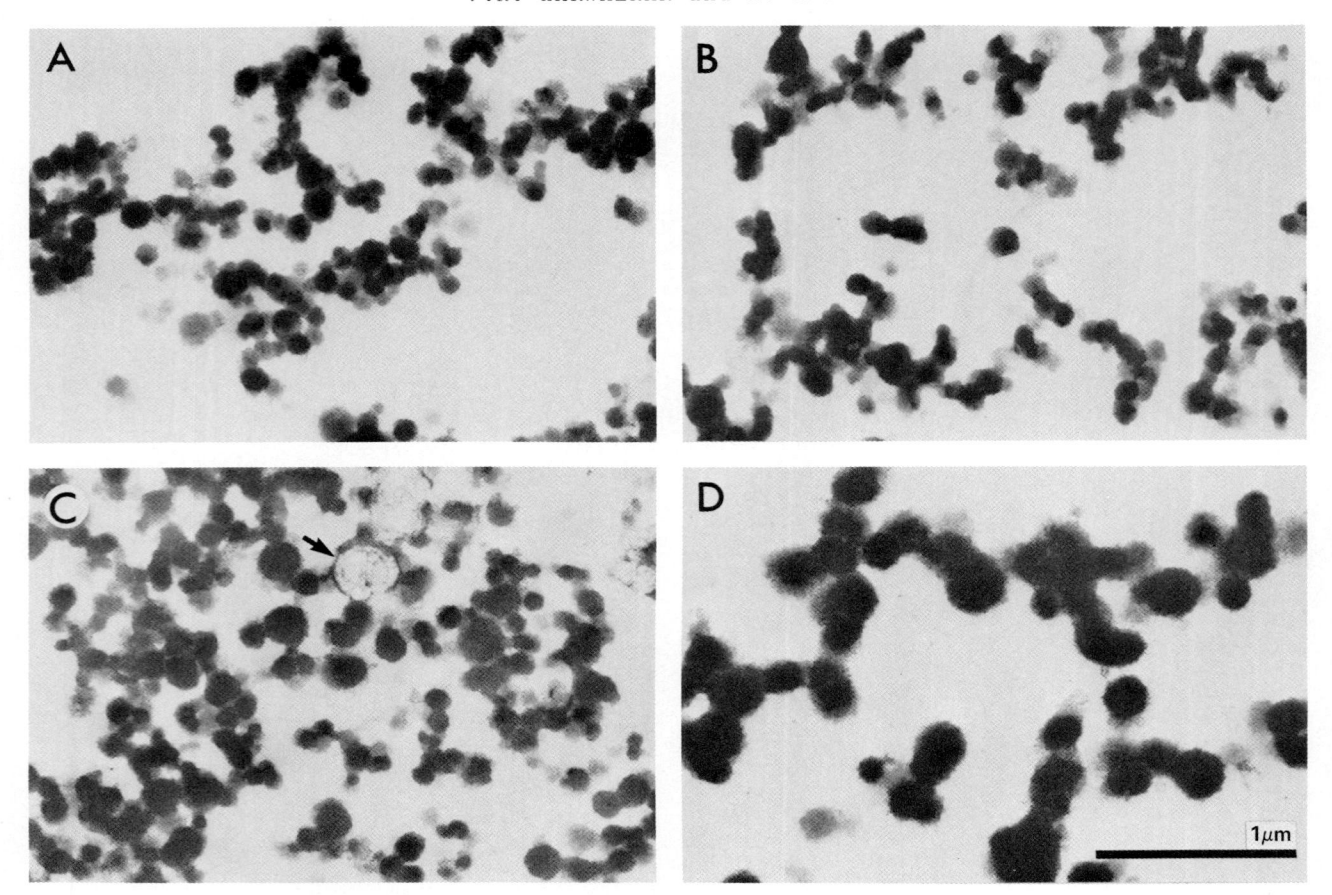

Fig. 6. Microstructure of floccules and gels obtained by heating acidulated milk (pH 4.6) to 40°C. The acidulants were: A: Hydrochloric acid; B: Citric acid; C: Oxalic acid; D: Glucono-δ-lactone. Arrow in C points to calcium oxalate crystals.

of casein particles in gels made with HCl. It is intriguing to observe such a great difference between the microstructures of gels made with HCl and citric acid although they are similar as far as firmness and syneresis are concerned. Furthermore, the gels made by incorporating GDL, particularly at high temperature, are considerably firmer than gels made with citric acid, yet the sizes of the particles are similar (e.g. Fig. 7 B vs. Fig. 8 B). The differences are presumably due to differences in the way in which these particles are joined into a network and in the strength of their linkages. Such differences in the strength of the linkages can not be demonstrated by electron microscopy.

The most interesting feature of the gels made at pH 5.5 is the observation of a unique structure of the casein particles. The thin lining surrounding the particle core was previously observed in GDL-induced skim milk gels and its existence was confirmed by several electron microscopical techniques[4,10]. The present work shows that all the acidulants produce this particular structure in milk gels at pH 5.5 (Fig. 7). There are some differences in the microstructure of these gels. For example, the core is not as distinctly dense in the floccules made with oxalic acid as in gels made with other acidulants although the lining is clearly visible. It appears that the micelle core is disintegrating, probably as a result of sequestering of calcium by oxalic acid[11]. Citric acid, also known to be a sequestering agent, does not show this behaviour.

The formation of a core-and-lining structure in gels at pH 5.5 seems to depend upon the temperature at which the gels are formed. Skim milk gels at pH 5.5 were produced by incorporating GDL at 70, 80, and 90°C. The microstructure of gels produced at 70 and 90°C are shown in Fig. 8. The core-and-lining structure is observed in gels made at 80°C (micrograph not shown) and 90°C. The 70°C gels indicate, though not as clearly, the beginning of formation of such structures. At present, there is no suitable explanation for the appearance of this structure. It is possible that the β-lactoglobulin-κ-casein complex interaction is involved in view of the requirements of temperatures higher than 70°C for the formation of the structure. The interaction between β-lactoglobulin and κ-casein is very extensive only above 70°C, being approximately 15% at 70°C[12]. Furthermore, the complex between β-lactoglobulin and κ-casein has been demonstrated to appear in the form of filamentous appendages on the surface of the casein particles[13]. Similar structures in heated milk have been frequently observed and variously described as *hairy*[14], *spiky* and *tendrils*[15], and *ragged* and *fuzzy*[9]. These filamentous appendages could interact in different

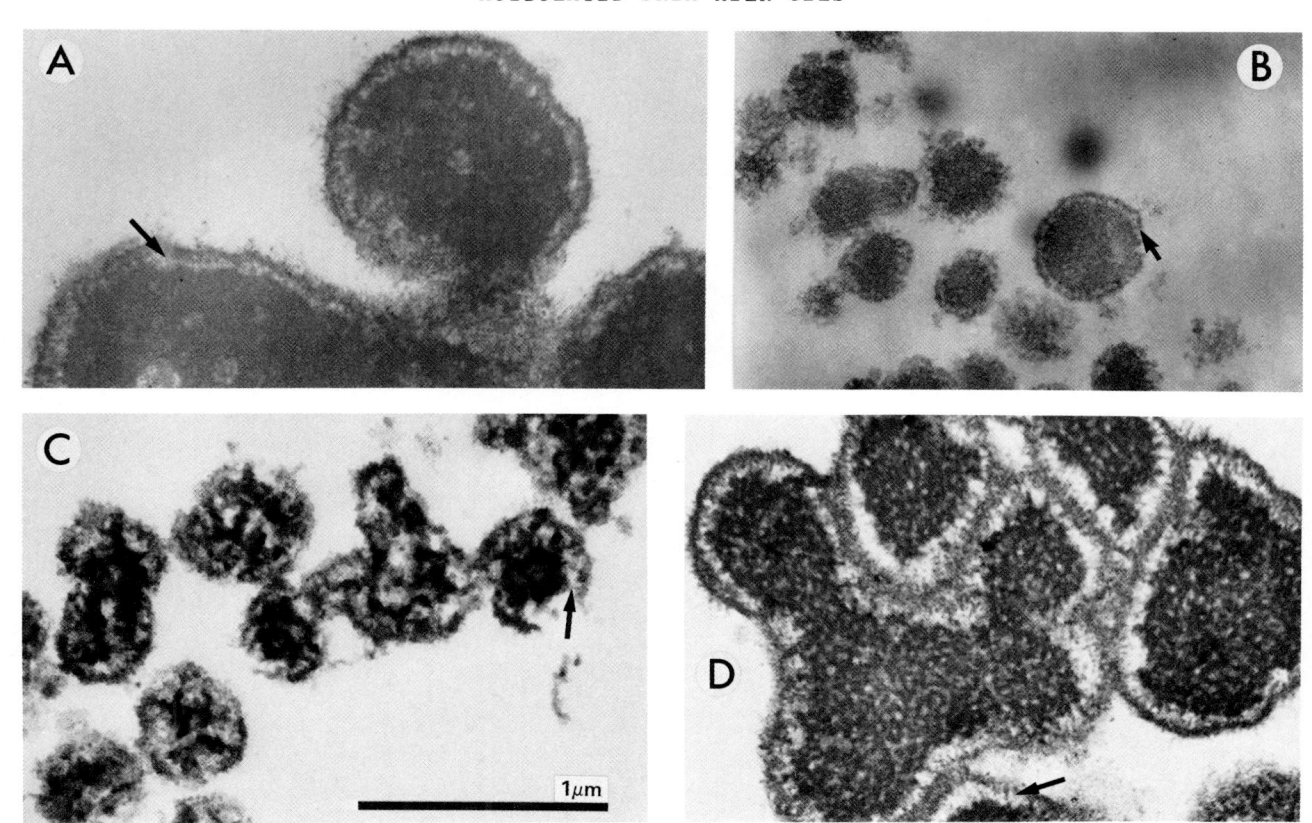

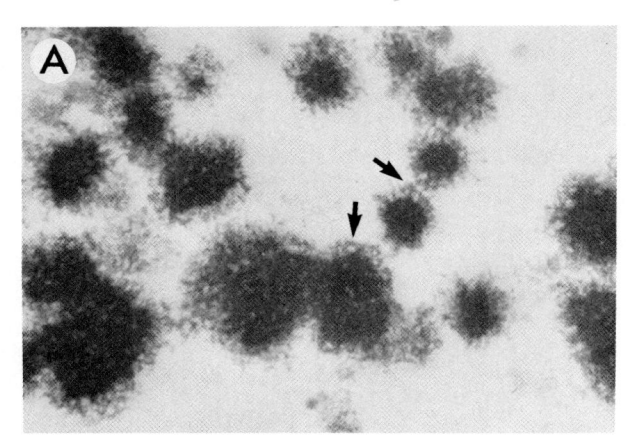

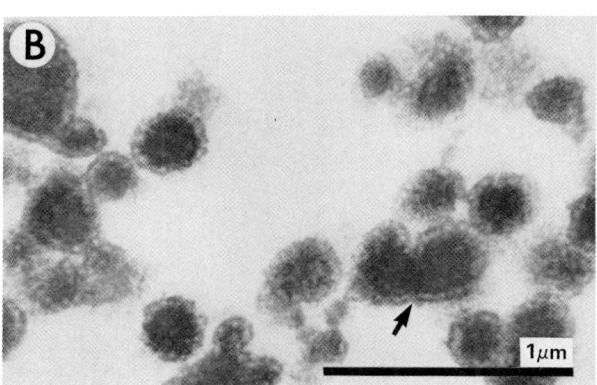

Fig. 7. Microstructure of floccules and gels obtained by heating acidulated skim milk (pH 5.5) to 90°C. The acidulants were: A: Hydrochloric acid; B: Citric acid; C: Oxalic acid; D: Glucono-δ-lactone. Arrows point to the free annular space between the lining and the micelle core. In C the cores are partly disintegrated.

Fig. 8. Microstructure of skim milk gels made with glucono-δ-lactone at pH 5.5 and at different temperatures. GDL was added and gelation took place at 70°C (A) or at 90°C (B).
In A, a lining starts to develop at the ends of appendages (arrows) on the surfaces of casein micelles. In B, the lining is developed around all the casein micelles (arrows).

stages culminating in the formation of a lining.
A model illustrating the mechanism of the formation of a lining is presented in Fig. 9. The different stages A, B, C, and D arbitrarily show the probable sequence of events as hypothesized above. The model presented is considered to be a realistic representation of changes taking place in the gels. This is illustrated in Fig. 10 by selected enlarged areas of sections from variously produced gels at pH 5.5. The micrographs of a GDL-induced gel made at 70°C (Fig. 10 A) shows mainly a ragged surface and some interaction between the appendages but the formation of the lining is not distinctly observed. The GDL-induced gel (Fig. 10 B) and floccules produced with oxalic acid (Fig. 10 C), both formed at 90°C, show a clear formation of the lining as an extension of

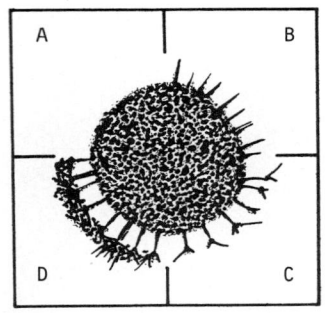

Fig. 9. A schematic diagram illustrating different stages in the formation of the core-and-lining structure in skim milk gels made at pH 5.5 and 90°C.
A: Surface of an unheated casein micelle. B: Formation of filamentous appendages resulting from a β-lactoglobulin-κ-casein interaction. C and D: Additional association of protein aggregates with the filamentous appendages and formation of a lining.

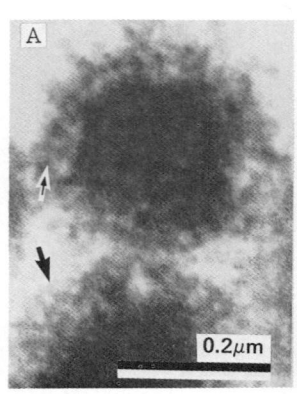

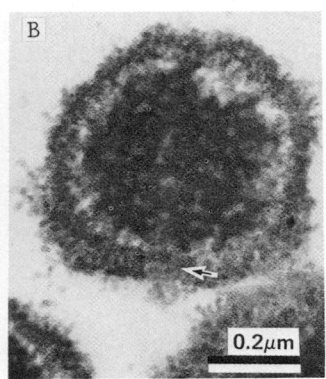

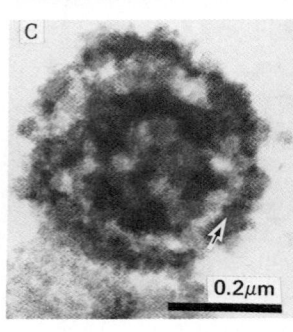

Fig. 10. Details of casein micelles in skim milk gels made at pH 5.5.
A: Glucono-δ-lactone-induced skim milk gel at 70°C; a black arrow points to protein aggregates corresponding to stage C in Fig. 9 and a white arrow points to the beginning of the formation of a lining. B: GDL-induced skim gel at 90°C; the lining is well developed around the entire casein micelle (arrow). C: Floccules produced by oxalic acid in skim milk at 90°C; the lining (arrow) is well developed and the core shows signs of disintegration.

an interaction between the appendages. High temperature and pH 5.5 seem to be conducive to the formation of this type of structure through the involvement of a complex between β-lactoglobulin and κ-casein. However, the proof of any definite involvement of such an interaction product in the formation of the core-and-lining structures requires further research.

Conclusions

This study has demonstrated that firmness and susceptibility to syneresis of milk gels made with acidulants such as hydrochloric and citric acids were different from such properties of milk gels made with glucono-δ-lactone. Microstructures of both kinds of gels were similar although the dimensions of the casein particles were different. A core-and-lining ultrastructure of casein micelles, previously observed in glucono-δ-lactone-induced skim milk gels, was shown to be present in milk gels made at pH 5.5 using the above acidulants, and in floccules produced with oxalic acid.

Acknowledgment

Skillful assistance provided by Miss Dorothy Sibbitt and Mr. J.A.G. Larose is acknowledged. Appreciation is expressed to Dr. H.W. Modler for useful suggestions. Electron Microscope Centre, Research Branch, Agriculture Canada in Ottawa provided facilities. This presentation is Contribution 455 from Food Research Institute, Agriculture Canada in Ottawa.

References

1. Harwalkar V.R., Kalab M. and Emmons D.B. Gels prepared by adding glucono-δ-lactone to milk at high temperature. Milchwissenschaft 32, 1977, 400-402.

2. McNurlin T.F. and Ernstrom C.A. Formation of curd by direct addition of acid to skim milk. J. Dairy Sci. 45, 1964, 647.

3. Breene W.M., Price W.V. and Ernstrom C.A. Changes in composition of Cheddar curd during manufacture as a guide to cheese making by direct acidification. J. Dairy Sci. 47, 1964, 840-848.

4. Harwalkar V.R. and Kalab M. Milk gel structure. XI. Electron microscopy of glucono-δ-lactone-induced skim milk gels. J. Texture Stud. 11, 1980, 35-49.

5. Kalab M, Voisey P.W. and Emmons D.B. Heat-induced milk gels. II. Preparation of gels and measurement of firmness. J. Dairy Sci. 54, 1971, 178-181.

6. Kalab M. Milk gel structure. VI. Cheese texture and microstructure. Milchwissenschaft 32, 1977, 449-458.

7. Spurr A.R. A low-viscosity epoxy resin embedding medium for electron microscopy. J. Ultrastruct. Res. 26, 1969, 31-43.

8. Kalab M. and Harwalkar V.R. Milk gel structure. II. Relation between firmness and ultrastructure of heat-induced skim milk gels containing 40-60% total solids. J. Dairy Res. 41, 1974, 131-135.

9. Kalab M., Emmons D.B. and Sargant A.G. Milk gel structure. V. Microstructure of yoghurt as related to heating of milk. Milchwissenschaft 31, 1976, 402-408.

10. Kalab M. Electron microscopy of milk products: A review of techniques. Scanning Electron Microsc. 1981: III, 453-472.

11. Odagiri S. and Nikerson T.A. Complexing of calcium by hexametaphosphate, oxalate, citrate and EDTA in milk. 1. Effect of complexing agents on turbidity and rennet coagulation. J. Dairy Sci. 47, 1964, 1306-1309.

12. Tumerman L. and Webb B.H. Coagulation of milk and protein denaturation. In: Fundamentals of Dairy Chemistry. Webb B.H. and Johnson A.H. (eds.),

Avi Publishing Co., Westport, Connecticut, 1965, 506-589.

13. Davies F.L., Shankar P.A., Brooker B.E. and Hobbs D.G. A heat-induced change in the ultrastructure of milk and its effect on gel formation in yoghurt. J. Dairy Res. 45, 1978, 53-58.

14. Harwalkar V.R. and Vreeman H.J. Effect of added phosphate and storage on changes in ultra-high temperature short-time sterilized concentrated skim milk. 2. Micelle structure. Neth. Milk Dairy J. 32, 1978, 204-216.

15. Andrews A.T., Brooker B.E. and Hobbs D.G. Properties of aseptically packed ultra-heat treated milk. Electron microscopic examination of changes occurring during storage. J. Dairy Res. 44, 1977, 283-292.

Discussion with Reviewers

M.V. Taranto: You describe a new syneresis test in Materials and Methods, yet you make no reference to the standard method which your test is replacing. What is the standard method your new test is replacing? On what basis was your centrifuge test compared to this standard method?
Authors: Several methods have been described in the literature for the measurement of syneresis in various systems. They are all as arbitrary as the test used in this study and none of them may be called *the standard method*. Thus, we have not replaced any standard method and, consequently, have not compared our test to other tests; such findings will be presented in the description of the test to be published separately. Emmons *et al.*[16] described a method for the measurement of syneresis in coagulated milk. The coagulum is cut into quarters in a 400-ml beaker and heated to 90°F for 1 h; the volume of separated whey is measured in millilitres and recorded as the *drainage test*. This test was modified and used by other authors[17,18]. Susceptibility to syneresis in milk gels made by acidulating skim milk with various acidulants under varying conditions of pH and temperature fluctuates within a wide range. Whey separation can be accelerated by centrifugation. However, centrifugation at a single g-force for an arbitrary period of time would certainly fail to reveal the susceptibility to syneresis in such diverse gels. Fig. 2 illustrates the separation of whey from the protein network over a range of centrifugal force from 1 to 1000 g. The advantage of this test is that it respects the peculiarities of various gels and presents the results in a graphic form. In fact, the slopes of the initial linear portions of the curves were of the greatest interest to us as the indicators of the susceptibility to syneresis.

M.V. Taranto: How do you know that the g-forces to which the gels are exposed in your centrifuge test do not disrupt the protein network and cause an increase in the loss of fluid?
Authors: We anticipate that centrifugation compacts rather than disrupts the protein network of the gels. A similar compaction may take place when whey is separated from the gels on prolonged standing or at a high temperature. The linear relationships obtained by centrifugation at low g-forces (up to 50-100 g), when the gels lost less than 10% of liquid, are indicative of the susceptibility to syneresis. The gels may collapse when the g-force is further increased but that does not affect the importance of the initial stages of centrifugation.

M.V. Taranto: You show a marked increase in gel firmness in Table 2 due to preheating yet no specific comments regarding the reason for this effect were given in the text. Would you comment on the mechanism by which preheating of milk leads to an increased firmness in the resultant gel?
Authors: The increased firmness of gels made from preheated (90°C for 5 min) milk is attributed to the formation of a network consisting of casein micelles arranged in chains[4]. It is known that preheating leads to the formation of a complex between β-lactoglobulin and κ-casein and that this complex protects the micelles from an excessive fusion[14,19,20]. A network composed of casein micelle chains is capable of immobilizing larger volumes of water and, thus, is less susceptible to syneresis than a network in which the micelles are fused in large clusters. A model explaining the different properties of gels made from preheated and non-preheated milk was published earlier[4].

C.A. Ernstrom: You found that skim milk gels made with GDL at 90°C were firmer than those made at 40°C. Could you attribute this to gel structure or was there more syneresis at 90°C than at 40°C?
Authors: There was no visible syneresis in either gels shortly after their preparation. However, the gels made with GDL at 90°C were firmer and less susceptible to syneresis than those made at 40°C when measured by the centrifugation method. This difference in firmness and syneresis is attributed, in our opinion, to the gel structure, *i.e.* to the arrangement of the protein particles in the network[4].

C.A. Ernstrom: Our experience with Cottage cheese made with HCl acidification was that it was difficult to get a firm, smooth, meaty texture. When cooked firm enough, it tended to be mealy. On the other hand, Cottage cheese acidified with GDL can be made into a firm, yet smooth curd. Do your results suggest an explanation for this difference?
Authors: The main difference between acidification with HCl and GDL is the rate at which pH is lowered, being considerably lower with GDL. Presumably, a lower rate leads to a network with fewer links between the casein particles whereas the higher rate leads to the formation of clusters or ring structures with a greater number of links between the particles. From the limited number of samples tested it appears that the cluster type structures would give a mealy texture whereas the chain type networks would give a firm, smooth, and meaty texture. This, however, needs to be confirmed by further experimental work.

M.V. Taranto: You mention the sequestering of calcium by oxalic acid as a possible cause for the disintegration of the casein micelles. These data suggest that Ca^{++} ions may play an important role in the gelation process. This role has been confirmed in natural cheese manufactured from a rennet curd (pH 5.5). Did you compare the Ca^{++} content of the gels produced at each pH-temperature combination? Have you tried adding additional Ca^{++}

to the oxalic acid sample to see if gelation could be induced? The treatment of the GDL and HCl gels with EDTA should provide the same evidence for the role of Ca^{++} in gelation. If the removal of Ca^{++} causes a disintegration of the protein network, this is a strong circumstantial evidence for the Ca^{++} requirement.

Authors: The role of Ca^{++} is perhaps more important in gels made at pH 5.5 than at pH 4.6. At the latter pH, the colloidal calcium phosphate is completely dissociated from the casein micelles. In this paper we have not examined the role of Ca^{++} in the gelation process but we believe that it would be useful area for further work. In related work, however, the sequestering of Ca^{++} with hexametaphosphate was found to disintegrate the casein micelles and, simultaneously, to prevent the formation of heat-induced milk gels[8].

D.J. Gallant: As your earlier published work showed, it is very important to study the influence of factors such as pH, preheating of milk, temperature of gelation, and type of acidulant used in skim milk processing. Concerning firmness, GDL-induced skim milk gels, especially those made at 90°C, were harder than gels obtained with other acidulants. Saio[21], in a study on tofu and on the relationship between texture and fine structure, noted that tofu coagulated with GDL was harder to penetrate but had a lower internal hardness and was more fragile and cohesive than tofu coagulated with $CaSO_4$. Did you find the same phenomenon for GDL-induced skim milk gels as described by Saio on tofu?

Authors: No, we did not. In our study we have compared the effects of various acidulants on the gelation of skim milk but we did not study the effect of $CaSO_4$. The firmness profiles of the skim milk gels were not similar to those of the soy protein gels; examples of such profiles were published earlier[1,5].

D.J. Gallant: As reported by the authors[4], the relation between rheological behaviour and microstructure of GDL-induced skim milk gels was more dependent upon the micellar arrangement than upon the size and shape of the casein micelles. Saio[21] stated in the paper mentioned above that the SEM images of GDL- and Ca-coagulated tofu showed different fine structures which were correlated with texture measurements. As the authors assume, it is intriguing to observe such differences between the microstructures of gels made with HCl and citric acids whereas firmness and syneresis are practically similar. Would it not have been more practical to use SEM rather than TEM for the aspect of correlations between texture and fine structure?

Authors: This question was asked before SEM micrographs were included in the revised version. Now it can be shown that SEM is suitable for revealing the overall microstructure of the gel network but is unsuitable for studies of the ultrastructure of casein micelles and their association in the network. Changes and differences at the micellar level were revealed by TEM (thin sections of embedded specimens and replicas of freeze-fractured specimens)[4]. SEM of these gels is quite difficult to carry out because the gels are very fluffy (approximately 5-6 g solids in 100 ml) and subject to charging artifacts in the microscope. Fig. 3 and 4 were obtained at a gun potential of 5 kV and a beam current of 40 µA; the scanning time was decreased to 20 sec from the usual 40 sec used for picture taking.

D.J. Gallant: As observed by the authors, differences in the strength of the linkages joining the casein particles cannot be shown in the present paper. It could certainly be demonstrated using freeze-etching which is a more accurate technique than most cytochemical techniques. I think that it would be interesting to follow by freeze-etching the steps of formation of the core-and-lining structures which seem to be linked to an effect of boiling or heating. At least the phenomenon is partly similar to the heating of yeasts in cooked loaf crumb (unpublished data). Would it not be interesting to use immuno-cytochemical techniques to study the formation of the β-lactoglobulin-κ-casein complex?

Authors: Freeze-fracturing was used to demonstrate[4] the formation of the core-and-lining structures of casein micelles in gels formed by heating skim milk at pH 5.5 to 90°C. The existence of these structures was also confirmed by replication of freeze-fractured and consecutively dried specimens[22] which corresponds, to some extent, to a very deep freeze-etching (Fig. 23 in the review by M. Kaláb[11] in this volume). We will probably disappoint the reviewer by saying that the images of freeze-fractured specimens showed less detail than did thin sections. We would not dare to comment on the apparent similarity between the core-and-lining structures of casein micelles and heated yeasts in cooked loaf crumb, as the materials are different both as far as the composition and the dimensions of the components are concerned. We agree that it would be interesting to study the formation of the β-lactoglobulin-κ-casein complex under the microscope.

M.V. Taranto: The lining around casein particles is observed in gels formed at pH 5.5 at temperatures above 70°C (Fig. 7). The gels formed at pH 4.6 and 90°C (Fig. 5) show no lining regardless of the type of acidulant, possibly because no β-lactoglobulin was incorporated into the protein network. (A) Have you performed any electrophoretic analyses to determine the different types of proteins that are found in the gels made at various pH and temperature combinations? (See C.V. Morr. J. Dairy Sci. 48(1), 1965, 8-13)[23].
(B) Have you ever made any gels from purified caseins (devoid of β-lactoglobulin) to see whether or not the lining structure is formed in gels made with the appropriate pH and temperature combinations?

Authors: (A) Polyacrylamide gel electrophoresis indicated that whey proteins, particularly β-lactoglobulin and serum albumin, were part of the insoluble protein network in gels made from milk heated to 90°C. In gels made from unheated skim milk the whey proteins were present in the liquid phase (whey).
(B) Casein micelles were isolated from fresh skim milk by centrifugation and were washed free of whey proteins. The casein micelles were suspended in a protein-free milk dialyzate and were gelled with GDL at 90°C at pH values ranging from 5.0 to 6.3. The core-and-lining structure was not observed

either in gels made at pH 5.5 or at any other pH value within the above range. However, when β-lactoglobulin was added to the casein micelle suspension, the core-and-lining structure was observed at pH 5.5. This investigation is still incomplete for publication.

S.M. Gaud: The authors state that *larger* particles in the gel resulted from the higher temperatures to which the gels were subjected. This is in contrast to the report in an earlier paper[9] that the particles in yoghurt gels prepared from heated milk were *smaller* than in unheated systems and that no change in micelle size was observed upon heating skim milk. Smietana et al.[24], on the other hand, did observe larger micelles in heated skim milk. This is not a major point since the authors are reporting and discussing their actual observations of the gels formed from acidulants but a comment could clarify these differences for the reader.

Authors: The study by Kalab et al.[9] was concerned with yoghurt made from milk which had been or had not been preheated to 90°C and was subsequently incubated with lactic cultures to produce yoghurt. Preheating leads to the formation of a β-lactoglobulin-κ-casein complex which prevents the casein micelles from excessive fusion during the coagulation of the milk[19,20] by the bacterial action (decrease of pH). Unheated milk does not contain the above complex and, hence, the unprotected casein micelles fuse into large particles. A behaviour similar to that in yoghurt was also observed in GDL-induced gels made from preheated milk[4]. The phenomenon observed by Smietana et al.[24] and by other authors, for example Creamer and Matheson, concerns the effect of temperature alone on the dimensions of the casein micelles: they become enlarged as the temperature is increased and while the milk remains liquid. Finally, this study deals with the dimensions of casein particles in gels made at different pH values and at different temperatures. If a certain amount of an acid is added to milk to bring the pH value to 5.5 (measured at 22°C) and the acidulated milk is heated to 90°C, the actual pH at that temperature is considerably lower. This means that the casein micelles are subjected not only to the effect of temperature but also, at the same time, to the effect of a low pH value (a high concentration of H^+ ions); the increase in the micelle dimensions and their aggregation take place at the same time. Thus, the casein micelles in gels made at 40°C are smaller than in gels made at 90°C.

D.N. Holcomb: Presumably disulfide bonding is involved (for example, [14]) in forming the filamentous appendages (Fig. 9 B). Can you make any statement about the type of bonding involved in formation of the lining - whether it is covalent or weaker (hydrophobic, electrostatic) in nature?

Authors: It is difficult to identify the precise nature of the bonds between the appendages from the experiments described in this study. Possibly a combination of both covalent (-S-S-) bond formation and noncovalent linkages, *e.g.* hydrophobic, or electrostatic forces, is involved. We did not study the contribution of the different types of bonding.

Additional References

16. Emmons D.B., Price W.V. and Swanson A.M. Tests to measure syneresis and firmness of Cottage cheese coagulum, and their application in the curd-making process. J. Dairy Sci. 42, 1952, 866-869.

17. Green M.L. Studies on the mechanisms of clotting of milk. Neth. Milk Dairy J. 27, 1973, 278-285.

18. Lelievre J. Rigidity modulus as a factor influencing the syneresis of renneted milk gels. J. Dairy Res. 44, 1977, 611-614.

19. Knoop A.-M. and Peters K.-H. Die Ausbildung der Gallertenstruktur bei der Labgerinnung und der Säuregerinnung der Milch. Kieler Milchwirt. Forsch. Ber. 27, 1975, 227-248.

20. Knoop A.-M. and Peters K.-H. Die submikroskopische Struktur der Labgallerte und des jungen Camembert-Käseteiges in Abhängigkeit von den Herstellungsbedingungen. Milchwissenschaft 27, 1972, 153-159.

21. Saio K. Tofu - Relationship between texture and fine structure. Cereal Foods World 24, 1979, 342-345 and 350-354.

22. Kalab M. Milk gel structure. XII. Replication of freeze-fractured and dried specimens for electron microscopy. Milchwissenschaft 35, 1980, 657-662.

23. Morr C.V. Effect of heat upon electrophoresis and ultracentrifugal sedimentation of skimmilk protein fractions. J. Dairy Sci. 48, 1965, 8-13.

24. Smietana Z., Jakubowski J., Poznanski S., Zuraw J. and Hosaja M. The influence of calcium ions and heat on size changes of casein micelles in milk. Milchwissenschaft 32, 1977, 464-467.

25. Creamer L.K. and Matheson A.R. Effect of heat treatment on the proteins of pasteurized skim milk. New Zealand J. Dairy Sci. Technol. 15, 1980, 37-49.

About SCANNING ELECTRON MICROSCOPY, Inc.

This not-for-profit organization was established with the following goals: **a.** *Promotion of advancement of science of SEM and related material characterization techniques;* **b.** *Promotion of application of these techniques in existing and new areas of applications;* **c.** *Promotion of these techniques so that their users obtain the best information of the highest quality from their instruments.*

SEM, Inc. publishes the journal "SCANNING ELECTRON MICROSCOPY" and sponsors the annual SEM meetings. Suggestions on activities that SEM may sponsor are invited.

SCANNING ELECTRON MICROSCOPY/1982

SEM/1982 will take place during **April 26-30** at **Disneyland Hotel in Anaheim** (southern California). A Call for Papers is available. In addition to a general session, programs of common interest include: *Analytical Electron Microscopy (including STEM); Microprobe Surface Analytical Techniques; Related Microscopy and Microanalytical Techniques; Forensic Applications* and nine hours of tutorials (for Sunday night and Monday) on **Introduction to Scanning Electron Microscopy.**

Comprehensive programs on SEM/STEM applications in: *Semiconductor Characterization and Failure Analysis, Material Characterization and Fractography, SEM in Geosciences and Particulate Characterization* are being planned.

Biological/Biomedical programs are being organized on *Cell Biology, Ultrastructural Effects of Radiation on Cells and Tissues, Muscle and Connective Tissue, Biological Microanalysis, Cell Surface Labeling, Cell Culture, Developmental Biology, Neurobiology, Sensory Organs, Cancer, Clinical Applications, Pathology of Particulate Related Diseases, Stones and Crystals in Diseases, Medical Microbiology/Virology, Parasitic Nematology, Skin Biology, Blood Forming and Immune Systems, Blood Vascular Systems, Digestive Systems, Male Reproductive System, Other Organ Systems, Mineralized Tissue, Plants and Their Environment, and Food Microstructure.*

REGISTRATION FOR SEM/1982:

The registration fees for SEM/1982 will be kept same as for 1981 & 1980. **Thus an early bird registration fee (available till Dec. 1, 1981) of: $25.00** (with no SEM/1982 published parts), or **$75.00** (with one SEM/1982 journal part), **or $124.00 (including complete set of SEM/1982), will admit the registrant to the entire 5 day meeting including all sessions, tutorials, and a comprehensive equipment exhibition.** Fees after Dec. 1 will be $15.00 higher (till March 20, 1982).

The subscription price for the 1982 issues of the SEM journal (if paid by March 20, 1982) is $99.00 (for U.S. delivery) and $109.00 (elsewhere).

TRAVEL SUPPORT FOR SEM/1982 MEETINGS:

Authors offering tutorial or review papers may apply for travel support. Other authors whose papers make **significant contributions to SEM/1982 program** may apply for **Presidential Scholarships** (up to 20 scholarships will be awarded). **A complete Letter of Intent** per the Call for Papers including the *extent of subsidy desired* (limited to $300 for travel within North America and $500 elsewhere) should be submitted. Scholarship applicants should also enclose a recommendation letter (preferably from someone associated with SEM Inc. activities). The decision to support travel is made in consultation with the **organizers of specific programs.**

The promise of travel support (to be confirmed in writing by Nov.-Dec.) will be automatically considered withdrawn, if **(a) the full paper was not submitted on time (January 15, 1982),** or (b) if the paper was not acceptable to reviewers or editors, or (c) if the paper was not presented by the person promised support.

SPECIAL PROGRAM ON ELECTRON BEAM INTERACTIONS WITH SOLIDS

Just prior to SEM/1982 meetings, a special program on **Electron-Beam Interactions with Solids for Microscopy, Microanalysis and Microlithography** will take place (during April 18-23 at Asilomar near Monterey in Northern California). **Dr. David F. Kyser** (Signetics Corp., Advanced Tech. Ctr., 811 E. Arques Ave., Sunnyvale, CA 94086, phone 408-746-1452, **as the general chairman**), and Dr. Dale E. Newbury (National Bureau of Standards, Washington, D.C.), Prof. H. Niedrig (Tech. Univ., Berlin, W. Germany), and Prof. R. Shimizu (Osaka Univ., Japan), are the organizers of this program. Twenty invited contributions of one hour each have already been planned. Interested contributors should contact one of the organizers. **Attendance at this program will be by application only.** Details are available on request.

PAPERS FOR THE SEM JOURNAL:

Papers for publication in the SEM journal may be offered at anytime per simple instructions available. In 1981, three parts containing papers organized by specific subjects, are being first published. The fourth part will contain all papers not included in earlier parts.

Papers for publications only are invited. Of maximum interest are papers emphasizing topics of general interest to all SEM users (SEM techniques, theory, instrumentation, interpretation, etc.; other related techniques; and novel or unusual applications of SEM). It is no longer necessary that a paper submitted to the SEM journal be presented at the SEM meetings. Application papers must conform to announced themes.

Oral presentation of a paper at some other meeting as well as publication in the form of summary or abstract (e.g., in proceedings, non-English publications, etc.) does not preclude consideration of a paper by SEM. In addition, review and tutorial papers can be also offered.

SEM journal enjoys a wide international circulation and is now abstracted/indexed by most leading services. Many libraries are subscribers on a standing order basis. Please recommend this publication to your colleagues and library.

ITEMS AVAILABLE AT NO CHARGE

—*Call for Papers and Hotel Information for SEM/1982.*
—*Ordering information for the SEM/1981 volume*
—*SEM Publications descriptions, price list, and ordering information.*
—*List of papers whose reprints are available from SEM/1978 1979 & 1980. (a) Physical Papers*
(b) Biological Papers
—*SEM Inc. mailing list form.*

For more information or suggestions, contact:

Dr. Om Johari, SEM, Inc.
P.O. Box 66507
AMF O'Hare (Chicago), IL 60666, USA
Phone 312-529-6677

STRUCTURES OF VARIOUS TYPES OF GELS AS REVEALED BY SCANNING ELECTRON MICROSCOPY (SEM)

V.E. Colombo, P.J. Späth[*]

Central Research Unit, F. Hoffmann-La Roche & Co., Ltd.,
CH-4002 Basel, Switzerland
*Institute of Hygiene and Medical Microbiology, University Berne,
Berne, Switzerland, present address:
Swiss Red Cross, Central Laboratory of the Transfusion Centre,
Berne, Switzerland

Abstract

This contribution presents a study of three different types of gels using scanning electron microscopy (SEM). The morphology of polyacrylamide, agarose and alginate gels was observed in relation to various dehydration and drying procedures. A main problem in the preparation of gels for SEM consists of specimen shrinkage. In this study the shrinkage was confined to only one dimension because thin slab gels which adhered firmly to a glass surface were used. The largest pores were found when polyacrylamide and agarose gels were freeze-dried. At the same time the gels underwent the least shrinkage. Chemical dehydration in 2,2-dimethoxypropane (DMP) preserved the details of the gel structures equally well as dehydration in ethanol. However, when compared with freeze-dried gels, chemically substituted and chemically dehydrated gel preparations showed pores which were smaller by more than one order of magnitude. Experiments indicated that the pores in agarose gels were of the same dimensions as pores in polyacrylamide gels, although the two different gel systems are capable of separating substances within considerably different molecular weight ranges. The apparent morphology of dehydrated and dried calcium alginate gels as revealed by SEM did not change according to the applied techniques.

KEY WORDS: Polyacrylamide gel, Agarose gel, Alginate gel, Gel morphology, Sample preparation, Dehydration, Drying, Artefacts, Scanning electron microscopy.

Introduction

The principles on which separation of macromolecules in polyacrylamide gels is based are still not well understood. The Ogston model might provide a good basis for explaining the empirical observations. It claims only minimal postulations towards pore architecture (1). This model was particularly applied to electrophoretic systems (2) and seems realistic in terms of polymer structures, such as polyacrylamide, cross-linked dextran, agarose or starch. According to Chrambach et al.(3), the retardation of proteins in polyacrylamide gels is believed to be based upon the ratio of the gel pore size vs. the size of the migrating proteins. Following these lines, the electron microscopical study of gel structures should provide a better understanding of the separation characteristics of polyacrylamide gels. Examination of polyacrylamide gel by transmission electron microscopy was initially not successful (4). Dehydration together with embedding of the gels failed to produce recognizable structures. Although early SEM studies on polyacrylamide gel frame work after critical-point drying did not show real structures (5), documentations of gel pores after freeze-drying were published (5, 6, 7). Therefore, in our initial work the freeze-drying method was used (8). In this work we attempted to determine whether the retardation of proteins during electrophoresis and the pore sizes as observed by SEM would show any correlation. The results obtained raised the question about the effect of dehydration and drying on apparent gel structures.

The present report deals with gel morphology after dehydration and drying of specimens to be subjected to SEM. The effect of a rapid chemical dehydration technique (9) was studied and was compared with both the ethanol dehydration and freeze-drying techniques. The present study also includes preparations of two other gel types. However, the results were intended to aim finally at a better understanding of the separation characteristics of polyacrylamide gels.

In contrast to the covalently cross-linked polyacrylamide gel, the agarose gel is a hydrocolloid of carbohydrate polymers which is stabilized by the formation of weak hydrogen and hydrophobic bonds (10). Alginate, the third gel type, is a derivative of polymannuronic and polyguluronic acid and is obtained from seaweed. It is widely used as a sizing agent and stabilizing colloid in food processing and cosmetic industries. Alginate structures can be stabilized and alginate gels can be rendered insoluble in water by the interaction of calcium ions with the acid residues (11).

Materials and Methods

All chemicals used throughout the study were of the highest commercial quality.

The Gels

Gel monomer concentrations (Definition of the terms T and C). The total concentration T (%, w/v) is the amount of monomer added to 100 ml water. C is the weight of the cross-linker compounds expressed in % of the total weight of the monomers (12).

Polyacrylamide Gels. Rod gels were prepared by polymerization of acrylamide (SERVA, Heidelberg, FRG) and of appropriate concentrations of N,N'-methylenebisacrylamide (BIS) and N,N'-diallyltartardiamide (DATD) with tetramethylethylenediamide (TEMED, 0.05 %) and ammoniumperoxysulfate (0.075%) in glass tubes (inner diameter 4 mm)(13). The gels were removed from the glass cylinders by injecting water between the glass and gel surface. The wet gels were sectioned with a razor blade and the sections (about 1 mm thick) were freeze-dried. Slab gels (thickness 0.105 mm) were polymerized in water in a slit-like compartment formed by the smooth surfaces of two microscopic slides which were separated by two cover slips. Slab gel monomer concentrations were T_{BIS} = 5.3 %, C_{BIS} = 5 %, and T_{DATD} = 6.5 %, C_{DATD} = 27 %. The thin gels remained mounted on one glass slide and were dehydrated according to one of the described methods.

Agarose Gels. 1 % agarose gels were formed in a manner similar to the above. The thin gels remained attached to the glass slide for further treatment.

Alginate Gels. A 2 % sodium alginate solution in distilled water was spread with a syringe into a 1 M $CaCl_2$-solution. The resulting alginate beadlets (size 2.5 to 3 mm in diameter) were hardened in the $CaCl_2$-solution for at least 2 hours and were then rinsed thoroughly with distilled water. The calcium alginate gels were sectioned with a razor blade into small segments prior to dehydration.

Sodium alginate beadlets were also frozen directly, without any further treatment, in melting nitrogen and were then freeze-dried.

Dehydration

Three procedures for the dehydration of the gel specimens were used and are outlined below:

Freeze-Drying. The preparations were frozen in melting nitrogen at -210° (14). Frozen specimens were transported in liquid nitrogen and then freeze-dried (freeze-drier Leybold-Heraeus, Köln, FRG) at -25° C. For the preparation of alginate gels, freeze-drying was carried out at a temperature below -45° C during 24 hours (freeze-drier WKF, Forschungs- und Laborgeräte GmbH, Modautal, FRG).

Chemical Dehydration with DMP (9). Dehydration by DMP (FLUKA, Buchs SG, Switzerland) was performed by immersing small gel particles into slightly acidified DMP (1 drop of 2M HCl per 20 ml) for at least 10 minutes. The length of time used for the chemical dehydration did not influence the preservation of any gel detail. The dehydrated material was transferred in acetone and kept there until critical-point drying. Alternatively, chemically dehydrated polyacrylamide gel preparations were air-dried.

Chemical Substitution with Ethanol and Freon[R] 113 (15). Samples were dehydrated by using a graded series of ethanol (20, 40, 60, 80, 95, 100 % ethanol for 20 minutes each), followed by a graded series of Freon 113 (30, 50, 70, 90, 100 % Freon 113 for 20 minutes each). The dehydrated material was critical-point dried.

Critical-Point Drying

Chemically substituted or chemically dehydrated samples were critical-point dried (apparatus POLARON Equipment Ltd., Watford, Great Britain) by using carbon dioxide or Freon 13 as the transitional fluid.

Air-Drying

Chemically dehydrated specimens were air-dried in a dust-protected area of the laboratory.

Surface Tension

The surface tension of DMP was measured by means of a interfacial tensiometer according to Lecomte du Noüy (A. Krüss, opt.-mech. Werkstätte, Hamburg, FRG) using a Pt-Ir-ring, 1.9 cm wide. The apparatus was calibrated with ethanol and the surface tension of DMP was determined at 22° C and at 7° C.

Scanning electron microscopy

The dry gel preparations were separated from the glass, fractured into pieces of appropriate size and immediately mounted on aluminum stubs with a conductive cement. The mounted specimens were coated with palladium/gold (40/60)(15 nm) and consecutively with gold (15 nm) by diode sputtering. By using two different targets, thermal stress during the sputtering process towards the fragile gel preparations should be minimized. Alternatively, the samples were vacuum-coated with carbon (5 nm) and with gold (25 nm). Preference was given to the former technique, because sputtered gel specimens seemed to be coated more uniformly. The consecutive observation and documentation was done on a JEOL JSM-35 SEM, operating at a acceleration voltage of 20 kV with a specimen tilt of 30°.

Results

At equimolar conditions with low concentrations of the cross-linkers BIS or DATD, almost identical dimensions and details of freeze-dried gels were observed (figures 1, 2). The retardation coefficients of a distinct protein are markedly different in such gels (13). In highly cross-linked equimolar BIS- and DATD-polyacrylamide gels, the differences in the retardation coefficients are less pronounced. In contrast to the situation at low concentrations of the cross-linkers, the gel pores are no more comparable (figures 3, 4).

From observations with SEM no correlation between the retardation coefficients and the apparent structures of freeze-dried gels could be concluded. Furthermore, the gel pores seemingly were too large compared with the components separated therein. Figure 5 shows a cut surface of a freeze-dried polyacrylamide gel, wherein trypsin inhibitor was electrophoretically separated. Freezing and sublimation of water might have altered the real gel structures. Otherwise, the freeze-drying method reduces other problems in gel preparation, such as shrinkage associated with air-drying (6) and critical-point drying (5).

During further work, prevention of freezing and shrinkage of gels during SEM preparation was attempted. The fast chemical dehydration with subsequent air- or critical-point drying of the gels was used. To avoid shrinkage at least in two dimensions, the slab gels remained mounted on one of the glass plates forming the compartment during gel polymerization. In chemically dehydrated slab gels, three zones were seen. Small pores could be detected in the inner region of the gels (figures 6, 7). Air-dried gels presented two sharply defined outer layers of a crust-like appearance (figure 6). When the DMP-dehydrated gels were critical-point dried, the two outer zones were inhomogeneous (figure 7).

Fig. 1. Typical structure of BIS-cross-linked polyacrylamide rod gels at low concentrations of the cross-linker; freeze-dried (T = 10.0%, C_{BIS} = 5.0%) (bar = 5 μm)

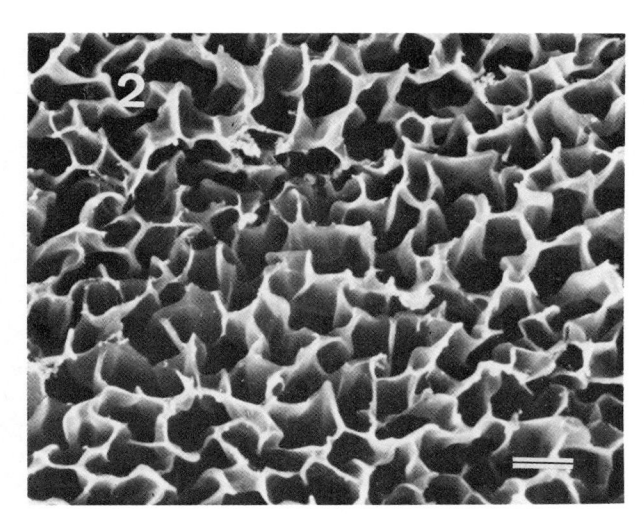

Fig. 2. Typical structure of equimolar DATD-cross-linked polyacrylamide rod gel (T = 10.2%, C_{DATD} = 7.2%) showing similar details of the pores; freeze-dried (bar = 5 μm)

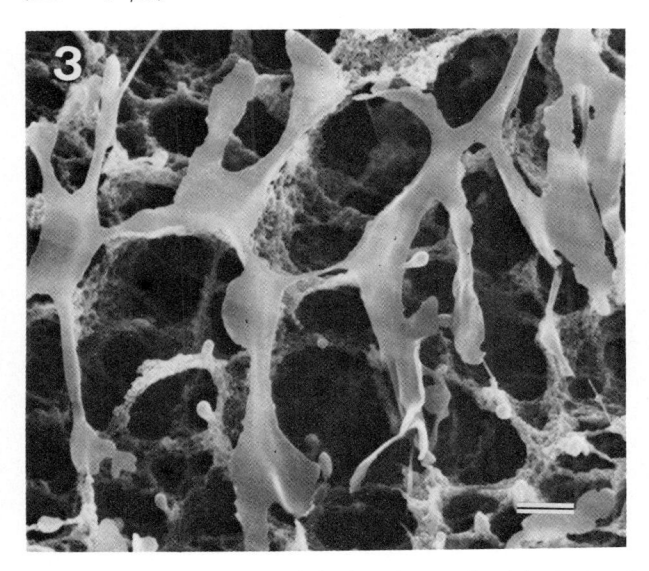

Fig. 3. Highly cross-linked polyacrylamide rod gel (T = 10.0%, C_{BIS} = 40.0%) demonstrating a significant alteration of the gel structure. The figure reveals a markedly decreased organization of the cellular structures; freeze-dried (bar = 5 μm)

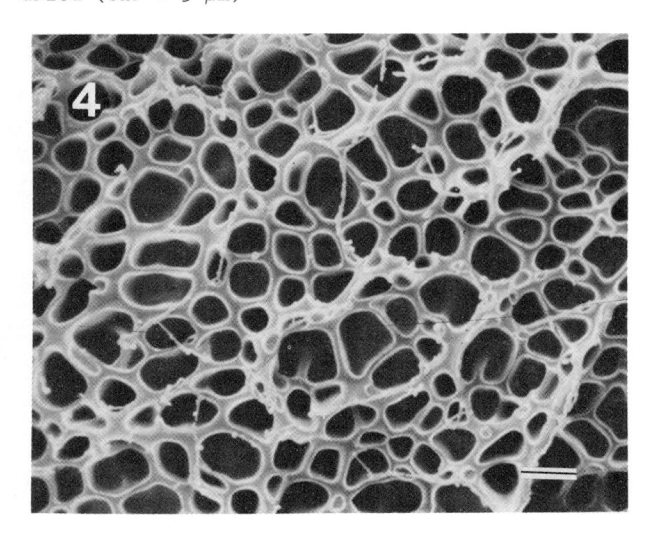

Fig. 4. Highly DATD-cross-linked polyacrylamide gel (T = 11.9 %, C_{DATD} = 49.7 %) showing a close relationship to the structure and dimension of the pores cross-linked with low concentrations of the cross-linker; the slightly thicker walls could be due to the applied high concentrations of DATD; freeze-dried (bar = 5 μm)

These gel structures correspond with the observed structures of ethanol-dehydrated preparations (figure 8), but were not similar to the structures seen in freeze-dried gels (figure 9). The results obtained with polyacrylamide gels were compared with agarose and alginate gel preparations. In freeze-dried agarose slab gels, fine details and dimensions of the pores were not substantially different from those of the freeze-dried polyacrylamide gels (figure 10). Ethanol-dehydrated agarose slabs were in all cases comparable to the corresponding polyacrylamide gel preparations (figure 11). However, chemically dehydrated agarose slab gels adhered tightly to the plate and could not be removed after dehydration. Therefore they were exempted from further examination. Calcium alginate gel showed round and regular pores without any variation in dimension and shape in regard to the applied dehydration and drying techniques (figures 12, 13). Sodium alginate showed an open network with structures similar in shape and dimensions to freeze-dried polyacrylamide and agarose gels (figure 14).

Table 1 summarizes the dominant features in the various gels, as well as typical features of the gels which occured during the different preparation techniques.

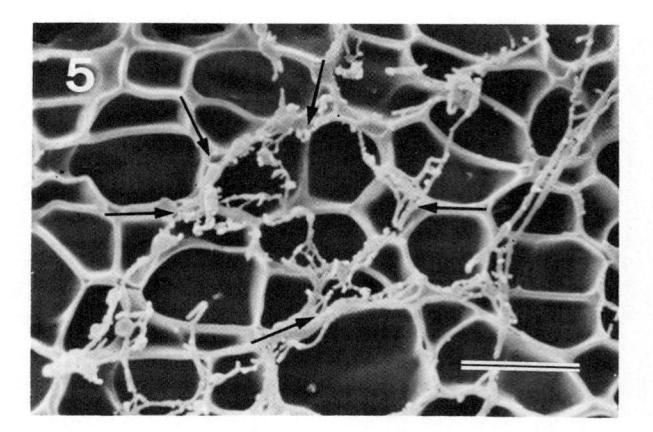

Fig. 5. BIS-cross-linked polyacrylamide gel (T =10%, C_{BIS} = 5 %) containing a protein. The slender network (arrows) is electrophoretically separated protein (trypsin inhibitor, molecular weight 21'000, precipitated with trichloroacetic acid) after staining with Coomassie blue. The sample shows a fractioned section of the stained area of a freeze-dried rod gel (bar = 5 μm)

Table 1. Some features of polyacrylamide, agarose and alginate gels in regard to different dehydration and drying techniques

Preparation / gel type	Freeze-drying	Critical-point drying DMP-dehydrated	Critical-point drying EtOH-dehydrated	Air-drying DMP-dehydrated
polyacrylamide gel	pores several μm wide	3 zones, 2 inhomogeneous layers, pores in submicron range, not comparable to freeze-dried gel	same as corresponding DMP-dehydrated gel, not comparable to freeze-dried gel	fairly good preservation, 2 homogeneous outer layers not comparable to freeze-dried gel
agarose slab gel	similar to polyacrylamide gel		similar to polyacrylamide gels, not comparable to freeze-dried gel	
sodium alginate gel	similar to polyacrylamide gel			
calcium alginate gel	regularly shaped pores, sizes below μm-range		regularly shaped pores, no differences from the corresponding freeze-dried gel	

Discussion

It was assumed on the basis of mathematical models that the gel retardation coefficients and the apparent pore dimensions were reciprocally correlated (3). The existence of an inverse proportion of the retardation coefficients and the apparent pore sizes in BIS-cross-linked polyacrylamide gels was confirmed in this study (figures 1, 3). The variation in the concentration of the cross-linking DATD component resulted in less pronounced differences in the retardation coefficients (13); the apparent microstructures in lowly and highly cross-linked DATD-polyacrylamide gels were very similar (figures 2, 4).

The coincidence of the relative values of retardation coefficients and the apparent changes in the gel pore size may mislead one to the conclusion that the structures of freeze-dried gels would not be correct and ought to be excluded on the basis of the following facts:

a) the observed structures of the gels were too large as compared to the proteins separated therein (figure 5),

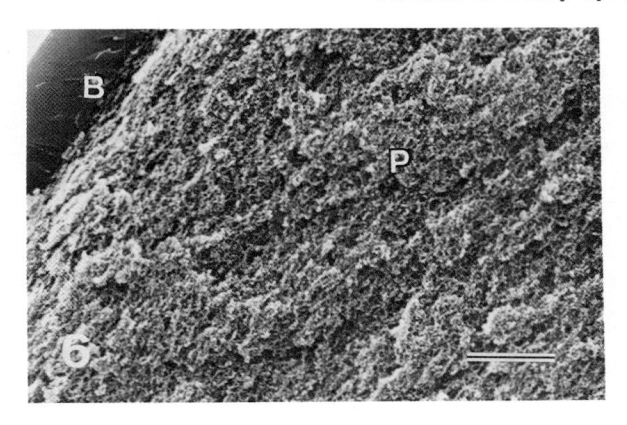

Fig. 6. DMP-dehydrated BIS-cross-linked slab polyacrylamide gel (T = 5.3 %, C_{BIS} = 5 %); air-dried. One of the two sharply bordered outer layers is shown (B). The central region displays fine pores of submicron dimension (P). Absence of greater pores in the region between the outer layer and the core (bar = 5 µm)

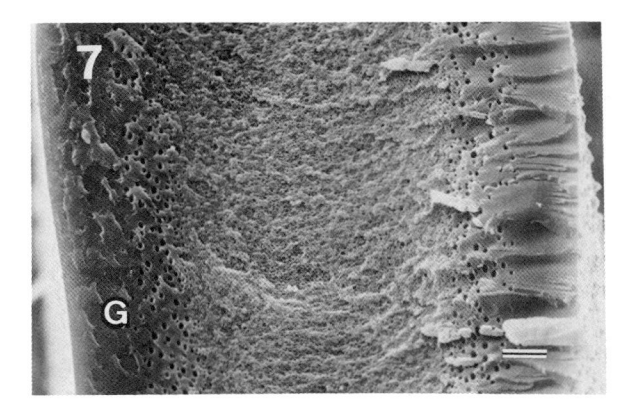

Fig. 7. DMP-dehydrated BIS-cross-linked slab polyacrylamide gel (T = 5.3 %, C_{BIS} = 5 %); critical-point dried. Two inhomogeneous identical layers are seen on both outer sides of the slab gel (G = side of the gel facing the glass during preparation). The region between the central pores and the outer borders displays greater pores (bar = 5 µm)

Fig. 8. Ethanol-dehydrated BIS-cross-linked slab polyacrylamide gel (T = 5.3 %, C_{BIS} = 5 %); critical-point dried, confirming fine details observed after chemical dehydration (bar = 5 µm)

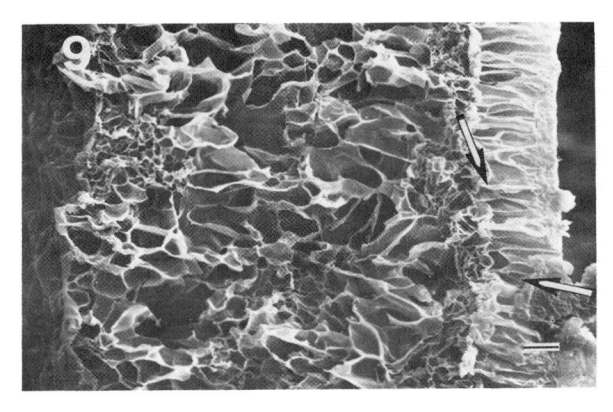

Fig. 9. BIS-cross-linked polyacrylamide gel (T = 5.3 %, C_{BIS} = 5 %); freeze-dried. The side of the thin gel facing the glass during preparation (arrow) shows elongated pores. The structures in the center of the slab gel display more or less round pores in the low µm range (bar = 10 µm)

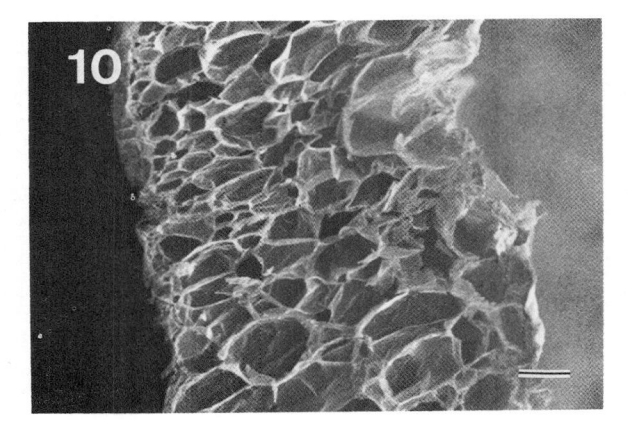

Fig. 10. Agarose slab gel (1%, w/v), freeze-dried, displaying typical gel pores (bar = 10 µm)

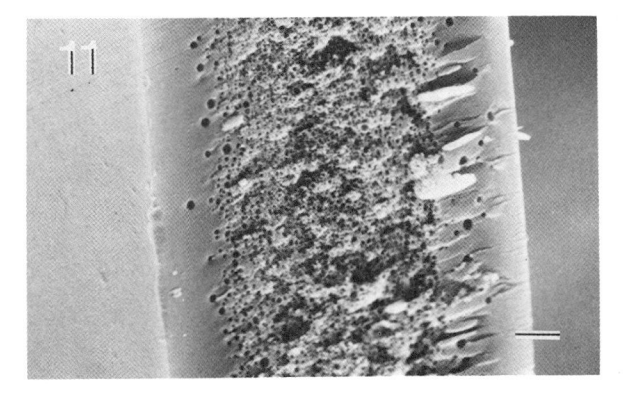

Fig. 11. Ethanol-dehydrated agarose slab gel (1%, w/v); critical-point dried. The structures show fine pores in the central region and two inhomogeneous outer layers (bar = 5 µm)

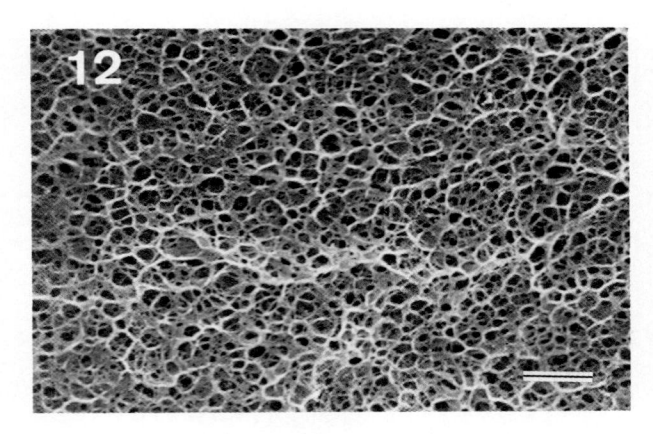

Fig. 12. Calcium alginate gel (2 %, w/v), freeze-dried, displays network structures with pores of sub-µm-dimensions (bar = 2 µm)

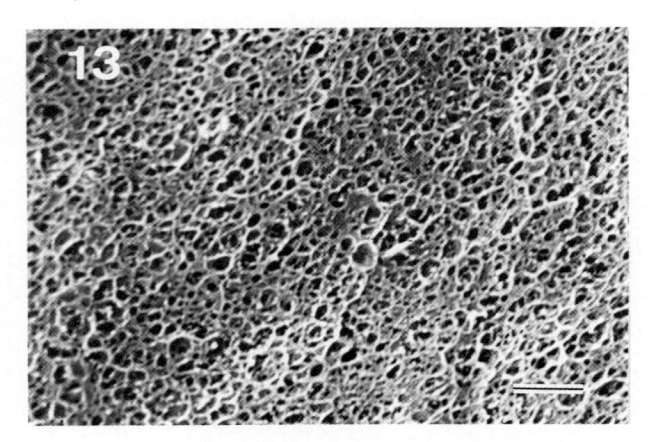

Fig. 13. Ethanol-dehydrated calcium alginate gel (2%, w/v), critical-point dried, demonstrates good preservation of fine details and structures of the small pores (bar = 2 µm)

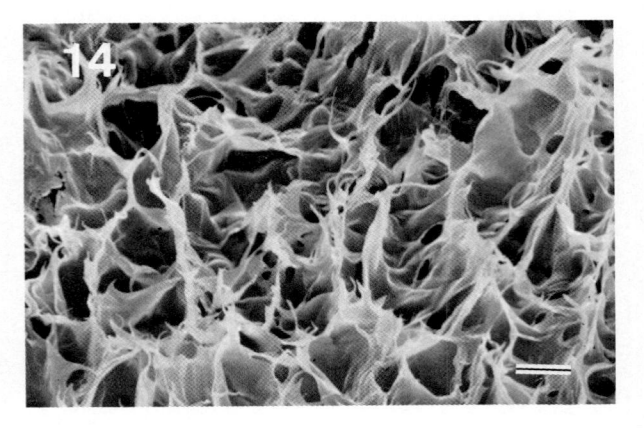

Fig. 14. Sodium alginate gel (2%, w/v), freeze-dried, displays network structure that is occasionally connected by fibrils (bar = 10 µm).

b) in equimolar BIS- and DATD-cross-linked polyacrylamide gels, the measured values of the retardation coefficients did not coincide with the apparent gel structures (figures 1-4, reference 13),

c) in 1%-agarose gels, molecules are separated within a considerably different molecular weight range than in polyacrylamide gels with T between 5 and 8 %; nevertheless, the observed pores of freeze-dried agarose gels were of the same dimension. The texture of freeze-dried polyacrylamide gels is, at least partially, the consequence of the freezing and the sublimation of the aqueous components. This is documented by the presence of oval pores in the border region that was facing the glass slide (figure 9). One might raise the question whether the freezing and subsequent sublimation process was retarded at this side of the thin gel.

The second part of this study was dedicated to the problem to what extent the gel structures were affected by dehydration and drying. In our approach, the freezing of the specimens was avoided and the inevitable shrinkage of the gels in DMP or ethanol was confined to one dimension, i.e. the thickness of the gels. An indication for only one-dimensional shrinkage of the gel preparation may be seen by the shape of the elliptical pores demonstrated in figures 7, 8, 11. The approved chemical dehydration technique was included in this work because acidified DMP reacts very rapidly and quantitatively with water (9). When chemical dehydration was followed by critical-point drying, the fine details observed were equivalent to those found after physical dehydration. The fact that the ultrastructure of DMP-dehydrated gels after air-drying was fairly well preserved may be attributable only to a small extent to the low surface tension of DMP, i.e. 20.4 mJ/m^2. Air-drying of chemically dehydrated SEM preparations may be essentially practicable owing to the permanent breakdown of the water as is present in ambient air, by the hydrolysis of DMP (16). We conclude that air-drying which follows DMP-dehydration may provide an acceptable alternative to the mostly dissatisfying technique of air-drying after dehydration by means of ethanol or Freon (17). Two border layers appeared in critical-point dried slab gels irrespective to the fact that only one side of the thin gel had direct contact to DMP or ethanol. The possibility that the drying step too may dispose of the apparent gel structures was indicated with DMP-dehydrated polyacrylamide gel preparations; comparatively larger pores were demonstrated to be located between the center and the outer borders only after critical-point drying, but not after air-drying. The apparent dimensions of the central pores in critical-point dried polyacrylamide slab gels are rather in accordance with the proportions claimed for the separation of macromolecules than the pores in freeze-dried gels.

So far, no interpretation can be induced by SEM in order to explain the empirically found sieving effect of polyacrylamide and agarose gels during electrophoresis. The present work suggests that dehydration and drying are serious impediments when preparing gels for electron microscopy, although artefacts may be introduced at almost any step of sample preparation. However, an extended SEM study upon the gel morphology of an additional type of a copolymer may mitigate this prejudice. The stability of calcium alginate originates first of all by virtue of the strong ionic binding sites of alginic acid with the divalent cation. Calcium alginate gels seemed to be rigid enough to overcome the stress of freeze-drying and critical-point drying (figures 12, 13). The observed details of calcium alginate gels were shown to be unrelated to the applied dehydration and drying techniques. In sodium alginate, there is obviously no ionic interaction for a stabilization of the gel structure. The stress which occurs during freeze-drying seemed to have similar effects as in freeze-dried polyacrylamide and agarose gels.

In our view, the controversy about the real structure of polyacrylamide and agarose gels is likely to be settled only when a preparatory treatment for these gels will be found that simultaneously prevents changes in size during preparation.

References

1. Ogston, A.G. The spaces in a uniform random suspension of fibres. Trans. Faraday Soc. 54 (1958) 1754-1757
2. Rodbard, D. and Chrambach, A. Unified theory for gel electrophoresis and gel filtration. Proc. Nat. Ac. Sci. 65 (1970) 970-977
3. Chrambach, A. Electrophoresis and electrofocussing on polyacrylamide gel in the study of native macromolecules. Molec. Cell. Biochem. 29 (1980) 23-46
4. Ruechel, R., Steere, R.L. and Erbe, E.F. TEM observations of freeze-etched polyacrylamide gels. J. Chromat. 166 (1978) 563-575
5. Gressel, J. and Robards, A.W. Polyacrylamide gel structure resolved ? J. Chromat. 114 (1975) 455-458
6. Blank, Z. and Reimschuessel, A.C. Structural studies of organic gels by SEM. J. Mater. Sci. 9 (1974) 1815-1822
7. Ruechel, R. and Brager, M.D. Scanning electron microscopic observations of polyacrylamide gels. Anal. Biochem. 68 (1975) 415-428
8. Colombo, V.E. and Spaeth, P.J. Scanning electron microscopic observations of differently cross-linked polyacrylamide gels. Experientia 35 (1979) 960
9. Muller, L.L. and Jacks, T.J. Rapid chemical dehydration of samples for electron microscopic examinations. J. Histochem. Cytochem. 23 (1975) 107-110
10. Arnott, S., Fulmer, A., Scott, W.E., Dea, I.C., Moorhouse, R. and Rees, D.A. The agarose double helix and its function in agarose gel structure. J. Mol. Biol. 90 (1974) 269-284
11. Mc Neely, W.H. and Pettitt, D.J. Algin; In: Industrial gums, Whistler, R.L. editor. Academic Press New York (1973) 49-81
12. Hjertén, S. "Molecular sieve" chromatography on polyacrylamide gels, prepared according to a simplified method. Arch. Biochem. Biophys. Suppl. 1 (1962) 147-151
13. Spaeth, P.J. and Koblet, H. Properties of SDS-polyacrylamide gels highly cross-linked with N,N'-diallyltartardiamide and the rapid isolation of macromolecules from the gel matrix. Anal. Biochem. 93 (1979) 275-285
14. Umrath, W. Cooling bath for rapid freezing in electron microscopy. J. Microsc. 101 (1974) 103-105
15. Cohen, A.L. Critical-point drying; In: Principles and Techniques of Scanning Electron Microscopy Vol 1, Hayat M.A. editor, Van Nostrand Reinhold Comp. New York (1974) 44-112
16. Erley, D.S. 2,2-Dimethoxypropane as a drying agent for preparation of infrared samples. Anal. Chem. 29 (1957) 1564
17. Albrecht, R.M., Rasmussen, D.H., Keller, C.S. and Hinsdill, R.D. Preparation of cultured cells for SEM: Air drying from organic solvents J. Microsc. 108 (1976) 21-29

Discussion with Reviewers

D.E. Carpenter: Can we use the lack of correlation in BIS vs DATD as a crosslinking agent to explain their relative virtues and shortcomings?
Authors: Not directly; in our view the question might possibly be answered on the molecular level. We anticipate that the introduction of two hydroxyl groups per molecule of the DATD-crosslinker agent is of primary importance. Furthermore, the changes in the molecular level could also influence the structures of the gels.

M.T. Postek: Most samples prepared for SEM are not attached to an impermeable barrier such as a glass slide. Please comment on the effect of this barrier on the reduction of surface area and dehydrant penetrability.
Authors: One of the reasons we studied extreme thin slab gels was to increase the surface/volume ratio in order to overcome diffusion problems which resulted from the effect of the non-permeable barrier.

D.E. Carpenter: Figures 1 and 5 should be nearly identical according to their description, yet the pore size in each is different. Is there some explanation for this?
Authors: The pressure during polymerization of gels is depending upon the volume of the polymerization mixtures (Ref. 7). The pores demonstrated in figures 1 and 5 are different due to this fact (figure 1 to 4: rod gel 4 mm wide; figure 5: rod gel 2 mm wide).

D.E. Carpenter: As pointed out there is an apparent discrepancy in the pore size of freeze-dried polyacrylamide and the denatured protein in the matrix. Is this just an artefact or perhaps are we missing something in the theory of electrophoretic separation?

Authors: We are sure that the shown protein material does not represent the real structure, since all proteins have a hydration layer. The observed structure is an aggregated protein which has been precipitated by trichloroacetic acid.

M. Kalab: The temperature of -25°C, at which some of your gels were freeze-dried, is within the range of severe ice crystal formation, and the temperature of -45°C is just at the lower limit of this range. Would the formation of ice crystals explain the markedly larger pores in freeze-dried gels?
Authors: We are aware of the fact that the formation of ice crystals is more likely at a temperature around -25°C than at considerably lower temperatures. However, when comparing the present freeze-dried gels with freeze-dried gels sublimated at -80°C (Ref. 5,6) we cannot discern any substantial differences in the size and structure of the gel pores.

D.E. Carpenter: The apparent pore size in the agarose gel and acrylamide gel were about the same, yet the agarose is used for the separation of higher molecular weight biopolymers. What is it about the structure of agarose that allows it to collapse to a greater extent that polyacrylamide? Can this tell us something about its structure?
Authors: We don't think that any precise answer to this point can be found yet. However, the agar gels are distinguished by the lack of membrane formation that must provide the major obstacle for migrating macroions in polyacrylamide gels of comparable concentrations (Ref. 4).

M. Kalab: What is the reason for transferring the gel samples, dehydrated in ethanol, into Freon 113?
Authors: For all kinds of SEM samples, it is often advantageous to interpose an intermediate fluid, even when the pure dehydration solvent is completely miscible with the transitional fluid. The use of Freon 113 is considered to be convenient, because its chemical composition is similar to that of transitional fluids and therefore, it does not require thorough removal as is the case with other critical-point methods (Ref. 15).

M. Kalab: How did the final thickness of the dried polyacrylamide gels compare with the initial thickness of 0.105 mm? Have you numerical data?
Authors: Table 2 summarizes the changes in thickness of thin polyacrylamide gels. The sizes of the slabs as seen with SEM were compared with the initial thickness.

M.T. Postek: Were the results between DMP-dried material (CPD out of acetone) and the ethanol dehydrated material (CPD out of Freon 113) comparable and were control samples run drying both sets using the same intermediate fluid?
Authors: For many SEM samples, the rapid DMP dehydration technique was found to work equally well as the ethanol dehydration method (Ref. 9; M.D. Maser and J.J. Trimble III, J. Histochem. Cytochem. 25, (1977) 247-251; L.E. Kahn et al., SEM/1977/I, 501-506). According to our experience with many kinds of samples, as e.g. fungal material, cosmetic preparations, mosquito cells and gels, we can fully confirm the usefulness and the equivalency of the DMP dehydration method to the ethanol or acetone dehydration method. We found it also reasonable to perform the critical-point drying of DMP dehydrated material out of acetone, since some acetone always originates from the hydrolysis of DMP by water. We could not detect any significant differences in drying DMP dehydrated material out of acetone or out of DMP.

M. Kalab: Were attempts made to prepare the gels on glass cover slips and to fracture them along with glass to which they adhered?
M.T. Postek: The chemically dehydrated agarose slab gels adhered tightly to the glass, according to your results. Is there any reason for this adherence and why did you not simply break the slab and view it as a cross section?
Authors: On the basis of the present results this extreme adherence of agarose films towards glass could not be explained. However, simple breaking of the glass plate with the agarose film still mounted was not successful because of many glass fragments.

Table 2. Changes of the thickness of dehydrated and dried polyacrylamide gel slabs compared with wet gel slabs (%)

	Polyacrylamide gels	
	BIS-crossl.	DATD-crossl.
Freeze-drying	18	< 10
Critical point drying		
Ethanol-dehydration	57	61
DMP-dehydration	57	58
Air-drying		
DMP-dehydration	--	77
not dehydrated	97	88

MICROSTRUCTURE OF MAYONNAISE AND SALAD DRESSING

M.A. Tung and L.J. Jones

Department of Food Science
University of British Columbia
Suite 248, 2357 Main Mall
Vancouver, B.C., Canada V6T 2A2

Abstract

Mayonnaise and salad dressing were examined by light microscopy (LM), transmission electron microscopy (TEM) and scanning electron microscopy (SEM). Sample preparation techniques were those of Becher (1965) for LM and Chang et al. (1972) for TEM. These methods require dilution of the emulsion, thus prohibiting the observation of the lipid droplets in their natural spacial configuration. Chemical fixation and critical point dehydration of the samples was attempted in an effort to examine undiluted emulsions using SEM. This technique was found to be successful in stabilizing the emulsions for microscopic examination.

Measurements obtained from SEM micrographs appeared to provide better determination of lipid droplet size distributions than did light micrograph measurements. Samples of mayonnaise were stored at 55°C for 3 days, and subsequently examined by SEM. A shift in droplet size distribution toward larger droplets was observed as a result of coalescence of lipid droplets. Centrifugation and turbidimetric studies confirmed that coalescence was occurring during this high temperature stress treatment.

The salad dressing, as expected, was observed using SEM to contain a lower concentration of lipid droplets than was observed in the mayonnaise samples. Amorphous material, assumed to be cooked starch paste, an ingredient in salad dressing, was observed between lipid droplets.

KEY WORDS: Food Microstructure, Food Emulsions, Mayonnaise, Salad Dressing, SEM, TEM, LM

Introduction

A wide variety of mayonnaise and salad dressing products are marketed commercially. These foods are oil-in-water emulsions containing vegetable oil, vinegar and an emulsifying agent (usually egg yolk). Salad dressing contains cooked starch paste as an additional ingredient. Due to the high volume of dispersed (oil) phase (65 - 80%), emulsion stability can become a problem in mayonnaise after prolonged storage. Although considerable research has been devoted to theoretical considerations of emulsion stability in general (Becher, 1965; Kitchener and Musselwhite, 1968), comparatively little research has been published concerning practical aspects of mayonnaise stability. One of the few papers of this type was published by Corran (1946), who reported on the effect of product formulation on the stability of mayonnaise.

Changes in emulsion stability may occur through the processes of creaming, flocculation and coalescence which represent physical changes in the dispersed droplets. Among other properties, the droplet size distribution and the nature of the stabilizing interfacial film influence the rate of destabilization of a food emulsion.

Microscopic examination of food emulsions has been utilized by other authors to study these properties (Chang et al. 1972; Walstra et al. 1969).

Light microscopy has been used widely to determine the droplet size distributions of emulsions (Becher, 1965). Some problems with this technique, identified by Walstra et al. (1969), include: 1) poor detection of very small globules, 2) mistaking of other particles for fat droplets, 3) non-uniform distribution of droplets over the counting area (larger droplets in one part of the slide and smaller droplets in other parts) and 4) flattening of larger droplets leading to overestimation of size. In spite of these limitations, light microscopic examination is generally considered to provide the best estimation of the size distribution of lipid droplets in an emulsion.

Scanning electron microscopy has not been employed in the study of high lipid content emulsions, due primarily to difficulties encountered in fixing the samples prior to observation.

However, examination of emulsions by SEM would overcome some of the problems of droplet size measurement identified by Walstra (1969), mentioned earlier.

Chang et al. (1972) studied mayonnaise using TEM of ultrathin sections. They observed the interfacial film around the fat droplets and speculated that electron dense particles at the oil-water interface were coalesced low-density lipoproteins and microparticles from egg yolk granules.

In the present study, samples of mayonnaise were examined by LM, SEM and TEM to compare the information provided by these techniques. Using SEM, two brands of mayonnaise and one salad dressing were examined before and after a high temperature stress treatment.

Materials and Methods

Light Microscopy

Samples (∼0.1g) of mayonnaise obtained from a local supermarket were diluted with 2.5 ml glycerol, 2.5 ml of a 0.1% sodium dodecylsulfate (SDS) solution (to inhibit droplet flocculation) and 1 ml Sudan IV stain, and shaken vigorously in a capped vial. After at least 1 h, smears of the samples were examined and photographed under bright field illumination and a 40X objective (N.A. 0.75), using a Wild M20 microscope and a Pentax 35mm camera loaded with Kodak Technical Pan 2415 film. Nine fields, covering a range of locations within the smear, were examined. Droplet size measurements were made from 13 x 18 cm photographic prints with a total magnification of 1150X. Three hundred droplets were measured representatively from the nine photographs of the sample. This procedure was duplicated with another mayonnaise sample from the same jar.

Scanning Electron Microscopy

Small samples of locally purchased mayonnaise or salad dressing (pH∼4), approximately (2mm)^3 in size, were dropped into vials containing 4% glutaraldehyde in 0.07 M phosphate buffer (pH 7) and left overnight at 4°C. The samples were rinsed in phosphate buffer three times for 5 min each. Osmium tetroxide (1%) in phosphate buffer for 4 h was used as a secondary fixative to stabilize lipids in the samples. Attempts were made to fix the samples at low pH (4) and in unbuffered fixatives, however, the rate of osmium tetroxide penetration into the sample was found to be unacceptably slow under these conditions. This poor penetration was observed under the microscope as a substantial loss of lipid material, after osmification of up to 4 h. A second set of rinses in phosphate buffer was followed by dehydration through a graded series of ethanol. Exhange of amyl acetate for ethanol was accomplished using a graded series of amyl acetate in 100% EtOH, followed by 1 h in 100% amyl acetate. The samples were critical point dried using liquid CO_2 in a Parr critical point drying bomb, mounted on aluminum stubs with silver paint, gold/palladium coated in a Technics sputter coating unit, and observed with an ETEC Autoscan SEM at an accelerating voltage of 20 kV.

Attempts to prepare the sample for SEM using freeze-drying were unsuccessful, as the emulsion broke during freezing, resulting in oil separation on warming the samples to room temperature.

Transmission Electron Microscopy

The techniques reported by Chang et al. (1972) for preparation of mayonnaise for examination by TEM were used in this study, with some modification. Samples of mayonnaise (∼ 1 g) were diluted with 5 ml of a warm aqueous solution containing 2% agar and 1% glycerol, the mixture vigorously shaken and allowed to solidify overnight. The emulsion-containing gel was disrupted with a glass rod, and small pieces (∼ 2 mm)^3 were transferred to vials. Fixation was accomplished using 5% glutaraldehyde in 0.05 M phosphate buffer (pH 7.2) for 2 h at room temperature. Following phosphate buffer rinses, the samples were post-fixed in 1% osmium tetroxide in phosphate buffer and rinsed again with the buffer. Dehydration was carried out using EtOH in graded concentrations, followed by block staining using a saturated solution of uranyl acetate in 70% EtOH for 3 h. Dehydration was completed through changes of 90 and 100% EtOH. Infiltration was carried out using propylene oxide and a 1:1 mixture of propylene oxide and Luft's Epon. Further infiltration in pure Epon for 90 minutes preceded embedding in Epon in Beem capsules and curing at 60°C overnight. Sections were cut using a Reichert Om-U3 ultramicrotome and were mounted on uncoated 200 mesh grids. A Zeiss EM-10, operated at an accelerating voltage of 80 kV was used to observe and photograph the sections.

Turbidity Measurements

Samples of mayonnaise and salad dressing were diluted with 0.1% SDS to achieve a final dilution of 1:4200. The absorbances of the diluted emulsions were measured at 500 nm using a Beckman Model DB spectrophotometer. This technique was employed by Pearce and Kinsella (1978) in the evaluation of protein emulsification properties. The measured absorbance is related to the interfacial area of the emulsion, and decreases in absorbance with time and/or temperature provide an indication of emulsion stability (Pearce and Kinsella, 1978).

Centrifugation

Emulsion stability is commonly measured in terms of the amounts of oil and/or cream separating from an emulsion during centrifugation (Becher, 1965; Wang and Kinsella, 1976). In the present study, samples of mayonnaise and salad dressing (∼ 25 g) were centrifuged in 50 ml polypropylene tubes in a Sorval Superspeed RC2-B centrifuge at 12,100 xg for 1h, both before and after high temperature stress treatment. Semi-quantitative measurements of the amounts of released oil and cream were made immediately after the tubes were removed from the centrifuge.

Results and Discussion

Composition

The proximate compositions of the three products used in this study are provided in Table 1. The data were supplied by the manufacturers.

Table 1. Proximate compositions of mayonnaise and salad dressing.

	Mayonnaise A	Mayonnaise B	Salad dressing
Fat, %	80	79	49
Moisture, %	17	17	36
Protein, %	1.5	1.1	0.69
Carbohydrate, %	0.74	1.7	12
Ash, %	1.5	1.6	2.1

Light microscopy

A typical light micrograph of diluted mayonnaise A is presented in Figure 1. Due to the non-uniform distribution of droplet sizes throughout the smear, the entire range of droplet sizes found in this product cannot be identified in this one micrograph. Measurements of 600 droplets, taken from 18 micrographs of duplicate diluted emulsions, resulted in the droplet size distribution presented in Figure 2. The sizes of the droplets were found to follow a log normal size distribution, typical of emulsion droplet distributions (Becher, 1965; Levius and Drommond, 1953; Groves and Freshwater, 1968). Difficulties were experienced in making droplet size measurements. Small droplets (< 1 µm) were difficult to identify and measure and the non-uniform spread of droplets throughout the smear created problems in obtaining a representative selection of droplets in the nine fields photographed for each sample.

Scanning Electron Microscopy

Observation of chemically fixed, dehydrated and critical point dried samples of mayonnaise, revealed that structure could be visualized only in regions of the sample that were both well-fixed by osmium tetroxide and undisturbed by physical actions during processing (Fig. 3). In interior regions of the samples, where the fixative had not penetrated adequately, the lipid was solubilized during dehydration, resulting in the empty network of continuous phase proteinaceous solids found throughout Figure 4. The low proportion of continuous phase in mayonnaise inhibited osmium tetroxide penetration, and thorough penetration could not be achieved, even after 4 h.

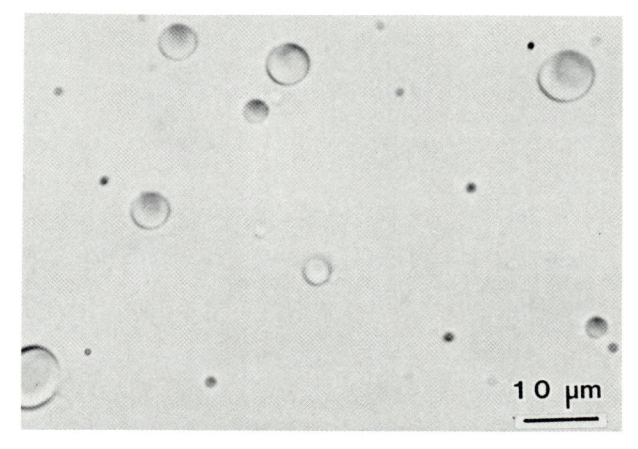

Figure 1. Light micrograph of mayonnaise.

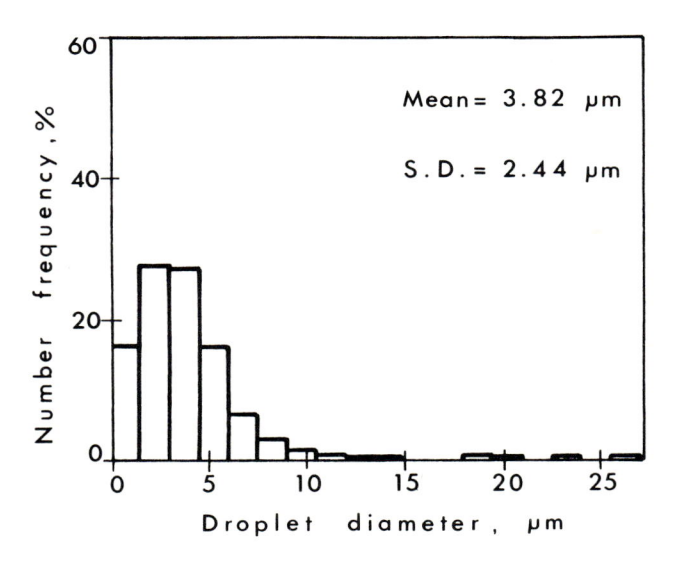

Figure 2. Frequency distribution of lipid droplet diameters determined from light micrographs of mayonnaise A.

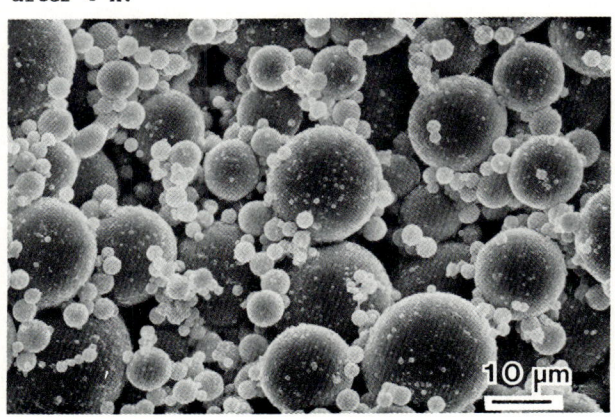

Figure 3. Scanning electron micrograph of mayonnaise.

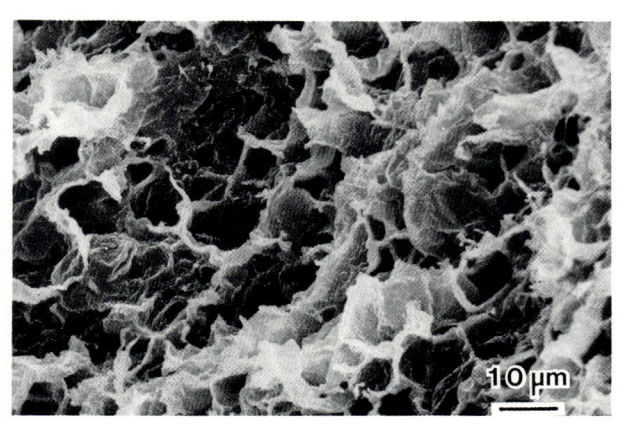

Figure 4. Scanning electron micrograph of mayonnaise.

Droplet size measurements of mayonnaise A, examined by light microscopy, were obtained from SEM micrographs similar to Figure 3 at a total magnification of 1000X. Due to the greater resolution and depth of field afforded by SEM, the droplets were easier to measure than those from light micrographs. However, the edges of some of the larger droplets in the SEM photographs were concealed by smaller droplets, possibly biasing the results toward a smaller mean droplet size. Three hundred droplets were measured from three well preserved regions of the sample, resulting in the distribution shown in Figure 5. The mean droplet diameter (number-average) determined for this sample from SEM micrographs was 2.23 µm, lower than 3.82 µm (p = 0.05) obtained by LM. It would appear from the histograms that a significant number of droplets, less than 1.5 µm in diameter, were missed in the LM measurements. Thus, SEM may be a more useful technique for determining droplet size distributions of emulsions.

Transmission Electron Microscopy

The wide size distribution of lipid droplets observed in mayonnaise samples using LM and SEM techniques was also observed in sections of diluted mayonnaise with TEM (Figure 6). The appearance of the emulsion with TEM agrees well with results obtained by Chang *et al.* (1972). At high magnification (Figure 7), parts of the interfacial film is visible and observations may be made relating to its composition and formation. Chang and co-workers (1972) speculated that the speckled, electron-dense material at the interface was low-density lipoproteins and microparticles of egg yolk granules.

Droplet size distributions were not determined from TEM micrographs, due to the large number of photographs that would be required, and the necessity for corrections to diameter measurements of particles from ultrathin sections.

Accelerated Aging Tests

Results obtained from turbidimetric and centrifugation studies indicated that during storage at 55°C for 3 days, two samples (different brands) of mayonnaise and one of salad dressing became increasingly unstable due to droplet coalescence. Storage of mayonnaise at 55°C would not commonly occur in commercial practice, however, these high temperature stress conditions were employed to accelerate the destabilization of the emulsion products. In all three products, measurements of absorbance, which is related to the interfacial area, were significantly higher in fresh than in stored samples (Table 2).

Table 2: Turbidity of diluted (1:4200) samples of mayonnaise and salad dressing in fresh and stored (3 days, 55°C) conditions.

	Absorbance (500 nm) n=2			
	fresh		stored	
	mean	S.D.	mean	S.D.
Mayonnaise				
brand A	0.618	0.018	0.508	0.004
brand B	.322	.002	.236	.016
Salad dressing	.613	.004	.510	.011

Means within each row are significantly different (p = 0.05) as determined by Student's t-test.

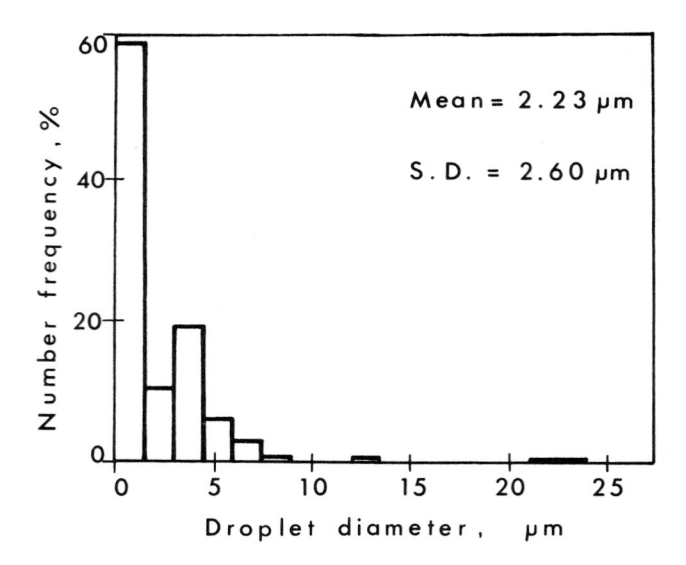

Figure 5. Frequency distribution of lipid droplet diameters determined from scanning electron micrographs of fresh mayonnaise A.

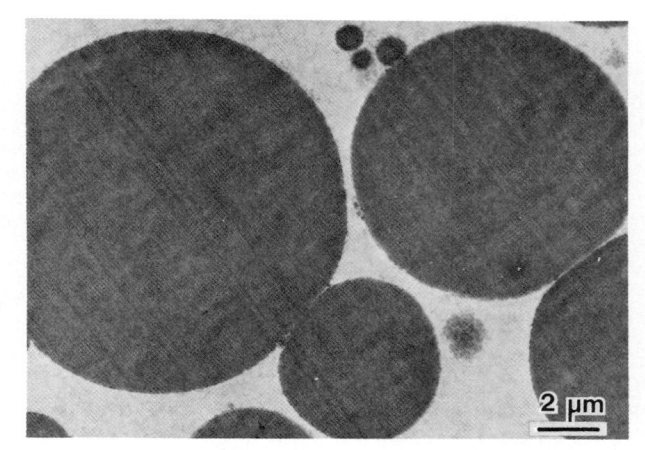

Figure 6. Transmission electron micrograph of mayonnaise.

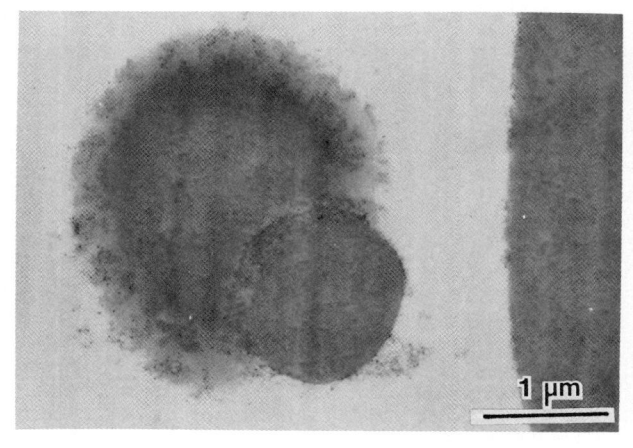

Figure 7. Transmission electron micrograph of mayonnaise.

As the droplets coalesce, the total interfacial area of a given weight of sample decreases, since larger droplets have less total surface area than an equivalent volume of smaller droplets.

The differences in absorbance values for mayonnaises A and B suggest that there is a larger interfacial area in mayonnaise A. Given that the products contain equal oil contents (Table 1), it would be expected from this result that mayonnaise A contains a higher proportion of oil in the form of small droplets with large surface area to volume ratios. However, other ingredients may also influence the turbidity of the diluted products - egg, spices and lemon juice. Therefore, conclusions regarding turbidity differences between brands cannot be interpreted from these data with any certainty.

Evidence that coalescence was occurring during storage was also obtained from centrifugation of fresh and stored samples. The amount of oil separating from the product during centrifugation is related to the degree of oil droplet coalescence. In fresh samples, only a thin film of oil formed on the top of all samples after 1 h of centrifugation at 12,100 xg. After 3 days at 55°C, however, up to 9% (w/w) of the emulsion separated out as oil, on centrifugation. The salad dressing was noticeably more stable to oil separation than the mayonnaises. There was no difference observed between mayonnaise A and mayonnaise B in the degree of oil separation.

Fresh samples of two different brands of mayonnaise were examined by SEM. Mayonnaise A (Figure 8) was observed to contain more very large and very small droplets as compared to mayonnaise B (Figure 9). Droplet size distributions were determined for both samples and the mean droplet diameters (Table 3) for the two brands were found to be not significantly different ($p = 0.05$).

Table 3: Droplet size distribution parameters of two brands of mayonnaise.

	droplet diameter (µm)	
	mean (number-average)	S.D.
Mayonnaise A	2.24	2.60
Mayonnaise B	2.21	2.18

Following high temperature stress treatment, changes in droplet size distributions occurred, resulting in more irregularly shaped, larger droplets (Figures 10 and 11).

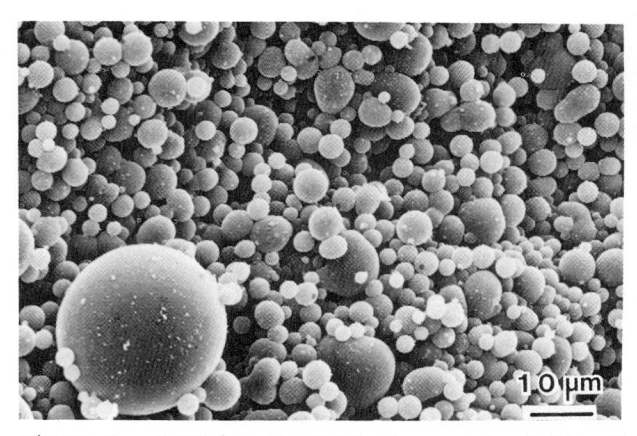

Figure 8. Scanning electron micrograph of "fresh" mayonnaise A.

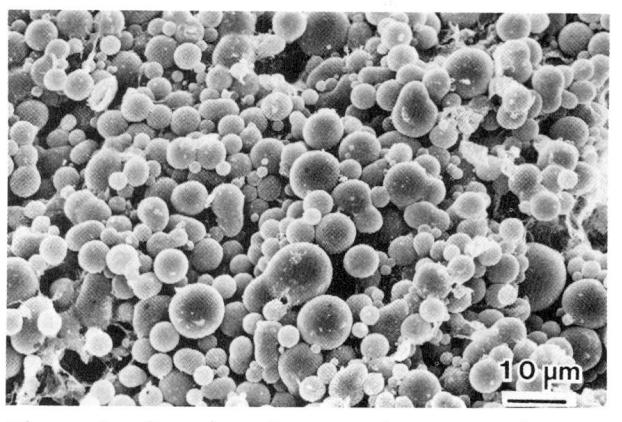

Figure 9. Scanning electron micrograph of "fresh" mayonnaise B.

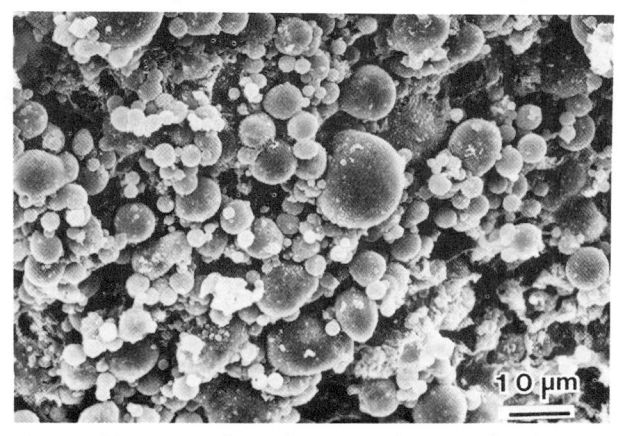

Figure 10. Scanning electron micrograph of "stored" mayonnaise A.

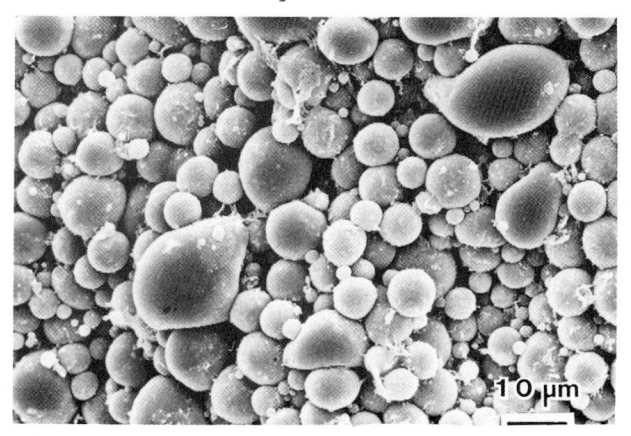

Figure 11. Scanning electron micrograph of "stored" mayonnaise B.

At an elevated temperature, increases in Brownian motion, decreases in continuous phase viscosity, increased solubilization of the surfactant and possibly changes in the electrical double layer around the lipid droplets, may contribute to increased rates of droplet coalescence. Droplet size measurements from SEM micrographs of mayonnaise A after storage at 55°C for 3 days, resulted in the distribution shown in Figure 12. A shift in droplet size toward larger droplets was observed, with an accompanying significant (p= 0.05) increase in the mean droplet diameter from 2.24 µm to 2.71 µm.

Salad dressing observed under the SEM, contains a lower concentration of lipid droplets, and a higher concentration of amorphous material between droplets (Figure 13). This material was assumed to be cooked starch paste, a stabilizing ingredient added to salad dresssing but not to mayonnaise. Mayonnaise must, by Canadian law, contain at least 65% vegetable oil, while salad dressing may contain as little as 35% oil. The salad dressing utilized in this study contained 49% oil. The appearance of salad dressing was not altered significantly as a result of high temperature storage. Due to the high concentration of non-lipid material, droplet distributions could not be accurately determined. The rate of droplet coalescence in the salad dressing may not be as high as in the mayonnaise, due to the stabilizing effect of the starch ingredients.

Conclusion

The microstructures of mayonnaise and salad dressing were examined by LM, TEM and SEM. These oil-in-water emulsions had a high concentration of dispersed phase, and much of the continuous phase was not visible under the microscope, or was removed during sample preparation. SEM was found to be an advantageous method of examining these products as compared to LM, providing a more representative view of the droplet distribution, as dilution of the sample was avoided. Also, small lipid droplets could be well resolved, distortion of lipid droplets was minimized, non-lipid material can be more readily identified, and Brownian motion was eliminated. Droplet size distributions obtained from SEM micrographs appeared to be more accurate than those determined from LM measurements, as evidenced by distributions which more closely approximated log normal distributions. TEM was confirmed as a useful technique for examining the interfacial surfaces of emulsions.

Two brands of mayonnaise were observed to differ in droplet size distribution, although the mean droplet diameters were found to be not significantly different. Salad dressing contains a significant amount of non-lipid material, which was observed between lipid droplets using SEM. Storage of the products for 3 days at 55°C was found to accelerate lipid droplet coalescence as determined by light scattering, centrifugation and microscopic techniques. Lipid droplet size measurements made from SEM micrographs, indicated a shift toward larger droplet sizes in samples which had been subjected to high temperature stress conditions.

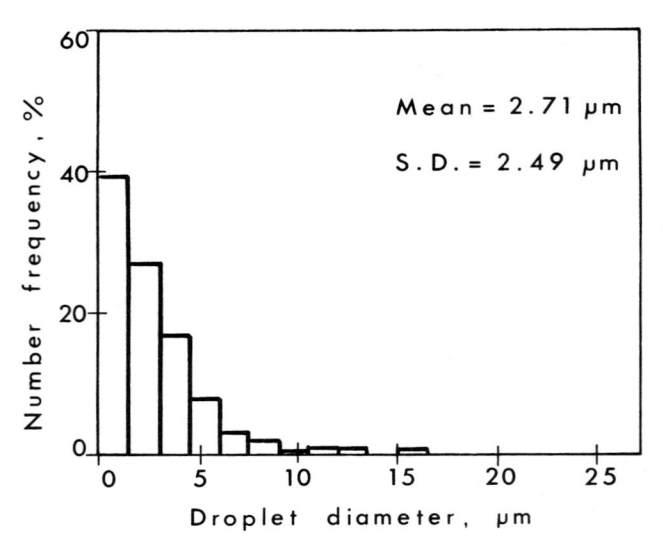

Figure 12. Frequency distribution of lipid droplet diameters determined from scanning electron micrographs of stored mayonnaise A.

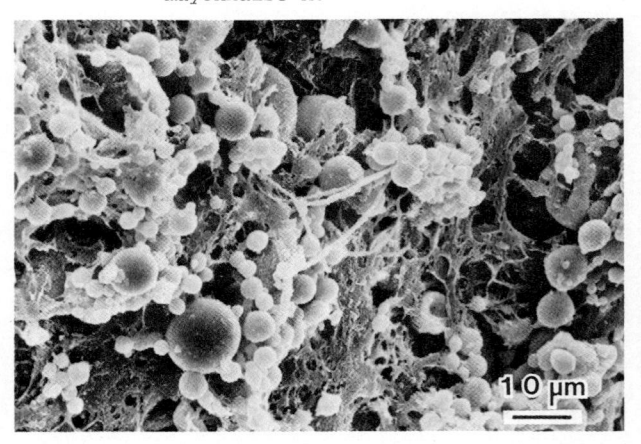

Figure 13. Scanning electron micrograph of salad dressing.

Acknowledgements

The authors are indebted to Mr. A Lacis, Department of Metallurgical Engineering, and Mr. L. Veto, Biosciences, U.B.C. for their suggestions regarding sample preparation and their assistance in the use of the microscopes.

References

Becher, P. Emulsions: Theory and Practice, 2nd ed., Reinhold Publishing Corporation, New York, NY, U.S.A. 1965 415-416, 95-149.

Chang, C.M., Powrie, W.D., Fennema, O. Electron microscopy of mayonnaise. Can. Inst. Food Sci. Technol. J. 5 1972 134-137.

Corran, J.W. Some observations on a typical food emulsion in Emulsion Technology. Chemical Publishing Co., Brooklyn, NY, U.S.A. 1946 176-192.

Groves, M.J., Freshwater, D.C. Particle-size analysis of emulsion systems. J. Pharm. Sci. 57 1968 1273-1291.

Kitchener, J.A., Mussellwhite, P.R. The theory of stability of emulsions in Emulsion Science, ed. Sherman P., Academic Press, Inc., London, England 1968 77-130.

Levius, H.P., Drommond, F.G. Elevated temperature as an artificial breakdown stress in the evaluation of emulsion stability. J. Pharm. Pharmacol. 5 1953 743-756.

Pearce, K.N., Kinsella, J.E. Emulsifying properties of proteins: Evaluation of a turbidimetric technique. J. Agric. Food Chem. 26 1978 716-723.

Walstra, P., Oortwijn, H., de Graaf, J.J. Studies on milk fat dispersion. I. Methods for determining globule-size distribution. Neth. Milk Dairy J. 23 1969 12-36.

Wang, J.C., Kinsella, J.E. Functional properties of novel proteins: Alfalfa leaf protein. J. Food Sci. 41 1966 286-292.

Discussion with Reviewers

R.J. Carroll: How did you determine the statistical significance of the difference in mean droplet diameters of fresh and stored mayonnaise A?

Authors: The emulsion droplets were observed to follow a log normal size distribution. Thus, in order to perform meaningful statistical analyses of the results, the droplet diameters must be transformed as their logarithms to obtain a distribution which approximates a normal distribution. Lipid droplets in fresh mayonnaise A formed a log distribution characterized by a mean of 0.1575 and a standard deviation of 0.3927, while stored sample droplets formed a log distribution with mean 0.2805 and standard deviation 0.3658. Using a Student's t-test for the comparison of the two means, the fresh mayonnaise mean droplet diameter was significantly different ($p = 0.05$) from the stored mayonnaise mean droplet diameter.

R.J. Carroll: How do the turbidity measurements of samples A and B relate to the SEM micrographs of Figures 8-11?

Authors: Coalescence is one process by which emulsion de-stabilization may occur, resulting in separation of oil and water phases. As coalescence of lipid droplets proceeds during storage, lipid droplets combine to form larger droplets, and an increase in mean droplet size is observed (Figure 8-11). Coalescence can also be monitored by following the reduction in interfacial area which occurs as a result of the fact that large lipid droplets have a lower surface area to volume ratio than an equivalent volume of small droplets. Thus, it would be anticipated that an increase in droplet size observed in SEM micrographs would be accompanied by a decrease in interfacial area, measured as a decrease in absorbance at 500 nm. This result was demonstrated in this study.

J. G. Oles: How do SEM/ TEM and LM compare with other particle size determination techniques, e.g. Coulter Counter?

D.N. Holcomb: Fat globule size distributions in mayonnaise can be determined using particle sizing equipment (such as "Coulter Counter"), but that requires dilutions of 10,000 times, or more. Have you observed, or would you expect differences in size distribution as obtained by a Coulter Counter and by your SEM technique?

Authors: Becher (1965) discussed several methods of determining emulsion droplet-size distributions. He identified microscopic measurements as providing the greatest amount of certainty in the results obtained. Light microscopy has been the most widely used technique for this purpose. The preparation time and complex analyses of droplet measurements made from thin sections made the use of TEM impractical for determining droplet size distributions. Scanning electron microscopy has not been used previously for emulsion distribution determinations; however, this study has demonstrated the merits of SEM for this purpose. Light-scattering techniques were also discussed by Becher (1965). This approach is only acceptable when an average particle size rather than a size distribution is required. Measurements of mayonnaise droplet size distributions using a Coulter Counter have not been attempted to date. Although theoretically, the droplet size distribution would not be affected by dilution of the oil-in-water emulsion, our limited experience with the technique has indicated that practically, the Coulter Counter is not a very satisfactory technique when a wide range of particle sizes is involved. Very small particles in relation to the aperture size are not adequately resolved, and large particles tend to clog the aperture.

R.J. Carroll: Were any droplets less than 1.0 µm found with the SEM? If so, how would this change mean droplet sizes?

Authors: Droplets as small as 0.35 µm in diameter were measured from SEM photographs. These small droplets were likely missed, or too poorly resolved to be accurately measured from LM photographs. Measurement of these small droplets, less than 1.0 µm in diameter in SEM photographs, would certainly result in a lower mean droplet diameter as compared to LM measurements. In this study, the mean droplet diameter of mayonnaise A was found by SEM to be 2.23 µm and by LM to be 3.82 µm. This difference was determined to be statistically significant ($p = 0.05$).

M. Kalab: Dilution with glycerol disrupts the original microstructure of mayonnaise. What was the reason for diluting the samples destined for

TEM? Were attempts made to embed undiluted samples?
Authors: Dilution of the samples was carried out according to the procedure of Chang et al. (1972). We assume that dilution was recommended by these authors to insure adequate penetration of fixatives and resins during preparation of the samples. Attempts were not made to embed undiluted samples, as previous experience in our laboratory with high-lipid materials has demonstrated that lipid-dense tissue blocks are extremely difficult to section. However, recently researchers in other laboratories have succeeded in embedding undiluted mayonnaise. We would certainly pursue this aspect in any future work.

D.N. Holcomb: Can you offer any explanation as to why OsO_4 would penetrate the samples at pH 7, but not at pH 4?
Authors: We cannot offer any explanation as to why the penetration of OsO_4 was so poor at pH 4. We have not found any information on the influence of pH on OsO_4 penetration in the microscopy literature. Other workers on this campus have experienced similar difficulties.

J.G. Oles: Does the fixing technique alter the membrane properties?
Authors: The emulsifying agents which form interfacial membranes in mayonnaise and salad dressing are phospholipids (lecithin) contributed by egg yolk. In areas of the specimen in which the fixatives have penetrated, the phospholipids will be well fixed by both glutaraldehyde and osmium tetroxide fixatives.

D.G. Schmidt: Is it not possible that melting of the oil phase in the electron microscope is responsible for the disappearance of the fat in Figure 4?
Authors: If melting of the oil phase was indeed occurring during observation under the electron microscope, this would have been detected as movement of the sample and a degradation of the image as the sample was moved around under the beam. It was evident that better preservation of oil droplet structure was attained near the edges of the sample, where OsO_4 had penetrated (as evidenced by the black color of the fragment). Therefore, we believe that the disappearance of fat in Figure 4 was caused by inadequate fixation and subsequent removal during dehydration.

Reviewer: How were the mean droplet diameters calculated from droplet size measurements?
Authors: Mean diameters of spherical particles can be calculated in a number of different ways (e.g. simple arithmetic mean, mean volume-surface diameter, etc.) (Becher, 1965). In this study, simple arithmetic means were calculated and were described as number-averages.

P. Walstra: How do you know that the sample of droplets is representative? The effective depth of the specimen may well be different for droplets of different sizes, because of screening or because of loss of droplets from the surface of the specimen. Moreover, how do you discriminate between very small droplets and other particles?
Authors: We can only assume that the sample of droplets is representative, based on the procedure of counting droplets from 3 different areas within the fractured surface of the specimen. Mayonnaise is an homogenous product, and we are confident that the droplets measured are representative of each product. The amount of non-lipid particulate matter in the prepared mayonnaise samples is low due to the removal of spices with the aqueous phase during ethanol dehydration. Therefore, there was very little material that could be confused with lipid droplets.

P. Walstra: The irregularly shaped droplets shown in Figs. 9 & 11, must be artefacts, as they cannot exist in the shapes shown in the real emulsion, because of the Laplace pressure.
Authors: The irregularly shaped droplets may result from fixation of the sample while droplets were coalescing. Droplet coalescence may not occur instantaneously, and as a result, irregularly shaped droplets may represent several lipid droplets in some intermediate stage of coalescence.

W. James Harper: Would you speculate on the predictable changes in the emulsion system by dilution and fixation procedures utilized and how these might differ for mayonnaise and salad dressing?
Authors: Artifacts may have been introduced in the sample preparation of salad dressing, as this product contains cooked starch paste which is known to be poorly stabilized by EM fixation techniques. (see also responses to questions by other reviewers).

SCANNING ELECTRON MICROSCOPY OF SOYBEANS AND SOYBEAN PROTEIN PRODUCTS

W. J. Wolf and F. L. Baker

Northern Regional Research Center, Agricultural Research
Science and Education Administration
U.S. Department of Agriculture
Peoria, Illinois 61604

Abstract

Use of scanning electron microscopy (SEM) to determine the structure of soybeans and how this native structure changes upon commercial processing of the seed into protein products is reviewed.

Soybean seed coat surfaces differ with variety; some are smooth and some are pitted. Seed coats may also be coated with a honeycomb-like material of unknown composition and biological function. In extreme cases it gives a bloom to the seeds. Seed coat damage in some seeds consists of missing layers of palisade and hourglass cells.

SEM reveals the protein bodies and lipid bodies in the cotyledonary cells. These structures can be detected to varying degrees after processing of soybeans into full-fat and defatted soy flours. Protein concentrate particles vary in size and shape depending on the process used in their preparation. Seed structure appears to be retained only in concentrates made by alcohol extraction. Isoelectric isolates are spherical with bumpy surfaces whereas neutralized isolates are smooth spheres. Some neutralized isolates are partially collapsed spheres.

SEM of extruded and steam textured soy flours shows fibrous structures. Spun isolate fibers are cyclindrical with surface striations running along their length. The fibers are observed after fabrication into ham or fried bacon bit analogs. Jet cooking of isolates forms fibers that are variable in size and less distinct than spun fibers.

Among oriental soybean foods, spray dried soy milk particles resemble those found in neutralized isolates. Yuba, the film formed when soy milk is heated, is a continuous protein matrix interspersed with lipid droplets. In tofu, a protein-lipid network forms the backbone of the gel; heat denaturation of the proteins appears necessary to form the three dimensional structure.

KEY WORDS: Soybean Structure; Soy Flours, Protein Concentrates, Protein Isolates, Oriental Soybean Foods

Introduction

Our use of scanning electron microscopy (SEM) in 1970 to examine protein bodies isolated from soybeans was probably the first application of this technique to soybeans (Wolf, 1970). Later, we applied SEM to a detailed examination of soybean seed structure and determined how structure is modified by commercial processing of this important oilseed into edible protein products (Wolf and Baker, 1972; 1975) that are now widely used as ingredients by the food industry. Since publication of our early papers, SEM has been applied extensively by many investigators both to extend our studies and to examine products and processes that we did not include in our work. We are, therefore, reviewing the literature to provide a summary of the current status of SEM studies in this research area.

Soybeans and derived protein products are comparatively easy to examine by SEM because they usually have a low moisture and do not need to be dried prior to examination. This eliminates drying artifacts and makes sample preparation simple. However, there are exceptions to these generalizations, and some of them will be discussed later. For most of our studies, we mounted the materials with plastic cement (large specimens, such as whole soybeans) or sprinkled them on double-coated cellulose tape attached to the specimen stub. The specimens were coated with gold-palladium (60:40) and examined in a Stereoscan Mark 2A scanning electron microscope (Cambridge Instrument Co., Ltd., London, England).

Seed Structure

Seed Coat

Although the seed coat or hull is removed in the preparation of food products, it is an important structural part of the soybean. It makes up about 8% of the seed weight and serves as a protective coating for the seed during storage and commercial handling. This is particularly important for the 37% of the U.S. soybean crop exported as beans to many parts of the world. Beans split during transportation and handling yield lower quality oil upon processing (Mounts et al., 1979).

The seed coat of commercial soybeans appears smooth to the naked eye, but in some varieties it is covered with pits (Fig. 1a). At higher magnification (Fig. 1b), it is apparent that these pits do not open into the interior of the seed coat, and one also sees the fibrous texture of the cuticular layer of the seed coat.

Seed coats of some commercial soybean varieties are dull in appearance with a distinct bloom, and SEM reveals material deposited on the surface. An extreme case of this is observed in Sooty soybeans (a black hay variety no longer grown commercially) as shown in Figures 2a and 2b. The nature of this honeycomb-like network is unknown. It is not removed by washing with water, hexane, or chloroform:methanol (2:1); hence, it is unlikely to be water-soluble proteins, water-soluble polysaccharides, lipids, or wax-like substances.

A view of a cross section of the seed coat reveals a variety of cells that make up this structure (Fig. 3). Although not apparent here, the outermost surface is covered with a cuticle (Saio et al., 1973) as seen in Figure 1b. Next is a layer of elongated palisade cells, followed by the distinctive hourglass cells. Following these are several layers of parenchyma cells, a single layer of rectangular aleurone cells, and finally, the compressed fiber-like cells that constitute the innermost surface of the seed coat.

Some soybeans have damaged areas in their seed coats in which the palisade cells and hourglass cells are missing (Fig. 4a). The edges of the damaged spots show a cross section of the seed coat layers that are absent within the damaged region, and the hourglass cells are very prominent (Fig. 4b). Similar cracks can be formed in intact seed coats by soaking the beans in water and then air drying them. Cracking probably is caused by the more rapid drying and shrinking of the seed coat as compared to the cotyledon. Because of the openness of the hourglass cell layer, damage of this type permits rapid penetration of moisture into the interior of the seed coat and is also a likely site for entry of insects and microorganisms. Such damage undoubtedly also weakens the seed coat and makes it more susceptible to cracking and formation of splits during commercial handling.

Soaking and cooking of soybeans in water swells the seed coat especially because of expansion of the parenchyma cells. Splitting between the parenchyma and the overlying hourglass cells is commonplace (Saio et al., 1973).

The role of seed coat structure in soaking and cooking of soybeans and other legumes has been investigated by SEM (Sefa-Dedeh and Stanley, 1979). Soybean seed coats are 2-4 times thicker than those of other legumes, yet soybeans hydrate more rapidly than pinto beans whose seed coat is only about one-half as thick.

Cotyledons

The cellular structure of the cotyledons was examined by transmission electron microscopy (TEM) in the 1960's (Bils and Howell, 1963; Tombs, 1967; Saio and Watanabe, 1968). These TEM studies revealed that the main subcellular structure consisted of protein bodies 5-20 μm in diameter surrounded by a cytoplasmic protein network in which were embedded the oil storage sites. These lipid bodies (sometimes referred to as spherosomes) are only 0.2-0.5 μm in diameter. Additional details on cotyledon cell structure are found in a recent dissertation (Bair, 1979) and in a study of dry and imbibed cotyledons (Webster and Leopold, 1977).

In our hands, freeze-fractured surfaces were more amenable to examination by SEM than mechanical fractures or sections (Wolf and Baker, 1975). A freeze-fractured surface (Fig. 5a) shows the cell wall and the protein bodies. Structural details, however, are obscured somewhat by the spongy network that covers the protein bodies (Fig. 5b). High magnification (Fig. 5c) indicates spherical particles within the spongy network, and these are the lipid bodies. This interpretation is strengthened by washing the fracture-surface with hexane (Fig. 5d). Hexane dissolves the oil from the lipid bodies and leaves only the cytoplasmic network, with depressions corresponding to the former location of the lipid bodies (Fig. 5e-f). The cytoplasmic network apparently is water soluble, because protein bodies isolated by sucrose density gradient centrifugation are smooth and the cytoplasmic network is no longer apparent.

The ease with which the cytoplasmic network and protein bodies dissolve in water is illustrated by simply washing an undefatted fracture surface in water for 1 min. Examination of the washed fracture surface reveals only the empty cells (Fig. 6a). High magnification shows the wrinkled surface of the cell wall (Fig. 6b). The water washings were noticeably turbid, but on centrifugation a fatty layer floated on top of the supernatant. The turbidity is, therefore, attributed to the lipid bodies. This rapid washing out of the cellular contents is undoubtedly of importance in the preparation of soy milk where beans are soaked, ground with water, and then cooked. Because the cell wall does not appear to rupture in the region below the fracture surface as a result of contact with water, mechanical rupture of cell walls by grinding is probably a key step in the efficient extraction of the proteins and dispersion of the oil in making soy milk.

Bair (1979) has recently isolated the lipid bodies from soybeans by three different methods. By TEM and SEM, the isolated lipid bodies appeared spherical and had an average diameter of 0.5 μm, but they ranged in size from 0.05 μm to 1.2 μm. Depending on the method of isolation, the lipid bodies had variable amounts of protein associated with them. The cleanest preparations contained 81% oil, 15% protein, and 4% unidentified substance.

We isolated the protein bodies from defatted soy flour by sucrose density gradient centrifugation at pH 5, the pH of minimum solubility of the proteins located within the protein bodies (Tombs, 1967). Examination of the isolated protein bodies by SEM revealed numerous spherical particles 0.5-4 μm in diameter plus amorphous material (Fig. 7). We did not find any protein bodies larger than 4 μm even though defatted flour contains numerous particles in the range of 5-10 μm in diameter. Apparently, the large protein bodies are not stable under these conditions and disrupt to form the amorphous material found in the preparation.

Studies using TEM indicate that cooking of soybeans at 115°C for 30 min results in rupturing

Scanning Electron Microscopy of Soybeans

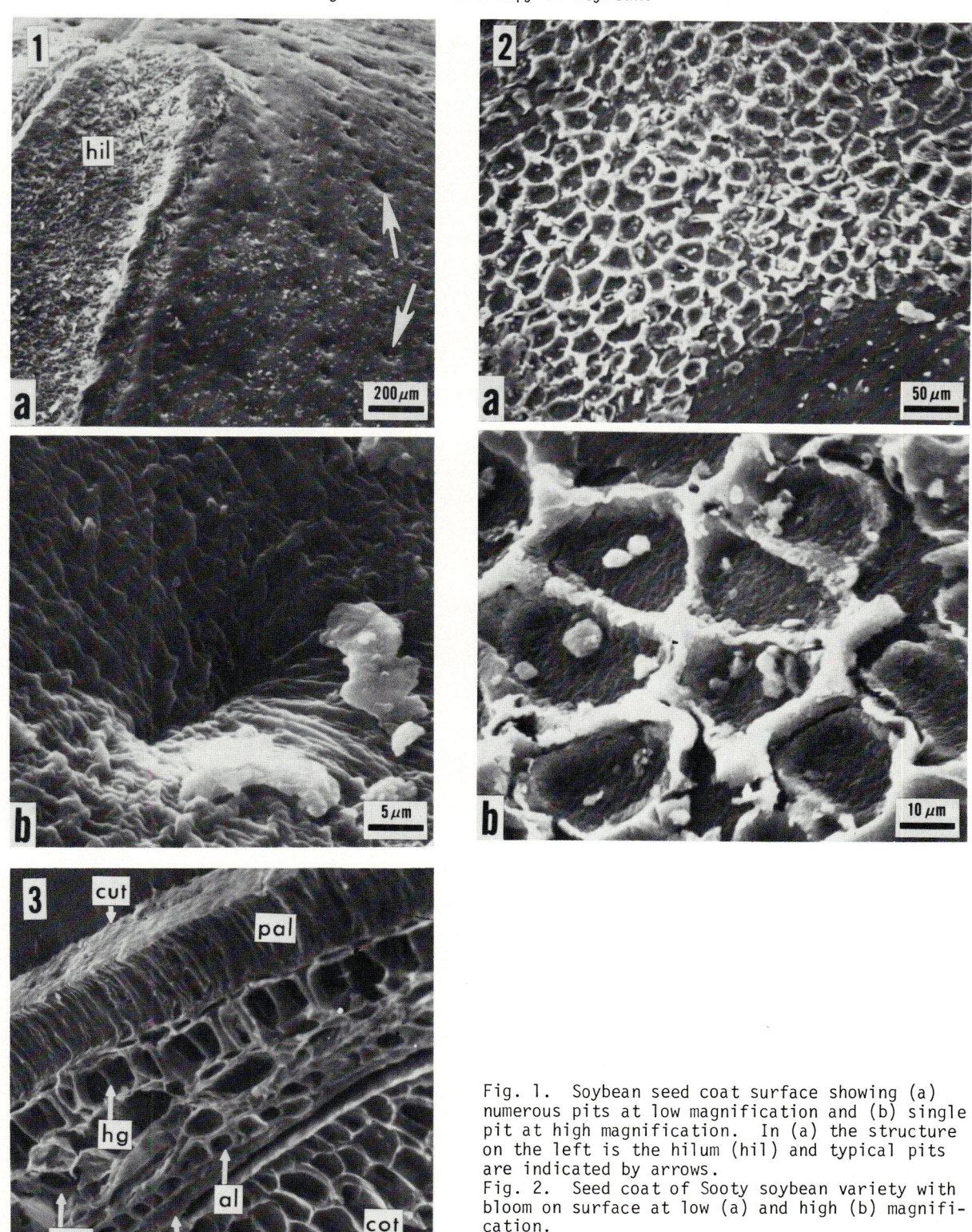

Fig. 1. Soybean seed coat surface showing (a) numerous pits at low magnification and (b) single pit at high magnification. In (a) the structure on the left is the hilum (hil) and typical pits are indicated by arrows.

Fig. 2. Seed coat of Sooty soybean variety with bloom on surface at low (a) and high (b) magnification.

Fig. 3. Section through the seed coat showing cuticle (cut), palisade layer (pal), hourglass cells (hg), parenchyma cells (par), aleurone cells (al), and compressed cells (com). A portion of the cotyledon (cot) is also shown.

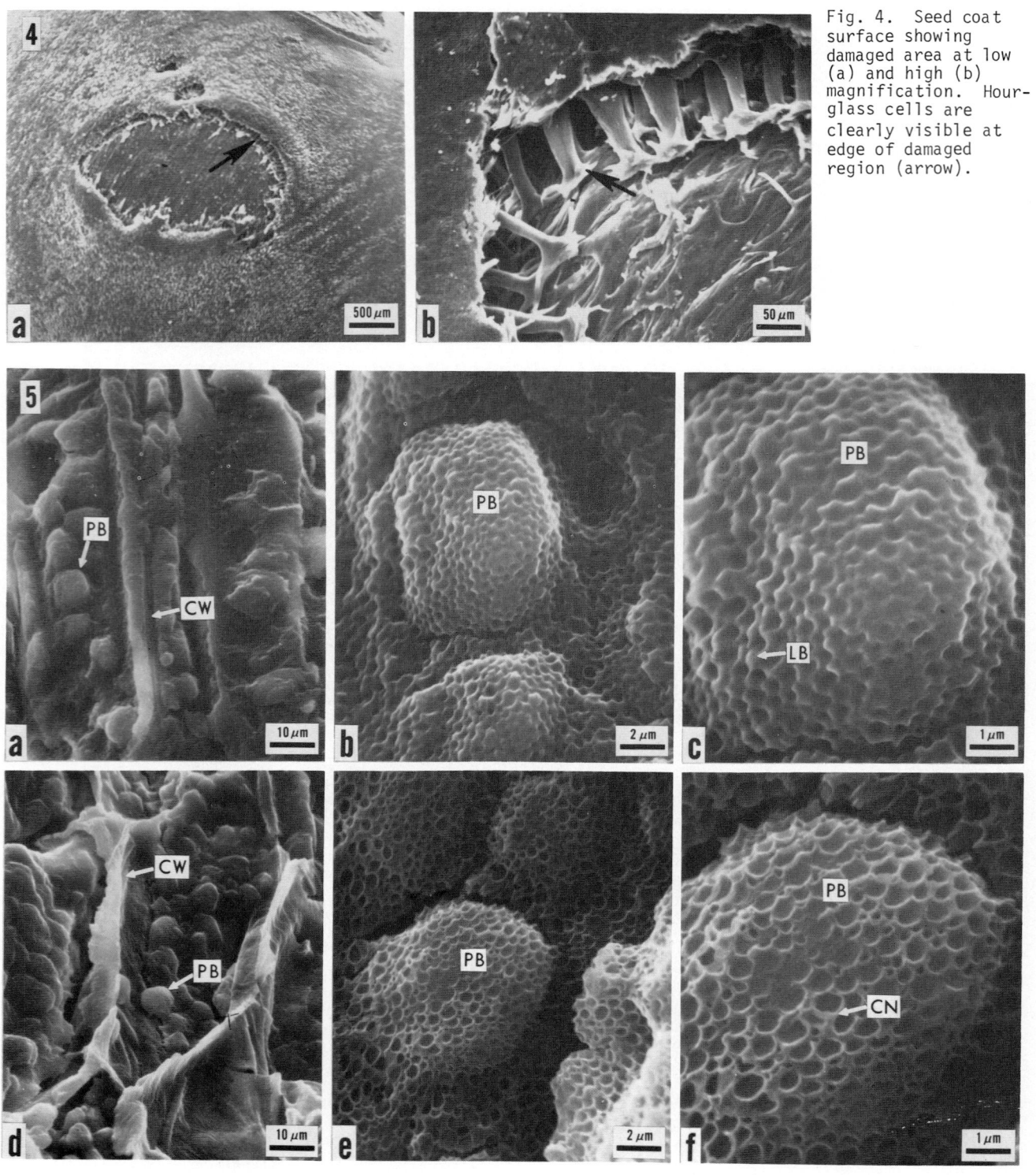

Fig. 4. Seed coat surface showing damaged area at low (a) and high (b) magnification. Hourglass cells are clearly visible at edge of damaged region (arrow).

Fig. 5. Freeze-fractured cotyledon (a); protein body covered with sponge-like cytoplasmic network with embedded lipid bodies (b, c); fracture surface after defatting with hexane (d); protein body in defatted fracture surface showing cytoplasmic network (e, f). Cell wall (CW), protein body (PB), lipid bodies (LB), and cytoplasmic network (CN) are identified.

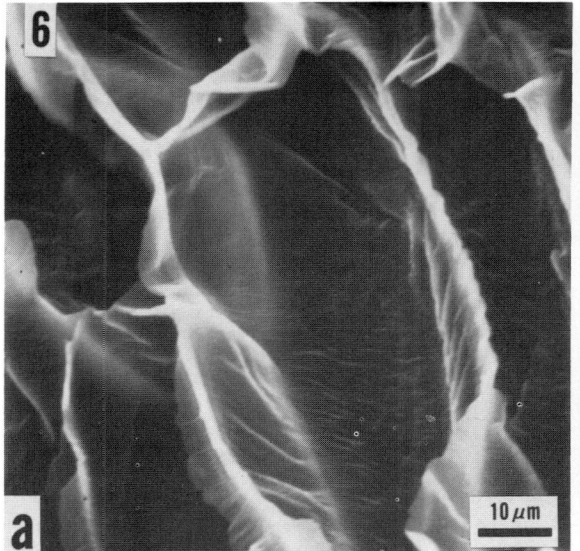

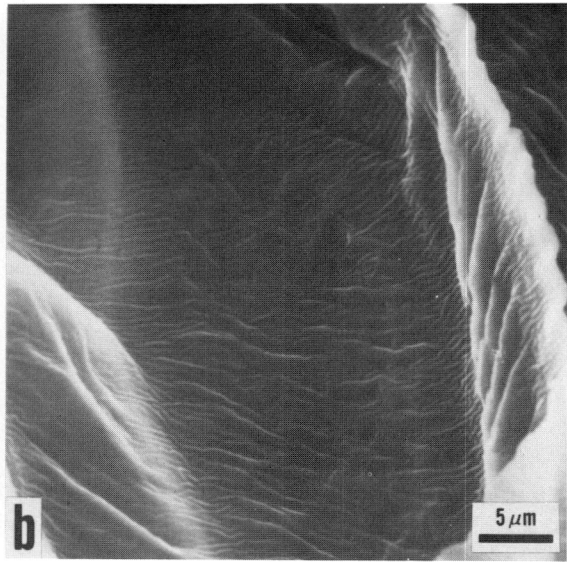

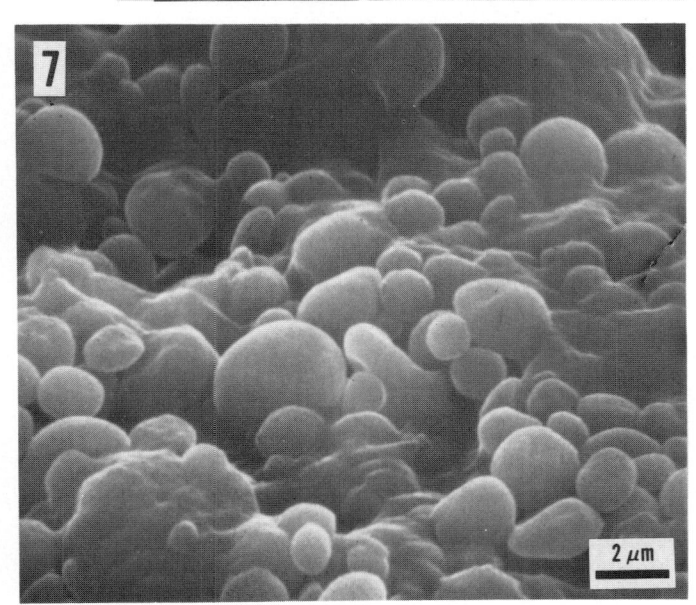

Fig. 6. Undefatted, freeze-fractured cotyledon surface after washing with water for 1 min. showing (a) empty cells and (b) wrinkled cell wall surface.
Fig. 7. Soybean protein bodies isolated by sucrose density gradient centrifugation.

of the lipid bodies and coalescing of the oil into large droplets, plus bursting of the protein bodies and curdling of the proteins (Saio and Watanabe, 1968). Cooking of soybeans at 100°C for 90 min caused separation of intact cells as seen in SEM studies by Sefa-Dedeh and Stanley (1979). Rupture of cell walls did not appear to take place.

In a study of the state of water in the seed, Okamura (1973) fractured soybean cotyledons at moistures ranging from 5.6 to 39.4% and then examined them by SEM after freeze drying. The cellular structure appeared to be covered with a film at moistures above 20%, which Okamura suggests is the transition zone between bound and free water in the seed. Possibly, fractures at these high moistures occurred by separation of adjacent cells at the middle lamella so that the cell wall is observed rather than the inside of the cells.

Structural Changes Caused by Processing

Full-Fat Soy Flour

This product is prepared by steaming soybeans to inactivate lipoxygenase and trypsin inhibitors and to modify the grassy/beany and bitter flavors of raw soybeans. After drying, the beans are dehulled and ground into a flour. The resulting product contains spherical or elliptical particles 6-10 μm in size, which are probably protein bodies that survived the processing (Fig. 8a). Lipid body-like particles are detectable on the surface of the protein bodies (Fig. 8b), but with greater difficulty than on unheated protein bodies (Fig. 5c). Modification of the lipid bodies and protein bodies as a result of steaming is likely; such changes occur on heating soybeans under more severe conditions (Saio and Watanabe, 1968). Further modification of soybean ultrastructure is suggested in full-fat soy flour made by extrusion cooking. The flour particles appear to be angular and to be aggregated into clusters (Mustakas et al., 1971). SEM micrographs for a germinated and a high-temperature soy flour (presumably both were full-fat products) reported by Pomeranz et al. (1977) resemble our results shown in Figure 8a.

For comparative purposes, Figure 8c shows a micrograph of an experimental full-fat flour that was prepared by cracking, dehulling, and pin milling (Wolf, 1970). Intact protein bodies are clearly seen along with some irregularly shaped particles.

Defatted Flours

The bulk of soybeans used in the United States is converted into oil and toasted meal. In this process, soybeans are cracked, dehulled, flaked, extracted with hexane, desolventized, and toasted to yield the meal that goes largely into animal feeds (Wolf and Cowan, 1975). Bair (1979) recently described the ultrastructural changes occurring during such processing.

For the preparation of edible flours, a modified procedure is used. After hexane extraction, a milder form of desolventizing is employed and the toasting step is varied depending on the intended end use. Consequently, defatted flours with minimum protein contents of 50% are available with cooking ranging from minimum heat treatment (white flours) to fully cooked (toasted flours). White flours contain numerous intact protein bodies (Fig. 9a), whereas toasted flours contain fewer particles that appear to be protein bodies (Fig. 9b). The cytoplasmic network observed on protein bodies in a fracture surface, as shown in Figures 5e-f, was not identifiable on the surface of protein bodies in either flour sample. SEM observations on defatted flours by others (Pomeranz et al., 1977) are in agreement with our results (Fig. 9).

Protein Concentrates

These commercial products containing a minimum of 70% protein can be prepared by (a) alcohol leaching (Mustakas et al., 1962), (b) dilute acid leaching (Sair, 1959), or (c) moist heat treatment followed by water leaching (McAnelly, 1964). Distinct differences in particle sizes and shapes occur in these products. The alcohol-leached concentrate consisted of particles ranging from about 5 to 80 μm in diameter (Fig. 10a). Some of the smaller particles appeared to be protein bodies that survived the leaching and drying steps involved in processing (Fig. 10b).

The acid-leached product contained partially collapsed spheres plus larger contorted particles covered with a smooth film (Fig. 10c-d). This product is neutralized after acid leaching and then spray dried. The neutralization step dissolves some of the proteins, and presumably the soluble proteins give rise to the collapsed particles and the continuous films during the spray drying operation.

The moist heat-treated and water-leached concentrate contained much larger particles and had a more granular appearance than either of the other two products (Fig. 10e). No residual structure of the cotyledon was evident, and a fibrous texture is suggested at higher magnification (Fig. 10f).

Protein Isolates

Isolates contain a minimum of 90% protein and are available in the isoelectric form and as neutralized proteinates. Both are spray dried. Isoelectric forms range in particle size from about 2 to 40 μm and have a bumpy appearance (Fig. 11a). Higher magnification suggests that the bumpy surface is the result of coalescence of smaller particles that probably represent aggregates formed when the protein is precipitated at pH 4.6 (Fig. 11b).

The proteinates have a different appearance than the isoelectric types. The particles are mainly spheres or collapsed spheres with a smooth surface (Fig. 11c-f). These are not the protein bodies that exist in defatted flour but they are formed as a result of spray drying the neutralized dispersions of the isolated proteins. The spherical particles likely are formed by drying of droplets of protein solution. The spheres are hollow and often collapse (Fig. 11e). Noncollapsed spheres nonetheless are hollow, as observed in a particle with a broken surface (Fig. 11f). Conditions during spray drying, such as protein concentration and temperature, may determine the extent of collapsed particles one observes.

A byproduct obtained in the preparation of protein isolates is the insoluble residue or spent flakes. This material consists primarily of cell walls plus residual protein. Empty cells are readily discernible in the spent flakes (Fig. 12a and b).

Textured Protein Products

Textured soybean proteins are manufactured from flours, concentrates, and isolates. The most widely used products are manufactured by thermoplastic extrusion of soy flour (Atkinson, 1970). Extrusion of wet flour at high temperature and pressure develops a chewy, fibrous structure (Fig. 13a and b). Flours also can be texturized by subjecting defatted soy flour to high-pressure steam in a rotary valve and expelling it through a gun barrel-like device (Strommer, 1975). The fibrous structure of the resulting product is readily observed by SEM (Fig. 13c and d).

SEM has been applied by a number of other workers to study extrusion of soy flours. Cumming et al. (1972) examined differences in structure of defatted flour before and after extrusion. A related study indicated that increasing process temperatures produced aligned fibers and a more porous structure. The pores apparently form as a result of flashing off of the steam as the material leaves the extruder die (Maurice et al., 1976). Aguilera and coworkers (1976) isolated soybean grits from seven preselected areas in an extruder and examined them to follow the progressive change from the cellular structure of the grits to the fibrous character of the textured grits. Taranto et al. (1978a) combined the techniques of light microscopy, TEM, and SEM to examine the structural relationship between the proteins and the insoluble carbohydrates in textured cottonseed and soy flours. In a second study, they compared textured flours made by extrusion and by a snack food press; they concluded that working and kneading of the flour by the extruder is unnecessary to develop texture (Taranto et al., 1978b).

The classical procedure for texturizing soy protein isolates is by spinning, which involves dissolving isolate in alkali and then pumping it through fine spinnerettes into an acid-salt bath that neutralizes the alkali and coagulates the protein into fibers. The fibers are then combined into bundles, stretched, washed, and formulated into meat analogs (Thulin and Kuramoto, 1967). The fibers appear relatively smooth at low magnification (Fig. 14a), but at higher magnification lengthwise striations are seen on the surface of the solid fibers (Fig. 14b). The structural identity of the fibers is readily observed in a ham analog (Fig. 14c) and in a fried

Scanning Electron Microscopy of Soybeans

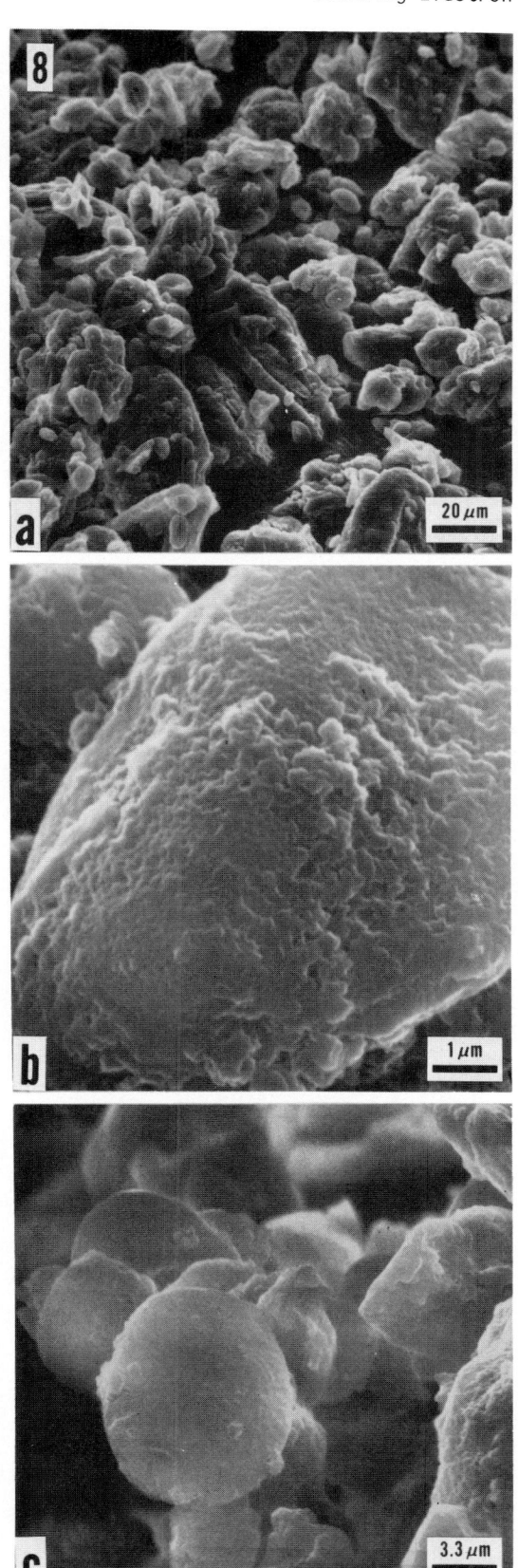

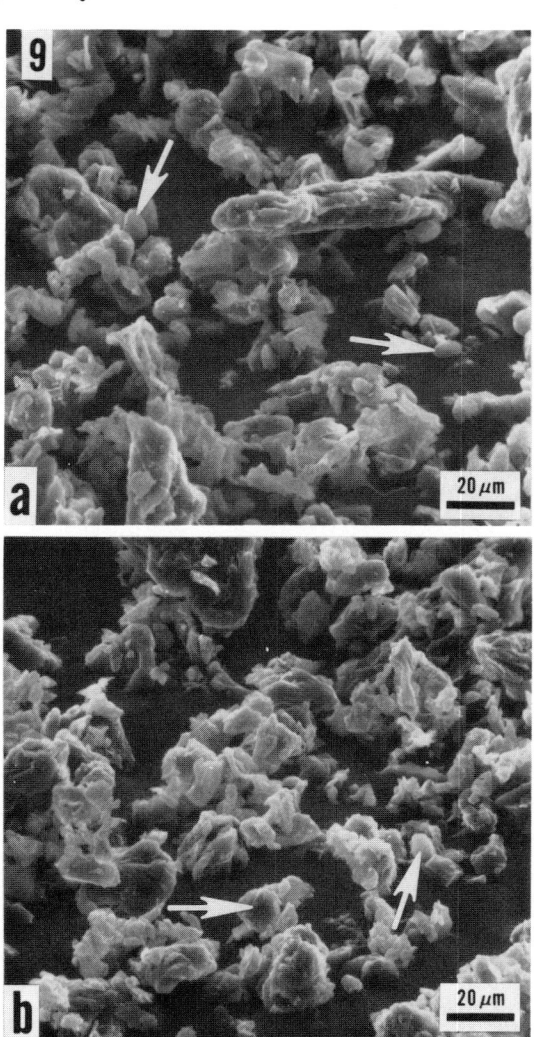

Fig. 8. Commercial full-fat soy flour showing (a) particle size and shape and (b) lipid bodies on surface of protein body. Experimental full-fat flour (c) clearly shows protein bodies.

Fig. 9. Electron micrographs of (a) "white" soy flour and (b) toasted soy flour. Protein body-like particles are indicated by arrows.

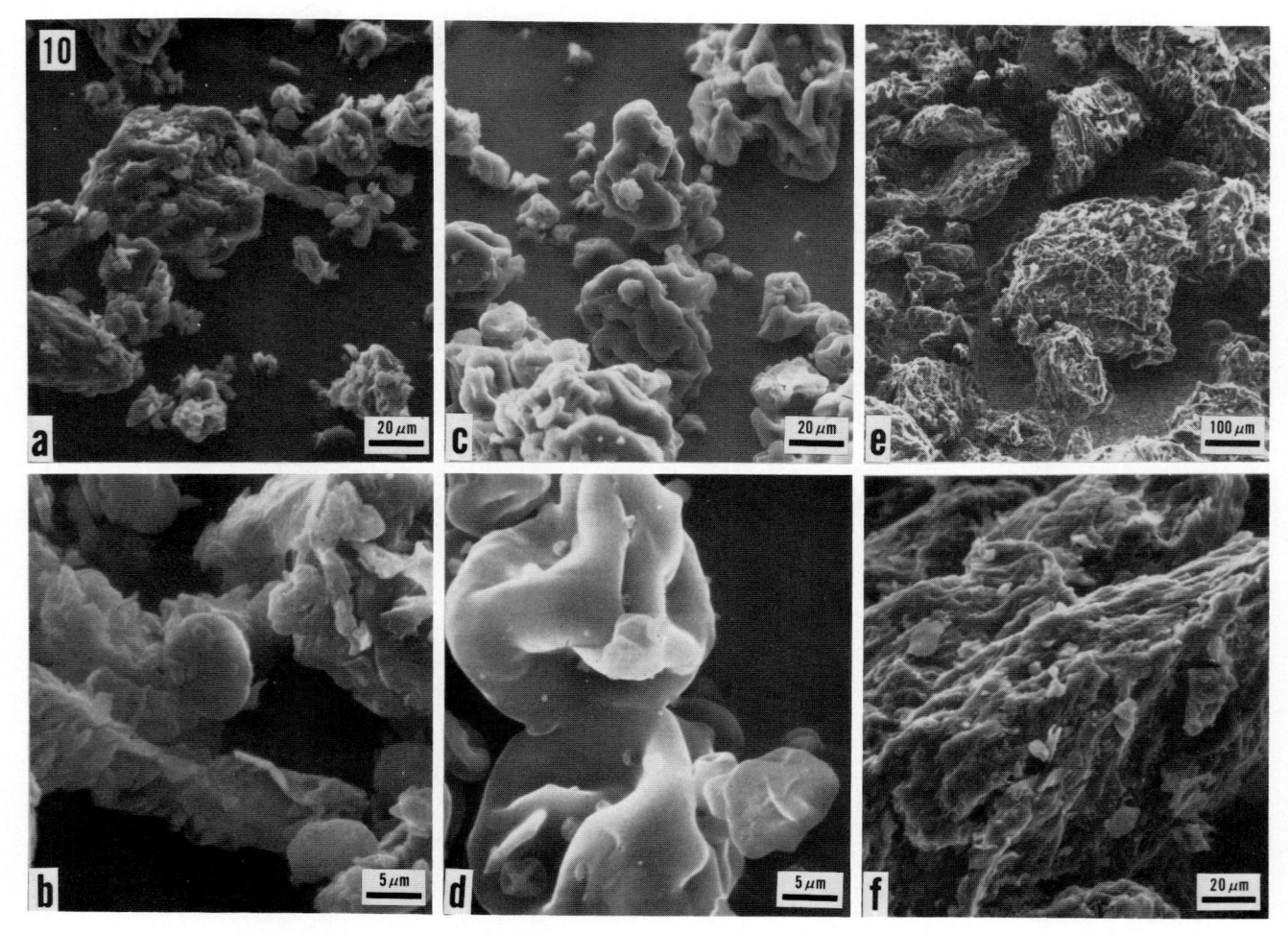

Fig. 10. Protein concentrates prepared by alcohol leaching (a, b); acid leaching (c, d); and by moist heat treatment-water leaching (e, f). Note the lower magnification in (e, f) as compared to (a-d).

bacon analog (Fig. 14d). Related SEM of spun fibers is described by Stanley et al. (1972).

Although the fibers are largely solid, internal cavities are sometimes seen; these are the points where the fibers rupture when they are stretched. Existence of a skin on the fiber surface, plus variations in texture of the fiber core ranging from smooth to fibrous, are also reported (Aguilera et al., 1975).

Another method for texturizing isolates consists of jet cooking a slurry of isolate, pumping it through a die under high pressure into air, and centrifuging to remove excess water (Hoer, 1972). The product obtained by this "dry spinning" method has a highly fibrous structure that simulates the chewiness of meat fibers (Fig. 15a and b).

Oriental Foods

In contrast to flours, concentrates, and isolates that are made from defatted flakes, many traditional Oriental soybean protein foods are made directly from soybeans. Although we have not studied the structures of these products, a brief summary of some SEM work on these materials is given here.

Spray-dried soy milk consists largely of spheres and collapsed spheres that resemble the particles found in soy proteinates (Fig. 11c-f). However, in contrast to the smooth surfaces on proteinate particles, the soy milk particles have a spongy appearance (Pomeranz et al., 1977).

When soy milk is heated, a surface film forms that can be lifted off and dried. The resulting product is known as yuba and consists primarily of protein and oil. Farnum et al. (1976) examined yuba films by SEM and TEM and concluded that they are a continuous protein matrix with dispersed droplets of lipid.

Saio (1979) has recently reviewed her studies on the structure of tofu, which is made by adding a coagulant such as calcium sulfate or glucono delta lactone to hot soy milk. TEM and SEM reveal a protein-lipid network, which is capable of retaining nine times its weight in water to give the characteristic gel structure of tofu. Because of the high water content of the tofu, samples had to be examined with a cryounit when using SEM. In a

Scanning Electron Microscopy of Soybeans

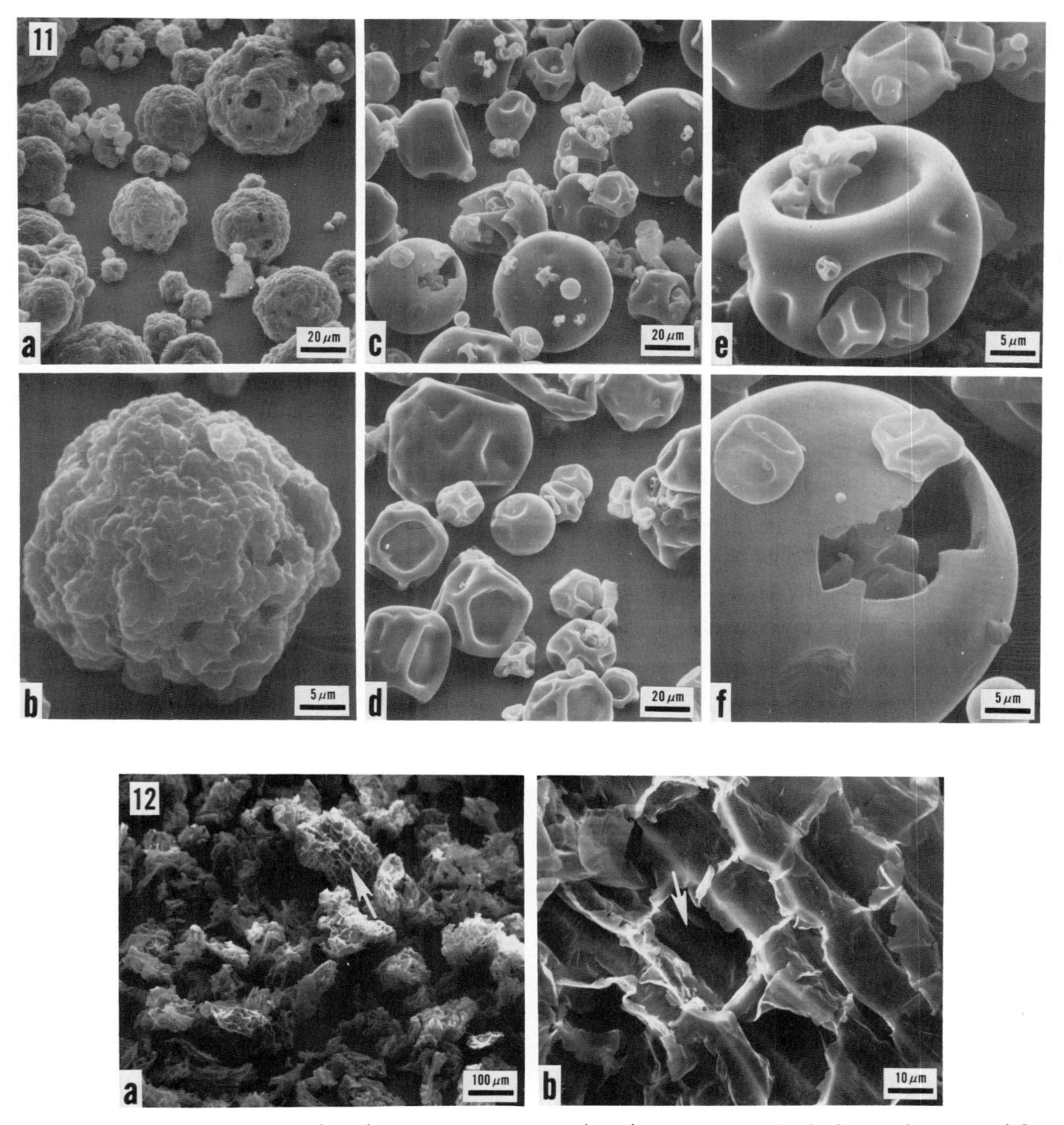

Fig. 11. Isoelectric isolate (a, b); sodium proteinates (c, d); collapsed spherical proteinate particles (e); and broken spherical proteinate particle (f).
Fig. 12. Spent flakes from protein isolate process (a) and empty cells in spent flake material (b). Arrows points to empty cells.

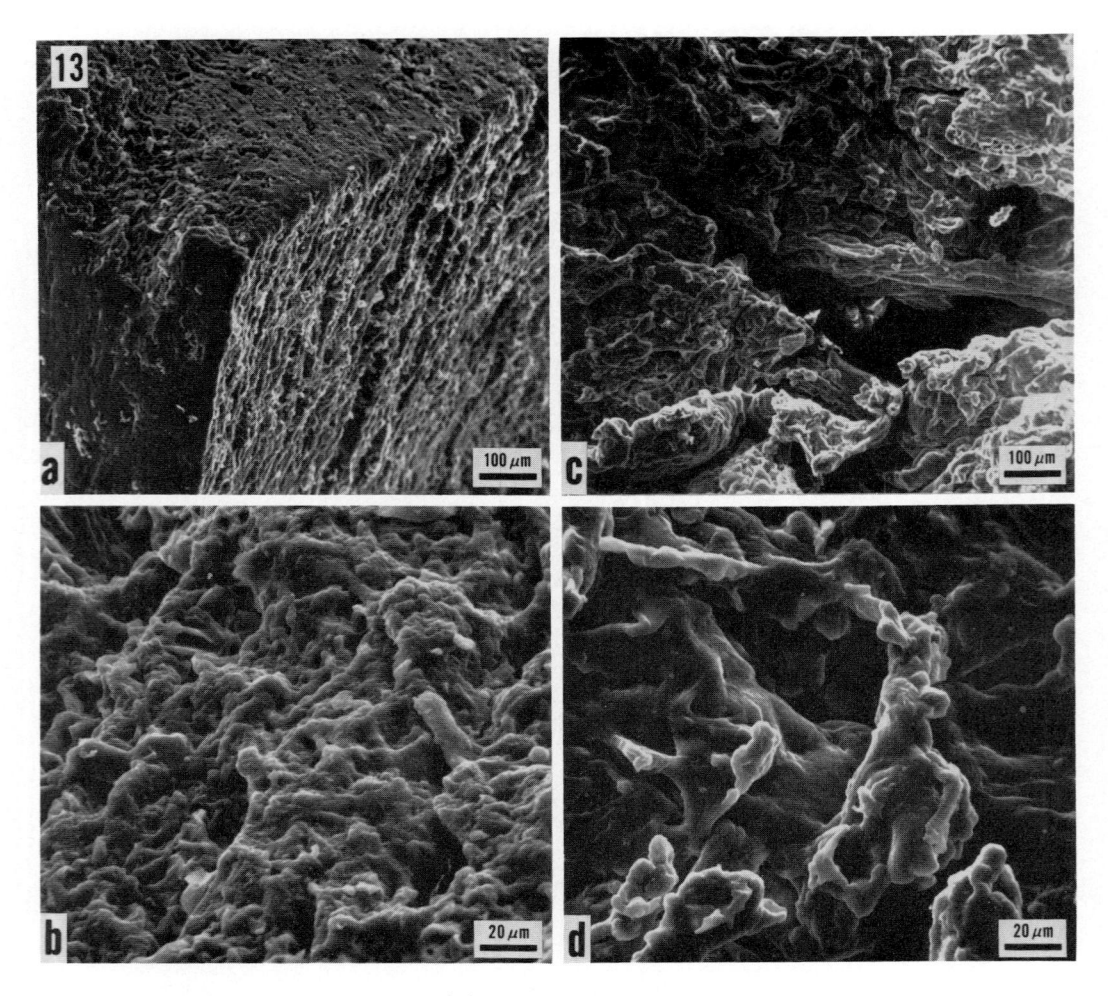

Fig. 13. Thermoplastic extruded soy flour (a) at low magnification showing three sides of a cube and, (b) at high magnification of right side of view in (a), and of steam texturized soy flour at low (c) and high (d) magnification.

related study, Lee and Rha (1978) examined the structure of protein aggregates prepared from a water extract of defatted meal using light microscopy and SEM. Hydrochloric acid and calcium chloride were used as coagulants, and samples for SEM were either freeze dried or fixed with glutaraldehyde and then freeze dried. Unheated protein solutions yielded globular aggregates, but these are not protein bodies as claimed. The protein bodies dissolve in the preparation of the original water extract. More likely these aggregates are artifacts of the precipitation process as seen in the commercial sample of isoelectric isolates (Fig. 11a-b). Network formation occurred only when the protein extracts were heated before coagulation of the protein; hence, denaturation is a prerequisite to development of the three-dimensional structure found in tofu.

Conclusions

Following the dictum that "the best treatment is no treatment" (Chabot, 1979), we have found SEM very useful in determining the structure of soybeans and soybean-derived products. The low moisture characteristic of many of the soybean products enables one to merely coat them with an electron-conducting layer and then examine them. SEM has not only confirmed previous work with TEM of soybean structure but also provided new information on how the native structure is modified during processing of the seed into flours, concentrates, and isolates. The technique is extremely useful in examining new structures formed from soy proteins, such as the textured products. Ideally, SEM studies should be complemented with light microscopy and TEM when time and facilities are available. A number of workers have followed this three-pronged approach.

References

Aguilera, J. M., F. V. Kosikowski, and L. F. Hood. Ultrastructure of soy protein fibers fractured by various texture measuring devices. J. Text. Stud. 6, 549-554. 1975.

Scanning Electron Microscopy of Soybeans

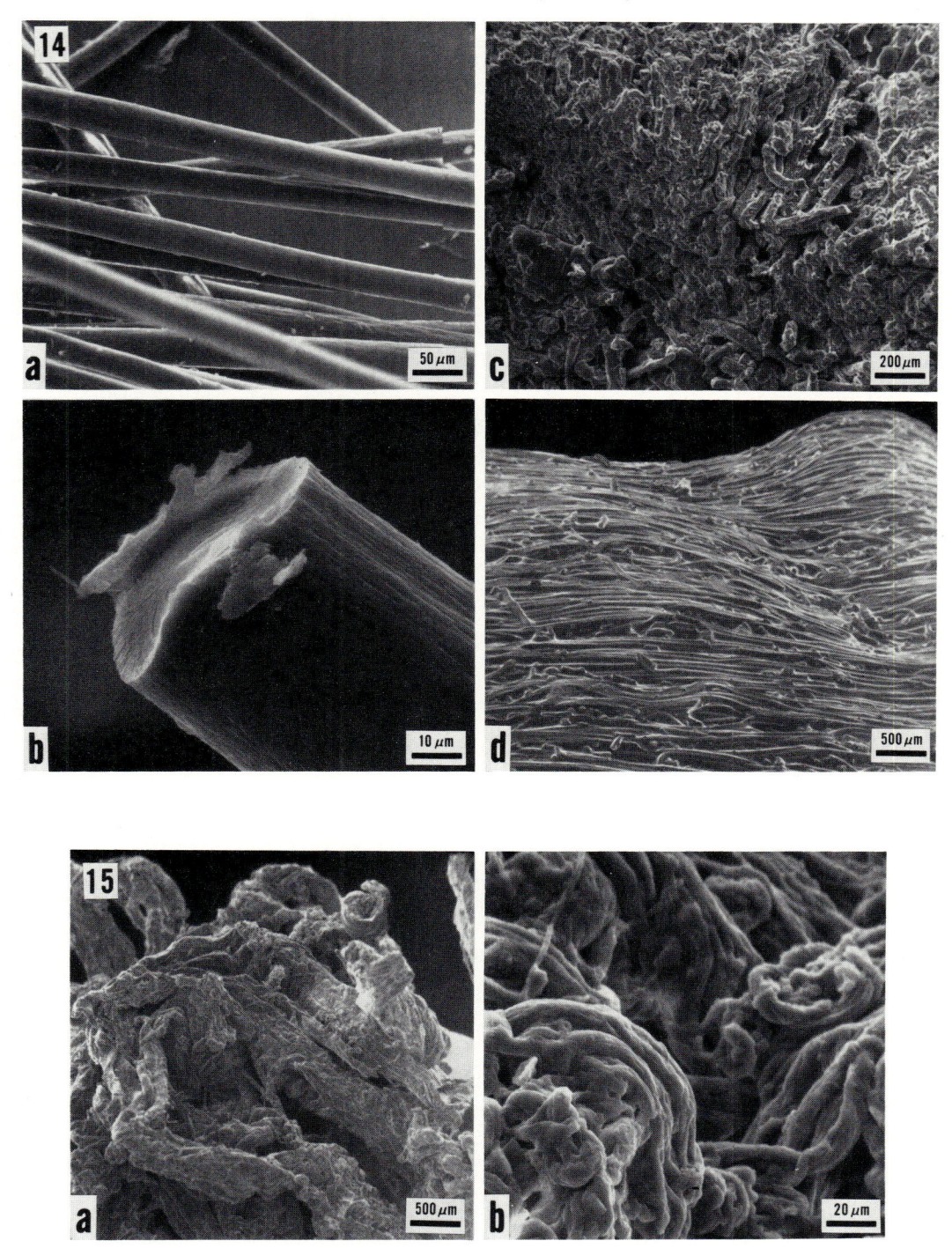

Fig. 14. Electron micrographs showing (a) spun isolate fibers; (b) cross section and striations on surface of individual fiber; (c) fibers in a ham analog; and (d) fibers in a fried bacon bit analog.
Fig. 15. Isolate texturized by jet cooking at low (a) and high (b) magnification.

Aguilera, J. M., F. V. Kosikowski, and L. F. Hood. Ultrastructural changes occurring during thermoplastic extrusion of soybean grits. J. Food Sci. 41, 1209-1213. 1976.

Atkinson, W. T. Meat-like protein food product. U.S. Patent 3,488,770. January 6, 1970.

Bair, C. W. Microscopy of soybean seeds: Cellular and subcellular structure during germination, development and processing with emphasis on lipid bodies. Ph.D. Dissertation, Iowa State University, Ames, Iowa. 1979.

Bils, R. F. and R. W. Howell. Biochemical and cytological changes in developing soybean cotyledons. Crop. Sci. 3, 304-308. 1963.

Chabot, J. F. Preparation of food science samples for SEM. In: SEM/1979/III, SEM, Inc., AMF O'Hare, Illinois, 279-286, 298. 1979.

Cumming, D. B., D. W. Stanley, and J. M. deMan. Texture-structure relationships in texturized soy protein. II. Textural properties and ultrastructure of an extruded soybean product. Can. Inst. Food Sci. Technol. J. 5, 124-128. 1972.

Farnum, C., D. W. Stanley, and J. I. Gray. Protein-lipid interactions in soy films. Can. Inst. Food Sci. Technol. J. 9, 201-206. 1976.

Hoer, R. A. Protein fiber forming. U.S. Patent 3,662,672. May 16, 1972.

Lee, C. H. and C. Rha. Microstructure of soybean protein aggregates and its relation to the physical and textural properties of the curd. J. Food Sci. 43, 79-84. 1978.

McAnelly, J. K. Method for producing a soybean protein product and the resulting product. U.S. Patent 3,142,571. July 28, 1964.

Maurice, T. J., L. D. Burgess, and D. W. Stanley. Texture-structure relationships in texturized soy protein. III. Textural evaluation of extruded products. Can. Inst. Food Sci. Technol. J. 9, 173-176. 1976.

Mounts, T. L., G. R. List, and A. J. Heakin. Postharvest handling of soybeans: Effects on oil quality. J. Am. Oil Chem. Soc. 56, 883-885. 1979.

Mustakas, G. C., W. J. Albrecht, G. N. Bookwalter, et al. New process for low-cost, high-protein beverage base. Food Technol. 25, 534-538, 540. 1971.

Mustakas, G. C., L. D. Kirk, and E. L. Griffin, Jr. Flash desolventizing defatted soybean meals washed with aqueous alcohols to yield a high protein product. J. Am. Oil Chem. Soc. 39, 222-226. 1962.

Okamura, T. Studies on the state of water in soybean seed in relation to moisture content. Res. Bull. Obihiro Zootech. Univ., Series I, 8, 261-317. 1973.

Pomeranz, Y., M. D. Shogren, and K. F. Finney. Flour from germinated soybeans in high-protein bread. J. Food Sci. 42, 824-827, 842. 1977.

Saio, K. Tofu-Relationships between texture and fine structure. Cereal Foods World 24, 342-345, 350-354. 1979.

Saio, K., K. Arai, and T. Watanabe. Fine structure of soybean seed coat and its changes on cooking. Cereal Sci. Today 18, 197-201, 205. 1973.

Saio, K. and T. Watanabe. Observation of soybean foods under electron microscope. Nippon Shokuhin Kogyo Gakkai-Shi (J. Food Sci. Technol.) 15, 290-296. 1968.

Sair, L. Proteinaceous soy composition and method of preparing. U.S. Patent 2,881,076. April 7, 1959.

Sefa-Dedeh, S. and D. W. Stanley. Textural implications of the microstructure of legumes. Food Technol. 33(10), 77-83. 1979.

Stanley, D. W., D. B. Cumming, and J. M. deMan. Texture-structure relationships in texturized soy protein. I. Textural properties and ultrastructure of rehydrated spun soy fibers. Can. Inst. Food Sci. Technol. J. 5, 118-123. 1972.

Strommer, P. K. Method for texturizing particulate protein material. U.S. Patent 3,883,676. May 13, 1975.

Taranto, M. V., G. F. Cegla, K. R. Bell, et al. Textured cottonseed and soy flours: A microscopic analysis. J. Food Sci. 43, 767-771. 1978a.

Taranto, M. V., G. F. Cegla, and K. C. Rhee. Morphological, ultrastructural and rheological evaluation of soy and cottonseed flours texturized by extrusion and non-extrusion processing. J. Food Sci. 43, 973-979, 984. 1978b.

Thulin, W. W. and S. Kuramoto. "Bontrae"- A new meat-like ingredient for convenience foods. Food Technol. 21, 168-171. 1967.

Tombs, M. P. Protein bodies of the soybean. Plant Physiol. 42, 797-813. 1967.

Webster, B. D. and A. C. Leopold. The ultrastructure of dry and imbibed cotyledons of soybean. Am. J. Bot. 64, 1286-1293. 1977.

Wolf, W. J. Scanning electron microscopy of soybean protein bodies. J. Am. Oil Chem. Soc. 47, 107-108. 1970.

Wolf, W. J. and F. L. Baker. Scanning electron microscopy of soybeans. Cereal Sci. Today 17, 124-126, 128-130, 147. 1972.

Wolf, W. J. and F. L. Baker. Scanning electron microscopy of soybeans, soy flours, protein concentrates, and protein isolates. Cereal Chem. 52, 387-396. 1975.

Wolf, W. J. and J. C. Cowan. Soybeans as a food source. CRC Press, Inc. Boca Raton, Florida, 101 pp. 1975.

Discussion with Reviewers

K. Saio: In discussing the seed coat, you stated that some varieties have pitted seed coats. Are there soybean varieties without such pits? The seed coats of Brazilian soybeans are often covered with a fine, red clay (Fig. 16) which cannot easily be removed by water washing and which has significant unfavorable effects on colors of traditional Japanese soybean foods. Because screening and breeding for varieties with fewer pits seemed like a useful approach to the problem, I checked several Brazilian and Japanese soybean varieties. However, all of them were

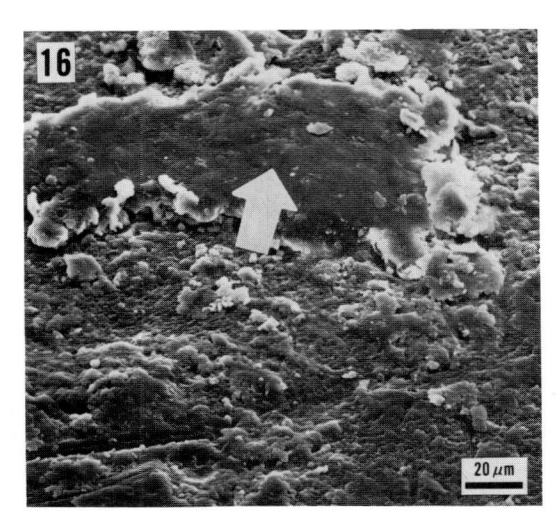

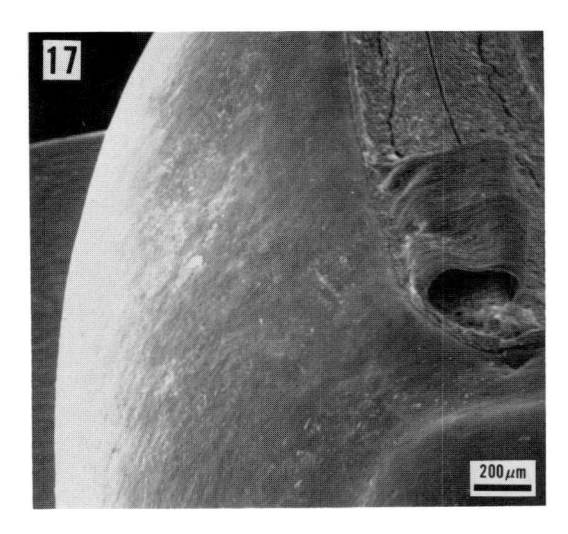

Fig. 16. Seed coat surface of Brazilian soybean showing adhering mass of fine, red clay indicated by arrow (figure provided by K. Saio).
Fig. 17. Seed coat surface of Peking variety soybean in the region of the hilum. Note absence of pits as observed in Figure 1a.

pitted to about the same degree with variations in numbers of pits in different areas of the seed coats.
Authors: On examining over 40 different varieties, plant introductions, and strains of soybeans, we found several samples that had seed coats that appeared very smooth. Two of these were Peking (Fig. 17) and Ilsoy varieties. Peking is a small seeded variety with a black seed coat while Ilsoy seed is larger but has a brown seed coat. Because of their seed coat colors and other characteristics these varieties are not grown commercially. However, they have the important characteristic of resistance to the cyst nematode and are maintained in the USDA germplasm collection. It would be interesting to determine whether Brazilian red clay adheres to these seeds as tenaciously as it does to seeds with pitted seed coats.

K. Saio: You have pointed out an interesting surface structure on the seed coat of Sooty soybeans (Fig. 2a-b). My research on this point is still limited, but I think it is noteworthy to call attention to this structural feature.
Authors: The bloom on Sooty soybeans is obvious to the naked eye and gives the seeds a dull, dusty, or moldy appearance. It also occurs, although to lesser degree, on commercial varieties such as Hark, Clark 63, and Provar. It is less noticeable on these varieties because of their yellowish seed coats. The significance of this bloom material to soybean food products is still unknown.

C. Bair: In view of the importance of the seed coat for permeability of the soybean, have "hard beans" (those which do not hydrate readily) been investigated microscopically and if so, how does the seed coat structure differ from that of normal seeds?
Authors: Saio [Soybeans resistant to water absorption, Cereal Foods World 21, 168-173 (1976)] found that "hard" soybeans had a thicker and denser seed coat than normal beans. She also found that seed coats from "hard" beans were high in calcium and silicon and that the micropyle of "hard" beans was closed but open in normal seeds. Duangpatra S. [Some characteristics of the impermeable seed coat in soybeans (Glycine max [L.] Merrill), Diss. Abstr. 37, 1061-B (1976)] observed that "hard" beans had a continuous layer of suberin in the inner layer of the palisade cells under the hilar region. However, in normal beans, the suberin occurred in the form of droplets with spaces between them that could permit the passage of water. It was hypothesized that in "hard" beans, the extra-hilar region of the seed coat contains an unidentified, water-impermeable substance in the cell walls. More work is necessary to confirm these observations and to test the hypothesis of a water impermeable barrier in "hard" seeds.

M. V. Taranto: In my work with soybean cotyledons, I had limited success in producing a "clean" freeze-fractured surface following your procedure as outlined in the 1975 Cereal Chemistry article. Would you briefly describe your technique for fracturing the frozen cotyledons?
Authors: We simply froze them in liquid nitrogen and dropped them on a hard surface to shatter them. This was clearly superior to mechanical fracturing or sectioning either with or without prior fixation. We have also noted variations in the quality of freeze-fractured surfaces, but we have examined too few samples to determine whether it is related to parameters such as variety, moisture content, or age of the sample.

C. Bair: What is the chemical nature of the spongy (cytoplasmic) network which often encompases the protein bodies?
Authors: Tombs (1967, text reference) isolated the protein bodies and found them to contain 97.5% protein. However, the isolated protein bodies accounted for only about 70% of the total protein of the bean. The remaining 30% of the proteins, including trypsin inhibitors, presumably is located in the cytoplasmic network. Soybeans also contain 9-11% sucrose, raffinose, plus stachyose. If the protein bodies are essentially pure protein, these sugars likely also occur in the cytoplasm.

C. Bair: The historical background of the spherosome, which is considered by some authors to be the direct precursor of the oil body, or the same as the oil body, has been clouded by confusion and much controversy. In light of my investigations, I believe that the oil containing particles found in soybeans should be referred to as lipid bodies, while the term spherosome should be reserved for those particles which are enzymatically active and possess other distinguishing characteristics.
Authors: We agree with this comment and have eliminated the term spherosome except where we refer to it parenthetically to indicate that it is sometimes used.

K. Saio: You have pointed out that mechanical rupture of cell walls is a key step in effective extraction of the proteins and oil with water. Recent results of ours on oil extraction from soybean tissues with various solvents, revealed that it took an incredibly long time without mechanical rupture of cell walls.
Authors: The need to flake soybeans (and presumably to mechanically rupture cell walls) prior to extraction of the oil with hexane is well known to the industry. However, data to substantiate the need for cell wall rupture are difficult to find. One report on this subject is that of Othmer and Agarwal [Extraction of soybeans: Theory and mechanism, Chem. Eng. Prog. 51, 372-378 (1955)]. They found that hexane extraction for a week removed less than 0.08% of the original oil in whole beans and less than 0.19% of the oil in beans cut in half.

M. V. Taranto: There is a growing concern about the phytic acid-phytate content of soy protein products. Phytate-free proteins exhibit different functional properties than the proteins produced by conventional technologies. Have you ever studied the structure of phytate-free proteins? If so, how does the structure of these protein systems compare with that of the conventional protein systems?
Authors: To the best of our knowledge, none of the commercial proteins are phytate-free; hence, we have not had the opportunity to compare phytate-free proteins with conventional ones. We, likewise, have not had access to experimental, phytate-free proteins processed under commercial conditions.

M. V. Taranto: Agglomeration is often used to improve the water-dispersibility of protein systems. Have you ever studied the structural differences between agglomerated and non-agglomerated protein isolates?
Authors: From a molecular viewpoint, all protein isolates are agglomerated systems, but we assume that you are referring to agglomeration in the sense applied to the process for instantizing dry milk. In milk, this involves moistening spray-dried powders and redrying to cause some of the lactose to crystallize and produce clustering of the particles in loose, spongy aggregates. Since soybeans do not contain lactose, the process is not applicable to soy isolates and consequently, agglomerated isolates are not available. The jet-cooked isolate shown in Figure 15 is an agglomerated product, but this product is insoluble and has been processed to induce texture rather than to improve water dispersibility.

K. Saio: Flours, concentrates, isolates, and textured products each show characteristic structures depending on the process used. The degree of change from the original soybean structure and fine structures of the products give an indication of their properties. For example, the spherical particles formed as a result of spray drying (Fig. 11c-f) are hollow and often collapsed which suggests low density and water dispersibility whereas in the texturized products, meat-like chewiness is indicated. I agree that the technique is extremely useful in the field of food technology.
Authors: In addition, SEM often provides information that would be difficult to obtain by other methods. For example, we found that SEM of isolates provides an explanation for the differences in the bulk densities of the two types of isolates. Isoelectric isolates have bulk densities of about 44 lb/cu. ft. whereas the sodium proteinates have densities of only 26-30 lb/cu. ft. The hollow particles found in the sodium proteinate forms (Fig. 11c-f) obviously account for much of the difference in bulk densities between the two isolate types.

SOYBEAN SEED-COAT STRUCTURAL FEATURES: PITS, DEPOSITS AND CRACKS

W. J. Wolf[1], F. L. Baker[1] and R. L. Bernard[2]

[1]Northern Regional Research Center, Agricultural Research
Science and Education Administration
U.S. Department of Agriculture
Peoria, Illinois 61604

[2]Agricultural Research, Science and Education Administration
U.S. Department of Agriculture, and Department of Agronomy
University of Illinois at Urbana-Champaign
Urbana, Illinois 61801

Abstract

Thirty-three soybean (*Glycine max*) cultivars plus a number of plant introductions and strains were surveyed for surface characteristics of their seed coats. Pitting of the seed-coat surface occurred in many cultivars, but a number of them were free of pits, and the degree of pitting ranged from light to heavy in those exhibiting this feature. Williams, a currently popular cultivar, was an example of one with a heavy degree of pitting. The different cultivars also showed a wide range in the amount of a deposit on the seed coat surface which appeared to be a "fingerprint" of the endocarp, the innermost layer of the pod wall. Some cultivars were free of this material whereas others were so densely covered that it was apparent to the naked eye. One cultivar also had deposits of crystal-like materials on its seed coat and on the inner surface of the seed pod. Seed coat cracks were prevalent in some cultivars. Similar cracks were formed in intact seeds by wetting and then drying them, thereby supporting previous work that cracking can occur in the field as a result of alternate wetting and drying of the mature seed. Pitting of the seed coat has no known importance in food processing of soybeans, but the material on the seed coat decreases luster, which is undesirable for some food uses. Seed-coat cracking is of economic significance because damaged seed coats can result in lowering of grade and poorer storage stability of the seed. Split seeds yield oil of lower quality than sound seeds and are discriminated against for the preparation of edible protein products such as flours.

KEY WORDS: Soybean; Cultivar; Seed-Coat Surface; Pits; Deposit; Endocarp Structure; Crystalline Residue; Cracks; Scanning Electron Microscopy

Introduction

The seed coat plays a significant role in determining the quality of soybean (*Glycine max*) seeds that is of concern to agronomists, farmers, soybean handlers, processors, and ultimately consumers of soybean foods. Intact soybeans have a low degree of fungal contamination in the interior of the seed as compared to split seeds (Hesseltine et al., 1978) and the seed coat is an effective barrier to fungi under normal storage conditions (Christensen and Kaufmann, 1972). It is therefore desirable to produce seed with intact seed coats and to avoid seed-coat damage in order to maintain good storage qualities.

Defects in the seed coat such as cracks may be naturally occuring or may result from harvesting or subsequent handling of the seeds. Artificial drying may also cause cracking of soybean seed coats (Ting et al., 1980). Seed-coat defects are undesirable because they permit rapid entry of moisture and provide easy access for microorganisms thereby leading to poor storage and to lower quality when the seeds are planted or processed. Defects likewise make the seed coat more susceptible to breakage during handling. Seed-coat breakage resulting in the formation of split seeds can lower the grade at which the beans are traded commercially and broken beans (splits) have been shown to yield oil of lower quality than whole beans (Mounts et al., 1979). Broken seeds also should be avoided for preparation of edible protein products. For example, specifications for edible soy flours state that they shall be processed from sound, clean beans (Smith and Circle, 1972).

Because of the importance of the seed coat in maintaining quality of soybeans, we have continued our earlier scanning electron microscope (SEM) studies of seed-coat structure (Wolf and Baker, 1972; Wolf and Baker, 1980). Previously we examined only a few cultivars, but we have now extended our SEM studies to over 30 cultivars to obtain information on the prevalence of some of the structural features we observed earlier. These structural characteristics include pits and deposits on the seed-coat surface. We also examined cracking of the seed coat in selected cultivars that are prone to this defect.

Materials and Methods

Cleaned samples of soybeans were obtained from the USDA experimental plots at the University of Illinois, and commercial samples of Beeson and Hawkeye soybeans were obtained from Pacific Grain Co., Farmer City, IL.

Seed-coat sections for light microscopy were prepared freehand and stained and mounted in Oil Red O in propylene glycol.

Hard soybeans (defined in Pitting in Hard Seeds) were separated as follows: A seed lot of the cultivar Altona, known to have an appreciable proportion of hard seeds, was hand selected to obtain the smallest seeds. This concentrated the hard seeds, since hardness is often associated with the smallest seeds in a batch. Next, the selected seeds were soaked in distilled water for 7 hr to distinguish the unswollen, hard seeds from the swollen, non-hard seeds. The hard seeds were then separated from the swollen seeds, blotted dry and mounted for SEM examination.

Specimens were glued onto aluminum stubs, coated with gold:palladium (60:40) and dabbed at the point of attachment with silver conducting paint to ensure a good conductive path to ground. Specimens were examined in a Steroscan Mark 2A scanning electron microscope (Cambridge Instrument Co., Ltd., London, England).

Pitting of Seed Coats

Degree of Pitting

Earlier, we and others (Wolf and Baker, 1972; Saio et al., 1973; Newell and Hymowitz, 1978) examined a limited number of cultivars and found that the seed coat surface was pitted. In surveying more than 30 cultivars, we have now found that not all soybeans have a pitted seed coat and that degree of pitting varies among cultivars. Figure 1 shows examples illustrating the range in pitting observed. Ilsoy soybeans (Fig. 1a) were smooth and free of pits, whereas Chippewa 64 (Fig. 1b), Disoy (Fig. 1c) and Hawkeye 63 (Fig. 1d) showed progressively increasing numbers of pits in the seed coat. Table 1 lists the 33 cultivars examined and our ratings of the degree of pitting found.

Williams, which is an example of extreme pitting, was of special interest because this is one of the most popular cultivars at present; it was grown on over 11 million acres (4.5 million hectares) in 1979. Examination of eight seeds produced in 1971 (at Urbana, Illinois) consistently revealed extensive pitting (Fig. 2a). Samples of Williams from the crop years 1977 to 1980 (also grown at Urbana) all revealed similar extents of pitting. These results suggest that this feature is characteristic of the cultivar. Samples of Williams seeds for 1979 and 1980 are shown in Figures 2b and 2c, respectively.

Characteristics of Pits

The pits frequently appear to be closed slots (Fig. 3a). They penetrate about 20 to 35% of the

Table 1

Seed-Coat Characteristics of 33 Soybean Cultivars

Cultivar	Pitting[a]	Surface deposit[a]	Seed-Coat Luster	Color
Adelphia	0	+	Shiny	Yellow
Altona	+	+	"	"
Amsoy 71	++	0	"	"
Beeson	++	++	"	"
Calland	++	++	Dull	"
Chestnut	0	+	"	Brown
Chippewa 64	+	0	Shiny	Yellow
Clark	+	+++	Dull	"
Clark 63	+	+++	"	"
Cloud	0	++	"	Black
Corsoy	0	++	"	Yellow
Cutler 71	+	++	Shiny	"
Disoy	++	+	Dull	"
Hark	++	+++	"	"
Harosoy	+++	++	"	"
Harosoy 63	++	+++	"	"
Hawkeye 63	+++	++	"	"
Ilsoy	0	0	Shiny	Brown
Kanrich	+	+++	Dull	Yellow
Kent	+++	++	"	"
Kim	++	++	"	Green
Lincoln	+++	+	Shiny	Yellow
Magna	+	+++	"	"
Merit	+	++	"	"
Midwest	+	+++	"	"
Peking	0	0	"	Black
Provar	0	+++	Dull	Yellow
Rampage	+++	0	Shiny	"
Sooty	0	++++	Bloom	Black
Verde	+	++	Dull	Green
Wayne	++	+	Shiny	Yellow
Williams	+++	0	"	"
Wisconsin Black	0	+++	Dull	Black

[a]Degree of pitting or surface deposit: 0, none to very light; +, light; ++, medium; +++, heavy; ++++, very heavy.

thickness of the palisade cells which make up the outermost cellular layer (Wallis, 1913). Figure 3b shows a profile of one of the deeper pits we observed in seed-coat sections. Light microscopy of a section through one of the pits suggests that there is an oval-shaped cavity below the pits (Fig. 3c). This observation was confirmed by SEM examination of another section through one of the pits (Fig. 3d). Failure to see a cavity in Figure 3b may have been caused by sectioning above the plane passing through the cavity.

The significance or function of the pits and their associated cavities is not known; they may be functional during seed development or may play a role during germination. Of the 33 cultivars listed in Table 1, nine have few or

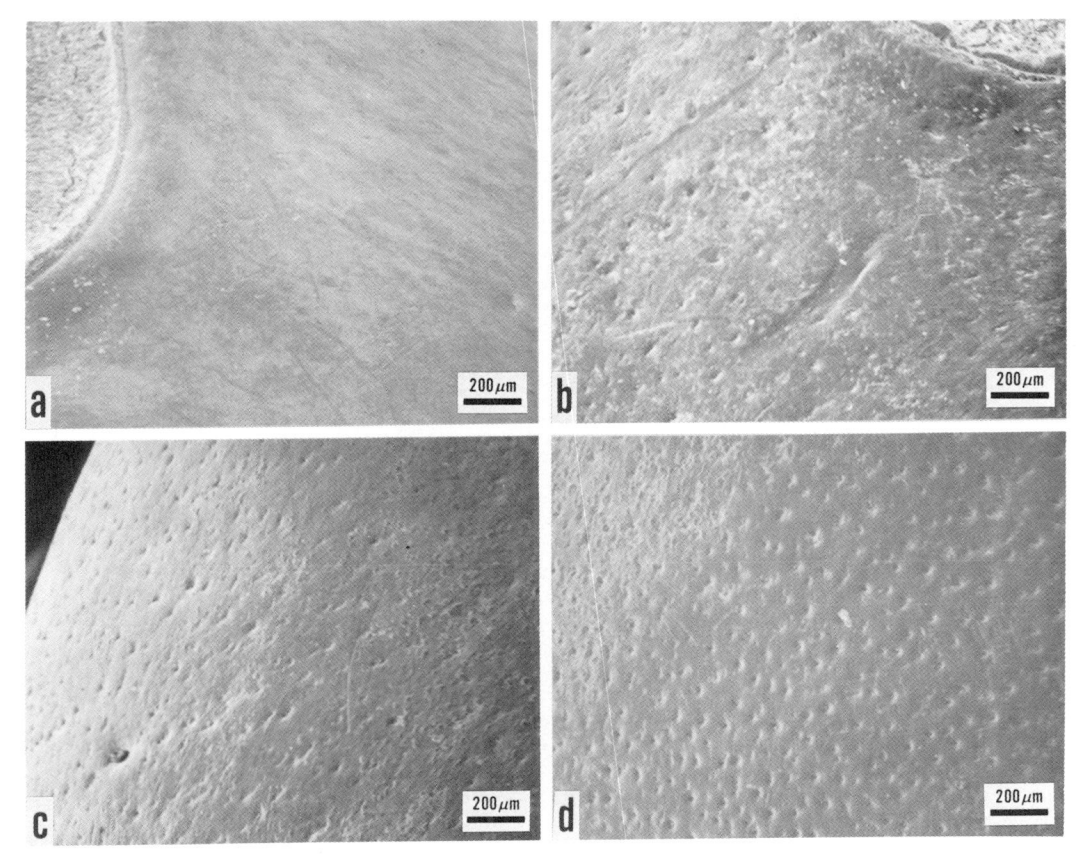

Fig. 1. Ilsoy (a); Chippewa 64 (b); Disoy (c); and Hawkeye 63 (d) seed coats illustrating the range in degree of pitting found in soybeans.

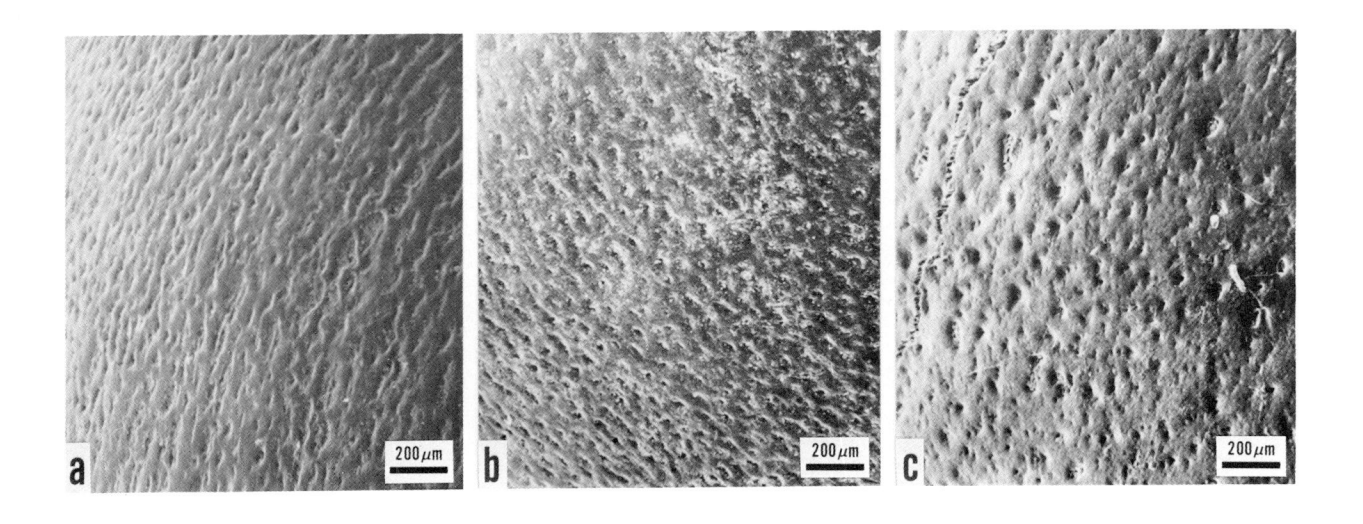

Fig. 2. Seed coats of Williams cultivar for crop years of 1971 (a); 1979 (b); and 1980 (c).

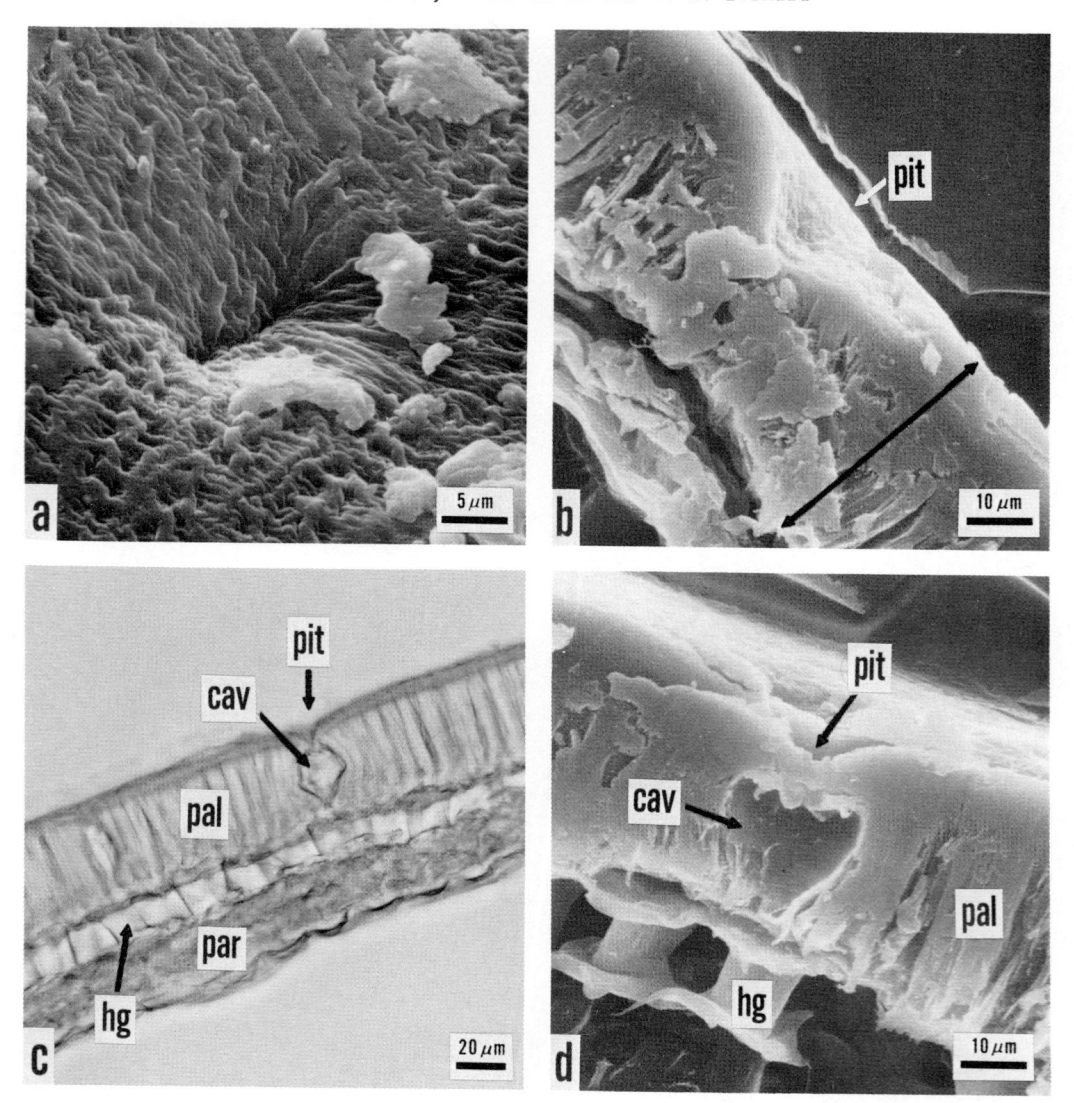

Fig. 3. Slot-shaped pit (a); section through a pit with double-headed arrow showing thickness of the palisade cellular layer (b); light micrograph of a section through a seed-coat pit showing cavity below pit (c); and scanning micrograph of section through pit showing cavity (d). Figure (a) obtained with Harosoy soybeans and all others with Hawkeye 63 cultivar.
Key: Palisade cells = pal; hourglass cells = hg; parenchyma cells = par; cavity = cav.

no pits hence they may not be essential for normal development of certain soybean cultivars.

Pitting in Hard Seeds

Soybeans that do not swell or swell very slowly when soaked in water are known as "hard" seeds (Carlson, 1973; Saio, 1976). This property is of concern to seed dealers and farmers because hard seeds are slow to germinate or may not germinate at all. Some hard seeds do not germinate when planted but germinate in the following growing season. This causes contamination of cultivars if different cultivars are planted in the succeeding season for producing seed stock. Hard seeds are also objectionable in food processing that involves soaking of the beans because of the non-uniformity that exists in size and texture at the end of the soaking. The hard seeds are much smaller than swollen seeds and have a hard texture typical of dry soybeans.

We have examined samples of hard seeds (see Materials and Methods) to determine whether "hardness" may be associated with the absence of pits on the seed coat surface. Hard Altona seeds clearly are pitted (Figs. 4a and 4b) and the degree of pitting is the same as in an unsoaked, normal Altona seed coat (Fig. 4c). Apparently the presence of pits does not make the seed coat of a hard seed permeable to water.

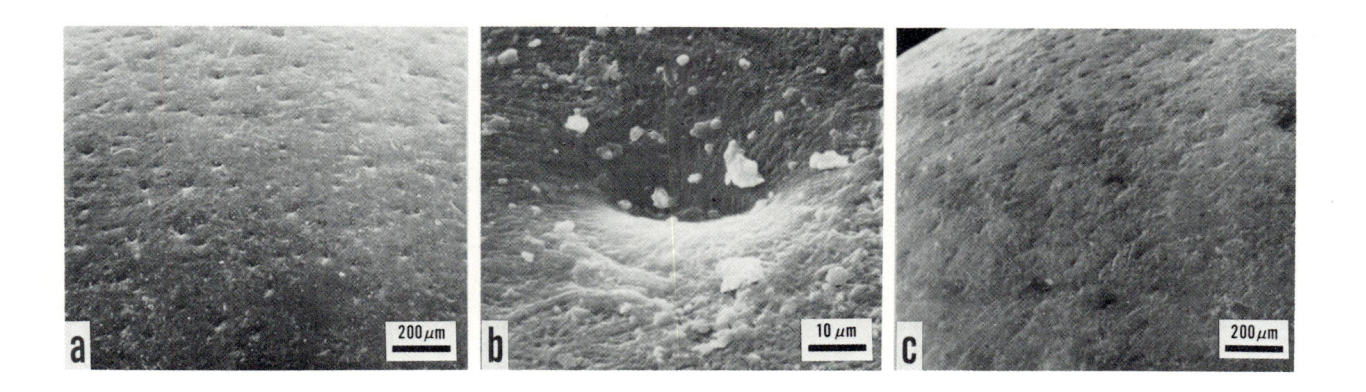

Fig. 4. Seed coats of Altona cultivar showing pits in a hard seed at low (a) and high magnification (b) and in an unsoaked, normal seed (c).

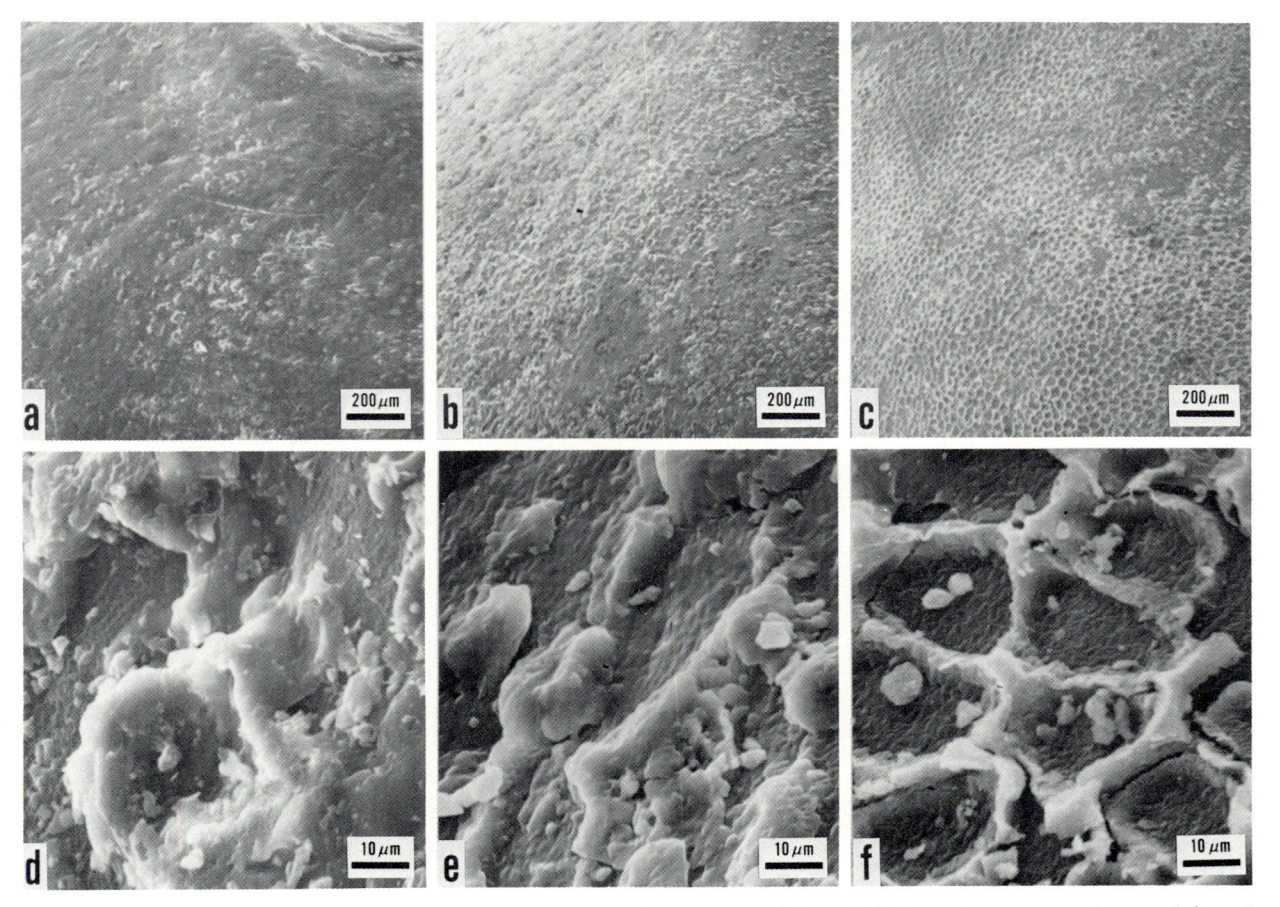

Fig. 5. Seed-coat surface deposit on Merit (a) and (d); Clark (b) and (e); and Sooty soybeans (c) and (f).

Deposits on Seed-Coat Surfaces

Degree of Deposits

In surveying the 33 cultivars listed in Table 1 for degree of pitting, we also found large variations in amounts of material deposited on the seed-coat surface. Figure 1a is illustrative of seed coats that were essentially clean of these deposits. Figure 5, in contrast, shows surface deposits on three different cultivars. Merit variety (Fig. 5a) has small amounts of surface deposits, whereas Clark (Fig. 5b) is more extensively covered and Sooty (Fig. 5c) is most heavily covered. The deposit on Sooty has a honeycomb-like appearance. At higher magnification (Figs. 5d-f), there is a similarity in the structure of the deposited material among the three varieties. Merit and Clark appear to be covered with the same material as that found on Sooty, but the honeycomb network is more disrupted than it is on the seed coat of Sooty soybeans.

Deposits and Seed-Coat Luster

Plant breeders have long noted that soybean cultivars vary in seed-coat luster and have classified them as shiny, dull or coated with bloom (Williams, 1950). Bloom is a whitish deposit on the surface similar in appearance to that on fruits such as grapes. It has been suggested that a dull luster is caused by a thin layer of the same material as that responsible for the bloom on cultivars such as Sooty and that soybeans having little or none of the surface coating are shiny (Bernard and Weiss, 1973). This interpretation is supported by the photomicrographs shown in Figure 5. In Table 1 we have indicated the extent of surface deposit on an arbitrary scale along with the classification for each cultivar. With the exception of Beeson and Midwest soybeans, all cultivars classified as shiny are free of surface deposit or only lightly coated. Cultivars rated as dull in appearance had medium to heavy deposits of material on the seed-coat surface.

Six strains from the soybean germplasm collection (PI 79648, PI 81766, PI 82278, PI 86046, PI 135589 and PI 157492) with black seed coats covered with bloom were examined in the SEM, and the surface deposits were identical to those of Sooty.

Nature of Seed-Coat Deposits

Williams (1950) stated that the bloom on Sooty cultivar seeds is waxy and easily scraped off. To determine the ease with which the surface deposit can be removed mechanically, we vigorously rubbed Clark and Sooty seeds with a clean cotton towel. Although the seeds had more sheen after rubbing, examination in the SEM revealed that much of the deposit remained (Fig. 6). Rubbing flattened the ridges of the deposits but did not appear to remove significant amounts of the material (compare with Figs. 5e and 5f).

Previously we had washed Clark and Sooty soybeans with water, hexane, or chloroform: methanol (2:1) for about five minutes, dried them and reexamined them. The different solvents did not appear to remove significant amounts of the surface deposits (Wolf and Baker, 1980). We now have soaked the soybeans in the same three solvents for 17 hr at room temperature, air-dried them for 24 hr and examined them in the SEM (Fig. 7). Water and hexane had little effect on the deposit on Clark seed coats (Figs. 7a and 7b). Chloroform-methanol gave the deposit ridges a more wrinkled appearance (Fig. 7c). With Sooty (Fig. 7d-7f), all solvent treatments may have removed some of the honeycomb-like deposits, giving them a more granular appearance than observed in the untreated sample (Fig. 5f). Cracking of the honeycomb structure in the water-soaked sample probably resulted from expansion and contraction of the seed coat during soaking and subsequent drying. Although some removal of the deposits is suggested as a result of prolonged soaking, especially with Sooty soybeans, it is clear that the bulk of the material remained on the seed coat. The deposits are obviously not water-soluble proteins, lipids or waxes but some of these materials may be present in the original structures.

Newell and Hymowitz (1978) reported that the seed coats of species of *Glycine* Willd. subgenus *Glycine* [relatives of the cultivated soybean, *Glycine max* (L.) Merr.,] typically exhibit a reticulate appearance on their surfaces. This network appears to result from adherence of the membraneous endocarp (Carlson, 1973) (the inner wall of the seed pod) to the seed coat. Newell and Hymowitz (1978) followed Hermann (1962) in calling this material "perisperm", but perisperm is a tissue located under the seed coat and therefore cannot be this material. Seed coats of *Glycine tomentella* examined by Newell and Hymowitz (1978) have a pattern similar to that found on Sooty cultivar. This similarity suggested that the surface deposit we have observed may be a residue of the endocarp.

Pods containing mature seeds were collected in the field, and their inner surfaces as well as the seed-coat surfaces were examined. Williams cultivar has a shiny seed-coat luster with little, if any, deposit on its seed coat. The inner surface or endocarp of Williams pods is a membranous layer with circular to hexagonal ridges giving the surface a network-like appearance (Fig. 8a-8b). These ridges presumably are walls of the parenchyma-type cells of this tissue (Carlson, 1973). Fibers lying below the membranous structure are apparent at both magnifications. Also noted in Fig. 8b are square to rectangular materials that appear to be crystals (further discussion of these "crystals" appears below).

Strain L72-1495 (closely related to Clark 63) has a dull seed coat luster and its seed coat has a fairly heavy deposit of network pattern on the surface with interspersed bare patches (Fig. 8c and 8d). The corresponding inner pod wall surface resembles that of Williams soybeans, but the ridge-like cell walls appear more prominent and more ragged on their upper edges (Figs. 8e and 8f). Similarities of size and shape between the seed-coat deposits and the cell wall ridges of the membranous endocarp of the pod wall suggest a common origin for the two structures.

Soybean seed-coat pits, deposits and cracks

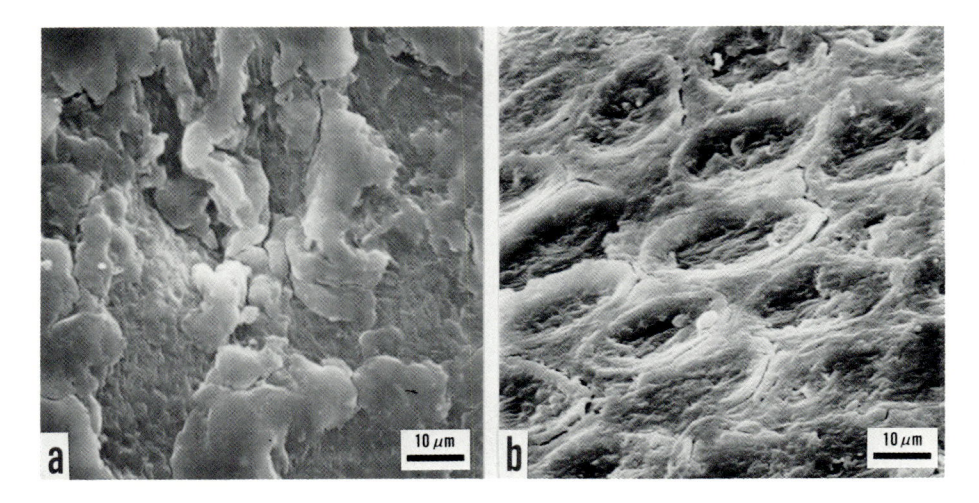

Fig. 6. Illustrations of the effect of mechanical rubbing on surface deposit of Clark (a) and Sooty (b) cultivars.

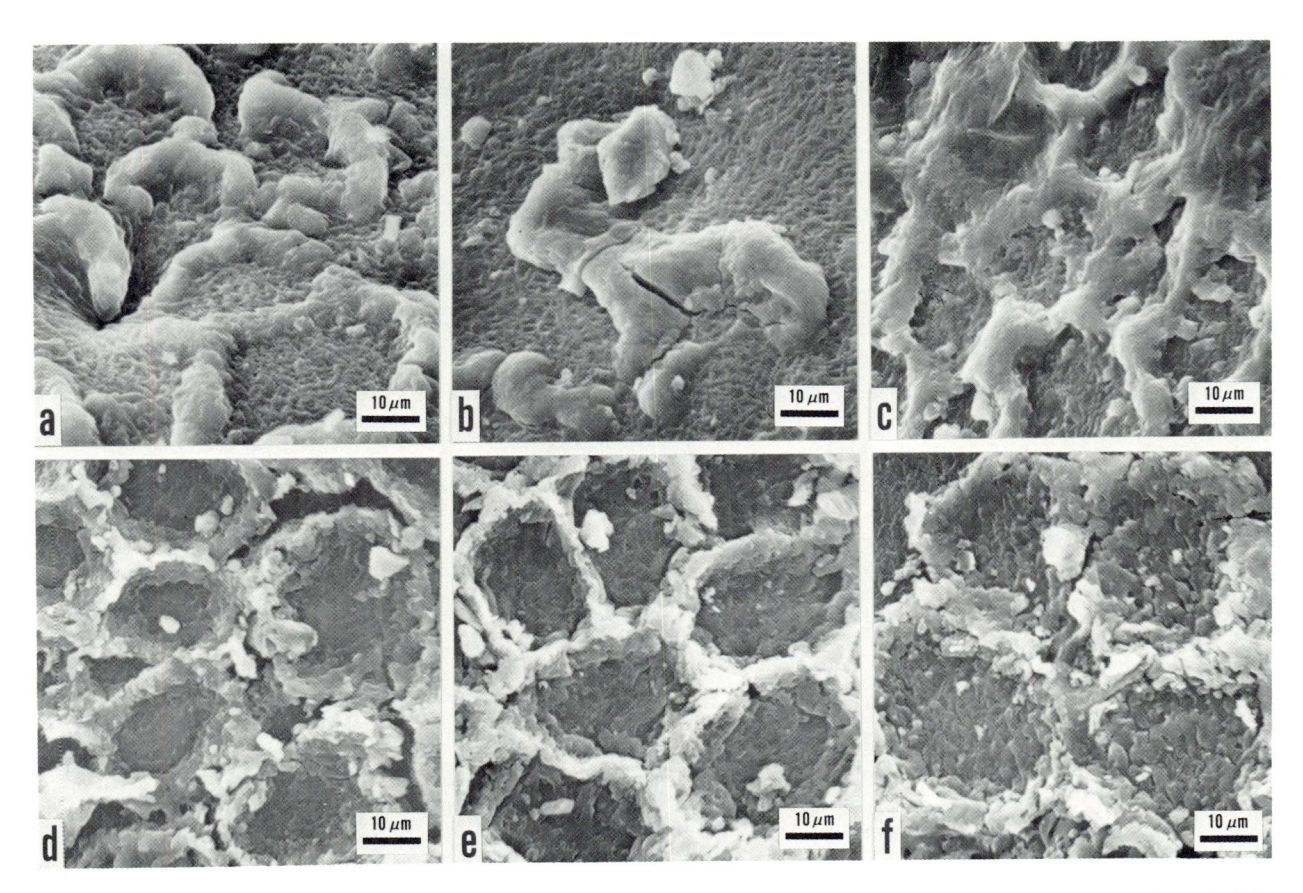

Fig. 7. Illustrations of effect of water (a) and (d); hexane (b) and (e); and chloroform:methanol (c) and (f) on seed-coat deposits of Clark (a) - (c) and Sooty cultivars (d) - (f).

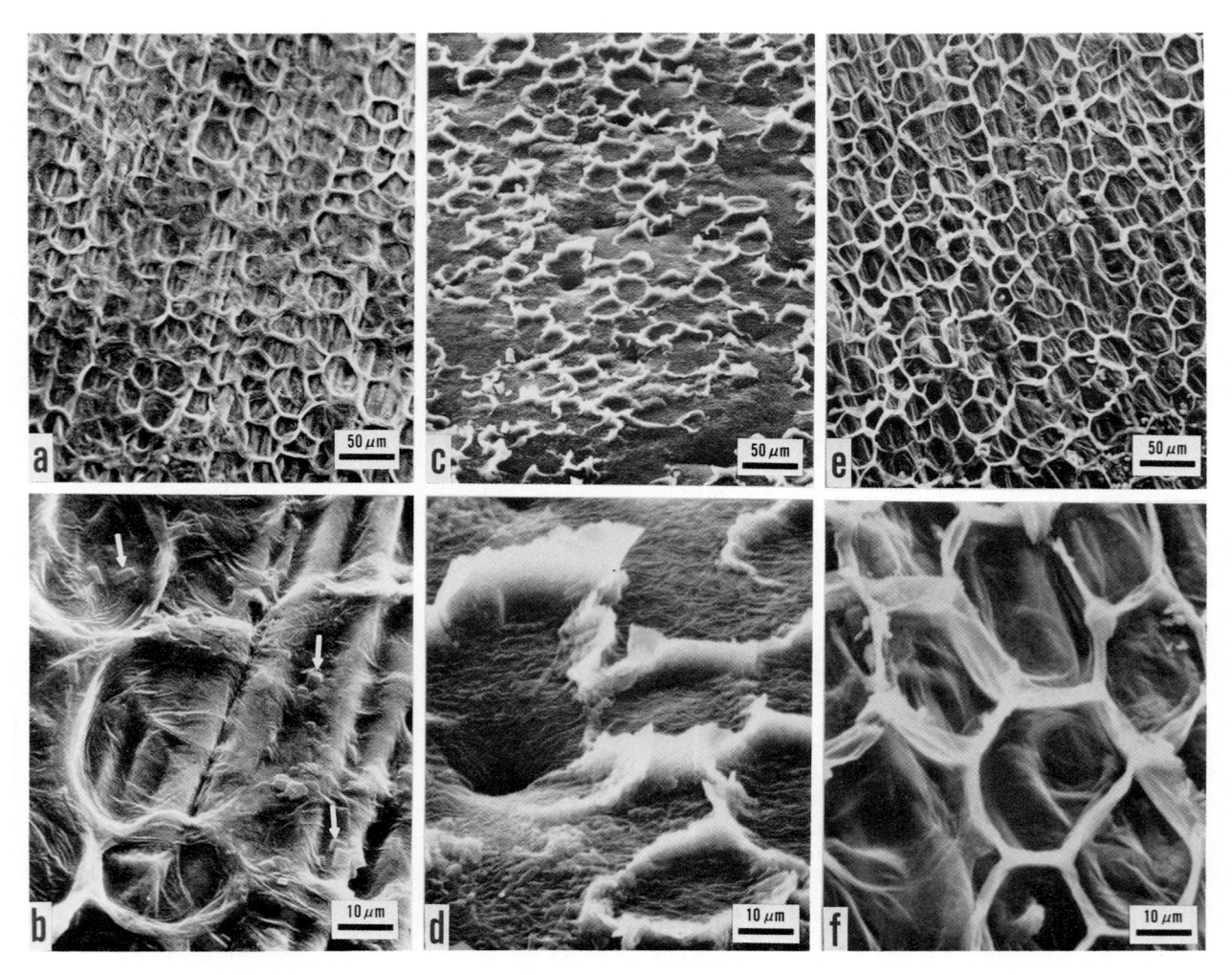

Fig. 8. Inner surface of Williams cultivar seed pod (a) and (b); seed coat surface of L72-1495 (c) and (d); and inner surface of L72-1495 seed pod (e) and (f). Arrows in (b) point to "crystals".

A closer correspondance between seed-coat residue and endocarp structure was observed in two semiwild soybeans originally from Korea, PI 339734 and PI 424078, plus a wild soybean (Glycine soja Sieb. and Zucc.) originally from China, PI 65549. All three have a distinct bloom on their seed coats. Figure 9 shows the seed coat and endocarp tissue of PI 339734. The seed-coat surface is nearly uniformly covered, with only an occasional bare spot (Fig. 9a), and the network is distinctly higher (Fig. 9b) than in strain L72-1495 (Fig. 8d). The seed-coat network of PI 339734 is like a "fingerprint" of the network on the pods (Fig. 9c and 9d), which fact suggests that the seed-coat deposits are residual endocarp. However, there are some distinct morphological differences between the two networks indicating that they are not identical. The ridges on the inner pod wall appear to be gathered membranes (probably cell walls), whereas the ridges on the seed coat are more amorphous with a slightly globular structure. Possibly the seed-coat ridges are extracellular material deposited along the ridges of the thickened cell walls of the membranous endocarp. In this regard, we have also noted a similarity of our micrographs of the seed-coat deposits to those of tomato fruit cutin (Kolattukudy, 1980). Cutin is an insoluble polyester that is the structural component of plant cuticle. If the seed-coat deposits are cutin, they may originate from the epidermal layer of the seed coat rather than the pod wall. Further work, including studies during seed development, is needed to identify the seed-coat deposits and their origin.

"Crystalline" Deposits

As pointed out earlier, the endocarp surface of Williams soybeans showed deposits of regularly shaped materials that appear to be crystalline (Fig. 8b). Similar rectangular, crystal-like material has been found on the surface of Williams seeds grown in 1980 (Fig. 10) obtained

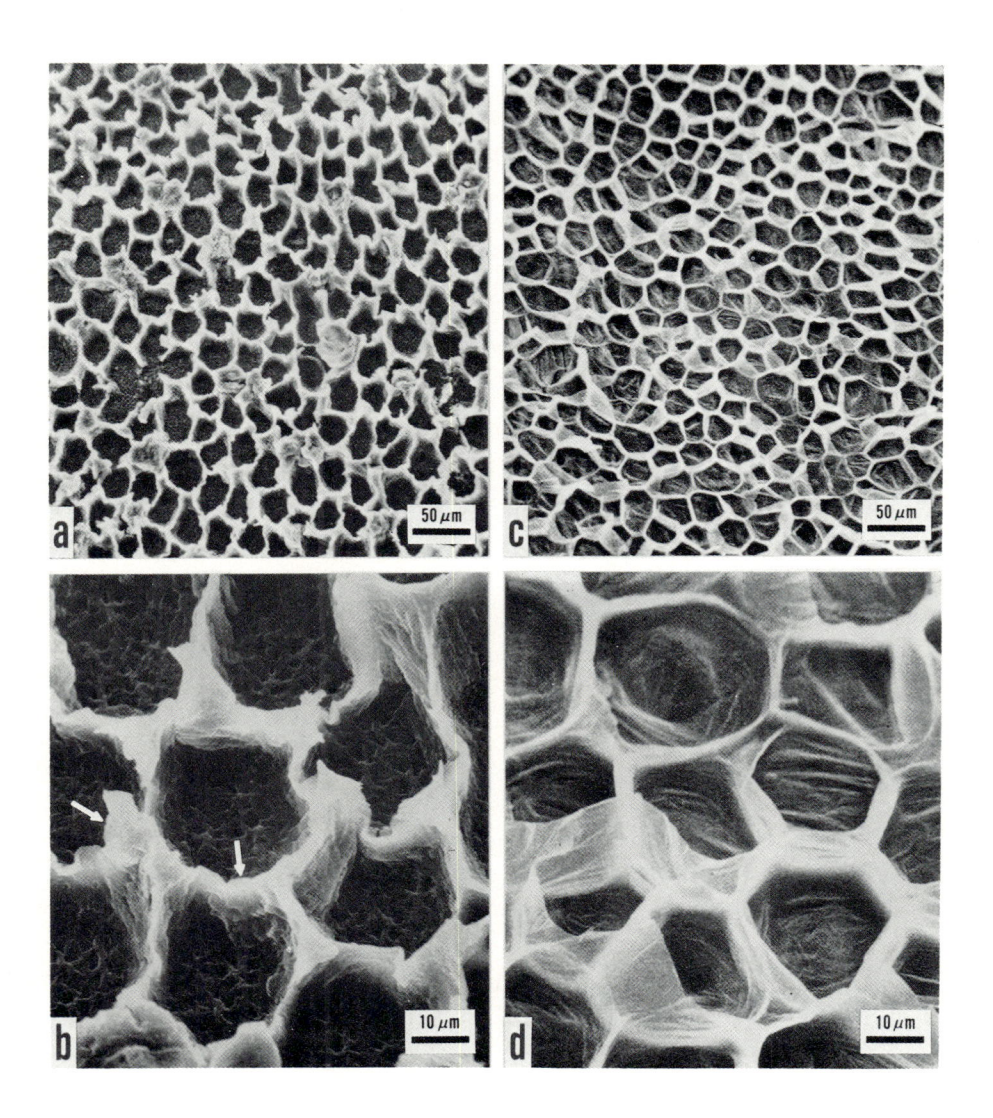

Fig. 9. Seed-coat surface (a) and (b) and inner seed pod surface (c) and (d) of PI 339734. Arrows in (b) indicate globular structure of seed-coat deposit.

from the pod used for Figure 8b. The "crystals" are about 3 X 12 µm. Because the seed was removed from the pod just before examination and the pod contained similar material, it is unlikely that the "crystals" are contaminants. The "crystals" appear to be quite fragile since Figure 10b shows several that are broken. We also observed "crystals" on the seed coat of Williams soybeans grown in 1971, hence the results shown in Figure 10b are not an isolated observation. We have not made a detailed search for these "crystals" on the seed coats of other cultivars, but we have not observed them in the course of our screening studies. The nature of these "crystals" is unknown. Wallis (1913) reported the occurrence of calcium oxalate crystals in cotyledonary cells, and Bair (1979) has confirmed Wallis' observation, but we have not found literature reports of their existance in or on the seed coat. Although the shapes of calcium oxalate crystals reported by Wallis and Bair are different than those we observed, crystal shapes of calcium oxalate vary as recently reviewed by Franceschi and Horner (1980). These authors also pointed out that calcium oxalate crystals occur on seed coats of certain species such as Oxalis stricta.

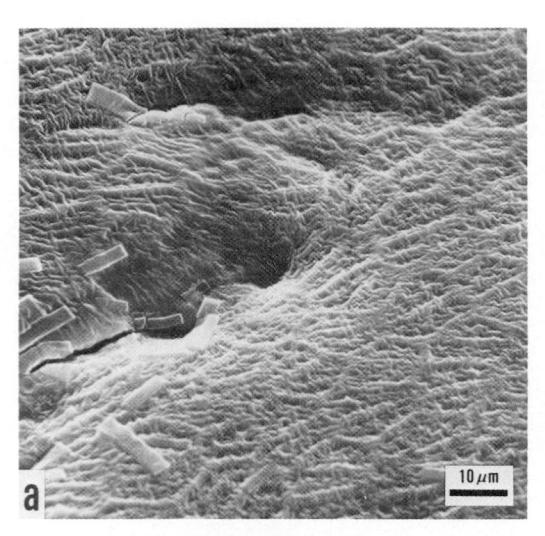

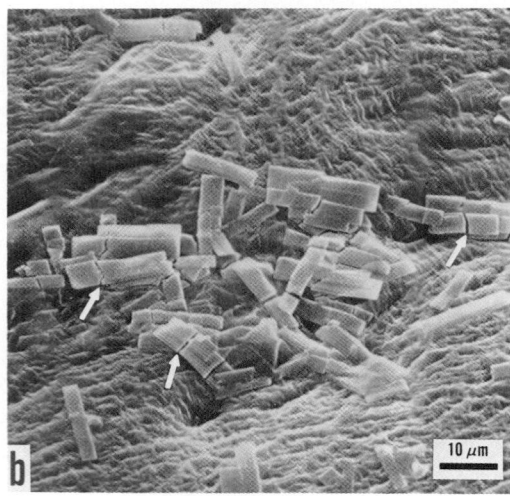

Fig. 10. Williams cultivar seed coat (a) and (b) showing "crystalline" deposits. Arrows in (b) point to broken "crystals".

Seed-Coat Cracking

Naturally Occurring Cracks

Seed-coat cracking is of at least two types. The first type is caused by mechanical impact during harvesting and handling and results in cracking of the seed coat through all of the cellular layers. For example, a single 30.5 m (100 ft) free-fall drop onto a concrete surface is capable of causing mechanical cracking resulting in 1.4 to 5.7% of seed breakage at normal, safe storage moisture contents (Foster and Holman, 1973). A second type of cracking involves splitting of only the outer layers of the seed coat. A variety of terms has been applied to these naturally occurring cracks: defective seed coat (Liu, 1949); growth cracks (Moore, 1972); and weather-induced coat fissures (Moore, 1972). Although morphologically similar, they may be formed under different conditions. One condition is during maturation where the cotyledons expand more rapidly than the seed coat, and seed coat rupturing occurs; certain varieties are prone to this type of cracking. A second set of conditions conducive to cracking of the seed coat consists of alternate wetting and drying of the mature seed while it is still in the field. Under these conditions the seed coat is believed to dry and shrink more rapidly than the cotyledons, and cracks develop. Both conditions set up stresses in the seed coat that result in rupture of the outer cellular layers. In the absence of clear-cut definitions, we refer to these cracks as internal stress cracks as opposed to mechanical cracks caused by external stresses. Cracking of seed coats during deep-bed drying of soybeans (Ting et al., 1980) likely is caused by internal stresses.

Strain T217 is a Korean cultivar that consistently has a large number of internal stress cracks in its black seed coat (Fig. 11a). It is not grown commercially in the U.S. Figure 11b is a section through the seed coat at the boundary between the intact seed coat and the surface exposed by a crack. The palisade and hourglass cellular layers end at the edge of the crack whereas the cellular layers below the hourglass cells are continuous. The seed-coat cracks thus penetrate only through the cuticle, palisade cells and the hourglass cells. The appearance of the cracks and orientation of the hourglass cells suggest that the initial separation occurs along a plane perpendicular to the seed coat and that lateral movement then occurs between the hourglass cells and the underlaying cellular layers. The exposed surface of interior cellular layers has a fibrous appearance (Fig. 11a).

Magna is a large-seeded, yellow soybean that occasionally is prone to internal stress cracks in its seed coat. Figure 11c shows one large and several small cracks in the seed coat of a Magna soybean. In contrast to T217 (Fig. 11a), the region between the edges of the large crack is comparatively smooth rather than fibrous. Higher magnification of the edge of the large crack reveals that cracking involved only the palisade and hourglass cellular layers (Fig. 11d) as noted in T217.

In the early 1960's Japanese tofu manufacturers began to import identity-preserved soybean cultivars from the United States. A preferred cultivar at that time was Hawkeye, but it was later displaced by Beeson which yields better. Beeson soybeans, however, readily form splits, which necessitates careful harvesting and handling of the seed (Paulsen et al., 1981).

Examination of commercial samples of Hawkeye and Beeson soybeans revealed that Beeson has more seed-coat cracks than Hawkeye seed. An internal stress crack and mechanical crack in a Beeson seed coat are seen in Figure 11e; the edges of the cotyledon faces are exposed as a result of the mechanical crack. A closeup view of the edge of an internal stress crack reveals exposed hourglass cells under the palisade cells (Fig. 11f).

Soybean seed-coat pits, deposits and cracks

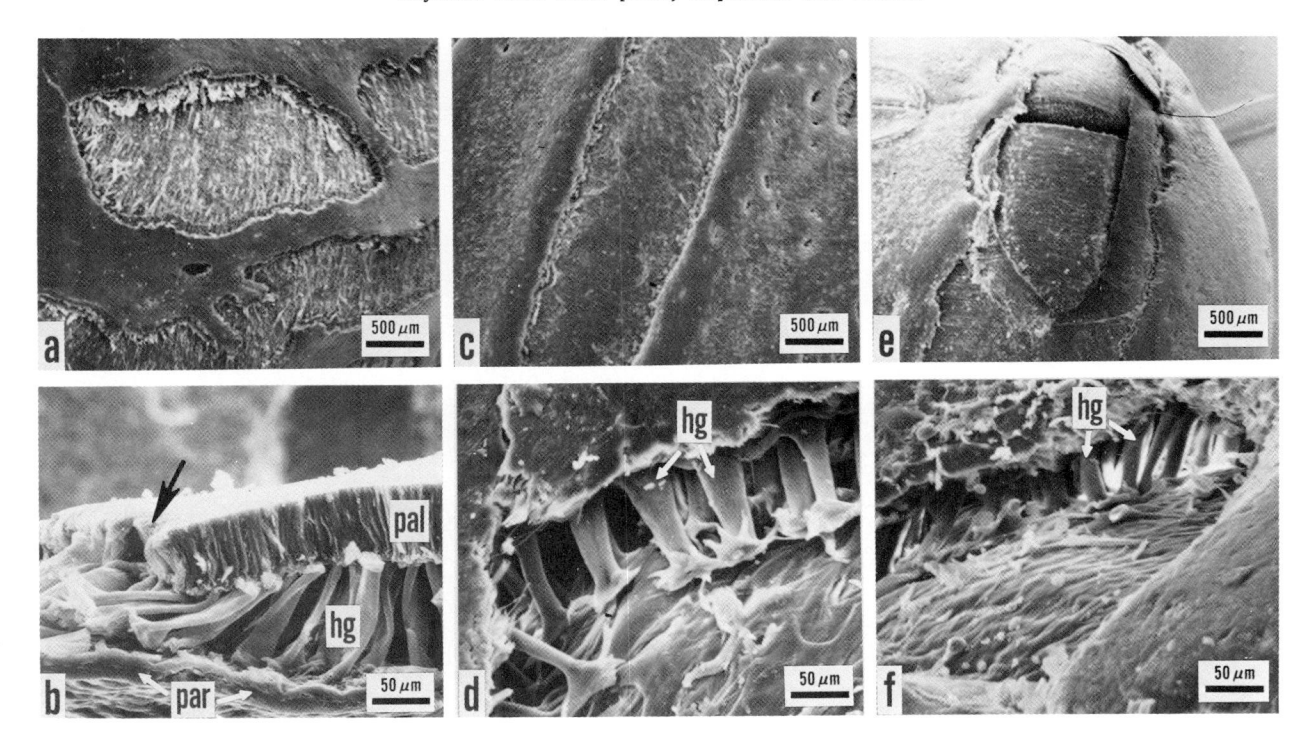

Fig. 11. Micrographs of seed-coat cracks in T217 (a); section of T217 seed coat showing termination of palisade and hourglass cells at boundary between crack and intact seed coat as indicated by arrow (b); seed-coat cracks in Magna (c); edge of crack in Magna (d); seed-coat cracks in Beeson (e); and edge of crack in Beeson showing exposed hourglass cells(f).
Key: Palisade cells = pal; hourglass cells = hg; parenchyma cells = par.

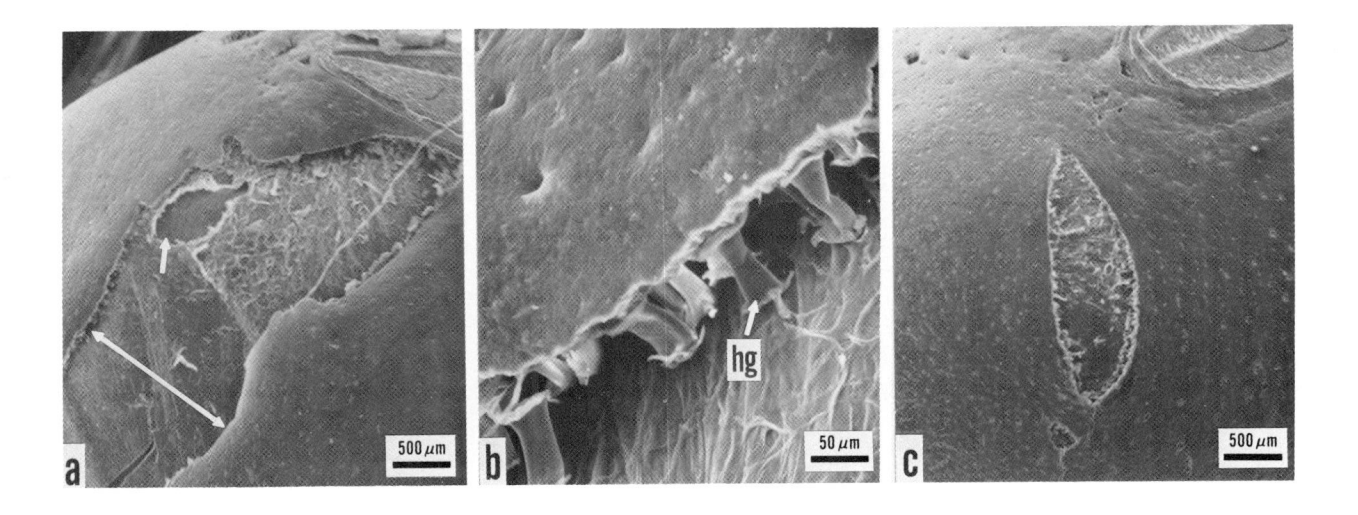

Fig. 12. Seed coat of Beeson cultivar after soaking in water and drying showing large induced crack through cuticle, palisade cells and hourglass cells (double-headed arrow) plus smaller crack (short arrow) through entire seed-coat (a), hourglass cells at edge of large crack [at left end of double-headed arrow in (a)] (b), and small crack near hilum (c). Key: Hourglass cells = hg.

Because of the "I"-shaped structure of hourglass cells, there is appreciable intercellular space in this layer; when exposed, it permits rapid entry of moisture and easy access for microorganisms.

Induced by Wetting and Drying

During development and maturation, moist soybean seeds are naturally protected against wide fluctuations in atmospheric moisture by the pod wall. The seed coats remain moist, swollen and turgid. As the seed matures the seed structures gradually dry, shrink and become firm. During early exposure of the mature seed in the field to dews and rains, the seed coat resists rapid uptake of water; but with repeated exposure, resistance is gradually reduced and seed quality deteriorates. Alternate cycles of wetting and drying progressively result in loosened, wrinkled seed coats and increased number of fissures (Moore, 1972).

To see whether cracking caused by field conditions could be duplicated in the laboratory, we soaked intact Beeson soybeans in water and then air-dried them. Soaking for as little as one hour followed by drying resulted in cracking that resembled internal stress cracks. Figure 12a shows a large induced crack in which the cuticle, plus the palisade and hourglass cellular layers have separated. Additional cracks through the remaining cellular layers are also visible. Higher magnification at the edge of the primary crack (near large arrow on left side of Fig. 12a) shows the exposed palisade and hourglass cells (Fig. 12b). Smaller induced stress cracks than shown in Figure 12a were also present (Fig. 12c). These results show that wetting followed by air-drying, where the seed coat dries more rapidly than the cotyledons, results in fissures that closely resemble those found in field-harvested seeds (Fig. 11e).

Conclusions

Our survey of more than thirty soybean cultivars has revealed a wide range in surface characteristics of their seed coats. These include the pits found in about two-thirds of the cultivars and various amounts of material on the seed coats. There is no obvious relationship between the two properties. For example, Rampage and Williams are highly pitted but have no surface deposits. Sooty, in contrast, has no pits but is an extreme case of a cultivar with deposited material on its seed coat. The significance of these properties to food processing of soybeans is unknown, but large amounts of material on the seed coat can affect luster of the seed. Lack of luster may be discriminated against by certain soybean users such as miso manufacturers who require soybeans with a light yellow, glossy seed coat (Hesseltine and Wang, 1972).

The "crystals" found on Williams soybeans apparently have not been reported before, and further study is needed to identify this material as well as to determine whether it occurs on the seed coats of other cultivars.

Probably of greater interest to food processors of soybeans than pitting or surface deposits is seed-coat cracking found, particularly in large-seeded soybeans. This defect is of demonstrated economic importance because it affects grading, handling, storage and quality of finished food products. It is also of concern to the producers of soybeans for seed, because it lowers germination and seedling vigor.

Acknowledgment

Dr. Uheng Khoo kindly prepared the seed-coat sections and performed the light microscopy.

References

Bair, C. W. Microscopy of soybean seeds: Cellular and subcellular structure during germination, development and processing with emphasis on lipid bodies. Ph.D. Dissertation, Iowa State University, Ames, Iowa. 1979, 276 pp.

Bernard, R. L., and M. G. Weiss. in Soybeans: Improvement, Production, and Uses, B. E. Caldwell, Ed., American Society of Agronomy, Inc., Madison, Wisc., U.S.A., 1973, Chap. 4.

Carlson, J. B. in Soybeans: Improvement, Production, and Uses, B. E. Caldwell, Ed., American Society of Agronomy, Inc., Madison, Wisc., U.S.A., 1973, Chap. 2.

Christensen, C. M., and H. H. Kaufmann. in Soybeans: Chemistry and Technology, Vol. 1, Proteins, A. K. Smith and S. J. Circle, Eds., Avi Publishing Co., Westport, Conn., U.S.A., 1972, Chap. 8.

Foster, G. H., and L. E. Holman. Grain breakage caused by commercial handling methods. Marketing Research Report No. 968, Agricultural Research Service, U.S. Department of Agriculture, 1973, 23 pp.

Franceschi, V. R., and H. T. Horner, Jr. Calcium oxalate crystals in plants. Bot. Rev. 46, 1980, 361-427.

Hermann, F. J. A revision of the genus Glycine and its immediate allies. Technical Bulletin No. 1268, Agricultural Research Service, U.S. Department of Agriculture, 1962, 82 pp.

Hesseltine, C. W., R. F. Rogers, and R. J. Bothast. Microbiological study of exported soybeans. Cereal Chem. 55, 1978, 332-340.

Hesseltine, C. W., and H. L. Wang. in Soybeans: Chemistry and Technology, Vol. 1, Proteins, A. K. Smith and S. J. Circle, Eds., Avi Publishing Co., Westport, Conn., U.S.A., 1972, Chap. 11.

Kolattukudy, P. E. Biopolyester membranes of plants: Cutin and suberin. Science 208, 1980, 990-1000.

Liu, H.-L. Inheritance of defective seed coat in soybeans. J. Hered. 40, 1949, 317-322.

Moore, R. P. Mechanisms of water damage in mature soybean seed. Proc. Official Seed Analysts 61, 1972, 112-118.

Mounts, T. L., G. R. List, and A. J. Heakin. Postharvest handling of soybeans: Effects on oil quality. J. Am. Oil Chem. Soc. 56, 1979, 883-885.

Newell, C. A., and T. Hymowitz. Seed coat variation in Glycine Willd. subgenus Glycine (Leguminosae) by SEM. Brittonia 30, 1978, 76-88.

Paulsen, M. R., W. R. Nave, T. L. Mounts, and L. E. Gray. Storability of harvest-damaged soybeans. Trans. ASAE 1981, Accepted for publication.

Saio, K. Soybeans resistant to water absorption. Cereal Foods World 21, 1976, 168-173.

Saio, K., K. Arai, and T. Watanabe. Fine structure of soybean seed coat and its changes on cooking. Cereal Sci. Today 18, 1973, 197-201, 205.

Smith, A. K., and S. J. Circle. Soybeans: Chemistry and Technology, Vol. 1, Proteins, Avi Publishing Co., Westport, Conn., U.S.A., 1972, Appendix, 448.

Ting, K. C., G. M. White, I. J. Ross, and O. J. Loewer. Seed coat damage in deep-bed drying of soybeans. Trans. ASAE 23, 1980, 1293-1296, 1300.

Wallis, T. E. The structure of the soya bean. Pharm. J. 37, 1913, (2597) 120-123.

Williams, L. F. in Soybeans and Soybean Products, Vol. I, K. S. Markley, Ed., Interscience Publishers, New York, U.S.A., 1950, Chap. 3.

Wolf, W. J., and F. L. Baker. Scanning electron microscopy of soybeans. Cereal Sci. Today 17, 1972, 124-126, 128-130, 147.

Wolf, W. J., and F. L. Baker. Scanning electron microscopy of soybeans and soybean protein products. Scanning Electron Microsc. 1980, III, 621-634.

Discussion with Reviewers

H. J. Arnott: Do you use a standard part of the seed for your assay or have you found that pitting is uniform so that standardization is not necessary?
K. Saio: In limited studies of the pits on the surface of soybean seed coats I found that the numbers of pits varied with the position on the seed-coat surface. Did you also observe this variation?
Authors: We have also noted that the pits are not uniformly distributed over the seed-coat surface. For our screening survey we routinely mounted the seeds with the hilum up and then examined both sides of the seed. Micrographs were taken at 50X magnification hence a relatively large area was recorded for arbitrarily classifying the degree of pitting (Table 1).

H. J. Arnott: It would seem possible to quantitate the number of pits to provide information as to the number per square μm or square mm.
Authors: We agree that such measurements could be made, but have not attempted to do so in these studies because of difficulties with cultivars where surface deposits are present. Some method for removing the surface deposits needs to be developed first.

L. B. Smith: Could the pits in the surface of the seed coats function to increase the surface area of the seed for more efficient imbibition of water? Is it really a fair assumption that the pits have no function and/or are not essential just because they do not occur in all varieties of soybean?
Authors: Although the pits increase the seed coat surface area, we have no evidence that they play a role in imbibition of water. In fact, in hard seeds of Altona cultivar the presence of pits did not make the seed coat permeable to water after soaking for 7 hr. Non-hard seeds take up water in 15 min or less.
We did not say that the pits have no function, but that they have no known function. We are not aware of any abnormalities associated with seeds of cultivars that lack pits. Hence there is no reason to think that the pits are necessary for normal development of the seeds.

J. D. Brisson: If the presence of pits does not make the seed coat permeable to water, what does?
Authors: It is conceivable that pits may increase seed-coat permeability to water in normal seeds, but we found that hard seeds of Altona cultivar with light seed coat pitting did not imbibe water in 7 hr. The problem of hard seeds is an intriguing one but we still do not have a good explanation of what causes hardness. Saio (1976, text reference) found that hard soybeans tend to have a thicker and denser seed coat plus a higher content of calcium and silicon in the seed coat than normal seeds. She also observed that the micropyle of hard beans was closed but open in normal seeds. Duangpatra S. [Some characteristics of the impermeable seed coat in soybean (Glycine max [L.] Merrill), Diss. Abstr. 37, 1976, 1061-B] found that hard beans had a continuous layer of suberin under the hilar region whereas in normal beans the suberin existed in the form of droplets with spaces between them that could permit passage of water. It was suggested that in hard beans, the extra-hilar region of the seed coat contains an unidentified, water-impermeable substance in the cell walls. These observations need to be confirmed and the hypothesis of a water-impermeable barrier has still to be tested.

J. D. Brisson: Since you suggest that "seed-coat deposits are residual endocarp," why have you tried to remove them with water, hexane or chloroform-methanol, and not with a cellulose solvent procedure?
Authors: We are still not completely convinced that the seed-coat deposits are residual endocarp because of the differences in morphology between the networks found on the seed coat and on the inner pod wall. Hexane and chloroform-methanol were chosen as solvents to remove the seed-coat deposits to test the statement of Williams (1950, text reference) that the deposits on Sooty are waxy. Your suggestion of using cellulose solvents is certainly worth trying.

L. B. Smith: Have you found any evidence of when this deposition of material onto the seed coat occurs? For example, examination of the seeds from the pods at various growth stages may suggest when in the development of the seed this material is deposited.
Authors: We have not examined seeds at various stages of development hence do not know when the deposition occurs.

C. W. Bair: The crystalline deposits on the endocarp surface are similar in size and shape to crystals I observed in cotyledonary tissue from an Amsoy 71 soybean variety. When sectioned material was viewed under polarized light, numerous, birefringent particles measuring 3.0 X 10.0 µm were observed. These particles were larger than starch granules and did not possess the characteristic centric polarized cross pattern typical of starch. Similar observations were reported by Wallis (1913, text reference) who described them as crystals of calcium oxalate. Although my studies did not include observations of the seed coat or endocarp surface, I found through examination of the hypocotyl and cotyledonary regions that distribution of the polarizing crystals was limited to cotyledonary tissues. A variety of shapes, druse (star-shaped), cylindrical, and prismatic forms with pyramid-shaped ends were observed in many preparations. Further work is definitely needed to chemically characterize and identify this crystalline material.
Authors: We agree with your comments, especially the need for further study of these crystalline substances.

L. B. Smith: Crystals found in the fruit coats of many plants have been reported throughout the literature. Did you find any evidence suggesting that these crystals could have been formed in the endocarp of the pod and deposited onto the surface of the seed coat?
Authors: No.

L. B. Smith: The interior layer illustrated in Figure 11a which is described as fibrous in nature, in the text, appears to be comprised of elongate cells. What evidence did you find which supports the fibrous rather than cellular nature of this area?
Authors: It is likely that the exposed layer consists of parenchyma cells that normally lie below the hourglass cells. We merely called attention to the fibrous appearance of the surface; we do not want to imply that this layer consists of fibers.

K. Saio: It is very interesting that soybean seed coats not only have pits, but also have deposits that appear to be fingerprints of the endocarp. Although you mentioned that the deposits are not protein, lipid or wax, if they are fibrous materials (cellulose, hemicellulose, lignin, cutin, etc.) which contain calcium or silicate, it is understandable why miso manufacturers prefer soybeans with glossy seed coats.

Authors: We are not aware of a scientific basis for the preference by miso manufacturers for soybeans with glossy seed coats, but your suggestion is an interesting speculation.

AN SEM STUDY OF THE EFFECTS OF AVIAN DIGESTION ON THE SEED COATS OF THREE COMMON ANGIOSPERMS

Lorraine B. Smith

Department of Biology
University of Texas at Arlington
Arlington, TX 76019

Abstract

A preliminary scanning electron microscope analysis of seed coat degradation in the avian digestive system has demonstrated surface alterations in the morphology of seed coats due to the digestive process. Seeds of Amaranthus palmeri S. Watson (Careless weed), Rhus glabra L. (Smooth sumac), and Abutilon theophrasti Medic (Velvetleaf), were collected and fed to captive starlings (Sturnus vulgaris vulgaris Linnaeus). The birds were sacrificed, and seeds were surgically removed from the digestive tracts for examination with the SEM. Changes in the seed coats due to the action of the muscular gizzard and digestive enzymes were studied. Results indicate that the most intense degradation occurred at different points along the tract, and the resistance of the seed coats varied with the species. The seeds of A. palmeri were degraded the most, whereas the seed and/or fruit coats of R. glabra and A. theophrasti were merely thinned. Passage through the digestive tract significantly decreased the viability of A. palmeri seeds. The digestive process did not alter the viability of R. glabra and A. theophrasti seeds. A marked decrease in germination occurred in the seeds of A. palmeri. However, germination in seeds of R. glabra and A. theophrasti was enhanced after passing through the birds.

KEY WORDS: Seed coat, degradation, avian digestion, scanning electron microscopy, seed germination

Introduction

Research done on the impact of avian digestive processes on seeds is scarce in the literature. A few studies have been reported on the effects of the avian alimentary tract on the seed and fruit coats of some common plants (De Vlaming and Proctor, 1968; Krefting and Roe, 1949; Proctor, 1968). However, details of the alterations in the surface of the seed and/or fruit coats are not clearly defined. With the aid of the scanning electron microscope (SEM), minute changes in the morphology of the seed and/or fruit coat as it passes through the digestive tract of an organism may be investigated. With the aid of viability and germination tests, one may relate the changes in the coats to the seeds' survival.

According to Wilson and Loomis (1967), a seed coat, by definition, is a protective outer covering of a seed which has developed from the integuments of the ovule. On the other hand, a fruit coat is an outer covering of a mature floral ovary which may contain one or more seeds. Seeds and/or fruits of three common angiosperms, Amaranthus palmeri S. Watson (Careless weed), Rhus glabra L. (Smooth sumac), and Abutilon theophrasti Medic (Velvetleaf), were fed to starlings (Sturnus vulgaris vulgaris Linnaeus). Alterations in the seed and/or fruit coats which were the result of the starlings' digestive processes were investigated in this study. These species were selected for this study because they are part of the starlings' normal diet (Robertson et al, 1978), and research accomplished in the past on seed-starling interactions has dealt with these species (Proctor, 1968).

The purpose of this investigation was to observe the changes which occurred in the surfaces of the seed and/or fruit coats at specific locations in the digestive tracts and upon removal from the feces. The relative resistance of the coats of each species of seed to degradation was noted. Viability and germination tests allowed comparison of untreated seeds (before digestion) with treated seeds (after digestion).

Materials and Methods

Mature seeds of *Amaranthus palmeri* S. Watson and *Rhus glabra* L. were collected in the fall of 1980, in Arlington, Texas. *Abutilon theophrasti* Medic seeds were collected in Lyon County, Kansas. The red, fleshy, outer coverings of the *R. glabra* fruits were mechanically removed. Fifteen starlings were captured with mist nets at the Village Creek Sludge Drying Beds in Fort Worth, Texas. The birds were housed separately, indoors, in metal cages. They were maintained on commercial canned cat food and were continually provided with water and grit for three days. This procedure allowed the birds to clear their digestive tracts of seeds prior to the start of the experiment. The birds were then starved for 24 hours, before large numbers of seeds mixed with canned cat food were introduced to the animals. Several hundred seeds were removed from the feces.

The birds were again starved for 24 hours, and the food mixture was reintroduced so that the seeds could be removed from the specific areas of the digestive tract. Using seed retention times for seeds mentioned by Proctor (1968), estimation of the location of the seed in the digestive tract was possible. The birds were sacrificed, their digestive tracts were removed, and seeds were then surgically excised. Approximately 50 seeds of each species were examined with the SEM.

All seeds used in this study were soaked in acetone for 45 seconds to remove superficial debris and any water which may have been present on the surface of the seeds. Some of the seeds were sectioned with a razor blade and extracted in acetone for 45 seconds to remove most of the internal lipid component for better viewing of the internal ultrastructure. Because of their small size (ca. 1.0 mm), the *A. palmeri* seeds were mounted on brass stubs with double stick cellophane tape. The *A. theophrasti* seeds and *R. glabra* fruits, which were much larger in size (ca. 3.0 mm x 2.5 mm), were mounted with colloidal graphite in isopropyl alcohol. The specimens were coated with approximately 200 A of gold in a vacuum sputter coater (Technics Hummer Jr.) for 2.5 minutes. Seeds were examined with a JEOL JSM 35-C scanning electron microscope.

Viability and germination tests were performed in triplicate on each of the three seed types before and after entering the digestive tract. For each of the viability tests, 100 seeds of each species were cut in half and placed in petri dishes lined with filter paper saturated with 10 mg tetrazolium chloride/100 ml water (Kozlowski, 1972). The seeds were allowed to soak in the aqueous solution overnight. They were examined for pink coloration of the embryo. The presence of pink coloration indicated that the seeds were viable (Kozlowski, 1972). Germination tests consisted of selecting 100 whole seeds of each species and placing them in petri dishes lined with filter paper moistened with distilled water. The dishes were placed in the dark and maintained at room temperature for 18 days. The appearance of a radicle was considered a positive indicator of germination capabilities. Untreated seeds (the control), and seeds subjected to the digestive processes of the birds were compared in both tests.

Results

The untreated surface of the seed coat of *A. palmeri* is shiny black, and relatively smooth. The seed is rounded, and a marginal rim with a single notch is usually discernible (Fig. 1). Two layers of the seed coat may be observed (Fig. 2). The outer layer is crustaceous (Kozlowski, 1972), and the inner layer is cellular in organization. The cells of the inner layer possess wall thickenings which are slightly visible in Figure 2 (arrows). The perisperm is very abundant and centrally located (Kozlowski, 1972), Fig. 2. Physical damage to the seed appeared to be initiated in the bird's muscular gizzard, where the seed coat was mechanically cracked (Fig. 3). In addition, the innermost layer of the seed coat was moderately disrupted as evidenced by disturbance of the cell walls (Fig. 4). The *A. palmeri* seeds from the intestine were similar in appearance to those found in the gizzard (Fig. 5). The outer layer of the seed coat was cracked, and the inner layer of the seed coat sustained considerable damage (Fig. 5). Almost all *A. palmeri* seeds taken from the excrement were in fragments, or they were severely cracked. Most of this species collected from the birds' excrement showed spherical structures associated with the outside of the seed surface (Fig. 6). These structures were often so densely aggregated around cracks in the seed surface that they obscured viewing of the cracks at lower magnifications. They appear either to have been derived from the outer layer of the seed coat (Fig. 7), or it is possible that they are microbial in origin.

The fruits of *Rhus glabra*, with the red, fleshy, outer coats removed, are kidney shaped. The surface of the fruit is relatively smooth in appearance prior to digestion (Fig. 8). The fruit coat of *R. glabra*, not inclusive of the red outer covering, is three-layered and encases one seed per fruit. The seed coat appears to be one cell layer thick (Fig. 9). Fruits which were removed from the gizzard demonstrated the presence of crystals in the fruit coat surface some of which were covered with a thin layer of endocarp material (Fig. 10). Fruits from the upper and lower intestine showed entire faces of crystals in the coat. The crystals occurred one per cell (Fig. 11). Examination of the surface of *R. glabra* fruits which had travelled the entire length of the digestive tract demonstrated that a very thin outer portion of the fruit coat was removed in digestion together with a fraction of the crystals which were present (Fig. 12).

The seeds of *Abutilon theophrasti* are cordate-shaped. The seed coat possesses a sculptured reticulate pattern with trichomes covering much of the surface (Fig. 13). The trichomes are epidermal extensions, each one originating from one cell of the seed coat (Fig. 14). *A. theophrasti* seeds removed from the proventriculus (glandular stomach) of the birds, showed

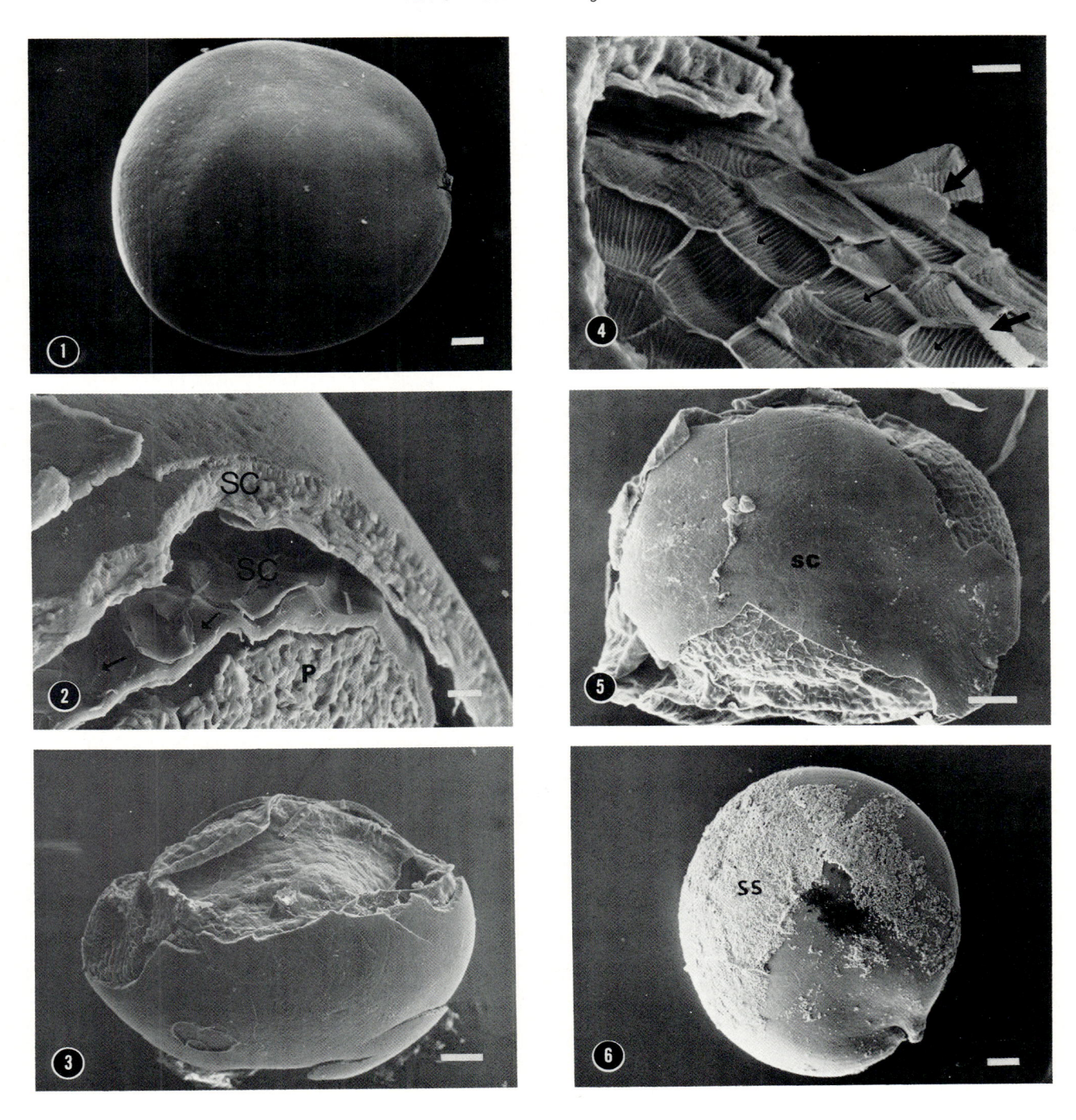

Fig. 1. Whole seed of Amaranthus palmeri. Bar = 100 μm.
Fig. 2. Sagittal view of A. palmeri seed showing the two-layered seed coat and perisperm. SC = seed coat; p = perisperm. Bar = 10 μm.
Fig. 3. A. palmeri seed from gizzard of bird. SC = seed coat. Bar = 100 μm.
Fig. 4. Disruption of the inner layer of the seed coat of A. palmeri (large arrows). Bar = 10 μm. Small arrows indicate cell wall thickenings in inner cellular layer.
Fig. 5. A. palmeri seed from intestine of bird. Bar = 100 μm.
Fig. 6. A. palmeri seed from excrement with spherical structures (SS) on surface. Bar = 100 μm.

remnants of the sculptured patterns but lacked trichomes (Fig. 15). The seeds taken from the gizzard illustrated that all superficial sculpturing and trichomes had been removed, and the thickenings in the seed coat surface which were present under the reticulate sculpture patterns were exposed. These thickenings, which appear to be accumulations of material in the center of the cells of the outer testal wall, had exhibited cracks (Fig. 16). The seed coat of A. theophrasti never cracked as extensively (Fig. 17) as did the seed coat of A. palmeri (Figs. 3-5, 7), even when subjected to the most intensive degrading processes of the digestive tract. The seed coats of A. theophrasti did undergo distinct alterations, however, because they were devoid of the sculpture patterns (Fig. 17) which were present before digestion (Fig. 13).

Seed viability and germination tests were performed on all three species of seeds (Table 1). These tests were conducted in triplicate to verify the results obtained. The viability of A. palmeri decreased considerably as did the number of seeds which germinated. The R. glabra and A. theophrasti seeds showed no reduction in viability, and both species demonstrated an increase in germination after passing through the birds (Table 1).

Table 1

Viability of seeds before and after subjection to avian digestive processes

Plant Species	Tetrazolium Chloride[a]	Germination[b]
Amaranthus palmeri		
Test 1	100 (56)	97 (12)
Test 2	98 (43)	92 (10)
Test 3	97 (58)	95 (10)
Rhus glabra		
Test 1	98 (96)	0 (12)
Test 2	96 (92)	0 (32)
Test 3	98 (96)	0 (18)
Abutilon theophrasti		
Test 1	100 (96)	0 (10)
Test 2	98 (96)	2 (8)
Test 3	99 (98)	0 (12)

[a] Number of seeds viable out of 100
[b] Number of seeds which germinated out of 100
Figure not in parenthesis = before digestion
Figure in parenthesis = after digestion

Discussion

In comparing digestive action on the seeds of the three species under investigation, it appears that the seeds of A. palmeri were affected the most while the seeds of R. glabra were affected the least. Proctor (1968) stated that smaller seeds are retained in the gizzards of birds longer than larger seeds. It follows that the smaller size of A. palmeri seeds partially dictated the length of time spent in the gizzard which subjected this species to grinding action for longer periods of time than would be expected for the other two species. This could account for the dramatic cracking of the seed coats in A. palmeri. Once the seed coats cracked, the internal tissues of the seeds probably were subject to enzymatic degradation.

The fruit coats of R. glabra were only slightly affected by digestive systems of the birds. The gizzard did not affect appreciably the outer fruit coat walls. By the time the R. glabra seeds had reached the intestine, enough of the fruit coat had been removed to expose entire faces of crystals which were present in the coat. The seed coat which lay under the three-layered fruit coat remained unaffected by the digestive process. As a result, the integrity of the internal tissues of the seeds was preserved.

In the case of A. theophrasti, the thick seed coat was merely thinned. Only slight cracking of the thickenings in the cells of the seed coat occurred while the seed was in the gizzard. Winter (1960) states that the seed coat of this species is very impervious to water. The digestive action of the avian system appears to have enhanced the ability of the A. theophrasti seeds to germinate (Table 1). The enzymes of the proventriculus had the most dramatic effect on the seed coats of this species. It was at this point that all trichomes were removed and only a few small remnants of the sculpture patterns on the surface of the seed coats remained.

The viability and germination tests which were conducted on the A. palmeri seeds before and after digestion suggest that this species was degraded the most. The tetrazolium chloride tests showed that the viability of this species had decreased markedly after passing through the birds. There was over 90% germination of the A. palmeri seeds before entering the birds (the control), but there was less than 15% germination of the seeds which had passed through the birds.

R. glabra seeds which were used as the control did not germinate before passing through the birds; however, this species did show greater than 10% germination after passing through the birds (Table 1). This implies that the digestive action of the birds removed the inhibition which prevented these seeds from germinating.

Furthermore, the A. theophrasti seeds which were used as the control exhibited no germination before passing through the birds. Enhancement of germination occurred in this species after the seeds passed through the birds (Table 1). Winter (1960) suggests that scarification of the seed coat in A. theophrasti interrupts the seeds' dormancy period. The tetrazolium chloride tests showed that the A. theophrasti seeds remained viable during the digestion process. Subjection of the seeds to muscular gizzard and digestive enzymes of the birds enabled the seeds to germinate. Scarification of the seed coats apparently allowed imbibition of water which triggered metabolic processes within the seed.

In summary, the effects that avian digestion have on a seed depends on the species of seed under investigation. More specifically, the thickness of the coat is very important in protecting the seed. However, it can act as an inhibitor to germination. Furthermore, the size of the seed is an important factor. Seed size

SEM of Seed Coat Degradation

Fig. 7. Relationship of spherical structures to surface of A. palmeri seed (from excrement). SC = seed coat; SS = spherical structures. Bar = 10 μm.
Fig. 8. Fruit of Rhus glabra lacking red outer covering. Bar = 1 mm.
Fig. 9. Sagittal view of R. glabra fruit showing the triple-layered fruit coat (FC), and single-layered seed coat (SC). Bar = 100 μm.
Fig. 10. Fruits of R. glabra with crystals covered by a thin layer of fruit coat material removed from gizzard of bird (arrows). Bar = 10 μm.
Fig. 11. Crystals in the fruit coat of R. glabra from the intestine. Bar = 10 μm.
Fig. 12. Disruption of crystals in the fruit coat of R. glabra (arrows). Bar = 10 μm.

influences the time it is retained in the bird. If a seed is held in the gizzard for longer periods of time, greater mechanical damage occurs. This was demonstrated in this study when observing the effects of the gizzard on *A. palmeri* and *A. theophrasti* seeds. The species of bird under study is also an important factor. De Vlaming and Proctor (1968) showed that certain species of birds degrade certain species of seeds to different degrees. For example, they showed that *A. palmeri* seeds have a high germination success after passing through the duck digestive system, whereas in this investigation, germination of this species of seed was very low after passing through starlings.

This study has endeavored to illustrate sites of seed coat degradation in avian digestion and the changes which have occurred in the morphology of the seed coat at these specific locations. The SEM has proven to be an excellent tool for studying these modifications.

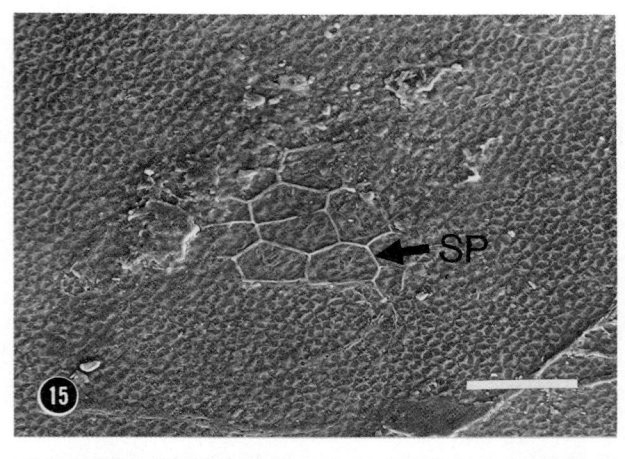

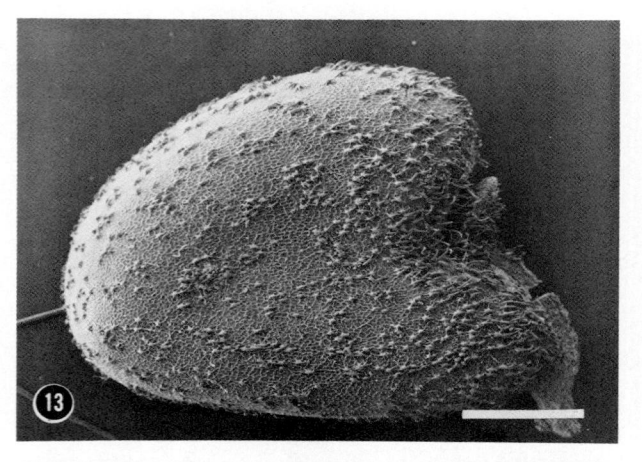

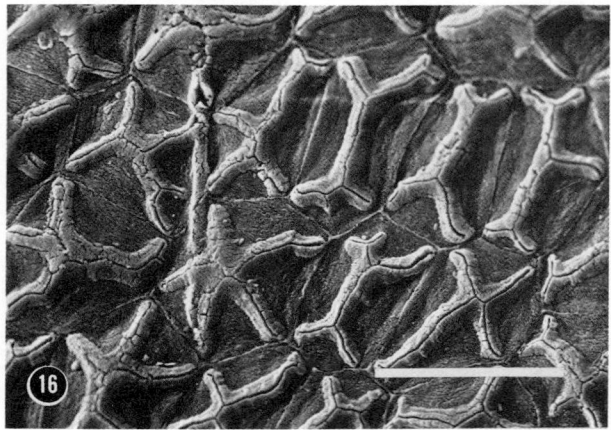

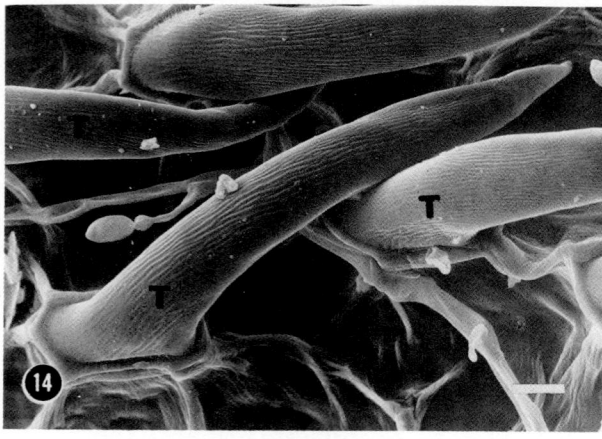

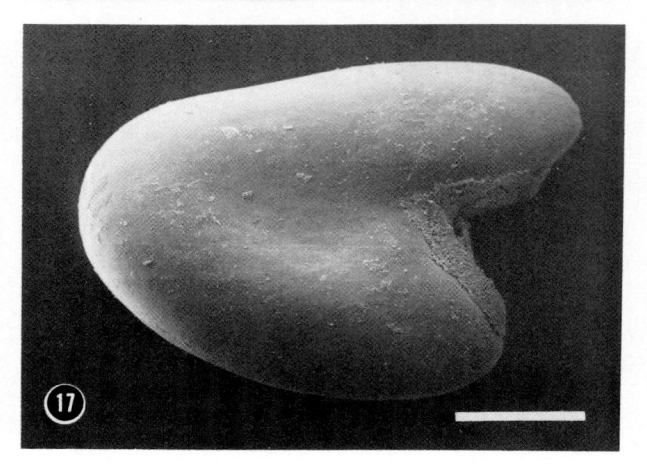

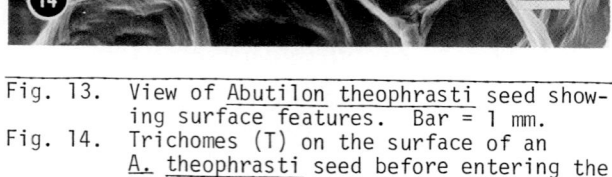

Fig. 13. View of *Abutilon theophrasti* seed showing surface features. Bar = 1 mm.
Fig. 14. Trichomes (T) on the surface of an *A. theophrasti* seed before entering the bird. Bar = 10 μm.
Fig. 15. Remnants of seed coat sculpture patterns (SP) in *A. theophrasti* (from proventriculus). Bar = 100 μm.
Fig. 16. Seed coat thickenings that occur under the sculpture patterns in *A. theophrasti* seeds (from gizzard). Bar = 10 μm.
Fig. 17. View of *A. theophrasti* seed after subjection to avian digestive processes. Bar = 1 mm.

Acknowledgements

I wish to thank Dr. Howard J. Arnott for the use of his laboratory and equipment, Dr. Robert L. Neill for aiding in the collection of the seed specimens and birds and for his encouragement, and John L. Darling for his many helpful suggestions. I also wish to express my appreciation to Mary Jane Goad for help with the manuscript.

References

De Vlaming, V. and V. W. Proctor. Dispersal of aquatic organisms; viability of seeds recovered from the droppings of captive killdeer and mallard ducks, Am. J. Bot., 55, (1), 1968, 20-26.

Kozlowski, T. T. Seed Biology, Academic Press, New York and London, 1972, Chapter 2.

Krefting, L. W. and Roe, E. I. The role of some birds and mammals in seed germination, Ecol. Monographs., 19, 1949, 271-286.

Proctor, V. W. Long-distance dispersal of seeds by retention in the digestive tracts of birds, Science., 160, 1968, 321-322.

Robertson, Raleigh J., Weatherhead, P. J., Phelam, F. J. S., Holroyd, G. L. and Lester, N. On assessing the economic and ecological impact of winter blackbird flocks, J. Wildl. Manage., 42, (1), 1978, 53-60.

Wilson, C. L. and Loomis, W. E. Botany, Holt, Rinehart and Winston, New York, Chicago, San Francisco, Toronto, London, 1967, Chapters 13 and 14.

Winter, D. M. The development of the seed of Abutilon theophrasti, II. Seed coat, Am. J. Bot., 47, 1960, 157-162.

Discussion with Reviewers

J. Croxdale: Can you be sure the crystals are being disrupted (Fig. 12)? The crystals appear to have fallen out during preparation.
Author: Firstly, very little preparation was performed on the seeds examined with the SEM. Immediately after removal from the digestive tracts, the seeds were extracted in acetone and placed on stubs for observation. Rhus glabra fruits were examined before and after extraction with acetone, and it was established that the acetone had no effect on the crystals.

Secondly, careful examination of the micrograph (Fig. 12), illustrates that portions of the crystals have been left behind in the cells of the fruit coat which suggests that the crystals have been dissolved within each of the cells by the digestive processes of the bird.

J. N. A. Lott: Since seeds of two of the species would not germinate under the conditions provided, was any attempt made to determine if some other factor other than impervious seed coats was involved? I am concerned, because other factors could be very important.
Author: The seeds that were selected from the stock which was being fed to the birds were used as control specimens. The seeds which were removed from the feces, acted as test samples. The control and test samples were subjected to the same temperatures, light exposures, etc. Therefore, the results obtained were relative. Since all conditions were the same for all the seeds tested, the only variable was that the control seeds had not been through the birds, while the test samples had been subjected to the digestive tracts.

J. N. A. Lott: Why were the seeds mixed with tinned cat food? It suggests that starlings would not eat the seeds unless they are tricked into eating them. Since birds would not normally be eating a diet of cat food plus seeds does this mixture reduce or increase the degree of damage to seeds? Since cat foods likely vary as to their oil content does the type of cat food make any difference?
Author: The seeds were mixed with tinned cat food so that large quantities of seeds could be fed to the birds without difficulty. If more seeds are fed to the birds at any given time, the more likely one will find large numbers of seeds in the digestive tracts. This greatly reduces the number of birds which must be killed. Also, as in the case of A. palmeri, the seeds were very small, and they would be difficult for the birds to obtain directly from their feeding dishes. In the field, starlings which are highly omnivorous, probably consume part of the plant together with the seeds which are present. Animal matter fed to the birds does influence the activity and size of the gizzard. Further studies are required to determine the effects imposed on the seeds by the meat content.

J. N. A. Lott: The differences between status in the proventriculus and gizzard were mentioned for one species (Abutilon theophrasti). Were any differences observed in other species between these two regions?
Author: There were no differences observed in A. palmeri or R. glabra from the proventriculi of the birds. One reason for this is that the seeds stay in this region a very short time. The trichomes of A. theophrasti were evidently very sensitive to the gastric secretions of this area.

H. T. Horner, Jr.: Based on your study, what part(s) of the digestive system were most degrading to the seeds?
Author: It appears that in A. palmeri seeds, degradation was initiated in the gizzards of the birds. In Rhus glabra fruits, the fruit coats appeared to be relatively unaffected until they reached the intestine. On the other hand, Abutilon theophrasti seeds appeared to be affected most by the proventriculi of the birds. This information suggests that the location of the greatest degree of degradation depends on the seed species in question.

H. T. Horner, Jr.: The fact that many seeds passed through the birds and remained viable raises a question as to their nutritive value to the birds. Are there other seed species that might be more appropriate to use?
Author: The seeds used in this experiment have been found in the stomachs of starlings. The total nutritive value gained by the birds may not be detectable by observations with the SEM. For instance, the birds may be extracting vitamins

and/or other chemicals from the seed and fruit coats.

J. N. A. Lott: Changes in the structure of the seed coat are noted but little is said about the internal tissues. Were any changes noted?
Author: The internal tissues were not studied during this work.

MICROSTRUCTURE OF TRADITIONAL JAPANESE SOYBEAN FOODS

K. Saio

National Food Research Institute, Ministry of Agriculture,
Forestry and Fisheries, 2-1-2 Kannondai, Yatabe, Tsukuba
Ibaraki 305, Japan

Abstract

Application of a cryounit on a scanning electron microscope (SEM) to examine the fine structure of traditional soybean products such as tofu and kori-tofu, is described. When examined in an SEM fitted with a cryounit, tofu showed a clear honeycomb-like structure including oil drops and the structural changes during the preparation of kori-tofu were clarified. The micrographs also indicated that chemical treatments used to fix the sample for a transmission electron microscope (TEM) did not produce any structural artifacts. Relationship bewteen texture and fine structure of tofu, which varies depending on the processing conditions, was also studied using SEM and TEM techniques.

The structure of seed coats and cotyledonary cells of soybeans were examined with a light microscope (LM), SEM and TEM. These techniques provided increased and diversified information about seed structure and chemical composition.

KEY WORDS: Soybean, tofu, kori-tofu, aburage, glucono-delta-lactone, cryounit, transmission electron microscope, light microscope, textural properties, seed coat, cotyledonary cells.

Introduction

Soybeans and soybean products are important as human foods in the Orient but in other countries they have been utilized mainly as a source of food oil, animal feeds and industrial polymers. Orientals have processed soybeans for centuries with sophisticated techniques into a wide variety of food products.

In the course of our investigation of these traditional soybean foods[1], electron microscopic techniques have been one of the important means used to study their morphology and ultrastructure. The data collected have provided valuable insights concerning the effect of processing procedures on the structure of the resultant products. This paper describes the use of microscopic techniques to study the structure of soybean foods. The detailed procedures should be applicable to the study of other natural and engineered food systems with minor modification.

Application of a cryounit fitted to SEM for the direct observation of fragile food systems

Tofu is a soft and smooth curd (protein gel) with about 90% moisture which is prepared by the coagulation of heated soybean milk. The fine structure of tofu was studied with TEM[2]. Pieces of tofu (5mm cube) were fixed with 5% glutaraldehyde in 0.1M phosphate buffer (pH 6.7) for 90-120 min and then fixed with 1% OsO$_4$ in phosphate buffer (pH 6.7) for 12-180 min after cutting into smaller pieces (1mm cube). The fixed pieces were carefully dehydrated stepwise with 30-100% acetone and embedded in Epon resin. The blocks thus prepared were cut with an LKB ultratome and the thin sections were stained with uranyl acetate. Drops of saturated uranyl acetate were placed on wax paper and the sections were floated on the surface of drops for 20-40 min and washed repeatedly with water. The stained sections were observed with a JEM 100-B. With TEM, a porous network composed of small protein granules with embedded oil drops was observed (Figure 1). This porous structure may explain why a large amount of water is retained in the gel.

Fixation of specimens always presents a major challenge as most commonly used fixatives attack one or more components with consequent damage to

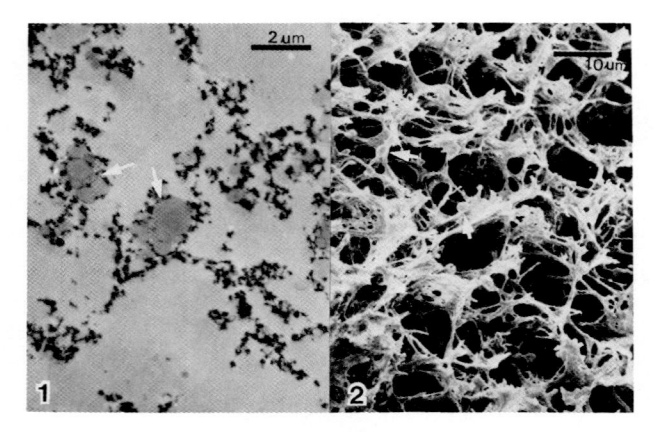

Figure 1. TEM - image of Momen tofu showing a network of protein granules and coalesced oil droplets (arrows).

Figure 2. SEM - image of Momen tofu showing oil droplets (arrows) scattered through a honeycomb-like structure.

the overall structure. Therefore, it is desirable to observe specimens in their native state to confirm the observations made on fixed samples. Utilization of a cryounit attached to a SEM provides a way to observe fragile specimens in their native state[3].

Pieces of tofu (5mm cube) were placed on specimen stubs, frozen promptly with liquid nitrogen, cryofractured with an installed knife and coated with gold in a cryounit after it was ascertained that no frost existed on the specimen. Samples prepared in this manner showed the mesh-like nature of the gel (Figure 2), with a size coinciding with that of the TEM-image (Figure 1). Oil drops shown in the TEM-image were recognized in the SEM-image as scattered droplets. These findings indicate that chemical treatments used to fix the sample did not produce any structural artifacts.

Kori-tofu is a tofu derivative and is prepared by freezing hard tofu at -20°C overnight, storing at 0 to -5°C for 2-3 weeks, thawing and then drying. Experimental results using kori-tofu at different stages during its manufacture are shown in Figure 3. Structural changes[4] can be observed at 3 stages: before freezing (fresh tofu); just after freezing; and after complete aging during frozen storage. The final product, kori-tofu, has a porous and spongy texture with longer shelf life than fresh tofu. The final gel network in the original tofu (1st stage) was collapsed and transformed into a coarse gel network after freezing, but the mesh-like nature of the gel was still observable (2nd stage). The protein network was severely compressed during frozen storage. Oil droplets were randomly distributed in the protein network but during freezing, those located near sites of the ice crystal formation were forced out of the network and aggregated with a subsequent increase in the size

Figure 3. Changes in SEM-images and TEM-images of kori-tofu during its preparation.
1st stage: before freezing, 2nd stage: just after freezing (frozen at -20°C overnight), 3rd stage: after aging (kept at 0 to -5°C for 18 days).
A: with a cryounit attached to the SEM, B: freeze-drying, C: with TEM.
Arrows identify oil droplets.

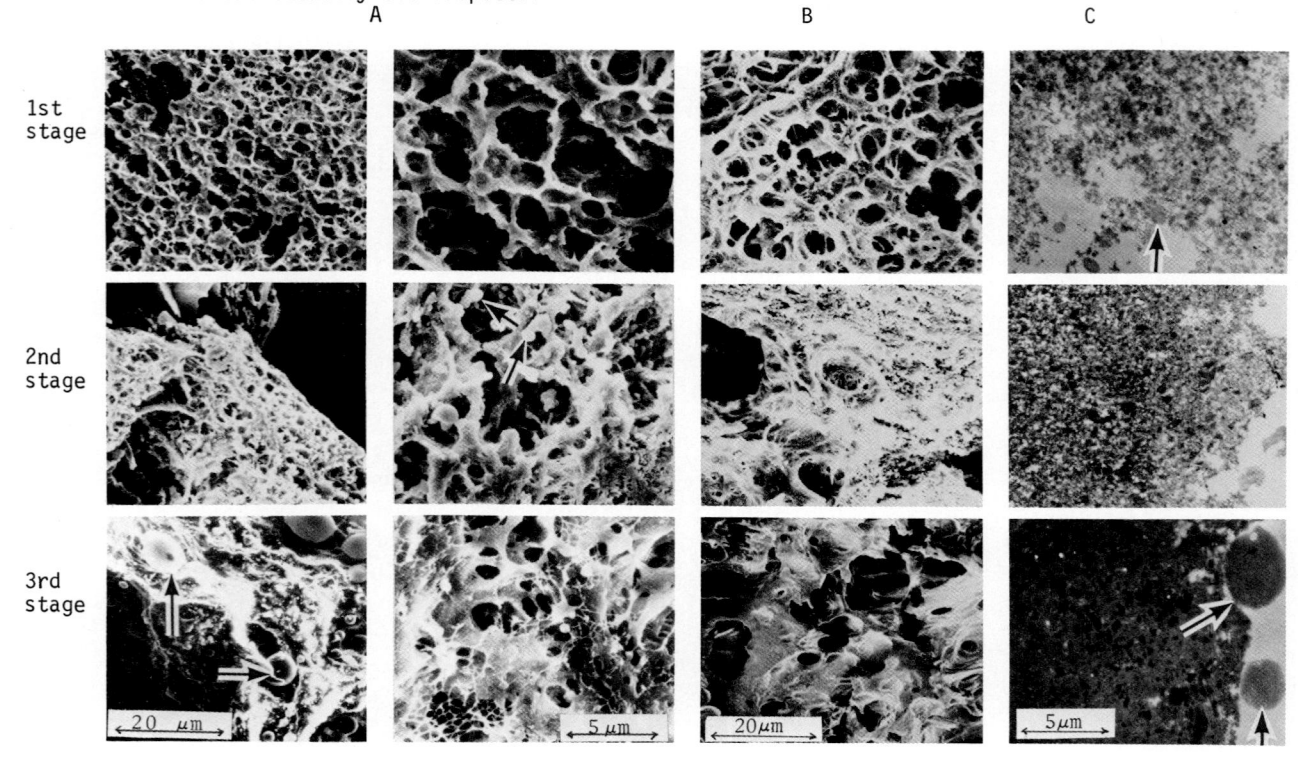

of the droplets (3rd stage).

As reference specimens, pieces of kori-tofu frozen with liquid nitrogen and immediately freeze dried overnight by use of a Virtis freeze drier, were observed with a JSM-50 after coating with gold. The results are also shown in Figure 3.

The morphologies and ultrastructures of tofu and kori-tofu were readily observed and distinct in the SEM-images of samples prepared with a cryounit and in the TEM-images of chemically fixed specimens. The structural features of freeze dried samples, however, were comparatively indistinct, especially the oil droplets in the respective SEM images (Figure 3-B). Chemical fixation followed by freeze drying should improve the quality of the SEM micrographs.

In conclusion, the cryotechnique in conjunction with SEM can be used to study the structure of food systems containing high contents of moisture, oil, starch and sugar which are difficult to chemically fix or sensitive to the electron beam. It also allows very quick observation of the foods in their native state. The cryotechnique was also applied to other food items like cooked spaghetti, bread dough and malt with successful results.

Relationship between texture and fine structure of tofu

Tofu is quite bland in taste and minute variations in its textural properties markedly influence its food quality and acceptance by consumers. Table 1 briefly shows the differences among various tofus and their derivatives. Factors affecting texture of tofu include: heating and coagulating temperature of soybean milk, solids and protein concentration of soybean milk; stirring rate before coagulation; amounts and varieties of coagulants; protein components and phosphorus content in soybean milk. Here I shall discuss only a few of the most significant factors.

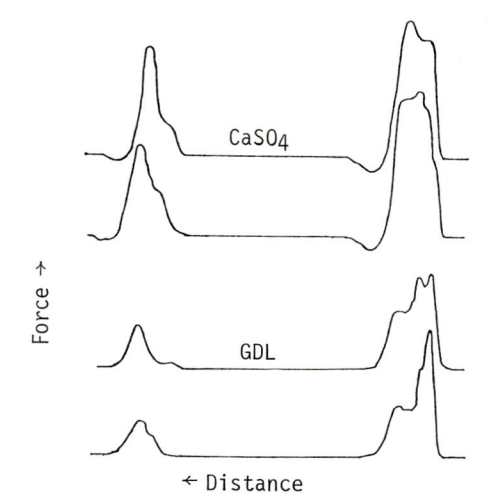

Figure 4. Texturometer profiles of tofu coagulated with $CaSO_4$ (upper curves) and glucono-delta-lactone (GDL, lower curves)

Coagulants: Historically the coagulant used was a concentrate of sea water, but now tofu is coagulated with calcium or magnesium salts or glucono-delta lactone (GDL). The latter causes soybean milk proteins to coagulate due to the liberation of protons during heating which lowers the pH of soybean milk. Figure 4 shows texturometer profiles of tofu coagulated with $CaSO_4$ or GDL. Tofu coagulated with GDL (GDL-tofu) was harder to penetrate, had a lower inner hardness, was more fragile and less cohesive than tofu coagulated with $CaSO_4$ (Ca-tofu). The SEM-images of Ca-tofu and GDL-tofu exhibited distinctly different fine structure (Figure 5). The protein network of GDL-tofu consisted of flocculent aggregates, and that of Ca-tofu showed a more honeycomb structure.

Table 1   Tofu and its Derivatives

	Fresh Tofu				Tofu Derivatives	
	Momen	Soft	Kinu	Packed	Aburage	Kori-tofu
Water added (no. of times)	10	7	5	5	10	15
Coagulant	$CaSO_4$	$CaSO_4$	$CaSO_4$ and/or GDL	GDL and/or $CaSO_4, CaCl_2$	$CaSO_4$	$CaSO_4$
processing	Thorough elimination of whey	Elimination of whey	Coagulation of whole soybean milk without elimination of whey	Cooked soybean milk is packed immediately after addition of coagulant and reheated. Whole soybean milk is coagulated	Soybean milk is moderately heated and coagulated, pressed and deep-fried in oil.	Soybean milk is coagulated with continuous stirring and whey eliminated. Tofu is frozen at -20°C overnight, kept at 0 to -5°C for 2-3 weeks, thawed and dried.
Chemical composition						
moisture(%)	86.8	88.9	89.4	90.0	44.0	8.1
crude protein(%)	6.8	5.7	5.0	4.5	18.6	50.2
crude fat(%)	5.0	3.8	3.3	3.2	33.1	33.4
ash(%)	0.6	0.6	0.6	0.6	1.4	2.8
Ca (mg%)	120	90	90	35	300	590
Texture	Hard, rough	Intermediate between momen and kinu tofus	Soft, smooth	Soft, smooth and fragile	Before frying, fresh tofu for aburage is harder than momen. Rough, coarse. After frying, aburage is texturized and chewy	Before freeze-drying, fresh tofu is harder than aburage-tofu. Coarse, lumpy. After freezing, tofu is spongy, elastic and chewy

Reproduced from CEREAL FOODS WORLD, 24, 1979, 342.

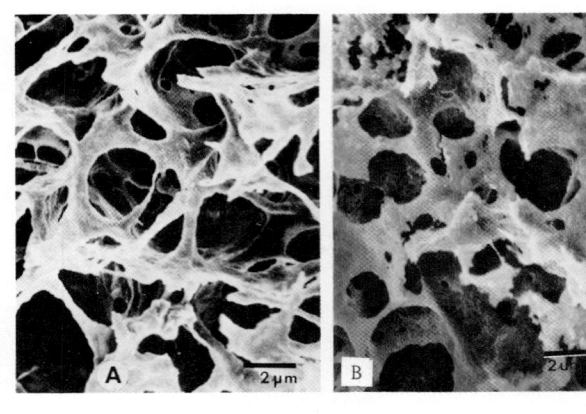

Figure 5. Comparison of SEM-images of tofu coagulated with CaSO4 (A) and GDL (B).

Protein components: The principal protein components of soybean storage protein are 7S and 11S globulins. The concentration of these two components depends on the soybean variety, and the 7S to 11S ratio in soybean milk distinctly affects the textural properties of tofu. Generally speaking, the lower the 7S to 11S ratio, the harder the tofu that is obtained. The hardness of tofu is also closely related to the amount of sulfhydryl groups in soybean milk. This fact suggests that sulfhydryl-disulfide exchange is involved in forming a three dimensional structure that gives rigidity to tofu. Figure 6 shows the relationships between amount of free SH groups and textural properties of Ca-tofu prepared from the protein fractions of 7S and 11S[5]. Hardness of 11S-tofu showed a strong dependence upon SH content whereas 7S-tofu did not. Presumably this behavior of the 11S protein accounts for the importance of the 7S to 11S ratio in soybean milk in determining the hardness of tofu. There is also a marked effect of SH content on adhesiveness of 7S and 11S tofus; both showed decreases in adhesiveness as SH content increased.

Figure 7 shows TEM-images of Ca-tofu prepared from either 7S or 11S. The protein granules in the network were larger in the hard 11S-tofu than in the soft 7S-tofu. In our experiment, it was also recognized that the larger protein granules were formed because of heating at higher temperatures and that they contributed to hardness of tofu too.

Structure of texturized soybean foods: The traditional Japanese foods of deep-fried tofu (aburage) and frozen, dried tofu (kori-tofu) can be considered types of texturized soybean products. Aburage is a puffed product that is expanded by two step-heating processes in oil at 120 and 180°C.[6] Kori-tofu is texturized during concentration of protein by formation of ice crystals and sulfhydryl-disulfide exchange between protein molecules on aging.[7,8] It is interesting to note that both were developed by monks in ancient times with the purpose of imparting textural properties like those of animal protein products to vegetable proteins. Modern techniques of preparing extruded and spun vegetable protein products also impart fibrous structure and chewiness with the same purpose. As shown in Figure 8, their fine structure[9] is in agreement with these properties.

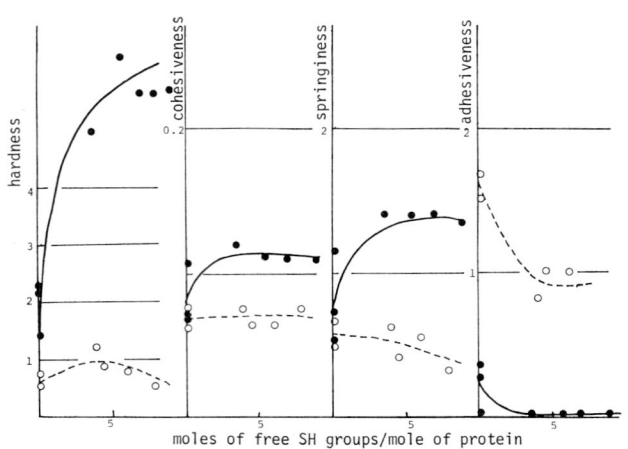

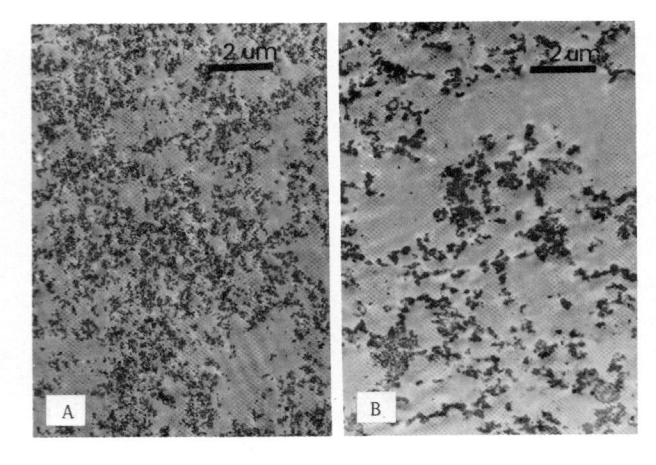

Figure 7. TEM-images of tofu prepared from 7S (A) or 11S (B) proteins.

Figure 6. Relationship between amount of free sulfhydryl groups and textural properties of tofu made with 7S (O--O) or 11S (●—●) proteins, using CaSO4. Sulfhydryl content was varied by treating with SH-blocking compounds or disulfide reducing agents.[5]

Seed structure by use of light microscope (LM) and TEM

Traditional Japanese soybean foods are pre-

Figure 8. SEM-images of extruded soy flour (A) and fiber type product from soy protein isolate (B).

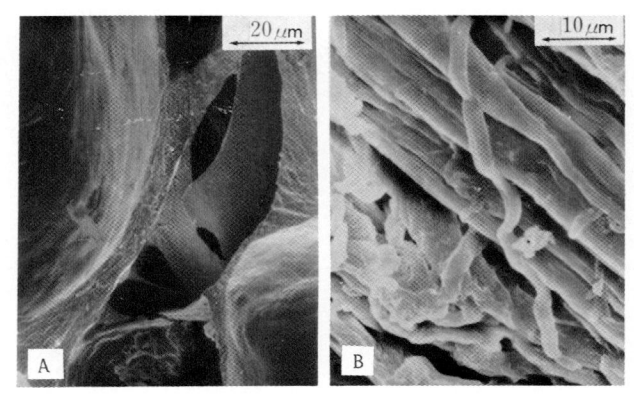

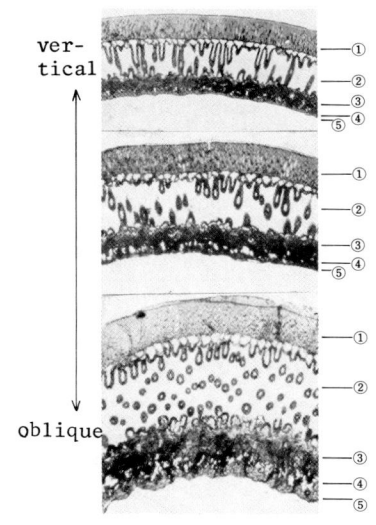

Figure 9. LM-images of cross sections of soybean seed coat.

Top to Bottom: Vertical to Oblique Sections.

1. palisade cells
2. hourglass cells
3. parenchyma
4. aleurone
5. compressed cells

Figure 11. LM-images of soybean cotyledonary cells.

Sudan Black B stained cytoplasmic part dark blue (A) Coomassie Brilliant Blue stained protein bodies violet (B). Schiff-periodic acid stained starch and cell walls red (C).

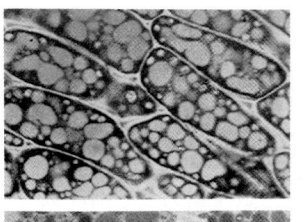

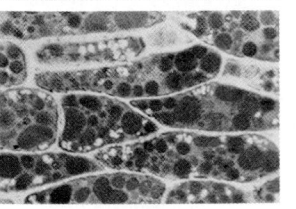

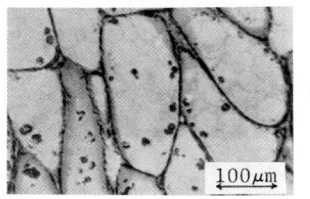

pared from whole beans which frequently are processed by water soaking followed by cooking or grinding. Information on seed structure is essential for understanding changes resulting from such processing. We have, therefore, examined the structure of whole soybeans.

Seed structure of soybeans was examined by use of LM, TEM and SEM (here, micrographs by SEM were not cited). Heterogeneity of seed tissues makes LM an excellent tool for locating the major constituents of plant cells because oil, protein and carbohydrates are often stored in distant organelles that are readily detected and identified by their interaction with specific stains.

The sample blocks prepared for TEM were sectioned (about 10μm) by an LKB ultratome and then examined using LM after affixing the sections on glass slides and staining components as follows:

Figure 10. TEM-images of soybean seed coat.

Samples were fixed with 0.6% potassium permanganate for 20-30 min., dehydrated with acetone and embedded in Epon resin. The thin sections were stained with uranyl acetate and lead nitrate. Seed coat consists of cuticular layer (cut, A), palisade cells (pal, A), hourglass cells (hgc, B), parenchyma (par, B, C, and D), aleurone cells (al, D) and compressed cells (en, D).

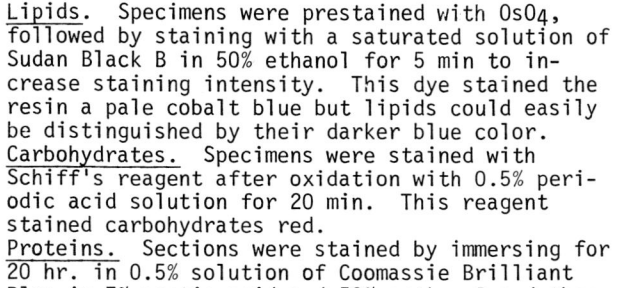

Lipids. Specimens were prestained with OsO_4, followed by staining with a saturated solution of Sudan Black B in 50% ethanol for 5 min to increase staining intensity. This dye stained the resin a pale cobalt blue but lipids could easily be distinguished by their darker blue color.

Carbohydrates. Specimens were stained with Schiff's reagent after oxidation with 0.5% periodic acid solution for 20 min. This reagent stained carbohydrates red.

Proteins. Sections were stained by immersing for 20 hr. in 0.5% solution of Coomassie Brilliant Blue in 7% acetic acid and 50% methanol and then by decolorizing with 7% acetic acid and 50% methanol. Proteins were stained violet.

It was possible to stain all components on one slide or on separate slides.

The observations with TEM were performed as described in figure captions.

Soybean Seed Coat

The structure of the soybean seed coat observed with LM is shown in Figure 9. As reported by Winton and Winton[10], the seed coat from outside to inside consists of a cuticular layer, palisade cells, hourglass cells, parenchyma cells, aleurone cells and compressed cells. The presence of pits, terminating in the cuticular layer or palisade cells was pointed out by Wolf and Baker[11].

The fine structures of each part as observed with TEM[12] are shown in Figure 10. The palisade cell walls were multilayered and almost completely lignified. The hourglass cells were more lignified than the palisade cells. The parenchyma contained no subcellular materials and were compressed into multilayers. Aleurone cells were recognized as living cells with thickened walls resembling collenchyma and contained intracellular substances such as protein bodies which were disrupted on cooking.

Soybean Cotyledonary Cells

The LM micrographs of soybean cotyledonary cells after staining[13] are shown in Figure 11.

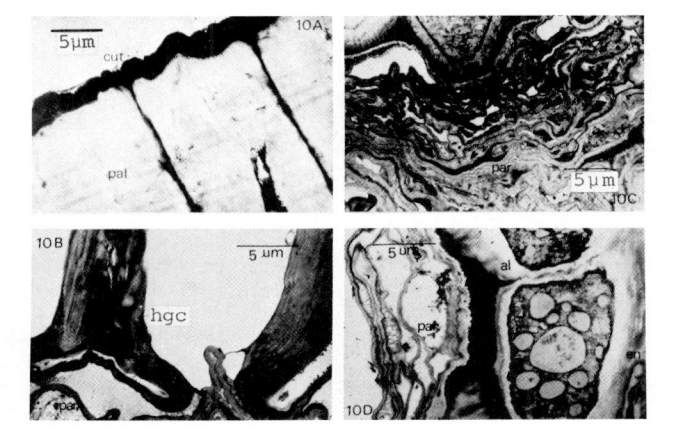

Sudan Black B stained the cytoplasmic network which is the intracellular site of oil storage but protein bodies are not stained (Fig. 11-A). Coomassie Brilliant Blue clearly stained the protein bodies (Fig. 11-B) whereas periodic acid-Schiff reagent stained cell walls and starch granules (Fig. 11-C).

On the other hand, the transmission electron micrograph shown in Figure 12 revealed distinct lipid bodies (spherosomes) in a cytoplasmic network. The lipid bodies could not be distinguished by Sudan Black B. The amyloplasts consisting of membranes contained small starch granules. The soybeans used here were a Chinese variety which contained a rather high content of carbohydrate.

Concluding Remarks

Microscopic techniques often provide different information than other methods. In addition, the combination of observations with LM, SEM and TEM provides increased and diversified knowledge. In our experiments, the information on the concept of fine structure in tofu was certainly expanded by comparison of the planar image (TEM) with the three dimensional image (SEM), and chemical compositions of subcellular structures in the TEM-image were verified by staining of LM samples. Using these techniques, many microscopy studies on kinds of gels, cooked spaghetti[14,15], noodles[16], heated dough[17], wheat gluten gel[18] and soybean protein gel[19,20] have been reported in recent years. Lee and Rha also reported SEM micrographs of soybean protein aggregates both calcium precipitated and isoelectric point precipitated[21,22]. All of this research revealed the intense relationships between textural properties and microstructure of gels. Methodological investigation on the observation of gels, which contain high content of water, seems to still continue with difficulties. Rapidly growing interest in this area, however, may result in development of better techniques for examining gels and emulsions which are important for preparing many fabricated foods.

References

1. Saio K. and T. Watanabe, Differences in functional properties of 7S and 11 S soybean proteins, J. of Texture Studies 9, 1978, 135-157
2. Saio K. and T. Watanabe, Observation of soybean foods under electron microscope, Nippon Shokuhin Kogyo Gakkaishi (Japan) 15, 1968 290-296
3. Nei T., H. Yotsumoto, Y. Hasegawa, Direct observation of frozen specimens with a scanning electron microscope, J. of Electron Microscopy (Japan), 20, 1971, 202-203
4. Saio K. and D. Gallant, Kori-tofu, the changes of fine structure during preparation, Japan Food Science (Japan), 15, 1974, 17-20
5. Saio K., M. Kajikawa and T. Watanabe, Food Processing characteristics of soybean proteins, Part II. Effect of sulfhydryl groups on physical properties of tofu-gel, Agr. Biol. Chem (Japan), 35, 1971, 890-898
6. Saio K., I. Sato and T. Watanabe, Food use of soybean 7S and 11S proteins, High temperature examination characteristics of gels. J. Food Sci., 39, 1974, 777-782
7. Hashizume K., K. Kakiuchi, E. Koyama and T. Watanabe, Denaturation of soybean protein by freezing, Agr. Biol. Chem (Japan), 35, 1971, 449-459
8. Hashizume K., Preparation of a new protein food material by freezing, JARQ, 12, 1978, 104-108
9. Taranto M.V., G.F. Cegla, K.R. Bell and K.C. Rhee, Textured cottonseed and soy flours: a microscopic analysis, J. of Food Sci. 43, 1978, 767-771
10. Winton A.L. and K.B. Winton, Structure and composition of foods, John Wiley & Sons, New York, U.S.A., 1932, Vol. I, 512-524
11. Wolf W.J. and F.L. Baker, Scanning electron microscopy of soybeans and soybean protein products, Scanning Electron Microsc. 1980; III: 621-634
12. Saio K., K. Arai and T. Watanabe, Fine structure of soybean seed coat and its changes on cooking, Cereal Science Today, 18, 1973, 197-201
13. Saio K. and K. Baba, Microscopic observation on soybean structure changes in storage, Nippon Shokuhin Kogyo Gakkaishi (Japan), 27, 1980, 343-347
14. Matsuo R.R., J.E. Dexter and B.L. Dronzec, Scanning electron microscopy study of spaghetti processing, Cereal Chem. 55, 1978, 744-753
15. Dexter J.E., B.L. Dronzec and R.R. Matsuo, Scanning electron microscopy of cooked spaghetti, Cereal Chem. 55, 1978, 23-30
16. Dexter, J.E., R.R. Matsuo and B.L. Dronzec, A scanning electron microscopy study of Japanese noodles, Cereal Chem., 56, 1979, 202-208
17. Tanaka K., S. Endo and S. Nagao, Effect of potassium iodate, and 1-ascorbic acid on the consistency of heated dough, Cereal Chem., 57, 1980, 169-174
18. Tu C.C. and C.C. Ysen, Effects of mixing and surfactants on microscopic structure of wheat glutenin, Cereal Chem., 55, 1978, 87-95
19. Siegel D.G., K.E. Cherch and G.R. Schmidt, Gel structure of nonmeat proteins as related to their ability to bind meat pieces, J. of Food Sci., 44, 1979, 1276-1279

Figure 12. TEM-image of soybean cotyledonary cells.
Samples were fixed with glutaraldehyde and OsO4, dehydrated with acetone and embedded in Epon resin. The thin sections were stained with uranyl acetate and lead nitrate.
PB: Protein body, LB: Lipid body, ST: Starch granule.

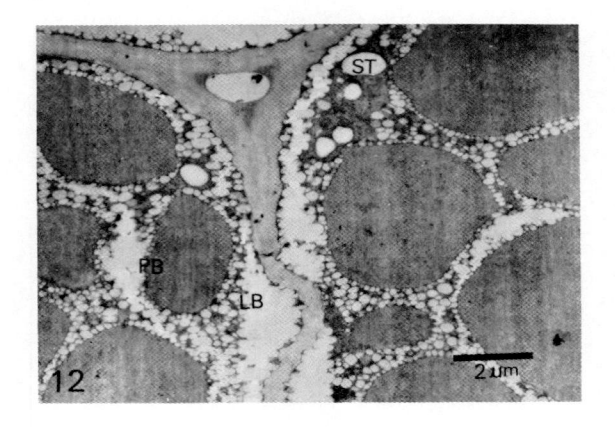

20. Siegel D.G., W.B. Tuley and G.R. Schmidt, Microstructure of isolated soy protein in combination ham, J. of Food Sci. 44, 1979, 1272-1275
21. Lee C.H. and C.K. Rha, Application of scanning electron microscopy for the development of materials for food, Scanning Electron Microsc. 1979; III: 465-472
22. Lee C.H. and C.K. Rha, Microstructure of soybean protein aggregates and its relation to the physical and textural properties of the curd, J. Food Sci. 43, 1978, 79-84

Acknowledgement

Sincere thanks to Dr. D. Galant, INRA, France and Mr. K. Ogura, JEOL Ltd., Tokyo for their cooperation in using cryotechniques with a SEM.

Discussion with Reviewers

M.V. Taranto: How did you determine whether the surface exposed by cryofracturing exhibited any frost damage?
Author: After putting the sample into the pre-cooled SEM, we cryofractured it with a chilled knife and observed the fracture surface. Because of the high vacuum, the surface was rarely covered with frost. After ascertaining that no frost was present, we coated the surface with gold using a mini-evaporator attached to the SEM pre-evacuation chamber [JEOL News, 14e, (2), 1976, 6-10].

M.V. Taranto: Have you ever applied the LM techniques you described to study the structure of tofu and its derivatives?
Author: I have not tried to apply LM techniques to study tofu. Although the network size of tofu looks a little bit too fine for LM, it would be interesting to try.

W.J. Wolf: I had difficulties in identifying the oil droplets in the SEM micrograph of momen tofu (Fig. 2). Have you examined samples before and after extraction with a lipid solvent to more clearly identify the sites of the oil droplets?
Author: With glossy photographs (Figs. 1 and 2), oil drops including the meshy structure of tofu are observed. When we actually worked on tofu with a cryounit in a SEM, they were clearly observed with high magnification. I did not, however, try to wash the surface of tofu with a lipid solvent as you did on soybean seeds. It would be interesting to ascertain the presence of oil drops, but I doubt whether lipid solvents can be effective on a material which contains more than 90% moisture. But, I just recently heard from a colleague that washing tofu after freeze-drying with a lipid solvent resulted in clearer honeycomb-like structures in SEM-image. It appears that the filaments of meshy structure might be covered with thick oily surfaces.

W.J. Wolf: Do you think that the irregular surface of the cuticular layer (Fig. 10A) corresponds to the pits that we have observed with the SEM? Would you expect shrinkage and rippling of the cuticle to occur as a result of sectioning and staining?

Author: I cannot conclude that the irregular surface of the cuticular layer corresponds to the pits which you pointed out. The pits were rarely sectioned in the sectioning process of LM. As I knew of the presence of the pits on the seed coats, I looked for such structures in the images of the sections, and I thought this irregular surface seemed to correspond to the pits. As I did not always examine continuous sections, the deepest size of the pits is uncertain.

C.K. Rha: You stated that "oil drops shown in the TEM image were recognized in the SEM-image with a high magnification. These findings indicate that chemical treatments used to fix the sample did not produce any structural artifacts". There is a distinct difference, however, in the protein portion which appears more or less homogeneously white in Fig. 1, while clear filamentous web-like morphology is evident in Fig. 2.
Author: As you know, only ultrathin sections can be examined in the TEM. Consequently, TEM micrographs show only a cross section of the meshy structure observed by SEM. In Fig. 1 the black granules are the protein aggregates. The white areas are not protein; they are the regions that were occupied by water before fixation. As I described in the answer to Dr. Wolf, some oil droplets seem to be removed during fixation. In this sense, I might have stated it too strongly when I indicated that no structural artifacts were produced.

C.K. Rha: Did the two samples (Figs. 5A and 5B) have the same original moisture content? The moisture content of the original samples would affect the structure.
Author: GDL-precipitated tofu generally contains 90-91% moisture, whereas, calcium-precipitated tofu (momen tofu) contains 86-88% moisture. Because different coagulants were used, the structural variations between Figs. 5A and 5B cannot be attributed solely to differences in moisture content. We have, however, observed structural dissimilarities between tofu samples prepared with calcium sulfate but with different moisture contents (Fig. 13). The sample with 90% moisture (Fig. 13A) was softer and less honeycombed than the one with 86% moisture (Fig. 13B).

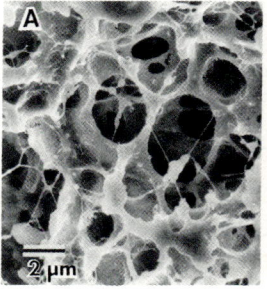

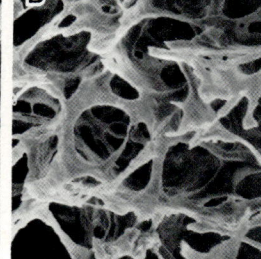

90% moisture   86% moisture

Fig. 13. Comparision of Microstructure of tofus with (A) 90% and (B) 86% moisture.

Davis, E. A. and Gordon, J.

Additional discussion with reviewers of the paper "Structural Studies of Carrots by SEM" continued from page 332.

S. B. Jones: Will you define the time frame implied by the different developmental stages?
S. H. Cohen: What are the average growth rates and size differences of the HiPak and Scarlet Nantes varieties?
Authors: The HiPak variety is of the Imperator variety type while Scarlet Nantes variety is of the Nantes variety type. The HiPak on the average develops to the earliest edible state in about 77 days, while the Scarlet Nantes arrives to that point in about 68 days. They continue growing until they are harvested after some 24 weeks from implantation. The data presented in this paper is primarily from 12 and 24 weeks of growth. At harvest (24 weeks), the length: diameter at crown for the HiPak is 5.37 and for the Scarlet Nantes is 4.27. Both are considered short carrots with their core cross sectional diameters about the same. At harvest, 10% HiPak and 5% of the Scarlet Nantes were infected with aster yellows.

S. B. Jones: What is unidirectional wall separation?
Authors: This was deduced from an article from Albersheim's group (Keegtra, K., Talmadge, K. W., Baver, W. D., Albersheim, P. The structure of plant cell walls III. A model of the walls of suspension-cultured sycamore cells based on the interconnections of the macromolecular components. Plant Physiol., 51:1973, 188-197.) They provide evidence for a model in which the cellulose component of the cell wall is embedded in a matrix of other polysaccharides such as rhamno-galacturonan, arabino-galactan, xyloglucan, and, perhaps, hydroxyproline-containing protein. These components are believed to be present in a specific pattern. The cell wall is probably achieved by polymerization of the repeated unit. This is important in terms of cell wall extension growth. Since walls grow throughout their length, the cellulose fibrils within the wall must be able to slide along their length relative to each other. The other polymer molecules then arrange themselves in their specific polymer unit structures to result in a stronger cell wall. It is this type of unidirectional extension that we believe is the source of the tubular formations of the phloem seen in heated tissue. The fibrillar-appearing material seen in Figure 6b results from some cleavage of the polymer connections in an organized manner. For further details on the cell elongation model see the article by Keegstra et al. (1973) referenced above.

EFFECTS OF EXOGENOUS ENZYMES ON OILSEED PROTEIN BODIES

R. D. Allen and H. J. Arnott

Department of Biology
University of Texas at Arlington
Arlington, TX 76019

Abstract

Scanning electron microscopy (SEM) was used to observe the nature and differences of the protein bodies of Yucca and sunflower seeds. Seeds were acetone extracted to remove the stored lipid which allows visualization of the protein bodies. Three proteolytic enzyme preparations were used to treat seed sections, and the protein body breakdown was observed.

Two basic schemes of breakdown are seen in Yucca protein bodies. The first involves a general removal of the outer proteinaceous matrix which reveals the protein body inclusions. In the second, deep holes appear in the protein body surface which suggest the protein body crystalloid inclusions have been digested. Sunflower protein bodies follow the second decomposition scheme. Protein body surface degradation also is seen in seeds of Yucca and sunflower treated with distilled water.

Parallels can be drawn between enzyme-treated protein decomposition and the digestion described during seed germination. While artificial protein body digestion cannot be considered a definitive model of natural protein body digestion during germination, it does serve to suggest a number of possible experiments during germination and helps to further establish the nature of the protein bodies in dormant seeds.

KEY WORDS: Yucca rupicola, Helianthus annuus, seed protein bodies, proteolytic enzymes, scanning electron microscopy, crystalloid, globoid, germination

Introduction

Protein is stored in both nutritive and embryonic tissues of many seeds. Stored protein is usually organized into distinct organelles called protein bodies. Protein bodies are reported to be enclosed by a single unit membrane and are composed of a homogenous protein matrix in which various inclusions may be imbedded (Ashton, 1976; Weber and Neuman, 1980; Webb and Arnott, 1980). Two types of protein body inclusions have been described: (a) spherical globoids which contain stored phosphates, particularly phytin, and (b) protein crystalloids composed mainly of crystalline protein (Weber and Neuman, 1980).

The inclusions present have been used to classify protein bodies into three categories: (1) protein bodies without inclusion, (2) protein bodies with globoid inclusions only, and (3) protein bodies with both globoid and crystalloid inclusions (Rost, 1972). In transmission electron microscope (TEM) studies perisperm protein bodies of several species of Yucca are reported to contain both globoid and crystalloid components (Horner and Arnott, 1965 and 1966). Sunflower cotyledon protein bodies have been investigated using TEM and were reported to contain only globoid inclusions (Buttrose and Lott, 1978), although Saio et al.(1977) previously reported a crystalloid-type inclusion within sunflower protein bodies.

Protein body digestion during germination is mediated by proteolytic enzymes within the seed (Young and Varner, 1958). Proteolytic agents may be produced de novo in the seed during germination, and the production of these agents is under hormonal control (Penner and Ashton, 1966 and 1967; Weber and Neuman, 1980). Proteolytic enzymes also have been demonstrated within the protein bodies of many seeds including sunflower (Schnarrenberger et al, 1972; Weber and Neuman, 1980). The endogenous proteases are associated with the water soluble albumin fraction of protein bodies which comprises the protein matrix (Weber and Neuman, 1980). Albumin fractions which have a sedimentation coefficient of 2S in sucrose gradient centrifugation have been demonstrated in the seeds of many genera including Yucca and Helianthus (Youle and Huang, 1981).

The pattern of protein body digestion during germination has been described (Horner and Arnott, 1966; Ashton, 1976; Weber and Neuman, 1980). Protein bodies undergo either internal or peripheral decomposition followed by fusion of the protein bodies into a large protein vacuole. A comprehensive bibliography of SEM studies of seeds, including germination, has been prepared (Brisson and Peterson, 1976). SEM was used to observe seeds experimentally treated with proteolytic enzymes. The purpose of these experiments was to determine the effects of exogenous proteases on protein bodies and to use this technique to further investigate protein body structure. Crude enzyme preparations were used to help assure nonspecific protein body digestion. Yucca rupicola and Helianthus annuus seeds were chosen based on availability and to provide a diversity of protein body types.

Materials and Methods

Seeds of Yucca rupicola Scheele (Agavaceae) were collected in their natural habitats near Austin, Texas, in 1980. Sunflower seeds, Helianthus annuus L. cv. Medium Black (Compositae) were obtained from a local retail supplier. Both seed types were sectioned with a razor blade. Seed sections were soaked in acetone to remove the stored lipid component. Yucca seeds require 15 to 30 second acetone extractions to allow adequate lipid removal from the cut surface to allow observation of the perisperm protein body. Sunflower seed sections require about 5 minutes acetone extraction for sufficient lipid removal from cut cotyledon surface. Cut seeds were air dried; those used for observations of normal protein bodies were mounted on stubs immediately, and seeds to be observed without acetone extractions were mounted following sectioning.

Aqueous solutions of three proteases were prepared: (1) papain crude (Pap) from papaya latex (Sigma Chemical No. P-3375), pH optima 6.2 solutions of 0.33% and 1.00%; (2) pancreatic crude (Pan) (Sigma Chemical No. P-4630), pH optima 7.5 solutions of 0.17% and 0.33%; and Adolf's meat tenderizer (MT), pH 6.8 -- salt, sugar, tri-calcium phosphate, papain, cottonseed and soybean oils -- solutions of 0.25% and 2.50%. Seed sections were soaked in enzyme solutions for periods ranging from 15 to 120 seconds, rinsed in acetone for 15 seconds, and air dried. Alternatively, seed sections were soaked in protease solutions prior to defatting with acetone and then soaked in acetone the required time for lipid removal.

Seed sections were soaked in distilled water for periods of 15 seconds to 15 minutes to show the effects of water alone on protein bodies. This water control was used to separate the effects of water from that of the three enzyme preparations. Both pre- and post-defatted seed sections were prepared and observed.

Variations in the length of time which seeds were soaked in water or enzyme solutions were found to have no effect on the pattern of protein body digestion and little effect on the extent of digestion. Therefore, a period of 60 seconds was arbitrarily established as a standard treatment time. Also, little difference was seen between seeds treated before or after defatting. Certain variations, particularly in the water-treated seeds, will be discussed later.

Prepared seed sections were mounted on brass or aluminum stubs with double-sided carpet tape and/or colloidal graphite. The prepared stubs were then sputter-coated with a gold target (Hummer Jr.) or with a gold-palladium target (Polaron SEM coating unit, E5100). Specimens were viewed with a JEOL JSM 35C SEM. Acceleration voltages of 15 kV or 25 kV were used, and the tilt angle was varied according to specimen demand. The diameter of objects (protein bodies, crystalloids, holes, etc.) was measured from the micrographs, and mean and standard deviations were computed.

Results

Yucca

Untreated perisperm cells contain large amounts of lipid-like material which is seen coating the other cell components (Figure 1). Protrusions in the lipid surface are assumed to be protein bodies. Rounded indentations on the embedded protein body surfaces suggest sites of possible removal of lipid bodies.

Acetone-extracted perisperm cells contain numerous spherical or ovoid protein bodies (Figure 2). The protein bodies range in diameter from 2.8 µm to 9.9 µm with a mean of 4.6 µm (± 1.1 µm). The surface of each protein body is marked by numerous indentations ranging in diameter from 0.1 µm to 1.0 µm with a mean of 0.6 µm (± 0.3 µm). Many protein bodies appear to be interconnected either by extensions of the protein body surface, or by thin strands extending between separate protein bodies or between protein bodies and the cell wall (Figure 2, arrows).

Water-treated (60 seconds) perisperm protein bodies exhibit extensive surface degradation (Figure 3). Polygonal areas of different sizes show where the protein matrix has been removed; the absence of surface indentations uniformly seen in acetone-extracted protein bodies suggests material is removed from the entire protein body surface. The protein bodies, which range in diameter from 0.3 µm to 6.2 µm with a mean of 3.8 µm (± 1.0 µm), are somewhat smaller than those in acetone-extracted seeds. Rounded protrusions are seen within the more heavily degraded surface areas which possibly represent embedded spherical inclusions such as globoids (Figure 3, arrows). Seed sections which have been water-treated for as little as 2 seconds showed an identical surface pattern which suggests an almost immediate dissolution of external protein body components. Many empty cells were seen in seeds soaked for 15 minutes; however, the remaining protein bodies, which ranged from 1.4 µm to 4.4 µm with a mean of 2.9 µm (± 0.7 µm), lacked surface indentations and were smaller than those treated for 60 seconds (Figure 4). The surface of the protein body in 15 minute-treated seeds had a granular appearance. Vigorous stirring during the 15 minute period increased the frequency of empty cells to nearly 100% which indicates that protein bodies are physically washed from the cells. In most of

Enzyme Effects on Seed Protein Bodies

Figure 1. Untreated Yucca perisperm cell. Note the thick lipid layer obscuring structural detail. Protrusions thought to be protein bodies are visible. Pb = protein body. Bar = 10 μm.
Figure 2. Acetone-extracted Yucca protein bodies. Note pattern of surface indentations and thin interconnecting strands. Pb = protein body; I = surface indentation; arrows = interconnecting strands. Bar = 1 μm.
Figure 3. Water-treated (60 seconds) Yucca protein bodies. Note extensive surface degradation and possible inclusion bodies. D = degraded surface area; arrows = possible inclusions. Bar = 10 μm.
Figure 4. Water-treated (15 minutes) Yucca protein bodies. Note granular appearance of protein body surface and protein body clumping. Bar = 10 μm.
Figure 5. Pap-treated Yucca protein bodies. Large protein body holes are present. Note the relatively smooth protein body surface and presence of free "globoid" bodies. H = protein body hole; G = globoid. Bar = 10 μm.
Figure 6. Pan-treated Yucca protein bodies. Protein body holes are visible and the protein body surfaces have a granular appearance. Possible embedded inclusions can be seen. H = Hole; arrows = inclusions. Bar = 10 μm.

the cells in water-treated seeds the protein bodies were clumped into a central mass, although not universal, it was more common in the 15 minute water-treated cells.

Pap-treated (1.00%) perisperm cells also contain protein bodies (2.5 µm to 5.0 µm; mean 3.5 µm, ±0.7 µm) which are somewhat smaller than those which are acetone extracted. These protein bodies are marked by one or more holes (2.1 µm to 2.9 µm; mean 2.5 µm, ± 0.3 µm) as shown in Figure 5. The protein bodies are clumped, and several objects can be seen on the outside of the protein bodies (Figure 5, G). These objects range in size from 1.1 µm to 1.8 µm with a mean of 1.4 µm (± 0.3 µm) and are most likely globoids which have been freed from the protein body interior. The holes seen in the protein body surface appear to widen internally and extend deeply into the protein body core. The protein body surfaces are relatively smooth and many small particles are seen on the surfaces.

Pan-treated (0.33%) perisperm cells contain protein bodies which range in size from 2.2 µm to 6.9 µm with a mean of 4.0 µm (± 0.9 µm), are somewhat smaller than acetone-extracted protein bodies, and also are marked with variously sized holes (Figure 6). The larger holes are very similar to those seen in Pap-treated seeds and range in size from 1.3 µm to 2.8 µm with a mean of 1.9 µm (± 0.5 µm) which give the protein bodies the appearance of hollow shells. The surfaces of the protein bodies are very rough which suggest a particulate nature, and numerous embedded spherical bodies, probably globoids, can be seen (Figure 6, arrows). The protein bodies also are clumped, and some fusion of protein bodies is evident.

MT-treated (0.25%) perisperm protein bodies range in size from 2.2 µm to 4.7 µm with a mean of 3.7 µm (± 0.7 µm) in diameter. It is evident that the outer portions of the protein bodies have been removed (Figure 7). Many globoid bodies (0.3 to 1.0 µm; mean 0.5 µm, ± 0.2 µm) can be seen in Figure 7, G. Both free and embedded globoids can be seen as well as rounded indentations in protein body surfaces where globoids once were (Figure 7, arrows). The protein body surfaces are rough and no clumping or protein body fusion is evident. The intimate association of globoid bodies in the protein matrix can be seen in Figure 8.

Protein matrix material is removed entirely from many of the protein bodies in 2.50% MT-treated seeds (Figure 9). Protein crystalloids which range in size from 2.3 µm to 4.0 µm with a mean of 2.9 µm (± 0.6 µm) clearly are visible along with numerous globoids. The globoids are similar in size to those seen in 0.25% MT-treated cells. Many small particles (approximately 0.2 µm) also are seen in the background and may be remnants of the protein matrix or small globoid-like particles.

Sunflower

Untreated cotyledon cells of sunflower cannot be distinguished clearly (Figure 10). Surface details appear to be obscured by a layer of lipid-like material. Rounded protrusions are seen which are thought to be protein bodies.

Acetone-extracted cotyledon cells contain protein bodies somewhat larger (3.5 to 9.8 µm; mean 5.8 µm, ± 1.8 µm) than those seen in Yucca (Figure 11). A regular pattern of surface indentations also is seen. Interconnections between separate protein bodies, and between protein bodies and the cell wall, are present as in Yucca, however, the cotyledon cell walls of sunflower are much thinner than the perisperm cell walls of Yucca. In certain cells the pronounced ridges which surround the protein body indentations appear membraneous in nature and separate from the protein body surface (Figure 12).

Water-treated protein bodies are slightly smaller than acetone-extracted protein bodies and range from 4.0 to 8.5 µm, mean 5.3 µm (± 1.4 µm) in diameter. Two distinct size groups of holes are seen in the surface of each protein body (Figure 13). Many small pits (approximately 0.3 µm in diameter) are seen along with several larger crater-like holes which range from 0.9 µm to 2.1 µm with a mean of 1.6 µm (± 1.3 µm). The holes do not appear to penetrate deeply into the protein body core. The indentation pattern as seen in acetone-extracted protein bodies is not present, and the protein body surface is smooth. Protein bodies soaked 15 minutes do not appear greatly different from those treated for 60 seconds (Figure 14). Protein body clumping and fusion are common in cells of both one minute and 15 minute treatment although only illustrated in the 15 minute cells (Figure 14). No free protein body inclusions are seen, however, possible embedded inclusions are visible in certain protein bodies.

Pap-treated (1.00%) protein bodies are considerably larger than in the acetone-extracted case and range in size from 3.0 to 15.0 µm with a mean of 8.3 µm (± 3.0 µm). Surface holes, 3.0 µm to 8.1 µm with a mean of 5.6 µm (± 1.6 µm), along with numerous smaller pits are shown in Figure 15. The protein body surfaces are fairly smooth and no regular indentation pattern is seen. No free protein body inclusions, such as globoids are visible. Clumping of protein bodies is seen; however, large scale protein body fusion is not evident.

Pan-treated (0.33%) protein bodies are degraded in a pattern similar to that seen in Pap-treated cells. Holes extending deeply into the protein body core, and in some cases completely through the protein body, are seen (Figure 16). Many smaller holes also are present. Some protein body clumping is apparent, but protein body fusion does not appear. The protein body surfaces are quite rough, and many small particles are present. Both Pan- and Pap-treated sunflower protein bodies appear quite similar, and variations in digestion extent and patterns may overlap. Both Pan- and Pap-treated protein bodies are larger than acetone-extracted bodies, and free globoids are not seen in either Pan- or Pap-treated cells.

MT-treated (0.25%) protein bodies (Figure 17) appear similar in some respects to those which are water-treated (Figures 13, 14). The surface holes are more extensive, and protein body fusion is always present. MT-treated protein bodies range in diameter from 3.6 µm to 8.0 µm with a mean of 4.9 µm (± 1.1 µm). Many small particles (approximately 0.3 µm) are seen on the protein body surfaces (Figure 17, arrows), and embedded globoids are evident in some cells. Progressive protein body fusion is evident in the 2.50% MT-treated

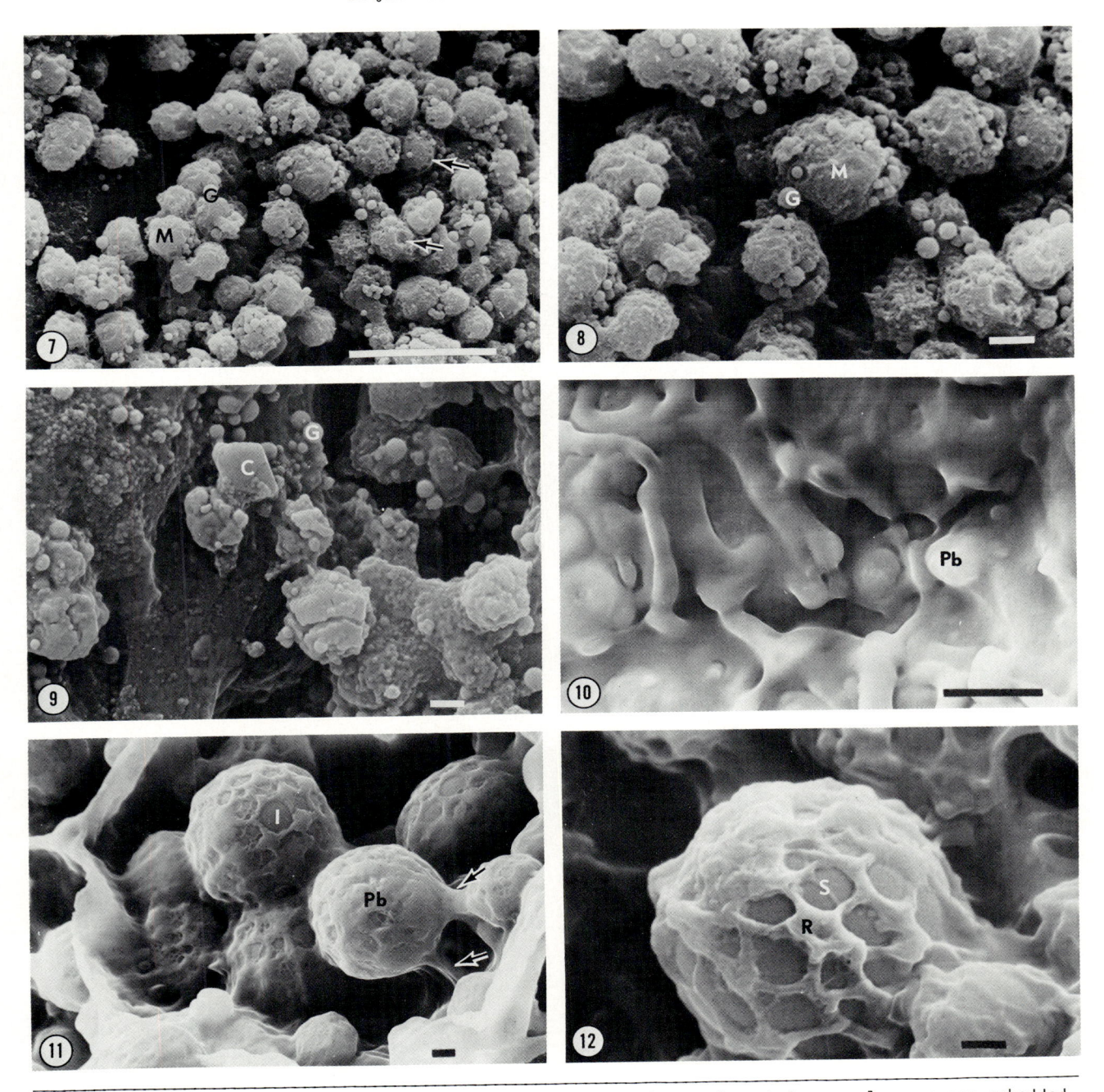

Figure 7. MT-treated (0.25%) Yucca protein bodies. Degraded protein matrix reveals numerous embedded globoids. Rounded pits where globoids have been removed are visible. M = protein matrix; G = globoid; arrows = pits. Bar = 10 μm.

Figure 8. MT-treated (0.25%) Yucca protein bodies. Close association of globoids in protein matrix is apparent. M = protein matrix; G = globoid. Bar = 1 μm.

Figure 9. MT-treated (2.50%) Yucca protein body remnants. Protein crystalloid inclusions are visible. Globoids and many small particles also are present. C = protein crystalloid; G = Globoid; Bar = 1 μm.

Figure 10. Untreated sunflower cotyledon surface. Thick lipid layer is present in which protrusions thought to be protein bodies are seen. Pb = protein body. Bar = 10 μm.

Figure 11. Acetone-extracted sunflower protein bodies. Note surface indentation pattern. Pb = protein body; I = indentations; arrows = interconnecting strands. Bar = 1 μm.

Figure 12. Acetone-extracted sunflower protein body. Note membraneous surface reticulum. R = surface reticulum; S = protein body surface. Bar = 1 μm.

cotyledon cells (Figure 18). Distinct protein bodies no longer can be seen, and the entire protein "vacuole" surface is covered with globoid bodies and particles ranging in size from 0.2 μm to 0.8 μm with a mean of 0.7 μm (± 0.2 μm). Remaining protein body holes are visible at the protein vacuole periphery.

Observations and measurements from all specimens are compiled in Table 1. This table will be used as reference in the Discussion.

Table 1
Results of Various Treatments on Yucca and Sunflower Protein Bodies

Yucca

Acetone-extracted -- Protein bodies (Pb) 2.8 to 9.9 μm, mean 4.6 μm ± 1.1 μm; surface indentations; Pb and cell wall interconnections
Water (60 seconds) -- Pb 1.3 to 6.2 μm, mean 3.8 μm ± 1.0 μm; polygonal depressions in surfaces, little Pb clumping; no Pb fusion
Water (15 minutes) -- Pb 1.4 to 4.4 μm, mean 2.9 μm ± 0.7 μm; no pits or holes; rough surface; Pb clumped; some fusion
Pap 1.00% (60 seconds) -- Pb 2.5 to 5.0 μm, mean 3.5 μm ± 0.7 μm; large holes (2.1 to 2.9 μm, mean 2.5 μm); Pb clumping; smooth surfaces; small particles on Pb surface; free "globoids"
Pan 0.33% (60 seconds) -- Pb 2.2 to 6.9 μm, mean 4.0 μm ± 0.9 μm); large holes (1.3 to 2.8 μm, mean 1.9 μm); Pb clumped; rough surfaces; embedded globoids
MT 0.25% (60 seconds) -- 2.2 to 4.7 μm, mean 3.7 μm ± 0.7 μm; rough Pb surface, many globoids both free and embedded; no fusion
MT 2.5% (60 seconds) -- No protein matrix; crystalloids visible, 2.3 to 3.9 μm, mean 2.9 μm ± 0.6 μm; numerous free globoids; no fusion

Sunflower

Acetone-extracted -- Pb 3.5 to 9.8 μm, mean 5.8 μm ± 1.8 μm; surface indentations; Pb and cell wall interconnections; cytoplasmic reticulum
Water (60 seconds) -- Pb 4.0 to 8.5 μm, mean 5.3 μm ± 1.4 μm; rounded surface holes and pits; smooth surface; Pb clumped; some Pb fusion
Water (15 minutes -- Pb fused 4.5 to 9.0 μm, mean 5.7 μm ± 1.5 μm; many holes and pits; smooth surface
Pap 1.00% (60 seconds) -- Pb 3.0 to 15.0 μm, mean 8.2 μm ± 3.0 μm; large holes (3.0 to 8.1 μm, mean 5.6 μm) and small pits present; smooth surface; Pb clumped; little Pb fusion
Pan 0.33% (60 seconds) -- Pb 3.3 to 12.5 μm, mean 7.8 μm ± 2.6 μm; large holes (1.5 to 2.7 μm, mean 1.8 μm); Pb clumped; rough Pb surface; many surface particles
MT 0.25% (60 seconds) -- Pb 3.6 to 8.0 μm, mean 4.9 μm ± 1.1 μm; large holes and surface particles; extensive Pb fusion
MT 2.50% (60 seconds) -- Protein "vacuole" (fused Pb), globoids and particles on surface; Pb shapes and holes visible at periphery

Discussion

Material soluble in acetone, which is assumed to be lipid, is removed from Yucca and sunflower seed sections to allow visualization of protein bodies within storage tissue cells. These protein bodies were found to vary considerably in size, averaging about 4.6 μm for Yucca and about 5.8 μm for sunflower (Table 1). Saio et al (1977) report sunflower protein bodies ranging from 0.5 to 10 μm with the highest frequency occurring at about 2μm. These measurements were taken from protein bodies isolated in a sucrose gradient and sonicated. The authors state that crystalloid-type structures were often found in the isolated protein body fractions which could cause an abnormally low average size. It also should be noted that the isolated protein bodies were hydrated which, as the present results indicate, has a marked effect on protein body morphology and size.

Protein body surface indentation patterns are seen in both seed types after the removal of lipids. Similar patterns have been reported in other seed species (Wolf and Baker, 1972; Swift and Buttrose, 1973; Webster and Leopold, 1977; Webb and Arnott, 1980). These indentations have been attributed to pressure on the protein bodies from surrounding lipid bodies or spherosomes.

The membraneous reticulum material observed on the surface of some sunflower protein bodies (Figure 12) is considered to be residual cytoplasm which in intact seeds surrounds the lipid bodies and remains following lipid removal. Similar cytoplasmic reticulum structures have been observed with SEM in zucchini cotyledon cells (Webb and Arnott, 1980).

Water-treated seeds of both species show surface characteristics indicating the removal of water soluble substances. Yucca protein bodies show a pattern of polygonal surface depressions and a reduction of protein body size when water treated for one minute (Figure 3). Further reductions in size and disappearance of the polygonal areas in specimens treated for 15 minutes indicate progressive dissolution of additional protein body material. Rounded holes are seen in sunflower protein bodies treated with water for one minute (Figure 13). These holes remain in specimens treated for 15 minutes, and a slight increase in protein body diameter is evident between specimens treated for 1 minute and those treated for 15 minutes. Apparently, specific areas of sunflower protein bodies are dissolved with the remainder being relatively water insoluble and apparently having a tendency to swell slightly. The effects of hydration, as well as the effects of various salt solutions on protein bodies of black beans (Phaseolus vulgaris), have been investigated (Varriano-Marston and DeOmana, 1979). Surface degradation is reported to occur and is more extensive in beans soaked in salt solutions. As previously stated, the isolated sunflower protein bodies studied by Saio et al (1977) were hydrated during preparation. SEM photographs of the latter protein bodies show surface patterns resembling some of the water-treated specimens in the present paper. Protein body clumping also is reported by Varriano-Marston and DeOmana (1979).

Enzyme Effects on Seed Protein Bodies

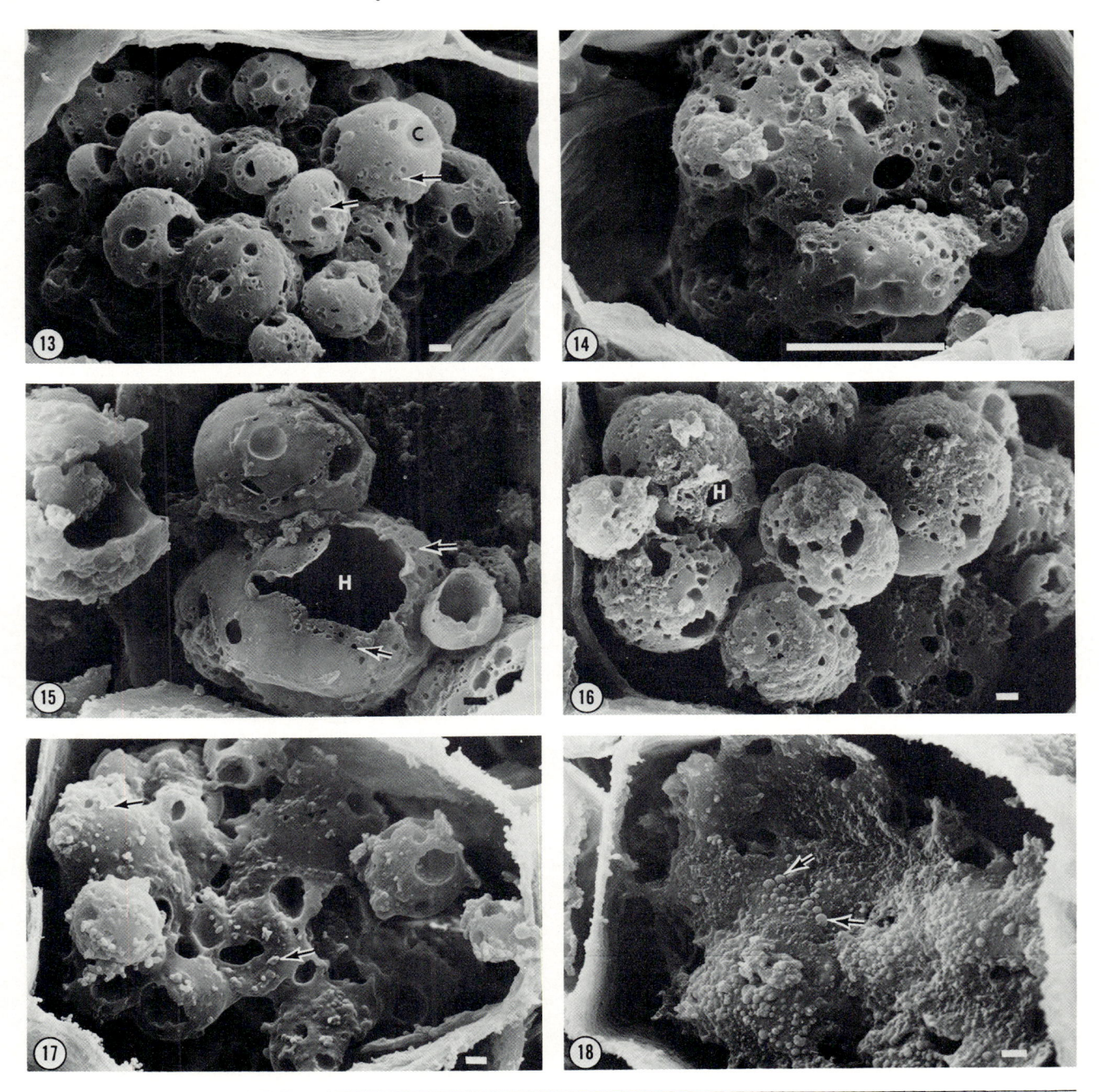

Figure 13. Water-treated (60 seconds) sunflower protein bodies. Crater-like surface holes and small pits are seen. Protein bodies are clumped. C = crater-like hole; arrows = pits. Bar = 1 μm.
Figure 14. Water-treated (15 minutes) sunflower protein bodies are fused into a mass. Surface holes and pits are present. Bar = 10 μm.
Figure 15. Pap-treated sunflower protein bodies. A large protein body hole is present along with smaller pits. Protein body surfaces are smooth. H = hole; arrows = pits. Bar = 1 μm.
Figure 16. Pan-treated sunflower protein bodies. Several holes are seen in each protein body and protein body surfaces are rough. H = hole. Bar = 1 μm.
Figure 17. MT-treated (0.25%) sunflower protein bodies are fused. Several holes are present and numerous particles are seen on the surface. Arrows = particles. Bar = 1 μm.
Figure 18. MT-treated (2.50%) sunflower protein bodies completely fused into a protein "vacuole." Particles and globoids cover entire surface. Arrows = globoids. Bar = 1 μm.

Water soluble proteins characterized as albumins have been shown to compose 27% of the total seed protein in a Yucca species, whereas sunflower seeds contained 62% albumin (Youle and Huang, 1981). Protein body albumins have been noted for their enzymatic activity (Aston, 1976). Release and activation of these enzymatic proteins by hydration is of considerable importance in germination (Weber and Neuman, 1980). Albumins also are especially rich in nitrogen and sulfur, and are important seed storage proteins (Youle and Huang, 1978 and 1981).

The present studies indicate that while large portions of Yucca protein bodies do appear to be water-soluble, only a relatively small amount of sunflower cotyledon protein bodies appears to be dissolved. It may be possible that protein removed from the surface holes combined with other protein dissolved from the remaining protein body mass account for 62% water-soluble protein. On the other hand, some factor that is released from the protein bodies during hydration may inhibit albumin solubility.

Yucca and sunflower protein bodies treated with either Pap or Pan enzyme solutions show similar patterns of decomposition (Figure 5, 6, 15, and 16). One or more large holes are usually seen in the protein body surface and are accompanied by numerous smaller pits. The holes extend deeply and widen into the protein body core. The remaining protein body structure appears as a hollow shell. Globulin storage proteins are associated with crystalloid protein body inclusions (Weber and Neuman, 1980). It is conceivable, in Yucca at least, that the crystalloid inclusion is more susceptible to Pap and Pan treatment than is the surrounding protein matrix and therefore is digested first. The result of this digestion leaves a hollow protein matrix shell. Since protein crystalloid type inclusions in sunflower protein bodies have been reported (Saio et al, 1977), it may be possible that a similar digestion pattern occurs. Further substantiation of the existence of crystalloid inclusions in sunflower is required. In any case, internal protein body portions of both Yucca and sunflower are most susceptible to digestion by these enzymes. In both seed types, holes produced by Pap treatment are somewhat larger than Pan-induced holes. Also, the protein body surfaces are smooth in Pap-treated cells and have a granular appearance in those treated with Pan. These differences may be due to differing enzyme strengths or relative concentration or perhaps to basic differences in the character and specificity of these plant and animal enzymes.

Yucca protein bodies treated with Pap and Pan are both somewhat smaller than acetone-extracted protein bodies. Sunflower protein bodies on the other hand increase in size when treated with Pap or Pan (Table 1). Protein body swelling accompanied by a decrease in protein body density has been reported during germination (Horner and Arnott, 1965; Ashton, 1976; Weber and Neuman, 1980). Perhaps the swelling that occurs in germination originates from a similar proteolytic enzyme which is released by the hydration.

Yucca and sunflower protein bodies do not show similar digestion patterns when treated with MT. Reductions in protein body size indicate that protein matrix material is removed progressively from Yucca protein bodies (Table 1). Spherical bodies embedded in remaining matrix material are seen in 0.25% MT-treated seed sections (Figure 7). Further removal of matrix material in 2.50% MT-treated cells reveals the protein crystalloid inclusion (Figure 8). The crystalloid structures that are revealed are slightly larger than the holes seen in Pap- or Pan-treated Yucca protein bodies. However, since these holes widen internally, a close correspondence between crystalloid size as revealed by MT treatment of Yucca sections and the size of the Pap- or Pan-induced protein body holes can be seen which further indicates crystalloid digestion in Pap- and Pan-treated protein bodies. Although Pap and MT contain the same basic enzyme, their effects are not similar. The reasons for this discrepancy are not clear. The pH of MT is somewhat higher than that of Pap, but if their pHs are matched, the differences in digestion patterns remain. The many extraneous ingredients in MT may be responsible for this difference in enzyme specificity.

The spherical bodies seen in both seed types are assumed to be globoids. The spheres are obviously embedded in the protein bodies and remain following acetone extraction and enzyme treatment; this seems to rule out identification as lipid and supports calling them globoids. Energy dispersive x-ray (EDX) analysis has not yet been performed to determine phosphorous content. Saio et al (1977) have reported EDX analysis of surface protuberances of isolated sunflower protein bodies indicating high phosphorus content. The protuberances appear to be embedded globoids which contain stored phosphates such as phytin (K, Mg, Ca, salt of myoinositol-hexaphosphoric acid) (Weber and Neuman, 1980). Saio et al (1977) also reported EDX analysis of sunflower protein bodies of various sizes. The amount of phosphorus was higher in the small protein bodies (1-2 μm), and the amount of sulfur was higher in the larger ones (6 μm). Since the protein matrix is composed of protein which is high in cysteine (Youle and Huang, 1981), the reported results may indicate that the small protein bodies contain a relatively small amount of protein matrix. It is also possible that much of the sulfur rich albumin fraction associated with the protein matrix was removed by hydration of the protein bodies during isolation procedures used by Saio et al (1977).

Protein body fusion and the presence of many globoid bodies are the outstanding features of MT-treated sunflower cotyledon cells (Figures 17 and 18). Since the protein vacuole remains after additional acetone extraction of enzyme-treated seeds, it does not appear that residual lipid is responsible for apparent protein body fusion. Fusion of protein bodies and formation of a large protein vacuole is reported during seed germination (Horner and Arnott, 1965 and 1966; Ashton, 1976; Weber and Neuman, 1980). Since "artificial" protein body digestion with exogenous proteases has not, as yet, been directly compared with actual protein body digestion during germination, comparisons between them are purely speculative.

While it is difficult to model the process of germination using exogenous enzymes and water, these experiments do provide some interesting comparisons. Certainly, some stages of germination can be, in part, duplicated by the present experiments. The common protein body swelling and fusion noted by many authors (Weber and Neuman, 1980) can be seen in certain treatments of both Yucca and sunflower. Horner and Arnott (1965) provide electron micrographs along with a schematic drawing of Yucca protein body digestion during germination. In three of the four Yucca species tested, protein body swelling and fusion were seen; holes in the protein body and in the protein vacuole surfaces also were reported. Peripheral decomposition and pitting of the protein body surfaces are apparent in the published micrographs (Horner and Arnott, 1965, Figures 28 and 30). The protein body digestion characteristics seen in the present study appear to resemble in many respects the germination protein body digestion reported in previous studies.

Additional experiments using these techniques on other seed types, and the use of additional enzymes, combined with EDX and TEM analysis may provide more information about protein body structure and protein body breakdown during germination.

Acknowledgements

The authors thank Ms. E. K. Flyger and Ms. M. J. Goad for their assistance in preparation of the manuscript. All work was carried out in the Electron Microscope Laboratory at The University of Texas at Arlington.

References

Ashton, FM. Mobilization of storage proteins of seeds, Ann. Rev. Plant. Physiol., 27, 1976, 95-117.

Brisson, JD, Peterson, RL. The scanning electron microscope and x-ray microanalysis in the study of seeds: A bibliography covering the period of 1967-1976, Scanning Electron Microsc. 1977; II, 697-711.

Buttrose, MS, Lott, JNA. Inclusions in seed protein bodies of the Compositae and Anacardiaceae: Comparison with other dicotyledon families, Can. J. Bot., 56, 1978, 2062-2071.

Horner, HT, Jr., Arnott, HJ. A histochemical and ultrastructural study of Yucca seed proteins, Am. J. Bot., 52, 1965, 1027-1038.

Horner, HT, Jr., Arnott, HJ. A histochemical and ultrastructural study of pre- and post-germinated Yucca seeds, Bot. Gaz., 127, 1966, 48-64.

Penner, D, Ashton, FM. Proteolytic enzyme control in squash cotyledons, Nature, 212, 1966, 935-936.

Penner, D, Ashton, FM. Hormonal control of Proteinase activity in squash cotyledons, Plant Physiol., 42, 1967, 791-796.

Rost, TL. The ultrastructure and physiology of protein bodies and lipids from hydrated dormant and non-dormant embryos of Setaria lutescens (Gramineae), Am. J. Bot., 59, 1972, 607-616.

Saio, K, Gallant, D, Pettit, L. Electron Microscope research on sunflower protein bodies, Cereal Chem., 54, 1977, 1171-1181.

Schnarrenberger, C, Oeser, A, Tolbert, NE. Isolation of protein bodies on sucrose gradients, Planta, 104, 1972, 185-194.

Swift, JG, Buttrose, MS. Protein bodies, lipid layers and amyloplasts in freeze etched pea cotydeons, Planta, 109, 1973, 61-72.

Varriano-Marson, E, DeOmana, E. Effects of sodium salt solutions on the chemical composition and morphology of black beans (Phaseolus vulgaris), J. Food Sci., 44, 1979, 531-536.

Webb, MA, Arnott, HJ. A scanning electron microscope study of zucchini seed cotyledons during germination, Scanning Electron Microsc. 1980; III, 581-590.

Weber, E, Neuman, D. Protein bodies, storage organelles in plant seeds, Biochem. Physiol. Pflanzen, 175, 1980, 279-306.

Webster, BD, Leopold, AC. The ultrastructure of dry and imbibed cotyledons of soybean, Am. J. Bot., 64, 1977, 1266-1293.

Wolf, WJ, Baker, FL. Scanning electron microscopy of soybeans, Cereal Sci. Today, 17, 1972, 124-130.

Youle, RJ, Huang, AHC. Protein bodies from the endosperm of castor bean. Subfractionation, protein contents, lectins and changes during germination, Plant Physiol. 61, 1978, 13-16.

Youle, RJ, Huang, AHC. Occurrence of low molecular weight and high cysteine containing albumin storage proteins in oilseeds of diverse species, Am. J. Bot. 68, 1981, 44-48.

Young, JL, Varner, JE. Enzyme synthesis in cotyledons of germinating seeds, Arch. Biochem. Biophys. 84, 1958, 71-78.

Discussion with Reviewers

H. T. Horner, Jr.: How can you be certain that the spheres shown in the MT-treated seeds are globoids? Have you used x-ray analysis to determine whether they contain calcium?

Authors: X-ray analysis has not yet been used; however, tentatively, we believe these bodies are best identified as globoids. These globoids have many characteristics similar to globoids described in published accounts (for Yucca, see Horner and Arnott, 1965 and 1966). Figure 19 shows an MT-treated Yucca protein body in which the

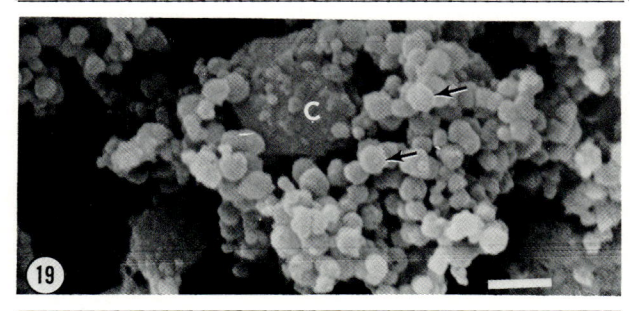

Figure 19. MT-treated Yucca protein body. Crystalloid (C) and associated globoids (arrows) are visible. Bar = 1 μm.

crystalloid and globoids are clearly seen. The globoids retain their associations with the original protein bodies.

W. J. Wolf: Can you explain why the globoids appear to dissolve in water but not in aqueous solutions of MT? Does Adolf's meat tenderizer contain constituents that may insolubilize the globoids? What happens if the surface is first treated with water, dried, and then treated with MT?
Authors: We do not feel that the globoids have been dissolved in water but rather that the protein bodies have not been sufficiently decomposed to allow the globoids to be viewed. If defatted seed sections are hydrated and allowed to air-dry, all the surface cells are completely empty probably due to the autolytic effects of water-activated enzymes within the protein bodies. This also occurs in seeds treated with enzymes for extended periods.

R. L. Peterson: Do you have any idea of the composition of each crude enzyme solution?
Authors: According to the Sigma Catalog, all crude proteases contain extraneous enzymes. Papain type II crude powder (P-3375) has an activity of 1.6 to 2.8 BAEE units (One unit will hydrolyze 1.0 µmole of α-N-Benzoyl-L-Arginine Ethyl Ester per minute.) per mg solid at pH 6.2 at 25°C and is not standardized with lactose. Pancreatic crude type I (P-4630) has an activity of 8 to 10 units per mg solid at pH 7.5 at 37°C. (One unit will hydrolyze casein to produce color equivalent to 1.0 µm, 18 µg, tyrosine per minute.) This preparation is substantially free of DNase but may contain a trace of RNase.

H. T. Horner, Jr.: Do you feel your approach to observing and characterizing protein bodies will be of value to seed analysts?
Authors: These techniques may be of value in future seed analyses. With the use of appropriate enzymes, specific protein body components may be selectively digested, and the relatively quick procedure allows many seed parameters to be tested in a short time.

I. B. Sachs: Did you at any time try the papainase of pineapple juice on the seed sections? If so, was its proteolytic activity higher or lower than the papain of papaya latex?
Authors: We have not used at this time any additional enzymes. It is not possible to speculate on the activity of pineapple papainase, but we do intend to run such an experiment soon.

TANNIN DEVELOPMENT AND LOCATION IN BIRD-RESISTANT SORGHUM GRAIN

P. Morrall[1], N. v.d. W. Liebenberg[2], C.W. Glennie[1]*

[1]Sorghum Beer Unit; [2]National Food Research Institute, Council for Scientific and Industrial Research, P.O. Box 395, PRETORIA, South Africa, 0001.

Abstract

Bird-resistant sorghum hybrids, SSK 52 and Red Nyoni, were collected at various stages of maturity from pre-flowering to time of harvest. All stages were examined using light and transmission electron microscopy.

Small tannin containing vesicles appeared at the inner integument cells just after fertilization and, at the milk stage of grain development, tannins were rapidly being deposited along the periphery of the central vacuole. The tannins developed until there was little evidence of cell structure and the testa consisted of a continuous layer of tannin. In the mature kernel the tannins extended from the testa through the cross and tube cells into the mesocarp, but they were not found in the aleurone layer. The tannins appeared very dense and amorphorous on electron micrographs. Besides being strongly osmiophilic, the tannins stained with vanillin-HCl reagent. Once deposited, the tannins did not appear to decrease as the grain matured.

*Author to whom correspondence should be addressed

KEY WORDS: *Sorghum bicolor*, tannin development, tannin location, testa development, inner integument.

Introduction

Condensed tannins occur as constituents of bird-resistant varieties of sorghum and little is known about their mode of deposition during grain development. Polyphenols may be located in both the pericarp and testa of sorghum kernels (Rooney et al., 1979; Blakely et al., 1979) while a testa must be present for a variety to contain appreciable amounts of tannin (Bullard and Ellis, 1979). Bird-resistant grains have a well-developed testa which contains varying amounts of tannin depending on the variety, while in non-bird-resistant varieties, which do not contain tannin, the testa can be absent (Wall and Blessin, 1969).

Other studies traced the morphological development of immature sorghum grain (Artschwager and McGuire, 1949; Saunders, 1955). They showed that the pericarp was derived from the ovary wall with no cell division occurring, only cell enlargement. The testa was found to develop from the inner integument and as the fertilized seed expanded the outer integument was crushed between the inner integument and the pericarp. The testa was highly coloured.

Because tannins inactivate enzymes required during the brewing of sorghum beer (Daiber, 1975) and reduce the nutritional quality of the grain, it is important to develop a better understanding of them. The structure of their monomeric building blocks has been established (Gupta and Haslam, 1978) and a biosynthetic sequence proposed (Gupta and Haslam, 1979). One aspect of sorghum tannins which has not been studied is the developmental morphology of their deposition. This paper examines the deposition of the tannins in the developing testa during growth of the grain.

The nomenclature for the structure of the mature caryopsis agrees with that of Doggett (1970) (Fig. 1). The pericarp is derived from the ovary wall and can be differentiated into a cutin covered epidermis on the exterior, subtended by a hypoderm (not always present), then comes the mesocarp which comprises starch containing cells in some varieties and the differentiated cross cell layer and tube cell layer. In bird-resistant sorghum, immediately beneath the cross and tube cells, there is a testa which is highly coloured and contains tannin. Below the testa is the aleurone layer of the endosperm (See Figs. 1, 2).

Materials and Methods

Bird-resistant sorghum varieties SSK 52 and Red Nyoni were collected at five-day intervals for the various stages of maturity, from the 1980 crop (Table 1). The SSK 52 grain was grown by the Plant and Seed Control Division, Roodeplaat Experimental Farm, South African Department of Agriculture and Fisheries, and the Red Nyoni grain was kindly supplied by Asgrow Ltd, Pretoria. The time lapse of 48 days should be noted between the physiological maturity of the grain and its time of harvest. The grain was collected in paper bags in which it was transported to the laboratory and fixed.

Table 1. Times of collection and stages of development of sorghum grain

Developmental stages	Days post-anthesis
Pre-flowering	-5
Flowering	0
Post-flowering	5
Milk stage	10
Soft dough stage	15
Hard dough stage	20
Physiological ripe	25
Time of harvest	73

For light microscopy, the grain was fixed in 10% formaldehyde and longitudinal sections were stained with safranin and counterstained with Fast Green. Material for electron microscopy was fixed in 5% glutaraldehyde, post-fixed in 1% OsO_4, embedded in an Epon-araldite mixture and the sections double-stained with uranyl acetate and lead citrate.

A widely-used method for tannin analysis is the reaction of condensed tannins with vanillin and HCl. To locate tannins in sections of sorghum grain, they were treated with a freshly prepared solution (1:1) of vanillin (4% in methanol) and HCl (8% in methanol) and then viewed under the light microscope.

Results and Discussion

A longitudinal section of sorghum grain is shown in Fig. 2. Only sufficient developmental morphology will be given to describe testa development, as seed morphology and cytology of fertilization have been ably described by Saunders (1955) and Artschwager and McGuire (1949). Before fertilization, the sorghum ovary consisted of an ovary wall enclosing a nucellus. Immediately interior to the ovary wall and covering about one-third of the area of the nucellus, was the outer integument. This integument varied in thickness between two and four cells. The nucellus was completely surrounded by the inner integument which was two cells thick, the inner cells being larger than the outer ones. It is the inner integument which develops into the testa by the accumulation of tannin. At this prefertilization stage, the integument cells were cuboidal and contained no detectable tannin.

Between the nucellus and the inner integument was a thin black layer which stained darkly with Sudan IV and, according to Saunders (1955), was composed of cutin. This cutin layer was electron dense and in the electron micrographs it had a well-defined sharp edge on the nucellus side, whilst the edge adjoining the inner integument was less sharp (Fig. 3). The appearance of this cutin layer suggested that it was deposited from the inner integument cells. This idea was supported by the occasional gaps which were observed between the nucellus and cutin layer while no gaps could be found between the inner integument and cutin layer. The cutin layer was continuous and completely covered the nucellus except near the hilum.

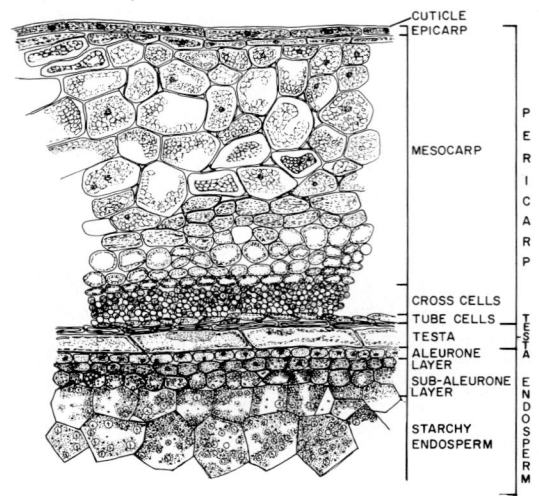

Fig. 1. A median longitudinal section of sorghum grain at the soft dough stage (15 days post anthesis) showing the outer layers of the dorsal surface of the caryopsis.

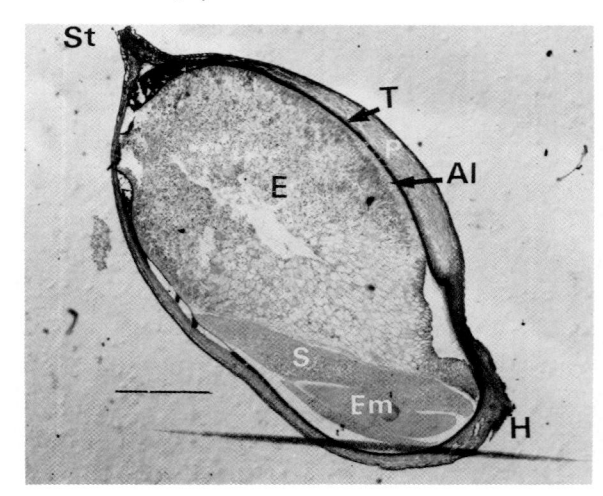

Fig. 2. A longitudinal section of SSK 52 sorghum grain at soft dough stage (15 days post anthesis) showing the pericarp (P), testa (T), aleurone layer (Al), endosperm (E), scutellum (S), embryo (Em), stylar end (St) and the hilar end (H). (Bar = 1 mm).

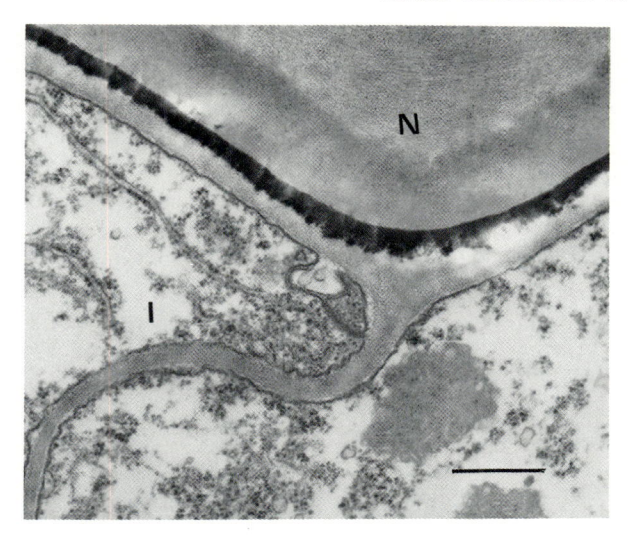

Fig. 3. SSK 52, soft dough stage (15 days post anthesis): the cell wall between the inner integument (I) and the nucellus (N). (Bar = 0.5 μm).

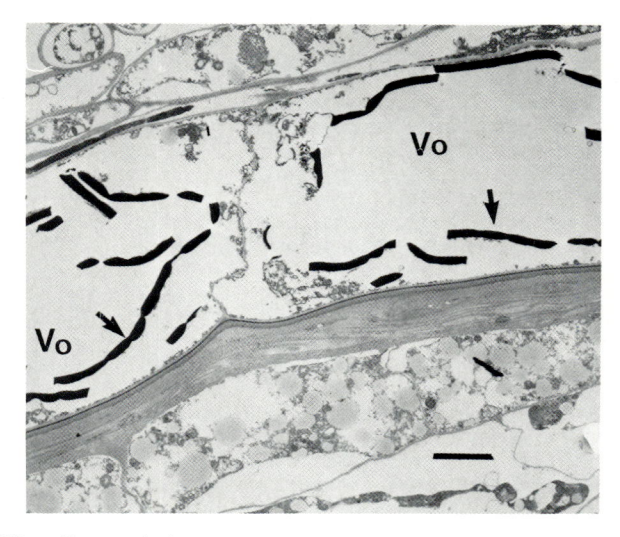

Fig. 5. Red Nyoni, milk-stage (10 days post anthesis): tannin deposition (arrows) around the periphery of the central vacuole (Vo) of the inner integument. (Bar = 5 μm).

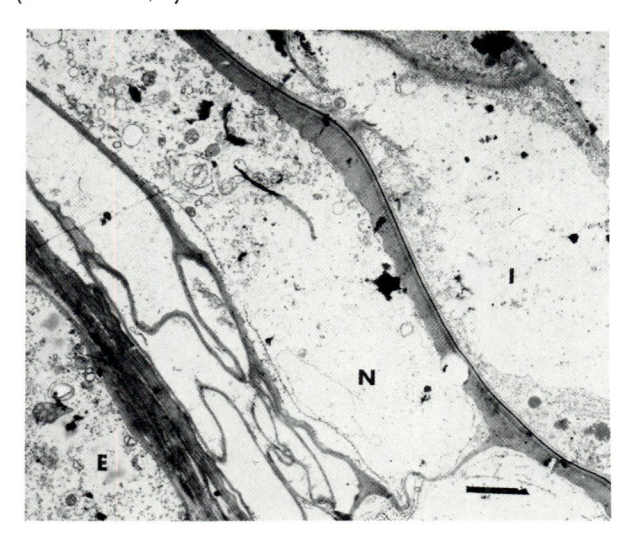

Fig. 4. SSK 52, milk-stage (10 days post anthesis): the outer cell layer of the nucellus (N) being compressed against the inner integument (I) by the developing endosperm (E). (Bar = 5 μm).

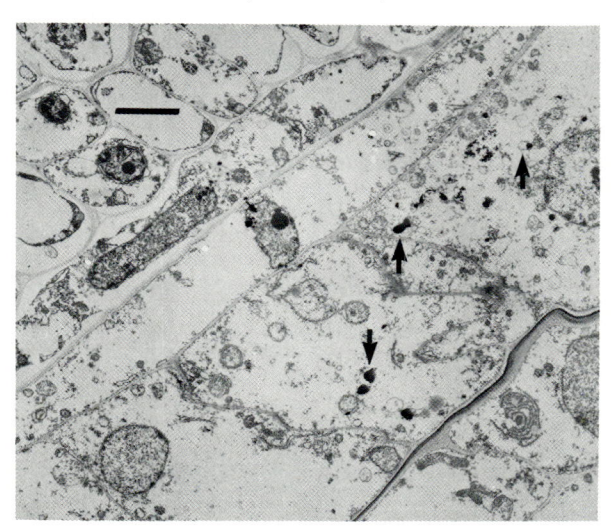

Fig. 6. Red Nyoni, post fertilization stage (5 days post anthesis): vesicles (arrows) in the cytoplasm of the inner integument filling with tannin (Bar = 5 μm).

During the pre-fertilization period, the nucellus was still intact. It was composed of large irregular shaped cells which contained a very diffuse cytoplasm. The exterior layer of nucellus cells were smaller and more cuboidal; they also had more cytoplasmic organization and more rigid cell walls than the interior cells.

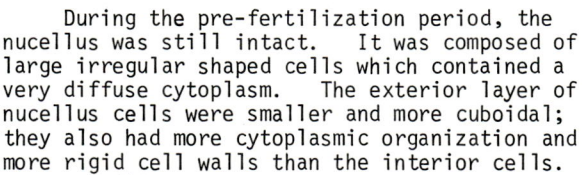

Fertilization introduced a period of rapid change. The caryopsis started to expand rapidly and the outer integument was crushed and lost. The disintegrating nucellus was crushed against the inner integument by the expanding endosperm but the outer layer of cells of the nucellus resisted this crushing and persisted until the milk stage (Fig. 4). This remaining nucellus appeared to still contain cytoplasm. The cell walls of the crushed nucellus cells accumulated on the exterior of the endosperm giving a thickened cell wall between the aleurone layer of the endosperm and the inner integument. Fig. 5 shows that this thickened wall had a striated appearance and was crowned by the layer of cutin from the inner integument.

After fertilization in Red Nyoni sorghum, small membrane-bound vesicles appeared in the cytoplasm of the inner integument cells (Fig. 6). These vesicles were filling with osmiophilic

295

substances, presumably tannin (Fig. 7). These cells contained organized cytoplasm and nuclei were still present. In cell suspension cultures of white spruce, tannin inclusions originated within cytoplasmic vacuoles and the tannin accumulated in the central vacuole through coalescence of the cytoplasmic vacuoles (Chafe and Durzan, 1973).

In this study, we found that during the milk stage there was a rapid deposition of tannin. The central vacuoles of the cells of both cell layers of the inner integument developed and enlarged to fill almost the entire cell (Fig. 8). Tannin had accumulated around the periphery of the vacuole of the cells of the inner cell layer (Fig. 5). There was very little cytoplasmic organization at this stage although the occasional cell still contained a nucleus (Fig. 8). Large quantities of tannin first accumulated in the inner cells of the integument while the outer layer of cells, which had more cytoplasm and were smaller, developed tannin more slowly. At this stage the pericarp had developed into its various well-defined layers of cross and tube cells, the starch-containing mesocarp and the heavily cutinized epidermal cells.

The integument cells were converted into a tannin-containing testa by the soft dough stage. The rapid increase in tannin content during this period was confirmed by measuring the amount of extractable tannin (Glennie, 1981). After the initial development of the testa, the rate of tannin deposition (apparent by microscopy) slowed but continued until, in the mature kernel, tannin was found in the inner layers of the mesocarp (Fig. 9).

At maturity, in SSK 52, the tannin obscured cell structure throughout the testa and no fine structure was evident (Fig. 10). The tannin penetrated into the cell walls and the testa appeared as a continuous amorphorous layer. While the tannin was present in both cell layers of the testa in Red Nyoni grain, it did not obliterate the cell walls and even small amounts of crushed cytoplasmic residue could be found (Fig. 11). The two cell layers often stained with different intensities. This is in agreement with Blakely *et al.*, (1979) who also found that the testa was composed of two overlapping layers often of two different colors. In a general comparison of micrographs of the two varieties, the SSK 52 grain appeared to contain heavier deposits of tannin than did Red Nyoni. When both grains were extracted and the extracts analysed for tannin content, the SSK 52 was found to contain approximately 20% more.

In all grains examined, the testae were continuous except near the hilum and were thicker around the crown end than at the basal region of the grain. Light micrographs (Fig. 2) showed that where the testa was thickest the pericarp was thinnest. As the grain matured there was no visible decrease in the testa but the amount of extractable tannin decreased. The amount of decrease varied with the sorghum variety being examined (Glennie, 1981; Price *et al.*, 1979). The decrease in extractable tannin was probably due to continued polymerization of the tannins to high molecular weight forms which were insoluble, and binding of the tannin to cell debris. Although the tannin penetrated well into the pericarp there was no evidence that it penetrated the thickened cell wall into the aleurone layer.

The tannin was strongly osmiophilic and also stained with the vanillin-HCl reagent. The vanillin reagent is specific for flavan-3-ols (Gupta and Haslam, 1979) and showed the pattern of tannin distribution in the sorghum grain. When sections of both grains were stained with vanillin-HCl and examined under the light microscope, tannins could be found mainly in the testa while small amounts appeared in the epidermis. Reichert *et al.* (1980) found a similar pattern with hand-dissected grains. We observed good staining of the testa from the soft dough stage onwards. Unfortunately, the vesicles which appeared after fertilization were too small to be observed under the light microscope so it was impossible to tell if they stained with vanillin or not.

It is probable that the testa began to protect the sorghum kernel at the very early stage when the endosperm had barely replaced the nucellus. By this stage the endosperm contained large amounts of starch and the grain was soft. During the final stages of grain maturation there was a decrease in the amount of extractable tannin but there was no visual evidence that there was a similar decrease in the tannin content of the testa. Besides protecting the grain against depredation by birds, the testa could also protect the seed against microbial attack (Swain, 1979). This protection could take the form of the inactivation of enzymes of any organism which tried to penetrate the testa.

References

Artschwager E, McGuire R.C. Cytology of reproduction in *Sorghum vulgare*. J Agric Res 78 1949 659-673.

Blakely M E, Rooney L W, Sullins R D, Miller F R. Microscopy of the pericarp and the testa of different genotypes of sorghum. Crop Science 19 1979 837-842.

Bullard R W, Elias D J. Sorghum polyphenols and bird resistance. In: Polyphenols in Cereals and Legumes Hulse J H (ed.). International Development Research Centre, Ottawa, Canada 1979 43-49.

Chafe S C, Durzan D J. Tannin inclusions in cell suspension cultures of white spruce. Planta 113 1973 251-262.

Daiber K H. Enzyme inhibition in sorghum grain and malt. J Sci Fd Agric 26 1975 1399-1411.

Doggett H. *Sorghum*. Longmans, Green and Co. Ltd, London 1970 p 69.

Glennie C W. Preharvest changes in polyphenols and the enzymes involved in their oxidation in sorghum grain. J Agric Fd Chem 29 1981 33-36.

Gupta R K, Haslam E. Plant proanthocyanidins, part 5. Sorghum polyphenols. J Chem Soc Perkin I 8 1978 892-896.

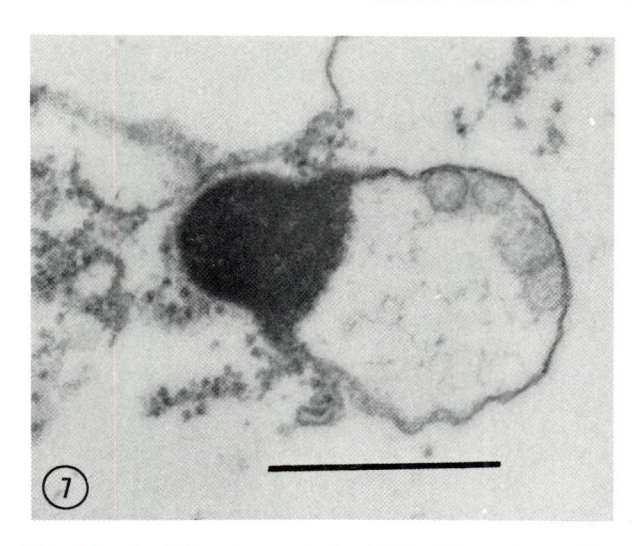

Fig. 7. Red Nyoni, post fertilization stage (5 days post anthesis): membrane-bound vesicle filling with tannin in the inner integument (see Fig. 6). (Bar = 0.5 μm).

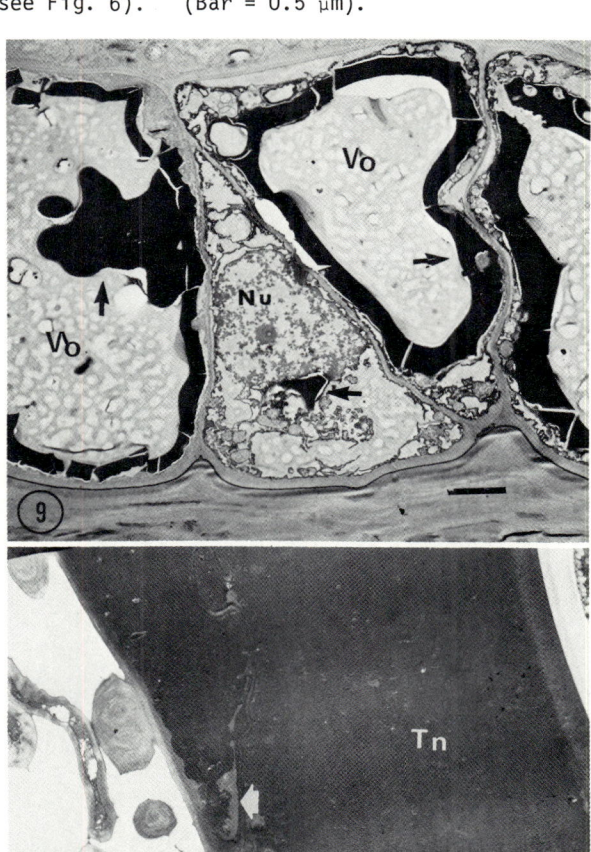

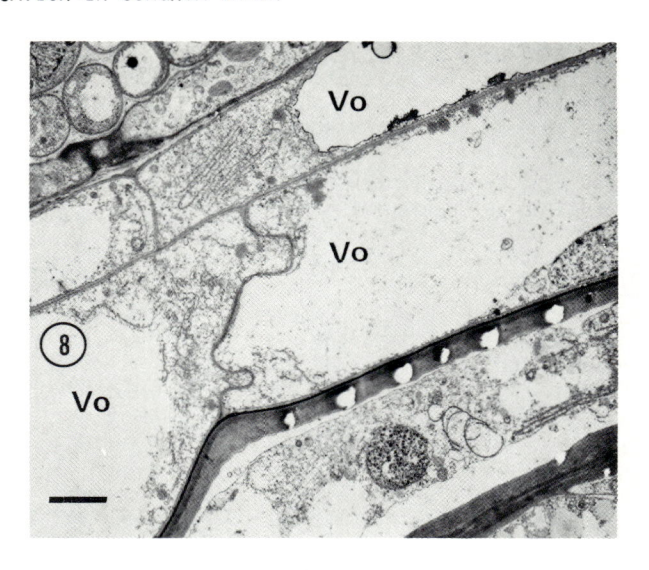

Fig. 8. SSK 52, milk stage (10 days post anthesis): the two cell layers of the inner integument showing large central vacuoles (Vo). (Bar = 5 μm).

Fig. 9. SSK 52, milk stage (10 days, post anthesis): the central vacuoles (Vo) of some cells contain tannin deposits (arrows) whilst other cells still contain nuclei (Nu). (Bar = 5 μm).

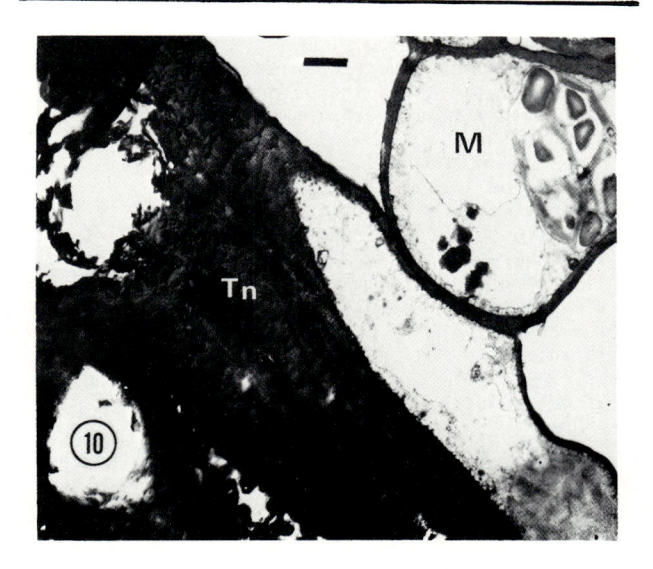

Fig. 10. SSK 52, mature grain (73 days post anthesis): tannin (Tn) in the testa and the inner layer of the mesocarp (M). (Bar = 0.5 μm).

Fig. 11. Red Nyoni, mature grain (73 days post anthesis): the cells of the testa still containing discernible cell walls (CW) and cytoplasmic residues (arrow). (Bar = 0.5 μm).

Gupta R K, Haslam E. Vegetable tannins - structure and biosynthesis. In: Polyphenols in Cereals and Legumes. Hulse J H (ed.). International Development Research Centre, Ottawa, Canada 1979 15-24.

Price M L, Stromberg A M, Butler L G. Tannin content as a function of grain maturity and drying conditions in several varieties of *Sorghum bicolor* (L.) Moench. J Agric Fd Chem 27 1979 1270-1274.

Reichert R D, Fleming S E, Schwab D J. Tannin deactivation and nutritional improvement of sorghum by anaerobic storage of H_2O-, HCl-, or NaOH-treated grain. J Agric Fd Chem 28 1980 824-829.

Rooney L W, Blakely M E, Miller F R, Rosenow D T. Factors affecting the polyphenols of sorghum and their development and location in the sorghum kernel. In: Polyphenols in Cereals and Legumes. Hulse J H (ed.). International Development Research Centre, Ottawa, Canada 1979 25-35.

Saunders E H. Developmental morphology of the kernel in grain sorghum. Cereal Chem 32 1955 12-25.

Swain T. Tannins and lignins. In: Herbivores - Their Interaction with Secondary Plant Metabolism. Rosenthal G A, Janzen D H (ed.). Academic Press, New York 1979 657-681.

Wall J S, Blessin C W. Composition and structure of sorghum grains. Cereal Sci Today 14 1969 264-270.

Discussion with Reviewers

M.L. Price: Collection of the grain at five-day intervals implies a very precise timed study was done. Were the heads individually tagged at certain stages of anthesis? Was grain only taken from the top 2 cm, middle 2 cm, lower 2 cm or some such standard location?
Authors: The taking of samples of sorghum grains from panicles of different stages of maturity is important and in practice quite difficult. The panicles were not individually tagged at anthesis. Samples of grain were taken from the middle 2-4 cm of the panicles of plants growing in a large stand. Grain was collected every five days and only grain which had reached the desired stage of maturity was collected. Thus, the grain samples obtained showed a uniformity of size, colour and moisture.

M.L. Price: Do varieties of sorghum which contain no tannin not show these membrane-bound vesicles with osmiophilic substances? This would give added credibility to the hypothesis that these contain tannin.
L.W. Rooney: Did the authors do any comparisons between bird-resistant sorghums and normal sorghum (non-resistant to birds)?
Authors: We are investigating the cytology of the development of low tannin grains.

M.L. Price: Do you have any explanation as to why the testa and pericarp thickness complemented each other, the one being thick when the other was thin?
Authors: No, it was simply an observation. Light enhances the accumulation of polyphenols and in this case the thin pericarp may allow in more light to encourage the accumulation of tannin. However, it should be remembered that the developing grain is partially covered by the glumes which are highly pigmented and dense but do not cover the stylar surface of the grain.

LIGHT MICROSCOPY OF PLANT CONSTITUENTS IN ANIMAL FEEDS

J.G. Vaughan

Biology Department
Queen Elizabeth College
Campden Hill Road
London W8 7AH, U.K.

Abstract

The use of the light microscope in the quality control of animal feeds is reviewed. Microscopy is important in establishing adulteration and in the control of safety and aesthetic properties. The technique must be regarded as complementary to chemical analysis. Facilities and organizations concerned with training and the standardization of feed microscopy techniques are described. Information is given on instrumentation and the microtechniques which are particularly relevant to feed microscopy. Examples of specific feed constituents are described to illustrate the microscopic features used in identification. Some areas for future research are suggested. A complete list of feed microscopy manuals is given.

KEY WORDS: Light microscopy, plant constituents, animal feeds, adulteration, admixture, toxicity, stimulants, flavour, organisations, training.

Introduction

The identification of plant (and animal) feeds by light microscopy is a well established technique. Since feed microscopy is complementary to chemical analysis, a comprehensive survey of an animal feed would include both techniques. Electron microscopy is not suitable for this type of identification because reasonably large quantities of feeds need to be examined and the limit of resolution of the light microscope is sufficient for identification. As a method of animal feed analysis, microscopy serves several purposes: (1) detection of adulteration or admixture; (2) identification of certain toxic and stimulatory plant materials; (3) recognition of materials which contribute to flavour characteristics; (4) inspection of compound manufactured feeds for compositional correctness.

In contrast to the chemical analysis of food, it has not been possible to automate analytical microscopy and consequently the microscopist is still the most important factor in the operation. This being the case, the operation of feed microscopy takes considerably longer than that of chemical analysis. However, the application of feed microscopy may be essential if problems are to be solved.

Chemical analysis may indicate discrepancies in the alleged make-up of a feed but it is only microscopy that can establish the physical structure of the material. In this situation, an adulterant is not necessarily a constituent of poor nutritive value but one which is hardly acceptable on the original contract. This being the case, it is not possible to produce a list of plant materials which are specifically adulterants although rice hulls are a notable exception.

There are two organizations concerned with the propagation of feed microscopy. In Europe, the 'Sektionen Futtermittel-Mikroscopie und Futtermittelmikrobiologie der Internationalen Arbeitsgemeinschaft fur Futtermitteluntersuchung' organizes an annual meeting at which papers are presented concerning contemporary problems in feed microscopy. These papers are published in a "Protokoll" but, unfortunately, its circulation is restricted to members of the associations. No training courses are given by this organization.

The American Association of Feed Microscopists also organizes an annual meeting and produces an official proceedings. This Association provides training courses in North America and, of late years, has been responsible for courses in other parts of the world.

Academic institutions can also be involved in this work, such as that found in the U.K., where a training course is given at Queen Elizabeth College, London University, on a biennial basis.

Many publications are available for the techniques and application in this area. However, some are presently out of print. The following is a comprehensive list of references and are listed in Appendix I. In English: Morris (1928), Parkinson and Fielding (1930), Winton and Winton (1932, 1935, 1937, 1939), Vaughan (1970), American Association of Feed Microscopists (1978), Huss (1976). In French: Collin and Perrot (1904) Bussard and Brioux (1925), Juillet et al. (1955), Ferrando and Henry (1966). In German: Moeller and Griebel (1928), Czaja (1971), Gassner (1973), Meszaros and Deutschmann, (1975).

When the wide range of plant constituents in animal feeds is studied, the complexity of this area is fully realized. Not only are some plant materials produced primarily for feeding purposes but also used are the by-products of many other industries (e.g. cereal milling, oil seed extraction, the production of beer, wine, soft drinks). However, in the context of the present tutorial paper, some special features as regards the training and importance of feed microscopists are emphasized.

Instrumentation

Two types of light microscope are used in this work - the stereomicroscope (magnifications x 5 to x 50) and the compound microscope (magnifications x 50 to 500). The stereomicroscope is used in the preliminary examination of the feed for the detection of large pieces of plant material. However, most of the plant materials in an animal feed are in a finely divided form and the compound microscope is required for identification.

Normally, the plant constituents of an animal feed are identified on the basis of the type and arrangement of cells present and therefore bright field light microscopy is employed. However, some identification is based on the morphology of starch grains and crystals. In these cases, polarized light microscopy is useful, starch grains and crystals being anisotropic objects. The polarizing system may be augmented with a Red I Compensator (Voehringer, 1979) which gives a distinct play of colours without staining.

Although the optical results of the latter method are attractive to the observer, the method does not often provide much extra diagnostic information. Phase contrast and fluorescence microscopy have not really contributed to the microscopic methods used in feed analysis.

Sample Preparation

Careful preparation of an animal feed is necessary before it can be investigated with the compound microscope. The plant fragments present are normally identified on the basis of arrangement of cells and tissues. Therefore, the fragments must be 'cleared' to remove starch and protein so that the cell wall pattern is apparent.

In most animal feeds, there is comparatively little oil or fat. However, if the plant cell pattern is masked by oil or fat then the offending substance can be removed with benzene or some similar organic solvent.

Most of the plant materials included in animal feeds have a high starch (e.g. cereals) or protein (e.g. oil seeds) content. Both these inclusions must be cleared and a variety of methods are available for this purpose (Wallis, 1965). In most cases, the production of a 'crude fibre' is useful. This involves alternate digestion of the feed with caustic soda (hydrolyzes protein) and nitric or sulphuric acid (hydrolyses starch). For future reference crude fibres may be preserved in formalin. Crude fibre production is a vigorous process and delicate plant tissues may be affected, also plant crystals of diagnostic importance, as in sesame, will be dissolved. If a more delicate treatment is required, then heating in chloral hydrate is recommended. In this method, plant crystals are not destroyed.

Pigmentation sometimes makes the interpretation of plant fragments difficult. Bleaching of such fragments may be carried out with commercial domestic bleach.

Once a feed has been cleared in the form of a crude fibre or by chloral hydrate then it may be mounted for microscopic examination in either water or a mixture (1:1) of water and glycerol. In the case of a plant starch alone, water is the best mountant.

It is rare that the plant fragments in feeds are stained prior to microscopic examination. The normal staining methods employed in plant histology are too time consuming and somewhat unnecessary. As far as the experienced feed microscopist is concerned, only simple histochemical tests are employed (iodine/potassium iodide solution for starch and protein; Sudan stains for fats; ferric chloride solution for tannins).

In normal feed microscopy, fairly rapid identification is required. This is because of the quick turnover of feedstuffs in the industry. Consequently, the microscopist attempts to base his identification of plant fragments on surface features. Nevertheless,

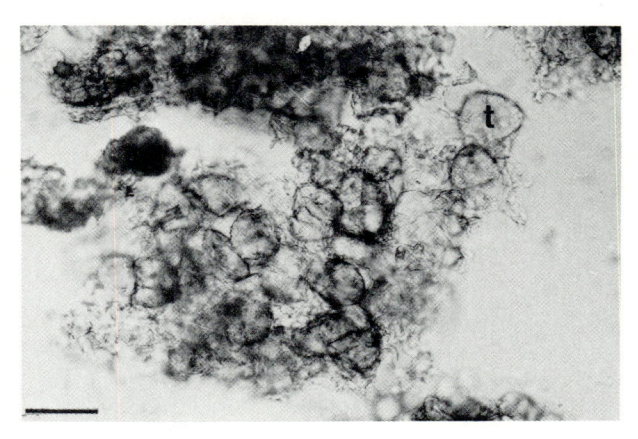

Fig.1. Illipe (Bassia) fragment showing tannin cells (t)
Bar = 100µm.

Fig.2. Sunflower (Helianthus) fragment showing phytomelan layer.
Bar = 100µm.

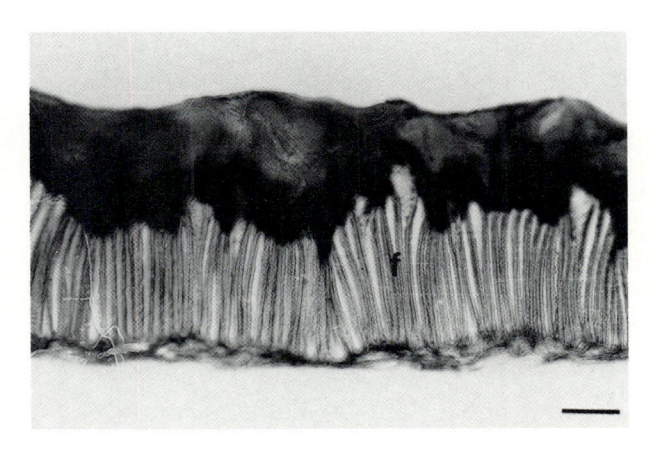

Fig.3. T.S. Sweet orange (Citrus) testa showing fibrous cells (f)
Bar = 10µm

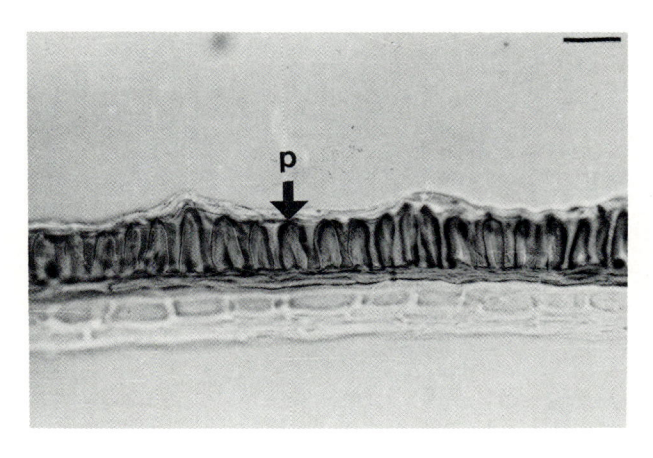

Fig.4. T.S. rape (Brassica campestris) testa showing palisade cells (p)
Bar = 10µm

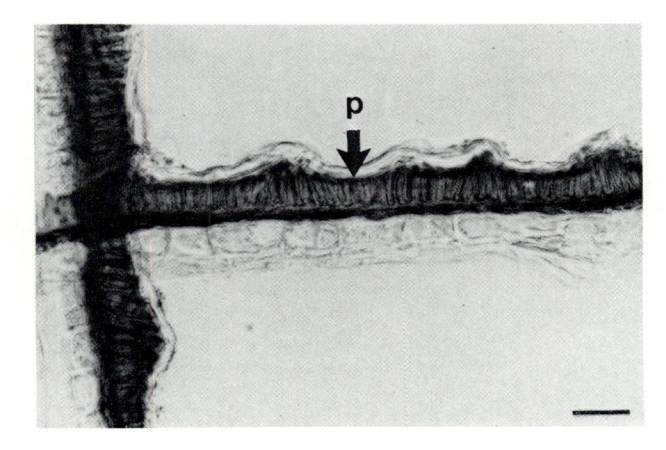

Fig.5. T.S. mustard (Brassica juncea) testa palisade cells (p)
Bar = 10µm

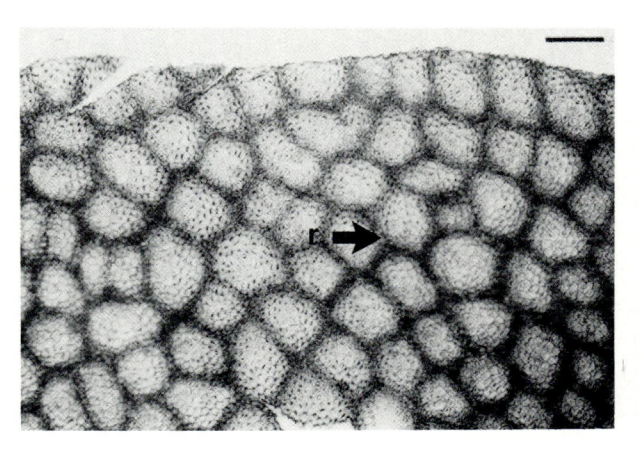

Fig.6. S.V. mustard (Brassica carinata) testa showing reticulation (r).
Bar = 100µm

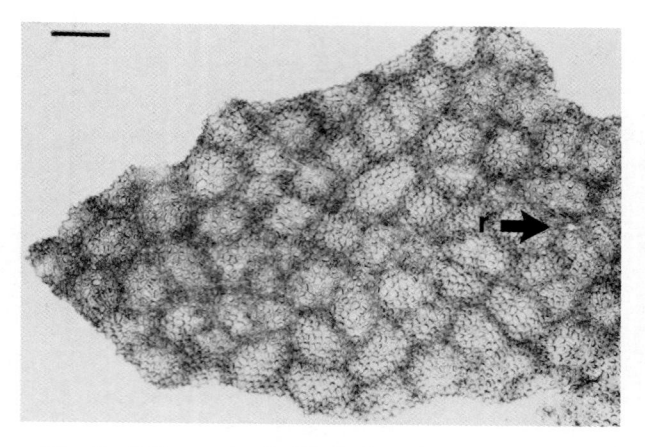

Fig.7. S.V. mustard (Brassica juncea) testa showing reticulation (r)
Bar = 100μm

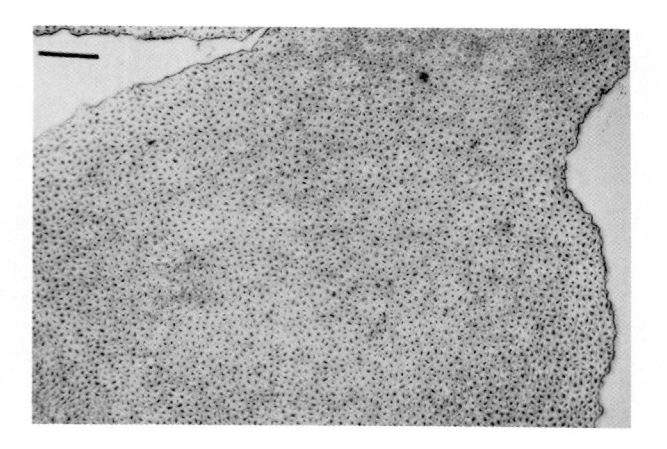

Fig.8. S.V. rape (Brassica campestris) testa
Bar = 100μm

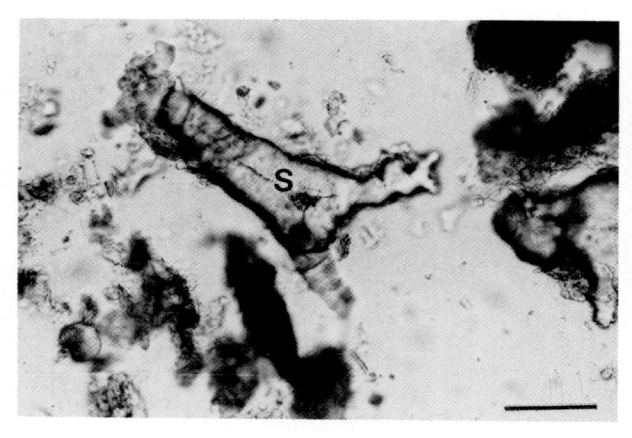

Fig.9. Olive (Olea) stone cell (S)
Bar = 100μm

Fig. 10. Wheat (Triticum) cross cells in pericarp
Bar = 100μm

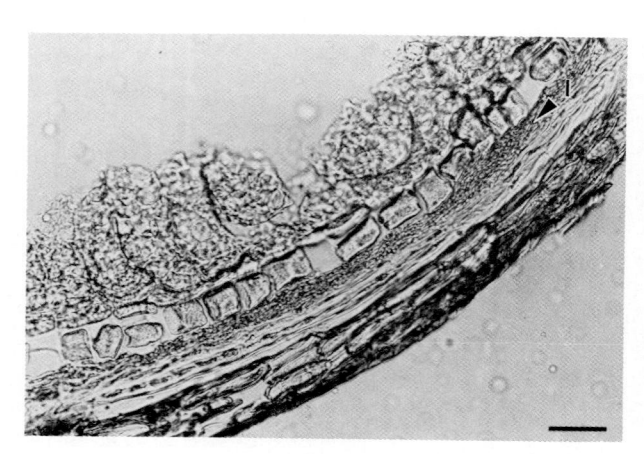

Fig. 11. T.S. darnel (Lolium temulentum) pericarp and endosperm showing fungus layer (l)
Bar = 10μm

Fig. 12. Wheat (Triticum) starch in polarized light.
Bar = 100μm

there are occasions when sections of the fragments must be prepared for accurate diagnosis. Various methods have been employed but in the author's experience the most satisfactory method is cutting the fragments, held between two pieces of cork or similar supporting material, with the Reichert sliding-type microtome (Vaughan, 1960).

Examples

Some examples of plant constituents, both in surface view and section, found in animal feeds are illustrated in Figs. 1-12.

Fig.1 is a fragment of illipe (Bassia). This is a member of the Sapotaceae and contains toxic saponins. It is identified by the presence of tannin cells. A common oil seed residue is sunflower (Helianthus). A diagnostic feature is the phytomelan layer (Fig.2). Citrus waste is sometimes included in animal feeds. Fig. 3 represents a section of the testa of sweet orange. This is identified by the fibrous cells.

In Europe, at least, rape seed residues are acceptable in feeds but, according to present regulations, mustard seed residues are forbidden. Fig. 4 is a section of a rape seed testa (Brassica campestris). The palisade cells, of more or less equal height, are the diagnostic feature. Fig. 5 is a section of a mustard (B.juncea) testa which shows palisade cells of unequal height. This difference in the palisade cells distinguishes rapes from mustards and can also be observed in surface views of the fragments. Mustards (B. carinata, Fig. 6; B. juncea, Fig. 7) show a distinct reticulation in surface view. A reticulation is not clear in rape fragments (Fig. 8).

Olive (Olea) pulp is identified by branched stone cells (Fig. 9). Fig. 10 represents the pitted cross cells of the wheat (Triticum) pericarp. Cross cell types may be used to distinguish the various cereals. Fig.11 is a section of the outer region of the darnel seed (Lolium temulentum) and shows the fungus layer with which is associated toxicity. Fig. 12 illustrates wheat starch grains as seen in polarized light.

Feed microscopy has always been used to detect plant constituents toxic to animals. As far as the European Economic Community is concerned (Vaughan and Stubbs, 1979) this is now part of the official legislation. The undesirable constituents include mustard (Brassica) seeds, castor (Ricinus), apricot and bitter almond, beech (Fagus sylvatica), rye ergot (Claviceps purpurea), Crotalaria, mowrah, Datura stramonium and darnel (Lolium temulentum). Some of these constituents are absolutely forbidden and consequently the feed microscopist is in a powerful position regarding the ultimate decision of acceptability.

In recent times in the U.K. there have been needs to detect stimulatory substances. Race horses are given 'dope' tests for the identification of alkaloids - theobromine and caffeine. A positive reaction means disqualification. These stimulants maybe introduced as chemicals but they are also associated with certain plant materials, namely coffee residues and cocoa shells (Theobroma). The naturally occurring stimulants may be part of the feed for certain animals. It is the responsibility for the feed microscopist to determine if these materials are present in rations for race horses.

Problems of palatability are sometimes encountered in the feed industry. These are notoriously difficult to deal with but the presence of high tannin materials such as sorghum or glucosinolate containing materials, such as mustard seed, as detected by microscopy, may be the cause.

Conclusions

In summary, it can be stated that feed microscopy is a well established technique for quality and identification work. If one were to look at specific work needed in this area, it would be to systematize evaluation of new feed sources such as Robinia and Manihot leaf meals, imported into Europe from China, also chemically treated wheat and barley straws. Also, research is needed in quantifying the plant constituents of animal feeds.

Appendix I

American Association of Feed Microscopists. Manual of Microscopic Analysis of Feeding Stuffs, 1978.

Bussard, L, and Brioux, C. Torteaux. Libraire Polytechnicque Ch. Berganger,Paris, 1925.

Collin, E and Perrot, Ed. Les Residues Industriels.A. Joanin, Paris, 1904.

Czaja A Th. Methoden der Lebensmittel-Mikroskopie und Lebensmittel Uberwachung. Umschau Verlag, Frankfurt am Main, 1971.

Ferrando R, and Henry N. Determination microscopique des Aliments du Betail. Vigot Freres, Paris, 1966.

Gassner G. Mikroskopische Untersuchungen Pflanzlicher Lebensmittel. G Fischer Verlag, Stuttgart, 1973.

Huss W. Microscopy and Quality Control in the Manufacture of Animal Feeds. Roche Information Service, 1976.

Juillet A, Susplugas J, and Courp J. Les Oleagineaux et les Tourteaux. Paul Lechevalier, Paris, 1955.

Meszaros L, and Deutschmann F. Atlas fur die Mikroskopie von Nahrungsgrundstoffen und Futtermitteln. Verlag J. Neumann-Neudamm, Berlin, 1975.

Moeller J, and Griebel C. Mikroskopie der Nahrungs und Genussmittel aus dem Pflanzenreich. Springer Verlag, Berlin, 1928.

Morris T N. Microscopic Analysis of Cattle-Foods. Cambridge University Press, London, 1928.

Parkinson S T, and Fielding W L. The Microscopic Examination of Cattle Foods. Headley Brothers, London, 1930

Vaughan J G. The Structure and Utilization of Oil Seeds. Chapman and Hall, London 1970

Winton A L, and Winton K B. The Structure and Composition of Foods. Vols. I, II, III, IV. John Wiley, New York, 1932-39

References

Vaughan J G. The preparation and staining of sections of cruciferous seeds coats. Stain Techn. 35, 1950, 229-231

Vaughan J G, and Stubbs J A. Animal feeds - plant constituents. In Food Microscopy. Academic Press, London and New York, 1979, p. 393-424

Voehringer H. Animal feeds - animal constituents. In Food Microscopy. Academic Press, London and New York, 1979, p. 425-443

Wallis T E. Analytical Microscopy. Edward Arnold, London 1965, p. 66

Discussion with Reviewer

M.H. Buri: How much sample is examined to determine if there is adulteration? Is it in proportion to the amount of that feed received? How much of various adulterants would be considered critical or have any effect on the animal?
Author: A 5.0g sample is normally satisfactory. The sample is obtained by the division technique practised and accepted in the country where the feed is being utilized. The extent to which the various adulterants would be considered critical or toxic depends on the regulations in operation in the country in question.
M.H. Buri: When you say that there is little oil or fat in most animal feeds, how much is considered acceptable before extraction of the fat would be necessary?
Author: The limit would be about 5% fat.
M.H. Buri: If benzene is not available, what other solvent is considered comparable? Could you recommend one that would extract the fat as completely as benzene? The use of benzene is controlled in the U.S.
Author: Light petroleum is often used for this purpose, also acetone or ether.
M.H. Buri: Would you recommend examination of the crude fibre of the feed rather than the feed as is except for sesame and similar substances? How much of the animal feed would have to be digested to obtain a representative sample?
Author: The process of digestion removes starch and protein and therefore microscopic examination of the crude fibre is more effective for the identification of plant tissues than in the case of an untreated feed. However, the original feed must be examined when crystal and starch characters are necessary for identification. The problem of a representative sample has already been discussed.
M.H. Buri: Is there a description somewhere on how to identify tannin cells, phytomelan layer, palisade cells, etc., or are these terms considered specific for animal feeds?
Author: These terms are derived from plant histology but they are explained in the books listed in Appendix I.
M.H. Buri: When any of the absolutely forbidden constituents are found in one sample of a specific size, is the animal feed checked against to verify this presence or is just finding the substance there in any number sufficient to reject a shipment? What level is considered unacceptable?
Author: Although the European Economic Community regulations indicate that some constituents are absolutely forbidden, in practice traces of such materials (up to about 3%) are probably allowed.

THE RELATIONSHIP BETWEEN WHEAT MICROSTRUCTURE AND FLOURMILLING

R. Moss, N.L. Stenvert, K. Kingswood,* and G. Pointing

Bread Research Institute of Australia, North Ryde, N.S.W., Australia
*Lord Rank Research Centre, High Wycombe, United Kingdom

Abstract

Six wheat cultivars and their corresponding mill brans were examined under the SEM. It was apparent that the extent of removal of endosperm (flour) from bran was related to the cleavage pattern of the grains. Good bran clean-up, and hence improved flour yield, was associated with inter-cellular cleavage. When fracture took place through the contents of the endosperm cells more endosperm adhered to the bran. The hardness of the endosperm cells determined the nature of the cleavage pattern. In a hard wheat, the presence of a continuous protein matrix around all the cell contents resulted in the boundary between the cell wall and cell contents becoming a zone of weakness. Hence cleavage was inter-cellular. The contents of the cell could then act as a single entity and the whole cell could be removed from the bran by the shear forces imparted during milling. It appears as if some of these forces were also redirected towards the bran, fragmenting it into relatively small pieces. In a soft wheat air spaces and discontinuities in the protein matrix made the cell fragile and hence the shear forces were not redirected, but passed through the cell rather than removing it cleanly from the bran. Increasing the water content of the wheat prior to milling favoured intra-cellular cleavage. Thus the relationship between bran clean-up and bran fracturing can be optimised by balancing intrinsic hardness and grain moisture content. This is the basis of wheat conditioning.

KEY WORDS: Bran, Conditioning, Flour, Hardness, Milling, Wheat

Introduction

There have been many papers published on the morphology of the wheat grain [1,2] some of which have also been related to the milling behaviour of wheat. Earlier papers used the light microscope exclusively [3-5]. This instrument can reveal much detailed information of the structure and chemical composition of the grain, but it is not suitable for studying hardness-related differences in the structure of the endosperm. This is because the protein, and to a lesser extent the starch, swell in the aqueous media always giving the impression of a very compact endosperm structure. The SEM (scanning electron microscope) does not require the use of aqueous media and hence is an extremely useful tool in studying wheat structure as it affects milling quality.

The aim of the milling process is to efficiently remove the endosperm of the grain from the germ and surrounding bran (Fig. 1) so that the maximum yield of white flour with the minimum contamination of non-endosperm material is achieved. Fluted rolls are used to break open the grain and scrape the endosperm from the bran; smooth rolls are used to reduce the endosperm particles into flour. The moisture content of the cleaned wheat is adjusted to between 14 and 17% prior to being fed to the fluted rolls. This process is called conditioning and the aim is to facilitate separation of bran and endosperm, and to toughen the bran thereby reducing the amount of bran powder in the flour.

Many factors influence the yield of clean white flour obtained from wheat and some lie solely within the realm of the milling engineer. Of the morphological factors to be considered, the shape of the grain and the amount of endosperm within the grain are obviously important. However due to the large area of contact between bran and endosperm the efficiency with which bran and endosperm can be separated (bran clean-up) is also a major factor influencing flour yield.

There have been several theories advanced to explain differences in the separation of bran from endosperm between wheat cultivars in terms of morphology. Larkin et al.[3] and

Bradbury et al.[4] reported that variation in the thickness of adjacent aleurone cells might play a role in trapping endosperm and thus reducing the amount of flour extracted. It has also been suggested by Larkin et al.[5] that the thickness of endosperm cell walls may serve as a criterion of millability. This study examines other factors which appear to have a more significant effect on milling yield and account for some other differences in milling behaviour.

Materials and Methods

Six wheat cultivars grown in New South Wales were selected for examination. Hardness was measured using the particle size index method[6], (P.S.I.) by grinding wheat (10g) on a Labconco mill and then sieving through a 15N sieve (85μm aperture). The percentage of material passing through this sieve is the P.S.I. The lower the P.S.I. value, the harder is the wheat.

Grain density was determined by the displacement of toluene by wheat (5g) in a specific gravity bottle (50ml) at 20°C. Protein content was determined by the Kjeldahl method using a factor of 5.7 to convert N to protein content[7]. All results are expressed on a 12% moisture basis[8]. The particle size distribution within the samples of mill bran was determined by sieving bran (5g) for one minute over 2 screens having apertures of 1.0mm and 0.5mm respectively. To observe the effect of very low moisture content on fracture patterns, wheats were dried to 6% moisture content by storing a thin layer of wheat in a desiccator containing concentrated sulphuric acid for up to five days.

Milling was performed with a Buhler experimental mill and wheats that were conditioned to 14.5% and 17.0% moisture content were allowed to equilibrate for 48 hours prior to milling. Only the three hardest cultivars and Egret were conditioned to 17% moisture content as, in commercial practice, soft wheats are not conditioned at this high level since they then perform badly on the mill.

Scanning electron microscopy was carried out using a Cambridge Stereoscan S600. Accelerating voltage was 15kV, which enabled adequate surface detail to be observed with minimal charging of the specimens. Grains were fractured transversely, care being taken that they were not cut open, as this destroys significant structural detail. Fractured grains and samples of mill bran were attached to aluminium stubs using double sided adhesive tape. To reduce charging to a minimum, samples were surrounded by silver conducting medium prior to coating with a thin layer of gold (300A°).

Results and Discussion

Cleavage pattern of grains at 12% moisture content (as received)

When the fractured grains are examined in order of decreasing hardness (Table 1) the cleavage pattern gradually changes from inter-cellular to intra-cellular as shown in the following photomicrographs.

Table 1. Grain density, hardness (PSI) and protein content of the selected wheats

Cultivar	Grain density (g/cc)	P.S.I. (%)	Protein (%)
Eagle	1.33	12.4	13.7
Cook	1.33	14.2	12.7
Halberd	1.33	17.6	12.5
ND131	1.32	26.8	11.8
Egret	1.30	28.4	11.3
ND36	1.28	31.7	12.4

Figure 2 shows a typical cleavage pattern of the hard wheat Eagle and the manner in which the fracturing changes from the sub-aleurone region to the central endosperm. In the sub-aleurone region (Fig. 3), the cleavage has taken place at the boundary between the cell walls and the cellular contents as can be seen by the fragments of cell wall. This boundary is a zone of weakness in hard wheats as the starch granules in the endosperm are firmly bound together by the continuous protein matrix. In the centre of the grain the endosperm is also vitreous due to the continuous protein matrix which surrounds and masks the starch granules (Fig. 4). However there is an increasing amount of intra-cellular fracturing. This is due in part to the presence of fissures within the central endosperm cells. In addition intra-cellular cleavage is more likely in the central endosperm as the cells are larger in this region[9].

The cleavage pattern of the next two wheats, Cook and Halberd, shows a slightly different pattern to that of Eagle. In the immediate sub-aleurone region (Fig. 5), the fracture has occurred between endosperm cells but intra-cellular fracture has taken place within 100μm of the aleurone layer. Although the central endosperm has a vitreous appearance the more clearly discernible shape of the starch granules indicates that the endosperm cells are not as hard as those of Eagle.

In ND131, Egret (Fig. 6) and ND36 there is very little inter-cellular cleavage in the sub-aleurone and in the central endosperm fracturing has taken place entirely through the cells. This is because the cellular contents are not firmly bound together by the protein matrix. There are also many air spaces in the matrix, as can be seen more clearly at higher magnification (Fig. 7). This contributes to the lower density of soft wheats (Table 1).

Effect of variation of grain moisture content on the cleavage pattern

Increasing the grain moisture content from 12% to 14.5% caused very little difference in the cleavage pattern of the grains but a further increase to 17% produced a marked change.

Even the hardest cultivar, Eagle, showed an appreciable amount of intra-cellular fracture in the sub-aleurone region (Fig. 8). Both Cook and Halberd fractured in a similar manner and showed only a narrow band of inter-cellular cleavage in the immediate sub-aleurone region (Fig. 9). At 17% moisture content Egret displayed the same

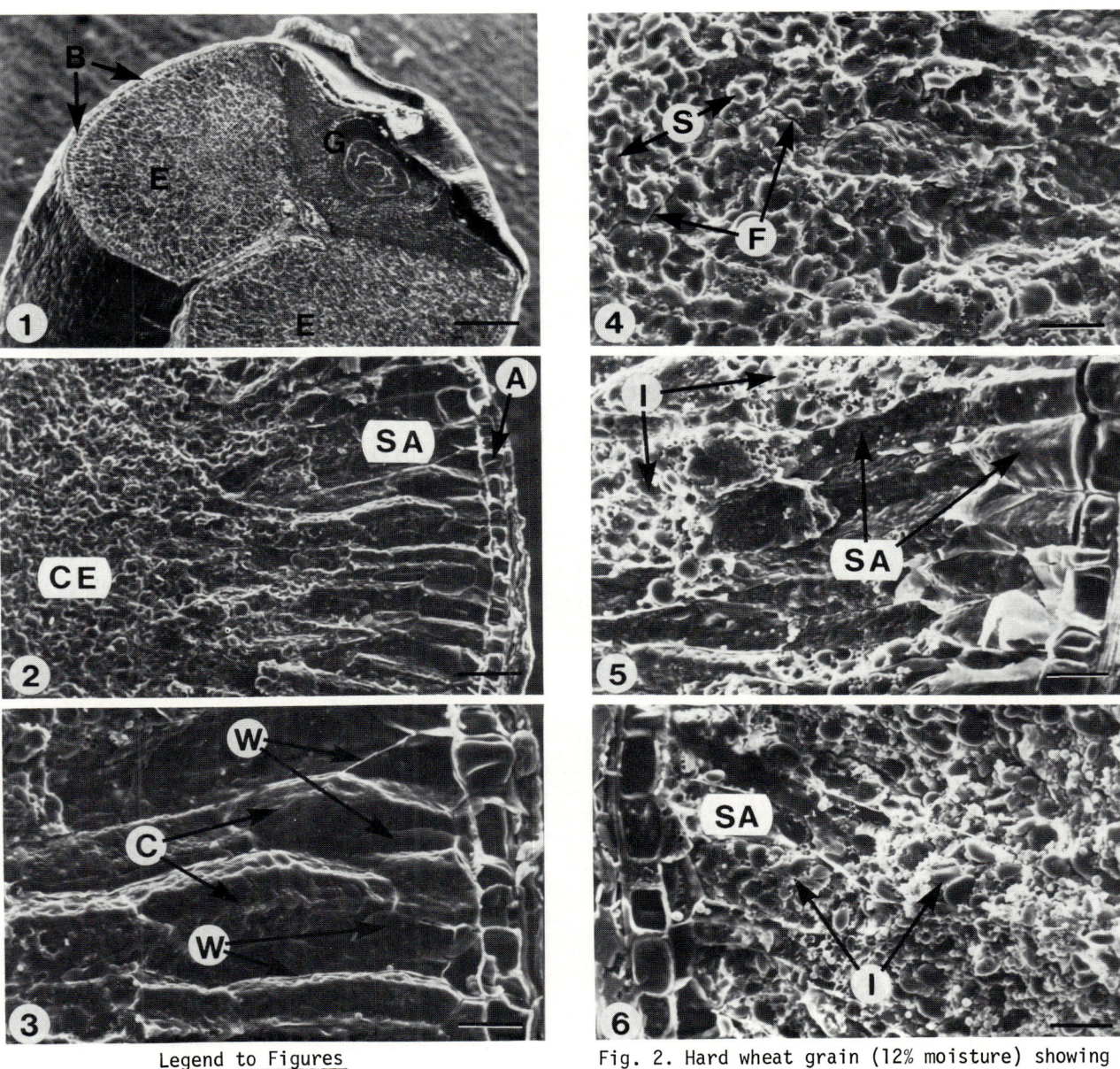

Legend to Figures

A	-	aleurone layer
B	-	bran
C	-	cell contents
CE	-	central endosperm
E	-	endosperm
F	-	fissure
G	-	germ
I	-	intra-cellular fracture
P	-	protein
S	-	starch
SA	-	sub-aleurone
SP	-	air spaces
W	-	cell wall

Fig. 1. Wheat grain fractured to show the germ, endosperm and bran. Bar = 400µm.

Fig. 2. Hard wheat grain (12% moisture) showing the different fracture patterns in the sub-aleurone and central endosperm. Bar = 100µm.

Fig. 3. Sub-aleurone region of cv. Eagle (12% moisture) showing that fracture has taken place at the interface between cell contents and endosperm cell walls. Bar = 40µm.

Fig. 4. Central endosperm of cv. Eagle (12% moisture) showing starch granules enclosed by the compact protein matrix nature of endosperm. Bar = 40µm.

Fig. 5. Sub-aleurone endosperm of cv. Cook (12% moisture) showing inter-cellular fracture in immediate sub-aleurone region and some intra-cellular fracture within 100µm of the aleurone. Bar = 40µm.

Fig. 6. Sub-aleurone region of cv. Egret (12% moisture) showing intra-cellular fracture in the immediate sub-aleurone region. Bar = 40µm.

fracture pattern as the softest wheat, ND36.

Decreasing the grain moisture content from 12% to 6% did not alter the manner in which the grains fractured, thus indicating that unlike other properties, (e.g. water penetration into grain[10]) the continuity and cohesion of the endosperm contents are not affected by reducing the moisture content to very low levels.

Effect of hardness and moisture level on bran size

It can be seen from Table 2 that the softer wheats contain a greater proportion of large (>1.0mm) bran flakes than the harder wheats. As this general relationship has been found to hold true for many hard and soft wheats examined over several years it seems to be associated with hardness rather than with the bran strength of specific cultivars. Furthermore variations in bran thickness do not appear to be an important factor in this regard[3,11].

Table 2. Varietal differences in the size of mill bran and the influence of conditioning moisture level

Cultivar	Moisture %	>1.0	0.5-1.0	<0.5
Eagle	12.0	42	55	3
	14.5	59	37	4
	17.0	72	25	3
Cook	12.0	43	55	2
	14.5	59	37	4
	17.0	76	22	2
Halberd	12.0	52	45	3
	14.5	54	43	3
	17.0	79	19	2
ND131	12.0	53	44	3
	14.5	69	28	3
Egret	12.0	68	31	1
	14.5	77	21	2
	17.0	78	20	2
ND36	12.0	55	43	2
	14.5	80	19	1

% of bran in stated particle size range (mm)

An explanation for the differences in bran particle size can be found in the manner in which hard and soft wheats fracture. In a hard wheat, the shear forces produced by the break rolls are deflected along the boundary between endosperm cells due to the hardness of the cell contents, and are thus directed radially towards the bran surface (Fig. 10). Some of these forces may be directed along the aleurone/endosperm interface but often the forces serve to fragment the bran into relatively small pieces. On the other hand, in soft wheats the shear forces tend to pass through, and be dissipated within, the endosperm cells because of the fragile nature of their contents (see dotted arrow in Fig. 10). Therefore the shear forces are not directed towards the bran.

All cultivars showed a trend towards larger bran fragments at higher conditioning moisture levels (Table 2), because the endosperm became softer and thus intra-cellular fracture was favoured. Additionally it has been reported that higher moisture levels serve to toughen the bran[12].

Effect of hardness and moisture content on bran clean-up

The amount of endosperm which adheres to the bran is also determined by the manner in which grain fracture occurs. When cleavage takes place at the interface between the cellular contents and the cell wall some of the shear forces are directed along the aleurone/endosperm interface as illustrated in Figure 10. This redirection of shear forces does not take place in a soft wheat as the contents of the endosperm cells are fragile and the forces are spent in disrupting the cellular contents. Thus at any one grain moisture level the softer wheats have more adhering endosperm than the harder wheats (compare Figs. 11 & 12).

It has been postulated that irregularities in the thickness of the aleurone cells[3,4] influence bran clean-up. However conditioning moisture level also affects bran clean-up but would not be expected to significantly alter the contour of the aleurone surface adjacent to the endosperm. It has also been suggested that endosperm cell walls can adversely affect bran clean-up[5]. However it can be seen from Figure 12 that the endosperm cell walls of Eagle do not appear to have trapped endosperm and the bran is very clean.

At higher moisture levels more endosperm adheres to the bran. Again an explanation for this behaviour can be found in terms of the change from inter-cellular cleavage to intra-cellular cleavage at higher moisture levels. The hardest wheat, Eagle, shows little difference in bran clean-up between 12% (Fig. 12) and 14.5% moisture, but some increase in adhering material occurred at 17% moisture (Fig. 13). This is because the endosperm of Eagle is particularly hard and compact and can tolerate higher grain moisture levels before bran clean-up is adversely affected. Cook and Halberd show a slight decrease in the bran clean-up between 12% and 14.5% and a more noticeable decrease in bran clean-up at 17% moisture. In contrast, the soft wheats have sufficient adhering endosperm to completely cover the aleurone surface even at 14.5% moisture (Fig. 14).

Effect of grain morphology on density and hardness

From Table 1 it can be seen that both grain density and P.S.I. rank the wheats in the same order, but in this case grain density does not discriminate between the three hardest varieties. The SEM reveals that variation in grain density is mainly due to differences in endosperm structure. The discontinuity of the protein matrix and air spaces within the endosperm are responsible for the low density of wheats like Egret and ND36. This is also the reason why such wheats are regarded as being soft - the discontinuities in the protein matrix allowing the endosperm to be easily shattered.

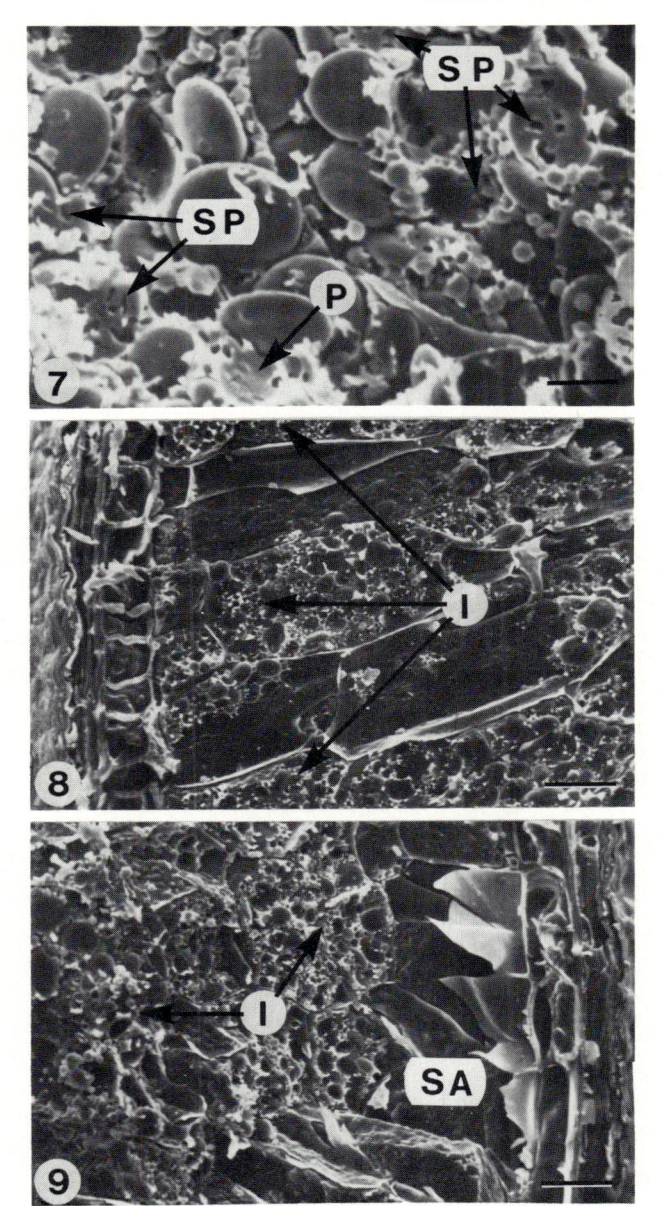

Fig. 7. Central endosperm of cv. Egret (12% moisture) showing discontinuity of the protein matrix and spaces within the endosperm. Bar = 10μm.

Fig. 8. Sub-aleurone region of cv. Eagle (17% moisture) demonstrating the occurrence of intra-cellular fracture due to the higher moisture content. Bar = 40μm.

Fig. 9. Sub-aleurone region of cv. Cook (17% moisture) where only a narrow (<100μm) region of inter-cellular fracturing has occurred. Bar = 40μm.

Fig. 10. Diagrammatic representation of the forces acting on one endosperm cell of a wheat grain. Large arrow at the base of the figure indicates the direction of the shear force due to the differential speed of the break rolls.

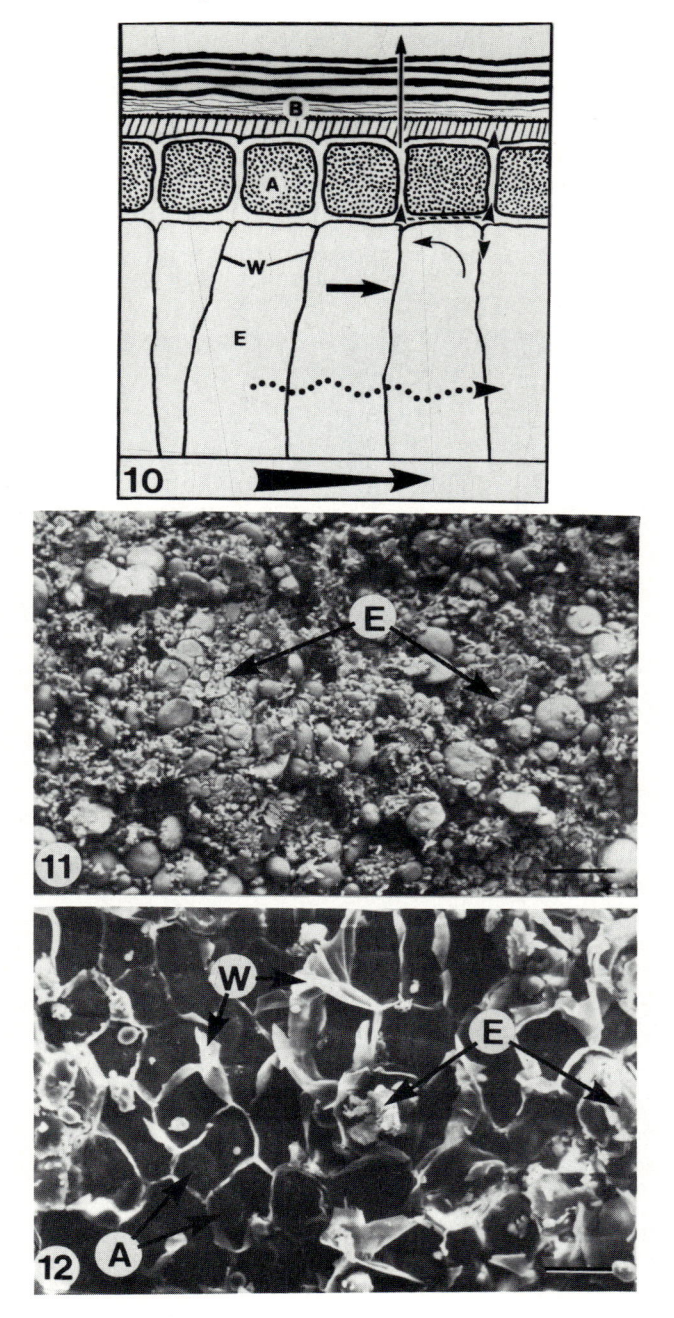

Fig. 10 (Cont'd). In a hard wheat the contents of the cell behave as one unit and hence the cell is clearly torn from the overlying aleurone layer (indicated by the dashed line). In a soft wheat the shear force passes through the contents of the cells as indicated by the dotted arrow.

Fig. 11. Endosperm face of bran from cv. ND36 milled at 12% moisture showing the large amount of adhering endosperm. Bar = 40μm.

Fig. 12. Endosperm face of bran from cv. Eagle milled at 12% moisture with only a few isolated fragments of adhering endosperm. Bar = 40μm.

It can also be seen from Table 1 that, for these samples, protein content per se does not have a marked effect on either grain hardness or density, but it has been shown that it can play a minor role[13].

Grain density and hardness are therefore interrelated and offer an explanation for differences in the efficiency of separation of bran and endosperm, and in the size of bran particles between different wheat cultivars. However grain density should not be considered as an absolute measure of grain hardness.

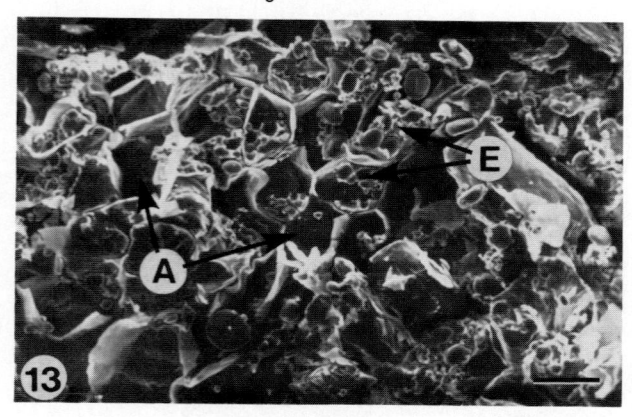

Fig. 13. Endosperm face of bran from cv. Eagle milled at 17% moisture illustrating the increased amount of adhering endosperm. Note separation of endosperm from bran still relatively efficient. Bar = 40μm.

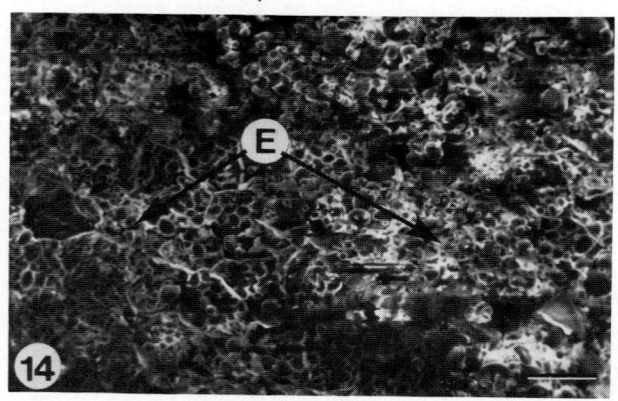

Fig. 14. Endosperm face of bran from cv. Egret milled at 14.5% moisture. Note the large amount of adhering endosperm, indicating the lack of tolerance to increased moisture content prior to milling. Bar = 40μm.

Conclusion

Two factors which influence milling yield and flour quality are, respectively, the efficiency with which the floury endosperm is removed from the overlying bran and the degree to which the bran is powdered or otherwise damaged. The present work demonstrates that both these factors are, in turn, influenced by the manner in which grain fractures during milling. If fracturing occurs at the boundary between the endosperm cell wall and cell contents, the endosperm is efficiently removed from the bran, but the bran is fractured into smaller pieces. When the grain is fractured by breaking open the contents of the endosperm cells, bran clean-up is poor but large pieces of bran are produced. The manner in which the grain is fractured is determined by both the hardness of the wheat grain and the moisture content of the grain. It is thus possible to optimise the relationship between grain hardness, bran clean-up and bran powdering by adjustment of moisture level and this is the basis of wheat conditioning.

Acknowledgements

The authors wish to thank Miss S. Berry for assistance with the SEM and the Fuel Geoscience Unit of the Institute of Earth Resources, CSIRO, for the use of the SEM.

References

1. Simmonds, D.H., Barlow, K.K., and Wrigley, C.W., 1973. The biochemical basis of grain hardness in wheat. Cereal Chem., 50, 553-562.
2. Hoseney, R.C., and Seib, P.A., 1974. Structural differences in hard and soft wheat. Bakers Digest, 47 (6), 26-28, 56.
3. Larkin, R.A., MacMasters, M.M., Wolf, M.J., and Rist, C.E., 1951. Studies on the relation of bran thickness to millability of some Pacific Northwest wheats. Cereal Chem., 28, 247-258.
4. Bradbury, D., MacMasters, M.M., and Cull, I.M., 1956. Structure of the mature wheat kernel. II. Microscopic structure of pericarp, seed coat, and other coverings of the endosperm and germ of hard red winter wheat. Cereal Chem., 33, 342-359.
5. Larkin, R.A., MacMasters, M.M., and Rist, C.E., 1952. Relation of endosperm cell wall thickness to the milling quality of seven Pacific Northwest wheats. Cereal Chem., 29, 407-413.
6. Symes, K.J., 1961. Classification of Australian wheat varieties based on the granularity of their wholemeal. Aust. J. Exp. Agric. Anim. Husbandry, 1, 18-23.
7. A.A.C.C. Cereal laboratory methods. Method No. 46-12. 7th Ed. American Association of Cereal Chemists, St. Paul, MN, 1976.
8. A.A.C.C. Approved methods of the American Association of Cereal Chemists. Method No. 44-15A. American Association of Cereal Chemists, St. Paul, MN, 1975.
9. Evers, A.D., 1970. Development of the endosperm of wheat. Ann. Bot., 34 (136), 547-555.
10. Moss, R., 1977. The influence of endosperm structure, protein content and grain moisture on the rate of water penetration into wheat during conditioning. J. Fd. Technol. 12, 275-283.
11. Crewe, J., and Jones, C.R., 1951. The thickness of wheat bran. Cereal Chem., 28, 40-49.
12. Lockwood, J.F. Flour Milling. Henry Simon Ltd., London, 1951, 168.
13. Stenvert, N.L., and Kingswood, K., 1977. The influence of the physical structure of the protein matrix on wheat hardness. J. Sci. Fd. Agric. 28, 11-19.

Discussion with Reviewers

D.R. Lineback: Do the air spaces and discontinuities in the protein matrix of soft wheat endosperm make the cell fragile or is it due to the low strength of the protein-starch interaction proposed in the papers by Simmonds? Simmonds indicates that the degree of adhesion between starch and protein is a more likely explanation for hardness and that in soft wheat this adhesion is lower than in hard wheat, resulting in ready fragmentation through the cells to release cell contents.

Authors: In one of the papers by Simmonds et al. (text reference 1) a relationship between P.S.I. and soluble material was shown whereby P.S.I. decreased (i.e. grain became harder) as the amount of extracted material increased. However there was a negligible difference in extracted material between Timgalen (a hard wheat) and soft Falcon (a very soft wheat) although there was a large difference in P.S.I. (11 units). An increase in extracted material only occurred at extremes of hardness.

Information is quoted from a previous paper by Barlow et al. to justify the adhesion theory of grain hardness. Barlow et al.[14] used a fluorescent antibody staining technique to demonstrate that, in one variety (Timgalen), there was a localisation of water soluble proteins around the starch granules. The authors observed "The concentration of fluorescence appears to be higher in the latter area (midendosperm), possibly because of the higher concentration of starch granules". They do not demonstrate that this material acted as an adhesive but speculated that it did since their testing demonstrated that there was no difference in the hardness of starch and isolated fragments of storage protein from hard and soft wheats. It is curious that the concentration of the proposed adhesive material was greater in the central endosperm as in the authors experience over many years this area tends to be softer than the sub-aleurone and is never harder. The authors also established that the chemical composition of the protein components which constitute the endosperm protein matrix did not differ. The combination of these observations led rise to the conclusion that "the adhesion between starch and storage protein is more important in determining grain hardness than is the composition of the protein matrix". These workers however did not take into account the continuity of the protein matrix as a factor in wheat hardness.

It is not necessary to postulate an adhesion theory to explain wheat hardness. When the protein matrix is discontinuous and air spaces are present in the endosperm the grain will be soft whether or not a special adhesive material is present around the starch. In addition not all starch granules are fractured when a hard wheat is split open, but many depressions, left by starch granules, can be seen in the protein matrix which would suggest that there is not necessarily an effective adhesive material surrounding the starch.

D.R. Lineback: Would the authors define "hardness" as used in this manuscript? Is it defined by the P.S.I. or does it have some other attributes which would be of importance for the reader to understand?

Authors: Hardness, as applied to wheat, means that the contents of the grain (i.e. the endosperm) are solid and compact and are not easily broken into fine particles during milling. The P.S.I. gives a reliable measure of hardness that is not erroneously affected by such factors as grain shape, and relates well to milling performance as it involves both grinding and sieving properties.

D.R. Lineback: Can grain hardness be altered or decreased by the adjustment of moisture level, i.e. can a hard wheat be altered to a soft wheat by the appropriate adjustment of moisture?

Authors: Moisture content does influence grain hardness but the extent to which hardness can be modified in this way is limited. Increasing the moisture content of wheat softens the endosperm but a hard wheat can only be made of intermediate hardness in this way.

D.R. Lineback: How does the larger size of cells in the central endosperm of a hard wheat relate to the ease of fracture. The papers by Simmonds and Hoseney indicate that there is very strong adhesion between the protein and starch in hard wheat resulting in fracture along the thin cell walls or across and through the cells, in which case starch granules are actually fractured, indicating the strength of the protein-starch interaction.

Authors: The larger cell size in the central endosperm could also tend to favour intracellular cleavage because there is consequently a larger area of inter-cellular contact. This would mean that a central endosperm cell is more firmly held in place by surrounding cells and it is therefore more difficult to achieve completely inter-cellular cleavage even though the endosperm is still hard.

D.R. Lineback: Can all endosperm be considered to be the same or is there some difference in properties between the central endosperm and that in the subaleurone area? The authors have mentioned larger cell size in the central endosperm area and Simmonds indicates that the concentration of starch granules associated with water-soluble protein is higher in the central endosperm. Is bran clean-up related to the properties of all endosperm in the grain or most closely to that in the subaleurone area?

Authors: All endosperm is not the same due to variations in starch and protein deposition during development. The protein content of the sub-aleurone region is higher than the central endosperm regions. This would favour the formation of a continuous protein matrix in the sub-aleurone endosperm resulting in increased hardness. Bran clean-up is most closely related to the properties of the endosperm with which it is immediately in contact i.e. the sub-aleurone region.

W.J. Wolf: What is the variability in the density for a single cultivar? Are all of the differences in density in Table 1 significantly different from a statistical standpoint? Hardness measured by the Particle Size Index is an indirect measurement of this physical property. Has anyone attempted to measure hardness more directly by using an instrument such as the Instron Universal Testing Machine?

Authors: Although the samples used in this study were uniform in grain density variability can occur within a single cultivar due to either agronomic or seasonal factors, as well as by position of the grain in the head. Statistically the grain densities of ND36 and Egret are different from those of the three hardest wheat varieties. Attempts have been made to use machines like the Instron but difficulty has been found due to the crease region of the grain and the hardness gradient which frequently exists within a grain. Barlow et al. have used a Leitz Miniload hardness tester to measure the hardness of starch and protein separately but this was done on material dispersed in a polyester-type resin.

P.S. Pescheek: Could you be more specific as to how the specimens were fractured prior to mounting. Since fracture paths might be expected to depend on the rate of specimen loading, I wonder how the shear rate applied during fracture for specimen preparation compares with that imposed during milling.

Authors: The grains were fractured by applying a downward pressure with a blunt, safety razor blade across the mid-dorsal region of the grain. A shear force was thus not applied during sample preparation, but the important point is that the grain was fractured so that the inherent weaknesses in the endosperm were exposed for examination. Care must be taken to avoid cutting the grain as this produces a smooth surface which masks structural detail.

The rate of shear in milling (as governed by the ratio of peripheral speeds of the mill rolls) does alter such flour characteristics as granularity and starch damage but the hardness of the grain is still a major factor governing the mill performance of wheat. The information gained by fracturing wheat in the manner described above does reveal the intrinsic structure responsible for hardness and relates well to milling behaviour.

References (Cont'd)

14. Barlow, K.K., Buttrose, M.S., Simmonds, D.H., and Vesk, M., 1973. The nature of the starch-protein interface in wheat endosperm. Cereal Chem., 50, 443-454.

SCANNING ELECTRON MICROSCOPY OF FLOUR-WATER DOUGHS TREATED WITH OXIDIZING AND REDUCING AGENTS

L.G. Evans,[1,2] A.M. Pearson,[1] and G.R. Hooper[3,4]

[1]Department of Food Science and Human Nutrition
[3]Center for Electron Optics
Michigan State University, East Lansing, MI 48824

Present address: [2]Pillsbury Company, Research and Development,
311 second Street S.E., Minneapolis, MN 55414

[4]Virginia Polytechnic Institute and State University,
Blacksburg, VA 24061

Abstract

Scanning Electron Microscopy (SEM) revealed that the gluten of undermixed flour-water doughs without oxidizing and reducing agents was not uniformly developed, with starch granules clustered together on the surface. Continued mixing distributed thin films of gluten throughout the mass. At optimum development, a strong association of starch and protein was apparent and when the dough was stretched, thin gluten fibrils formed and were oriented in the direction of stress. The ability to form gluten fibrils increased during proofing. It decreased upon overmixing, and the gluten film disintegrated into a spindly weblike structure that failed to form fibrils on stretching. Unstretched doughs containing oxidizing and reducing agents did not differ from the control. After proofing, the gluten of doughs containing potassium-bromate appeared stronger than controls as indicated by the formation of thicker fibers upon stretching. Optimally mixed potassium-iodate containing doughs formed thin fibrils upon stretching. Azodicarbamide (ADA) restricted thin fibril formation so that the gluten formed wide flat strips upon stretching. Even after overmixing of the ADA-treated dough, the starch-protein bond was maintained and the gluten resisted breakdown. The disulfide reducing agents, cysteine and glutathione, had similar effects on gluten but differed in the extent of reactivity. Cysteine severely damaged gluten strength and prevented fibril formation at all stages, whereas, glutathione destroyed gluten integrity only after overmixing. The reducing agents allowed extensibility to increase at the expense of elasticity. Upon loss of elasticity, fibrils could no longer form. Gluten became discontinuous and stretched irregularly.

KEY WORDS: Flour-Water Doughs, Wheat Dough, Dough Conditioners, Oxidizing Agents, Reducing Agents, Oxidants, Reductants, Dough Structure, Mixing Conditions, Dough Development.

Introduction

Oxidizing and reducing agents are widely used in commercial breadmaking. Since they develop a balance between elasticity and extensibility in the dough, long periods of bulk fermentation are not necessary[1,2]. Oxidizing agents strengthen the dough during mixing, proofing and initial baking stages. They act by inhibiting disulfide exchange reactions, and perhaps by forming intermolecular disulfide bonds which improve baking qualities without long aging periods.[3] Reducing agents, on the other hand, immediately accelerate dough development during mixing[2], and decrease development time[4], making the dough more extensible and easier to process.

Although a great deal is known about the microstructure of dough [5,6,7,8,9], scanning electron microscopy (SEM) and transmission electron microscopy (TEM) have offered new approaches to understanding the changes occurring in the ultrastructure during dough mixing and development[10,11,12,13,14,15,16]. Little is known however, about the effects of oxidizing and reducing agents on the ultrastructure of dough. Therefore, the present investigation was undertaken to determine their effects on doughs during the following stages of development: (1) undermixing, (2) optimal development, (3) after proofing, and (4) overmixing.

Materials and Methods

Flour-Water Dough

In order to simplify the system and facilitate interpretation, a flour-water dough was used throughout the study rather than a complete bread dough formulation. Untreated white flour (Pillsbury Company, St. Louis, MO) containing 11.3% protein, 14.0% moisture and 0.46% ash was used for making the dough. The oxidizing agents and concentrations used were: potassium bromate (75 ppm), potassium iodate (75 ppm) and azodicarbamide (ADA) (45 ppm). The reducing agents and concentrations used were: cysteine (75 ppm) and glutathione (75 ppm). Levels were selected at the legal limit for baked goods under regulations set forth by the Food and Drug Administration, 1976. The farinograph was used to determine the water absorption of each flour,

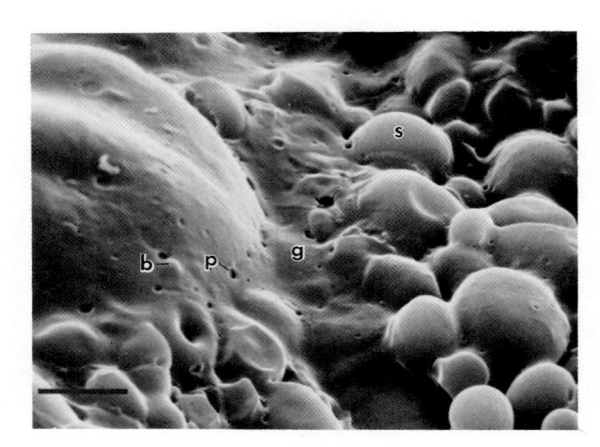

Fig. 1 Optimally mixed flour-water dough. b = bumps; g = gluten film; p = pores; and s = starch granules. Bar is 5μm.

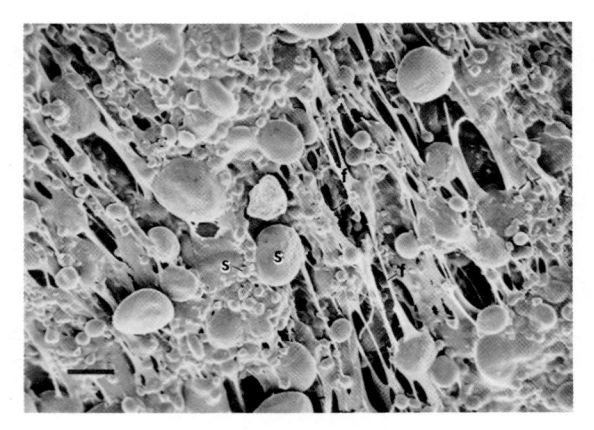

Fig. 2 Incubated optimally mixed flour-water dough after stretching. f = gluten fibrils formed on stretching; s = starch granules. Bar is 20μm.

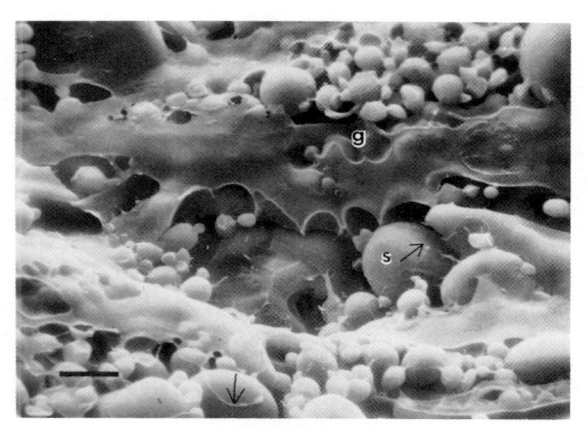

Fig. 3 Undermixed flour-water control dough showing lack of uniform development. g = gluten sheet incomplete; arrows point to film-like residues capping the starch granules (s). Bar is 10μm.

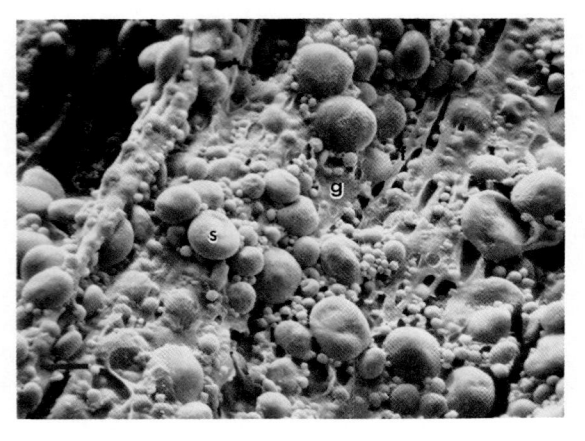

Fig. 4 Stretched undermixed flour-water dough. s = starch granules; g = gluten; note lack of association between starch granules and gluten; f = fibrils formed on stretching the gluten. Bar is 20μm.

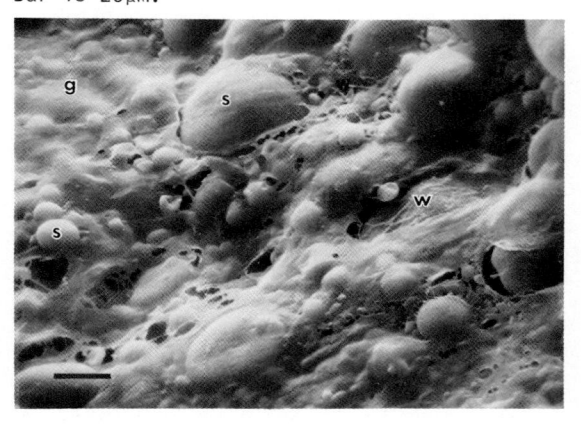

Fig. 5 Overmixed flour-water control dough. g = gluten; s = starch granules; w = spindly web of gluten; note that the gluten has pulled away from the starch granules and collapsed into a spindly web. Bar is 10μm.

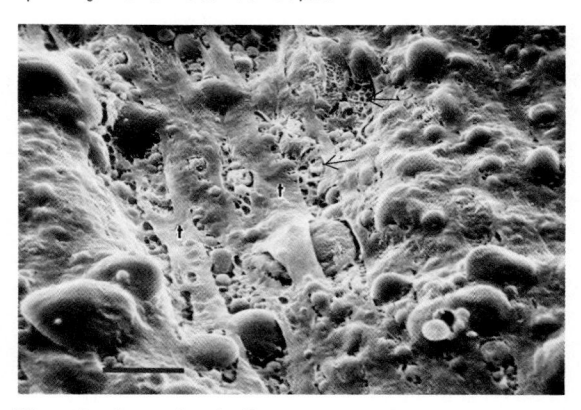

Fig. 6 Overmixed flour-water dough control dough after stretching. t = strips of thick gluten; arrows point to disintegrated protein made up of fibrillar webs. Bar is 20μm.

Table 1, and to mix each sample. The procedure for a constant dough weight was followed according to the standard farinograph methods as outlined by the American Association of Cereal Chemists, 1962. Water absorption was adjusted so that each dough was mixed to the same consistency, 500 Brabender Units. Optimum mixing was considered to be the point at which the dough offered maximum resistance against the mixing blades of the farinograph.

Each oxidizing and reducing agent was mixed with the dry flour for nine minutes in the farinograph and then hyrated with the required amount of water, Table 1. The doughs were then mixed to one of four stages of development: (1) undermixed (2.5 minutes), the control only; (2) optimally mixed, time being dependent on the additive as shown in Table 1; (3) optimally mixed and incubated (analogous to "proofing" in yeasted doughs) at 40°C for 90 minutes; and (4) overmixed (11.0 minutes).

Scanning Electron Microscopy

Immediately after mixing, two small samples (approximately 2 mm in diameter and 10-15 mm in length) were removed at each stage of development. One sample was maintained in its original shape so as not to disturb the surface structure, whereas, the other was stretched lengthwise to a point just short of breaking. Both samples were frozen rapidly by placing in contact with aluminum foil coated with dry ice. They were then transferred to previously chilled (-70°C) sample vials for storage in ethanol-dry ice slush until lyophilized in a Virtis freeze dryer for at least 12 hours.

The freeze-dried dough was divided into small segments and mounted on aluminum stubs with conductive adhesive for easy examination of the dough surface. The specimens were coated with gold using a sputter coating apparatus and examined with an ISI-Super-Mini scanning electron microscope at 10 kV. Duplicate samples from each treatment were prepared and examined, then representative sections were photographed.

Results and Discussion

Characteristics of the Dough

The water absorptions, peak mixing times and stabilities of the doughs as determined by the farinograph are presented in Table 1. The water absorption for the control dough (without oxidizing or reducing agents) was 61.6%, the peak time was 7.0 min and stability was 10 min. Addition of potassium-bromate did not alter any of these properties from those of the control. The other oxidizing and reducing agents increased water absorption of the flour, reduced mixing times and reduced the stability of the dough. Apparently, the additives enhanced disaggregation of the hydrated proteins, allowing them to absorb more water during mixing. Reducing agents are reported to expedite the disaggregation process by the scission of disulfide bonds. On the other hand, the action of oxidizing agents is more complex involving sulfhydryl-disulfide interchange. Once the SH groups are oxidized, the interchange mechanism in the dough becomes slower. Since there is less disulfide interchange to release the stress, the dough becomes more resistant to deformation. With the continued shear and tearing of mixing, more protein bonds are forcibly cleaved so as to disaggregate the protein particles[1].

Potassium-bromate had no effect on the farinograph in this study since the dough temperature was about 30°C. This is in agreement with previous studies which showed potassium-bromate lacks activity below 40-50°C.[17] Thus, bromate normally affects the dough during the later stages of proofing and in early baking. Since bromate and cysteine differ in reaction time, they are commonly combined as the oxidizing and reducing agents, respectively, in chemical dough development systems[1]. The action of bromate is delayed so that it does not interfere with the rapid action of cysteine[1,23], and thus, there is sequential development and maturation of the dough.

Of the oxidants tested, ADA exerted the greatest effect upon water absorption and on the mixing properties of the doughs, even though it was tested at the lowest level (45 ppm vs 75 ppm). Cysteine was a more reactive reducing agent than glutathione according to farinograph time and stability measurements, Table 1, and had a greater influence on dough ultrastructure. In support of this effect, cysteine has been demonstrated to be more effective than glutathione for extraction of gluten proteins from water-flour suspensions[1].

SEM Observations

Control Doughs. SEM examination of optimally developed flour-water doughs without any additives revealed a pebbly mass of starch granules bound by a continuous gluten matrix, Fig. 1. In some areas, the gluten formed a thin film that conformed with the shape of the starch granules, whereas, in other areas it was thicker and had a sheet-like character. The dough was characterized by a continuous and strongly adhering gluten layer covering the starch granules. The stretched samples at optimal mixing, Fig. 2, formed a multitude of long thin gluten fibrils oriented in the direction of stretching and extending over and between the large starch granules.

The gluten surface was pitted with small round pores, Fig. 1, ranging from less than 0.2μm to about 1.5μm in diameter. They were comparable in size to the lipid-rich inclusions reported in the protein matrix of wheat endosperm and dough by other workers[12,13,18] using TEM. Lipid-rich material was shown to originate from remnants of cytoplasmic organelles surviving grain maturation[12]. It seems probable that the pores are artifacts of preparation due to volatilization in SEM column and/or migration of lipid inclusions to dough matrix after freeze drying. In addition to the pores, some small bumps were also observed in the gluten film, Fig. 1.

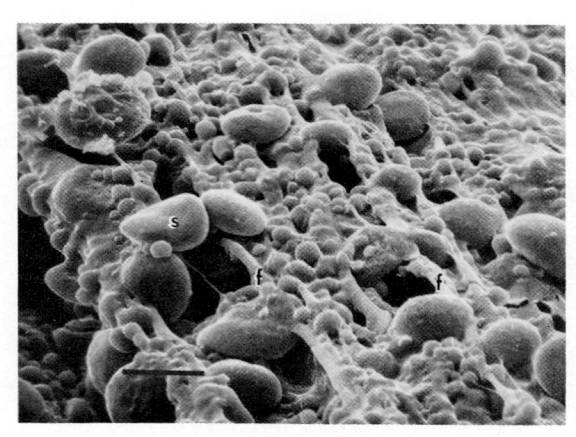

Fig. 7 Incubated optimally mixed flour-water dough containing potassium-bromate after stretching. f = thick fibrils of gluten; s = starch granules. Bar is 20μm.

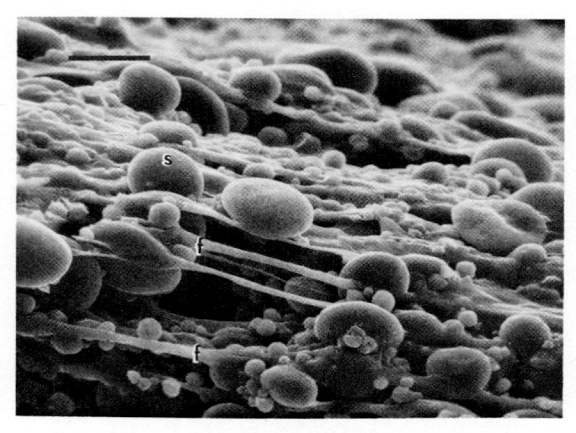

Fig. 8 Optimally mixed flour-water dough containing potassium-bromate upon stretching (unincubated). f = long thin fibrils of gluten; s = starch granules. Bar is 20μm.

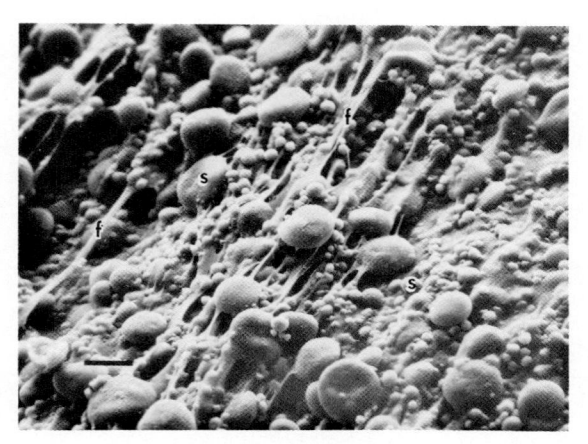

Fig. 9 Optimally mixed flour-water dough containing potassium-iodate upon stretching. f = long, thread-like gluten fibrils formed on stretching; s = starch granules. Bar is 20μm.

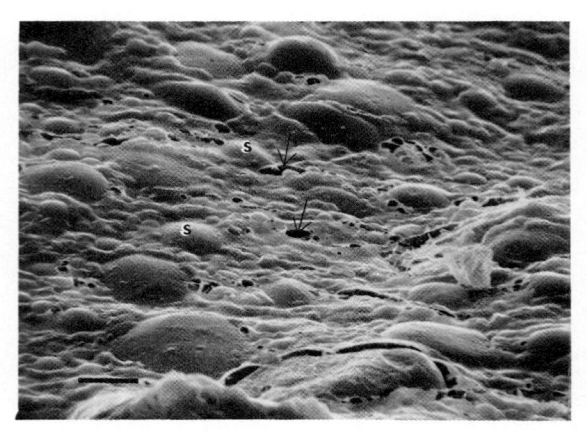

Fig. 10 Overmixed flour-water dough containing iodate. Note the obscured shape of the starch granules (s); arrows point to breaks in gluten sheet, which is very thin. Bar is 10μm.

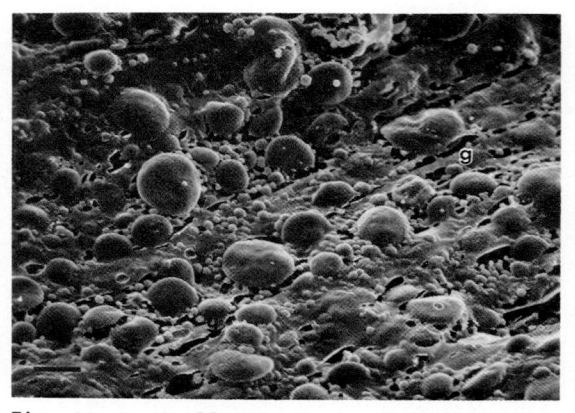

Fig. 11 Optimally mixed flour-water dough containing ADA after stretching. g = flat strips of gluten oriented in direction of stretching. Bar is 20μm.

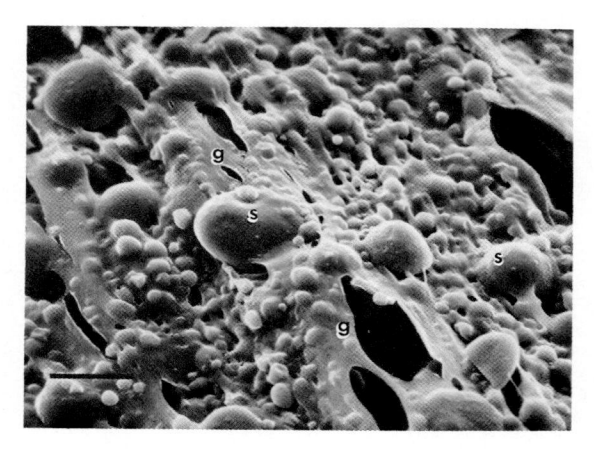

Fig. 12 Overmixed flour-water dough containing ADA after stretching; note elongated holes in gluten surrounded by thick gluten (g) strips; s = starch granules; Note how the gluten (g) drapes itself around the starch granules (s). Bar is 20μm.

Although the origin of the bumps is not certain, they may represent another form of protein, such as the "adhering protein"[19]. The bumps were rarely noticed in the gluten covering the small starch granules and were not observed in highly stretched fibrils.

In undermixed doughs (2.5 min), the gluten was not uniformly developed. For the most part, the starch granules tended to be clustered together, crowded and overlapped as shown in Fig. 3. In completely underdeveloped areas, the protein and starch were not closely associated and thin film-like residues capped the starch granules. With gluten development, the long axis of the starch granules tended to become oriented with the dough-surface. Stretched samples of the undermixed dough demonstrated that the starch-protein association was weak, as illustrated by Fig. 4. It can be seen that the undermixed dough did not form long thin fibrils like the control, Fig. 2. Results suggest that the starch granules in the optimally mixed control act as anchor points and facilitate formation of gluten fibrils upon stretching; whereas, in the undermixed dough, the starch-protein association is not developed so that the starch granules are obstacles to gluten continuity.

In overmixed control doughs, the gluten sheet became unusually thin and lost the smooth appearance of the control, as can be seen by comparing Figs. 5 and 1. Overmixing resulted in areas where the gluten had pulled away from the starch granules and collapsed into a spindly web, Fig. 5. Breakdown of the gluten sheet and loss of its integrity were characteristic of overmixing, consistent with earlier reports[8]. Stretched overmixed flour-water dough formed thick gluten strips connected by fibrillar webs of disintegrating protein, and fibril formation was absent, Fig. 6.

Oxidizing Agents.

Potassium-bromate - As already pointed out, bromate had no measurable effect on water absorption, peak mixing time or stability of the dough, Table 1. Observations by SEM further verified that optimally and overmixed dough containing bromate behaved like the corresponding controls. However, stretched samples of the incubated dough containing bromate formed thick fibrils, Fig. 7, in comparison to the optimally mixed fresh (unincubated) sample, Fig. 8. Thus, the bromate-treated, incubated dough formed wide strips or thick fibers upon stretching, but resisted extension into thin thread-like fibrils, Fig. 7. Changes in the bromated dough during incubation are in agreement with rheological data [17,20], which demonstrates that doughs containing bromate behaved normally during relaxation tests at 25 to 30°C but became less relaxed above 35°C. The dough consistency at 40°C can be explained by the increased activity of bromate, which restricts the reactivity of SH-groups of the gluten[17].

Potassium-iodate - Optimally mixed unstretched dough treated with iodate did not appear to be different from the control. Although the iodate-treated dough formed some fibrils upon stretching, Fig. 9, fibril formation was less abundant than for the stretched flour-water control, Fig. 2. The fibrils which formed while stretching the iodate-treated dough, however, were long and thread-like, corresponding to a more extensible dough. The unexpected increase in extensibility may have resulted from the high level of iodate used where the reverse in the expected strengthening effect has been observed[21]. The increased extensibility results from a sudden decrease in SH-groups, thus restricting the thiol-disulfide interchange during dough mixing. The stress of mixing may then lead to breakage of cross-links[21]. After incubation of the optimally mixed iodate-treated sample, however, it did not show any apparent differences from the control.

Overmixing the dough containing iodate caused breakdown of the gluten, as illustrated in Fig. 10. The gluten appeared thin and failed to maintain its continuity on the surface. Many starch granules were coated with a weblike film, apparently from collapse and shrivelling of the gluten. The contracted protein left large gaps around the starch granules and irregular holes in the gluten sheet similar to that observed in the overmixed control, Fig. 5.

ADA - No effects from addition of ADA were observed in unstretched samples of either optimally mixed or proofed doughs. Upon stretching of optimally mixed and proofed samples, however, the character of the gluten was changed, resulting in thick strips oriented in the direction of stretching, Fig. 11. The thick strips contrast to the rounded fibril formation observed in the stretched optimally mixed control, Fig. 2.

Overmixing of dough containing ADA did not result in the gluten deterioration that was characteristically observed in overmixed control samples. This is illustrated in Fig. 12, which shows the stretched, overmixed dough containing ADA. The stretched dough shows elongated holes that seem to initiate the division of the gluten into flat strips. The gluten is draped snugly over the starch granules revealing their prominent contours. This is in contrast to the stretched, overmixed, control or iodate-containing doughs, Fig. 5 and 10, in which gluten breakdown is evident. The integrity of the ADA sample shown by SEM does not reflect the farinograph data, Table 1, which indicate a relatively short stability compared to control. It is possible that accumulation of biurea (the reaction product of ADA) during extended mixing reverses the stiffening effect of the oxident upon the dough. The interaction of biurea may have relieved the intermolecular stresses in the gluten and allowed the gluten to extend by thinning of the sheets rather than by forced cleavage of disulfide bonds as discussed earlier for iodate treated sample.

Reducing Agents.

Cysteine - Immediately after mixing,

cysteine-treated dough was very extensible, and it became even more so following incubation or overmixing. Although optimally mixed dough containing cysteine was similar in appearance to the optimally mixed control dough, it stretched irregularly, was inelastic and was characterized by damaged areas as shown in Fig. 13. The gluten sheet was completely disrupted around many of the starch granules, although it still covered them with smooth protein caps. Generally, the gluten sheet did not form fibrils on stretching, but instead disintegrated. Incubation of the cysteine-treated dough gave results similar to those for that of the optimally mixed sample, with battered irregular holes being widely distributed throughout the gluten. Furthermore, the fragile sheet did not form fibrils on stretching, Fig. 14.

On overmixing, the cysteine-treated dough revealed scattered patches of breakdown, with some areas being relatively intact and others being almost completely disrupted, Fig. 15. Although the gluten adhered to the surface of the starch granules and did not shrivel and disintegrate as it did for the control dough, Fig. 5, the dough was characterized by breaks between the granules. The increased extensibility and fragility of the cysteine-reduced dough may be caused by a reduction of the glutenin fraction through the scission of disulfide bonds to a molecular size corresponding to that of gliadin[1,22].

Glutathione - SEM showed that optimally mixed dough containing glutathione resulted in a thin gluten film, covering and conforming to the shape of the starch granules, Fig. 16. Upon stretching of the freshly mixed dough some fibrils were formed, Fig. 17, but fewer than for the stretched optimally mixed control, Fig. 2. After incubation of the optimally mixed glutathione treated dough, stretching resulted in more fibril formation, Fig. 18, than was the case for freshly mixed samples, Fig. 17. The starch-protein association remained strong and cohesive.

Overmixed glutathione-treated dough did not exhibit breakdown, until stretched. The unstretched gluten sheet appeared flat and partially obscured the starch granules. On stretching of the overmixed glutathione-treated dough, severely disrupted areas were observed, with large irregularly shaped holes destroying the continuity of the gluten sheet, Fig. 19. Although results suggest that glutathione contributed to the decrease in gluten strength in this sample, the starch-protein association[24-26] was still partially maintained.

In contrast to cysteine, glutathione was not detrimental to gluten continuity until the dough was overmixed. The lesser reactivity of glutathione is expected due to its lower SH-content per gram. In addition, the effects of glutathione were more evenly distributed throughout the dough than was true for cysteine, with the latter causing more breakdown of the gluten in localized areas. It has been demonstrated that glutathione causes an increase in the gliadin-like fraction of gluten, with the increase being less than that caused by cysteine[1,24]. The discontinuity of overmixed glutathione-treated dough may be associated with such an increase in the gliadin-like properties of gluten.

Control of gluten's unique film-forming properties is a continual challenge of modern baking science. Oxidizing and reducing agents have been widely employed to modify its characteristics and yield consistent high-quality performance from variable wheat sources. The ultimate goal is to develop a gluten structure which is cohesive, resilient and has maximized gas holding ability. The observations made in this study provide new insights to the physical changes in dough structure and function as induced by mechanical action and chemical dough improvers. The general strengthening effects of oxidizing agents and the weakening effects of reducing agents support and extend results of previous physical and chemical studies on dough behavior. The primary basis of these effects is best described through the thiol-disulfide interchange theory[23]. An additional hypothesis bears significantly on the stretching behavior of the experimental doughs in this study[22]. In this theory, gluten is viewed as a random linear polymer of polypeptide chains. It was reasoned that if highly asymmetrical (linear) molecules were dominant in gluten, the orientation caused by stretching would improve intermolecular adhesion and tensile strength in the direction of stress. In contrast, if the molecule tended to be symmetrical, the intermolecular forces per unit volume would not increase appreciably on stretching and the gluten would lack viscoelasticity. In the linear model, elasticity arises from the tendency of extended or unfolded polypeptide chains to return to their contracted conformations of lowest free energy[22]. It follows that if gluten's linear polymers are diminished to a size corresponding to that of gliadin by a reducing agent, then there would be an apparent loss in elasticity as well as a reduction in tensile strength upon stretching. This characterization of intermolecular behavior helps explain the disintegration of the gluten film and the loss of fibril-forming ability in stretched cysteine- and glutathione-treated doughs and the strengthening effect of bromate and ADA-treated doughs, such that thick fibers tended to dominate over the thin fibrils of the control upon stretching. In this study the control dough was best balanced in rheological properties since the additives were added at high levels specifically to observe their effects.

Proper dough moulding and the ability to retain gas while rising is highly dependent on a balance of elasticity and extensibility so that gluten is maintained in a cohesive continuous film. An imbalance may lead to doughs which are too bucky or too slack, either of which yields low volume poor quality products and results in severe processing difficulties. This study using SEM forms a basis for explaining the action of dough improvers and should ultimately contribute to controlling the balance between elasticity and extensibility in doughs.

Acknowledgment

Michigan Agricultural Experiment Station Journal Article No. 9912.
Thanks to Mary Ann Iliff for typing the manuscript.

Table 1
Dough Properties Determined by the Farinograph

Additive	Absorption (%)	Peak Time (min)	Stability (min)
None	61.6	7.0	10.0
Potassium-bromate	61.6	7.0	10.0
Potassium-iodate	62.5	6.5	4.0
ADA*	62.6	6.0	3.5
L-cysteine	62.5	2.5	2.5
Glutathione	62.5	3.0	4.0

* ADA, Azodicarbamide

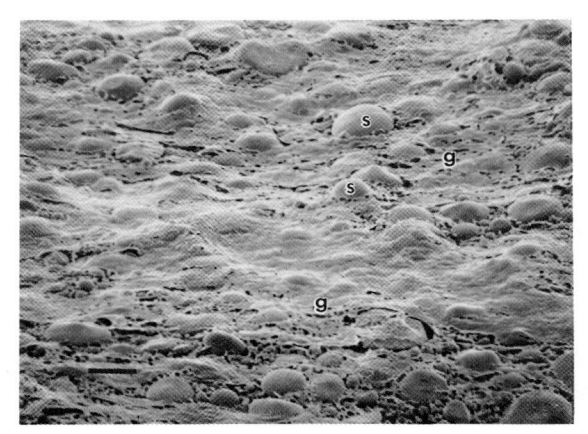

Fig. 13 Optimally mixed flour-water dough containing cysteine after stretching. Note the numerous holes in the gluten (g) sheet; s = starch granules. Bar is 20μm.

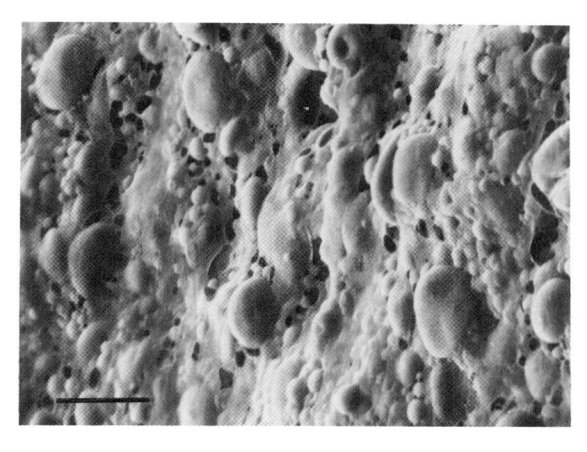

Fig. 14 Incubated and stretched optimally mixed flour-water dough containing cysteine. Note holes in gluten sheet showing the lack of continuity. Bar is 20μm.

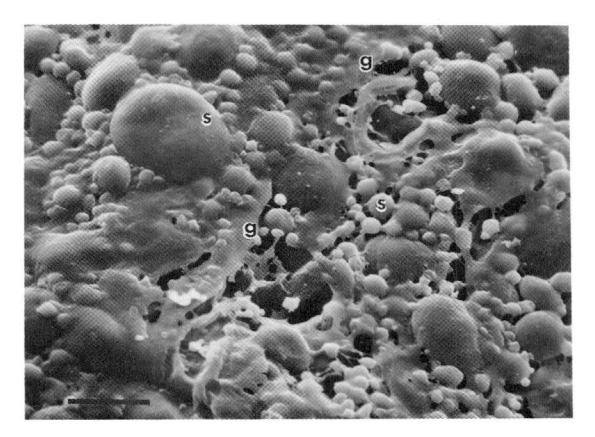

Fig. 15 Overmixed flour-water dough containing cysteine. s = starch granules; note large holes in the gluten sheet (g). Note many free starch granules on surface of the gluten sheet. Bar is 20μm.

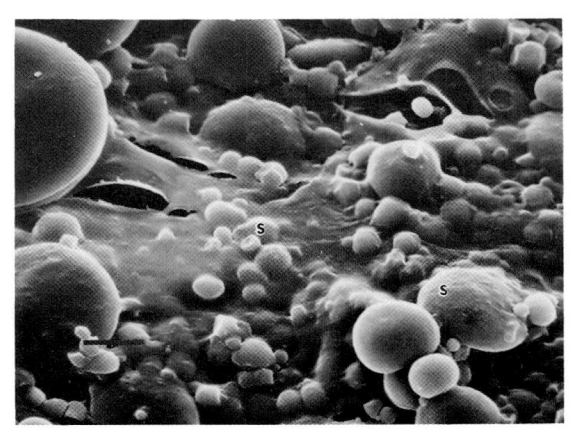

Fig. 16 Optimally mixed flour-water dough containing glutathione. Note thin film of gluten covering the starch granules (s). Bar is 10μm.

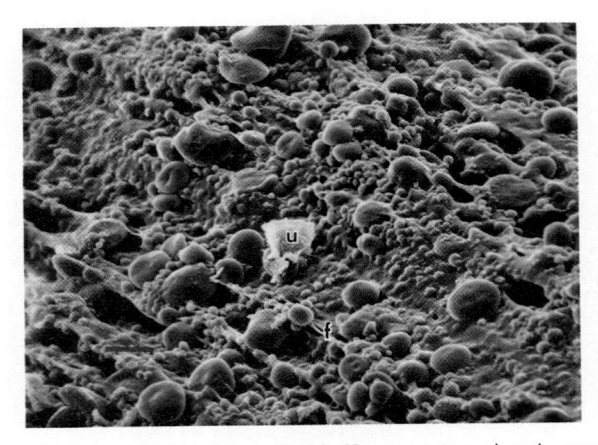

Fig. 17 Optimally mixed flour-water dough containing glutathione after stretching. f = stretched fibrils; u = unidentified material. Bar is 20µm.

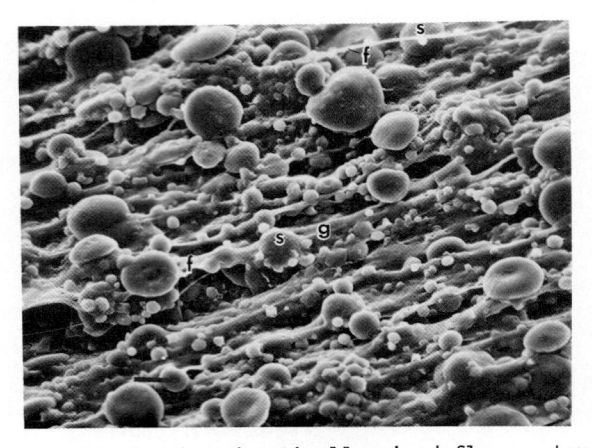

Fig. 18 Incubated optimally mixed flour-water dough containing glutathione after stretching. f = long, thin gluten fibrils. Note the gluten (g) and starch granules (s) seem to be strongly associated. Bar is 20µm.

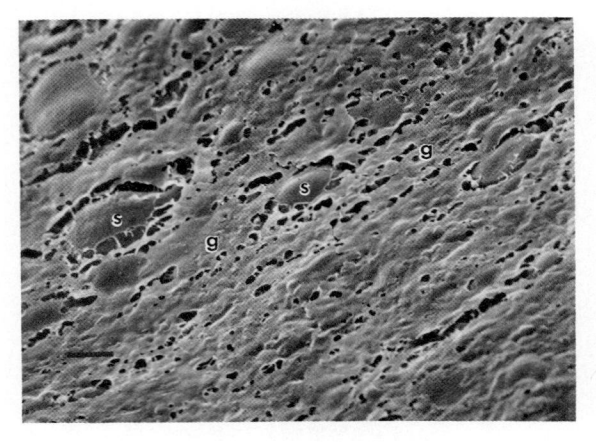

Fig. 19 Overmixed flour-water dough containing glutathione after stretching. Note extensive holes in the gluten (g). s = starch granules. Bar is 20µm.

References

1. Tsen, C.C. Effects of oxidizing and reducing agents on changes of flour proteins during dough mixing. Cereal Chem. 46, 1969, 435-442.

2. Pomeranz, Y. and Shellenberger, J.A. Bread Science and Technology AVI Publ. Co., Inc., Westport, USA, 44-54.

3. Frater, R., Hird, F.J.R., Moss, H.J., and Yates, J.R. A role for thiol and disulfide groups in determining the rheological properties of dough made from wheaten flour. Nature. 186, 1960, 451-454.

4. Henika, R.G. and Rodgers, N.E. Reactions of cysteine, bromate and whey in a rapid break making process. Cereal Chem. 42, 1965, 397-408.

5. Butterworth, S.W., and Colbeck, W.J. Some photomicrographic studies of dough and bread structure. Cereal Chem. 15, 1938, 475-488.

6. Burhans, M.E. and Clapp, J. A microscopic study of bread dough. Cereal Chem. 19, 1942, 196-216.

7. Sandstedt. R.M., Schaumburg, L. and Fleming, J. The microscopic structure of bread and dough. Cereal Chem. 31, 1954, 43-49.

8. Moss, R. A study of the microstructure of bread dough. CSIRO Food Res. Quarterly. 32, 1972, 50-56.

9. Moss, R. Dough microstructure as affected by the addition of cysteine, bromate and ascorbic acid. Cereal Sci. Today. 19, 1974, 557-561.

10. Aranyi, C. and Hawrylewicz, E.J. A note on scanning electron microscopy of flours and doughs. Cereal Chem. 45, 1968, 500-502.

11. Simmonds, D.H. Wheat grain morphology and its relationship to dough structure. Cereal Chem. 49, 1972a, 324-335.

12. Simmonds, D.H. The ultrastructure of mature wheat endosperm. Cereal Chem. 49, 1972b, 212-222.

13. Khoo, U., Christianson, D.D. and Inglett, G.E. Scanning and transmission electron microscopy of dough and bread. Bakers Digest. 49, 1975, 24-31.

14. Evans, L.G., Volpe, T. and Zabik, M.E. Ultrastructure of bread dough with yeast single cell protein and/or emulsifier. J. Food Sci. 42, 1977, 70-74.

15. Pomeranz, Y., Shogren, M.D., Finney, K.F. and Bechtel, D.B. Fiber in breadmaking - Effects on functional properties. Cereal Chem. 54, 1977, 25-41.

16. Bechtel, D.B., Pomeranz, Y. and de Francisco, A. Bread making studied by light transmission electron microscopy. Cereal Chem. 55, 1978, 392-401.

17. Jelaca, S. and Dodds, N.J.H. Studies of some improver effects at high dough temperatures. J. Sci. Food Agric. 20, 1969, 540-545.

18. Seckinger, H.L. and Wolf, M.J. Lipid distribution in the protein matrix of wheat endosperm as observed by electon microscopy. Cereal Chem. 44, 1967, 669-674.

19. Hess, K. Protein, Kleber und Lipoide in Weizenkorn und Mehl. Kolloid A. 136, 1954, 84-99.

20. Dempster, C.J., Cunningham, C.K., Fischer, M.H., Hlynka, I. and Anderson, J.A. Comparative study of the improving action of bromate and iodate by baking data, rheological measurements and chemical analysis. Cereal Chem. 33, 1956, 221-239.

21. Bloksma, A.H. Rheology and chemistry of dough. In Pomeranz, Y. (ed.) Wheat Chemistry and Technology. American Association of Cereal Chemists, Inc., St. Paul, USA, 1971, 523-584.

22. Ewart, J.A.D. Recent research and dough viscoelasticity. Bakers Digest, 46, (7), 1972, 22-28.

23. Goldstein, S. Sulfhydryl- und Disulfidgruppen der Klebereiweisse und ihre Bezienhung zur Bäckfähigkeit der Brotmehle. Mitt. Lebensm. Hyg. Bern. 48, 1957, 87-93.

24. Finney, K.F., Tsen, C.C. and Shogren, M.D. Cysteine's effect on mixing time, water absorption, oxidation requirement and loaf volume of Red River 68. Cereal Chem. 48, 1971, 540-544.

25. Tsen, C.C. Oxidation of sulfhydryl groups of flour by bromate under various conditions and during the breadmaking process. Cereal Chem. 45, 1968, 531-538.

26. Villegas, E., Pomeranz, Y. and Shellenberger, J.A. The effects of thiolated gelatins and glutathione on rheological properties of wheat doughs. Cereal Chem. 40, 1964, 694-703.

Discussion with Reviewers

Reviewer II: In Fig. 5, how extensive were the areas where gluten had pulled away from the starch granules and collapsed into a spindly web compared to those where no pulling away had occurred? Could the breakdown have occurred during sample preparation because of the fragile nature of the protein film?
W. J. Wolf: Do you have pictures at higher magnification of the areas in overmixed dough, Fig. 6, where the film was disintegrated? They appear to be honeycomb-like and appear to show no orientation in the direction of the stretching.
D. R. Lineback: Do the authors feel confident that their sample preparation did not result in any artifacts in the SEM photomicrographs, particularly in the preparation of the "stretched samples"?
Authors: Freeze drying has been demonstrated to be the method of sample preparation which offers minimal, if any, distortion in dough, (Varriano-Marston, E. A comparison of dough preparation procedures for SEM., October, 1977, 32-36). Specific artifacts common to this method have not been identified to the knowledge of these authors. The collapsed areas seen in Fig. 5 were sufficient to characterize the dough appearance when compared to controls. The effect was not uniform across the dough surface and some areas, though not a majority, of the gluten sheet appeared intact. In the stretched samples the stress on the dough undoubtedly increased the fragility of the gluten film. It is possible that the collapsed, web-like appearance is an artifact of drying in very fragile areas and may only be indicative of variations in the strength or cohesiveness of the fresh sample. It remains that distinctively different behavior was observed between the control, oxidized and reduced doughs.

F. MacRitchie: Can you provide a brief description of the farinograph and the parameters it is used to measure?
Authors: The Brabender Farinograph is a recording dough mixer and measures power required to mix a dough at a constant speed. Most commonly, this instrument is used to determine water absorption and to characterize a flour by various defined indices. The water absorption is that amount of water required to yield a dough of predetermined consistency (e.g. 500 B.U.), and it generally increases with the protein quality and content of the flour. High absorption is desirable because of the impact on bakery production yields. The peak time relates to the mixing time required to develop maximum resistance of the dough against the mixer blades. Resistance increases as the gluten is hydrated and develops into an elastic cohesive matrix in the dough. The stability is the time for which the mixing curve is maintained within a given range of maximum dough consistency. A complete reference can be found in The Farinograph Handbook, by L. Locken, et.al. American Association of Cereal Chemists, St. Paul, Minnesota, 1962.

Reviewer II & VI: Would the author describe the sampling procedure in more detail as dough microstructure is very easily altered during sampling, particularly when the sample dimensions are as small as those described (2mm diameter, 10-15mm in length)?
Authors: A sample area was measured on the dough surface and excised using a fine razor blade and tweezers. For stretched samples, the dough strips were pulled by hand in a gentle steady motion using blunted tweezers until the dough was very near the breaking point. They were immediately frozen on dry ice. An improved method may incorporate an apparatus which is capable of pulling at a constant force/unit time and freeze the sample by submerging in liquid nitrogen for even faster freezing.

Reviewer II: Were the small round pores in the gluten surface of the optimally mixed control dough present in all the other doughs?
Authors: Yes, the pores were observed in all doughs to varying degrees depending on the development of the gluten. They were not visible where gluten had been stretched extensively and were only seen at magnifications over about 2000X.

Reviewer II: Re amyloplast residues. I have some doubt as to whether these films are amyloplast residues. In some cases the films appear to be continuous with the gluten matrix. It is also surprising that the author and co-workers have not seen evidence of such residues when wheat grains have been fractured. Is it possible that the films could be due to water soluble protein, and thus have been formed during freeze drying?

W. J. Wolf: Can you rule out that the "Amyloplast membrane residues" in Fig. 3 aren't thin gluten films? The work of Simmonds (Ref. 12) indicates that the membrane residues are very thin and perhaps difficult to see at 1000X. Have you observed them in flours before mixing with water?

Authors: The authors could not demonstrate that the films observed are the amyloplast membrane residues and agree that the reference be changed to film-like residues instead. It was considered that the turned or ruffled edges of a membrane would be visible at 1000X even though it would not if viewed in a cross-section. No, the authors have not seen them in flour before mixing with water.

W. J. Wolf: Appearance of the stretched, incubated dough containing bromate, Fig. 7, closely resembles the stretched underdeveloped dough, Fig. 4. Are there any rheological similarities between these two types of doughs?

Authors: Unfortunately, this study was not coupled with rheological analysis of the doughs. This is a very relevant question and would require further study.

W. J. Wolf: Using equal weights of cysteine and glutathione results in a molar concentration of cysteine that is 2.5 times that of glutathione. Can the lower reactivity of glutathione be attributed to its lower molar concentration?

Authors: Yes, since there is only one sulfhydryl group in either the cysteine or glutathione molecule the -SH reactivity should be proportional on a mole for mole basis.

D. R. Lineback: Do the authors regard the starch-protein interaction as being a mechanical one or is there some type of chemical association between the gluten fibrils or strands and the starch granules?

Authors: It is thought to be primarily physical although the exact mechanism of interaction is not known. The starch-protein interface appears to be affected considerably upon hydration and through additions of various ingredients.

D. R. Lineback: Do the authors see or detect any differences between the behavior of the large and small wheat starch granules in the different doughs or in the interactions with the gluten?

Authors: No, the granules seem similar.

Reviewer II: In reference to Fig. 4, the undermixed dough, compared to the control in Fig. 2, there is an alternative explanation for the observations regarding starch granules as anchor points. The ability to form gluten fibrils in an optimally mixed dough is due to the gluten matrix having the desired degree of interconnection and balance of extensibility and elasticity. As a consequence of these properties the starch granules in such a dough do not act as points of weakness, but they do not have to be anchor points - the continuous phase is the gluten matrix. Therefore when the dough is stretched the gluten is also stretched, to a degree dependent on its properties.

W. J. Wolf: In the undermixed dough, Figs. 3 and 4, the starch granules are relatively free of gluten and the protein matrix appears fibrous and more continuous than in the optimally mixed sample, Fig. 2. Are the starch granules really obstacles to gluten continuity in an undermixed dough as suggested? It seems more likely that they disrupt the continuity of an optimally mixed rather than an undermixed dough. Subsequently, when the optimally mixed dough is stretched, the discontinuities introduced by the granules result in the formation of fibrils.

Authors: The reviewers offer reasonable explanations for the observations. The gluten continuity exists in both undermixed and optimally mixed doughs and it is the continuous phase. The most significant change is the distribution of gluten in a thin film-like matrix surrounding all the starch granules as mixing proceeds. Where starch granules are referred to as points of weakness or anchor points in the continuous phase this description could simply be interpreted as disruptive obstacles in the film.

Reviewer II: Why was no mention made of ascorbic acid? In chemical dough development systems bromate is generally combined with ascorbic acid, and the reducing agents cysteine or sodium metabisulphite are frequently added.

Authors: Ascorbic acid and many other agents are widely used in chemical dough development, usually in combination. Ascorbic acid functions as an oxidizing agent in the presence of oxygen and as a reducing agent when oxygen is excluded during mixing. Because this work was not a comprehensive study or review of oxidizing and reducing agents, only the additives included in the experiments were discussed. These were selected somewhat arbitrarily but certainly represent clear examples of the effects of extreme oxidation or reduction in the dough. Ascorbic acid undoubtedly represents a worthwhile candidate for future study.

STRUCTURAL STUDIES OF CARROTS BY SEM

E. A. Davis and J. Gordon

Department of Food Science and Nutrition
University of Minnesota
1334 Eckles Avenue
St. Paul, Minnesota 55108

Abstract

A review of carrot ultrastructure using scanning electron microscopy has been presented, with dual emphasis being placed on (1) choice of fixation method and (2) by examination of carrot tissue whose cellular structures are expected to differ as functions of variety, maturity, infection, and heat treatment. Unique structural differences were observed, clearly making it a useful approach in studies of this type. More specifically, four different methods of sample preparation were compared in different combinations of mechanical excision and cryofracture. It was concluded that of the methods investigated, cryofracturing the tissue after immersions in liquid nitrogen, subsequently fixing with glutaraldehyde and osmium tetroxide and critical point drying after dehydrating with acetone, gave the most information. Morphological and textural differences were found between two varieties of carrots (Scarlet Nantes and HiPak), and these were attributed to differences in growth rates. Diseased carrots in both varieties had unique cellular development but, except for the hardness parameter, showed few differences in textural properties when compared to healthy carrots. Although carrots cooked by steaming, boiling, and pressure cooking resulted in varying degrees of cellular disruption, no statistically significant differences due to cooking methods were found. The integrated results from these studies provided confirmation of cell wall structure that can be related to existing theories of cellular development and breakdown.

KEY WORDS: Carrots, Food Microstructure, SEM Carrots, Carrot Morphology, Carrot Growth, Carrot Varieties, Carrot Disease, Aster Yellows, Aster Yellows Carrots, Carrots Cooked

Introduction

Food scientists are often interested in the relationship between the morphological changes that food components undergo during processing and cooking and the final textural or sensory characteristics. Scanning electron microscopy (SEM) can be useful in the study of plant source foods such as carrots and potatoes, in order to see cellular morphology changes that take place at various stages of growth, storage, infection and cooking. Two things need to be investigated prior to use of SEM for this purpose in such high moisture-containing tissues. First, the choice of procedure for sample preparation has to be made. Second, the manner in which one arrives at conclusions that depend on cellular differentiation as functions of developing or damaged plant cellular physiology has to be established.

A major consideration in the choice of a preparative method is whether artifacts are introduced during sample handling. Since it is very difficult to eradicate artifacts completely, choices must be made as to the best method of sample preparation depending on the information that is being sought and on how that information can be interpreted.

Carrots have not been extensively studied by SEM in the past. For this reason, this review will mainly serve to integrate the information from three intensive investigations on carrot phloem and xylem tissue by SEM in our laboratory. Other information from the literature will be included. First, choice of fixation method will be discussed in terms of cellular morphologies observed. Then, this fixation method will be used to study cellular differentiation patterns observed in carrots that were of different varieties, different levels of maturity, healthy and aster yellows infected, and subjected to different heat treatments. In this way, the sensitivity of SEM in demonstrating structural changes, or differences, in tissues that appeared to be the same on the basis of instrumental texture profile analysis (TPA) illustrated the potential of SEM techniques as a powerful tool in further understanding the physicochemical interactions that might be occurring during processing and cooking.

Fixation Methods

The use of SEM for studying surfaces of plant tissues is well-documented (Heslop-Harrison and Heslop-Harrison, 1969; Falk et al., 1971; Pomeranz, 1976; Panessa and Gennaro, 1972, 1974; Robards, 1978; Vaughan, 1979). Methods of fixation and dehydration often result in a negative surface charge on the sample when subjected to the electron beam, even when complete infiltration is ensured by prolonged fixation and dehydration techniques (Heslop-Harrison, 1970; Panessa and Gennaro, 1972).

In the case of processed and cooked foods, the cellular components are often softer than the raw sample and, therefore, more easily damaged during excision, fixation and handling. The best solution would be to leave the water in the tissue and view the surface directly without any fixation. However, SEM, which operates under a high vacuum system and results in the sample being bombarded by electrons under high energy for imaging, must be used with dehydrated materials at room temperature. Several groups have used SEM to view frozen samples subjected to ultra-low temperature freezing. These techniques were reviewed by Echlin and Moreton (1976), Robards (1974), Rebhun (1972), Echlin (1978), Bullivant (1970), and Van Harrevold et al. (1974). Specific applications of low temperature microscopy in the food area were made with potato tissue by Davis and Gordon (1978) and with muscle by Varriano-Marston et al. (1978). Freezing, freeze-drying, and freeze substitution techniques have also been considered by MacKenzie (1972) and Boyde and Echlin (1973), as well as Boyde (1974). Since these are highly specialized techniques, requiring equipment that is not available in most foods research facilities and their corresponding electron microscopy support facilities, the remainder of this review will focus on the use of more conventional methods of fixation.

A review of the few published SEM studies of carrot xylem and phloem tissue showed that the methods used to fix the tissue involved mainly glutaraldehyde and osmium tetroxide (OsO_4) fixation, followed by dehydration in acetone or ethanol, followed by critical point drying. For details of these studies see the work of Colvin and Leppard (1973), Davis et al. (1976a) and Grote and Fromme (1978) as specifically applied to carrots.

A study by Mohr and Stein (1969) using delicate tomato parenchyma tissue seemed applicable for use in the processed food area. They used liquid nitrogen in a cryofracture technique before fixation, followed by conventional fixation techniques. It was concluded that when the rapid-freeze, "moderate-thaw" (freezing rate > $10^{o}C$/min; thawing still air in 1 to 5 min) procedure was used on that tissue, only a small amount of damage was observed when the tissue was viewed by TEM. Because SEM studies on surface morphology are done at lower magnification and resolution than TEM studies, we felt we could adapt this procedure for our SEM carrot studies.

Humphreys et al. (1974), on the other hand, successfully used cryofracture after the plant sample had undergone fixation, ethanol dehydration and critical point drying. This procedure was also evaluated in our laboratory for applicability to carrot studies.

An additional criterion for selection of fixation methods for carrots undergoing processing is related to the structural role of both the xylem and phloem parenchymous tissue and xylem tracheary elements. The fixation method needs to be such that cellular differentiation observed in the micrographs can be related to carrot growth, disease, storage and processing conditions.

On the basis of the review of studies just summarized and the special requirements of carrot tissue and its response to processing, the following methods were evaluated in our laboratory:

(1) Mechanically excised, glutaraldehyde and OsO_4 fixed, acetone dehydrated, and critical point dried (Davis et al., 1976a).
(2) Mechanically excised, glutaraldehyde and OsO_4 fixed, ethanol dehydrated and critical point dried (Grote et al., 1978).
(3) Cryofractured in liquid nitrogen, glutaraldehyde and OsO_4 fixed, acetone dehydrated and critical point dried (Davis et al., 1976a).
(4) Mechanically excised, glutaraldehyde and OsO_4 fixed, ethanol dehydrated, cryofractured in liquid nitrogen and critical point dried (Humphreys et al., 1974).

More specifically, the following procedures were used: During excision and/or cryofracture; (a) whenever mechanical excision took place in the procedure, small pieces of carrot about 0.2 cm on any one side were cut and placed in glutaraldehyde. (b) whenever liquid N_2 cryofracture occurred prior to fixation and dehydration, small pieces of carrot were excised no larger than 0.4 cm on any one side, immersed in liquid N_2 and cryofractured with a hammer between two pieces of plexiglass under liquid N_2. The tap of the hammer was hard enough to shatter the sample into small but not powder-like pieces. These were placed immediately into cacodylate-glutaraldehyde buffer. (c) whenever liquid N_2 cryofracture occurred after ethanol dehydration, samples were cryofractured as in (b) but after the solvent dehydration series and prior to critical point drying with amylacetate-ethanol-carbon dioxide. During fixation and dehydration, the small samples across all procedures were placed into 2% glutaraldehyde in 0.1 M cacodylate buffer (pH 7), for 4 days at 3°C to allow complete infiltration into the hard cellular areas of the tissues. Next, samples were washed 3 times, 10 min each, in a 1:1 solution of 0.5 M sucrose and 0.2 M cacodylate buffer (pH 7). They were then placed in a 1:1 solution of 2% OsO_4 and 0.5 M cacodylate buffer (pH 7) for 4 days. After this treatment, the samples were washed 3 times, 10 min each washing, with 0.12 M sucrose. Then the samples were dehydrated by successively washing each for 15 min in 15, 25, 50, 75, and 95% acetone or ethanol and, finally, by washing 3 times in 100% acetone or ethanol for 15 minutes. The samples were critical point dried in a Sorvall critical point drying system with acetone-carbon dioxide for the acetone dehydrated samples and in amylacetate-ethanol-carbon dioxide for the ethanol dehydrated sample. The samples were glued to SEM target stubs with conducting silver paint, coated 2 times each with carbon and then gold in a Denton Vacuum DV-502 gold and carbon evaporator, and viewed in a Cambridge S600 scanning electron microscope operated at 15 kV.

Representative micrographs of carrot xylem parenchymous tissue prepared by the above methods of fixation are found in Figures 1-3 as well as previous work on phloem and xylem tracheary elements by Davis et al. (1976a). Raw tissues were found to have a similar appearance for all preparative methods studied. An example is seen in Figure 1 for xylem parenchymous tissue prepared by mechanical excision-ethanol dehydration (Method 2). However, the cooked samples were affected by method of preparation. For example, steam cooked, mechanically excised samples prepared by both mechanical excision methods (Methods 1 and 2), had an appearance similar to that seen in Figure 2a where the cells appear somewhat compressed and damaged. When prepared by the cryofracture-acetone dehydration procedure (Method 3) the cell structures were clearly and smoothly fractured (Figure 2b). Furthermore, when the samples were prepared by the procedure of cryofracture after fixation and ethanol dehydration (Method 4) the cells appeared torn and irregular in shape as seen in Figure 2c.

Steam-heated parenchymous tissue appeared more compressed and folded when prepared by Methods 1 and 2, as opposed to Method 3. Figure 3a is typical of what we see using Methods 1 and 2, and Figure 3b is typical of what we see by Method 3. Also Method 4 (Figure 3c), as for the xylem above, resulted in torn and ruptured appearing cell walls (Figure 3c).

It was concluded that, of the methods evaluated, the cryofracture-acetone dehydration procedure (Method 3) would offer the least damage and the most information in subsequent studies involving carrot growth, storage, disease and cooking.

Morphology During Growth, Disease and Storage

Carrots purchased from the retail market may differ somewhat in surface morphology of their cellular structure due to variety, growth level, disease and storage. These factors are recognized as sources of variability in quality characteristics. However, comparative SEM studies of carrot structure have not been reported. Such baseline data are important for structural studies when using these relatively new methods of preparation and observation. In the discussion which follows, particular emphasis will be focused on structures most susceptible to change during cooking and storage (Davis et al. 1976a, b, 1977) such as the phloem and xylem areas. Two varieties of carrots were chosen: Scarlet Nantes and HiPak. These are a fast and a slow growing variety, respectively. Furthermore, these were grown by the University of Minnesota Agricultural Extension Service on experimental plots and, therefore, had a known planting and harvesting history. Healthy and aster yellows infected types of the above varieties were available for study during growth and post-harvest storage. Details of these variety, growth, and storage studies are given by Davis et al. (1977). Textural measurements were made to determine whether differences observed by SEM were also detectable in the instrumental texture profile analysis with Instron UTM-M using a compression head (Davis et al., 1977). The data which emerged from these studies will be given after a brief description of aster yellows disease.

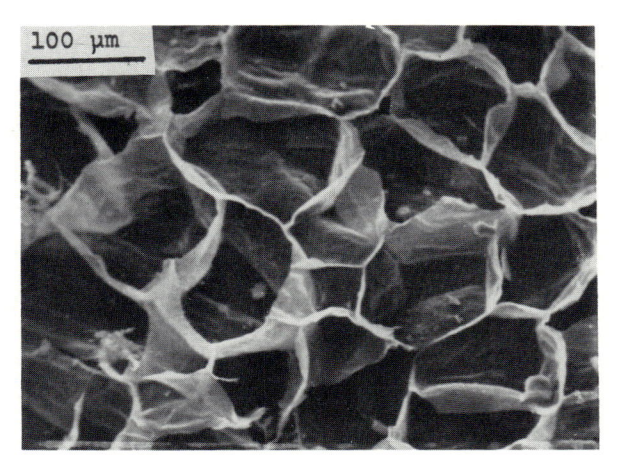

Figure 1. Raw carrot, xylem. Mechanically excised-ethanol dehydrated (Method 2).

Aster yellows, which is transmitted by mycoplasma, is the most common disease in carrots and many other vegetables (Haggis and Sinha, 1978; Maramorosch, Granados, and Hirumi, 1970; Schultz, 1973; Henne, 1970; Hervey and Schroeder, 1949; Petzold et al., 1977; Esau, 1977). Mycoplasma-like organisms are believed to have been observed by Haggis and Sinha (1978) in other plant tissue but others have not observed this. We did not observe any in our studies either. It has been established that 85-100% of a carrot crop can be infected by aster yellows (Schultz, 1973). In advanced stages of infection, the carrots have a characteristic stunted and hairy appearance and a bitter taste (Hervey and Schroeder, 1949). Doi et al. (1967) have looked at the ultrastructure of healthy and infected carrots by TEM but no SEM studies have been reported.

The phloem in early developmental stages of growth contains cell wall fibrous material such as that seen in Figure 4a, and granular material which can be seen in greater detail in the earlier work by Davis et al. (1977). As it reaches fuller maturity, the cell walls appear firmer with no fibrous material as seen in Figure 4b. Also, the granular material appears less frequently in the cells. HiPak carrots which were harvested at the same time as the Scarlet Nantes carrots did not contain as much granular material as the Scarlet Nantes carrots at the early stages of development (Figure 4c). The aster yellows infected HiPak phloem (Figure 4d) shows cellular malformation in that the cell walls no longer have a honeycombed compartmentalized appearance typical of the healthy cells in Figure 2b as well as no apparent fibrous cell wall material. This lack of compartmentalization persists throughout the growth period for both varieties of diseased carrots although somewhat more intracellular material is apparent in the diseased HiPak as in the healthy HiPak carrots. Thus, there is incomplete cell development, probably as a result of the absence of the intracellular fibrillar material which is necessary for cell wall development. This is not surprising, since aster yellows is considered a phloem disease.

Models of cell wall structure proposed by Albersheim et al. (1973, 1976) and the occurrence of the fibrillar material within the cell walls of suspension cultures of carrot cells observed by Colvin and Leppard (1973), are consistent with the findings of these studies. For example, we can interpret the "cotton-like" appearance of the cell walls in Figure 4a as being due to cellulose fibers upon which are laid other polysaccharide fibrils composed of rhamno-galacturonans, arabino-galactan, xyloglucan, and, perhaps, hydroxyproline-containing proteins to form the cell wall structure.

Storage for 10 weeks did not appear to affect the cellular appearance of the healthy Scarlet Nantes or HiPak phloem. However, diseased Scarlet Nantes phloem exhibited cell wall separation (Figure 4e) in a unidirectional (or tubular direction) manner. This cell wall separation in a unidirectional manner was also observed during cooking of healthy carrots (Davis et al., 1976b) and will be reviewed in the next section. These observations are in accord with the cell wall extension and cell elongation theory of Keegstra et al. (1973) that during growth other polysaccharides bond with the cellulose fibers to form an organized structure. Our interpretation is that incomplete cell wall formation could result in defects manifested in a unidirectional manner giving rise to the tubular-like areas seen in the SEM micrographs.

Parenchymous tissue of the xylem had similar properties to that of the phloem areas both within the same varieties and across the varieties irrespective of carrot treatment or growth. For example, the HiPak xylem parenchymous tissue also contained more intracellular material at each stage of development. These areas were not as affected by the aster yellows infection as the phloem, further substantiating this as a phloem infection. There is some evidence that tracheary elements developed more slowly in the HiPak, as seen in Figures 5a, 5b, and 5c, which are Scarlet Nantes early in the growth cycle, at harvest, and HiPak at harvest, respectively. For example, Scarlet Nantes, Figure 5a, has the tracheids exposed with parenchymous cells adhering to their surface, while in Figure 5b the cells appear firmly implanted against the tracheary elements. In Figure 5c, for HiPak however, the appearance is similar to that seen in Figure 5a, although the growth level is the same as that seen in Figure 5b. Tracheary elements in the Scarlet Nantes and HiPak infected xylem (Figure 5d) looked like those early in the growth cycle (Figure 5a).

Texture profile analysis for hardness, brittleness (sometimes referred to as fracturability), and total work of compression were supportive of the morphological observations. The Scarlet Nantes had significantly higher values at the 1% level than the HiPak (Davis et al., 1977). Hardness values for healthy samples were higher than those for the diseased samples at the 5% level, but the other TPA parameters were not significantly different. Possibly enough structural support is present in the xylem so that most of the TPA parameters show no disease effects, even though the disease may delay the overall development of the xylem. This would imply that the parenchymous phloem tissue has a different composition from the xylem parenchymous tissue since the cell walls in diseased specimens had appeared rigid and fewer textural differences were expected. This is in contrast to varietal differences in the rate of cell wall development, which one would expect to result in greater textural differences.

Morphology of Cooked Carrots

Since we now have baseline information on the cellular structure of carrots in the unheated stage, let us review the physical transformations observed by SEM of phloem and xylem carrot tissue after it is boiled, steamed or pressure cooked. Details of this study can be found in an article by Davis et al. (1976b) in which the observations of changes induced by normal home cooking techniques were reported. Specific cooking conditions are found in Table 1.

Table 1. Preparation and Cooking methods for carrots.

Cooking method[a]	Cooking water (ml)	Cooking time (min)
Steamed	160	31
Boiled (covered)	700	22
Pressure-cooked (15 lbs. pressure)	60	3

[a]All carrots were cut lengthwise and quartered in 2-1/2 in. lengths. Raw weight for each sample was 325 g.

Previous attempts to explain how the cell walls separate and lose their rigidity during heating have been based on light microscopy studies. Simpson and Halliday (1941) observed cell wall fracture in steamed carrot and parsnip xylem and phloem. Prior to the studies of Davis et al. (1976a, b; 1977) SEM had not been used for this purpose.

In the studies by Davis et al. (1976a, b) SEM was used to characterize the surface morphology of xylem and phloem tissue of mature carrots when cooked by steaming, boiling and pressure cooking. It was found that although all were brought to the same degree of cooking the morphology of the phloem and xylem parenchymous tissue appeared to be different. In contrast, the TPA for hardness, brittleness and total work of compression did

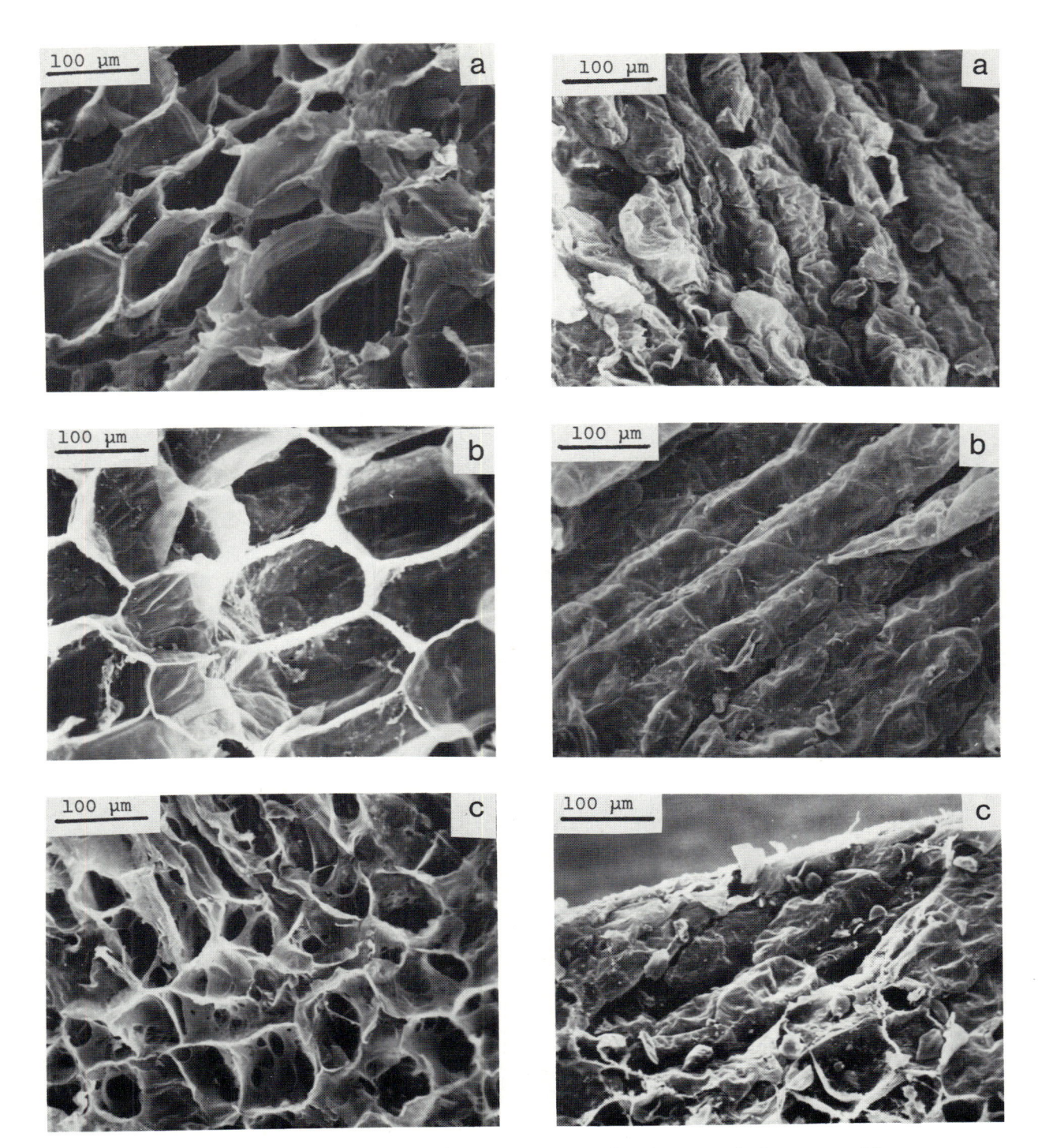

Figure 2. Steam cooked carrot, xylem parenchymous tissue.
a. Mechanically excised-acetone dehydrated (Method 1).
b. Cryofractured-acetone dehydrated (Method 3).
c. Ethanol dehydrated-cryofractured (Method 4).

Figure 3. Steam cooked carrot, phloem.
a. Mechanically excised-acetone dehydrated (Method 1).
b. Cryofractured-acetone dehydrated (Method 3).
c. Ethanol dehydrated-cryofractured (Method 4).

not yield statistically significant differences due to cooking method (Davis et al., 1977). In establishing cooking times for the three methods we attempted to cook to the same degree of subjective "doneness"; the instrumental measurements confirmed this subjective judgment where no significant differences due to cooking methods were found using analysis of variance in which we considered method cooking, variety, and disease.

The different methods of cooking resulted in differences of phloem morphology (Davis et al., 1976b). For example, unheated phloem has a regular honeycombed cellular appearance similar to that seen in Figure 1 for xylem parenchymous tissue. However, heating resulted in long tubular cellular separation (Figure 3b for steam cooked phloem) or collapse (Figure 6a for boiled phloem). However, the integrity of the cell walls were still retained in the sense that holes did not appear in them. Pressure cooked and boiled phloem tissue had similar appearing tubules. The tubes even appeared involuted (Figure 6b, arrow) into double walled structures, with the pressure cooked samples appearing most collapsed. Upon closer examination of the steam cooked phloem tubular areas (Figure 6b) we saw some fibrillar material (arrow) that appeared similar to that seen by Colvin and Leppard (1973). Although there is controversy as to the exact nature of these fibrils or microfibrils (Leppard and Colvin, 1978) we are not able to specifically evaluate this with the data presented in this review. However, these observations are further physical support of the theoretical models of cell wall structure theory as proposed by Albersheim et al. (1973) in which cellulose is believed to be embedded in the complex polysaccharides such as rhamno-galacturonan. As cooking proceeds, these polysaccharides change conformation and extrude, leaving cellulose fibers freestanding. In this way, we have gone full circle in that one can postulate the development of a fibrillar material, as proposed by Colvin and Leppard (1973) in their carrot cell culture studies and as observed in our growth studies observations, its disappearance at maturity (embedded in matrix) and its eventual reappearance after exposure to heat and water has weakened other structures. Colvin and Leppard documented the presence of this material with cultured carrot cells, but they could not see these fibrils in mature carrot cells due to the tightness of the walls in these cells. Because the cell walls separated in tubular fashion for the phloem upon heating, further support for the cell wall extension and cell elongation theory described by Keegstra et al. (1973) was also obtained. This can mean that the highly developed directional bonding with the cellulose fibers develops during the growth process but it is difficult to see when growth is proceeding normally. In the growth studies of the infected carrots, cell wall separations that were unidirectional could be observed, however.

Heating of the xylem resulted in separation and collapse of the parenchymous tissue as noted in the phloem. The pressure cooked samples had structures that were the most disrupted, steam cooked samples, the least disrupted. The main difference in the cooked xylem parenchymous tissue was the lack of tubular formations seen in the phloem. The tracheary elements of the xylem retained their integrity and rigidity upon heating. The unheated xylem (Figure 7a) appeared to have its parenchymous tissue firmly attached to the tracheary elements but the cooked xylem, seen for example, in steam cooked xylem (Figure 7b), contained parenchymous tissue that was either partially or totally pulled away from it. For a more detailed discussion, see Davis et al. (1976b).

The manner in which the parenchymous tissue breaks down in the heated xylem implies that its composition is different from the phloem. This is probably due to the fact that the xylem is mainly non-living material and the phloem is mainly living. Because the tracheary elements survive the heat treatment, the texture of the cooked sample is probably most influenced by these elements.

Conclusions

It can be concluded from the studies reviewed here that visualization of differentiation is possible by SEM for carrot phloem and xylem, once a fixation method is chosen, as a function of growth level, variety type, health state, storage period, and cooking method. Also, this information when integrated can be related to existing theories of cellular development and breakdown. Furthermore, this review demonstrates that SEM is a sensitive tool which is useful in studies of the physico-chemical characteristics of carrot tissue and can add new information that may be usefully applied to the study of other plant food systems before, during, and after processing.

Acknowledgements

The assistance of Ms. Su-In Hsieh is greatly acknowledged. The cooperation of Dr. O. C. Turnquist, Department of Horticultural Science and Landscape Architecture, is also much appreciated. This project was supported by the University of Minnesota Agricultural Experiment Station, Projects 18-27 and 18-63 and Scientific Journal Series No. 11,166.

References

Albersheim, P., Bauer, W. D., Keegstra, K., et al. The structure of the wall of suspension-cultured sycamore cells. In: Biogenesis of Plant Cell Polysaccharides. F. Loewus (ed.), Academic Press, New York, U.S.A., 1973, 117-147.

Albersheim, P. The Primary Cell Wall. In: Plant Biochemistry, J. Bonner and J. E. Varner (eds.), Academic Press, New York, U.S.A., 1976, 225-274.

Boyde, A. Freezing, freeze-fracturing and freeze-drying in biological specimen preparation for the SEM. In: SEM/1974, IIT Research Institute, Chicago, IL, 60616, 1043-1046.

Boyde, A. and Echlin, P. Freezing and freeze-drying - a preparative technique for SEM. In: SEM/1973, IIT Research Institute, Chicago, IL, 60616, 759-766.

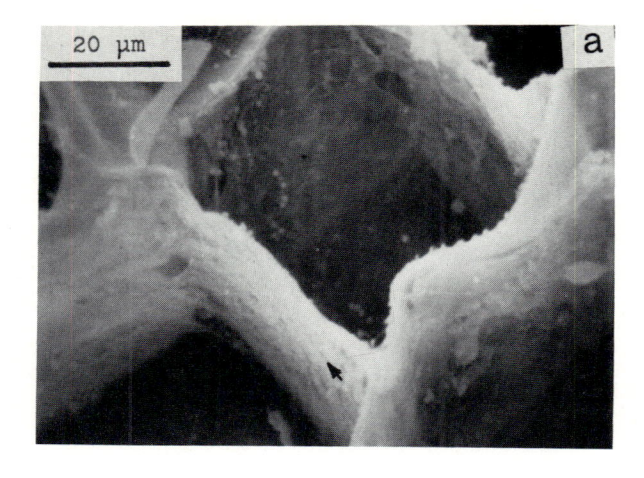

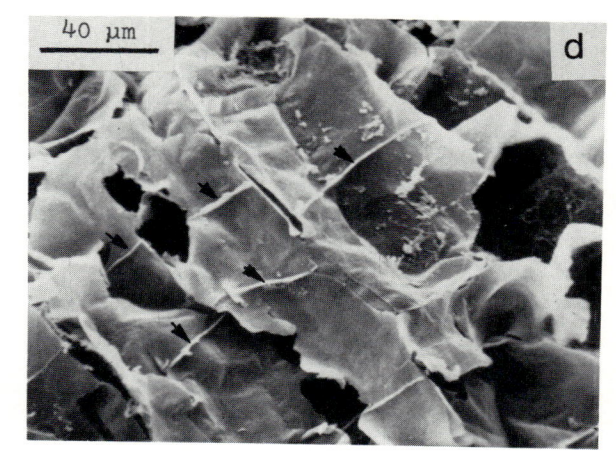

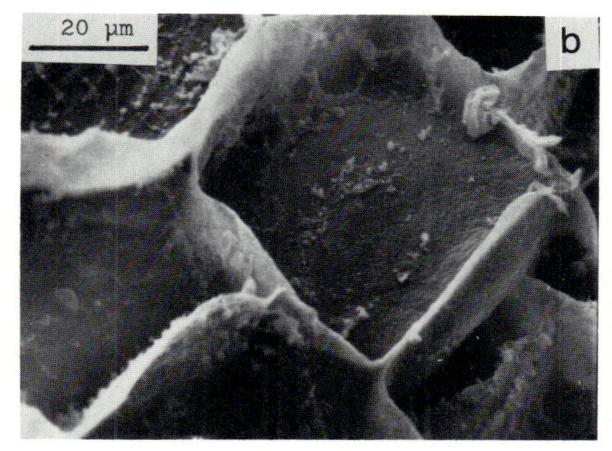

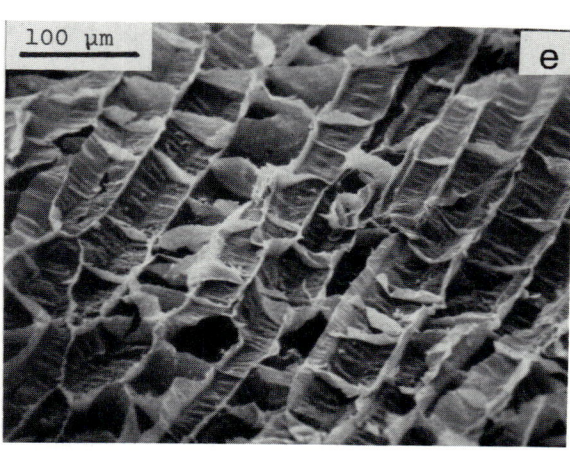

Figure 4. Carrot phloem during growth and disease.
   a. Scarlet Nantes, early development. Contains fibrous material (arrow). (Reprinted by permission from the Home Econ. Res. J.)
   b. Scarlet Nantes, mature.
   c. HiPak, same growth period as 4b. Note intracellular material (arrows). (Reprinted by permission from the Home Econ. Res. J.)
   d. HiPak, same growth period as 4b, aster yellows infected. Note incomplete cell wall formations (arrows).
   e. Scarlet Nantes, mature, aster yellows infected, after storage. Note unidirectional cell separation.

Bullivant, S. Present status of freezing techniques. In: Some Biological Techniques in Electron Microscopy. D. F. Parsons (ed.), Academic Press, New York, 1970, 101-146.

Colvin, J. R. and Leppard, G. G. Fibrillar modified polygalacturonic acid in, on, and between plant cell walls. In: Biogenesis of Plant Cell Polysa charides, F. Loewus (ed.), Academic Press, New York, U.S.A., 1973, 315-331.

Davis, E. A. and Gordon, J. Application of low temperature microscopy to food systems. J. Microsc. 112:1978, 205-214.

Davis, E. A., Gordon, J., and Hutchinson, T. E. Specimen preparation of raw and cooked carrot phloem and xylem for the scanning electron microscope. Home Ec. Res. J. 4:1976a, 163-166.

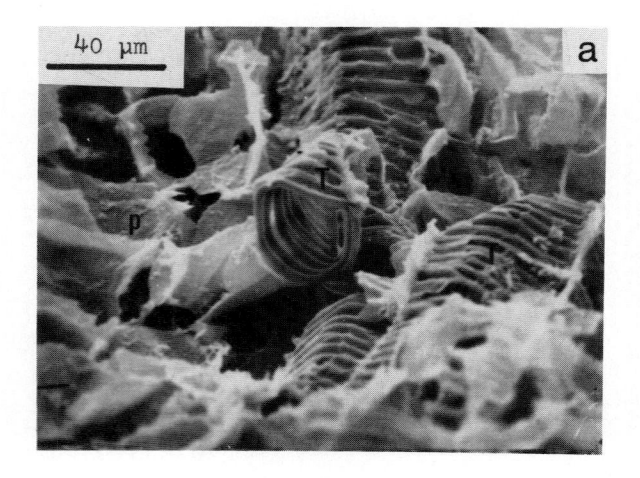

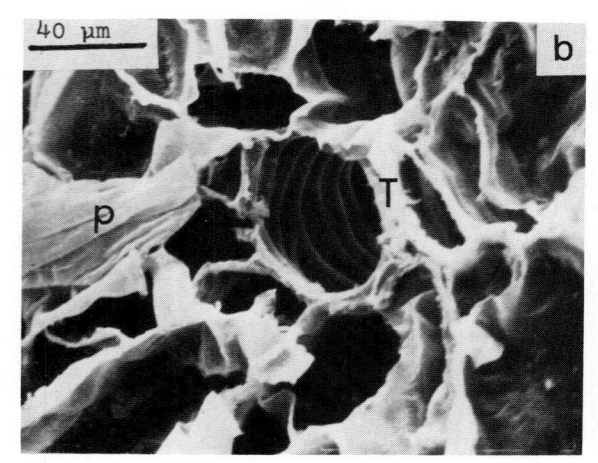

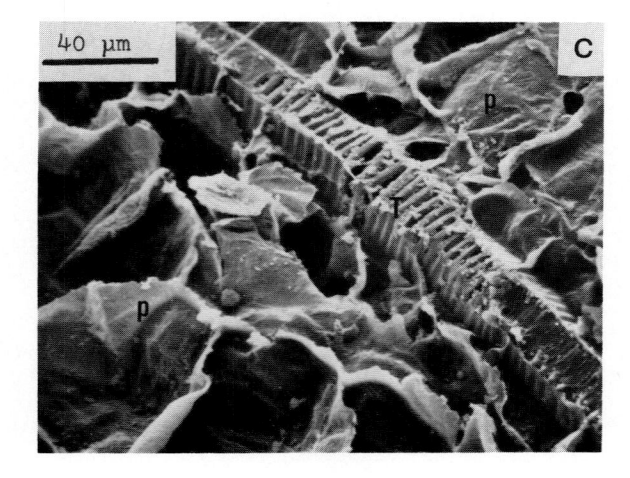

Figure 5. Carrots, xylem, tracheary elements (T) and parenchymous tissue (P).
  a. Scarlet Nantes, early development.
  b. Scarlet Nantes, mature.
  c. HiPak, same growth period as 5b.
  d. Scarlet Nantes, mature, aster yellows diseased.
  e. HiPak, same growth period as 5b, aster yellows diseased.

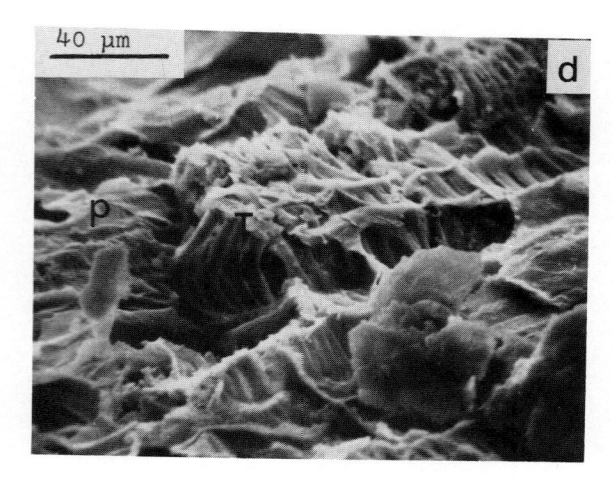

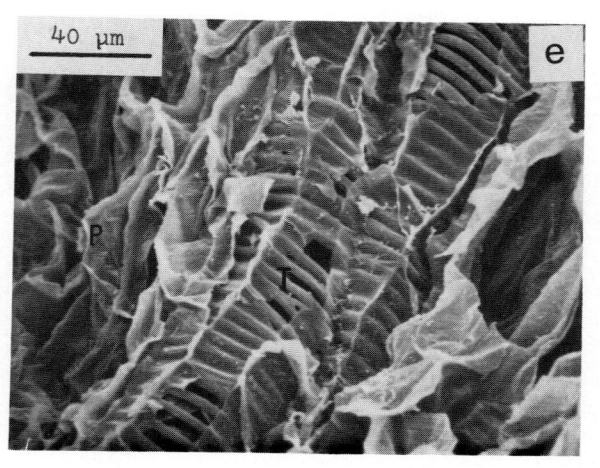

Davis, E. A., Gordon, J., and Hutchinson, T. E. Scanning electron microscope studies on carrots: Effects of cooking on the xylem and phloem. Home Ec. Res. J. 4:1976b, 214-224.

Davis, E. A., Gordon, J., and Hutchinson, T. E. Morphological comparison of two varieties of carrots during growth and storage: Scanning electron microscopy. Home Ec. Res. J. 6:1977, 15-23.

Doi, Y., Teranaka, M., Yora, K., et al. Mycoplasma - or PLT group-like microorganisms found in the phloem elements of plants infected with mulberry dwarf, potato witches broom, aster yellows, or paulownia witches broom. Ann. Phytopath. Soc. Japan, 33:1967, 259-266.

Echlin, P. Low temperature electron microscopy: A review. J. Microsc. 112(1): 1978, 47-61.

Echlin, P. and Moreton, R. Low temperature techniques for scanning electron microscopy. In: SEM/1976/I, IIT Research Institute, Chicago, IL, 60616, 753-761.

Esau, K. Membranous modifications in sieve element plastids of spinach affected by the aster yellows disease. J. Ultrastruct. Res. 59:1977, 87-100.

Falk, R., Gifford, E., and Cutter, E. The effects of various fixation schedules on the scanning electron microscopic image of Trapaeolum majus. Amer. J. Botany, 58:1971, 676-680.

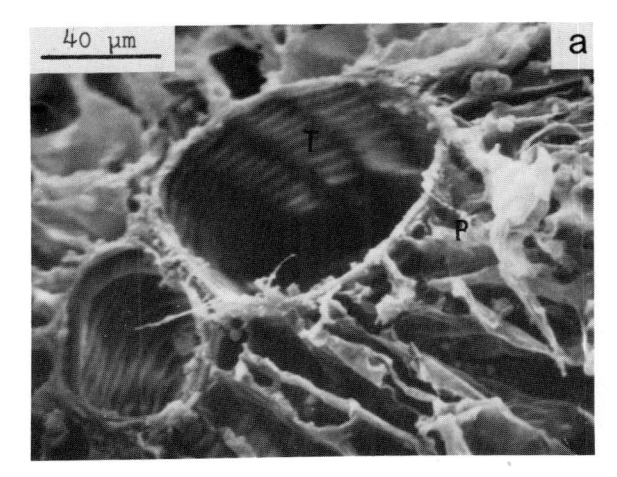

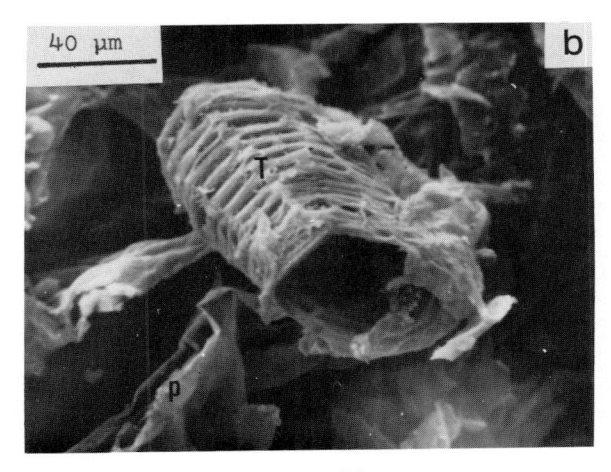

Figure 6. Cooked carrot, phloem.
  a. Boiled. Note tubules which appear collapsed.
  b. Steam cooked tubules with fibrous material between them (arrow). (Level of magnification and the nature of our material caused the micrograph to appear blurry.)

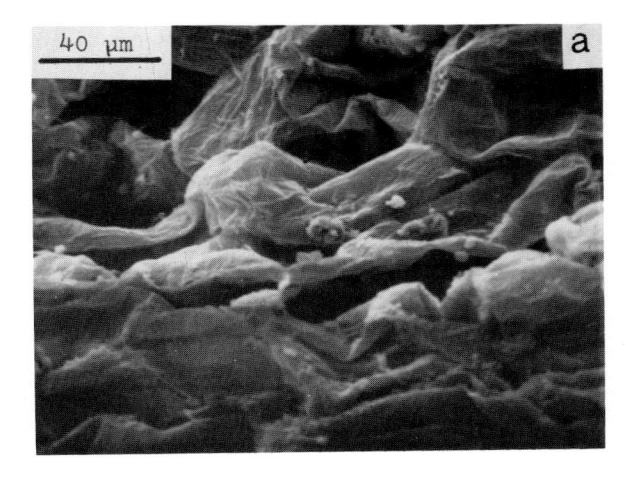

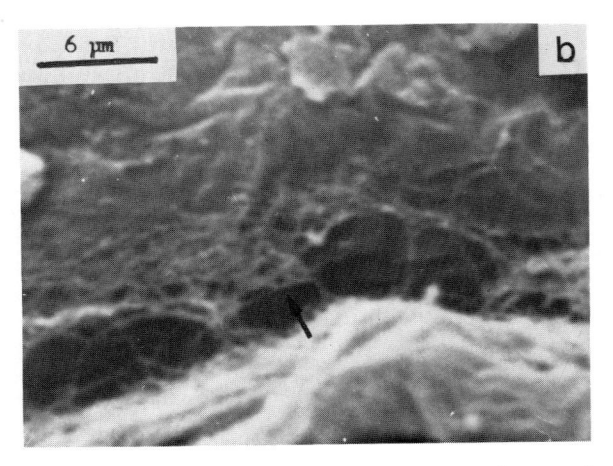

Figure 7. Carrots, xylem, cooked, tracheary elements (T) and parenchymous tissue (P).
  a. Raw.
  b. Steam cooked.

Grote, M. and Fromme, H. G. Electron microscopic studies in cultivated plants. II. Fresh and stored roots of daucus carota L. Z. Lebensm. Unters.-Forsch. 166:1978, 74-79.

Haggis, G. H. and Sinha, R. C. Scanning electron microscopy of mycoplasma organisms after freeze fracture of plant tissues affected with clover phyllody and aster yellows. Phytopath. 68:1978, 677-680.

Henne, R. C. Effect of five insecticides on populations of the six-spotted leafhopper and the incidence of aster yellows in carrots. Canad. J. Plant Sci. 50:1970, 169-174.

Hervey, G. E. R. and Schroeder, W. T. The Yellows Disease of Carrots, Geneva Agric. Exper. Sta. Bull. 737, New York, U.S.A., 1949.

Heslop-Harrison, Y. Scanning electron microscopy of fresh leaves of Pinguicula. Science, 167:1970, 172-174.

Heslop-Harrison, Y. and Heslop-Harrison, J. Scanning electron microscopy of leaf surfaces. In: SEM/1969, IIT Research Institute, Chicago, IL, 60616, 119-126.

Humphreys, W. J., Spurlock, B. O., Johnson, J. S. Critical point drying of ethanol-infiltrated, cryofractured biological specimens for scanning electron microscopy. In: SEM/1974, IIT Research Institute, Chicago, IL, 60616, 275-282.

Keegstra, K., Talmadge, K. W., Bauer, W. D., et al. The structure of plant cell walls, III. A model of the walls of suspension-cultured sycamore cells based on the interconnections of the macromolecular components. Plant Physiol., 51:1973, 188-197.

Leppard, G. G. and Colvin, J. R. Nascent cellulose fibrils in green plants. J. Microsc. 113(2):1978, 181-184.

MacKenzie, A. P. Freezing, freeze-drying and freeze-substitution. In: SEM/1972, IIT Research Institute, Chicago, IL, 60616, 273-280.

Maramorosch, K., Granados, R. R., and Hirumi, H. Mycoplasma diseases of plants and insects. Adv. Virus Res., 16:1970, 135-193.

Mohr, W. P. and Stein, M. Effect of different freeze thaw regimes on ice formation and ultrastructural changes in tomato fruit parenchyma tissue. Cryobiology, 6:1969, 15-31.

Panessa, B. J. and Genarro, J. Preparation of fragile botanical tissues and examination of intracellular contents by SEM. In: SEM/1972/II, IIT Research Institute, Chicago, IL, 60616, 327-334.

Panessa, B. J. and Genarro, J. Intracellular structures. In: Principles and techniques of scanning electron microscopy: Vol. 1, M. A. Hayat (ed.), Van Nostrand Reinhold Co., New York, U.S.A., 1974, 226-241.

Petzold, H., Marwitz, R., Özel, M., et al. Versuche Zum rasterelecktronenmikrosckopischen Nachweis von mykoplasmähnlicken Organismen. Phytopath. Z. 89:1977, 237-248.

Pomeranz, Y. Scanning electron microscopy in food science and technology. Adv. Food Res. 22:1976, 205-307.

Rebhun, L. I. Freeze-substitution and freeze-drying. In: Principles and Techniques of Electron Microscopy, Vol. 2, M. A. Hyaat (ed.), Van Nostrand Reinhold Co., New York, U.S.A., 1972, 3-49.

Robards, A. W. Ultrastructural methods for looking frozen cells. Sci. Prog. (Oxf.), 61: 1974, 1-40.

Robards, A. W. Scanning electron microscopy. In: Electron Microscopy and Cytochemistry of Plant Cells, J. L. Hall (ed.), Elseiver/North-Holland Publishers, Amsterdam, The Netherlands, 1978, 343-415.

Schultz, G. H. Plant resistance to aster yellows. Proc. North Central Branch Entomol. Soc. Amer., 28:1973, 93-99.

Simpson, J. L. and Halliday, E. G. Chemical and histological studies of the disintegration of cell-membrane materials in vegetables during cooking. Food Res. 6:1941, 189-198.

Van Harrevold, A., Trubatch, J., and Steiner, J. Rapid freezing and electron microscopy for the arrest of physiological processes. J. Microsc. (Oxf.) 100:1974, 189-198.

Varriano-Marston, E., Davis, E. A., Hutchinson, T. E., and Gordon, J. Postmortem aging of bovine muscle: A comparison of two preparation techniques for electron microscopy. J. Food Sci. 43:1978, 680-683.

Vaughan, J. G. Food Microscopy. Academic Press, New York, U.S.A., 1979, 1-650.

Discussion with Reviewers

S. B. Jones: Evaluation of "doneness" would seem to be important in relation to amount of collapse observed in cell walls, etc. Would the authors comment on how these evaluations were made and what SEM shows of "overcooked" tissue?

Authors: The reference to "doneness" which was initially described in the text of the paper most of the time has been changed to cooked. "Doneness" perhaps was a poor choice of word since we used specific conditions for cooking the samples as seen in Table 1 of the text. These conditions conformed to traditional modes of steaming, boiling, or pressure cooking carrots for home consumption (Halliday, E. G. and Noble, I. T., Food Chemistry and Cookery, University of Chicago Press, Chicago, 1943, 57-63). Specific cooking times were established in preliminary experiments using trained observers, and these cooking times, once chosen, were used in subsequent experiments. Carrot tissue may well be "overcooked" within certain cellular components such as parenchymous tissue. We did not study this rigorously, but expect greater cellular collapse and distortion upon "overcooking." This would make an interesting study in the future. An important point in the studies discussed here is that differences in the disruption of cellular integrity can be observed by SEM in tissue brought to the same stage of cooking, as defined by instrumental or sensory methods, when the cooking conditions (temperature, time, and amount of water) are different. Thus, similar TPA's do not necessarily imply the same structure.

S. B. Jones: What are the amounts of xylem and phloem relative to the total carrot?

Authors: The xylem and phloem have long been recognized as important structure elements of plants. Thus, both can be expected to contribute significantly to the textural properties of carrot, since it is composed of approximately 28% phloem and 20% xylem tissue, with tracheary elements making up 15% of the xylem. The remainder is composed of cambium, cortex, endodermis, and pericycle. This was estimated from information in Esau's paper on carrot anatomy (Esau, K., Developmental anatomy of the fleshy storage organ of Daucus carota. Hilgardia, 13:1940, 175-226).

For additional discussion see page 282.

THE MICROSTRUCTURE OF ORANGE JUICE

G. G. Jewell

Group Research (A39)
Cadbury Schweppes Ltd
Bournville
Birmingham
United Kingdom
B30 2LU

Abstract

In order to elucidate the microstructure present in orange juice, samples of juices and fruits were examined by thin-sectioning, freeze-etching and negative staining. From thin-sectioning the following structures could be recognised, cellular debris consisting primarily of cell wall material, numerous vesicles, membrane bound bodies, chromoplasts and lipid droplets. From comparison with the fresh orange tissue, certain components could be specifically identified namely cell wall material from the juice sac and chromoplasts. The freeze-etching technique revealed the material described above but in addition crystalline inclusions. Negative staining also revealed the crystalline inclusions, which were subsequently identified as the flavanone hesperidin.

KEY WORDS: Food, Orange Juice, Sectioning, Freeze Etching, Negative Staining, Cell Walls, Lipids, Flavanone Crystals, Membranes

Introduction

For most consumers in the USA and Western Europe the second most important attribute of an orange beverage, after flavour, is appearance. A product which has good colour and body or cloud, will be rated more highly than a watery looking product which lacks cloud. Since orange juice is prepared by expressing or squeezing an intact fruit, the cloud will obviously be comprised of a range of disintegrated cellular constituents, suspended in a solution of cellular contents (water, sugar, oil, acid, protein, etc.) together with any enzymes which may have been liberated. In the commercial production of orange juice, a heating stage is employed both to inactivate enzymes and to destroy any microbial contaminants.

One of the major problems facing the producer of orange juice is to produce a product with a consistent and stable level of cloud. Much work has been done on the enzymatic aspects of juice processing yet it is still true that fruit grown in different areas, or the fruit grown in the same area, but processed on different extraction plants can give products of widely varying cloudiness. A detailed knowledge of the types of particles present in orange juice should be useful in understanding this problem. Surprisingly little microscopy has been undertaken on either oranges as fruit or as juice. Resch and Schara (1970) have employed light microscopy to study the anatomical elements of the orange, and also made some preliminary studies of juice particulates. Mizrahi and Berk (1970) using a combination of light and electron microscopy, on the cloud from Shamouti oranges identified chromoplastids, fragments of pulp and rag, oil droplets and needle like particles which they identified as the flavanone hesperidin. The method of specimen preparation adopted by Mizrahi and Berk was to dry down a suspension of juice particles onto a carbon coated support grid. Although this is a perfectly valid technique, it can only yield a somewhat limited amount of information about particles which may have been subjected to gross distortion during the drying operation. It was the aim of this study to apply a wider range of electron microscopy techniques such as thin sectioning, freeze-etching and negative staining, to the study of the particles in orange juice.

Materials and Methods

Two cultivars of oranges and a range of juices (concentrated to various levels) have been examined during this work. The oranges were purchased from a local store and the observations were made on either Spanish Navel's or Israeli Jaffa's. The juices were obtained from several commercial sources. The juices are classified according to their country of origin and their soluble solids content. The soluble solids are expressed as degrees Brix, where a 10% w/v aqueous solution of sucrose would be $10°$ Brix. Freshly expressed orange juice is $10 - 11°$ Brix. The following range of samples has been examined:

    Israeli $11°$ Brix
    Israeli $44°$ Brix
    Israeli $66°$ Brix
    Spanish $55°$ Brix
    South African $55°$ Brix

Microscopy Methods

Thin Sectioning

In order to embed and thin-section the juice, it was first necessary to encapsulate them in agar using the method introduced by Salyaev (1968) see Fig.1. The anatomical portions of the fruit, with the exception of the juice sac, were sliced into cubes of less than 1 mm dimensions. The sliced tissue, juice sacs and agar capsules containing juice were fixed in phosphate buffered 3% glutaraldehyde pH 7.2 for 4 hrs at $+4°C$ (Sabatini, Bensch and Barnett, 1963) rinsed in distilled water, and then postfixed in a 1% osmium tetroxide solution pH 7.2 for 24 hrs at $+4°C$ (Palade, 1952). The fixed material was rinsed, dehydrated in alcohol series and finally embedded in Maraglas (Freeman and Spurlock 1962). Ultra thin sections were obtained and stained with aqueous uranyl acetate (1% w/v) and lead citrate (Reynolds 1963).

Negative Staining

The cloud from juices was isolated by centrifuging at 30,000 x g for 30 minutes, washed twice and stained by suspending in 2% phosphotungstic acid (PTA) at pH 5.0. This negative stain was chosen from a range of preliminary studies, since it gave adequate staining, and the pH was nearest to that of the natural juices, i.e. pH 4.5 - 5.5

Freeze Etching

The single strength juices ($11°$ Brix) were treated with glycerol prior to freeze etching; this was achieved by making the sample 30% with respect to glycerol. The concentrated juices did not need cryoprotection. A small (2 mm diameter) drop of the sample was placed onto a copper freeze etching stub and rapidly cooled by plunging into liquid Freon^{12} at $-150°C$. Stubs were stored in liquid nitrogen until required. Freeze etching was performed using an NGN FE 600 machine (manufactured by the NGN Co, Accrington, Lancs, England). The following method was employed. The stage was cooled to $-140°C$ and the sample placed on the stage. A vacuum was drawn and when a pressure of 2×10^{-6} torr had been achieved, the sample was warmed to $-100°C$. The sample was then planed/fractured using the microtome knife which had been precooled to $-190°C$. Etching was achieved by leaving the knife over the sample for between 2 and 6 minutes. A carbon-platinum replica of the freshly etched surface was prepared by evaporating on a layer of carbon/platinum and then carbon. The replica was cleaned by passing through a graded series of hydrochloric acid solutions.

All of the samples prepared for transmission electron microscopy were examined with an Hitachi HS7S microscope operating at 50 kV.

Results

A schematic representation of the structure of an orange is given in Fig. 2.

1. Thin Sectioning

a) Natural orange tissues

The cells of the flavedo (Fig.3 and 4) were 8 - 10 μm in diameter, and bounded by cell walls 1 - 2 μm thick which were traversed by plasmadesmata. A centrally placed nucleus was obvious in most cells, as were densely staining droplets which were presumed to be lipid because of their reaction with osmium tetroxide. Those lipid droplets which were between 0.5 and 1.0 μm diameter were not bounded by a discernible membrane. Mitochondria were also evident, frequently with swollen cristae. Whether this was related to fruit being purchased from a store and not fresh could not be determined.

The cells of the albedo were 20 - 50 μm diameter and had extremely thick cell walls up to 20 μm across. The cytoplasm was normally a thin layer compressed against the cell wall (Fig.5). The cells were separated by massive intercellular spaces 20 - 40 μm wide. The combination of very thick cell walls and large intercellular spaces made it very difficult to obtain satisfactory thin sections, where walls did not exhibit some compression or folding (Fig.6).

The outermost cells of the juice sac were columnar and covered by a cuticle (Fig.7). These cells contained a tonoplast of variable size, nuclei, chromoplasts and densely staining droplets presumed to be lipid. Fine detail of these outer cells is shown in Fig.8, and detail of chromoplasts in Fig.9. The walls of the outer cells were 0.5 - 1.0 μm thick, but those around the cells at the centre of the juice sac were exceptionally thin being 10 - 30 nm in diameter (Fig.10). There was little evidence of gradation in wall thickness between these two extremes, but by tracing the thin walls back to their branching point from a relatively thick wall, it was evident that they were of similar structure and function.

b) Juices

The sections obtained from the micro encapsulated single strength juice revealed a range of cell walls and cellular debris (Fig.11). The cell walls were of two distinct types. Relatively thick (0.5 - 1.5 μm thick) some of which exhibited an intensely staining cuticle, and very much thinner walls (10 - 20 nm in width). There was also evidence for aggregates of vesicles and intensely staining droplets (Fig.12) in the size range 0.5 - 2.0 μm.

The appearance of the concentrated juices

THE MICROSTRUCTURE OF ORANGE JUICE

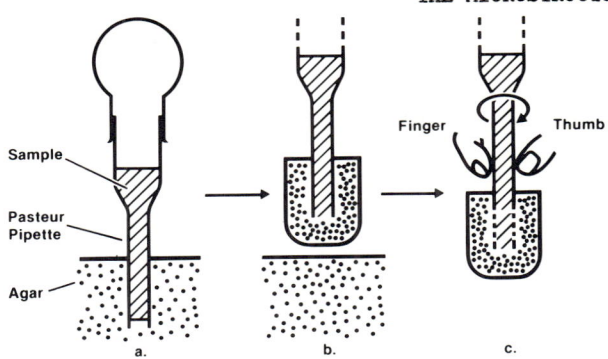

a. Pasteur pipette full of juice dipped in molten agar
b. Pipette removed and agar allowed to set
c. Microcapsule manipulated off pipette by use of finger and thumb. The microcapsule is filled by discharge from the pipette. Then fully removed and sealed with agar.

Fig.1 Schematic representation of the preparation of agar microcapsules to the method of Selyaev (1968).

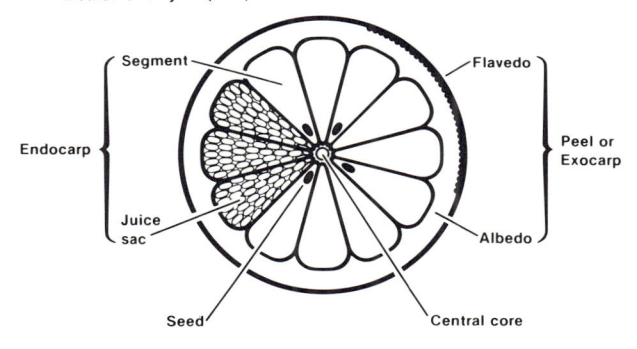

Fig 2. Schematic view of a cross section of an orange

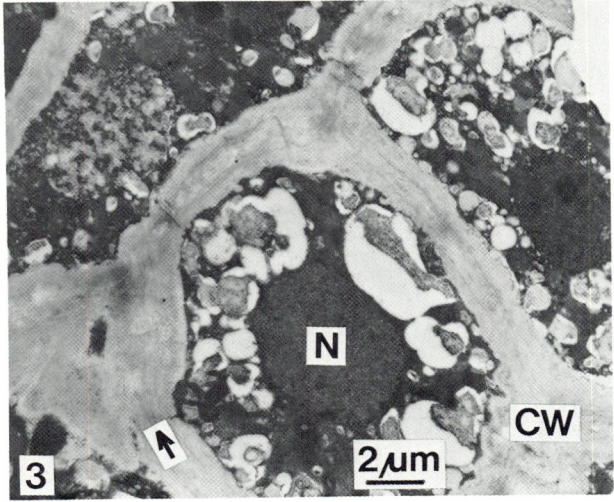

Fig. 3. A thin section of flavedo, showing nucleus (n), densely staining cytoplasm and plasmadesmata (arrow) crossing the cell wall (cw).

Fig. 4. A thin section of flavedo, showing lipid bodies (arrow).

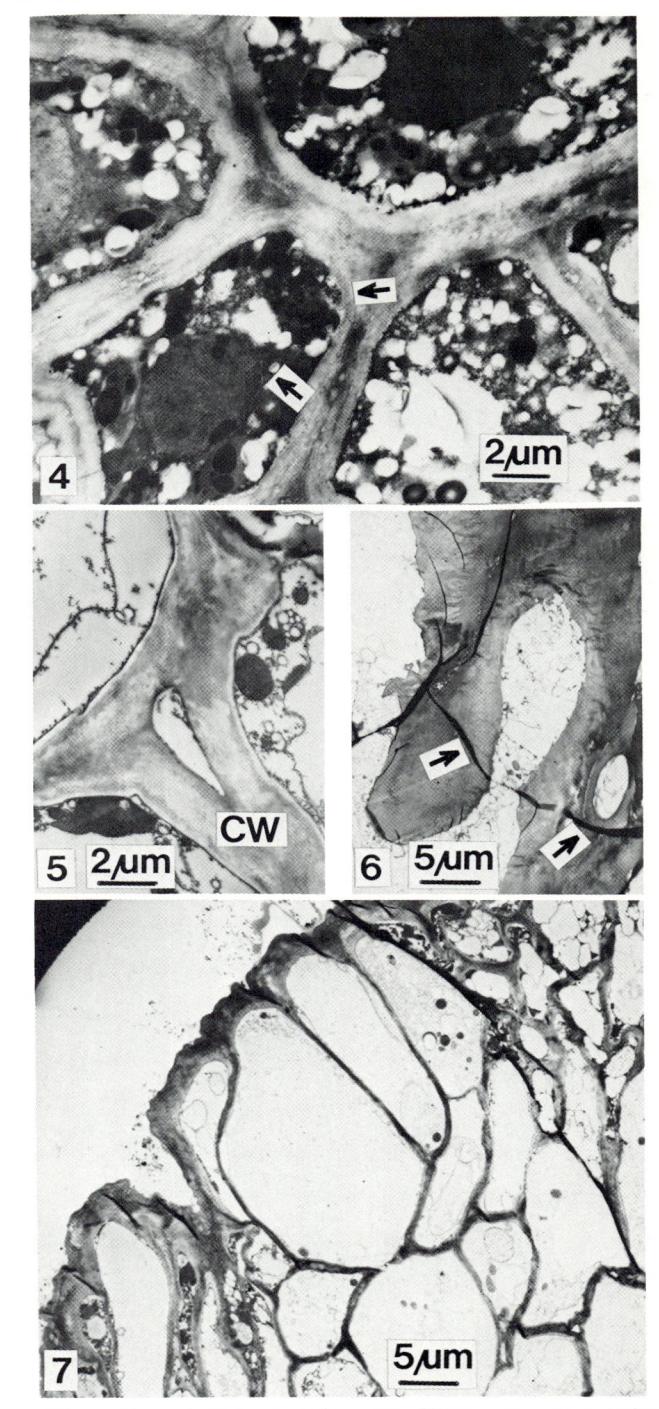

Fig. 5. A thin section of albedo, showing thin layer of cytoplasm against the cell wall (cw).

Fig. 6. A thin section of albedo, showing thick cell walls and cytoplasm. Folds are visible in the walls (arrow).

Fig. 7. A thin section of the outermost cells of a juice sac.

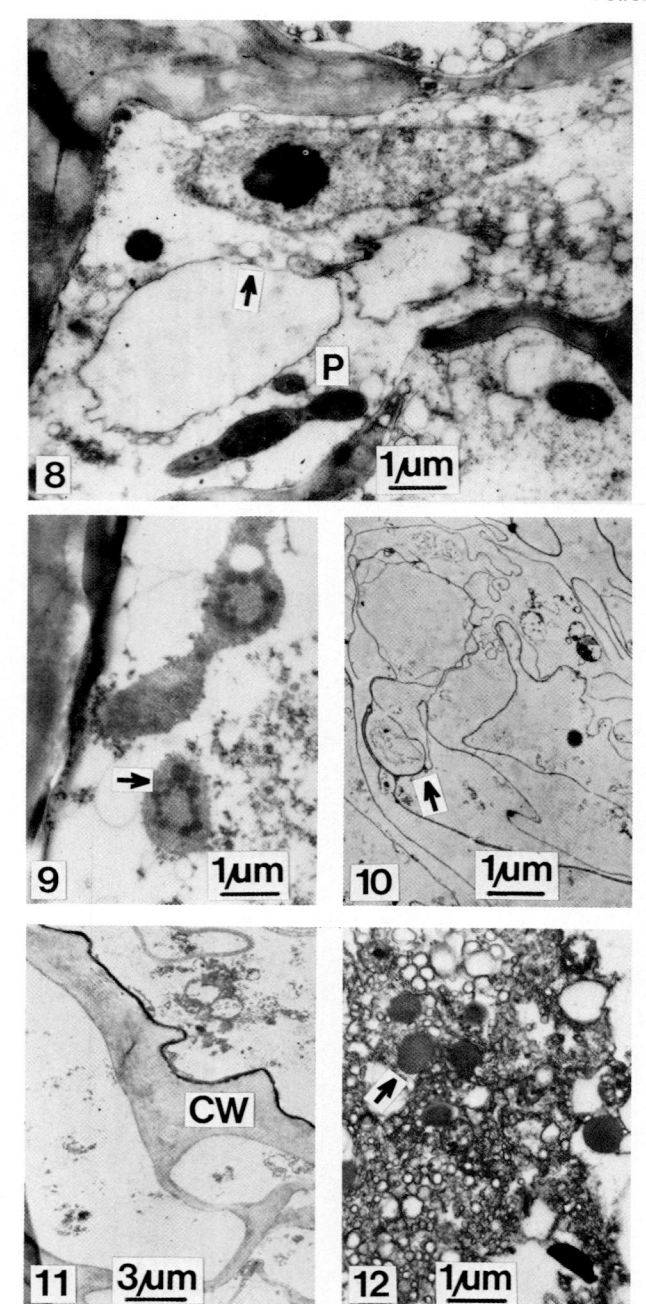

Fig. 8. Detail of juice sac, showing vesicles (arrow) and chromoplast (p).

Fig. 9. Detail of chromoplasts with lipid droplets (arrow).

Fig. 10. A thin section of inner cells of juice sac, showing very thin cell walls, which may be traced back to junction with thicker walls (arrow)

Fig. 11. A thin section of single strength juice, showing walls (cw) and cellular debris.

Fig. 12. A thin section of single strength juice, showing a complex array of vesicles and lipid droplets (arrow).

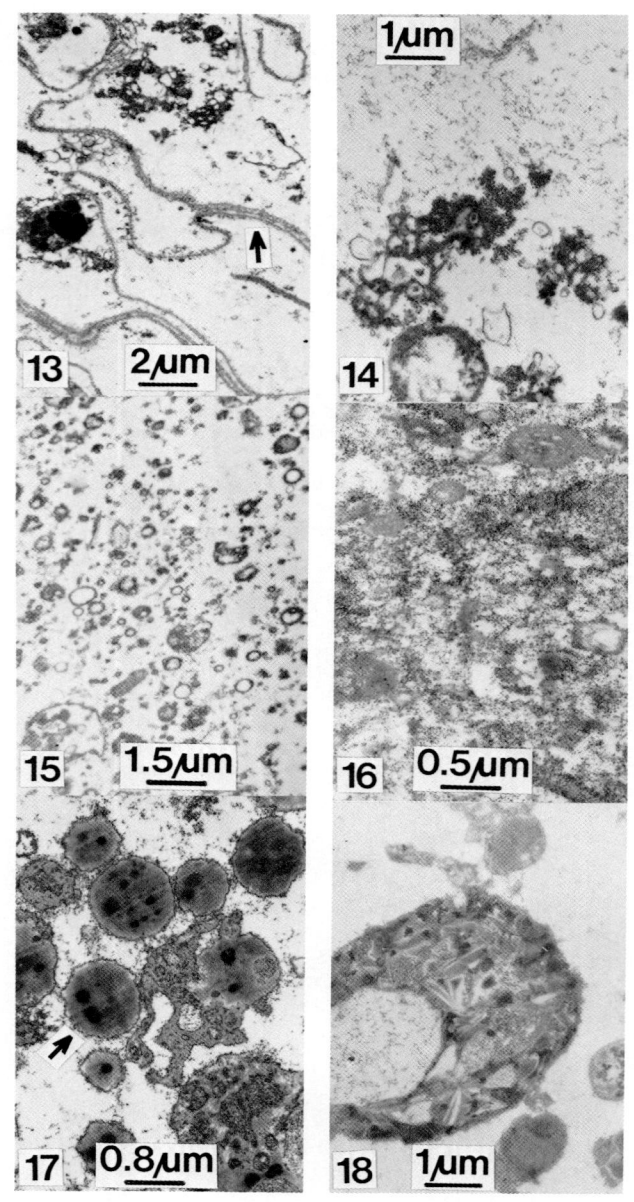

Fig. 13. A thin section of concentrated juice, showing cell walls (arrow) and cellular debris.

Fig. 14. Details of concentrated juice, showing clusters of coagulated debris.

Fig. 15. A section through isolated cloud showing numerous vesicles.

Fig. 16. A section through isolated cloud showing granular debris.

Fig. 17. A section through neck ring material, showing droplets with lipid inclusions (arrow).

Fig. 18. A section through neck ring material, showing inclusions within the droplet.

(Figs.13 and 14) was very similar to the single strength except for a greater extent of coagulated intensely staining material.

   c) Isolated cloud and neck rings

Two distinct types of particles were found in the cloud obtained by centrifugation. Fig.15, shows the numerous membrane-and vesicle-like bodies, whilst Fig.16 shows cell wall material and intensely staining granular material. The material from the neck rings (Figs.17 and 18) was comprised of numerous roughly spherical particles with diameters ranging from 0.5 - 5.0 µm. Many of the particles were bounded by fine granular material, and contained densely staining droplets, granular material and crystal-like inclusions.

2. Freeze Etching

The single strength juices with cryo protectant, and concentrated juices contained a very similar range of particles. Cell wall material was visible in some preparations (Fig.19) and was generally about 0.1 µm wide. Numerous aggregates of material were present (Fig.20), the aggregates which were 0.5 to 1.5 µm diameter contained numerous droplets. The droplets (Fig.21) were in the range 40 - 250 nm diameter. Crystalline inclusions were also observed (Fig.22), the dimensions of the crystals as seen in the fracture was of the order of 0.1 µm. None of the aggregates of droplets and crystals appeared to be bounded by a limiting membrane.

3. Negative Staining

The dominant structures observed in this form of specimen preparation were needle-like crystals (Figs.23, 24, 25). The needles were in the range 15 nm wide 300 nm long to 60 nm wide and 600 nm long. Fine fibrillar material was occasionally observed with aggregates of crystals (Fig.23). In addition, small circular particles about 15 - 60 nm in diameter were present as well as particulate material with a single range of 20 - 200 nm (Fig.25). Occasionally structures were seen which resembled cellular organelles, an example is given in Fig.26 which may represent a golgi apparatus or dictysome.

Discussion

From the results described above, it is evident that orange juice contains numerous types of cellular fragments. It is also clear that no one single preparation technique allows the clear identification of all the particles, e.g. Freeze etching and negative staining reveal numerous crystalline inclusions, whereas these are not visible in the thin sections. Other work (Jewell 1975) confirm that the crystals were the flavanone hesperidin. This material is readily soluble in alcohol, and this would not survive the embedding regime.

The size range of the particles observed (20 - 10,000 nm) indicated that much smaller particles were present than had been observed by light microscopy, and the objects identified as droplets under the light microscope were found to be aggregates of numerous types of particles when studied by electron microscopy.

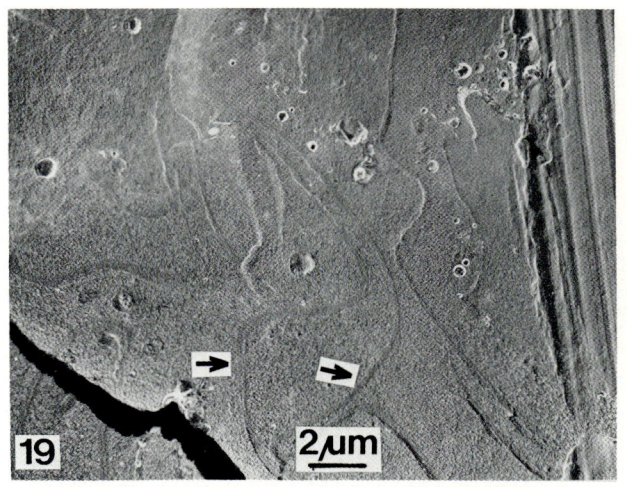

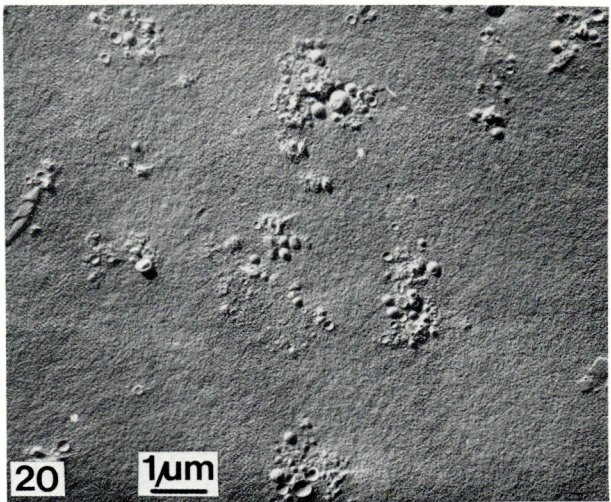

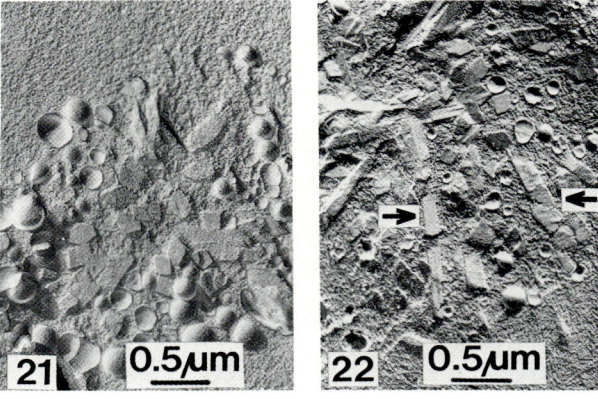

Fig. 19. A freeze etch replica of single strength juice, showing cell walls (arrow).

Fig. 20. A freeze etch replica of single strength juice showing aggregates of material.

Fig. 21. A freeze etch replica of concentrated juice, showing detail within aggregate.

Fig. 22. A freeze etch replica of single strength juice, showing aggregate with crystalline inclusions (arrow).

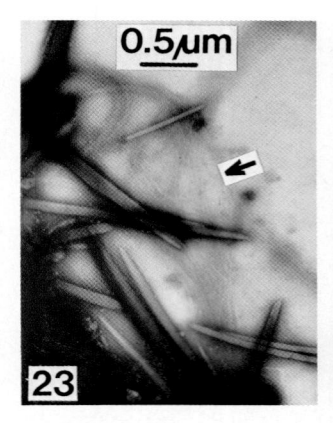

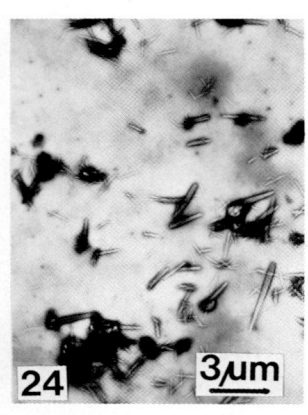

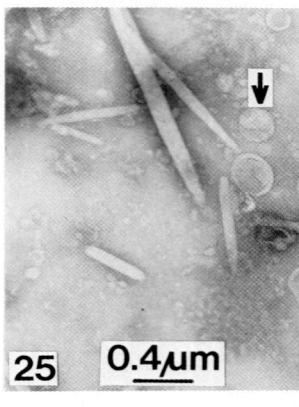

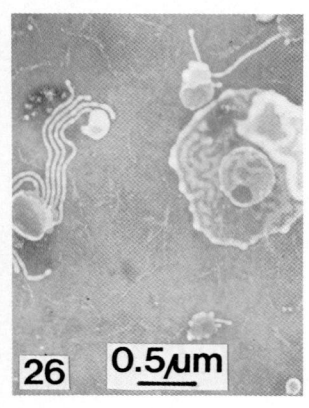

Fig. 23. A negatively stained preparation of juice, showing crystals and fine fibrils. (arrow).

Fig. 24. Negatively stained crystals from single strength juice.

Fig. 25. A negatively stained preparation of concentrated juice, showing crystals and vesicles (arrow).

Fig. 26. A negatively stained preparation of juice, showing cellular organelles.

Acknowledgements

This project was undertaken at the Leatherhead Food RA as part of the requirements for a collaborative PhD study with the University of Surrey, England.

References

Freeman, J.A., and Spurlock, B.D., (1962) "A new epoxy embedment for electron microscopy". J. Cell. Biol. 13 437 - 44.

Jewell, G.G., (1975) "An electron microscopy study of the particulate material in citrus juices" - PhD Thesis - University of Surrey-England.

Mizrahi, S., and Berk, Z., (1970) "Physico-chemical characteristics of orange juice cloud". J. Sci. Fd. Agric. 21 (May) 250 - 253.

Palade, G.E., (1952) "A study of fixation for electron microscopy". J. Exptl. Med. (95) (3) 285 - 295.

Resch, A., and Schara, A., (1970) "The anatomical elements of orange juice concentrates". Confructa 15 (3) 166 - 172.

Reynolds, E.S., (1963) "The use of lead citrate at high pH as an electron-opaque stain in electron microscopy". J. Cell. Biol. 17 208 - 212.

Sabatini, D.D., Bensch, K., and Barnett, R.J., (1963) "Cytochemistry and electron microscopy. The preservation of cellular ultrastructure and enzymatic activity by aldehyde fixation". J. Cell Biol. 17 19 - 58.

Salyaev, R.K., (1968) "A method for the examination of liquid samples". Proc. 4th Eur. Reg. Conf. Electr. Micr. - Tipographia Poliglotta Vaticana, Roma. 2 37.

Discussion with Reviewers

J. F. Chabot: Can you tell what properties of oranges and which processing conditions will result in increased "cloud"?
Author: The important factors in cloud are the size, number and type of particles present and their stability. High levels of crystalline hesperidin impart good cloud, but the problem is that hesperidin has a high density and so tends to settle out. Other aspects to be considered are the role of pectin material and lipids in stabilising the cloud particles.

M. Kalab: Were any microstructural differences found between the Spanish and Israeli oranges?
Author: The Israeli fruit had a much thicker albedo than the Spanish fruit. Otherwise the detailed morphology was similar.

M. Kalab: What is meant by "neck rings"?
Author: Occasionally, when concentrated orange juice, or a beverage containing orange juice has prolonged storage in a glass bottle, a ring of highly coloured material forms in the top of the neck of the container. This material has been called the neck ring.

Reviewer 1: Is there another type of negative stain better than PTA, but with slightly different pH values, which might work better?
Author: A range of stains and pH's were evaluated before selecting the system given. The stains were PTA, uranyl acetate and ammonium molybdate over a pH range from 4.0 - 9.0.

Reviewer 1: Why don't concentrated juices need cryo protection?
Author: The high level of sugar (66° Brix) present in the concentrated juices, acted as a very efficient cryoprotectant.

J. F. Chabot: What do you think is in the vacuoles illustrated in Fig. 3?
Author: It is very difficult to tell, they appear to contain a mixture of membranous and granular material. I am unable to make a positive assignment. The problem might be resolved by studies on fresh fruit harvested from the tree, rather than a store-purchased fruit.

SUBJECT INDEX

Aburage	275
Adulteration	179, 299
Aginosperm	267
Albedo	333
Amylose	25
Asbestos	39
Avian Digestion	267
Bacteria	111
Bloom, Chocolate	25
Bloom, Gelatine	25
Bovine, Beef	1, 51, 73, 79
Bran	305
Bread	9
Buttermilk	123, 179, 193
Candida Cipolytica	17
Candy	25
Carcass	51, 79, 87
Carrots	323
Casein	111, 123, 169, 193, 211
Cheese	111, 123, 143, 153, 163, 169, 211
Brick	111
Cheddar	163
Cottage	111, 123
Cream	123, 153
Crystals In	111, 123, 143
Gouda	111
Imitation	17, 153, 169
Mozzarella	163, 169
Neufchatel	153
Process	143
Cherry, Cracking	1
Chocolate	25
Chromoplast	333
Chrysotile	39
Cocoa Butter	25
Cold Stage	123, 275
Collagen	1
Compression Forces	105, 153, 163, 169, 211, 275
Connective Tissue	1
Contaminants	39
Cooking, Food	25, 51, 323
Corn Starch	25
Cotyledonary	275, 283
Cream	187, 193
Critical Point Drying	9, 61, 123, 223
Cryo-fracture	9, 61, 123, 187, 275, 323
Crystals in Milk	179, 193, 203
Cultivar	252, 305
Curd	111, 143, 153, 163
Cytoplasm	333
Dairy Products	See Milk Products
Digestion, Avian	267
Dough	313
Dressing	231
Drying, of Specimens	9, 61, 123, 223
Electron Microscopy	See Scanning EM and Transmission EM
Emulsions	1, 51, 99, 105, 143, 187, 231

Enzymes	39, 51, 73, 87, 203, 267, 283
Extraneous Matter	39
Fat, Meat	25, 99, 105, 193
Fat, Milk	123, 153, 163, 179, 187
Fat, Vegetable	231
Feed, Animal	299
Filteration	39
Fixation	9, 25, 61, 105, 123, 323
Flavanone Crystals	333
Flavedo	333
Flour, Soy	239
Flour, Wheat	305, 313
Fluorescence	39
Frankfurters	51, 105
Freeze Drying	9, 17, 123, 223
Freeze Etching	25, 123, 187, 333
Freeze Fracturing	9, 61, 123, 153, 187, 193
Fungus, Phytophthora Infestans	1
Gelatin	25, 169
Gels	25, 169, 211, 223
Glass, As Contaminant	39
Glucono-Delta-Lactone (GDL)	275
Gluten	313
Gold, Colloidal (Marker)	123
Gum Arabic	169
Ham	51
Hardness of Wheat	305
Insect Fragments	39
Juice, Orange	333
Kori-tofu	275
Lactose	179, 193, 203
Laser, Diffraction	73
Lead	39
Leatherhead Food Res. Assoc.	25
Light Microscopy	39, 61, 99, 105, 111, 163, 231, 275, 293, 299
Lipids	See Fat
Low-Temperature Microscopy	123, 187, 275
Mailard Reaction	203
Maize-Starch	25
Mayonnaise	231
Meat	1, 9, 25, 51, 61, 73, 79, 87, 99, 105
Cold Shortening	73
Contraction	79
Electrical Stimulation	79
Emulsion	1, 51, 99, 105
Specimen Preparation	9, 25, 61, 99
Stability	105
Tenderness	1, 25, 51
Membrane Fragments (Milk)	179
Micelles	123, 193, 211
Microanalysis	39
Microscopy	See Light Microscopy, Scanning EM and Transmission EM

339

Milk Products	9, 25, 111, 123, 143, 153, 163, 169, 179, 187, 193, 203, 211	Soy Flour	239
		Specimen Preparation	
Buttermilk	123, 179, 193	Coatings, Conductive	9, 61, 123
Cheese	See Cheese	Drying	9, 61, 123, 223
Cream	187, 193	Filteration	39
Curd	111, 143, 153, 163	Fixation	9, 25, 61, 105, 123, 323
Powder	111, 123, 193	Freeze Drying	9. 17, 123, 223
Quarg Powder	193	Freeze Etching	25, 123, 187, 333
Skim Milk	179, 193, 211	Freeze-Thaw Effects	87
Specimen Preparation	9, 123, 153	Freezefracture/cryofracture	9, 61, 123, 193
Whey	111, 193, 203	O-T-O Method	61
Yoghurt	111, 193	Replication	25, 123, 193
Milling of Wheat	305	Shadowing Metal	123
Mitochondria	51, 61, 87	Staining	25, 61, 99, 123, 333
Momen Tofu	275	Thin-Sectioning	25, 105, 123, 179, 333
Muscle	1, 9, 25, 51, 61, 73, 79, 87, 99, 105	Spray Dried Foods	17, 111, 179, 193, 239
		Stage, Tensile	1
Mustard	299	Staining	25, 61, 99, 123, 333
		Starch	9, 25, 313
Oil-Seed	283 (See also Soybean)	Sunflower Seed	283, 299
Orange Juice	333	Syneresis	211
Oriental Foods	239, 275		
		Tannin	293
Paint, as Contaminant	39	Tensile Stress	1
Parenchymous	323	Tensile, Stage	1
Pectin	25	Texturized Protein	9, 239
Pest, Fragments	39	Tofu	275
Phloem, Carrot	323	Transmission Electron Microscopy	25, 39, 51, 61, 73, 79, 87, 105, 111, 123, 153, 179, 193, 211, 231, 275, 293, 333
Polyacrylamide	223		
Pork, Porcine	25, 61, 87		
Potato	1		
Preparation	See Specimen Preparation	Vesicles	293
Precess Cheese	143	Viscosity	17
Protein	17, 111, 239, 275, 283 See also Meat		
		Water Activity	203
Protein, Texturized	9, 239	Weissenberg Test	169
PSE Muscle	79, 87	Wheat	299, 305, 313
		Whey	111, 193, 203
Quarg Powder	193	Whipping of Cream	187
Replicas (Specimen Preparation)	25, 123, 153, 193	Xylem	323
Rheology	1, 17, 105, 143, 153, 163, 169, 211, 275	Yeast	17
		Yoghurt	111, 193
Rodent Excrement	39	*Yucca Rupicola*	287
Rodent, Hair	39		
Sacromere	See Meat		
Salad Dressing	See Dressing		
Sausage	99		
Scanning Electron Microscopy	1, 9, 17, 25, 39, 51, 61, 73, 79, 87, 99, 105, 111, 123, 143, 153, 163, 169, 179, 187, 203, 211, 223, 231, 239, 253, 267, 275, 283, 305, 313, 323		
Sedimentation	17		
Seed Coats	239, 253, 267, 275, 283, 293, 299, 305		
Seeds	9, 239, 253, 267, 283, 293, 299, 305		
Shear Forces	105		
Silicates, As Contaminants	39		
Skim Milk	179, 193, 211		
Sorghum Grain	293		
Soy Bean	253, 275		
Foods	239, 275		
Curd	17, 169		
Isolates	17, 169, 239		

AUTHOR INDEX

Aberle, E.D.	99
Addis, P.B.	87
Albright, F.R.	39
Allen, R.D.	283
Arnott, H.J.	283
Baker, F.L.	239, 253
Bernard, R.L.	253
Buchheim, W.	193
Carroll, R.J.	1, 105
Chabot, J.F.	9
Chen, S.L.	163
Cloke, J.D.	87
Cohen, S.H.	73
Colombo, V.E.	223
Davis, E.A.	87, 323
Ernstrom, C.A.	143
Evans, L.G.	313
Forrest, J.C.	99
Froehlich, D.A.	153
Geissinger, H.D.	61
Glennie, C.W.	293
Gordon, J.	87, 323
Graham, R.	39
Grider, J.	87
Harwalkar, V.R.	211
Hooper, G.R.	313
Hsieh, S-I.	87
Jewell, G.G.	333
Jones, L.J.	231
Jones, S.B.	1
Judge, M.D.	99
Kalab, M.	111, 123, 143, 153, 179, 211
Kingswood, K.	305
Labuza, T.P.	203
Lee, C.-H.	17
Lee, C.M.	105
Lewis, D.F.	25
Liebenberg, N.V.D.W.	293
McGrath, C.J.	87
Miller, B.G.	99
Morrall, P.	293
Moss, R.	305
Pearson, A.M.	313
Pointing, G.	305
Ray, F.K.	99
Rayan, A.A.	143
Rha, C.K.	17
Rhee, K.C.	163
Saio, K.	275
Saltmarch, M.	203
Sargant, A.G.	153
Schmidt, D.G.	187
Smith, L.B.	267
Spath, P.J.	223
Stanley, D.W.	61
Stasny, J.T.	39
Stenvert, N.L.	305
Taranto, M.V.	163, 169
Trusal, L.R.	73
Tung, M.A.	231
Van Hooydonk, A.C.M.	187
Van Sickle, D.C.	99
Vaughan, J.G.	299
Voyle, C.A.	51, 79
Wan, P.J.	163
Wolf, W.J.	239, 253
Yang, C.S.T.	169

FOOD MICROSTRUCTURE

An international Journal on the microstructure and microanalysis of foods, feeds and their ingredients.

Microscopic examination of food stuffs is invaluable. The era in which food product development and improvement was as likely done by a chef as by a food scientist ended decades ago; solution of present day problems in the food industry requires increasingly sophisticated tools and approaches. The new journal, **Food Microstructure** has been established to publish papers dealing with modern approaches including microscopic examination of foods, feeds, and their ingredients, the efforts to understand the factors which lead to the microstructures observed, etc. The promptness of publication, thorough reviewing of manuscripts and high-quality reproduction of micrographs make this journal unique in this field.

FOOD MICROSTRUCTURE will follow the tradition of other high-quality publications for which its publisher (the not-for-profit organization: SEM, Inc.) has been well-known. Reviewers are encouraged to ask questions which a reader might have; these questions, along with author's answers will form an integral part ("Discussion with Reviewers") of each paper.

FOOD MICROSTRUCTURE will contain papers on microscopy and microanalysis of beverages, fruits, vegetables, cereals, meat, seafood, milk products, edible oils, fats, etc. Techniques will include any type of microscopy (scanning electron, transmission electron or light microscopy), x-ray microanalysis or other related microscopic and microanalytical methods.

PUBLICATION: Two (2) issues of the Journal will be published in 1982.

CALL FOR PAPERS

The Food Microstructure journal invites original papers, review articles and short notes on topics listed on the reverse. In addition, books for reviews, as well as announcements and news items related to the scope of the Journal, are invited. A complete Call for Papers is available.

All manuscripts and communications regarding the Food Microstructure journal should be sent either to one of the editors listed above or directly to Dr. Om Johari, P.O. Box 66507, AMF O'Hare, IL 60666 U.S.A. Authors may also submit manuscripts through an appropriate *member of the Editorial Board*.

EDITORS

Dr. Miloslav Kalab
Food Research Inst.
Agriculture Canada
Ottawa, Canada K1A 0C6
613-995-3700 X275

Dr. Samuel H. Cohen
Food Science Lab
Army Natick R&D Lab.
Natick, MA 01760 USA
617-653-1000 X2578

Prof. Eugenia A. Davis
Dept. Food Sci. & Nutri.
Univ. of Minnesota
St. Paul, MN 55108 USA
612-373-1158

Dr. David N. Holcomb
Kraft Inc., R&D
801 Waukegan Rd.
Glenview, IL 60025 USA
312-998-3724

EDITORIAL BOARD

W. Buchheim	Bund. Milchforschung, Kiel, W. Germany
R.J. Carroll	Eastern Reg. Res. Ctr., USDA, Philadelphia, PA
C.L. Davey	Meat Ind. Res. Inst., Hamilton, New Zealand
R.G. Fulcher	Agriculture Canada, Ottawa
D.J. Gallant	Ministry of Agriculture, Nantes, France
H.D. Geissinger	Univ. of Guelph, Ontario, Canada
A.M. Hermansson	The Swedish Food Inst., Goteborg, Sweden
R. Moss	Bread Res. Inst., North Ryde, Australia
Y. Pomeranz	U.S. Grain Marketing Res. Ctr., Manhattan, KS
M.W. Rüegg	Fed. Dairy Res. Inst., Liebefeld, Switzerland
K. Saio	National Food Res. Inst., Ibaraki, Japan
M.V. Taranto	ITT Continental Baking Co., Rye, NY
M.A. Tung	Univ. of British Columbia, Vancouver, Canada
E. Varriano-Marston	Univ. Delaware, Newark, DE
J.G. Vaughan	Queen Elizabeth College, London, U.K.
C.A. Voyle	Meat Research Inst., Bristol, U.K.
W.J. Wolf	USDA Northern Reg. Res. Lab., Peoria, IL